THE ELEMENTS

	IIIA	IVA	VA	VIA	VIIA	O
						2 **He** 4.0026 ± 0.00005
	5 **B** 10.811 ± 0.003	6 **C** 12.01115 ± 0.00005	7 **N** 14.0067 ± 0.00005	8 **O** 15.9994 ± 0.0001	9 **F** 18.9984 ± 0.00005	10 **Ne** 20.183 ± 0.0005

→ IB   IIB

13 **Al** 26.9815 ± 0.00005	14 **Si** 28.086 ± 0.001	15 **P** 30.9738 ± 0.00005	16 **S** 32.064 ± 0.003	17 **Cl** 35.453 ± 0.001	18 **Ar** 39.948 ± 0.0005

28 **Ni** 58.71 ± 0.005	29 **Cu** 63.54 ± 0.005	30 **Zn** 65.37 ± 0.005	31 **Ga** 69.72 ± 0.005	32 **Ge** 72.59 ± 0.005	33 **As** 74.9216 ± 0.00005	34 **Se** 78.96 ± 0.005	35 **Br** 79.909 ± 0.002	36 **Kr** 83.80 ± 0.005
46 **Pd** 106.4 ± 0.05	47 **Ag** 107.870 ± 0.003	48 **Cd** 112.40 ± 0.005	49 **In** 114.82 ± 0.005	50 **Sn** 118.69 ± 0.005	51 **Sb** 121.75 ± 0.005	52 **Te** 127.60 ± 0.005	53 **I** 126.9044 ± 0.00005	54 **Xe** 131.30 ± 0.005
78 **Pt** 195.09 ± 0.005	79 **Au** 196.967 ± 0.0005	80 **Hg** 200.59 ± 0.005	81 **Tl** 204.37 ± 0.005	82 **Pb** 207.19 ± 0.005	83 **Bi** 208.980 ± 0.0005	84 **Po** (210)	85 **At** (210)	86 **Rn** (222)

63 **Eu** 151.96 ± 0.005	64 **Gd** 157.25 ± 0.005	65 **Tb** 158.924 ± 0.0005	66 **Dy** 162.50 ± 0.005	67 **Ho** 164.930 ± 0.0005	68 **Er** 167.26 ± 0.005	69 **Tm** 168.934 ± 0.0005	70 **Yb** 173.04 ± 0.005	71 **Lu** 174.97 ± 0.005
95 **Am** (243)	96 **Cm** (247)	97 **Bk** (247)	98 **Cf** (249)	99 **Es** (254)	100 **Fm** (253)	101 **Md** (256)	102 **No** (253)	103 **Lr** (257)

Atomic Weights are based on C^{12}—12.0000 and Conform to the 1961 Values

INTRODUCTION TO
Organic Laboratory Techniques

A MICROSCALE APPROACH

INTRODUCTION TO
Organic Laboratory Techniques

A MICROSCALE APPROACH

DONALD L. PAVIA   **GARY M. LAMPMAN**   **GEORGE S. KRIZ**

Western Washington University, Bellingham, Washington

RANDALL G. ENGEL

Green River Community College, Auburn, Washington

SAUNDERS GOLDEN SUNBURST SERIES
SAUNDERS COLLEGE PUBLISHING

Harcourt Brace Jovanovich College Publishers
Fort Worth   Philadelphia   San Diego
New York   Orlando   Austin   San Antonio
Toronto   Montreal   London   Sydney   Tokyo

Text Typeface: Times Roman
Compositor: York Graphic Services, Inc.
Acquisitions Editor: John Vondeling
Copy Editor: Linda Kesselring
Manager of Art and Design: Carol C. Bleistine
Art Coordinator: Doris Roessner
Text Designer: Edward A. Butler
Cover Designer: Lawrence R. Didona
Production: York Production Services
Production Coordinator: Joanne Cassetti
Director of EDP: Tim Frelick

Cover Credit: © Phil Schofield/Allstock, Inc.

Printed in the United States of America

INTRODUCTION TO ORGANIC LABORATORY TECHNIQUES, A Microscale Approach, First Edition

ISBN 0-03-025418-3

Library of Congress Catalog Card Number: 89-10336

2 061 987654

To All of Our Organic Chemistry Students

Preface

Introduction to Organic Laboratory Techniques: A Microscale Approach is dedicated to the microscale approach to the teaching of the organic laboratory. The experiments are, for the most part, based on or have evolved from experiments that have been included in our previous three editions of the "macroscale" textbook. Some new experiments have also been added. These include:

Experiment 1, *Isolation of the Active Ingredient in an Analgesic Drug*

Experiment 14, *Elimination Reactions: 2-Butanol and 2-Bromobutane*

Experiment 23, *Chiral Reduction of Ethyl Acetoacetate*

Experiment 24, *NMR Determination of the Optical Purity of (S)-(+)-Ethyl 3-Hydroxybutanoate*

Experiment 36, *1,4-Diphenyl-1,3-butadiene*

Experiment 37, *Relative Reactivities of Several Aromatic Compounds*

Experiment 49, *The Diels-Alder Reaction of Cyclopentadiene with Maleic Anhydride*

Experiment 55, *"Pet Molecule" Project*

The organization of this textbook has been changed significantly from our earlier textbooks. This book is divided into five parts, plus appendices. Part One contains the experiments that we feel are essential for any student wishing to learn the important microscale techniques. We recommend that all of these experiments be assigned before experiments are selected from the rest of the textbook.

Part Two contains approximately 45 experiments from which the instructor may select. These experiments cover a wide range of techniques and reactions. We have continued to include essays at various points throughout the textbook, in order to provide the student with enrichment material. Part Three is devoted to the identification of organic compounds, including carbohydrates and amino acids.

Many people feel that the students need some experience with standard scale, or "macroscale" procedures and equipment in their general organic chemistry laboratory course. For this reason, Part Four contains five macroscale experiments. We have added an essay on how to scale up a reaction, and we have included proven experiments from the third edition of our macroscale laboratory textbook. The experiments have been chosen to provide experience in macroscale filtration, crystallization, and extraction, plus use of a separatory funnel, simple distillation, and fractional distillation.

Part Five contains the chapters on techniques. These chapters have been reorganized from the previous books, and they have been rewritten to focus on microscale laboratory techniques. Detailed descriptions of the macroscale methods have been retained, however, in order that this section might continue to serve as a valuable reference work. Finally, the five appendices include tables of unknowns, procedures for the preparation of derivatives, and coverage of infrared, proton NMR, and carbon-13 NMR spectroscopies.

We have continued to place special emphasis on laboratory safety, and every attempt to make the experiments as safe and reliable as possible has been exerted. Most of the experiments have been developed with student help, and nearly all of the experiments have been thoroughly class-tested. A few experiments have been student-tested or author-tested.

Additional information on safety is included in the instructor's manual that accompanies this textbook. The instructor's manual is also intended to assist persons preparing materials for organic laboratory classes. The time required for each experiment is given and possible laboratory schedules are described. Finally, answers to all questions and problems are included in the instructor's manual.

The development of a new textbook is built upon the encouragement and contributions of many people, and we would be remiss if we did not acknowledge them. Certainly we must salute those pioneers in the development of practical microscale experiments for the organic laboratory, particularly Professors Dana W. Mayo, Ronald M. Pike, Samuel S. Butcher, and Kenneth L. Williamson. Locally, our efforts have been encouraged by our colleagues, particularly Joseph Crook, Edward Fohn, Bruce Haulman, Siegfried Lodwig, Richard Rutkowski, Ruth Schoonover, Harold Taylor, and Mark Wicholas. Our institutions, Western Washington University and Green River Community College, as well as The Western Foundation, have provided us with a great deal of financial support, including released time, photocopying, telephone, postage, and travel. Steve Ware, of Chemglass, Inc., provided us with glassware kits and other items of equipment.

We received many offers from colleagues in the Pacific Northwest to assist in the development of our manuscript and in the class-testing of experiments. We are especially grateful to Nancy Howe (Everett Community College), James Patterson (University of Washington), and Jack Surendranath (Bellevue Community College), for their efforts. Production of this textbook was capably handled by W. B. Saunders Company and York Production Services. We thank all who contributed, with special thanks to our production coordinator, Marilyn James. We have also been most impressed with the art work executed by Larry Ward and the speed with which he produced it.

The manuscript was reviewed and helpful criticism was provided by Professors Jay Bardole (Vincennes University), Gary Carroll (Santa Barbara City College), Joseph Landesberg (Adelphi University), Donna Nelson (University of Oklahoma), Mary O'Brien (Edmonds Community College), Cornelius Steelink (University of Arizona), and Anthony Winston (West Virginia University). We thank them for their helpful comments. We are grateful to Professor Bruce P. Ronald (Idaho State University), who developed the procedure for determining the refractive index on small quantities of liquid and gave us permission to describe this technique and to Professor Ronald Starkey (University of Wisconsin-Green Bay), who allowed us to describe his method for centrifuging samples collected from a gas chromatograph.

We are especially grateful to the students and friends who volunteered to participate in the development of microscale experiments or who offered help and criticism. We thank Robert Anderson, John Arthur, Chris Bacon, Douglas Bryan, Jan Bussey, Debi Dorton, Sean Ebnet, Nancy McEhleran, Keith Frisbee, Dennis Garrison, Don Hamlin, John Harris, Kristi Huling, Joe Kane, Steve Kouri, Erik Leininger, Kim Lofgren, Tuan Nguyen, Raymond Peterson, Kim Splett, Jim Sutton, Kevin Uttech, Jim Verheyden, Anita Wahler, and Lisa Wanttaja. We give special thanks to Michael McVay, who provided valuable help in the development of many experiments and excellent technical assistance. DeeDee Lombard and Denice Hougen provided important help in a number of areas.

Finally, we must thank our families and special friends for their encouragement, support and patience.

Donald L. Pavia
Gary M. Lampman
George S. Kriz
Randall G. Engel

Table of Contents

xi

FOREWORD TO THE STUDENT AND WORDS OF ADVICE

Welcome to organic chemistry! Organic chemistry can be fun, and we hope to prove it to you. The organic chemistry laboratory, using microscale experiments, is also a pleasant place to work. The laboratory environment is cleaner and safer than it has been true with traditional laboratories, and the level of skills that you will develop will be higher. The personal satisfaction that comes with performing a sophisticated experiment skillfully and successfully will be great.

For you to get the most out of the laboratory course, you should strive to do several things. First, you need to understand the organization of this laboratory manual and how to use the manual effectively. It is your guide to learning. Second, you must try to understand both the purpose and the principles behind each experiment you do. Third, you must try to organize your time effectively **before** each laboratory period.

ORGANIZATION OF THE TEXTBOOK

Consider briefly how this textbook is organized. There are five introductory sections, of which this Foreword is the first; a section on laboratory safety is second; dispensing reagents is third; advance preparation and laboratory records make up the fourth; and laboratory glassware is the fifth. Beyond these introductory sections, the textbook is divided into five parts. Part One contains nine experiments which are intended to introduce the student to most of the essential techniques of microscale organic chemistry. Part One also contains an introductory essay which outlines the basic methods of microscale laboratory. This essay contains several laboratory exercises designed to provide practical experience in these methods. We recommend that each student perform all or most of the experiments in Part One. Part Two consists of 46 experiments, which may be assigned as part of your laboratory course. Your instructor will choose a set of these experiments. Part Three is devoted to the identification of organic compounds, and it contains three experiments which provide experience in the analytical aspects of organic chemistry. Interspersed within these first three parts of the textbook are numerous covering essays that provide background information related to the experiments. Part Four contains five experiments intended to provide experience with important "macroscale" methods. These include macroscale filtration, crystallization, and extraction, plus use of a separatory funnel, simple distillation, and fractional distillation. There is also an essay which describes how to scale up a reaction. Part Five comprises a series of detailed instructions and explanations dealing with the techniques of organic chemistry, with particular reference to microscale methods. The techniques are extensively developed and used, and you will become familiar with them in the context of the experiments. Within each experiment, you will find a section, "Required Reading," that indicates which techniques should be studied to do that experiment. Extensive cross-referencing to the techniques chapters in Part Five is included in each experiment. Each experiment also contains a section, "Special Instructions," which lists special safety precautions and specific instructions to the student. The Appendices to this textbook contain sections dealing with infrared spectroscopy, nu-

1

clear magnetic resonance spectroscopy, and carbon-13 nuclear magnetic resonance. Many of the experiments included in Parts One, Two, Three, and Four utilize these spectroscopic techniques, and your instructor may choose to add them to other experiments.

ADVANCE PREPARATION

It is essential to plan carefully for each laboratory period so that you will be able to keep abreast of the material you will learn in your organic chemistry laboratory course. You should not treat these experiments as a novice cook would treat *The Good Housekeeping Cookbook*. You should come to the laboratory with a plan for the use of your time and some understanding of what you are about to do. A really good cook does not follow the recipe line by line with a finger, nor does a good mechanic fix your car with the instruction manual in one hand and a wrench in the other. In addition, it is unlikely that you will learn much if you try to follow the instructions blindly, without understanding them. It can't be emphasized strongly enough that you should come to the lab **prepared.**

If there are items or techniques you do not understand, you should not hesitate to ask questions. You will learn more, however, if you figure things out on your own. Don't rely on others to do your thinking for you.

You should read the section entitled ''Advance Preparation and Laboratory Records'' right away. Although your instructor will undoubtedly have a preferred format for keeping records, much of the material here will help you in learning to think constructively about laboratory experiments in advance. It would also save time if, as soon as possible, you read the first six techniques sections in Part Five. These techniques are basic to all the experiments in this textbook. You should also read the essay, ''Introduction to Microscale Laboratory,'' on pp 35–48. The laboratory class will begin with experiments almost immediately, and a thorough familiarity with this particular material will save you much valuable laboratory time. You should also read the sections ''Laboratory Safety'' and ''Dispensing Reagents.'' Knowing what to do and what not to do in the laboratory is of paramount importance, since the laboratory has many potential hazards associated with it.

BUDGETING TIME

As mentioned in the Advance Preparation section, you should have read several chapters of this book even before your first laboratory class meeting. You should also read the assigned experiment carefully before every class meeting. Having read the experiment will allow you to schedule your time wisely. Often you will be doing more than one experiment at a time. Experiments like the fermentation of sugar or the chiral reduction of ethyl acetoacetate require a few minutes of advance preparation **one week** ahead of the actual experiment. At other times you will have to catch up on some unfinished details of a previous experiment. For instance, usually it is not possible to determine accurately a yield or a melting point of a product immediately after you first obtain the product. Products must be free of solvent to give an accurate weight or melting-point range; they have to be "dried." This drying is done usually by leaving the product in an open container on your desk. Then when you have a pause in your schedule during the subsequent experiment, you can determine these missing data using a sample that is dry.

THE PURPOSE

The main purpose of an organic chemistry laboratory course is to teach you the techniques necessary for a person dealing with organic chemicals. You will learn how to handle equipment that is becoming increasingly common in many laboratories. You will also learn the techniques needed for separating and purifying organic compounds. If the appropriate experiments are included in your course, you may also learn how to identify unknown compounds. The experiments themselves are only the vehicle for learning these techniques. The technique chapters in Part Five are the heart of this textbook, and you should learn them thoroughly. Your instructor may provide laboratory lectures and demonstrations explaining the techniques, but the burden is on you to master them fully by familiarizing yourself with the material in the technique chapters.

Besides good laboratory technique and the methods of carrying out basic laboratory procedures, other things you will learn from this laboratory course are

1. How to take data carefully.
2. How to record relevant observations.
3. How to use your time effectively.
4. How to assess the efficiency of your experimental method.
5. How to plan for the isolation and purification of the substance you prepare.
6. How to work safely.

In choosing experiments, we have tried whenever possible to make them relevant, and, more importantly, interesting. To that end, we have tried to make them a learning experience of a different kind. Most experiments are prefaced by a background essay to place things in context and provide you with some new information. We hope to show you that organic chemistry pervades your lives (drugs, foods, plastics, perfumes, and so on). Furthermore, you should leave your course well trained in organic laboratory techniques. We are enthusiastic about our subject and hope you will receive it with the same spirit.

Donald L. Pavia
Gary M. Lampman
George S. Kriz
Randall G. Engel

LABORATORY SAFETY

In any laboratory course, familiarity with the fundamentals of laboratory safety is critical. Any chemistry laboratory, particularly an organic chemistry laboratory, can be a dangerous place in which to work. Understanding potential hazards will serve well in minimizing that danger for you. Remember that if you have a serious accident, it will not be reversible. You won't get a second chance!

EYE SAFETY

> **Always wear approved
> safety glasses or goggles.**

First and foremost, ALWAYS WEAR APPROVED SAFETY GLASSES OR GOGGLES. This sort of eye protection must be worn whenever you are in the laboratory. Even if you are not actually carrying out an experiment, a person near you might have an accident that could endanger your eyes, so eye protection is essential. Even dishwashing may be hazardous. Cases are known in which a person has been cleaning glassware only to have an undetected piece of reactive material explode, sending fragments into the person's eyes. To avoid this sort of accident, it is necessary to wear your safety glasses at all times.

> **Learn location of
> eyewash facilities.**

If there are eyewash fountains in your laboratory, you should determine which one is nearest to you. In case any chemical enters your eyes, go immediately to the eyewash fountain and flush your eyes and face with large amounts of water. If an eyewash fountain is not available, the laboratory will usually have at least one sink fitted with a piece of flexible hose. When the water is turned on, this hose can be aimed upward and directly into the face, thus working much like an eyewash fountain. Care should be taken not to set the water flow rate too high, or damage to the eyes can result.

FIRES

> **Use care with open flames
> in the laboratory.**
>
> **No smoking.**

Equally important is the need to stress caution about fire. Because an organic chemistry laboratory course deals with flammable organic solvents at all times, the danger of fire is always present. Because of this danger, DO NOT SMOKE IN THE LABORA-TORY. Furthermore, exercise supreme caution when you light matches or use any open flame. Always check to see whether your neighbors on either side, across the bench, and behind you are using flammable solvents. If so, either delay your use of a flame or move to a safe location, such as a fume hood, to use your open flame. Many flammable organic substances are the source of dense vapors that can travel for some distance down a bench. These vapors present a fire danger, and you should be careful, since the source of those vapors may be far away from you. Do not use the bench sinks to dispose of flammable solvents. If your bench has a trough running along it, only **water** (No flammable solvents!) should be poured into it. The trough is designed to carry the water from the condenser hoses and aspirators—not flammable materials.

> **Learn location of fire extinguishers, fire showers, and fire blankets.**

For your own protection in case of a fire, you should learn immediately where the nearest fire extinguisher, fire shower, and fire blanket are. You should learn how these safety devices are operated, particularly the fire extinguisher. Your instructor can demonstrate how to operate it.

If you have a fire, the best advice is to get away from it and let the instructor or laboratory assistant take care of it. DON'T PANIC! Time spent in thought before action is never wasted. If it is a small fire in a container, it usually can be extinguished quickly by placing a wire gauze screen with a ceramic fiber center or, possibly, a watch glass, over the mouth of the container. It is good practice to have a wire screen or watch glass handy whenever you are using a flame. If this method does not take care of the fire and if help from an experienced person is not readily available, then extinguish the fire yourself with a fire extinguisher.

Should your clothing catch on fire, DO NOT RUN. Walk *purposefully* toward the nearest fire blanket or fire shower station. Running will fan the flames and intensify them. Wrapping yourself in the fire blanket will smother the flames quickly.

ORGANIC SOLVENTS: THEIR HAZARDS

> **Avoid contact with organic solvents.**

It is essential to remember that most organic solvents are **flammable** and will burn if they are exposed to an open flame or a match. Remember also that many are toxic or carcinogenic, or both. For example, many chlorocarbon solvents, when accumulated in the body, cause liver deterioration similar to the cirrhosis caused by the excessive use of ethanol. The body does not rid itself easily of chlorocarbons nor does it **detoxify** them; thus, they build up over time and may cause illness in the future. Some chlorocarbons are also suspected to be cancer-causing agents (carcinogens). MINIMIZE YOUR EXPOSURE. Constant and excessive exposure to benzene may cause a form of leukemia. Don't sniff benzene, and avoid spilling it on yourself. Many other solvents, such as chloroform and ether, are good anesthetics and will put you to sleep if you breathe too much of them. They subsequently cause nausea. Many of these solvents have a synergistic effect with ethanol, meaning that they enhance its effect. Pyridine causes temporary impotence. In other words, organic solvents are just as dangerous as corrosive chemicals, such as sulfuric acid, but manifest their hazardous nature in other, more subtle ways. Minimize any direct exposure to solvents, treating them with respect. The laboratory room should be well ventilated. Normal cautious handling of solvents should not cause any health problem. If you are trying to evaporate a solution in an open container, you must do the evaporation in the hood. Excess solvents should be discarded in a container specifically intended for waste solvents, rather than down the drain at the laboratory bench.

> **Do not breathe
> solvent vapors.**

If you want to check the odor of a substance (not a solvent), you should be careful not to inhale very much of the material. The technique for smelling flowers is **not** advisable here; you could inhale dangerous amounts of the compound. Rather, a technique for smelling minute amounts of a substance is used. You should pass a stopper or spatula moistened with the substance (if it is a liquid) under your nose. Alternatively, you may hold the substance away from you and waft the vapors toward you with your hand. But you should never hold your nose over the container and inhale deeply!

> **Solvent hazards.**
> ———————
> **Learn these!**

A list of common organic solvents follows, with a discussion of toxicity, possible carcinogenic properties, and precautions that should be used when these solvents are being handled. A tabulation of the compounds currently suspected of being carcinogens can be found at the end of this chapter.

Acetic acid: Glacial acetic acid is corrosive enough to cause serious acid burns on the skin. Its vapors can irritate the eyes and nasal passages. Care should be exercised not to breathe the vapors and not to allow them to escape into the laboratory.

Acetone: Relative to other organic solvents, acetone is not very toxic. It is flammable, however. It should not be used near open flames.

Benzene: Benzene can cause damage to bone marrow, it is a cause of various blood disorders, and its effects may lead to leukemia. Benzene is considered a serious carcinogenic hazard. Benzene is absorbed rapidly through the skin. It also poisons the liver and kidneys. In addition, benzene is flammable. Because of its toxicity and its carcinogenic properties, benzene should not be used in the laboratory; some less dangerous solvent should be used instead. In this textbook, **there are no experiments that call for benzene.** Toluene is considered a safer alternative solvent in procedures that specify benzene.

Carbon tetrachloride: Carbon tetrachloride can cause serious liver and kidney damage as well as skin irritation and other problems. It is absorbed rapidly through the skin. In high concentrations it can cause death, owing to respiratory failure. Moreover, carbon tetrachloride is suspected of being a carcinogenic material. Although this solvent has the advantage of being nonflammable (in the past, it was used on occasion as a fire extinguisher), it should not be used routinely in the laboratory since it causes health problems. If no reasonable substitute exists, however, it must be used in small quantities, as in preparing samples for infrared (IR) and nuclear magnetic resonance (NMR) spectroscopy. In such cases, it must be used in a hood.

Chloroform: Chloroform is like carbon tetrachloride in its toxicity. It has been used as an anesthetic. However, chloroform is currently on the list of suspect carcinogens. Because of this, chloroform should not be used routinely as a solvent in the laboratory. Occasionally it may be necessary to use chloroform as a solvent for special samples. Then, it must be used in a hood. Methylene chloride is usually found to be a safer substitute in procedures that specify chloroform as a solvent. Deuterochloroform, $CDCl_3$, is a common solvent for NMR spectroscopy. Caution dictates that it should be treated with the same respect as chloroform.

1,2-Dimethoxyethane (Ethylene glycol dimethyl ether or Monoglyme): This is a relatively nontoxic solvent. Because it is miscible with water, it is a useful alternative to solvents such as dioxane and tetrahydrofuran, which are more hazardous. 1,2-Dimethoxyethane is flammable and should not be handled near open flames. On long exposure to light and oxygen, explosive peroxides may form.

Dioxane: Dioxane has been used widely in the past because it is a convenient, water-miscible solvent. It is now suspected, however, of being carcinogenic. Additionally, it is toxic, affecting the central nervous system, liver, kidneys, skin, lungs, and mucous membranes. Dioxane is also flammable and tends to form explosive peroxides when it is exposed to light and air. Because of its carcinogenic properties, it is no longer used in the laboratory unless absolutely necessary. Either 1,2-dimethoxyethane (also known as ethylene glycol dimethyl ether or Monoglyme) or tetrahydrofuran is a suitable, water-miscible alternative solvent.

Ethanol: Ethanol has well-known properties as an intoxicant. In the laboratory, the principal danger arises from fires, since ethanol is a flammable solvent. When using ethanol, take care to work where there are no open flames.

Diethyl ether (Ether): The principal hazard associated with diethyl ether is fire or explosion. Ether is probably the most flammable solvent one is likely to find in the laboratory. Because the vapors are much more dense than air, they may travel along a laboratory bench for a considerable distance from their source before being ignited. Before you begin to use ether, it is very important to be certain that no one is working with matches or any open flame. Ether is not a particularly toxic solvent, although in high enough concentrations it can cause drowsiness and perhaps nausea. It is used as a general anesthetic. Ether can form highly explosive peroxides when exposed to air. Consequently, it should never be distilled to dryness.

Hexane: Hexane may be irritating to the respiratory tract. It can also act as an intoxicant and a depressant of the central nervous system. It can cause skin irritation since it is an excellent solvent for skin oils. The most serious hazard, however, comes from its flammable nature. The precautions recommended for the use of diethyl ether in the presence of open flames apply equally to hexane.

Ligroin: See Hexane.

Methanol: Much of the material outlining the hazards of ethanol apply to methanol. Methanol is more toxic than ethanol; ingestion can cause blindness and even death. Because methanol is more volatile, the danger of fires is more acute.

Methylene chloride (Dichloromethane): Methylene chloride is not flammable. Unlike other members of the class of chlorocarbons, it is not currently considered as a serious carcinogenic hazard. Recently, however, it has been the subject of much serious investigation and there have been proposals to regulate it in industrial situations where workers have high levels of exposure on a day-to-day basis. Methylene chloride is less toxic than chloroform and carbon tetrachloride. It can cause liver damage when ingested, however, and its vapors may cause drowsiness or nausea.

Pentane: See Hexane.

Petroleum ether: See Hexane.

Pyridine: There is some fire hazard associated with pyridine. The most serious hazard arises from its toxicity, however. Pyridine may cause depression of the central nervous system; irritation of the skin and respiratory tract; damage to the liver, kidneys, and gastrointestinal system; and even temporary sterility. Pyridine should be treated as a highly toxic solvent. It should be handled only in the fume hood.

Tetrahydrofuran: Tetrahydrofuran may cause irritation of the skin, eyes, and respiratory tract. It should never be distilled to dryness, since it tends to form potentially explosive peroxides on exposure to air. Tetrahydrofuran does present a fire hazard.

Toluene: Unlike benzene, toluene is not considered a carcinogen. However, it is at least as toxic as benzene. It can act as an anesthetic and also cause damage to the central nervous system. If benzene is present as an impurity in toluene, then one must expect the hazards associated with benzene to manifest themselves. Toluene is also a flammable solvent, and the usual precautions about working near open flames should be applied.

Certain solvents, because of their carcinogenic properties, should not be used in the laboratory. Benzene, carbon tetrachloride, chloroform, and dioxane are among these solvents. For certain applications, however, notably as solvents for IR or NMR spectroscopy, there may be no suitable alternative solvent. When it is necessary to use one of these solvents, safety precautions will be recommended, or you will be referred to the discussion in Technique 18.

Because relatively large amounts of solvents may be used in a large organic laboratory class, one must consider safe means of storing these substances. Only the amount of solvent that is needed for a particular experiment being conducted should be kept in the laboratory room. The preferred location for bottles of solvents being used during a class period is in a hood. When the solvents are not being used, they should be stored in a fireproof storage cabinet for solvents. If possible, this cabinet should be ventilated into the fume hood system.

WASTE-SOLVENT DISPOSAL

> **Do not pour flammable solvents into troughs or sinks.**
>
> ---
>
> **Use waste containers.**

Because of the toxicity and flammability hazards, it is not acceptable to dispose of solvents by pouring them down the sink. Municipal sewage-treatment plants are not equipped to remove these materials from sewage. Furthermore, with volatile and flammable materials, a spark or an open flame can cause an explosion in the sink or further down the drains.

The appropriate disposal method for waste solvents is to pour them into appropriately labeled waste-solvent containers. These containers should be placed in the hoods in the laboratory. When these containers are filled, they should be disposed of safely either by incineration or by burial in a designated hazardous-waste dump.

USE OF FLAMES

Even though organic solvents are frequently flammable (for example, hexane, ether, methanol, acetone, and petroleum ether), there are certain laboratory procedures for which a flame may be used. Most often these procedures involve an aqueous solution. In fact, as a general rule, a flame should be used to heat only aqueous solutions. Most organic solvents boil below 100 °C, and a steam bath, water bath, or sand bath may be used to heat these solvents safely. A listing of common organic solvents is given in Table 3–1, p 547, of Technique 3. Solvents marked in that table with boldface type will burn. Ether, pentane, and hexane are especially dangerous, since in combination with the correct amount of air, they explode.

Some common sense rules apply to using a flame in the presence of flammable solvents. Again, we stress that you should check to see whether anyone in your vicinity is using flammable solvents before you ignite any open flame. If someone is using a flammable solvent, move to a safer location before you light your flame. Your laboratory should have an area set aside for using a burner to prepare micropipets or other pieces of glassware. If it is not prudent to use a flame at your bench, move to a safer location for your operations. Methods for heating solvents safely are discussed in detail in Technique 3, starting on p 547.

The drainage troughs or sinks should never be used to dispose of flammable organic solvents. They will vaporize if they are low boiling and may encounter a flame further down the bench on their way to the sink.

INADVERTENTLY MIXED CHEMICALS

To avoid unnecessary hazards of fire and explosion, **never pour any reagent back** into a stock bottle. There is always the chance that you may accidentally pour back some foreign substance that will react explosively with the chemical in the stock bottle. Of course, by pouring reagents back into stock bottles you may introduce impurities that could spoil the experiment for the person using the stock reagent after you. Pouring things back into bottles is thus not only a dangerous practice, it is also inconsiderate.

UNAUTHORIZED EXPERIMENTS

You should never undertake any unauthorized experiments. The risk of an accident is high, particularly with an experiment that has not been completely checked to reduce the hazard. You should never work alone in the laboratory. Minimal safety considerations require that another person should be present also.

FOOD IN THE LABORATORY

Because all chemicals are toxic to some extent, you should avoid accidentally ingesting toxic substances; therefore, never eat or drink any food in the laboratory. There is always the possibility that whatever you are eating or drinking may become contaminated with a potentially hazardous material.

SHOES

You should always wear shoes in the laboratory. Even open-toed shoes or sandals offer inadequate protection against spilled chemicals or broken glass.

FIRST AID: CUTS, MINOR BURNS, AND ACID OR BASE BURNS

> **Any injury, no matter how small, must be reported to your instructor as soon as practicable.**

If any chemical enters your eyes, immediately irrigate the eyes with copious quantities of water. Slightly warm water, if it is available, is preferable. Be sure that the eyelids are kept open. Continue flushing the eyes in this way for 15 minutes.

In case of a cut, wash the wound well with water, unless you are specifically instructed to do otherwise. If necessary, apply pressure to the wound to stop the flow of blood.

Minor burns caused by flames or contact with hot objects may be soothed by immediately immersing the burned area in cold water or cracked ice for about five minutes. Applying salves to burns is discouraged. Severe burns must be examined and treated by a physician.

For chemical acid or base burns, the burned area should be rinsed first with copious quantities of water. After thorough rinsing, as an option, one of the following neutralizing solutions may be applied. For acid burns, apply a dilute solution of sodium bicarbonate (no substitutes). For burns by strong bases, a dilute solution of a weak acid, such as 2% acetic acid solution, a 25% vinegar solution, or a 1% boric acid solution, may be applied. Follow the use of an optional neutralizing solution by more rinsing with water. Most well-equipped laboratories always have these safety solutions available in containers clearly marked ''FOR ACID BURNS'' and ''FOR BASE BURNS.'' You should learn where these safety solutions are. After applying either sodium bicarbonate or dilute acid solution, flush the affected area with water for 10 to 15 minutes. Do not use the safety solutions for chemical purposes. If you do, they may not be available when they are really needed.

If you accidentally ingest a chemical, immediately begin drinking large quantities of water while proceeding immediately to the nearest medical assistance. It is important that the examining physician be informed of the exact nature of the substance ingested.

CARCINOGENIC SUBSTANCES

The accompanying table is taken from a list published by the Occupational Safety and Health Administration (OSHA) of substances suspected of being carcinogens. The table is not complete but merely lists substances likely to be found in an organic chemistry laboratory. Complete listings can be found in *Chemical and Engineering News,* July 31, 1978, p 20 and *NIH Guidelines for the Laboratory Use of Chemical Carcinogens,* May 1981, p 11. An additional list of suspected carcinogens has been compiled by the U.S. Department of Labor. Some substances from this list are also included in the table.

Acetamide	Hydrazine and its salts
Acrylonitrile	Lead(II) acetate
4-Aminobiphenyl	Methyl chloromethyl ether
Asbestos	Methyl methanesulfonate
Aziridine (ethyleneimine)	N-Methyl-N-nitrosourea
Benzene	4-Methyl-2-oxetanone
Benzidine	(β-butyrolactone)
Bis(2-chloroethyl) sulfide	1-Naphthylamine
Bis(chloromethyl) ether	2-Naphthylamine
Carbon tetrachloride	4-Nitrobiphenyl
Chloroform	N-Nitrosodimethylamine
Chromic oxide	2-Oxetanone (β-propiolactone)
Coumarin	Phenylhydrazine and its salts
Diazomethane	Polychlorinated biphenyl
1,2-Dibromo-3-chloropropane	Progesterone
1,2-Dibromoethane	Tannic acid
Dimethyl sulfate	Tannins
p-Dioxane	Testosterone
Ethyl carbamate	Thioacetamide
Ethyl diazoacetate	Thiourea
Ethyl methanesulfonate	o-Toluidine
1,2,3,4,5,6-Hexachlorocyclohexane	Trichloroethylene
	Vinyl chloride

Besides the specific substances listed, there are some general classes of compounds cited by OSHA that tend to have carcinogenic behavior. They are listed below.

Alkylating reagents
Androgens
Arsenic and its compounds
Azo and diazo compounds
Beryllium and its compounds
Cadmium and its compounds
Chromium and its compounds
Estrogens
Hydrazine derivatives
Lead(II) compounds
Nickel and its compounds
Nitrogen mustards (β-halo amines)
N-Nitroso compounds
Polyhalogenated compounds
Polycyclic aromatic amines and hydrocarbons
Sulfur mustards (β-halo sulfides)

REFERENCES

Fire Protection Guide on Hazardous Materials. 9th ed. Quincy, MA: National Fire Protection Association, 1986.

Gosselin, R. E., Smith, R. P., and Hodge, H. C. *Clinical Toxicology of Commercial Products*. 5th ed. Baltimore: Williams and Wilkins, 1984.

Green, M. E., and Turk, A. *Safety in Working with Chemicals*. New York: Macmillan, 1978.

Lenga, R. E., ed. *The Sigma-Aldrich Library of Chemical Safety Data*. Milwaukee, WI: Sigma-Aldrich Corp., 1985.

McKusick, B. C. ''Prudent Practices for Handling Hazardous Chemicals in Laboratories.'' *Science,* 211 (February 20, 1981): 777.

Merck Index. 10th ed. Rahway, N.J.: Merck and Co., 1983.

NIH Guidelines for the Laboratory Use of Chemical Carcinogens. Washington, D.C.: National Institutes of Health, U.S. Department of Health and Human Services, 1981.

''OSHA Issues Tentative Carcinogen List.'' *Chemical and Engineering News,* 56 (July 31, 1978): 20.

Prudent Practices for Disposal of Chemicals from Laboratories. Washington, D.C.: Committee on Hazardous Substances in the Laboratory, National Research Council, National Academy Press, 1983.

Prudent Practices for Handling Hazardous Chemicals in Laboratories. Washington, D.C.: Committee on Hazardous Substances in the Laboratory, National Research Council, National Academy Press, 1981.

Renfrew, M. M., ed. *Safety in the Chemical Laboratory*. Easton, PA: Division of Chemical Education, American Chemical Society, 1967–1981.

Safety in Academic Chemistry Laboratories. 4th ed. Washington, D.C.: Committee on Chemical Safety, American Chemical Society, 1985.

Sax, N. I. *Dangerous Properties of Industrial Materials*. New York: Van Nostrand Reinhold, 1975.

Sax, N. I. and Lewis, R. J., ed. *Rapid Guide to Hazardous Chemicals in the Work Place*. New York: Van Nostrand Reinhold Co., 1986.

Searle, C.E., ed. *Chemical Carcinogens*. ACS Monograph No. 173. Washington, D.C.: American Chemical Society, 1976.

DISPENSING REAGENTS

Although safety is much improved in a microscale organic laboratory, careless dispensing of reagents can still be a hazard in any laboratory. When reagents are spilled you may be subjected to an unnecessary health or fire hazard. In addition, you may waste expensive chemicals or destroy balance pans and clothing.

Generally speaking, it is not a good idea to pour small amounts of chemicals from large bottles. If you pour materials from containers, you are likely to spill some solid or liquid. Instead, you should use a spatula for transferring solids. For liquids, you should use a graduated pipet, automatic pipet, or Pasteur pipet to transfer a liquid from a larger container to a smaller one. When using a pipet to dispense liquids, you should never attempt to fill the pipet by mouth suction. If mouth suction is used, there is a danger of filling the mouth with toxic or corrosive liquids or harmful vapors. Always use a pipet pump or rubber bulb with a pipet. Spatulas and pipets are used in both macroscale or microscale experiments. When reading the procedures given in this book, you will notice that amounts of solids are expressed in units of mass (g), while liquids are given in volume (mL).

WEIGHING AND TRANSFERRING SOLIDS

Your instructor may store solids in small containers near the balance. Place your conical vial in a small beaker and take these with you to the balance. Use the larger of your two spatulas (p 48) to aid the transfer of the solid to a piece of weighing paper that has been folded once in the middle (never weigh directly into a conical vial). It is not necessary to obtain the exact amount specified in the experimental procedure, and trying to be exact requires too much time at the balance (Other students will need the balance, too!). For example, if you obtained 0.140 g of a solid, rather than the 0.136 g specified in a procedure, the actual amount weighed would be recorded in your notebook. Use the amount you weighed to calculate the theoretical yield, if this solid is the limiting reagent.

While still at the balance, carefully transfer the solid to your vial (or flask). The beaker acts as a trap for any solid that fails to make it into the vial. If a spill occurs, carefully remove the material from the beaker and place it in the vial. Do not attempt to transfer the solid when holding the vial in your hand. Otherwise, you may place most of the solid on the floor.

MEASURING AND TRANSFERRING LIQUIDS WITH AUTOMATIC AND GRADUATED PIPETS

Liquid reagents should be stored in small containers in a hood. Usually, an automatic pipet or a graduated pipet with pipet pump is placed near the reagent for your use. Automatic pipets must not be used with corrosive liquids, such as sulfuric or hydrochloric acids. With corrosive liquids, you will be provided with a graduated pipet and pipet pump. Use the pipet provided to remove the required volume of liquid and transfer it to a conical vial or flask. In some cases, you may need to preweigh the container before dispensing the liquid into the container.

Before using the pipets, you should read the essay Introduction to the Microscale Laboratory beginning on p 35 and practice using the pipets. The automatic pipet may be damaged if improperly used. Ask the instructor or assistant for help.

Make sure that pipets are returned to their proper place, near the reagent, when you are finished with them. If an accurate weight is required, the preweighed container should be weighed after the liquid has been transferred. Weighing is especially important if the liquid is a limiting reagent (see Advanced Preparation and Laboratory Records, in the next section). The laboratory procedure usually specifies when you should weigh the liquid after it has been measured with a pipet. With non-limiting reagents, you will not need to weigh the liquid. You must, however, be careful in measuring the volume accurately. Masses of these liquids may be calculated from the density of the liquid, as described in Technique 1.

MEASURING AND TRANSFERRING LIQUIDS WITH PASTEUR PIPETS

Less precise measurements may be made with calibrated Pasteur pipets (see the essay "Introduction to the Microscale Laboratory," beginning on p 35). Your instructor may provide you with a calibrated Pasteur pipet and bulb for transferring liquids where an accurate volume is not required. The pipet may be used to transfer a volume of 1.5 mL or less. You may find that the instructor has taped a test tube to the side of the storage bottle. The pipet is stored in the test tube with that particular reagent. The calibrated Pasteur pipet is widely used in transferring solvents. In some cases, you can use the pipet to add a solvent directly to your conical vial or flask. In some cases, you may use the pipet to dispense solvent into a container to be taken to your bench for use in crystallization.

A Pasteur pipet may also be supplied for dropwise addition of a particular reagent to a reaction mixture. For example, concentrated sulfuric acid is often dispensed in this way. When sulfuric acid is transferred, you should take care to avoid getting the acid into the dropper bulb.

MEASURING AND TRANSFERRING LIQUIDS WITH GRADUATED CYLINDERS

With larger volumes, you may want to use a graduated cylinder to measure a liquid. This device is widely employed with larger volumes of liquids employed in macroscale experiments, but it may also be used in microscale procedures. Use a **clean and dry** Pasteur pipet to transfer the liquid from the storage container into the graduated cylinder. Do not attempt to pour the liquid directly into the cylinder from the storage bottle or you may spill the fluid. Some instructors may want you to first pour some of the liquid into a beaker, and then use a Pasteur pipet to transfer the liquid to a graduated cylinder. Remember, you should not take more than you need. Excess material should never be returned to the storage bottle. Unless you can convince someone else to take it, it must be poured into the appropriate waste container. You should be frugal in your estimation of amounts needed. The environment does not need any more of our junk!

ADVANCED PREPARATION AND LABORATORY RECORDS

In the Foreword, mention was made of the importance of advance preparation for laboratory work. Presented here are some suggestions about what specific information you should try to get in your advance studying. Since much of this information must be obtained while preparing your laboratory notebook, the two subjects, advance study and notebook preparation, will be developed simultaneously.

An important part of any laboratory experience is learning to maintain very complete records of every experiment undertaken and every item of data obtained. Far too often, careless recording of data and observations has resulted in mistakes, frustration, and lost time due to needless repetition of experiments. If reports are required, you will find that proper collection and recording of data can make your report writing much easier.

Because organic reactions are seldom quantitative, special problems result. Frequently reagents have to be used in large excess to increase the amount of product. Some reagents are expensive, and therefore care in the amounts of these substances used is necessary. Very often, many more reactions take place than you desire. These extra reactions, or **side reactions,** may form other products besides the desired product. These are called **side products.** For all of these reasons, you must plan your experimental procedure carefully before undertaking the actual experiment.

THE NOTEBOOK

For recording data and observations during experiments, a BOUND NOTEBOOK SHOULD BE USED. The notebook should have consecutively numbered pages. If it does not, you should number the pages immediately. A spiral-bound notebook or any other type from which the pages can be removed easily is not acceptable, since the possibility of losing the pages is great.

All data and observations must be recorded in the notebook. Paper towels, napkins, toilet tissue, or scratch paper have a tendency to become lost or destroyed. It is bad laboratory practice to record information on such random and perishable pieces of paper. All entries in your notebook must be recorded in **permanent** ink. It can be frustrating to have important information disappear from the notebook because it was recorded in washable ink and could not survive a flood caused by the student at the next position on the bench. Because you will be using your notebook in the laboratory, it is quite likely that the book will become soiled or stained by chemicals, filled with scratched-out entries, or even slightly burned.

Your instructor may wish to check your notebook at any time, so you should always have it up to date. If your instructor requires reports, you can prepare them expeditiously from the material recorded in the laboratory notebook.

NOTEBOOK FORMAT: ADVANCE PREPARATION

Individual instructors vary greatly in the type of notebook format they prefer; such variation stems from differences in philosophies and experience. You will have to obtain specific directions in preparing a notebook from your own instructor. Certain features, however, are common to most notebook formats. The following discussion presents what might be included in a typical notebook.

You can save much time in the laboratory if you understand fully the procedure of the experiment and the theory underlying it. It will be very helpful if you know the main reactions, the potential side reactions, the mechanism, the stoichiometry, and the procedure before you come to the laboratory. Understanding the procedure by which the desired product is to be separated from undesired materials is also very important. If you have examined each of these topics before coming to class, you will be prepared to do the experiment efficiently. You will have your equipment and reagents already prepared when they are to be used. Your reference material will be at hand when you need it. Finally, with your time efficiently organized, you will be able to take advantage of long reaction or reflux periods to perform other tasks, such as doing shorter experiments or finishing previous ones.

For experiments in which a compound is synthesized from other reagents, that is, with **preparative** experiments, it is essential to know the main reaction. In order to perform stoichiometric calculations, the equation for the main reaction should be balanced. Therefore, before you begin the experiment, your notebook should contain the balanced equation for the pertinent reaction. Using the preparation of isopentyl acetate, or banana oil, (Experiment 6) as an example, you would write:

$$\underset{\text{Acetic acid}}{CH_3\overset{\overset{\displaystyle O}{\|}}{C}-OH} + \underset{\text{Isopentyl alcohol}}{CH_3\overset{\overset{\displaystyle CH_3}{|}}{C}HCH_2CH_2-OH} \xrightarrow{H^+} \underset{\text{Isopentyl acetate}}{CH_3\overset{\overset{\displaystyle O}{\|}}{C}-O-CH_2CH_2\overset{\overset{\displaystyle CH_3}{|}}{C}HCH_3} + H_2O$$

The possible side reactions that divert reagents into contaminants (side products) should, if they are known, also be entered in the notebook before beginning the experiment. These side products will have to be separated from the major product during purification. In the preparation of isopentyl acetate, there are no significant side reactions.

Physical constants such as melting points, boiling points, densities, and molecular weights should be listed in the notebook when this information is needed to perform an experiment or to do calculations. These data are located in such sources as the *Handbook of Chemistry and Physics,* the *Merck Index,* or Lange's *Handbook of Chemistry.* In many of the experiments in this textbook, some of this information is given within the experimental procedure. Physical constants required for an experiment should be written in your notebook before you come to class.

　　　Advance preparation may also include examination of some subjects, information not necessarily recorded in the notebook, that would prove useful in understanding the experiment. Included among these subjects are an understanding of the mechanism of the reaction, an examination of other methods by which the same compound might be prepared, and a detailed study of the experimental procedure. Many students find that an outline of the procedure, prepared **before** they come to class, helps them use their time more efficiently once they begin the experiment. Such an outline could very well be prepared on some loose sheet of paper, rather than in the notebook itself.

　　　Once the reaction has been completed, the desired product does not magically appear as purified material; it must be isolated from a frequently complex mixture of side products, unreacted starting materials, solvents, and catalysts. You should try to outline a **separation scheme** in your notebook for isolating the product from its contaminants. At each stage you should try to understand the reason for the particular instruction given in the experimental procedure. This not only will familiarize you with the basic separation and purification techniques used in organic chemistry but also will help you understand when to use these techniques. Such an outline might take the form of a flowchart. For example, see the separation scheme for isopentyl acetate. Careful

Separation scheme for isopentyl acetate

attention to understanding the separation may, besides familiarizing you with the procedure by which the desired product is separated from impurities in your particular experiments, also prepare you for original research, where no experimental procedure exists.

In designing a separation scheme, you should note that the scheme outlines those steps undertaken once the reaction period has been concluded. For this reason, the represented scheme did not include such steps as the addition of the reactants (isopentyl alcohol and acetic acid) and the catalyst (sulfuric acid), or the heating of the reaction mixture.

For experiments in which a compound is isolated from a particular source and not prepared from other reagents, some of the information described in this section will not be appropriate. Such experiments are called **isolation experiments.** A typical isolation experiment involves isolating a pure compound from a natural source. Some examples include the isolation of caffeine from tea or the isolation of cinnamaldehyde from cinnamon. Although isolation experiments require somewhat different advance preparation, this advance study may include looking up physical constants for the compound isolated and outlining the isolation procedure. A detailed examination of the separation scheme is very important here since this is the heart of such an experiment.

NOTEBOOK FORMAT: LABORATORY RECORDS

When you begin the actual experiment, your notebook should be kept nearby so that you will be able to record in it those operations you perform. When you are working in the laboratory, the notebook serves as a place where a rough transcript of your experimental method is recorded. Data from actual weighings, volume measurements, and determinations of physical constants are also noted. This section of your notebook should **not** be prepared in advance. The pupose here is not to write a recipe, but rather to provide a record of what you did and what you **observed.** These observations will help you to write reports without having to resort to memory. They will also help you or other workers to repeat the experiment in as nearly as possible the same way. The sample notebook pages found in this section illustrate the type of data and observations which should be written in your notebook.

When your product has been prepared and purified, or isolated if it is an isolation experiment, you should record such pertinent data as the melting point or boiling point of the substance, its density, its index of refraction, and the conditions under which spectra were determined.

NOTEBOOK FORMAT: CALCULATIONS

A chemical equation for the overall conversion of the starting materials to products is written on the assumption of simple ideal stoichiometry. Actually this assumption is seldom realized. Side reactions or competing reactions will also occur, giving other products. For some synthetic reactions, an equilibrium state will be reached in which an appreciable amount of starting material still is present and can be recovered. Some of the reactant may also remain if it is present in excess or if the reaction was incomplete. A reaction involving an expensive reagent illustrates another need for knowing how far a particular type of reaction converts reactants to products. In such a case, it is preferable to use the most efficient method for this conversion. Thus, information about the efficiency of conversion for various reactions is of interest to the person contemplating the use of these reactions.

The quantitative expression for the efficiency of a reaction is given by a calculation of the **yield** for the reaction. The **theoretical yield** is the number of grams of the product expected from the reaction on the basis of ideal stoichiometry, with side reactions, reversibility, and losses ignored. In order to calculate the theoretical yield, it is first necessary to determine the **limiting reagent.** The limiting reagent is the reagent that is not present in excess and on which the overall yield of product depends. The method for determining the limiting reagent in the isopentyl acetate experiment is illustrated in the sample notebook on pp 20–21. You should consult your general chemistry textbook for more complicated examples. The theoretical yield is then calculated from the expression,

$$\text{Theoretical yield} = (\text{moles of limiting reagent})(\text{ratio})(\text{MW of product})$$

The ratio here is the stoichiometric ratio of product to limiting reagent. In the preparation of isopentyl acetate, that ratio is 1:1. One mole of isopentyl alcohol, under ideal circumstances, should yield one mole of isopentyl acetate.

The **actual yield** is simply the number of grams of desired product obtained. The **percentage yield** describes the efficiency of the reaction and is determined by

$$\text{Percentage yield} = \frac{\text{actual yield}}{\text{theoretical yield}} \times 100$$

Calculation of the theoretical yield and percentage yield can be illustrated using hypothetical data for the isopentyl acetate preparation:

$$\text{Theoretical yield} = (6.45 \times 10^{-3} \, \cancel{\text{mol isopentyl alcohol}}) \left(\frac{1 \, \cancel{\text{mol isopentyl acetate}}}{1 \, \cancel{\text{mol isopentyl alcohol}}} \right)$$

$$\times \left(\frac{130.2 \text{ g isopentyl acetate}}{1 \, \cancel{\text{mol isopentyl acetate}}} \right) = 0.840 \text{ g isopentyl acetate}$$

$$\text{Actual yield} = 0.354 \text{ g isopentyl acetate}$$

$$\text{Percentage yield} = \frac{0.354 \text{ g}}{0.840 \text{ g}} \times 100 = 42.1\%$$

For experiments that have the principal objective of isolating a substance such as a natural product, rather than preparing and purifying some reaction product, the **weight percentage recovery** and not the percentage yield, is calculated. This value is determined by

$$\text{Weight percentage recovery} = \frac{\text{weight of substance isolated}}{\text{weight of original material}} \times 100$$

THE PREPARATION OF ISOPENTYL ACETATE (BANANA OIL)

MAIN REACTION

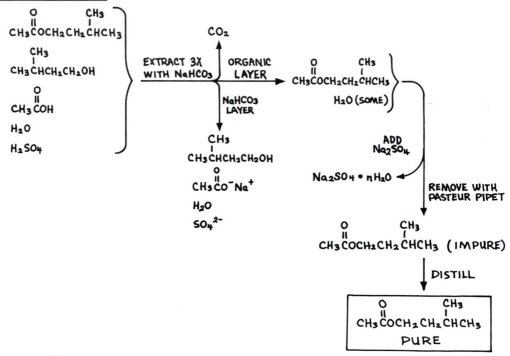

$$CH_3COH \quad + \quad CH_3CHCH_2CH_2OH \xrightarrow{H^+} CH_3COCH_2CH_2CHCH_3 + H_2O$$

ACETIC ACID   ISOPENTYL ALCOHOL   ISOPENTYL ACETATE

TABLE OF PHYSICAL CONSTANTS

	MW	BP	DENSITY
ISOPENTYL ALCOHOL	88.2	132°C	0.813 g/mL
ACETIC ACID	60.1	118	1.06
ISOPENTYL ACETATE	130.2	142	0.876

SEPARATION SCHEME

DATA AND OBSERVATIONS

0.70 mL OF ISOPENTYL ACETATE WAS ADDED TO A PRE WEIGHED 5-mL CONICAL VIAL:

VIAL + ALCOHOL	25.524 g
VIAL	24.955 g
	0.569 g ISOPENTYL ALCOHOL

ACETIC ACID (1.4 mL) AND THREE DROPS OF CONCENTRATED H_2SO_4 (USING A PASTEUR PIPET) WERE ALSO ADDED TO THE CONICAL VIAL ALONG WITH A SMALL BOILING STONE. A WATER-COOLED CONDENSER TOPPED WITH A DRYING TUBE CONTAINING A LOOSE PLUG OF GLASS WOOL WAS ATTACHED TO THE VIAL. THE REACTION MIXTURE WAS REFLUXED IN A SAND BATH (ABOUT 155°) FOR 75 MIN. AND THEN COOLED TO ROOM TEMPERATURE. THE COLOR OF THE REACTION MIXTURE WAS BROWNISH-YELLOW.

A sample notebook, page 1

THE BOILING STONE WAS REMOVED AND THE REACTION MIXTURE WAS EXTRACTED THREE TIMES WITH 1.0 mL OF 5% $NaHCO_3$. THE BOTTOM AQUEOUS LAYER WAS REMOVED AND DISCARDED AFTER EACH EXTRACTION. DURING THE FIRST TWO EXTRACTIONS, MUCH CO_2 GAS WAS GIVEN OFF. THE ORGANIC LAYER WAS A LIGHT YELLOW COLOR. IT WAS TRANSFERRED TO A DRY CONICAL VIAL, AND TWO FULL MICROSPATULAS OF ANHYDROUS Na_2SO_4 WERE ADDED TO DRY THE CRUDE PRODUCT. IT WAS ALLOWED TO SET WITH OCCASIONAL STIRRING FOR 10 MINS.

THE DRY PRODUCT WAS TRANSFERRED TO A 3-mL CONICAL VIAL AND A BOILING STONE WAS ADDED. A DISTILLATION APPARATUS USING A HICKMAN STILL, A WATER-COOLED CONDENSER, AND A DRYING TUBE PACKED WITH $CaCl_2$ WAS ASSEMBLED. THE SAMPLE WAS HEATED IN A SAND BATH (COVERED WITH ALUMINUM FOIL) AT ABOUT 180°C. THE LIQUID BEGAN BOILING AFTER ABOUT FIVE MINS, BUT NO DISTILLATE APPEARED IN THE HICKMAN STILL UNTIL ABOUT 20 MINS. LATER. ONCE THE PRODUCT BEGAN COLLECTING IN THE HICKMAN STILL, THE DISTILLATION REQUIRED ONLY ABOUT TWO MINS. TO COMPLETE. ABOUT 1-2 DROPS REMAINED IN THE DISTILLING VIAL. THE ISOPENTYL ACETATE WAS TRANSFERRED TO A PRE-WEIGHED 3-mL CONICAL VIAL.

$$\begin{array}{rl} \text{VIAL + PRODUCT} & 20.428g \\ \text{VIAL} & \underline{20.074g} \\ & 0.354g \quad \text{ISOPENTYL ACETATE} \end{array}$$

THE PRODUCT WAS COLORLESS AND CLEAR. BP (MICRO TECHNIQUE): 140°C. THE IR SPECTRUM WAS OBTAINED.

<u>CALCULATIONS</u>

DETERMINE LIMITING REAGENT:

ISOPENTYL ALCOHOL $0.569g \left(\dfrac{1 \text{ MOL ISOPENTYL ALCOHOL}}{88.2g} \right) = 6.45 \times 10^{-3}$ MOL

ACETIC ACID $1.40 \text{ mL} \left(\dfrac{1.06g}{\text{mL}} \right) \left(\dfrac{1 \text{ MOL ACETIC ACID}}{60.1g} \right) = 2.47 \times 10^{-2}$ MOL

SINCE THEY REACT IN A 1:1 RATIO, ISOPENTYL ALCOHOL IS THE LIMITING REAGENT.

THEORETICAL YIELD =

6.45×10^{-3} MOL ISOPENTYL ALCOHOL $\left(\dfrac{1 \text{ MOL ISOPENTYL ACETATE}}{1 \text{ MOL ISOPENTYL ALCOHOL}} \right) \left(\dfrac{130.2g \text{ ISOPENTYL ACETATE}}{1 \text{ MOL ISOPENTYL ACETATE}} \right)$

$= 0.840g$ ISOPENTYL ACETATE

PERCENTAGE YIELD $= \dfrac{0.354g}{0.840g} \times 100 = 42.1\%$

A sample notebook, page 2

Thus, for instance, if 0.014 g of caffeine were obtained from 2.3 g of tea, the weight percentage recovery of caffeine would be

$$\text{Weight percentage recovery} = \frac{0.014 \text{ g caffeine}}{2.3 \text{ g tea}} \times 100 = 0.61\%$$

LABORATORY REPORTS

Various formats for reporting the results of the laboratory experiments may be used. The report may be written directly into your notebook in a format similar to the sample notebook pages included in this section. Alternatively, your instructor may require a more formal report which is written separately from your notebook. When original research is performed, these reports should include a detailed description of all the experimental steps undertaken. Frequently the style used in scientific periodicals such as **Journal of the American Chemical Society** is applied to writing laboratory reports. Your instructor is likely to have his or her own requirements for laboratory reports and should detail the requirements to you.

SUBMISSION OF SAMPLES

In all preparative experiments, and in some of the isolation experiments, you will be required to submit the sample of the substance you prepared or isolated to your instructor. How this sample is labeled is very important. Again, learning a correct method of labeling bottles and vials can save time in the laboratory, because fewer mistakes will be made. More importantly, learning to label properly can decrease the danger inherent in having samples of material which cannot be identified correctly at a later date.

Solid materials should be stored and submitted in containers that permit the substance to be removed easily. For this reason, narrow-mouthed bottles or vials are not used for solid substances. Liquids should be stored in containers that will not let them escape through leakage. You should be careful not to store volatile liquids in containers that have plastic caps, unless the cap is lined with an inert material such as Teflon. Otherwise, the vapors from the liquid are likely to come in contact with the plastic and dissolve some of it, thus contaminating the substance being stored.

On the label, print the name of the substance, its melting or boiling point, the actual and percentage yields, and your name. An illustration of a properly prepared label follows:

> **Isopentyl Acetate**
> **BP 140 °C**
> **Yield 0.354 g (42.1%)**
> **Joe Schmedlock**

LABORATORY GLASSWARE

Since your glassware is expensive and you are responsible for it, you will want to give it proper care and respect. If you read this section carefully and follow the procedures presented here, you may be able to avoid some unnecessary expense. You may also save time since cleaning problems and replacing broken glassware are time-consuming.

For those of you who are unfamiliar with the equipment found in an organic laboratory, or who are uncertain about how such equipment should be treated, this section will provide some useful information. Topics such as cleaning glassware, caring for glassware when using corrosive or caustic reagents, and assembling components from your organic laboratory kit are included. At the end of this section are illustrations and names of most of the equipment you are likely to find in your drawer or locker.

CLEANING GLASSWARE

Glassware can be cleaned easily if you clean it immediately. It is good practice to do your ''dishwashing'' right away. With time, the organic tarry materials left in a container begin to attack the surface of the glass. The longer you wait to clean glassware, the more extensively this interaction will have progressed. Cleaning is then more difficult, because water will no longer wet the surface of the glass as effectively. If you are not able to wash your glassware immediately after use, you should soak the dirty pieces in soapy water. A ½ gallon plastic container provides a convenient vessel in which to soak and wash your glassware. The use of a plastic container also helps to prevent the loss of small pieces of equipment used in microscale techniques.

Various soaps and detergents are available for washing glassware. They should be tried first when washing dirty glassware. Organic solvents can also be used, since the residue remaining in dirty glassware is likely to be soluble in some organic solvent. After the solvent has been used, the conical vial or flask probably will have to be washed with soap and water to remove the residual solvent. When you use solvents in cleaning glassware, use caution since the solvents are hazardous (see section entitled Laboratory Safety). You should try to use fairly small amounts of a solvent for cleaning purposes. Usually 1–2 mL will be sufficient. Acetone is commonly used, but it is expensive. Your **wash acetone** can be used effectively several times before it is ''spent.'' Once your acetone is ''spent,'' dispose of it as directed by your instructor.

Acetone is very flammable. Do not use it around flames.

If acetone does not work, other organic solvents such as methylene chloride or toluene can be used in the same way as acetone.

For troublesome stains and residues that insist on adhering to the glass in spite of your best efforts, a mixture of sulfuric acid and nitric acid can be used. Cautiously add about 20 drops of concentrated sulfuric acid and five drops of concentrated nitric acid to the flask or vial.

Safety glasses must be worn when you are using this cleaning solution. Do not allow the solution to come into contact with your skin or your clothing. It will cause severe burns and create holes in your clothing. It is also possible that the acids will react with the residue in the container.

Swirl the acid mixture in the container for a few minutes. If necessary, place the glassware in a warm sand bath and heat cautiously to accelerate the cleaning process. Continue heating until any sign of a reaction ceases. When the cleaning procedure is completed, decant the mixture into an appropriate waste container.

Do not pour the acid solution into a waste container that is intended for organic wastes.

Rinse the piece of glassware thoroughly with water and then wash with soap and water. For most common organic chemistry applications, any stains that survive this treatment are not likely to cause difficulty in subsequent laboratory procedures.

If the glassware is contaminated with stopcock grease (unlikely with the glassware recommended in this book), rinse the glassware with a small amount (1–2 mL) of methylene chloride. Discard the rinse solution into an appropriate waste container. Once the grease is removed, wash the glassware with soap or detergent and water.

DRYING GLASSWARE

The easiest way to dry glassware is to allow it to stand overnight. Conical vials, flasks, and beakers should be stored upside down on a piece of paper towel to permit the water to drain from them. Drying ovens can be used to dry glassware if they are available, and if they are not being used for other purposes. Rapid drying can be achieved by rinsing the glassware with acetone and air-drying it or placing it in an oven. First, the glassware is thoroughly drained of water. Then it is rinsed with one or two **small** portions (1–2 mL) of acetone. You should not use any more acetone than is suggested here. The used acetone is returned to a waste acetone container for recycling. After you rinse the glassware with acetone, dry it by placing it in a drying oven for a few minutes or allow it to air-dry at room temperature. The acetone can also be removed by aspirator suction. In some laboratories, it may be possible to dry the glassware by blowing a **gentle** stream of dry air into the container. (Your laboratory instructor will indicate if you should do this.) Before drying the glassware with air, you should make sure that the air line is not filled with oil. Otherwise the oil will be blown into the container, and you will have to clean it over again. It is not necessary to blast the acetone out of the glassware with a wide-open stream of air; a gentle stream of air is just as effective and will not startle other people in the room.

You should not dry your glassware with a paper towel unless the towel is lint-free. Most paper will leave lint on the glass that can interfere with subsequent procedures performed in the equipment. Sometimes it is not necessary to dry a piece of equipment thoroughly. For example, if you are going to place water or an aqueous solution in a container, it does not need to be completely dry.

GROUND-GLASS JOINTS

It is likely that the glassware in your organic kit has **standard-taper ground-glass joints.** For example, the air condenser in the figure consists of an inner (male) ground-glass joint at the bottom and an outer (female) joint at the top. Each end is ground to a precise size which is designated by the symbol ⧲ followed by two numbers. A common joint size in microscale glassware is ⧲ 14/10. The first number indicates the diameter (in mm) of the joint at its widest point, and the second number refers to its length (see figure). One advantage of standard-taper joints is that the pieces fit together snugly and form a good seal. In addition, standard-taper joints allow all glassware components with the same joint size to be connected together, thus permitting the assembly of a wide variety of apparatus. One disadvantage of glassware with ground-glass joints, however, is that it is very expensive.

Some pieces of glassware with ground-glass joints also have threads cast into the outside surface of the outer joints (see top of air condenser in figure). The threaded joint allows the use of a plastic screw cap with a hole in the top to fasten two pieces of glassware together securely. The plastic cap is slipped over the inner joint of the upper piece of glassware, followed by a rubber O-ring (see figure on page 26). The O-ring should be pushed down so that it fits snugly on top of the ground-glass joint. The inner ground-glass joint is then fitted into the outer joint of the bottom piece of glassware. The screw cap is tightened without excessive force to attach the entire apparatus firmly together. The O-ring provides an additional seal that makes this joint air-tight. With this connecting system, it is unnecessary to use any type of grease to seal the joint. The

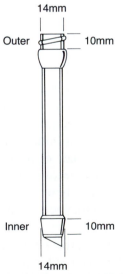

Illustration of ⧲ 14/10 inner and outer joints showing dimensions

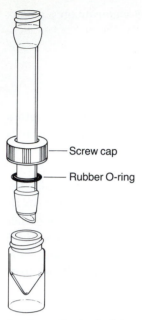

Screw cap

Rubber O-ring

A microscale standard-taper joint assembly

O-ring **must be used** to obtain a good seal and to lessen the chances of breaking the glassware when you tighten the plastic cap.

It is important to make sure no solid or liquid is on the joint surfaces. Such material will lessen the efficiency of the seal, and the joints may leak. The presence of solid particles could cause the ground-glass joints to break when the plastic cap is tightened. Also, if the apparatus is to be heated, material caught between the joint surfaces will increase the tendency for the joints to stick. If the joint surfaces are coated with liquid or adhering solid, they should be wiped with a cloth or lint-free paper towel before assembling.

SEPARATING GROUND-GLASS JOINTS

The most important thing you can do to prevent ground-glass joints from becoming "frozen" or stuck together is to disassemble the glassware as soon as possible after a procedure is completed. Even when this precaution is followed, ground-glass joints may become stuck tightly together. The same is true of glass stoppers in bottles or conical vials. Since microscale glassware is small and very fragile, it is relatively easy to break a piece of glassware when trying to pull two pieces apart. If the pieces do not separate easily, you must be careful when you try to pull them apart. The best way is to hold the two pieces, with both hands touching, as close as possible to the joint. With a firm grasp, try to loosen the joint with a slight twisting motion (do not twist very hard). If this does not work, try to pull your hands apart without pushing sideways on the glassware.

If it is not possible to pull the pieces apart, the following methods may help. A frozen joint can sometimes be loosened if you tap it **gently** with the wooden handle of a spatula. Then try to pull it apart as described above. If this procedure fails, you may try heating the joint in hot water or a steam bath. If this heating fails, the instructor may be able to advise you. As a last resort, you may try heating the joint in a flame. You should not try this unless the apparatus is hopelessly stuck, because heating by flame often causes the joint to expand rapidly and crack or break. If you use a flame, make sure the joint is clean and dry. Heat the outer part of the joint slowly, in the yellow portion of a low flame, until it expands and breaks away from the inner section. The joint should be heated very slowly and carefully, or it may break.

ETCHING GLASSWARE

Glassware that has been used for reactions involving strong bases such as sodium hydroxide or sodium alkoxides must be cleaned thoroughly **immediately** after use. If these caustic materials are allowed to remain in contact with the glass, they will etch the glass permanently. The etching makes later cleaning more difficult, since dirt particles may become trapped within the microscopic surface irregularities of the etched glass. Furthermore, the glass is weakened, so the lifetime of the glassware is decreased. If caustic materials are allowed to come into contact with ground-glass joints without being removed promptly, the joints will become fused or "frozen." It is extremely difficult to separate fused joints without breaking them.

ASSEMBLING THE APPARATUS

Care must be taken when assembling the glass components into the desired apparatus. You should always remember that Newtonian physics applies to chemical apparatus, and unsecured pieces of glassware are certain to respond to gravity. You should always clamp the glassware securely to a ring stand. Throughout this textbook, the illustrations of the various glassware arrangements include the clamps that attach the apparatus to a ring stand. You should assemble your apparatus using the clamps as shown in the illustrations.

CAPPING CONICAL VIALS OR OPENINGS

The plastic screwcaps used to join two pieces of glassware together can also be used to cap conical vials (see figure) or other openings. A Teflon insert, or liner, fits inside the cap to cover the hole when the cap is used to seal a vial. Only one side of the liner is coated with Teflon. This side should always face towards the inside of the vial. (Note that the O-ring is not used when the cap is used to seal a vial.) To seal a vial, it is necessary to tighten the cap firmly, but not too tight. It is possible to crack the vial if you apply too much force. Some Teflon liners have a soft backing material (silicone rubber) that allows the liner to compress slightly when the cap is screwed down. It is easier to cap a vial securely with these liners without breaking the vial than with liners which have a harder backing material.

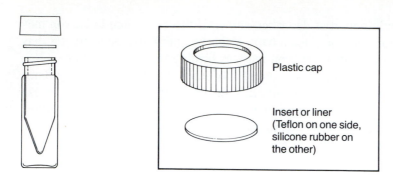

ATTACHING RUBBER TUBING TO EQUIPMENT

When you are attaching rubber tubing to the glass apparatus or when you are inserting glass tubing into rubber stoppers, it helps to lubricate the rubber tubing or the rubber stopper with either water or glycerin beforehand. Without such lubrication, it can be difficult to attach rubber tubing to the sidearms of items of glassware such as condensers and filter flasks. Furthermore, glass tubing may break when it is being inserted into rubber stoppers. Water is a good lubricant for most purposes. Water should not be used as a lubricant when it might contaminate the reaction. Glycerin is a better lubricant than water and should be used when there is considerable friction between the glass and rubber. If glycerin is the lubricant, be careful not to use too much.

DESCRIPTION OF EQUIPMENT

The components of the organic kit recommended for use in this textbook are given in the figure. Notice that most of the joints in these pieces of glassware are ⊺ 14/10 and all the outer joints are threaded. The organic kits used in your laboratory may have different joint sizes or some of the outer joints may not be threaded. In particular, many organic kits contain a number of pieces of glassware with ⊺ 7/10 joints. These kits will work as well with the experiments in this book as the glassware recommended in the figure. In addition, there are microscale kits containing glassware which is connected together without the use of ground-glass joints. The experiments in this book can also be performed with these glassware kits. Modifications with organic kits not containing the recommended glassware are discussed in the Technique chapters and in some of the experiments.

The last two figures include glassware and equipment which are commonly used in the organic laboratory. Your glassware and equipment may vary slightly from the pieces shown here. Another useful item not included in these figures is a sheet of glass or Teflon, approximately 14 inches square. This provides a clean work area on which procedures are performed. This is particularly important when working with the small amounts of materials characteristic of microscale laboratory experiments.

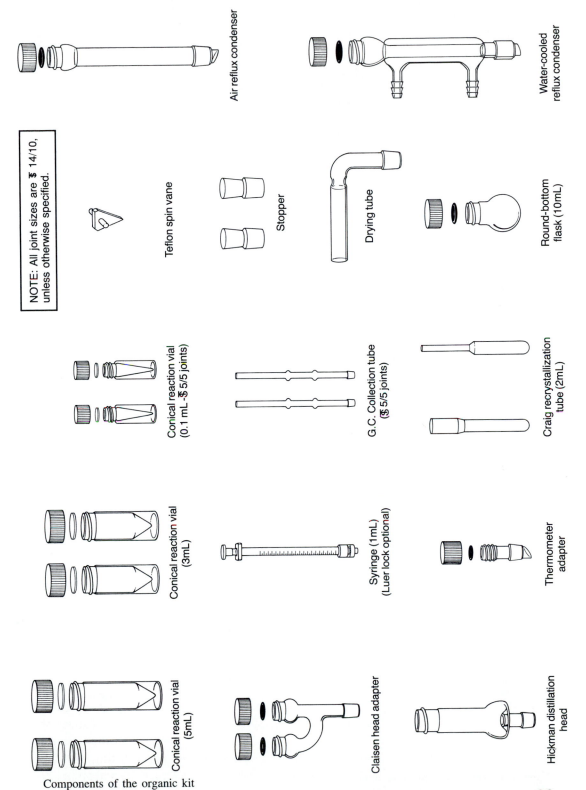

Air reflux condenser

Water-cooled reflux condenser

NOTE: All joint sizes are $ 14/10, unless otherwise specified.

Teflon spin vane

Stopper

Drying tube

Round-bottom flask (10mL)

Conical reaction vial (0.1 mL–$ 5/5 joints)

G. C. Collection tube ($ 5/5 joints)

Craig recrystallization tube (2mL)

Conical reaction vial (3mL)

Syringe (1mL) (Luer lock optional)

Thermometer adapter

Conical reaction vial (5mL)

Claisen head adapter

Hickman distillation head

Components of the organic kit

29

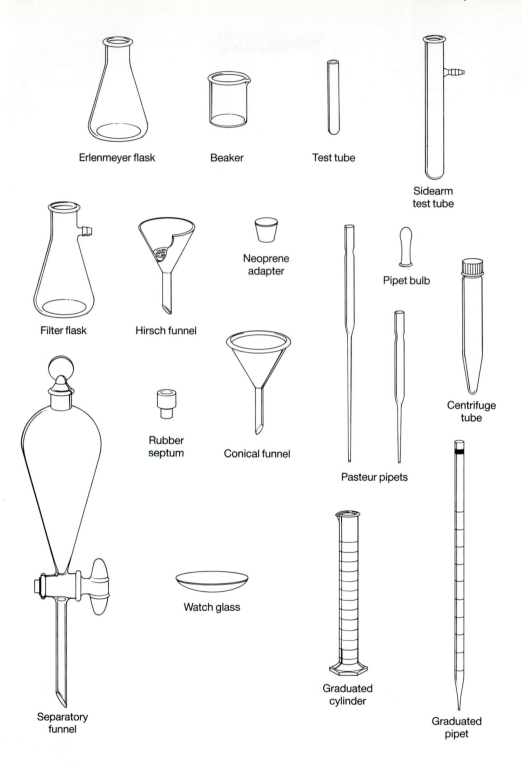

Erlenmeyer flask

Beaker

Test tube

Sidearm
test tube

Filter flask

Hirsch funnel

Neoprene
adapter

Pipet bulb

Centrifuge
tube

Rubber
septum

Conical funnel

Pasteur pipets

Separatory
funnel

Watch glass

Graduated
cylinder

Graduated
pipet

Equipment commonly used in the organic laboratory

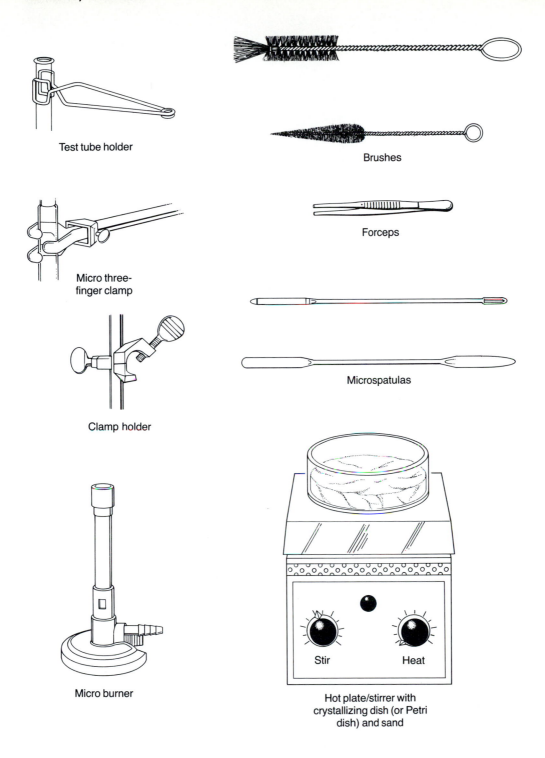

Test tube holder

Brushes

Micro three-finger clamp

Forceps

Clamp holder

Microspatulas

Micro burner

Hot plate/stirrer with crystallizing dish (or Petri dish) and sand

Equipment commonly used in the organic laboratory

Part One

Introduction to Microscale Methods

Essay
INTRODUCTION TO MICROSCALE LABORATORY

This textbook teaches students the important laboratory techniques of organic chemistry and illustrates many important reactions and concepts. In the traditional approach to teaching this subject, the quantities of chemicals used were on the order of 5 to 100 grams, and glassware was designed to contain up to 500 mL of liquids. This scale of experiment we might call a **macroscale** experiment. The approach used here, a **microscale** approach, differs from the traditional laboratory course in that nearly all of the experiments use very small amounts of chemicals. Quantities of chemicals used range from about 5 to 500 **milli**grams (0.005 to 0.500 grams), and glassware is designed to contain less than 10 mL of liquids. The advantages include improved safety in the laboratory, reduced risk of fire and explosion, and reduced exposure to hazardous vapors. This approach decreases the need for hazardous waste disposal, leading to reduced contamination of the environment. You will learn to work with the same level of care and neatness that has previously been confined to courses in analytical chemistry.

This essay introduces the equipment and shows how to construct some of the apparatus needed to carry out the first few experiments. Detailed discussion of how to assemble apparatus and how to practice the important techniques is found in Part Five (The Techniques) of this textbook. This essay provides only a brief introduction, sufficient to allow you to begin working. You will need to read the Techniques chapters for more complete discussions.

Microscale organic experiments require you to develop careful laboratory techniques and to become familiar with apparatuses which is somewhat unusual when compared with traditional glassware. We strongly recommend that each student do Laboratory Exercises 1 through 5 contained within this essay. These exercises will acquaint you with the most basic microscale techniques. To provide a strong foundation, we recommend that each student complete Experiments 1 through 9 in Part One of this textbook before attempting any other experiments in the textbook.

> **READ:** "Forword to the Student and Words of Advice," pp 1–3
> "Laboratory Safety," pp 4–12

A. Heating Baths

Sand Bath. The most convenient means of heating chemical reactions on a small scale is to use a **sand bath.** The sand bath consists of a Petri dish or a small

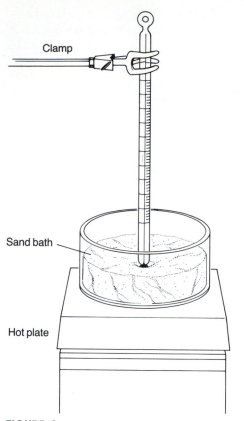

Clamp

Sand bath

Hot plate

FIGURE 1. Sand bath, with hot plate and thermometer

crystallizing dish that has been filled to a depth of about one centimeter with sand. The sand bath is heated by placing it on a hot plate. The temperature of the sand bath may be monitored by clamping a thermometer in position so that the bulb of the thermometer is buried in the sand. A sand bath, with thermometer, is shown in Figure 1.

A common problem with sand baths is in knowing which setting to use with the hot plate in order to arrive at a desired temperature. It is good practice, therefore, to calibrate the sand bath before using it. Calibration will give an approximate idea of how hot plate settings correspond to sand bath temperatures. Provided that you always use the same sand bath and hot plate, your calibration will be a very useful guide in doing your experiments.

Laboratory Exercise 1

> **NOTE:** This exercise involves some lengthy heating periods. You can move on to other laboratory exercises in this essay during these heating periods.

Prepare a sand bath by pouring sand into a small Petri or crystallizing dish until the depth of the sand reaches about one centimeter. Place the sand bath on a hot plate, and clamp a thermometer into a vertical position over the sand bath so that the bulb of the thermometer is buried in the sand near the center of the sand bath.

Select five equally spaced temperature settings, including the lowest and the highest settings, on the heating control of the hot plate. Set the dial to the first of these settings and monitor the temperature recorded on the thermometer. When the thermometer reading arrives at a constant value, record this final temperature, along with the dial setting.

Repeat this procedure with the remaining four settings. For each trial, record the dial setting and the final temperature. Pay special attention when the hot plate is on its highest setting. If the temperature rises above 250 °C, discontinue the calibration. Prepare a graph of the data by plotting the final temperature on the vertical scale of a piece of graph paper and the dial setting on the horizontal scale. Draw a calibration line through the points on the graph. Place this calibration curve in your notebook to serve as a reference for future experiments.

Water Bath. When precise control at lower temperatures (below about 80 °C) is desired, a suitable alternative is to prepare a **water bath.** The water bath consists of a beaker filled to the required depth with water. The hot plate is used to heat the water bath to the desired temperature. The water in the water bath can evaporate during heating. It is useful to cover the top of the beaker with aluminum foil to diminish this problem.

B. Conical Reaction Vials

One of the most versatile pieces of glassware contained in the microscale organic glassware kit is the **conical reaction vial.** This vial is used as a vessel in which organic reactions are performed. It may serve as a storage container. It is also used for extractions (see Technique 7). A reaction vial is shown in Figure 2.

The flat base of the vial allows it to stand upright on the laboratory bench. The interior of the vial tapers to a narrow bottom. This shape makes it possible to withdraw liquids completely from the vial, using a disposable Pasteur pipet. The vial has a screw

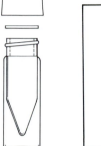
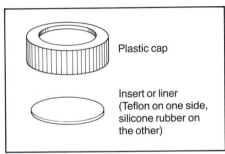

Plastic cap

Insert or liner
(Teflon on one side,
silicone rubber on
the other)

FIGURE 2. A conical reaction vial. (The inset shows an expanded view of the cap, with its Teflon insert.)

cap, which tightens by means of threads cast into the top of the vial. The top also has a ground-glass inner surface. This ground-glass joint allows one to assemble components of glassware tightly.

The plastic cap which fits the top of the conical vial has a hole in the top. This hole is large enough to permit the cap to fit over the inner joints of other components of the glassware kit (see below). A Teflon insert, or liner, fits inside the cap to cover the hole when the cap is used to seal a vial tightly. Notice that only one side of the liner is coated with Teflon. This side should always face toward the inside of the vial. An O-ring fits inside the cap when the cap is used to fasten pieces of glassware together. The cap and its Teflon insert are shown in the expanded view in Figure 2.

Do not use the O-ring when the cap is used to seal the vial.

The components of the glassware kit can be assembled into one unit that holds together firmly and clamps easily to a ring stand. The cap from the conical vial is slipped over the inner (male) joint of the upper piece of glassware, and a rubber O-ring is also fitted over the inner joint. The apparatus is then assembled by fitting the inner ground-glass joint into the outer (female) joint of the reaction vial. The screw cap is then tightened to attach the entire apparatus firmly together. The assembly is illustrated in Figure 3.

The walls of the conical vials are made of thick glass. Heat does not transfer through these walls very quickly. This means that if the vial is subjected to rapid

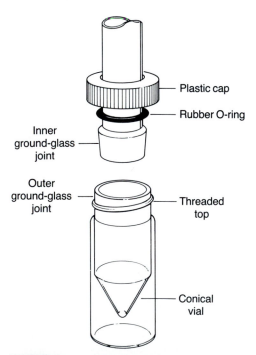

FIGURE 3.   Assembling glassware components

changes in temperature, strain set up within the glass walls of the vial may cause the glass to crack. For this reason, do not attempt to cool these vials quickly by running cold water on them. It is safer to allow them to cool naturally by allowing them to stand.

While the conical vials have flat bottoms, intended to allow them to stand up on the laboratory bench, this does not prevent them from falling over.

**It is good practice to store the vials standing upright
inside of small beakers.**

The vials are somewhat top-heavy, and it is very easy to upset them. The beaker will prevent the vial from falling over onto its side.

C. Measurement of Solids

Weighing substances to the nearest milligram requires that the weighings be done on a sensitive top-loading balance or an analytical balance.

You must not weigh chemicals directly on balance pans.

Many chemicals can react with the metal surface of the balance pan and thus ruin it. All weighings must be made into a container which has been weighed previously **(tared).** This tare weight is subtracted from the total weight of container plus sample to give the weight of sample. Some balances have a built-in compensating feature which allows the user to subtract the tare weight of the container automatically, thus giving the weight of the sample directly. A top-loading and an analytical balance are shown in Figure 4.

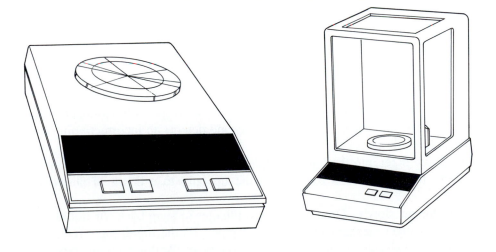

A. Top-loading balance   B. Analytical balance

FIGURE 4.   Laboratory balances

Balances of this type are quite sensitive and expensive. Care must be taken not to spill chemicals on the balance. It is also important to make certain that any spilled materials are cleaned up immediately.

D. Measurement of Liquids

In microscale experiments, liquid samples are measured using a pipet. When small quantities are used, graduated cylinders do not provide the accuracy needed to give good results. There are two common methods of delivering known amounts of liquid samples, **automatic pipets** and **graduated pipets.** When accurate quantities of liquid reagents are required, the best technique is to deliver the desired amount of liquid reagent from the pipet into a container whose tare weight has been determined previously. The container, with sample, is then weighed a second time in order to obtain a precise value for the amount of reagent.

Automatic Pipets. Automatic pipets may vary in design, according to the manufacturer. The following description, however, should apply in a general sense to most models. The automatic pipet consists of a handle which contains a spring-loaded plunger and a micrometer dial. The dial controls the travel of the plunger and is the means used to select the amount of liquid which the pipet is intended to dispense. Automatic pipets are designed to deliver liquids within a particular range of volumes. For example, a pipet may be designed to cover the range from 10 to 100 μL (0.010 to 0.100 mL) or from 100 to 1000 μL (0.100 to 1.000 mL).

Automatic pipets must never be dipped directly into the liquid sample without a plastic tip. The pipet is designed so that the liquid is drawn only into the tip. The liquids are never allowed to come in contact with the internal parts of the pipet. The plunger has two detent, or ''stop,'' positions used to control the filling and dispensing steps. Most automatic pipets have a stiffer spring that controls the movement of the plunger from the first to the second detent position. The user will find a greater resistance as the plunger is pressed past the first detent.

To use the automatic pipet, follow the steps outlined below. These steps are also illustrated in Figure 5.

1. Select the desired volume by adjusting the micrometer control on the pipet handle.
2. Place a plastic tip on the pipet. Be certain that the tip is attached securely.
3. Push the plunger down to the first detent position. Do not press the plunger to the second position. If the plunger is pressed to the second detent, an incorrect volume of liquid will be delivered.
4. Dip the tip of the pipet into the liquid sample. Do not immerse the entire length of the plastic tip in the liquid. It is best to dip the tip only to a depth of about one centimeter.
5. Release the plunger **slowly.** Do not allow the plunger to snap back, or liquid may splash up into the plunger mechanism and ruin the pipet. Furthermore, rapid release of the plunger may cause air bubbles to be drawn into the pipet. At this point the pipet has been filled.
6. Move the pipet to the receiving vessel. Touch the tip of the pipet to an interior wall of the container.

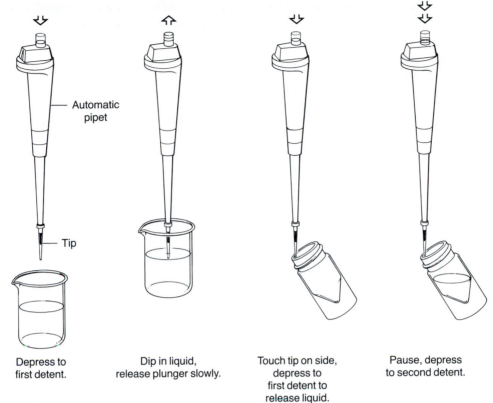

FIGURE 5.   Use of an automatic pipet

7. Slowly push the plunger down to the first detent. This action will dispense the liquid into the container.

8. Pause one or two seconds and then depress the plunger to its second detent position to expel the last drop of liquid. The action of the plunger may be stiffer in this range than it was up to the first detent.

9. Withdraw the pipet from the receiver. If the pipet is to be used with a different liquid, remove the pipet tip and discard it.

Automatic pipets are designed to deliver aqueous solutions with an accuracy of within a few percent. The amount of liquid actually dispensed varies, however, depending upon the viscosity, surface tension, and vapor pressure of the liquid. The typical automatic pipet is very accurate with aqueous solutions, but it is not always as accurate with other liquids.

Laboratory Exercise 2

Accurately weigh a 3-mL conical reaction vial, with cap and Teflon insert, on a balance. Determine its weight to the nearest milligram (nearest 0.001 g). Using the automatic pipet, dispense 0.500 mL of water into the vial, replace the cap assembly, and weigh the vial a second time. Determine the weight of water dispensed. Calculate the density of water from your results. Repeat the experiment using 0.500 mL of methylene chloride (dichloromethane). Dispose of any excess methylene chloride in a designated waste container. Calculate the density of methylene chloride from your data. Record your results in your notebook, along with your comments on any deviations from literature values which you may have noticed. At room temperature, the density of water is 0.997 g/mL, and the density of methylene chloride is 1.33 g/mL.

Graduated Pipets. A less-expensive means of delivering known quantities of liquids is to use a graduated pipet. Graduated pipets should be familiar to students who have taken general chemistry or quantitative analysis courses. Since they are made of glass, they are inert to most organic solvents and reagents. Disposable serological pipets may be an attractive alternative to standard graduated pipets.

Liquids should never be drawn into the pipets using mouth suction. A pipet bulb or a pipet pump, not a rubber dropper bulb, must be used to fill pipets. We recommend the use of a pipet pump. A pipet fits snugly into the pipet pump, and the pump can be controlled to deliver precise volumes of liquids. Control of the pipet pump is accomplished by rotating a knob on the pump. Suction created when the knob is turned draws the liquid into the pipet. Liquid is expelled from the pipet by turning the knob in the opposite direction. The pump will work satisfactorily with organic, as well as aqueous, solutions.

An alternative, and less expensive, approach is to use a rubber pipet bulb. Use of the pipet bulb is made more convenient by inserting a plastic automatic pipet tip into a rubber pipet bulb.[1] The tapered end of the pipet tip fits snugly into the end of a pipet. Drawing the liquid into the pipet is made easy, and it is also convenient to remove the pipet bulb and place a finger over the pipet opening to control the flow of liquid.

The calibrations printed on graduated pipets are reasonably accurate, but you should practice using the pipets in order to achieve this accuracy. When accurate quantities of liquids are required, the best technique is to weigh the reagent which has been delivered from the pipet. The use of a graduated pipet with a pipet pump is shown in Figure 6.

The following description, along with Figure 6, illustrates how to use a graduated pipet. Insert the end of the pipet firmly into the pipet pump. Rotate the knob of the pipet pump in the correct direction (counter-clockwise or up) to fill the pipet. Fill the pipet to a point just above the uppermost mark, and then reverse the direction of rotation of the knob to allow the liquid to drain from the pipet until the meniscus is adjusted to the 0.00-mL mark. Move the pipet to the receiving vessel. Rotate the knob of the pipet pump (clockwise or down) to force the liquid from the pipet. Allow the liquid to drain from the pipet until the meniscus arrives at the mark corresponding to the volume which you wish to dispense. Remove the pipet and drain the remaining liquid into a waste receiver. Avoid transferring the entire contents of the pipet when measuring volumes with a pipet. Remember that in order to achieve the greatest possible accuracy with this method, you should deliver volumes as a **difference** between two marked calibrations.

---

[1] This technique has been described in: Deckey, G. ''A Versatile and Inexpensive Pipet Bulb.'' *Journal of Chemical Education, 57* (July 1980): 526.

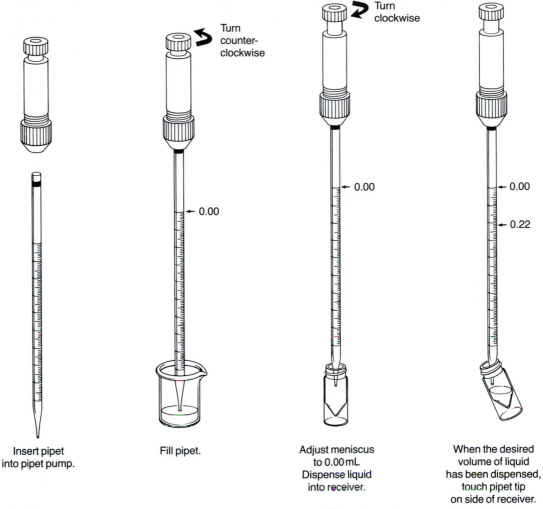

Turn counter-clockwise

Turn clockwise

← 0.00

← 0.00

← 0.00

← 0.22

| Insert pipet into pipet pump. | Fill pipet. | Adjust meniscus to 0.00 mL
Dispense liquid into receiver. | When the desired volume of liquid has been dispensed, touch pipet tip on side of receiver. |

FIGURE 6. Use of a graduated pipet. (The figure shows, as an illustration, the technique required to deliver a volume of 0.22 mL from a 1.00-mL pipet.)

Laboratory Exercise 3

Accurately weigh a 3-mL conical reaction vial, with cap and Teflon insert, on a balance. Determine its weight to the nearest milligram (0.001 g). Using a 1.0-mL graduated pipet, dispense 0.50 mL of water into the vial, replace the cap assembly, and weigh the vial a second time. Determine the weight of water dispensed. Calculate the density of water from your results. Repeat the experiment using 0.50 mL of methylene chloride (dichloromethane). Dispose of any excess methylene chloride in a designated waste container. Calculate the density of methylene chloride from your data. Record your results in your notebook, along with your comments on any deviations from literature values which you may have noticed. At room temperature, the density of water is 0.997 g/mL, and the density of methylene chloride is 1.33 g/mL.

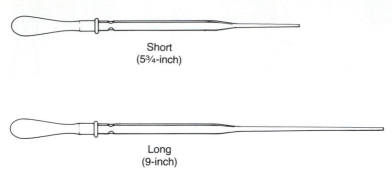

Short
(5¾-inch)

Long
(9-inch)

FIGURE 7.   Disposable Pasteur pipets

Disposable (Pasteur) Pipets.   A very convenient way of dispensing liquids when a great deal of accuracy is not required is to use a **disposable pipet,** or **Pasteur pipet.** Two sizes of Pasteur pipets are shown in Figure 7. Even though accurate calibration may not be required when these pipets are used, it is nevertheless handy to have some idea of the volume contained in the pipet. A crude calibration is, therefore, recommended.

Laboratory Exercise 4

On a balance, weigh 0.5 grams (0.5 mL) of water into a 3-mL conical vial. Select a short (5¾-inch) Pasteur pipet and attach a rubber bulb. Squeeze the rubber bulb before inserting the tip of the pipet into the water. Try to control how much you depress the bulb so that when the pipet is placed into the water and the bulb is completely released, only the desired amount of liquid will be drawn into the pipet. (This skill may take some time to acquire, but it will facilitate your use of a Pasteur pipet.) When the water has been drawn up, place a mark with an indelible marking pen at the position of the meniscus. A more durable mark can be made by scoring the pipet with a file. Repeat this procedure with 1.0 grams of water, and make a 1-mL mark on the same pipet.

Additional Pasteur pipets can be calibrated easily by holding them next to the pipet which has been calibrated in Laboratory Exercise 4 and scoring a new mark on each pipet at the same level as the mark placed on the calibrated pipet. We recommend that several Pasteur pipets be calibrated at one time for use in future experiments.

E. Extraction

A technique which is applied frequently in the purification of organic reaction products is **extraction.** In this method a solution is mixed thoroughly with a second solvent. The second solvent is not miscible with the first solvent. When the two solvents are mixed, the dissolved substances (solutes) distribute themselves between the two solvents until an equilibrium is established. When the mixing is stopped, the two immiscible solvents separate into two distinct layers. The solutes will be distributed between the two solvents so that each solute will be found in greater concentration in that solvent in which it is more soluble. Separation of the two immiscible solvent layers thus becomes a means of separating solutes from one another based on their relative solubilities in the two solvents.

In a common application, an aqueous solution may contain both inorganic and organic products. An organic solvent which is immiscible with water is added, and the mixture is shaken thoroughly. When the two solvent layers are allowed to form again, upon standing, the organic solutes will be transferred to the organic solvent, while the inorganic solutes will remain in the aqueous layer. When the two layers are separated, the organic and inorganic products will be separated from one another. The separation, as described here, may not be complete. The inorganic materials may be somewhat soluble in the organic solvent and the organic products may retain some water-solubility. Nevertheless, reasonably complete separations of reaction products can be achieved by the extraction method.

For microscale experiments, the conical reaction vial is the glassware item used for extractions. The two immiscible liquid layers are placed in the vial, and the top is sealed with a cap and a Teflon insert (Teflon side toward the inside of the vial). The vial is shaken to provide thorough mixing between the two liquid phases. As the shaking continues, the vial is vented periodically by loosening the cap and then tightening it again. After about 5 or 10 seconds of shaking, the cap is loosened to vent the vial, retightened, and the vial is allowed to stand upright in a beaker until the two liquid layers separate completely.

Separation of the two liquid layers is accomplished by withdrawing the **lower** layer using a disposable Pasteur pipet. This separation technique is illustrated in Figure 8. Care must be taken not to disturb the liquid layers by allowing bubbles to issue

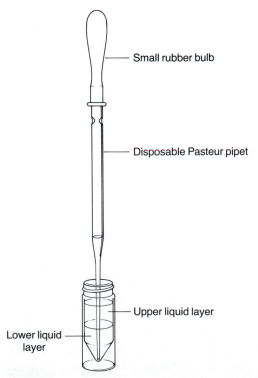

Small rubber bulb

Disposable Pasteur pipet

Upper liquid layer

Lower liquid layer

FIGURE 8.   Separation of immiscible liquid layers in a conical vial

from the pipet. The pipet bulb should be squeezed to the required amount before the pipet is introduced into the vial. Care should also be taken not to allow any of the upper liquid layer to enter the pipet. The pointed shape of the interior of the conical vial makes it easy to remove all of the lower layer without allowing it to be contaminated by some of the upper liquid layer. More precise control in the separation can be achieved by using a filter tip pipet (see Technique 4, Section 4.6, p 574).

Laboratory Exercise 5

Place approximately 1.0 mL of a saturated aqueous solution of iodine and potassium iodide in a 3.0 mL conical reaction vial. Add 1.0 mL of methylene chloride to the vial. Use your calibrated Pasteur pipet for this addition of methylene chloride. Be careful not to squeeze the rubber dropper bulb too firmly as you fill the Pasteur pipet. Seal the vial firmly using the cap and a Teflon insert. Shake the vial for a few seconds, and vent the vial by loosening the cap carefully. Tighten the cap and shake the vial again. Vent the vial and repeat the shaking process a third time. Loosen the cap to vent the vial, re-tighten it, and allow the vial to stand in a beaker on the desk top until the two layers have separated completely. Tap the sides of the vial to force the methylene chloride layer to the bottom of the vial.

    Open the vial and withdraw the **lower** layer (methylene chloride solution) from the vial using a short disposable Pasteur pipet. A better alternative is to use a filter tip pipet (Technique 4, Section 4.6, pp 574–575). Make sure to squeeze the rubber bulb before inserting the pipet tip into the solution. Draw the lower layer carefully into the pipet without allowing any of the upper layer to enter the pipet. Withdraw the pipet, and dispense the liquid into a small test tube. Note the color of the organic and aqueous phases. How have they changed?

    Repeat this process using a second, fresh 1.0-mL portion of methylene chloride. Again note the colors of the two liquid phases and how they may have changed during this second extraction. After comparing the colors, combine this methylene chloride solution with that obtained in the first extraction.

    If desired, a third extraction may be done. When the experiment has been completed, discard all organic solutions in the appropriate waste container. In your laboratory notebook, record all observations.

F. Heating under Reflux

A frequent technique in organic chemistry is to carry out a reaction which is heated to the boiling point of the solvent. A difficulty arises in that one cannot safely heat a closed system, and yet one does not wish to allow the solvent to escape during the period of heating. The technique which is applied is **heating under reflux.** In this technique, a condenser is attached vertically to the reaction vessel. The vapors of solvent rise up into the condenser. The condenser removes heat from the vapors, returning them to liquid. The liquid falls back into the reaction vessel. In this way the system remains open to the atmosphere, but solvent vapors are not allowed to escape.

    The condenser which is used can be either an air condenser or a water-jacketed condenser. An air condenser is adequate for most applications, but a water-jacketed

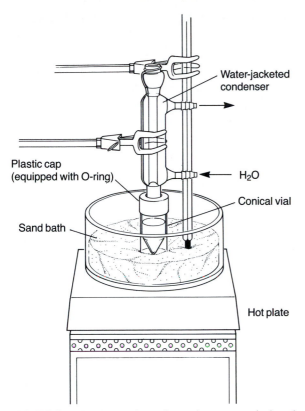

FIGURE 9. Heating under reflux using a water-jacketed condenser

condenser may be used in cases where the solvent is very volatile or where the ambient air temperature is high. A typical assembly for heating under reflux, using a water-jacketed condenser, is shown in Figure 9. A similar assembly would be used with an air condenser.

G. Other Useful Techniques

The practice of organic chemistry requires you to master many more techniques than the ones described in this essay. Those techniques included here are only the most elementary ones, those which are needed to get you started in the laboratory. Additional techniques are described fully in Part Five of this textbook, and Experiments 1 through 9 are intended to give you a good exposure to the most important of them.

There are some other practical hints which ought to be introduced at this point. The first of these involves manipulation of small amounts of solid substances. The efficient transfer of solids requires a spatula which is very small. We recommend that each student have two **microspatulas,** similar to those shown in Figure 10, included as part of their standard desk stock. The design of these spatulas permits the handling of milligram quantities of substances without undue spillage or waste. The larger style (see Figure 10) is more useful when relatively large quantities of solid must be dispensed.

FIGURE 10.   Microspatulas

A clean work area is of utmost importance when working in the laboratory. The need for cleanliness is particularly great when working with the small amounts of materials characteristic of microscale laboratory experiments. We recommend that each student work space be equipped with a sheet of clean glass on which the procedures are performed. A sheet of glass, approximately 14 inches square, is quite suitable. All of the edges of the glass should be wrapped with heavy tape to prevent chipping and to protect against injury. It is important to keep the glass surface **CLEAN!** Materials spilled on the glass surface can then be recovered easily without undue contamination.

> **It must be stressed that each student read the chapter**
> **"Laboratory Safety." There is no substitute for preparation**
> **and care in the prevention of accidents.**

With this final word of caution and advice, we hope that you will enjoy the learning experience which you are about to begin. Learning the care and precision which microscale experiments require may seem to be difficult at first, but it will not be long before you will be comfortable with the scale of the experiments. You will develop much better laboratory technique as a result of microscale practice, and this added skill will serve you well in future courses and endeavors.

Experiment 1
Isolation of the Active Ingredient in an Analgesic Drug

Extraction
Filtration
Melting point

The analgesic (pain relieving) drugs found on the shelves of any drug or grocery store generally fall into one of four categories. These drugs may contain **acetylsalicylic acid,** **acetaminophen,** or **ibuprofen** as the active ingredient; or, some **combination** of these compounds may be used in a single preparation. All tablets, regardless of type, contain a large amount of starch or other inert substance. This material acts as a binder to keep the tablet from falling apart and to make it large enough to handle. Some analgesic drugs also contain caffeine or buffering agents. In addition, many tablets are coated to make them easier to swallow and to prevent one from experiencing the unpleasant taste of the drugs.

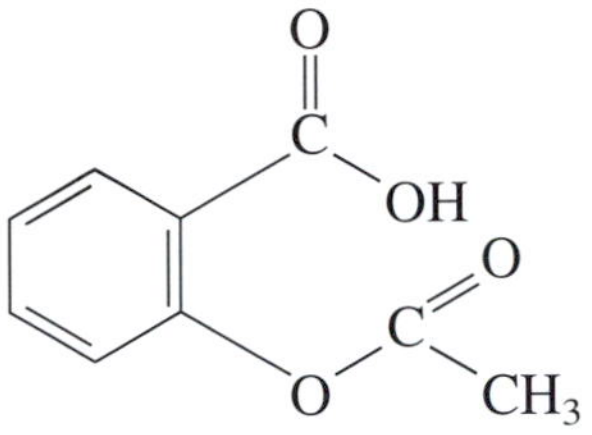

Acetylsalicylic acid

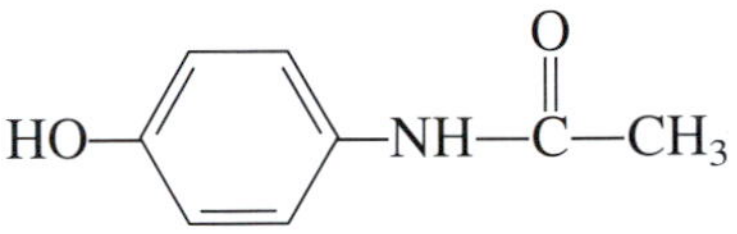

Acetaminophen

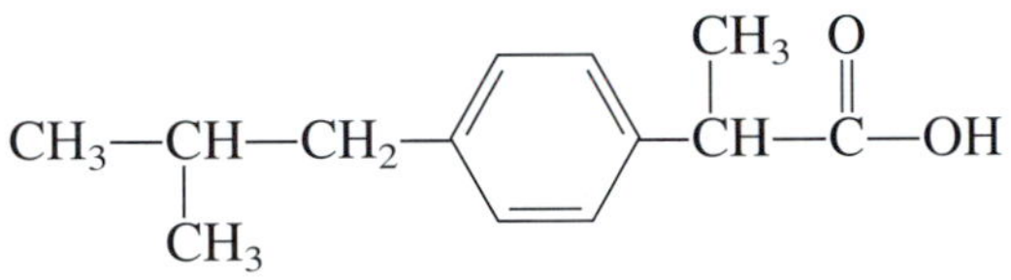

Ibuprofen

The three drugs, along with their melting points and common brand names, are listed below.

DRUG	MP	BRAND NAMES
Acetylsalicylic acid	135–136 °C	Aspirin, ASA, Acetylsalicylic acid, Generic aspirin, Empirin
Acetaminophen	169–170.5 °C	Tylenol, Datril, Panadol, Non-aspirin pain reliever (various brands)
Ibuprofen	75–77 °C	Advil, Brufen, Motrin, Nuprin

The purpose of this experiment is to demonstrate some important techniques which are applied throughout this textbook and to allow the student to become accustomed to working in the laboratory at the microscale level. More specifically, you will extract (dissolve) the active ingredient of an analgesic drug by mixing the powdered tablet with a solvent, methanol. Two steps are required to remove the fine particles of binder, which remain suspended in the solvent. First, centrifugation will be used to remove most of the binder. The second step will be a filtration technique using a Pasteur pipet packed with alumina (finely ground aluminum oxide). The solvent will then be evaporated to yield the solid analgesic, which will be collected by filtration on a Hirsch funnel. Finally, you will test the purity of the drug by doing a melting point determination.

REQUIRED READING

Review: Introduction to Microscale Laboratory (pp 35–48)

New: Technique 3   Reaction Methods, Section 3.9
           Technique 4   Filtration, Sections 4.1–4.6
           Technique 6   Physical Constants, Part A, Melting Points

SPECIAL INSTRUCTIONS

You will be allowed to select an analgesic which is a member of one of the classes described above. You should use an uncoated tablet which contains only a single ingredient analgesic and binder. If it is necessary to use a coated tablet, try to remove the coating when the tablet is crushed. To avoid decomposition of aspirin, it is essential to minimize the length of time that it remains dissolved in methanol. This experiment should not be stopped until after the drug is dried on the Hirsch funnel.

PROCEDURE

Crush one tablet into a powder using a mortar and pestle. If the tablet is coated, try to remove fragments of the coating material with forceps after the tablet is first crushed. Add the powdered tablet to a 3-mL conical vial. Using a calibrated Pasteur pipet (p 44), add about 2 mL of methanol to the vial. Cap the vial and mix thoroughly by shaking. The cap should be loosened at least once during the mixing process in order to release any pressure which may build up in the vial.

Allow the undissolved portion of the powder to settle in the vial. A cloudy suspension may remain even after five minutes or more. You should wait only until it is obvious that the larger particles have settled completely. Using a filter tip pipet (Figure 4–9, p 575), transfer the liquid phase to a centrifuge tube. Add a second 2-mL portion of methanol to the conical vial and repeat the shaking process described above. After the solid has settled, transfer the liquid phase to the centrifuge tube containing the first extract.

Place the tube in a centrifuge along with another centrifuge tube of equal weight on the opposite side. Centrifuge the mixture for two to three minutes. The suspended solids should collect on the bottom of the tube leaving a clear or nearly clear **supernatant liquid,** the liquid above the solid. If the liquid is still quite cloudy, repeat the centrifugation for a longer period of time or at a higher speed. Being careful not to disturb the solid at the bottom of the tube, transfer the supernatant liquid with a Pasteur pipet to a test tube or small beaker.

Prepare an alumina column using a Pasteur pipet, as shown in the figure. Insert a small ball of cotton into the top of the column. Using a long thin object such as a glass stirring rod or a wooden applicator stick, push the cotton down so that it fits into the Pasteur pipet where the constriction begins. Add about 0.5 g of alumina to the pipet, and

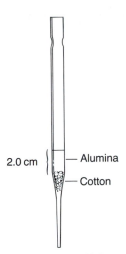

2.0 cm — Alumina

— Cotton

Column for purifying analgesic drug

tap the column with your finger to pack the alumina. Clamp the pipet in a vertical position so the liquid can drain from the column into a small beaker or a 5-mL conical vial. Place a small beaker under the column. With a calibrated Pasteur pipet, add about 2.0 mL of methanol to the column and allow the liquid to drain to the top of the alumina. Once methanol has been added to the alumina, the column should not be allowed to run dry. If necessary, add more methanol.

It is essential that the methanol not be allowed to drain below the surface of the alumina.

When the level of the methanol reaches the surface of the alumina, transfer the solution containing the drug from the beaker or test tube to the column using a Pasteur pipet. Collect the liquid which passes through the column into a 5-mL conical vial. When all the liquid from the beaker has been added to the column and has penetrated the alumina, add an additional 1.0 mL of methanol to the column and allow to drain. This will ensure that all of the analgesic drug has been eluted from the column.

If you are isolating aspirin, it is essential that the following evaporation procedure be completed in 10–15 minutes. Otherwise, the aspirin may partially decompose. Using a Pasteur pipet, transfer about half the liquid in the 5-mL conical vial to another small container. Evaporate the methanol in the 5-mL conical vial using a sand bath or a water bath at about 50 °C. To speed up evaporation, direct a gentle stream of dry air or nitrogen into the vial containing the liquid (Figure 3–12A, p 561). Evaporate the solvent until the volume is less than about 1 mL. Then add the remainder of the liquid and continue evaporation.

When the solvent has completely evaporated or it is apparent that the remaining liquid is no longer evaporating, remove the vial from the sand bath (or water bath) and allow it to cool to room temperature. (The volume of liquid should be less than 0.5 mL when you discontinue evaporation.) If liquid remains, which is likely with the ibuprofen- or acetaminophen-containing analgesics, place the cool vial in an ice-water bath for 10–15 minutes. Prepare the ice-water bath in a small beaker using both ice and water. Be sure that the vial cannot tip over. Crystallization of the product may occur more readily if you scrape the inside of the vial with a microspatula or a glass rod (not fire-polished). If the solid is very hard and clumped together, you should use a microspatula to break up the solid as much as possible before going on to the next step.

Set up a Hirsch funnel for vacuum filtration (see Technique 4, Section 4.3, and Figure 4–6, p 571). Moisten the filter paper with a few drops of methanol, and turn on the aspirator to the fullest extent. Use a microspatula to transfer the material in the conical vial to the Hirsch funnel. The vacuum will draw any remaining solvent from the crystals. Allow the crystals to dry for 5–10 minutes while air is drawn through the crystals in the Hirsch funnel.

Carefully scrape the dried crystals from the filter paper onto a tared (previously weighed) watch glass. If necessary, use a spatula to break up any remaining large pieces of solid. Allow the crystals to air-dry on the watch glass. To determine when the crystals are dry, move them around with a dry spatula. When the crystals no longer clump together or cling to the spatula, they should be dry. If you are working with ibuprofen, the solid will be slightly sticky even when it is completely dried. Weigh the watch glass with the crystals to determine the weight of analgesic drug which you have isolated. Use the weight of the active ingredient specified on the label of the container as a basis for calculating the weight percentage recovery.

Use a small sample of the crystals to determine the melting point (see Technique 6, Sections 6.5–6.8, p 600). Crush the crystals into a powder, using a stirring rod, in order to determine their melting point. You may observe some "sweating" or shrinkage (see Technique 6, Section 6.8, p 603) before the substance actually begins to melt. The beginning of the melting point range is when actual melting is observed, not when the solid takes on a slightly wet or shiny appearance or when shrinkage occurs. If you have isolated ibuprofen, the melting point may be somewhat lower than that given on p 49.

At the instructor's option, place your product in a small vial, label it properly (p 22), and submit it to your instructor.

QUESTIONS

1. Why was the percentage recovery less than 100%? Give several reasons.

2. Why was the tablet crushed?

3. What was the purpose of the centrifugation step?

4. What was the purpose of the alumina column?

5. If 185 mg of acetaminophen were obtained from a tablet containing 350 mg of acetaminophen, what would be the weight percentage recovery?

6. A student, who was isolating aspirin, stopped the experiment after the filtration step with alumina. One week later, the methanol was evaporated and the experiment was completed. The melting point of the aspirin was found to be 110–115 °C. Explain why the melting point was very low and why the melting range was so wide.

Essay
ASPIRIN

Aspirin is one of the most popular cure-alls of modern life. Even though its curious history began over 200 years ago, we still have much to learn about this enigmatic remedy. No one yet knows exactly how or why it works, yet more than 20 million pounds of aspirin are consumed each year in the United States.

The history of aspirin began on June 2, 1763, when Edward Stone, a clergyman, read a paper to the Royal Society of London entitled "An Account of the Success of the Bark of the Willow in the Cure of Agues." By ague, Stone was referring to what we now call malaria, but his use of the word *cure* was optimistic; what his extract of willow bark actually did was to reduce the feverish symptoms of the disease. Almost a century later, a Scottish physician was to find that extracts of willow bark would also alleviate the symptoms of acute rheumatism. This extract was ultimately found to be a powerful **analgesic** (pain reliever), **antipyretic** (fever reducer), and **anti-inflamma-tory** (reduces swelling) drug.

Soon thereafter, organic chemists working with willow bark extract and flowers of the meadowsweet plant (which gave a similar principle) isolated and identified the active ingredient as salicylic acid (from *salix*, the Latin name for the willow tree). The substance could then be chemically produced in large quantity for medical use. It soon became apparent that using salicylic acid as a remedy was severely limited by its acidic properties. The substance caused severe irritation of the mucous membranes lining the mouth, gullet, and stomach. The first attempts at circumventing this problem by using the less acidic sodium salt (sodium salicylate) were only partially successful. This substance gave less irritation but had such an objectionable sweetish taste that most

Salicylic acid   **Sodium salicylate**   **Acetylsalicylic acid (Aspirin)**

people could not be induced to take it. The breakthrough came at the turn of the century (1893) when Felix Hofmann, a chemist for the German firm of Bayer, devised a practical route for synthesizing acetylsalicylic acid, which was found to have all the same medicinal properties without the highly objectionable taste or the high degree of mucosal-membrane irritation. Bayer called its new product "aspirin," a name derived from *a* for acetyl, and the root *-spir,* from the Latin name for the meadowsweet plant, *spirea.*

53

The history of aspirin is typical of many of the medicinal substances in current use. Many began as a crude plant extract or folk remedy whose active ingredient was isolated and structure was determined by chemists, who then improved on the original.

In the last few years the mode of action of aspirin has just begun to unfold. A whole new class of compounds, called **prostaglandins,** has been found to be involved in the body's immune responses. Their synthesis is provoked by interference with the body's normal functioning by foreign substances or unaccustomed stimuli.

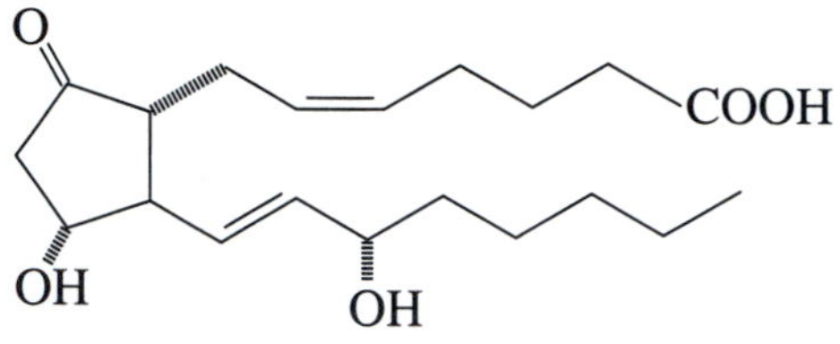

Prostaglandin E₂   **Prostaglandin F₂ₐ**

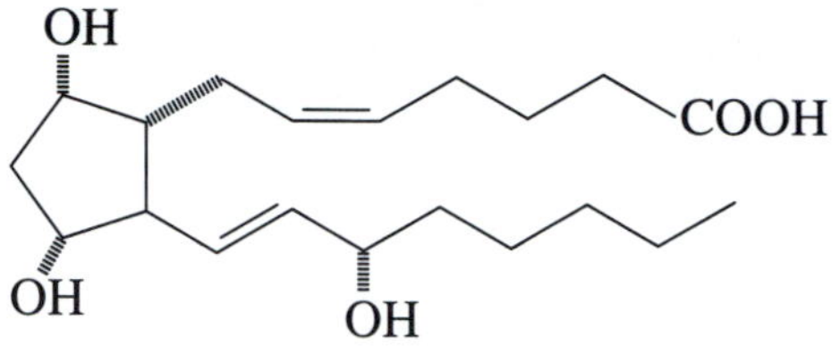

These substances are involved in a wide variety of physiological processes and are thought to be responsible for evoking pain, fever, and local inflammation. Aspirin has recently been shown to prevent bodily synthesis of prostaglandins and thus to alleviate the symptomatic portion (fever, pain, inflammation, menstrual cramps) of the body's immune responses (that is, the ones that let you know something is wrong). A recent report suggests that the role of aspirin may be to inactivate one of the enzymes responsible for the synthesis of prostaglandins. The natural precursor for prostaglandin synthesis is **arachidonic acid.** This substance is converted to a peroxide intermediate by

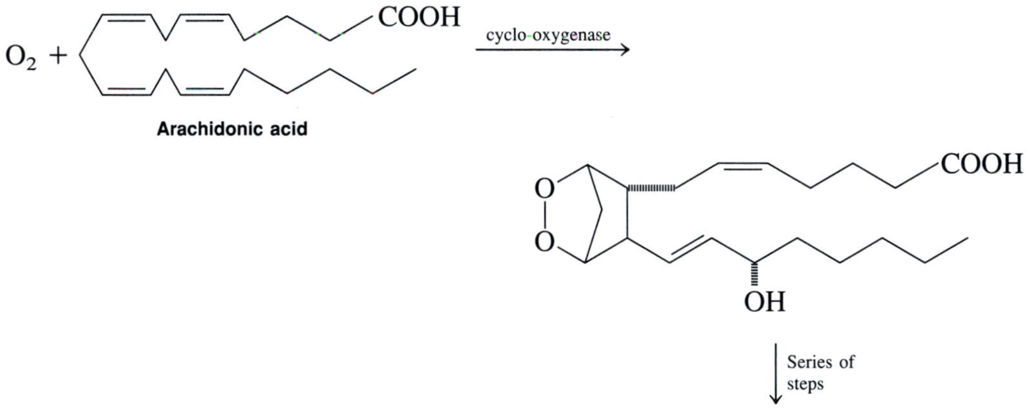

Arachidonic acid

Series of
steps

Prostaglandins

an enzyme called **cyclo-oxygenase,** or prostaglandin synthase. This intermediate is converted further to prostaglandin. The apparent role of aspirin is to attach an acetyl group to the active site of cyclo-oxygenase, thus rendering it unable to convert arachidonic acid to the peroxide intermediate. In this way, prostaglandin synthesis is blocked.

Aspirin tablets (5-grain) are usually compounded of about 0.32 g of acetylsalicylic acid pressed together with a small amount of starch, which binds the ingredients. Buffered aspirin usually contains a basic buffering agent to reduce the acidic irritation of mucous membranes in the stomach, since the acetylated product is not totally free of this irritating effect. Bufferin contains 5 grains of aspirin (1 grain = 0.0648 g), 0.75 grain of aluminum dihydroxyaminoacetate, and 1.5 grains of magnesium carbonate. Combination pain relievers usually contain aspirin, acetaminophen, and caffeine. Excedrin, for instance, contains 0.750 g aspirin, 0.250 g acetaminophen, and 0.065 g caffeine.

REFERENCES

"Aspirin Action Linked to Prostaglandins." *Chemical and Engineering News* (August 7, 1972): 4.

"Aspirin—Good News, Bad News." *Saturday Review* (November 25, 1972): 60.

Collier, H. O. J. "Aspirin." *Scientific American, 209* (November 1963): 96.

Collier, H. O. J. "Prostaglandins and Aspirin." *Nature, 232* (July 2, 1971): 17.

Pike, J. E. "Prostaglandins." *Scientific American, 225* (November 1971): 84.

Roth, G. J., Stanford, N., and Majerus, P. W. "Acetylation of Prostaglandin Synthase by Aspirin." *Proceedings of the National Academy of Science of the U.S.A., 72* (1975): 3073.

"The Uses of Aspirin." *The Medicine Show*. Mt. Vernon, N.Y.: Consumers Union, 1980.

Vane, J. R. "Inhibition of Prostaglandin Synthesis as a Mechanism of Action for Aspirin-like Drugs." *Nature-New Biology, 231* (June 23, 1971): 232.

Experiment 2
Acetylsalicylic Acid

Crystallization
Vacuum filtration
Melting point
Esterification

Aspirin (acetylsalicylic acid) can be prepared by the reaction between salicylic acid and acetic anhydride:

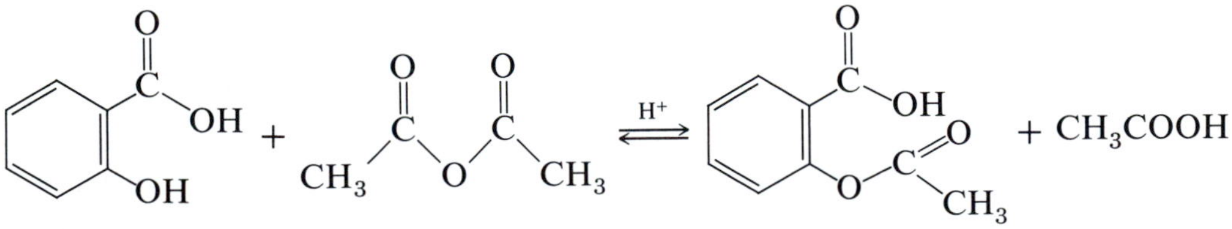

| Salicylic acid | Acetic anhydride | Acetylsalicylic acid | Acetic acid |

In the above reaction, the **hydroxyl group** (—OH), on the benzene ring in salicylic acid reacts with acetic anhydride to form an **ester** functional group. Thus, the formation of acetylsalicylic acid is referred to as an **esterification** reaction. This reaction requires the presence of an acid catalyst, indicated by the H^+ above the equilibrium arrows.

When the reaction is complete, some unreacted salicylic acid and acetic anhydride will be present along with acetylsalicylic acid, acetic acid, and the catalyst. The technique used to purify the acetylsalicylic acid from the other substances is called **crystallization.** This technique will be studied in more detail in Experiment 3. The basic principle is quite simple. At the end of this reaction, the reaction mixture will be hot, and all substances will be in solution. As the solution is allowed to cool, the solubility of acetylsalicylic acid will decrease, and it will gradually come out of solution, or crystallize. Since the other substances are either liquids at room temperature or are present in much smaller amounts, the crystals formed will be composed mainly of acetylsalicylic acid. Thus, a separation of acetylsalicylic acid from the other materials will have been accomplished. The purification process is facilitated by the addition of water after the crystals have formed. The water decreases the solubility of acetylsalicylic acid and dissolves some of the impurities.

The most likely impurity in the final product is salicylic acid itself, which can arise from incomplete reaction of the starting materials or from **hydrolysis** (reaction with water) of the product during the isolation steps. The hydrolysis reaction of acetylsalicylic acid produces salicylic acid. Salicylic acid and other compounds which contain a hydroxyl group on the benzene ring are referred to as **phenols.** Phenols form a highly colored complex with ferric chloride (Fe^{3+} ion). Aspirin is not a phenol, because it does not possess a hydroxyl group directly attached to the ring. Since aspirin will not give the color reaction with ferric chloride, the presence of salicylic acid in the final product is easily detected. The purity of your product will also be determined by obtaining the melting point.

REQUIRED READING

Review: Introduction to Microscale Laboratory (pp 35–48)

Technique 4	Filtration, Sections 4.1–4.6
Technique 6	Physical Constants, Part A, Melting Points

New:

Technique 1	Measurement of Volume and Weight
Technique 2	Heating and Cooling Methods
Technique 3	Reaction Methods, Sections 3.1–3.4
Essay	Aspirin

SPECIAL INSTRUCTIONS

This experiment involves concentrated phosphoric acid, which is highly corrosive. It will cause burns if it is spilled on the skin. Care should be exercised in handling it. The acetylsalicylic acid crystals should be allowed to air-dry overnight after filtration on the Hirsch funnel.

PROCEDURE

Prepare a hot water bath using a 250-mL beaker and a hot plate. Use about 100 mL of water and adjust the temperature to about 50 °C. Weigh 0.210 g of salicylic acid (MW = 138.1) and place this in a 5-mL conical vial. It is not necessary for you to weigh exactly 0.210 g of salicylic acid. Try to obtain a weight within about 0.005 g of the indicated weight without spending excessive time at the balance. Record the actual weight in your notebook, and use this weight in any subsequent calculations. Using an automatic pipet (or a graduated pipet and pipet pump), add 0.480 mL of acetic anhydride (MW = 102.1, d = 1.08 g/mL), followed by exactly one drop of concentrated phosphoric acid from a Pasteur pipet.

> **CAUTION:** Concentrated phosphoric acid is highly corrosive. You must handle it with great care.

Add a magnetic spin vane (Figure 3–4A, p 552) and attach an air condenser to the vial. Clamp this assembly so that the vial is partially submerged in the hot water bath, as shown in Figure 2–5 (p 544). Stir the mixture with the spin vane until the salicylic acid dissolves. (If the spin vane becomes stuck in the solid salicylic acid, insert a microspatula through the air condenser into the conical vial and gently push the spin vane until it begins spinning.) Heat the mixture for three to four minutes after the solid dissolves to complete the reaction.

Remove the vial from the water bath and allow it to cool. After the vial has cooled enough for you to handle it, detach the air condenser and remove the spin vane with forceps or a magnetic stirring bar. (If you use forceps, be sure to clean them.) Place the conical vial in a small beaker and allow the vial to cool to room temperature, during which time the acetylsalicylic acid should begin to crystallize from the reaction mixture. If it does not crystallize, scratch the walls of the vial with a glass rod (not fire-polished) and cool the mixture slightly in an ice-water bath (Technique 2, Section 2.6, p 545) until crystallization has occurred. (Scratching the inside walls of the container often helps to initiate crystallization.) After crystal formation is complete (usually when the product appears as a solid mass), add 3.0 mL of water (measured with a 10-mL graduated cylinder) and stir thoroughly with a microspatula.

Set up a Hirsch funnel for vacuum filtration (see Technique 4, Section 4.3, and Figure 4–6, p 571). Moisten the filter paper with a few drops of water and turn on the aspirator to the fullest extent. Transfer the mixture in the conical vial to the Hirsch funnel. When you have removed as much product as possible from the vial, add about 1 mL of cold water to the vial using a calibrated Pasteur pipet (p 44). Stir the mixture and transfer the remaining crystals and water to the Hirsch funnel. When all the crystals have been collected in the funnel, rinse them with several 0.5-mL portions of cold water. Continue drawing air through the crystals on the Hirsch funnel by suction until the crystals are nearly dry (5–10 minutes). Remove the crystals for air-drying on a watch glass or clay plate. It is convenient to hold the filter paper disc with forceps while **gently** scraping the crystals off the filter paper with a microspatula. If the paper is scraped too hard, small pieces of paper will be removed along with the crystals. To completely dry the crystals, it will be necessary to set the crystals aside overnight. Weigh the dry product, and calculate the percentage yield of acetylsalicylic acid (MW = 180.2).

TEST FOR PURITY

This test can be performed on a sample of your product which is not completely dry. To determine if there is any salicylic acid remaining in your product, carry out the following procedure. Obtain three small test tubes. Add 0.5 mL of water to each test tube. Dissolve a small amount of salicylic acid in the first tube. Add a similar amount of your product to the second tube. The third test tube, which contains only solvent, will serve as the control. Add one drop of 1% ferric chloride solution to each tube and note the color after shaking. Formation of an iron-phenol complex with Fe(III) gives a definite color ranging from red to violet, depending on the particular phenol present.

As an additional test for purity, determine the melting point of your product (see Technique 6, Sections 6.5–6.8, p 600). The melting point must be obtained with a completely dried sample. Pure aspirin has a melting point of 135–136 °C.

Place your product in a small vial, label it properly (p 22), and submit it to your instructor.

ASPIRIN TABLETS

Aspirin tablets are acetylsalicylic acid pressed together with a small amount of inert binding material (usually starch). One can test for the presence of starch by boiling approximately one-fourth of an aspirin tablet with 2 mL of water. The liquid is cooled, and a drop of iodine solution is added. If starch is present, it will form a complex with the iodine. The starch-iodine complex is deep blue violet. Repeat this test with a commercial aspirin tablet and with the acetylsalicylic acid prepared in this experiment.

QUESTIONS

1. What is the purpose of the concentrated phosphoric acid used in the first step?

2. What would happen if the phosphoric acid were left out?

3. If one were to use 250 mg of salicylic acid and excess acetic anhydride in the above synthesis of aspirin, what would be the theoretical yield of acetylsalicylic acid in moles? In mg?

4. What is the equation for the decomposition reaction which can occur with aspirin?

5. Most aspirin tablets contain five grains of acetylsalicylic acid. How many mg is this?

6. A student performed the reaction in this experiment using a water bath at 90 °C instead of 50 °C. The final product was tested for the presence of phenols with ferric chloride. This test was negative (no color observed); however, the melting point of the dry product was 122–125 °C. Explain these results as completely as possible. (It may be helpful to read the introduction to Experiment 3 on p 62.)

7. If the aspirin crystals were not completely dried before the melting point was determined, what effect would this have on the observed melting point?

Essay
ANALGESICS

Acylated aromatic amines (those having an acyl group, $R-\overset{\overset{\displaystyle O}{\|}}{C}-$, substituted on nitrogen) are important in over-the-counter headache remedies. Over-the-counter drugs are those you may buy without a prescription. Acetanilide, phenacetin, and acetaminophen are mild analgesics (relieve pain) and antipyretics (reduce fever) and are important, along with aspirin, in many nonprescription drugs.

Acetanilide   **Phenacetin**   **Acetaminophen**

The discovery that acetanilide was an effective antipyretic came about by accident in 1886. Two doctors, Cahn and Hepp, had been testing naphthalene as a possible vermifuge (an agent that expels worms). Their early results on simple worm cases were very discouraging, so Dr. Hepp decided to test the compound on a patient with a larger variety of complaints, including worms—a sort of "shotgun" approach. A short time later, Dr. Hepp excitedly reported to his colleague, Dr. Cahn, that naphthalene had miraculous fever-reducing properties.

In trying to verify this observation, the doctors discovered that the bottle they thought contained naphthalene had apparently been mislabeled. In fact, the bottle brought to them by their assistant had a label so faint as to be illegible. They were sure that the sample was not naphthalene since it had no odor. Naphthalene has a strong odor reminiscent of mothballs. So close to an important discovery, the doctors were nevertheless stymied; they appealed to a cousin of Hepp, who was a chemist in a nearby dye factory, to help them identify the unknown compound. This compound turned out to be acetanilide, a compound with a structure not at all like that of naphtha-

Naphthalene

lene. Certainly, Hepp's unscientific and risky approach would be frowned on by doctors today; and to be sure, the Food and Drug Administration (FDA) would never allow human testing before extensive animal testing (consumer protection has progressed). Nevertheless, Cahn and Hepp made an important discovery.

In another instance of serendipity, the publication of Cahn and Hepp, describing their experiments with acetanilide, caught the attention of Carl Duisberg, director of research at the Bayer Company in Germany. Duisberg was confronted with the problem of profitably getting rid of nearly 50 tons of *p*-aminophenol, a by-product of the synthesis of one of Bayer's other commercial products. He immediately saw the possibility of converting *p*-aminophenol to a compound similar in structure to acetanilide, by putting an acyl group on the nitrogen. It was then believed, however, that all compounds having a hydroxyl group on a benzene ring (that is, phenols) were toxic. Duisberg devised a scheme of structural modification of *p*-aminophenol to get the compound phenacetin. The reaction scheme is shown here.

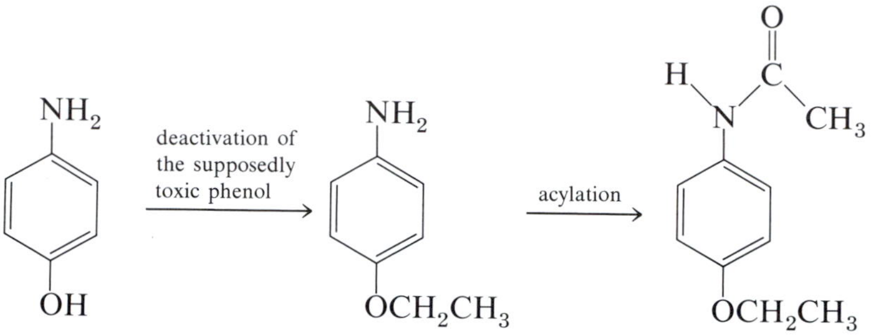

p-Aminophenol   Phenacetin

Phenacetin turned out to be a highly effective analgesic and antipyretic. A common form of combination pain reliever, called an APC tablet, was once available. An APC tablet contained **A**spirin, **P**henacetin, and **C**affeine (hence, **APC**). Phenacetin is no longer used in commercial pain-relief preparations. It was later found that not all aromatic hydroxyl groups lead to toxic compounds, and today the compound acetaminophen is very widely used as an analgesic in place of phenacetin.

Another analgesic, structurally similar to aspirin, that has found some application is **salicylamide.** Salicylamide is found as an ingredient in some pain-relief preparations, although its use is declining.

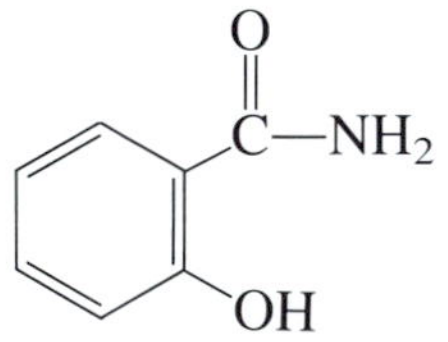

Salicylamide

On continued or excessive use, acetanilide can cause a serious blood disorder called **methemoglobinemia.** In this disorder, the central iron atom in hemoglobin is converted from Fe(II) to Fe(III) to give methemoglobin. Methemoglobin will not function as an oxygen carrier in the bloodstream. The result is a type of anemia (deficiency of hemoglobin or lack of red blood cells). Phenacetin and acetaminophen cause the same disorder, but to a much lesser degree. Since they are also more effective as antipyretic and analgesic drugs than acetanilide, they are preferred remedies. Acetaminophen is marketed under the trade names Tylenol and Datril and is often successfully used by persons who are allergic to aspirin.

Heme portion of blood-oxygen carrier, hemoglobin

Recently a new drug has appeared in over-the-counter preparations. This drug is **ibuprofen,** which is marketed as a prescription drug in the United States under the name Motrin. Ibuprofen was first developed in England in 1964. United States marketing rights were obtained in 1974. Ibuprofen is now sold without prescription under brand names, which include Advil and Nuprin. Ibuprofen is principally an anti-inflammatory drug, but it is also effective as an analgesic and an antipyretic. It is particularly effective in the treatment of the symptoms of rheumatoid arthritis and menstrual cramps. Ibuprofen appears to control the production of prostaglandins, which parallels the mode of action of aspirin. An important advantage of ibuprofen is that it is a very powerful pain reliever. One 200-mg tablet is as effective as two tablets (650 mg) of aspirin. Furthermore, ibuprofen has a more advantageous dose-response curve, which means that taking two tablets of this drug is approximately twice as effective as one tablet for certain types of pain. Aspirin and acetaminophen reach their maximum effective dose at two tablets. Little additional relief is gained at doses above that level. Ibuprofen, however, continues to increase its effectiveness up to the 400-mg level (the equivalent of four tablets of aspirin or acetaminophen). Ibuprofen is a relatively safe drug, but its use should be avoided in cases of aspirin allergy, kidney problems, ulcers, asthma, hypertension, or heart disease.

Ibuprofen

Analgesics and Caffeine in Some Common Preparations

	ASPIRIN	ACETAMINOPHEN	CAFFEINE	SALICYLAMIDE
Aspirin*	0.325 g	—	—	—
Anacin	0.400 g	—	0.032 g	—
Bufferin	0.324 g	—	—	—
Cope	0.421 g	—	0.032 g	—
Excedrin (Extra-Strength)	0.250 g	0.250 g	0.065 g	—
Tylenol	—	0.325 g	—	—
B. C. Powder	0.650 g	—	0.032 g	0.195 g

NOTE: Nonanalgesic ingredients, e.g., buffers, are not listed.
* 5-grain tablet (1 grain = 0.0648 g).

REFERENCES

Barr, W. H., and Penna, R. P. "O-T-C Internal Analgesics." In *Handbook of Non-Prescription Drugs,* 7th ed., edited by G. B. Griffenhagen. American Pharmaceutical Association, 1982.

Consumers Union of the United States. "Aspirin and its Competitors." *The Medicine Show.* New York: Pantheon Books, 1980. pp. 15–33.

Flower, R. J., Moncada, S., and Vane, J. R. "Analgesic-Antipyretics and Anti-inflammatory Agents; Drugs Employed in the Treatment of Gout." In A. G. Gilman, L. S. Goodman, T. W. Rall, and F. Murad, *The Pharmacological Basis of Therapeutics.* 7th ed. New York: Macmillan, 1985.

Hansch, C. "Drug Research or the Luck of the Draw." *Journal of Chemical Education, 51* (1974): 360.

Ray, O. S. "Internal Analgesics." *Drugs, Society, and Human Behavior,* 2nd ed. St. Louis: C. V. Mosby, 1978.

Senozan, N. M. "Methemoglobinemia: An Illness Caused by the Ferric State." *Journal of Chemical Education, 62* (March 1985): 181.

"The New Pain Relievers." *Consumer Reports, 49* (November 1984): 636–638.

Experiment 3
Acetaminophen

Crystallization
Filtration
Use of a Craig tube
Preparation of an amide

Preparation of acetaminophen involves treating an amine with an acid anhydride to form an amide. In this case, *p*-aminophenol, the amine, is treated with acetic anhydride to form acetaminophen (*p*-acetamidophenol), the amide.

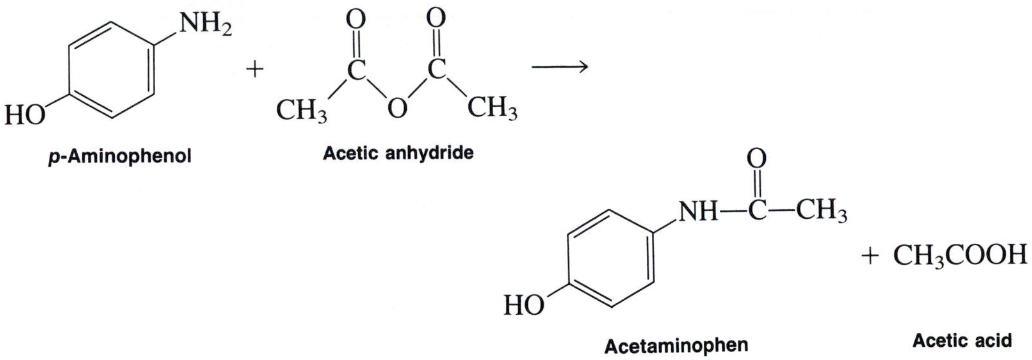

The first part of this experiment is the purification of *p*-aminophenol, which is commercially available only in an impure form. After dissolving the *p*-aminophenol in hot water, the solution is allowed to cool, yielding a crystalline solid. The mixture is filtered on a Hirsch funnel to isolate a material which is sufficiently pure for the preparation of acetaminophen. The purified *p*-aminophenol undergoes the reaction shown above to produce acetaminophen. This product will be impure mainly because of the presence of unreacted *p*-aminophenol and the products of various **side reactions.** Side reactions are those which occur in addition to the main reaction of interest. Because the products of these side reactions are different compounds than the desired substance, it is common to obtain a crude product contaminated with the products of side reactions. The crude acetaminophen will be purified by a microscale crystallization technique utilizing a Craig tube.

REQUIRED READING

Review: Introduction to Microscale Laboratory (pp 35–48)
         Techniques 1 and 2
         Technique 3   Reaction Methods, Sections 3.1–3.3
         Technique 4   Filtration, Sections 4.1–4.6
         Technique 6   Physical Constants, Part A, Melting Points

New: Technique 4   Filtration, Section 4.7
         Technique 5   Crystallization
         Essay   Analgesics

SPECIAL INSTRUCTIONS

p-Aminophenol is a skin irritant and is toxic. It is best to use a fresh bottle (not darkly colored) of **p-Aminophenol**.

PROCEDURE

PURIFICATION OF *p*-AMINOPHENOL

Adjust your sand bath for a temperature of about 120 °C. Place about 0.250 g (record the actual weight in your notebook) of *p*-aminophenol in a 25-mL Erlenmeyer flask. Using a 10-mL graduated cylinder, add 7.5 mL of water to the flask.

Avoid spills or contact with the skin.

With occasional swirling, heat the mixture by partially submerging the flask in the sand bath. Heat until the solid has completely dissolved.

Remove the flask from the sand bath and allow the solution to cool slowly to room temperature. Crystallization should occur during this cooling period. If it does not, scratch the inside surface of the flask with a glass rod (not fire-polished) to induce crystallization (Technique 5, Section 5.7, Part A, p 591). Then place the flask in an ice-water bath (Technique 2, Section 2.6, p 545). Be sure that both water and ice are present and that the beaker is small enough so there is no chance that the flask can tip over.

When crystallization is complete, vacuum filter the crystals using a Hirsch funnel (see Technique 4, Section 4.3, and Figure 4–6, p 571). Moisten the filter paper with a few drops of water and turn on the aspirator. Swirl the mixture in the flask and pour about one-third of the mixture into the funnel. When the liquid has passed through the filter, repeat this procedure until you have transferred all the liquid to the Hirsch funnel. Using your spatula, scrape out as many of the crystals as possible from the flask. Add about 1 mL of ice-cold water (measured with a calibrated Pasteur pipet) to the flask. Swirl the liquid in the flask and then pour the remaining crystals and water into the Hirsch funnel. Not only does this help transfer all the crystals to the funnel, but the water also rinses the crystals on the funnel. If necessary, repeat with another 1-mL portion of ice-cold water. Wash the crystals on the funnel with 1 mL of cold **methanol,** making sure that all the crystals are rinsed by the cold solvent. Methanol is used instead of water because methanol is more volatile, and the crystals will dry more rapidly. Some impurities are also washed away by the methanol. Continue drawing air through the crystals on the Hirsch funnel by suction for about five minutes. Transfer the crystals onto a watch glass or clay plate for air-drying. Separate the crystals as much as possible with a spatula. The crystals will dry in about 10 minutes. Weigh the crystals and calculate the percentage recovery from the crystallization. Compare the color of your purified *p*-aminophenol with the starting material.

PREPARATION OF ACETAMINOPHEN

Weigh about 0.100 g of purified *p*-aminophenol (MW = 109.1) and place this in a 3-mL conical vial. Using an automatic pipet (or a graduated pipet and pipet pump), add 0.300 mL of water and 0.110 mL of acetic anhydride (MW = 102.1, d = 1.08 g/mL). Place a spin vane in the conical vial, and attach an air condenser. Heat the reaction mixture in a sand bath at about 115 °C (see inset in Figure 3–2A, p 550) with gentle stirring. The conical vial should be partially buried in the sand so that the vial is nearly at the bottom of the sand bath. After the solid has dissolved (it may dissolve, precipitate, and redissolve), heat the mixture for an additional 10 minutes to complete the reaction.

Remove the vial from the sand bath and allow it to cool. When the vial has cooled to the touch, detach the air condenser and remove the spin vane with clean forceps or a magnetic stirring bar. Rinse the spin vane with two to three drops of warm water, allowing the water to drop into the conical vial. Place the conical vial in a small beaker and allow it to cool to room temperature. If crystallization has not occurred, scratch the inside of the vial with a glass stirring rod to initiate crystallization. Cool the mixture in an ice bath for 15–20 minutes and collect the crystals by vacuum filtration on a Hirsch funnel (see Technique 4, Section 4.3, and Figure 4–6, p 571). Rinse the vial with about 1 mL of ice water and transfer this mixture to the Hirsch funnel. Wash the crystals on the funnel with two additional 0.5-mL portions of ice water. Dry the crystals for 5–10 minutes by allowing air to be drawn through them while they remain on the Hirsch funnel. Transfer the product to a watch glass or clay plate and allow the crystals to dry in air. It may take several hours for the crystals to dry completely, but you may go on to the next step before they are totally dry. Weigh the crude product and set aside a small sample for a melting point determination. Record the appearance of the crystals in your notebook.

PURIFICATION OF ACETAMINOPHEN

Crystallize the crude acetaminophen from a solvent mixture composed of 50% water and 50% methanol by volume. The crystallization is performed in a Craig tube (Technique 5, Section 5.4, and Figure 5–5, p 586). The solubility of acetaminophen in this hot (nearly boiling) solvent is about 0.2 g/mL. Although you can use this as a rough indication of how much solvent is required to dissolve the solid, you should still use the technique in Figure 5–5. Add small portions (several drops) of hot solvent until the solid dissolves. Step 2 in Figure 5–5 (removal of insoluble impurities) should not be required in this crystallization. When the solid has dissolved, place the Craig tube in a 10-mL Erlenmeyer flask, insert the inner plug, and allow the solution to cool. If you want the solution to cool more slowly and possibly to obtain better (purer) crystals, place about 8 mL of hot water (about 50 °C) in the Erlenmeyer flask before setting the Craig tube in the flask. Be sure the Craig tube doesn't float in the water.

When the mixture has cooled to room temperature, place the Craig tube in an ice-water bath for several minutes. If necessary, induce crystallization using the methods described in Technique 5, Section 5.7B, p 592. Collect the crystals using the apparatus shown in Figure 4–11 on p 576. Place the assembly in a centrifuge (be sure it is balanced by a centrifuge tube filled with water so that both tubes contain the same weight), and turn on the centrifuge for several minutes. Collect the crystals on a watch glass or piece of smooth paper, as shown in Figure 5–5 on p 586. Set the crystals aside to air-dry. Very little additional time should be required to complete the drying.

Weigh the purified acetaminophen (MW = 151.2) and calculate the percentage yield. This calculation should be based on the weight of purified p-aminophenol used to prepare acetaminophen. Determine the melting point. Compare the melting point and the color of the final product with that of the crude acetaminophen. Pure acetaminophen melts at 169–170.5 °C. Place your product in a properly labeled vial and submit it to your instructor.

QUESTIONS

1. During the crystallization of *p*-aminophenol, why was the mixture cooled in an ice bath?

2. In the reaction between *p*-aminophenol and acetic anhydride to form acetaminophen, 0.300 mL of water was added. What was the purpose of the water?

3. If 0.130 g of *p*-aminophenol is allowed to react with excess acetic anhydride, what is the theoretical yield of acetaminophen in moles? In grams?

4. Give two reasons, discussed in Experiments 2 and 3, why the crude product in most reactions is not pure.

5. Phenacetin has the structure shown. How might it be prepared?

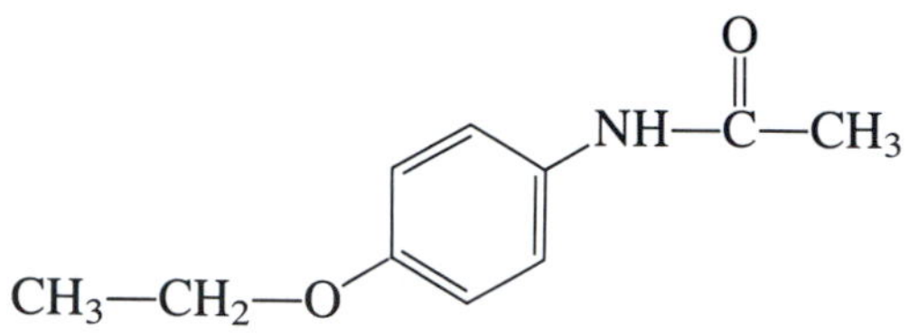

$$CH_3-CH_2-O-\underset{}{\bigcirc}-NH-\overset{\overset{\textstyle O}{\|}}{C}-CH_3$$

Essay
IDENTIFICATION OF DRUGS

Frequently a chemist is called on to identify a particular unknown substance. If there is no prior information to work from, this can be a formidable task. There are several million known compounds, both inorganic and organic. For a completely unknown substance, the chemist must often use every available method. If the unknown substance is a mixture, then the mixture must be separated into its components and each component identified separately. A pure compound can often be identified from its physical properties (melting point, boiling point, density, refractive index, and so on) and a knowledge of its functional groups. These can be identified by the reactions that the compound is observed to undergo or by spectroscopy (IR, ultraviolet, NMR, and mass spectroscopy). The techniques necessary for this type of identification are introduced in a later section.

A somewhat simpler situation often arises in drug identification. The scope of drug identification is more limited, and the chemist working in a hospital trying to identify the source of a drug overdose, or the law enforcement officer trying to identify a suspected illicit drug or a poison, usually has some prior clues to work from. So does the medicinal chemist working for a pharmaceutical manufacturer who might be trying to discover why a competitor's product is better than his.

Consider a drug overdose case as an example. The patient is brought into the emergency ward of a hospital. This person may be in a coma or in a hyperexcited state, have an allergic rash, or clearly be hallucinating. These physiological symptoms are themselves a clue to the nature of the drug. Samples of the drug may be found in the patient's possession. Correct medical treatment may require a rapid and accurate identification of a drug powder or capsule. If the patient is conscious, the necessary information can be elicited orally; if not, the drug must be examined. If the drug is a tablet or a capsule, the process is often simple, since many drugs are coded by a manufacturer's trademark or logo, by shape (round, oval, bullet shape), by formulation (tablet, gelatin capsule, time-release microcapsules), and by color.

It is more difficult to identify a powder, but under some circumstances such identification may be easy. Plant drugs are often easily identified since they contain microscopic bits and pieces of the plant from which they are obtained. This cellular debris is often characteristic for certain types of drugs, and they can be identified on this basis alone. A microscope is all that is needed. Sometimes chemical color tests can be used as a confirmation. Certain drugs give rise to characteristic colors when treated with special reagents. Other drugs form crystalline precipitates of characteristic color and crystal structure when treated with appropriate reagents.

If the drug itself is not available and the patient is unconscious (or dead), identification may be more difficult. It may be necessary to pump the stomach or bladder contents of the patient (or corpse), or to obtain a blood sample, and work on these. These samples of stomach fluid, urine, or blood would be extracted with an appropriate organic solvent, and the extract would be analyzed.

Often the final identification of a drug, as an extract of urine, serum, or stomach fluid, hinges on some type of **chromatography.** Thin-layer chromatography (TLC) is often used. Under specified conditions, many drug substances can be identified by their R_f values and by the colors that their TLC spots turn when treated with various reagents or when they are observed under certain visualization methods. In the experiment that follows, TLC is applied to the analysis of an unknown analgesic drug.

REFERENCES

Keller, E. "Forensic Toxicology: Poison Detection and Homicide." *Chemistry, 43* (1970): 14.

Keller, E. "Origin of Modern Criminology." *Chemistry, 42* (1969): 8.

Lieu, V. T. "Analysis of APC Tablets." *Journal of Chemical Education, 48* (1971): 478.

Neman, R. L. "Thin Layer Chromatography of Drugs." *Journal of Chemical Education, 49* (1972): 834.

Rodgers, S. S. "Some Analytical Methods Used in Crime Laboratories." *Chemistry, 42* (1969): 29.

Tietz, N. W. *Fundamentals of Clinical Chemistry.* Philadelphia: W. B. Saunders, 1970.

Walls, H. J. *Forensic Science.* New York: Praeger, 1968.

A collection of articles on forensic chemistry can be found in:

Berry, K., and Outlaw, H. E., eds. "Forensic Chemistry—A Symposium Collection." *Journal of Chemical Education, 62* (December 1985): 1043–1065.

Experiment 4
TLC Analysis of Analgesic Drugs

Thin-layer chromatography

In this experiment, thin-layer chromatography (TLC) will be used to determine the composition of various over-the-counter analgesics. If the instructor chooses, you may also be required to identify the components and actual identity (trade name) of an unknown analgesic. You will be given two commercially prepared TLC plates with a flexible backing and a silica gel coating with a fluorescent indicator. On one TLC plate, you will spot five standard compounds often used in analgesic formulations. In addition, a standard reference mixture containing four of these same compounds will be spotted. Ibuprofen is omitted from the standard mixture because it would overlap with the aspirin after the plate is developed. The reference substances are

Acetaminophen	(Ac)
Aspirin	(Asp)
Caffeine	(Cf)
Ibuprofen	(Ibu)
Salicylamide	(Sal)

They will all be available as solutions of 1 g of each dissolved in 20 mL of a 50:50 mixture of methylene chloride and ethanol. The purpose of the first plate is to determine the order of elution (R_f values) of the known substances and to index the standard reference mixture. On the second plate, the standard reference mixture will be spotted along with several solutions prepared from commercial analgesic tablets. The crushed tablets will also be dissolved in a 50:50 methylene chloride–ethanol mixture. At your instructor's option, one of the analgesics to be spotted on the second plate may be an unknown.

Two methods of visualization will be used to observe the positions of the spots on the developed TLC plates. First, the plates will be observed while under illumination from a short-wavelength ultraviolet (UV) lamp. This is done best in a darkened room or in a fume hood that has been darkened by taping butcher paper or aluminum foil over the lowered glass cover. Under these conditions, some of the spots will appear as dark areas on the plate, while others will fluoresce brightly. This difference in appearance under UV illumination will help to distinguish the substances from one another. You will find it convenient to outline very lightly in pencil the spots observed and to place a small **x** inside those spots that fluoresce. For a second means of visualization, iodine vapor will be used. Not all the spots will become visible when treated with iodine, but at least two will develop a deep brown color. The differences in the behaviors of the various spots with iodine can be used to further differentiate among them.

It is possible to use several developing solvents for this experiment, but ethyl acetate with 0.5% glacial acetic acid added is preferred. The small amount of glacial acetic acid supresses ionization of both the ibuprofen and the aspirin, allowing them to travel upward on the plates. Without the acid, these compounds do not move.

In some analgesics you may find other ingredients besides the five mentioned above. Some include an antihistamine and some a mild sedative. For instance, Midol contains N-cinnamylephedrine (cinnamedrine), an antihistamine, while Excedrin PM contains the sedative methapyrilene hydrochloride. Cope contains the related sedative methapyrilene fumarate.

REQUIRED READING

Review: Essay   Analgesics

New: Technique 12   Column Chromatography, Sections 12.1–12.3
         Technique 13   Thin Layer Chromatography
         Essay   Identification of Drugs

SPECIAL INSTRUCTIONS

Note that the developed plates must be examined under ultraviolet light first, and with iodine vapor second. The iodine permanently affects some of the spots. Aspirin presents some special problems since it is present in a large amount in many of the analgesics, and since it hydrolyzes easily. For these reasons, the aspirin spots often show excessive tailing. Take special care to notice that, although they have similar R_f values, the aspirin and ibuprofen spots each have a different appearance when viewed under UV illumination.

PROCEDURE

You will need at least 12 capillary micropipets to spot the plates. The preparation of these pipets is described and illustrated in Technique 13, Section 13.4, p 728.

After preparing the micropipets, obtain two 10-cm × 6.6-cm TLC plates (Eastman Chromagram Sheet, No. 13181) from your instructor. These plates have a flexible backing, but they should not be bent excessively. They should be handled carefully or the adsorbent may flake off them. Also, they should be handled only by the edges; the surface should not be touched. Using a lead pencil (not a pen) **lightly** draw a line across the plates (short dimension) about 1 cm from the bottom. Using a centimeter ruler, move its index about 0.6 cm in from the edge of the plate and lightly mark off six 1-cm intervals on the line (see figure). These are the points at which the samples will be spotted.

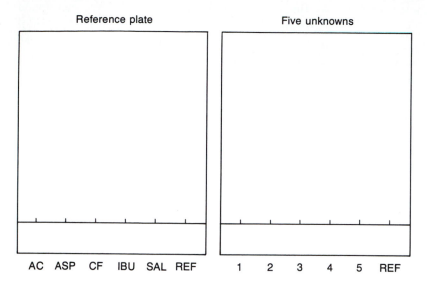

On the first plate, starting from left to right, spot first acetaminophen, then aspirin, caffeine, ibuprofen, and salicylamide. This order is alphabetic and will avoid any further memory problems or confusion. Solutions of these compounds will be found in small bottles on the side shelf. The standard reference mixture, also found on the side shelf, is spotted in the last position. The correct method of spotting a TLC plate is described in Technique 13, Section 13.4, p 728. It is important that the spots be made as small as possible and that the plates not be overloaded. If these cautions are disregarded, the spots will tail and will overlap one another after development. The applied spot should be about 1–2 mm (1/16 in.) in diameter. If scrap pieces of the TLC plates are available, it would be a good idea to practice spotting on these before preparing the actual sample plates.

When the first plate has been spotted, obtain a 16-oz wide-mouthed screw-cap jar for use as a development chamber. The preparation of a development chamber is described in Technique 13, Section 13.5, p 730. Since the backing on the TLC plates is very thin, if they touch the filter paper liner of the development chamber **at any point,** solvent will begin to diffuse onto the absorbent surface at that point. To avoid this, you may either omit the liner or make the modification described below.

If you wish to use a liner, use a very narrow strip of filter paper (approximately 5 cm wide). It should be folded into an L shape and should be long enough to traverse the bottom of the jar and extend up the side to the top of the jar. TLC plates placed in the jar for development should **straddle** this liner strip but not touch it.

When the development chamber has been prepared, obtain a small amount of the development solvent (0.5% glacial acetic acid in ethyl acetate). This mixture should have been prepared by your instructor; it contains such a small amount of acetic acid that small individual portions are difficult to prepare. Fill the chamber with the development solvent to a depth of about 0.5 cm. If you are using a liner, be sure it is saturated with the solvent. Recall that the solvent level must not be above the spots on the plate or the samples will dissolve off the plate into the reservoir instead of developing. Place the spotted plate in the chamber (straddling the liner if one is present) and allow the plate to develop.

When the solvent has risen to a level about 0.5 cm from the top of the plate, remove the plate from the chamber (in the hood) and, using a lead pencil, mark the position of the solvent front. Set the plate on a piece of paper towel to dry. It may be helpful to place a small object under one end to allow optimum air flow around the drying plate. When the plate is dry, observe it under a short-wavelength UV lamp, preferably in a darkened hood or a darkened room. Lightly outline all of the observed spots with a pencil. Carefully notice any differences in behavior between the aspirin and the ibuprofen. Both compounds have similar R_f values, but the spots have a different appearance under UV illumination. Most analgesics do not contain both aspirin and ibuprofen in the same preparation, but you will need to be able to distinguish them from one another to identify which one is present. Before proceeding, make a sketch of the plate in your notebook and note the differences in appearance that you observed. Next, place the plate in a jar containing a few iodine crystals, cap the jar, and warm it **gently** on a steam bath or hot plate until the spots begin to appear. Notice which spots become visible and note their relative colors. Remove the plate from the jar and record your observations in your notebook. Using a ruler marked in millimeters, measure the distance that each spot has travelled relative to the solvent front. Calculate R_f values for each spot (Technique 13, Section 13.9, p 733).

Next, obtain half a tablet of each of the analgesics to be analyzed. If you were issued an unknown, you may analyze four other analgesics of your choice; if not, you may analyze five. The experiment will be most interesting if you make your choices to give a wide spectrum of results. Try to pick at least one analgesic each containing aspirin, acetaminophen, ibuprofen, and, if available, salicylamide. If you have a favorite analgesic, you may wish to include it among your samples. Take each analgesic half-tablet, place it on a smooth piece of notebook paper, and crush it well with a spatula. Transfer each crushed half-tablet to a small, labelled test tube or Erlenmeyer flask. Using a graduated cylinder, mix together 15 mL of absolute ethanol and 15 mL of methylene chloride. Mix the solution well. Add 5 mL of this solvent to each of the crushed half-tablets and then heat each of them **gently** for a few minutes on a steam bath or sand bath at about 100 °C. Not all the tablet will dissolve, since the analgesics usually contain an insoluble binder. In addition, many of them contain inorganic buffering agents or coatings that are insoluble in this solvent mixture. After heating the samples, allow them to settle and then spot the clear liquid extracts on the second plate. At the sixth position, spot the standard reference solution. Develop the plate in 0.5% glacial acetic acid-ethyl acetate as before. Observe the plate under UV illumination and mark the visible spots as you did for the first plate. Repeat the visualization using iodine. Sketch the plates in your notebook and record your conclusions about the contents of each tablet. If you were issued an unknown, try to determine its identity (trade name).

Essay
CAFFEINE

The origins of coffee and tea as beverages are so old that they are lost in legend. Coffee is said to have been discovered by an Abyssinian goatherd who noticed an unusual friskiness in his goats when they consumed a certain little plant with red berries. He decided to try the berries himself and discovered coffee. The Arabs soon cultivated the coffee plant, and one of the earliest descriptions of its use is found in an Arabian medical book circa 900 A.D. The great systematic botanist, Linnaeus, named the tree *Coffea arabica*.

One legend of the discovery of tea—from the Orient, as one might expect—attributes the discovery to Daruma, the founder of Zen. Legend has it that he inadvertently fell asleep one day during his customary meditations. To be assured that this indiscretion would not recur, he cut off both eyelids. Where they fell to the ground, a new plant took root that had the power to keep a person awake. Although some experts assert that the medical use of tea was reported as early as 2737 B.C. in the pharmacopeia of Shen Nung, an emperor of China, the first indisputable reference is from the Chinese dictionary of Kuo P'o, which appeared in 350 A.D. The nonmedical, or popular, use of tea appears to have spread slowly. Not until about 700 A.D. was tea widely cultivated in China. Since tea is native to upper Indochina and upper India, it must have been cultivated in these places before its introduction to China. Linnaeus named the tea shrub *Thea sinensis;* however, tea is more properly a relative of the camellia, and botanists have renamed it *Camellia thea*.

The active ingredient that makes tea and coffee valuable to humans is **caffeine.** Caffeine is an **alkaloid,** a class of naturally occurring compounds containing nitrogen and having the properties of an organic amine base (alkaline, hence, **alkaloid**). Tea and coffee are not the only plant sources of caffeine. Others include kola nuts, maté leaves, guarana seeds, and in small amount, cocoa beans. The pure alkaloid was first isolated from coffee in 1821 by the French chemist Pierre Jean Robiquet.

XANTHINES
Xanthine R = R′ = R″ = H
Caffeine R = R′ = R″ = CH_3
Theophylline R = R″ = CH_3, R′ = H
Theobromine R = H, R′ = R″ = CH_3

Caffeine belongs to a family of naturally occurring compounds called **xanthines.** The xanthines, in the form of their plant progenitors, are possibly the oldest known stimulants. They all, to varying extents, stimulate the central nervous system and the skeletal muscles. This stimulation results in an increased alertness, the ability to put off sleep, and an increased capacity for thinking. Caffeine is the most powerful xanthine in this respect. It is the main ingredient of the popular No-Doz keep-alert

72

tablets. While caffeine has a powerful effect on the central nervous system, not all xanthines are as effective. Thus theobromine, a xanthine found in cocoa, has fewer central nervous system effects. It is, however, a strong **diuretic** (induces urination) and is useful to doctors in treating patients with severe water-retention problems. Theophylline, a second xanthine found along with caffeine in tea, also has few central nervous system effects but is a strong myocardial (heart muscle) stimulant; it dilates (relaxes) the coronary artery that supplies blood to the heart. Theophylline, also called aminophylline, is often used in treating congestive heart failure. It is also used to alleviate and to reduce the frequency of attacks of angina pectoris (severe chest pains). In addition, it is a more powerful diuretic than theobromine. Since it is also a vasodilator (relaxes blood vessels), it is often used in treating hypertensive headaches and bronchial asthmas.

One can develop both a tolerance for the xanthines and a dependence on them, particularly caffeine. The dependence is real, and a heavy user (>5 cups of coffee per day) will experience lethargy, headache, and perhaps nausea after about 18 hours of abstinence. An excessive intake of caffeine may lead to restlessness, irritability, insomnia, and muscular tremor. Caffeine can be toxic; but to achieve a lethal dose of caffeine, one would have to drink about 100 cups of coffee over a relatively short period.

Caffeine is a natural constituent of coffee, tea, and kola nuts (Kola nitida). Theophylline is found as a minor constituent of tea. The chief constituent of cocoa is theobromine. The amount of caffeine in tea varies from 2 to 5%. In one analysis of black tea, the following compounds were found: caffeine, 2.5%; theobromine, 0.17%; theophylline, 0.013%; adenine, 0.014%; and guanine and xanthine, traces. Coffee beans can contain up to 5% by weight of caffeine, and cocoa contains around 5% theobromine. Commercial cola is a beverage based on a kola nut extract. We cannot easily get kola nuts in this country, but we can get the ubiquitous commercial extract as a syrup. The syrup can be converted into "cola." The syrup contains caffeine, tannins, pigments, and sugar. Phosphoric acid is added, and caramel is also added to give the syrup a deep color. The final drink is prepared by adding water and carbon dioxide under pressure, to give the bubbly mixture. The Food and Drug Administration currently requires that a "cola" contain **some** caffeine but limits this amount to a maximum of 5 milligrams per ounce. To achieve a regulated level of caffeine, most manufacturers remove all caffeine from the kola extract and then re-add the correct amount to the syrup. The caffeine content of various beverages is listed in the accompanying table.

Amount of Caffeine (mg/oz) Found in Beverages

Brewed coffee	18–25	Tea	5–15
Instant coffee	12–16	Cocoa	1
		(but 20 mg/oz theobromine)	
Decaffeinated coffee	5–10	Coca-Cola	3.5

NOTE: The average cup of coffee or tea contains about 5 oz of liquid. The average bottle of cola contains about 12 oz of liquid.

Because of the central nervous system effects from caffeine, many persons prefer to use **decaffeinated** coffee. The caffeine is removed from coffee by extracting the whole beans with an organic solvent. Then the solvent is drained off, and the beans are steamed to remove any residual solvent. The beans are dried and roasted to bring out the flavor. Decaffeination reduces the caffeine content of coffee to the range of 0.03 to 1.2% caffeine. The extracted caffeine is used in various pharmaceutical products, such as APC tablets.

Adenine   Guanine   Caffeine

Caffeine has always been a controversial compound. Some religions forbid the use of beverages containing caffeine, which they consider an addictive drug. In fact, many people consider caffeine an addictive drug. Recently there has been concern because caffeine is structurally similar to the purine bases adenine and guanine, which are two of the five principal bases organisms use to form the nucleic acids DNA and RNA. It is feared that the substitution of caffeine for adenine or guanine in either of these genetically important substances could lead to chromosome defects.

A portion of the structure of a DNA molecule is shown on the facing page. The typical mode of incorporation of both adenine and guanine is specifically shown. If caffeine were substituted for either of these, the hydrogen bonding necessary to link the two chains would be disturbed. Although caffeine is most similar to guanine, it could not form the central hydrogen bond, since it has a methyl group instead of a hydrogen in the necessary position. Hence, the genetic information would be garbled, and there would be a **break** in the chain. Fortunately, little evidence exists of chromosome breaks due to caffeine. Many cultures have been using tea and coffee for centuries without any apparent genetic problems.

Recent work by researchers at Ohio State University, however, suggests that consumption of caffeine and other xanthine compounds is related to the development of cystic breast disease, which is a nonmalignant but often painful condition characterized by a fibrous growth in the breast. When all coffee, tea, colas, and chocolate were eliminated from the diet of women suffering from cystic breast disease, most women experienced a resolution of this disease.

Another problem, not related to caffeine but rather to the beverage tea, is that in some cases persons who consume high quantities of tea may show symptoms of Vitamin B_1 (thiamine) deficiency. It is suggested that the tannins in the tea may complex with the thiamine, rendering it unavailable for use. An alternative suggestion is that caffeine may reduce the levels of the enzyme transketolase, which depends on the presence of thiamine for its activity. Lowered levels of transketolase would produce the same symptoms as lowered levels of thiamine.

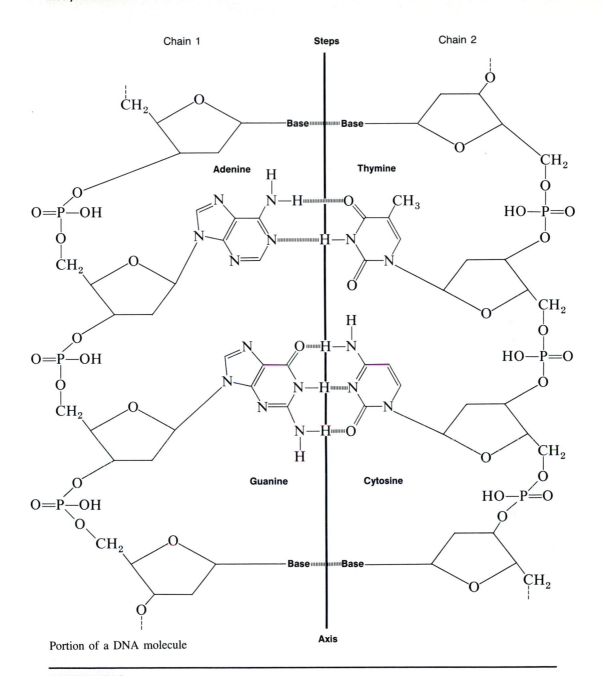

Portion of a DNA molecule

REFERENCES

"Caffeine May Damage Genes by Inhibiting DNA Polymerase." *Chemical and Engineering News, 45* (November 20, 1967): 19.

Emboden, W. "The Stimulants." *Narcotic Plants.* New York: Macmillan, 1972.

Minton, J. P., Foecking, M. K., Webster, D. J. T., and Matthews, R. H. "Caffeine, Cyclic Nucleotides, and Breast Disease." *Surgery, 86* (1979): 105.

Ray, O. S. "Caffeine." *Drugs, Society and Human Behavior.* 2nd ed. St. Louis: C. V. Mosby, 1978.

Ritchie, J. M. "Central Nervous System Stimulants. II: The Xanthines." In L. S. Goodman and A. Gilman, *The Pharmacological Basis of Therapeutics.* 7th ed. New York: Macmillan, 1985.

Taylor, N. *Plant Drugs that Changed the World*. New York: Dodd, Mead, 1965. Pp. 54–56.

Taylor, N. ''Three Habit-Forming Nondangerous Beverages.'' In *Narcotics—Nature's Dangerous Gifts*. New York: Dell, 1970. (Paperbound revision of *Flight from Reality*.)

''Tea, Powdered Milk May Harm Health.'' *Chemical and Engineering News, 54* (May 3, 1976): 27.

Experiment 5
Isolation of Caffeine from Tea

Isolation of a natural product
Extraction
Sublimation

In this experiment, caffeine is isolated from tea leaves. The chief problem with the isolation is that caffeine does not exist alone in tea leaves, but it is accompanied by other natural substances from which it must be separated. The main component of tea leaves is cellulose, which is the principal structural material of all plant cells. Cellulose is a polymer of glucose. Since cellulose is virtually insoluble in water, it presents no problems in the isolation procedure. Caffeine, on the other hand, is water soluble and is one of the main substances extracted into the solution called tea. Caffeine constitutes as much as 5% by weight of the leaf material in tea plants.

Tannins also dissolve in the hot water used to extract tea leaves. The term *tannin* does not refer to a single homogeneous compound, or even to substances that have similar chemical structure. It refers to a class of compounds that have certain properties in common. Tannins are phenolic compounds having molecular weights between 500 and 3000. They are widely used to ''tan'' leather. They precipitate alkaloids and proteins from aqueous solutions. Tannins are usually divided into two classes: those that can be **hydrolyzed** (react with water) and those that cannot. Tannins of the first type that are found in tea generally yield glucose and gallic acid when they are hydrolyzed. These tannins are esters of gallic acid and glucose. They represent structures in which some of the hydroxyl groups in glucose have been esterified by digalloyl groups. The nonhydrolyzable tannins found in tea are condensation polymers of catechin. These polymers are not uniform in structure; catechin molecules are usually linked at ring positions 4 and 8.

Glucose if R = H
A tannin if some R = Digalloyl

A digalloyl group

Catechin

When tannins are extracted into hot water, some of these compounds are partially hydrolyzed to form free gallic acid. The tannins, because of their phenolic groups, and gallic acid, because of its carboxyl groups, are both acidic. If sodium carbonate, a base, is added to tea water, these acids are converted to their sodium salts that are highly soluble in water.

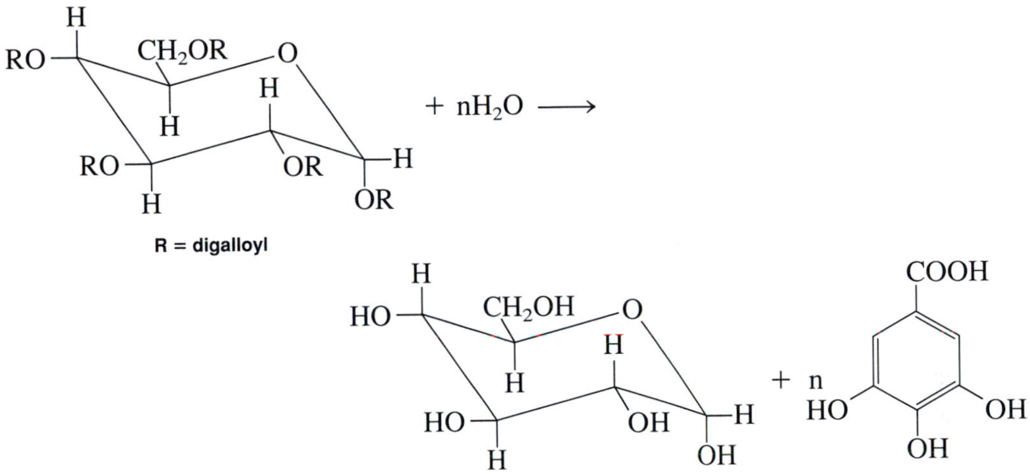

R = digalloyl

Glucose   Gallic acid

Although caffeine is soluble in water, it is much more soluble in the organic solvent, methylene chloride. Caffeine can be extracted from the basic tea solution with methylene chloride, but the sodium salts of gallic acid and the tannins remain in the aqueous layer.

The brown color of a tea solution is due to flavonoid pigments and chlorophylls and to their respective oxidation products. Although chlorophylls are soluble in methylene chloride, most other substances in tea are not. Thus, the methylene chloride extraction of the basic tea solution removes nearly pure caffeine. The methylene chloride is easily removed by evaporation (bp 40 °C) to leave the crude caffeine. The caffeine is then purified by sublimation.

REQUIRED READING

Review: Introduction to Microscale Laboratory (pp 35–48)
          Techniques 1 and 2
          Technique 3   Reaction Methods, Section 3.9
          Technique 6   Physical Constants, Part A, Melting Points

New: Technique 7   Extraction, Sections 7.1–7.6, 7.8, and 7.9
          Technique 15   Sublimation
          Essay   Caffeine

SPECIAL INSTRUCTIONS

Be careful when handling methylene chloride. It is a toxic solvent, and you should not breathe it excessively or spill it on yourself. Dispose of the tea bags in a waste container, not in the sink. The extraction procedure with methylene chloride calls for a centrifuge tube with a screw cap. A cork can also be used to seal the tube; however, the cork will absorb a small amount of the liquid. Rather than shaking the centrifuge tube, agitation can be accomplished conveniently with a vortex mixer.

PROCEDURE

Place 10 mL of water in a 30-mL beaker. Cover the beaker with a watch glass and heat the water in a sand bath (about 130 °C) until it is almost boiling. Place a tea bag[1] into the hot water so that it lies flat on the bottom of the beaker and is covered as completely as possible with water. Replace the watch glass and continue heating for about 15 minutes. During this heating period, it is helpful to push down **gently** on the tea bag with a test tube several times. This will help to ensure that all the tea leaves make good contact with water.

Using a Pasteur pipet, transfer the concentrated tea solution to a centrifuge tube with a screw cap. To squeeze additional liquid out of the tea bag, hold the tea bag on the inside wall of the beaker and roll a test tube back and forth, while exerting **gentle** pressure on the tea bag. Press out as much liquid as possible without breaking the bag. Combine this liquid with the solution in the centrifuge tube. Place the tea bag on the bottom of the beaker again, and pour 2 mL of hot water over the bag. Squeeze the liquid out, as just described, and transfer this liquid to the centrifuge tube. Add 1.0 g of sodium carbonate to the hot liquid. Cap the tube and shake the mixture until the solid dissolves.

Cool the tea solution to room temperature. Using a calibrated Pasteur pipet (p 44), add 3 mL of methylene chloride to extract the caffeine (Technique 7, Section 7.4, p 622). Cap the centrifuge tube and gently shake the mixture for several seconds. Vent the tube to release the pressure, being careful that the liquid does not squirt out toward you. Shake the mixture more vigorously for about 30 seconds with occasional venting. To separate the layers and to break the emulsion (see Technique 7, Section 7.9, p 631), centrifuge the mixture for several minutes (be sure to balance the centrifuge by placing a tube of equal weight on the opposite side). If an emulsion still remains (indicated by a green-brown layer between the clear methylene chloride layer and the top aqueous layer), centrifuge the mixture again.

Remove the lower organic layer with a filter tip pipet (Figure 4–9, p 575) and transfer it to a test tube. Be sure to squeeze the bulb before placing the tip of the Pasteur pipet into the liquid, and try not to transfer any of the dark aqueous solution along with

---

[1] The weight of tea in the bag will be given to you by your instructor. This can be determined by opening several bags of tea and determining the average weight. If this is done carefully, the tea can be returned to the bags, which can be restapled.

the methylene chloride layer. Repeat the extraction one more time, just as above, using a fresh 3-mL portion of methylene chloride. Combine both organic layers in the test tube. If there are visible drops of the dark aqueous solution in the test tube, transfer the methylene chloride solution to another test tube using a clean, dry Pasteur pipet. If necessary, leave a small amount of the methylene chloride solution behind in order to avoid transferring any of the aqueous mixture. Add a small amount (three to four microspatulafuls, using the V-grooved end) of granular anhydrous sodium sulfate to dry the organic layer (Technique 7, Section 7.8, p 629). If all the sodium sulfate clumps together when the mixture is stirred with a spatula, add some additional drying agent. Allow the mixture to stand for 10–15 minutes. Stir occasionally with a spatula.

Transfer the dry methylene chloride solution with a Pasteur pipet to a dry 20 × 150-mm side arm test tube while leaving the drying agent behind. Evaporate the methylene chloride by heating the test tube in a sand bath at 40–55 °C (Technique 3, Section 3.9, p 560). This should be done in a hood and can be accomplished more rapidly if a stream of dry air or nitrogen gas is directed at the surface of the liquid. When the solvent is evaporated, the crude caffeine will coat the bottom of the test tube. Do not heat the test tube after the solvent has evaporated, or you may sublime some of the caffeine.

SUBLIMATION OF CAFFEINE

Caffeine can be purified by sublimation (Technique 15, Section 15.4, p 756). Assemble a sublimation apparatus as shown in Figure 15–2C, p 756. Insert the inner glass tube into a #1 neoprene adapter using a drop of glycerin as lubricant. The tube should be about 1 cm from the bottom of the side arm test tube when the neoprene adapter is inserted into the test tube. Wipe off any glycerin remaining on the inner tube with soft tissue paper soaked in a small amount of acetone. Be sure the tube is clean and dry. Install a trap (Figure 4–6, p 571) between the aspirator and the sublimation apparatus. Turn on the aspirator and press the inner tube and the adapter into the test tube until a good seal is obtained. At the point at which a good seal is achieved, you should hear or observe a change in the water velocity in the aspirator. Place **ice cold** water in the inner tube. Heat the sample gently and carefully with a microburner to sublime the caffeine. Hold the burner in your hand (hold it at its base, **not** by the hot barrel) and apply heat by moving the flame back and forth under the test tube and up the sides. If the sample begins to melt, remove the flame for a few seconds before you resume heating. When sublimation is complete, discontinue heating. Remove the cold water and remaining ice from the inner tube and allow the apparatus to cool with the aspirator still running.

When the apparatus is at room temperature, stop the aspirator and **carefully** remove the inner tube with the neoprene adapter still attached. If this operation is done carelessly, the sublimed crystals may be dislodged from the inner tube and fall back into the test tube. Scrape the sublimed caffeine onto a tared piece of smooth paper and determine the weight of caffeine recovered. Calculate the weight percentage recovery of caffeine from the tea leaves using the weight given to you by your instructor. Determine the melting point of the purified caffeine. The melting point of pure caffeine is 236 °C; however, the observed melting point will be lower. Submit the sample to the instructor in a labeled vial.

QUESTIONS

1. Outline a separation scheme for isolating caffeine from tea. Use a flowchart similar in format to that shown in the **Advance Preparation and Laboratory Records** section (see p 17).

2. Why was the sodium carbonate added?

3. The crude caffeine isolated from tea has a green tinge. Why?

4. What are some possible explanations for why the melting point of your isolated caffeine was lower than the literature value (236 °C)?

5. An alternative procedure for removing the tannins and gallic acid is to heat the tea leaves in an aqueous mixture containing calcium carbonate. Calcium carbonate reacts with the tannins and gallic acid to form insoluble calcium salts of these acids. If this procedure were used, what additional step (not done in this experiment) would be needed in order to obtain an aqueous tea solution?

6. What would happen to the caffeine if the sublimation step were performed at atmospheric pressure?

Essay
ESTERS—FLAVORS AND FRAGRANCES

Esters are a class of compounds widely distributed in nature. They have the general formula

$$R-\overset{\overset{\displaystyle O}{\displaystyle \|}}{C}-OR'$$

The simple esters tend to have pleasant odors. In many cases, although not exclusively so, the characteristic flavors and fragrances of flowers and fruits are due to compounds with the ester functional group. An exception is the case of the essential oils. The organoleptic qualities (odors and flavors) of fruits and flowers may often be due to a single ester, but more often, the flavor or the aroma is due to a complex mixture in which a single ester predominates. Some common flavor principles are listed in Table 1. Food and beverage manufacturers are thoroughly familiar with these esters and often use them as additives to spruce up the flavor or odor of a dessert or a beverage. Many times such flavors or odors do not even have a natural basis, as is the case with the ''juicy fruit'' principle, isopentenyl acetate. An instant pudding that has the flavor of rum may never have seen its alcoholic namesake—this flavor can be duplicated by the proper admixture, along with other minor components, of ethyl formate and iso-butyl propionate. The natural flavor and odor are not exactly duplicated, but most

TABLE 1. Ester Flavors and Fragrances

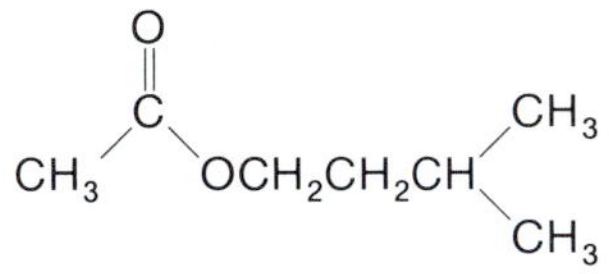

Isoamyl acetate
Banana
(Alarm pheromone of honeybee)

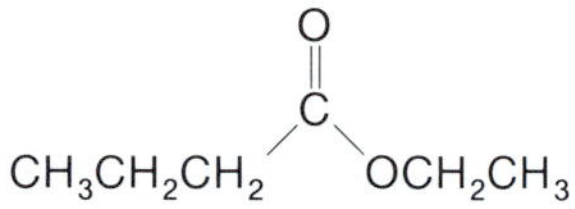

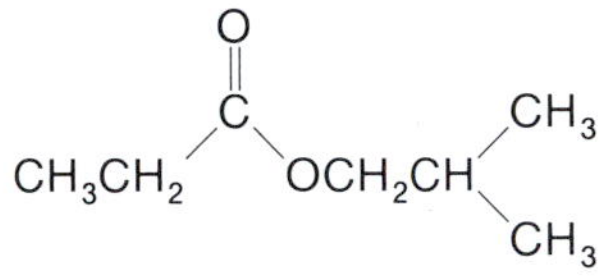

Isobutyl propionate
Rum

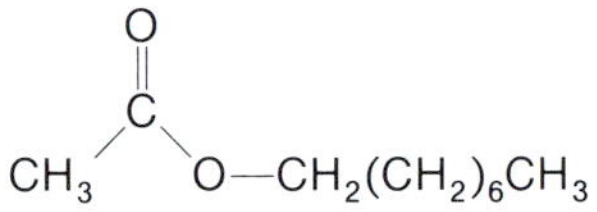

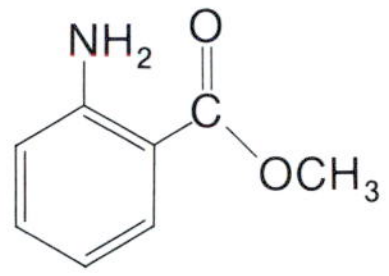

Methyl anthranilate
Grape

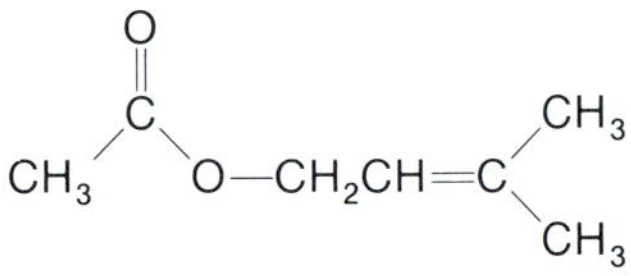

Benzyl acetate
Peach

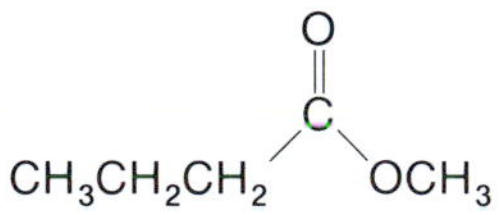

Methyl butyrate
Apple

Ethyl butyrate
Pineapple

Octyl acetate
Oranges

Isopentenyl acetate
"Juicy fruit"

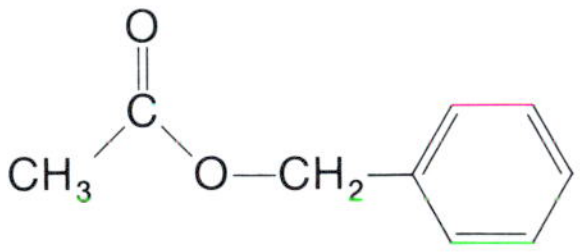

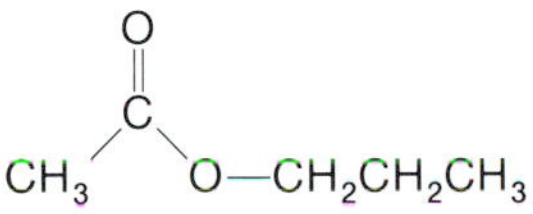

n-Propyl acetate
Pear

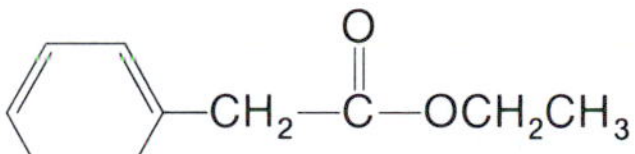

Ethyl phenylacetate
Honey

people can be fooled. Often only a trained person with a high degree of gustatory perception, a professional taster, can tell the difference.

A single compound is rarely used in good-quality imitation flavoring agents. A formula for an imitation pineapple flavor that might fool an expert is listed in Table 2. The formula includes 10 esters and carboxylic acids that can easily be synthesized in the laboratory. The remaining seven oils are isolated from natural sources.

Flavor is a combination of taste, sensation, and odor transmitted by receptors in the mouth (taste buds) and nose (olfactory receptors). The stereochemical theory of odor is discussed in the essay that precedes Experiment 8. The four basic tastes (sweet, sour, salty, and bitter) are perceived in specific areas of the tongue. The sides of the tongue perceive sour and salty tastes, the tip is most sensitive to sweet tastes, and the back of the tongue detects bitter tastes. The perception of flavor, however, is not so

TABLE 2. Artificial Pineapple Flavor

PURE COMPOUNDS	%	ESSENTIAL OILS	%
Allyl caproate	5	Oil of sweet birch	1
Isoamyl acetate	3	Oil of spruce	2
Isoamyl isovalerate	3	Balsam Peru	4
Ethyl acetate	15	Volatile mustard oil	1
Ethyl butyrate	22	Oil cognac	5
Terpinyl propionate	3	Concentrated orange oil	4
Ethyl crotonate	5	Distilled oil of lime	2
Caproic acid	8		19
Butyric acid	12		
Acetic acid	5		
	81		

simple. If it were, it would require only the formulation of various combinations of four basic substances: a bitter substance (a base), a sour substance (an acid), a salty substance (sodium chloride), and a sweet substance (sugar), to duplicate any flavor! In fact, we cannot duplicate flavors in this way. The human actually possesses 9000 taste buds. The combined response of these taste buds is what allows perception of a particular flavor.

Although the "fruity" tastes and odors of esters are pleasant, they are seldom used in perfumes or scents that are applied to the body. The reason for this is chemical. The ester group is not as stable to perspiration as the ingredients of the more expensive essential-oil perfumes. The latter are usually hydrocarbons (terpenes), ketones, and ethers extracted from natural sources. Esters, however, are used only for the cheapest toilet waters, since on contact with sweat, they undergo hydrolysis, giving organic acids. These acids, unlike their precursor esters, generally do not have a pleasant odor.

$$R-\underset{\underset{OR'}{}}{\overset{\overset{O}{\|}}{C}} + H_2O \longrightarrow R-\underset{\underset{OH}{}}{\overset{\overset{O}{\|}}{C}} + R'OH$$

Butyric acid, for instance, has a strong odor like that of rancid butter (of which it is an ingredient) and is a component of what we normally call body odor. It is this substance that makes foul-smelling humans so easy for an animal to detect when he is downwind of them. It is also of great help to the bloodhound, which is trained to follow small traces of this odor. Ethyl butyrate and methyl butyrate, however, which are the **esters** of butyric acid, smell like pineapple and apple, respectively.

A sweet fruity odor also has the disadvantage of possibly attracting fruit flies and other insects in search of food. Isoamyl acetate, the familiar solvent called banana oil, is particularly interesting. It is identical to the alarm **pheromone** of the honeybee. *Pheromone* is the name applied to a chemical secreted by an organism that evokes a specific response in another member of the same species. This kind of communication is common between insects who otherwise lack means of intercourse. When a honeybee worker stings an intruder, an alarm pheromone, composed partly of isoamyl acetate, is secreted along with the sting venom. This chemical causes aggressive attack on the intruder by other bees, who swarm after the intruder. Obviously it wouldn't be wise to wear a perfume compounded of isoamyl acetate near a beehive. Pheromones are discussed in more detail in the essay preceding Experiment 46.

REFERENCES

Benarde, M. A. *The Chemicals We Eat*. New York: American Heritage Press, 1971. Pp. 68–75.

Gould, R. F., ed. *Flavor Chemistry, Advances in Chemistry,* No. 56. Washington: American Chemical Society, 1966.

Pyke, M. *Synthetic Food*. London: John Murray, 1970.

Rasmussen, P. W. "Qualitative Analysis by Gas Chromatography—G. C. versus the Nose in Formulation of Artificial Fruit Flavors. *Journal of Chemical Education, 61* (January 1984): 62.

Shreve, R. N., and Brink, J. *Chemical Process Industries*. 4th ed. New York: McGraw-Hill, 1977.

The Givaudan Index. New York: Givaudan-Delawanna, 1949. (Gives specifications of synthetics and isolates for perfumery.)

Experiment 6
Isopentyl Acetate (Banana Oil)

Esterification
Heating under reflux
Extraction
Simple distillation
Microscale boiling point

In this experiment, we prepare an ester, isopentyl acetate (isoamyl acetate). This ester is often referred to as banana oil, since it has the familiar odor of this fruit.

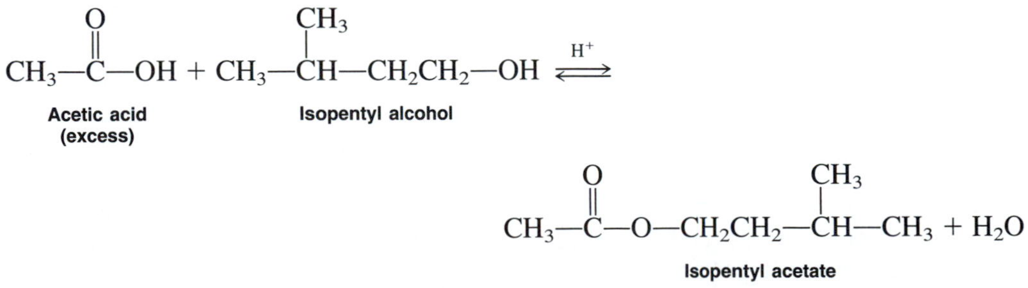

$$CH_3-\overset{\overset{O}{\|}}{C}-OH + CH_3-\overset{\overset{CH_3}{|}}{CH}-CH_2CH_2-OH \xrightarrow{H^+} $$

Acetic acid   **Isopentyl alcohol**
(excess)

$$CH_3-\overset{\overset{O}{\|}}{C}-O-CH_2CH_2-\overset{\overset{CH_3}{|}}{CH}-CH_3 + H_2O$$

Isopentyl acetate

Isopentyl acetate is prepared by the direct esterification of acetic acid with isopentyl alcohol. Since the equilibrium does not favor the formation of the ester, it must be shifted to the right, in favor of the product, by using an excess of one of the starting materials. Acetic acid is used in excess because it is less expensive than isopentyl alcohol and more easily removed from the reaction mixture.

In the isolation procedure, much of the excess acetic acid and the remaining isopentyl alcohol are removed by extraction with sodium and bicarbonate and water. After drying with anhydrous sodium sulfate, the ester is purified by distillation. The purity of the liquid product is analyzed by performing a microscale boiling point determination.

REQUIRED READING

Review: Introduction to Microscale Laboratory (pp 35–48)
         Techniques 1 and 2

New: Technique 3   Reaction Methods, Sections 3.2–3.4, and 3.6
         Technique 6   Physical Constants, Part B, Boiling Points
         Technique 7   Extractions, Separations, and Drying Agents
         Technique 8   Simple Distillation
         Essay   Esters—Flavors and Fragrances

If performing the optional infrared spectroscopy, also read:

         Technique 18   Preparation of Samples for Spectroscopy, Part A

SPECIAL INSTRUCTIONS

Since a one-hour reflux is required, the experiment should be started at the very beginning of the laboratory period. During the reflux period, other experimental work may be performed. Be careful when measuring the concentrated sulfuric acid. It will cause extreme burns if it is spilled on the skin.

PROCEDURE

Weigh (tare) an empty 5-mL conical vial and record its weight. Place approximately 0.7 mL of isopentyl alcohol (MW = 88.2, d = 0.813 g/mL) in the vial using an automatic pipet (or a graduated pipet and pipet pump). Reweigh the vial containing the alcohol and subtract the tare weight to obtain an accurate weight for the alcohol. Add 1.4 mL of glacial acetic acid (MW = 60.1, d = 1.06 g/mL) using an automatic pipet (or a graduated pipet and pipet pump). Using a disposable Pasteur pipet, add two to three drops of concentrated sulfuric acid. Then add a small boiling stone (or a magnetic spin vane).

Assemble a reflux apparatus using a water-cooled condenser (Figure 3–2A, p 550) and top it with a drying tube (Figure 3–6A, p 554) containing a loose plug of glass wool. Using a hot plate and a sand bath, bring the mixture to a boil (sand bath about 150–160 °C). Be sure to stir the mixture if you are using a spin vane instead of a boiling stone. Continue heating under reflux for 60–75 minutes. Remove the heating source, and allow the mixture to cool to room temperature.

Disassemble the apparatus and, using a forceps, remove the boiling stone (or spin vane). Using a calibrated Pasteur pipet (p 44), slowly add 1.0 mL of 5% aqueous sodium bicarbonate to the cooled mixture in the conical vial. Stir the mixture in the vial with a microspatula until carbon dioxide evolution is no longer vigorous. Then cap the vial and shake gently with venting until the evolution of gas is complete. Using a filter tip pipet (Figure 4–9, p 575), remove the lower aqueous layer and discard it. Repeat the extraction two more times, just as above, using a fresh 1.0-mL portion of 5% sodium bicarbonate solution each time.

If droplets of water are evident in the vial containing the ester, transfer the ester to a dry conical vial using a Pasteur pipet. Add a small amount (one to two microspatulafuls, using the V-grooved end) of granular anhydrous sodium sulfate to dry the ester. Allow the capped solution to stand for 10–15 minutes. Transfer the dry ester with a filter tip pipet into a 3-mL conical vial while leaving the drying agent behind. If necessary, pick out any granules of sodium sulfate with the end of a spatula.

Add a boiling stone (or a magnetic spin vane) to the dry ester. Clamping the glassware above the sand bath, assemble a distillation apparatus using a Hickman still and a water-cooled condenser (Figure 8–5, p 643). Top the apparatus with a drying tube packed loosely with a small amount of calcium chloride held in place by plugs of cotton. Begin the distillation by lowering the distillation assembly into the sand bath (180 °C).[1] Continue the distillation until only one or two drops of liquid remain in the distilling vial. If the Hickman head fills before the distillation is complete, it may be necessary to empty it using a Pasteur pipet (see Figure 8–6A, p 644), and transfer the distillate to a tared (pre-weighed) conical vial. Unless you have a side-ported Hickman still, it will be necessary to remove the condenser in order to perform the transfer. When the distillation is complete, transfer the final portion of distillate to this same vial.

Weigh the product and calculate the percentage yield of the ester. Determine its boiling point (bp 142 °C) using a microscale boiling point determination (Technique 6, Section 6.10, p 607).

INFRARED SPECTROSCOPY

At your instructor's option, obtain an infrared spectrum using salt plates (Technique 18, Section 18.2, p 771). Compare the spectrum with the one reproduced in this experiment and include it with your report to the instructor. If any of your sample remains after performing the determination of the infrared spectrum, submit it in a properly labeled vial along with your report.

---

[1] It may be a good idea to cover the top of the sand bath with a square of aluminum foil with a tear from the center of one edge to the middle. This will keep heat from radiating upward and heating the walls of the Hickman still. It will also increase the temperature of the sand bath.

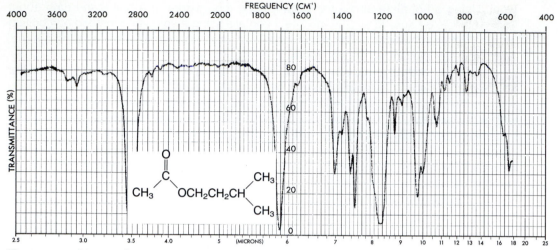

IR spectrum of isopentyl acetate, neat

QUESTIONS

1. One method for favoring the formation of an ester is to add excess acetic acid. Suggest another method, involving the right-hand side of the equation, that will favor the formation of the ester.

2. Why is it easier to remove excess acetic acid from the products than excess isopentyl alcohol?

3. Why is the reaction mixture extracted with sodium bicarbonate? Give an equation and explain its relevance.

4. Which starting material is the **limiting reagent** in this procedure? Which reagent is used in excess? How great is the molar excess (how many times greater)?

5. How many grams are there in 1.00 mL of isopentyl acetate? Look up the density of isopentyl acetate in a handbook.

6. How many moles of isopentyl acetate are there in 1.00 g of isopentyl acetate? You will need to calculate the molecular weight of isopentyl acetate.

7. Suppose that 1.00 mL of isopentyl alcohol was reacted with excess acetic acid and that 1.00 g of isopentyl acetate was obtained as product. Calculate the percentage yield.

8. Outline a separation scheme for isolating pure isopentyl acetate from the reaction mixture.

9. Interpret the principal absorption bands in the infrared spectrum of isopentyl acetate. (Appendix 3 may be of some help in answering this question.)

10. Write a mechanism for the acid-catalyzed esterification of acetic acid with isopentyl alcohol. You may need to consult the chapter on carboxylic acids in your lecture textbook.

Essay
TERPENES AND PHENYLPROPANOIDS

Anyone who has walked through a pine or a cedar forest, or anyone who loves flowers and spices, knows that many plants and trees have distinctively pleasant odors. The essences or aromas of plants are due to volatile or **essential oils,** many of which have been valued since antiquity for their characteristic odors (frankincense and myrrh, for example). A list of the commercially important essential oils would run to over 200 entries. Allspice, almond, anise, basil, bayberry, caraway, cinnamon, clove, cumin, dill, eucalyptus, garlic, jasmine, juniper, orange, peppermint, rose, sandalwood, sassafras, spearmint, thyme, violet, and wintergreen are but a few familiar examples of such valuable essential oils. Essential oils are used for their pleasant odors in perfumes and incense. They are also used for their taste appeal as spices and flavoring agents in foods. A few are valued for antibacterial and antifungal action. Some are used medicinally (camphor and eucalyptus) and others as insect repellents (citronella). Chaulmoogra oil represents one of the few known curative agents for leprosy. Turpentine is used as a solvent for many paint products.

Essential oil components are often found in the glands or intercellular spaces in plant tissue. They may exist in all parts of the plant but are often concentrated in the seeds or flowers. Many components of essential oils are steam-volatile and can be isolated by steam distillation. Other methods of isolating essential oils include solvent extraction and pressing (expression) methods. Esters (see essay, p 80) are frequently responsible for the characteristic odors and flavors of fruits and flowers, but other types of substances may also be important components of odor or flavor principles. Besides the esters, the ingredients of essential oils may be complex mixtures of hydrocarbons, alcohols, and carbonyl compounds. These other components usually belong to one of the two groups of natural products called **terpenes** or **phenylpropanoids.**

TERPENES

Chemical investigations of essential oils in the nineteenth century found that many of the compounds responsible for the pleasant odors contained exactly ten carbon atoms. These ten-carbon compounds came to be known as **terpenes** if they were hydrocarbons, and as **terpenoids** if they contained oxygen and were alcohols, ketones, or aldehydes. Eventually it was found that there are also minor and less-volatile plant constituents containing 15, 20, 30, and 40 carbon atoms. Since compounds of ten carbons were originally called ''terpenes,'' they came to be called **monoterpenes.** The other terpenes were classified in the following way.

87

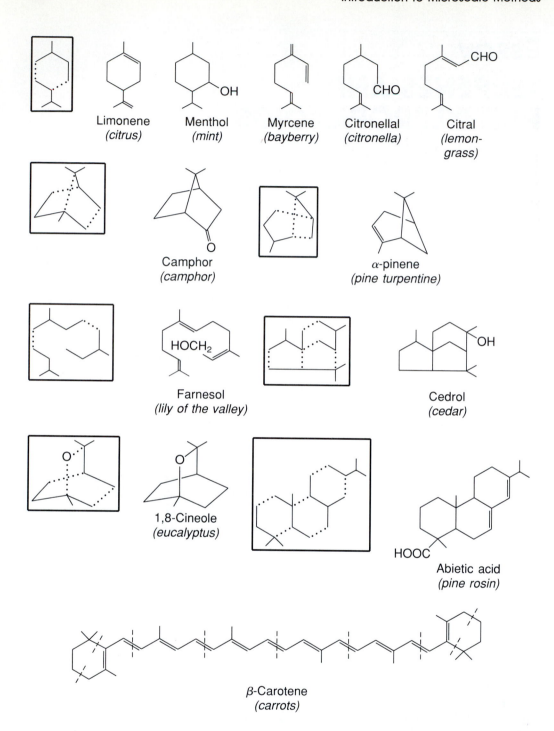

Limonene
(citrus)

Menthol
(mint)

Myrcene
(bayberry)

Citronellal
(citronella)

Citral
(lemon-grass)

Camphor
(camphor)

α-pinene
(pine turpentine)

Farnesol
(lily of the valley)

Cedrol
(cedar)

1,8-Cineole
(eucalyptus)

Abietic acid
(pine rosin)

β-Carotene
(carrots)

Selected terpenes

CLASS	NO. OF CARBONS	CLASS	NO. OF CARBONS
Hemiterpenes	5	Diterpenes	20
Monoterpenes	10	Triterpenes	30
Sesquiterpenes	15	Tetraterpenes	40

Further chemical investigations of the terpenes, all of which contain multiples of five carbons, showed them to have a repeating structural unit which was based on a five-carbon pattern. This structural pattern corresponds to the arrangement of atoms in the simple five-carbon compound isoprene. Isoprene was first obtained by the thermal "cracking" of natural rubber.

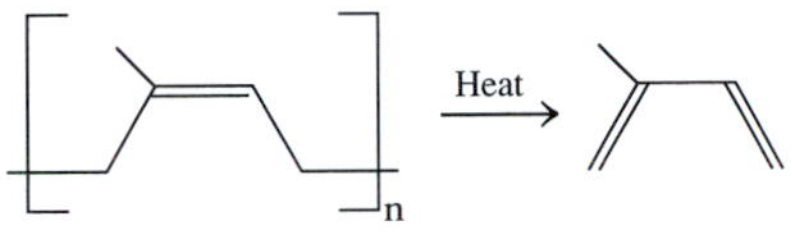

Natural rubber   Heat →   **Isoprene**

As a result of this structural similarity, a diagnostic rule for terpenes, called the **iso-prene rule,** was formulated. This rule states that a terpene should be divisible, at least formally, into **isoprene units.** The structures of a number of terpenes, along with a diagrammatic division of their structures into isoprene units, is shown in the figure which accompanies this essay. Many of these compounds represent odors or flavors which should be very familiar to you.

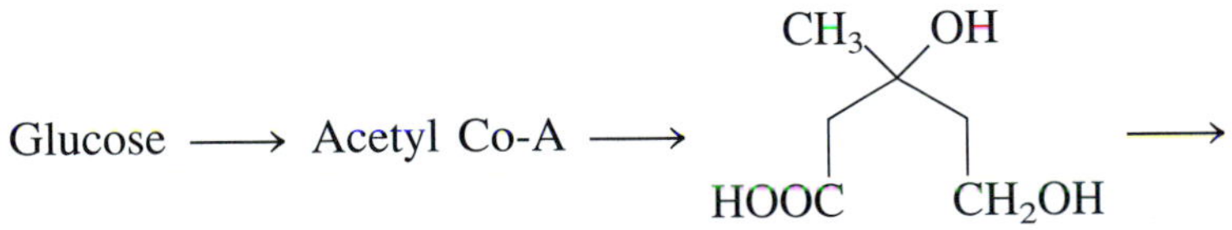

Glucose ⟶ Acetyl Co-A ⟶

$$CH_3 \quad OH$$
$$HOOC \quad CH_2OH$$

Mevalonic acid

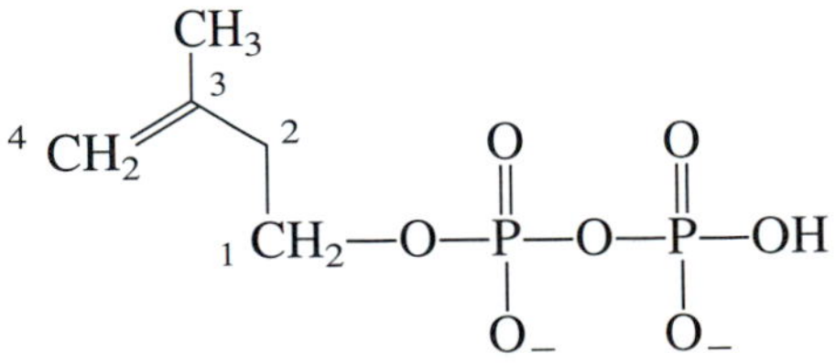

$$CH_3$$
$${}^4 CH_2 \quad {}_2 \quad {}_1 CH_2 - O - P - O - P - OH$$

Isopentenyl pyrophosphate

Modern research has shown that terpenes do not arise from isoprene; it has never been detected as a natural product. Instead, the terpenes arise from an important biochemical precursor compound called **mevalonic acid.** This compound arises from acetyl coenzyme A, a product of the biological degradation of glucose (glycolysis), and is converted to a compound called isopentenyl pyrophosphate. Isopentyl pyrophosphate and its isomer 3,3-dimethylallyl pyrophosphate (double bond moved to the second position) are the five-carbon building blocks used by nature to construct all of the terpene compounds.

PHENYLPROPANOIDS

Aromatic compounds, those containing a benzene ring, are also a major type of compound found in essential oils. Some of these compounds, like *p*-cymene, are actually cyclic terpenes that have been aromatized (had their ring converted to a benzene ring), but most are of a different origin.

Benzene	**p-Cymene**	**Phenylpropane**	**Phenylalanine & Tyrosine** (R = H) (R = OH)

Many of these aromatic compounds are **phenylpropanoids,** compounds based on a phenylpropane skeleton. Phenylpropanoids are related in structure to the common amino acids phenylalanine and tyrosine, and many are derived from a biochemical pathway called the **shikimic acid pathway.**

Caffeic acid (coffee)	**Eugenol (cloves)**	**Vanillin (vanilla)**

It is also common to find compounds of phenylpropanoid origin that have had the three-carbon side chain cleaved. As a result, phenylmethane derivatives, such as vanillin, are also quite common in plants.

REFERENCES

Cornforth, J. W. ''Terpene Biosynthesis.'' *Chemistry in Britain, 4* (1968): 102.
Geissman, T. A., and Crout, D. H. G. *Organic Chemistry of Secondary Plant Metabolism.* San Francisco: Freeman, Cooper and Co., 1969.
Hendickson, J. B. *The Molecules of Nature.* New York: W. A. Benjamin, 1965.

Pinder, A. R. *The Terpenes*. New York: John Wiley and Sons, 1960.

Ruzicka, L. "History of the Isoprene Rule." *Proceedings of the Chemical Society* (London), 1959: 341.

Sterret, F. S. "The Nature of Essential Oils. Part I. Production." *Journal of Chemical Education, 39* (1962): 203.

Sterret, F. S. "The Nature of Essential Oils. Part II. Chemical Constituents. Analysis." *Journal of Chemical Education, 39* (1962): 246.

Experiment 7
Essential Oils from Spices: Oil of Cloves or Cinnamon

Isolation of a natural product
Steam distillation

The technique of steam distillation permits separating volatile components from non-volatile materials without the need for raising the temperature of the distillation above 100 °C. Steam distillation provides a means of isolating natural products such as **essential oils** without the risk of decomposing them thermally. Essential oils are volatile compounds responsible for the aromas commonly associated with many plants (see essay, "Terpenes and Phenylpropanoids").

The chief constituents of essential oils from cloves and cinnamon are aromatic and volatile with steam. In this experiment we isolate the main component of the essential oil derived from these spices by steam distillation. Identification and characterization of the essential oils will be accomplished by infrared spectroscopy.

Oil of cloves (from *Eugenia caryophyllata*) is rich in **eugenol** (4-allyl-2-methoxyphenol). Caryophyllene is present in small amounts, along with other terpenes. Eugenol (bp 250 °C) is a phenol, or an aromatic hydroxy compound.

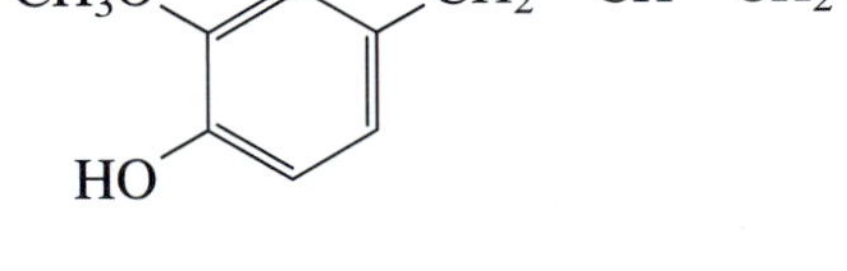

Eugenol

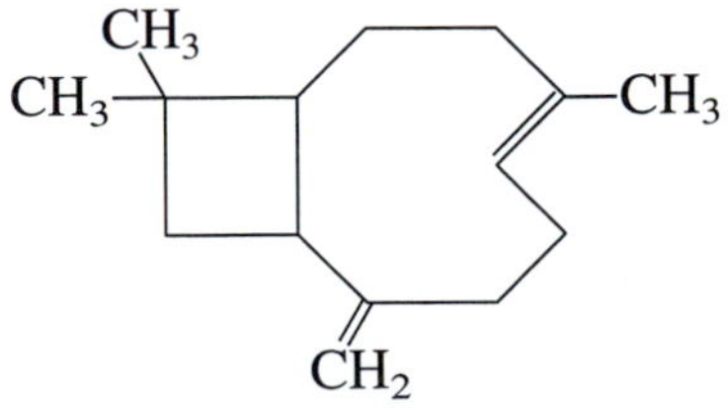

Caryophyllene

The principal component of cinnamon oil (from *Cinnamomum zeylanicum*) is cinnamaldehyde (*trans*-3-phenylpropenal). Cinnamaldehyde, which has a boiling point of 252 °C, is an example of a phenylpropanoid.

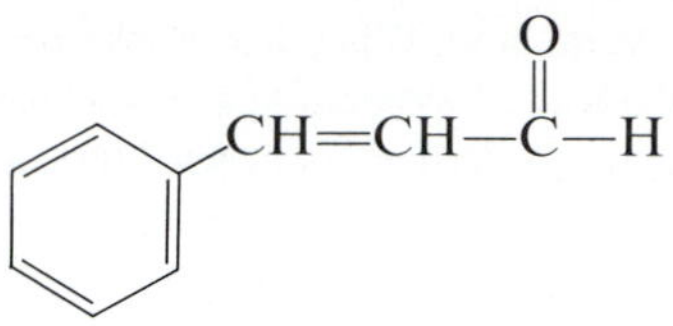

$$CH=CH-\overset{\overset{\displaystyle O}{\|}}{C}-H$$

Cinnamaldehyde

REQUIRED READING

Review: Introduction to Microscale Laboratory (pp 35–48)
          Techniques 1 and 2
          Technique 3   Reaction Methods, Section 3.9
          Technique 7   Extractions, Sections 7.4 and 7.8
          Technique 18   Preparation of Samples for Spectroscopy, Part A
          Appendix 3   Infrared Spectroscopy

New: Technique 11   Steam Distillation
          Essay   Terpenes and Phenylpropanoids

          If performing the optional NMR analysis, also read:

          Technique 18   Preparation of Samples for Spectroscopy, Part B
          Appendix 4   Nuclear Magnetic Resonance Spectroscopy

---

SPECIAL INSTRUCTIONS

In order to complete the distillation in a reasonable length of time, it will be necessary to boil the mixture as rapidly as possible without allowing the boiling mixture to rise above the neck of the Hickman head. This will require that you work with very careful attention during the distillation procedure. The distillation will require one to two hours for either spice.

PROCEDURE

Assemble an apparatus using a 10-mL round-bottom flask, a Hickman head, and a water cooled reflux condenser, as shown in Figure 11–3, p 693. It is not necessary to use the stainless steel sponge which is indicated in this figure. Adjust the temperature of the sand bath initially to about 130 °C. During the distillation, you should be able to increase the sand bath temperature to about 150 °C, as frothing becomes less of a problem.

   If you are steam distilling oil of cloves, place about 0.30 g (record actual weight) of ground cloves into the round-bottom flask and add 4.0 mL of water and a boiling stone. For the distillation with cinnamon, use about 0.50 g of ground cinnamon and 6.0 mL of water and add a small magnetic stirring bar instead of a boiling stone. In order to prevent the cinnamon from clumping, it will be necessary to stir the mixture during the distillation. Note that with cinnamon, you will be instructed to add a fresh portion of ground cinnamon and water halfway through the distillation.

Begin heating the mixture to provide a steady rate of distillation (see Technique 11, Section 11.3, Method A, p 691). The temperature of the mixture can be controlled by raising or lowering the round-bottom flask in the sand. You may also push the sand towards the flask with a spatula when a higher temperature is needed, or the temperature of the sand bath may be increased. With cinnamon it is essential to control the temperature so that the boiling mixture does not rise above the neck of the Hickman head. While being careful, remember that in order to complete the distillation in a reasonable length of time, you will have to boil the mixture as rapidly as possible without allowing the boiling mixture to splash into the Hickman head. The distillation with cloves can be done with more vigorous boiling; however, some care should be exercised to ensure that the mixture does not froth too much.

To perform the following operations, it will be necessary to remove the reflux condenser. During the distillation, replace the water lost through distillation by adding water to the boiling mixture using a Pasteur pipet which is passed through the Hickman head into the round-bottom flask. As the distillation proceeds, collect the distillate with a second Pasteur pipet (5¾-inch) and use it to rinse the walls of the Hickman head at 5 to 10 minute intervals. As the distillate continues to collect in the reservoir of the Hickman head, use the Pasteur pipet to transfer it to a 5-mL conical vial. The rinsing and transferring operations are best accomplished if the Pasteur pipet is bent slightly at the end.

Continue the steam distillation with frequent rinsing of the Hickman head and collection of the distillate in the 5-mL conical vial until you have obtained 2.5 mL of distillate **if you are distilling oil of clove.** Then proceed to the next paragraph. **If you are working with cinnamon,** stop the distillation when you have collected about 1.5 mL of distillate. Raise the assembly and allow the round-bottom flask to cool until you can touch it with your fingers. Remove the flask and discard the mixture of ground cinnamon and water, but save the stirring bar! It is not necessary to clean the round-bottom flask or the Hickman head. Add 0.50 g of cinnamon (record actual weight), 6.0 mL of water, and a magnetic stirring bar to the flask. Reassemble the apparatus and continue the distillation. Rinse the Hickman head as before and transfer the distillate to the same 5-mL conical vial. When you have collected another 1.5 mL of distillate (for a total of 3.0 mL), the distillation may be stopped.

The essential oil is now extracted with methylene chloride (see Technique 7, Section 7.4, p 622). Using a calibrated Pasteur pipet (p 44), add 1.0 mL of methylene chloride to the 5-mL conical vial containing the distillate. Cap the vial securely and shake it vigorously with frequent venting. Allow the layers to separate. Using a filter tip pipet (see Figure 4–9), transfer the lower methylene chloride layer to a second dry 5-mL conical vial. Repeat this procedure with two more 1.0-mL portions of methylene chloride and combine all the methylene chloride extracts in the second 5-mL conical vial. If there are visible drops of water in the vial, it will be necessary to transfer the methylene chloride solution with a dry Pasteur pipet to another dry conical vial. Dry the methylene chloride solution by adding three to four microspatulafuls (measured with the V-groove end) of granular anhydrous sodium sulfate to the conical vial (see Technique 7, Section 7.8, p 629). Let set for 10–15 minutes with occasional stirring.

While the organic solution is being dried, clean and dry the first 5-mL conical vial and weigh it accurately. With a clean, dry filter tip pipet transfer the dried organic layer to this tared vial. Use small amounts of clean methylene chloride to rinse the solution completely into the tared vial. Be careful to keep the sodium sulfate from being transferred to the vial. Remove the methylene chloride by leaving the vial in the hood until the solvent has evaporated and nothing but an oily residue remains. As an alternative, you may evaporate the methylene chloride in the hood with a gentle stream of dry air or nitrogen while heating the conical vial in a warm sand bath at 40–50 °C (see Technique 3, Section 3.9, p 560). Tap the vial with your finger to provide agitation in order to facilitate removal of the remaining solvent.

When the solvent has been removed, weigh the conical vial. Calculate the weight percentage recovery of the oil from the original amount of spice used.

Obtain the infrared spectrum of the oil as a pure liquid sample (Technique 18, Section 18.2, p 771). It may be necessary to use a micro syringe or a Pasteur pipet with a narrow tip to transfer a sufficient amount of liquid from the vial to the salt plates. Include the spectrum in your laboratory report, along with an interpretation of the principal peaks. At the instructor's option, determine the nuclear magnetic spectrum of the oil (Technique 18, Section 18.9, p 782).

QUESTIONS

1. Why are eugenol and cinnamaldehyde steam-distilled rather than purified by simple distillation?

2. A natural product (MW = 150) distills with steam at a boiling temperature of 99 °C at atmospheric pressure. The vapor pressure of water at 99 °C is 733 mmHg.

    (a) Calculate the weight of the natural product that co-distills with each gram of water at 99 °C.

    (b) How much water must be removed by steam distillation to recover this natural product from 0.5 g of a spice which contains 10% of the desired substance?

3. In a steam distillation, the amount of water actually distilled is usually greater than the amount calculated, assuming that both water and organic substance exert the same vapor pressure when they are mixed that they exert when each is pure. Why does one recover more water in the steam distillation than was calculated?

4. Explain how caryophyllene fits the isoprene rule (see essay, p 87).

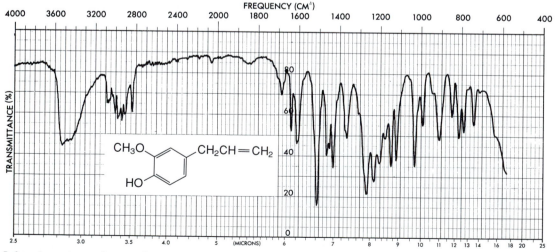

Infrared spectrum of eugenol, neat

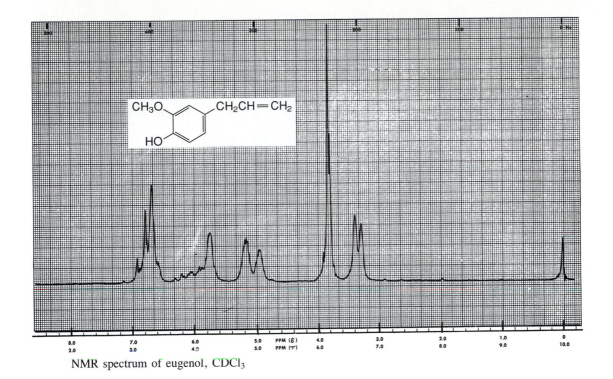

NMR spectrum of eugenol, CDCl₃

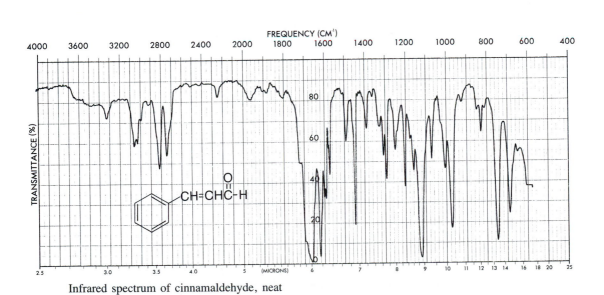

Infrared spectrum of cinnamaldehyde, neat

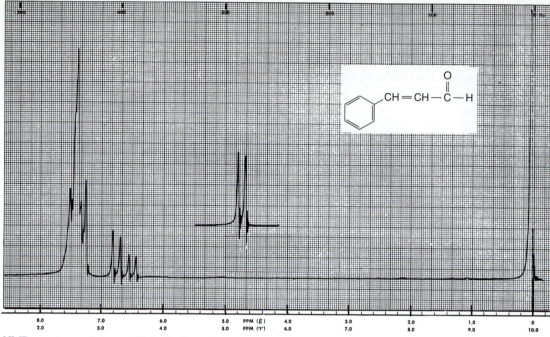

NMR spectrum of cinnamaldehyde (Doublet offset by 300 Hz)

Essay
STEREOCHEMICAL THEORY OF ODOR

The human nose has an almost unbelievable ability to distinguish odors. Just consider for a few moments the different substances you are able to recognize by odor alone. Your list should be a very long one. A person with a trained nose, a perfumer, for instance, can often recognize even individual components in a mixture. Who has not met at least one cook who could sniff almost any culinary dish and identify the seasonings and spices that were used? The olfactory centers in the nose can identify odorous substances even in a very small amount. With some substances, studies have shown that as little as one ten-millionth of a gram (10^{-7} g) can be perceived. Many animals, for example, dogs and insects, have an even lower threshold of smell than humans (see essay on pheromones that precedes Experiment 46).

There have been many theories of odor, but few have persisted very long. Strangely enough, one of the oldest theories, although in modern dress, is still the most current theory. Lucretius, one of the early Greek atomists, suggested that substances having odor gave off a vapor of tiny "atoms," all of the same shape and size, and that they gave rise to the perception of odor when they entered pores in the nose. The pores would have to be of various shapes and the odor perceived would depend on which pores the atoms were able to enter. We now have many similar theories regarding the action of drugs (receptor-site theory) and the interaction of enzymes with their substrates (the lock-and-key hypothesis).

A substance must have certain physical characteristics to have the property of odor. First, it must be volatile enough to give off a vapor that can reach the nostrils. Second, once it reaches the nostrils, it must be somewhat water-soluble, even if only to a small degree, so that it can pass through the layer of moisture (mucus) that covers the nerve endings in the olfactory area. Third, it must also have lipid solubility to allow it to penetrate the lipid (fat) layers that form the surface membranes of the nerve cell endings.

Once we pass these criteria, we come to the heart of the question. Why do substances have different odors? In 1949, R. W. Moncrieff, a Scotsman, resurrected Lucretius' hypothesis. He proposed that in the olfactory area of the nose there is a system of receptor cells of several different types and shapes. He further suggested that each receptor site corresponded to a different type of primary odor. Molecules that would fit these receptor sites would display the characteristics of that primary odor. It would not be necessary for the entire molecule to fit into the receptor, so that for larger molecules, any portion might fit into the receptor and activate it. Molecules having complex odors would presumably be able to activate several different types of receptors.

Moncrieff's hypothesis has been strengthened substantially by the work of J. E. Amoore, who began studying the subject as an undergraduate at Oxford in 1952. After an extensive search of the chemical literature, Amoore concluded that there were only seven basic primary odors. By sorting molecules with similar odor types, he even formulated possible shapes for the seven necessary receptors. For instance, from the literature, he culled more than 100 compounds that were described as having a "camphoraceous" odor. Comparing the sizes and shapes of all these molecules, he postulated a three-dimensional shape for a camphoraceous receptor site. Similarly, he derived shapes for the other six receptor sites. The seven primary receptor sites he formulated are shown in the figure (p 98), along with a typical prototype molecule of the appropriate shape to fit the receptor. The shapes of the sites are shown in perspective. Pungent and putrid odors are not thought to require a particular shape in the odorous molecular but rather to need a particular type of charge distribution.

You can verify quickly that compounds with molecules of roughly similar shape have similar odors if you compare nitrobenzene and acetophenone with benzaldehyde or *d*-camphor and hexachloroethane with cyclooctane. Each group of substances has the same basic odor **type** (primary), but the individual molecules differ in the **quality** of the odor. Some of the odors are sharp, some pungent, others sweet, and so on. The second group of substances all have a camphoraceous odor, and the molecules of these substances all have approximately the same shape.

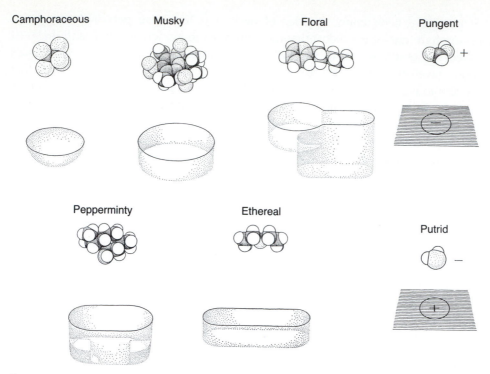

An interesting corollary to the Amoore theory would be the postulate that if the receptor sites are chiral, then optical isomers (enantiomers) of a given substance might have **different** odors. This circumstance proves true in several cases. It is true for (+)- and (−)-carvone; we investigate the idea in Experiment 8 in this textbook.

Several workers have tested Amoore's hypothesis by experiment. The results of these studies are generally favorable to the hypothesis—so favorable that some chemists now elevate the hypothesis to the level of theory. In several cases, researchers have been able to ''synthesize'' odors almost indistinguishable from the real thing by properly blending primary odor substances. The primary odor substances used were unrelated to the chemical substances composing the natural odor. These experiments, and others, are described in the articles listed at the end of this essay.

REFERENCES

Amoore, J. E. *The Molecular Basis of Odor*. American Lecture Series Publication, No. 773. Springfield, IL. Thomas, 1970.

Amoore, J. E., Johnston, J. W., Jr., and Rubin, M. ''The Stereochemical Theory of Odor.'' *Scientific American, 210* (February 1964): 1.

Amoore, J. E., Rubin, M., and Johnston, J. W., Jr. ''The Stereochemical Theory of Olfaction.'' *Proceedings of the Scientific Section of the Toilet Goods Association* (Special Supplement to No. 37) (October 1962): 1–47.

Burton, R. *The Language of Smell*. London: Routledge & Kegan Paul, 1976.
Moncrieff, R. W. *The Chemical Senses*. London: Leonard Hill, 1951.
Theimer, E. T., ed. *Fragrance Chemistry*. New York: Academic Press, 1982.

Experiment 8
Spearmint and Caraway Oil: (−)- and (+)-Carvones

Stereochemistry   Spectroscopy
Gas chromatography   Refractometry
Optical rotation

(R)-(−)-Carvone
from spearmint oil

(S)-(+)-Carvone
from caraway-seed oil

In this experiment, we isolate (+)-carvone from caraway-seed oil or (−)-carvone from spearmint oil using preparative gas chromatography. The odors of these optical isomers are distinctly different from each other. The presence of one or the other of these isomers is responsible for the characteristic odors of each of the two oils. The difference in their odors is to be expected, since the odor receptors in the nose are chiral (see essay, ''Stereochemical Theory of Odor''). This phenomenon in which a chiral receptor interacts differently with each of the enantiomers of a chiral compound is called **chiral recognition.** Another example of chiral recognition can be found in the effect that these two carvone isomers have on rats. The toxicity of (+)-carvone in rats is 400 times greater than that of (−)-carvone.

Although we should expect the optical rotations of the isomers (enantiomers) to be of opposite sign, the other physical properties should be identical. Thus, for both (+)- and (−)-carvone, we predict that the infrared and nuclear magnetic resonance spectra, the gas chromatographic retention times, the refractive indices, and the boiling points should all be identical. Hence, the only difference in properties one will observe for the two carvones are the odors and the signs of rotation in a polarimeter.

Caraway-seed oil contains mainly limonene and (+)-carvone. The gas chromatogram for this oil is shown in the figure. The (+)-carvone (bp 230 °C) can easily be separated from the lower-boiling limonene (bp 177 °C) by gas chromatography, as shown in the figure. Using gas collection tubes, the (+)-carvone and limonene can be collected separately as they elute from the gas chromatography column. **Spearmint oil** contains mainly (−)-carvone with a smaller amount of limonene and very small amounts of the lower boiling terpenes, α- and β-phellandrene. The gas chromatogram for this oil is also shown in the figure. The (−)-carvone can easily be collected as it exits the column. It is more difficult, however, to collect limonene in a pure form. It is likely to be contaminated with the other terpenes, since they all have similar boiling points.

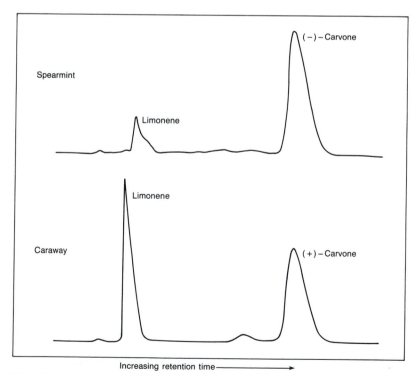

Gas chromatograms of caraway and spearmint oil

REQUIRED READING

Review: Introduction to Microscale Laboratory (pp 35–48)

   Technique 18   Preparation of Samples for Spectroscopy, Part A, Infrared

   Appendix 3   Infrared Spectroscopy

New: Technique 14   Gas Chromatography
   Technique 16   Polarimetry
   Essay   Stereochemical Theory of Odor

If performing any of the optional procedures, also read:

   Technique 6   Physical Constants, Part B, Boiling Points
   Technique 17   Refractometry
   Technique 18   Preparation of Samples for Spectroscopy, Part B, Nuclear Magnetic Resonance

   Appendix 4   Nuclear Magnetic Resonance Spectroscopy
   Appendix 5   Carbon-13 Nuclear Magnetic Resonance Spectroscopy

SPECIAL INSTRUCTIONS

Your instructor will either assign you spearmint or caraway oil, or have you choose one of the oils. You will also be given instructions on which optional procedures, if any, you are to perform. Your instructor may have you isolate both the carvone and limonene components or only carvone. Avoid contact with (+)-carvone which is very toxic.

NOTE TO THE INSTRUCTOR: This experiment may be scheduled along with another experiment. An appointment schedule for using the gas chromatograph should be arranged so that students are able to make efficient use of their time.

The gas chromatograph should be prepared as follows: column temperature, 200 °C; injection and detector temperature, 210 °C; carrier gas flow rate, 20 mL/min. The recommended column is 8 feet long with a stationary phase such as Carbowax 20M. It is convenient to use a Gow-Mac 69–350 instrument with the preparatory accessory system for this experiment.

The polarimeter cells should be filled in advance with (+)-carvone or (−)-carvone.

PROCEDURE

PREPARATION AND MEASUREMENT OF STANDARDS
(Procedures for instructor)

The instructor should prepare gas chromatograms using both carvone isomers and limonene as reference standards. Appropriate reference standards include a mixture of (+)-carvone and limonene and a second mixture of (−)-carvone and limonene. The chromatograms should be posted with retention times, or each student should be provided with a copy of the appropriate chromatogram.

In addition, the polarimeter cells should be filled by the instructor. One cell is filled with commercially available (+)-carvone and the other cell with (−)-carvone.

SEPARATION BY GAS CHROMATOGRAPHY

The instructor may prefer to perform the sample injections or have a laboratory assistant perform them. The sample injection procedure requires careful technique, and the special microliter syringes that are required are very delicate and expensive. If students are to perform the sample injections themselves, they must have adequate instruction beforehand.

Inject 50 μL of caraway-seed or spearmint oil onto the gas chromatography column. Just before a component of the oil (limonene or carvone) elutes from the column, install a gas collection tube at the exit port, as described in Technique 14, Section 14.10, p 748. To determine when to connect the gas collection tube, refer to the chromatograms prepared by your instructor. These chromatograms have been run on the same instrument you are using under the same conditions. Ideally, you should connect the gas collection tube just before the limonene or carvone elutes from the column and remove the tube as soon as all the component has been collected, but before any other compound begins to elute from the column. This can be accomplished most easily by watching the recorder as your sample passes through the column. The collection tube is connected (if possible) just before a peak is produced, or as soon as a deflection in the pen is observed. When the pen has returned to the base line, the gas collection tube is removed.

This procedure is relatively easy for collecting the carvone component of both oils and for collecting limonene in caraway-seed oil. Because of the presence of several terpenes in spearmint oil, it is somewhat more difficult to isolate a pure sample of limonene from spearmint oil (see chromatogram in figure). In this case, you must try to collect only the limonene component and not any other compounds, such as the terpene which produces a shoulder on the limonene peak in the chromatogram for spearmint oil.

After collecting the samples, insert the ground joint of the collection tube into a 0.1-mL conical vial, using an O-ring and cap to fasten the two pieces together securely. Insert the top of the collection tube through the hole in the rubber septum cap and place this assembly into a test tube, as shown in Figure 14–10, p 749. Put cotton on the bottom of the test tube to prevent breakage. Balance the centrifuge by placing a tube of equal weight on the opposite side. During centrifugation, the sample is forced into the bottom of the conical vial. Disassemble the apparatus, cap the vial, and perform the analyses described on page 103.

ANALYSIS OF THE CARVONES

The samples obtained by gas chromatography and centrifugation should be analyzed by the methods below, and the instructor will indicate which methods to use. Compare your results with those obtained by someone who used a different oil. In addition, measure the observed rotation of the commercial samples of (+)-carvone and (−)-carvone.

Odor. About 8 to 10% of the population cannot detect the difference in the odors of the optical isomers. Most people, however, find the difference quite obvious.

Gas Chromatography. Determine the retention times of the components (see Technique 14, Section 14.7, p 747). Calculate the percentage composition of the carvone sample by the method explained in Technique 14, Section 14.11, p 750.

Infrared Spectroscopy. Obtain the infrared spectrum of the (−)-carvone sample from spearmint or of the (+)-carvone sample from caraway (see Technique 18, Section 18.2, p 771). At the option of the instructor, obtain the infrared spectrum of the (+)-limonene, which is found in both oils. If possible, determine all spectra on neat samples. It may be necessary in some cases, however, to add one to two drops of carbon tetrachloride to the sample. Thoroughly mix the liquids by drawing the mixture into a Pasteur pipet and expelling several times. It may be helpful to draw the end of the pipet to a narrow tip in order to withdraw all the liquid in the conical vial. As an alternative, a microsyringe may be used. Obtain a spectrum of this solution, as described in Technique 18, Section 18.5, p 777.

Polarimetry. With the help of the instructor or assistant, obtain the observed optical rotation α of the (+)-carvone and (−)-carvone samples. The specific rotation is calculated from the relation given on p 761 of Technique 16. The concentration c will equal the density of the substances analyzed at 20 °C. The values are 0.9608 g/mL for (+)-carvone and 0.9593 g/mL for (−)-carvone. The literature values for the specific rotations are as follows: $[\alpha]_D^{20} = +61.7°$ for (+)-carvone and $-62.5°$ for (−)-carvone. These values are not identical because trace amounts of impurities are present.

ANALYSIS PROCEDURES (optional)

Some of these procedures may require more sample than provided by following the instructions given above. Several students may need to combine their samples, or you may perform the procedures on the commercial samples.

Refractive Index. Use the technique for obtaining the refractive index on a small volume of liquid, as described in Technique 17, Section 17.2, p 766. Obtain the refractive index for the carvone sample from spearmint or caraway-seed oil. At 20 °C, the (+)- and (−)-carvones have the same refractive index, equal to 1.4989.

Boiling Point. Determine the boiling point on the carvone you isolated using the micro boiling point technique (Technique 6, Section 6.10, p 607). The boiling points for both carvones are 230 °C at atmospheric pressure.

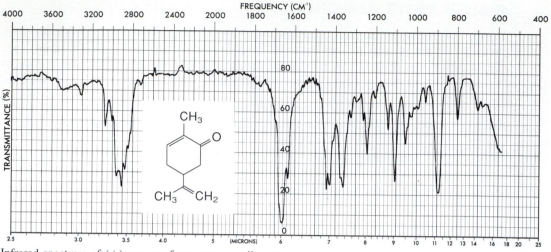

Infrared spectrum of (+)-carvone from caraway oil, neat

Nuclear Magnetic Resonance Spectroscopy.

Using an NMR instrument, obtain a proton NMR spectrum of the carvone. Compare your spectrum with the NMR spectra for (−)-carvone and (+)-limonene shown in this experiment. If your NMR instrument is capable of obtaining a carbon-13 NMR spectrum, determine a carbon-13 spectrum. Compare your spectrum of carvone with the carbon-13 NMR spectrum shown in this experiment.

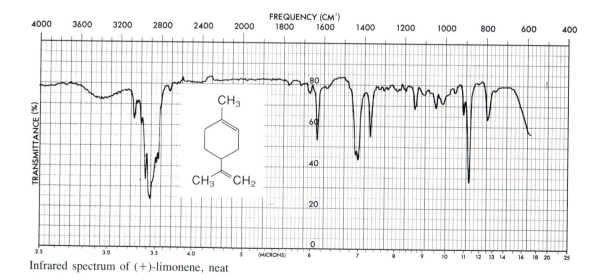

Infrared spectrum of (+)-limonene, neat

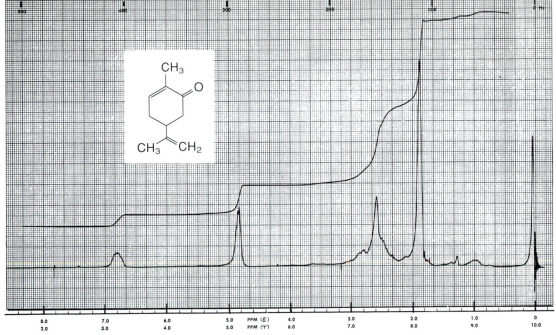

NMR spectrum of (−)-carvone from spearmint oil

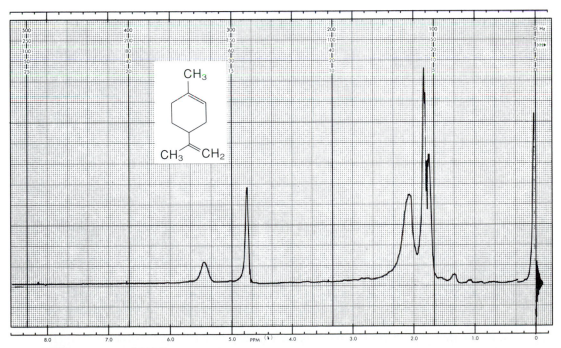

NMR spectrum of (+)-limonene

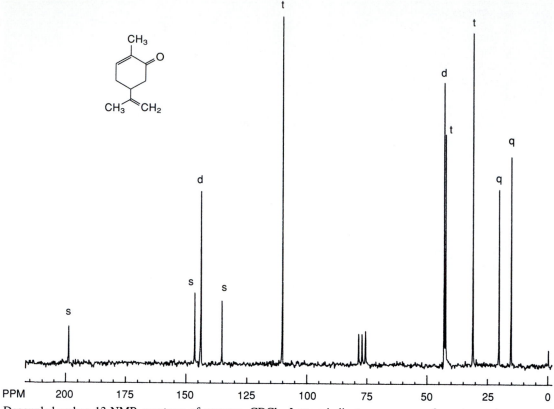

Decoupled carbon-13 NMR spectrum of carvone, $CDCl_3$. Letters indicate appearance of spectrum when carbons are coupled to protons (s = singlet, d = doublet, t = triplet, q = quartet).

REFERENCES

Friedman, L., and Miller, J. G. "Odor Incongruity and Chirality." *Science, 172* (1971): 1044.

Murov, S. L., and Pickering, M. "The Odor of Optical Isomers." *Journal of Chemical Education, 50* (1973): 74.

Russell, G. F., and Hills, J. I. "Odor Differences between Enantiomeric Isomers." *Science, 172* (1971): 1043.

QUESTIONS

1. Interpret the IR spectra for carvone and limonene and the NMR and carbon-13 spectra of carvone.

2. Identify the chiral centers in α-phellandrene, β-phellandrene, and limonene.

3. Explain how carvone fits the isoprene rule (see essay, p 87).

4. Using the Cahn-Ingold-Prelog sequence rules, assign priorities to the groups around the chiral carbon in carvone. Draw structural formulas for (+)- and (−)-carvone with the molecules oriented in the correct position to show the R and S configurations.

5. Explain why limonene elutes from the column before either (+)- or (−)-carvone.

6. Explain why the retention times for both carvone isomers are the same.

Essay
THE CHEMISTRY OF VISION

An interesting and challenging topic for chemists to investigate is how the eye functions. What chemistry is involved in detection of light and transmission of that information to the brain? The first definitive studies on how the eye functions were begun in 1877 by Franz Boll. Boll demonstrated that the red color of the retina of a frog's eye could be bleached yellow by strong light. If the frog was then kept in the dark, the red color of the retina slowly returned. Boll recognized that a bleachable substance had to be somehow connected with the ability of the frog to perceive light.

Most of what is now known about the chemistry of vision is the result of the elegant work of George Wald, Harvard University; his studies, which began in 1933, ultimately brought about his receiving the Nobel Prize in biology. Wald identified the sequence of chemical events during which light is converted into some form of electrical information that can be transmitted to the brain. Here is a brief outline of that process.

The retina of the eye is made up of two types of photoreceptor cells: **rods** and **cones.** The rods are responsible for vision in dim light, and the cones are responsible for color vision in bright light. The same principles apply to the chemical functioning of the rods and of the cones; however, the details of functioning are less well under-stood for the cones than for the rods.

Each rod contains several million molecules of **rhodopsin.** Rhodopsin is a complex of a protein, **opsin,** and a molecule derived from Vitamin A, 11-*cis*-retinal (sometimes called **retinene**). Very little is known about the structure of opsin. The structure of 11-*cis*-retinal is shown below.

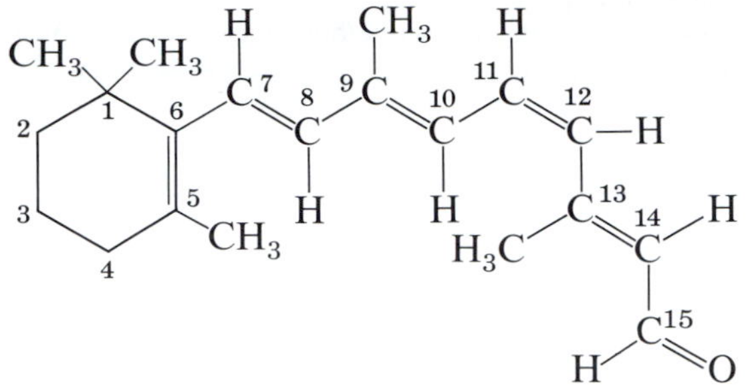

11-*cis*-retinal

The detection of light involves the initial conversion of 11-*cis*-retinal to its all-*trans* isomer. This is the only obvious role of light in this process. The high energy of a quantum of visible light promotes the fission of the π bond between carbons 11 and 12. When the π bond breaks, free rotation about the σ bond in the resulting radical is possible. When the π bond re-forms after such rotation, all-*trans*-retinal results. All-*trans*-retinal is more stable than 11-*cis*-retinal, which is why the isomerization proceeds spontaneously in the direction shown.

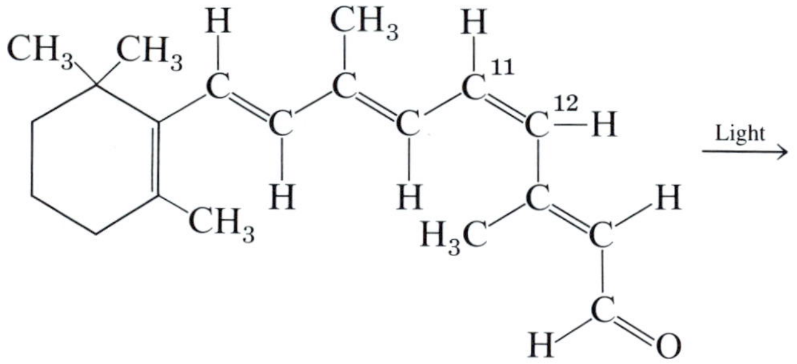

11-*cis*-retinal

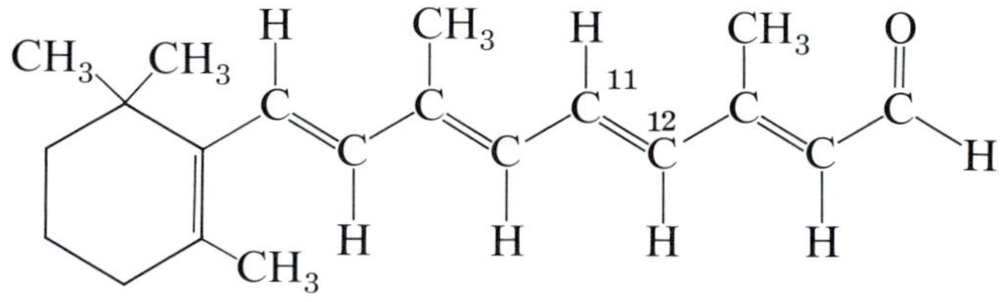

All-*trans*-retinal

The two molecules have different shapes due to their different structures. The 11-*cis*-retinal has a fairly curved shape, and the parts of the molecule on either side of the *cis* double bond tend to lie in different planes. Because proteins have very complex and specific three-dimensional shapes (tertiary structures), 11-*cis*-retinal will associate

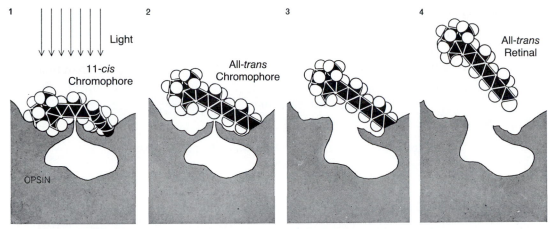

with the protein opsin in a particular manner. All-*trans*-retinal has an elongated shape, and the entire molecule tends to lie in a single plane. This different shape for the molecule, compared with the 11-*cis* isomer, means that all-*trans*-retinal will have a different association with the protein opsin.

In fact, all-*trans*-retinal associates very weakly with opsin because its shape does not fit the protein. Consequently, the next step after the isomerization of retinal is the dissociation of all-*trans*-retinal from opsin. The opsin protein undergoes a simultaneous change in conformation as the all-*trans*-retinal dissociates.

At some time after the 11-*cis*-retinal–opsin complex receives a photon, a message is received by the brain. It was originally thought that either the isomerization of 11-*cis*-retinal to all-*trans*-retinal or the conformational change of the opsin protein was an event that generated the electrical message sent to the brain. Current research, however, indicates that both of these events occur too slowly relative to the speed with which the brain receives the message. Current hypotheses invoke involved quantum mechanical explanations, which hold it significant that the chromophores (light-absorbing groups) are arranged in a very precise geometrical pattern in the rods and cones, allowing the signal to be transmitted rapidly through space. The main physical and chemical events Wald discovered are illustrated in the figure for easy visualization. The question of how the electrical signal is transmitted still remains unsolved.

Wald was also able to explain the sequence of events by which the rhodopsin molecules are regenerated. After dissociation of all-*trans*-retinal from the protein, the following enzyme-mediated changes occur. All-*trans*-retinal is reduced to the alcohol all-*trans*-retinol, also called all-*trans*-Vitamin A.

All-*trans*-Vitamin A

All-*trans*-Vitamin A is then isomerized to its 11-*cis*-Vitamin A isomer. Following the isomerization, the 11-*cis*-Vitamin A is oxidized back to 11-*cis*-retinal, which forthwith recombines with the opsin protein to form rhodopsin. The regenerated rhodopsin is then ready to begin the cycle anew, as illustrated in the accompanying diagram.

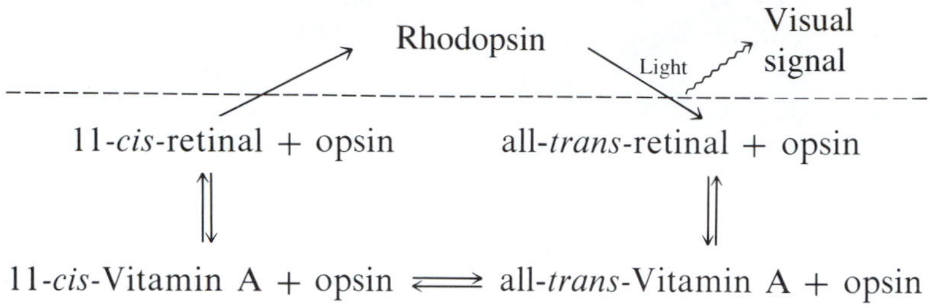

By this process, as little light as 10^{-14} of the number of photons emitted from a typical flashlight bulb can be detected. The conversion of light into isomerized retinal exhibits an extraordinarily high quantum efficiency. Virtually every quantum of light absorbed by a molecule of rhodopsin causes the isomerization of 11-*cis*-retinal to all-*trans*-retinal.

As you can see from the reaction scheme, the retinal derives from Vitamin A, which merely requires the oxidation of a —CH$_2$OH group to a —CHO group to be converted to retinal. The precursor in the diet that is transformed to Vitamin A is β-carotene. The β-carotene is the yellow pigment of carrots and is an example of a family of long-chain polyenes called **carotenoids.**

In 1907 Willstätter established the structure of carotene, but it was not known until 1931 to 1933 that there were actually three isomers of carotene. The α-carotene differs from β-carotene in that the α isomer has a double bond between C$_4$ and C$_5$ rather than between C$_5$ and C$_6$, as in the β isomer. The γ isomer has only one ring, identical to the ring in the β isomer, while the other ring is opened in the γ form between C$_{1'}$ and C$_{6'}$. The β isomer is by far the most common of the three.

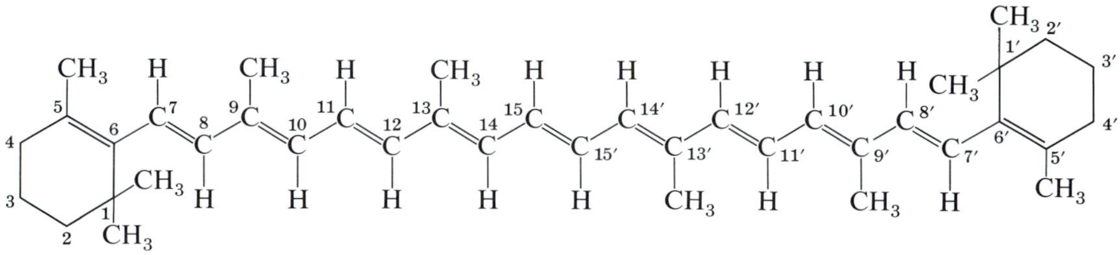

β-Carotene

The substance β-carotene is converted to Vitamin A in the liver. Theoretically, one molecule of β-carotene should give rise to two molecules of the vitamin by cleavage of the C$_{15}$–C$_{15'}$ double bond, but actually only one molecule of Vitamin A is produced from each molecule of the carotene. The Vitamin A thus produced is converted to 11-*cis*-retinal within the eye.

Along with the problem of how the electrical signal is transmitted, color perception is also currently under study. In the human eye there are three kinds of cone cells, which absorb light at 440, 535, and 575 nm, respectively. These cells discriminate among the primary colors. When combinations of them are stimulated, full color vision is the message received in the brain.

Since all these cone cells use 11-*cis*-retinal as a substrate-trigger, it has long been suspected that there must be three different opsin proteins. Recent work has begun to establish how the opsins vary the spectral sensitivity of the cone cells even though all of them have the same kind of light-absorbing chromophore.

Retinal is an aldehyde and it binds to the terminal amino group of a lysine residue in the opsin protein to form a Schiff base, or imine linkage ($>C=N-$). This imine linkage is believed to be protonated (with a plus charge) and to be stabilized by being located near a negatively charged amino acid residue of the protein chain. A second negatively charged group is thought to be located near the 11-*cis* double bond. Researchers have recently shown, from synthetic models that use a simpler protein than opsin itself, that forcing these negatively charged groups to be located at different distances from the imine linkage causes the absorption maximum of the 11-*cis*-retinal chromophore to be varied over a wide enough range to explain color vision.

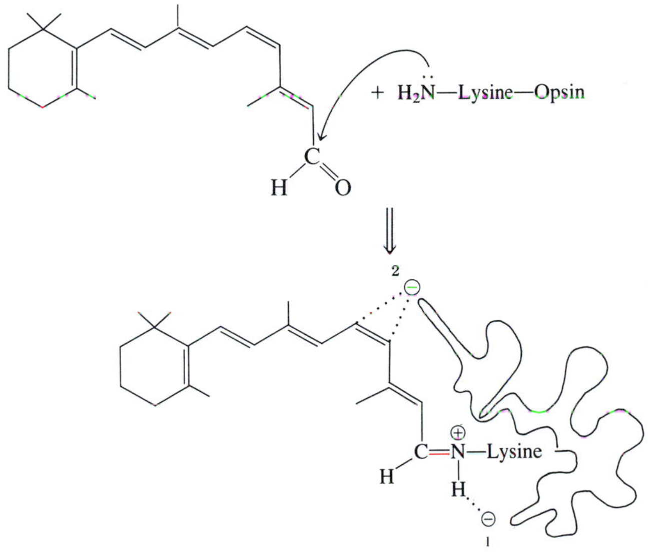

Rhodopsin

Whether there are actually three different opsin proteins, or whether there are just three different conformations of the same protein in the three types of cone cells, will not be known until further work is completed on the structure of the opsin or opsins.

REFERENCES

Clayton, R. K. *Light and Living Matter.* Vol. 2: *The Biological Part.* New York: McGraw-Hill, 1971.
Fox, J. L. ''Chemical Model for Color Vision Resolved.'' *Chemical and Engineering News, 57* (46) (November 12, 1979): 25.

A review of articles by Honig and Nakanishi in the *Journal of the American Chemical Society* (*101*, [1979]: 7082, 7084, 7086).

Hubbard, R., and Kropf, A. "Molecular Isomers in Vision." *Scientific American, 216* (June 1967): 64.

Hubbard, R., and Wald, G. "Pauling and Carotenoid Stereochemistry." In *Structural Chemistry and Molecular Biology*. Edited by A. Rich and N. Davidson. San Francisco: W. H. Freeman, 1968.

MacNichol, E. F., Jr. "Three Pigment Color Vision." *Scientific American, 211* (December 1964): 48.

Rushton, W. A. H. "Visual Pigments in Man." *Scientific American, 207* (November 1962): 120.

Wald, G. "Life and Light." *Scientific American, 201* (October 1959): 92.

Wolken, J. J. *Photobiology*. New York: Reinhold Book Co., 1968.

Zurer, P. S. "The Chemistry of Vision." *Chemical and Engineering News, 61* (November 28, 1983): 24.

Experiment 9
Isolation of Chlorophyll and Carotenoid Pigments from Spinach

Isolation of a natural product
Extraction
Column chromatography
Thin-layer chromatography

Photosynthesis in plants takes place in organelles called **chloroplasts.** Chloroplasts contain a number of colored compounds (pigments) which fall into two categories, **chlorophylls** and **carotenoids.**

Chlorophyll *a*

$$\text{Phytyl} = -CH_2-CH=C(CH_3)-CH_2-(CH_2-CH_2-CH(CH_3)-CH_2)_2-CH_2-CH_2-CH(CH_3)-CH_3$$

Chlorophylls are the green pigments which act as the principal photoreceptor molecules of plants. They are capable of absorbing certain wavelengths of visible light that are then converted by plants into chemical energy. Two different forms of these pigments found in plants are **chlorophyll *a*** and **chlorophyll *b*.** The two forms are identical, except that the methyl group that is shaded in the structural formula of chlorophyll *a* is replaced by a —CHO group in chlorophyll *b*. **Pheophytin *a*** and **pheophytin *b*** are identical to chlorophyll *a* and chlorophyll *b*, respectively, except that in each case the magnesium ion, Mg^{2+}, has been replaced by two hydrogen ions, $2H^+$.

Carotenoids are yellow pigments which are also involved in the photosynthetic process. The structures of **α-** and **β-carotene** are given in the essay preceeding this experiment. In addition, chloroplasts also contain several oxygen-containing derivatives of carotenes, called **xanthophylls.**

In this experiment you will extract the chlorophyll and carotenoid pigments from spinach leaves using acetone as the solvent. The pigments will be separated by column chromatography using alumina as the adsorbent. Increasingly more polar solvents will be used to elute the various components from the column. The colored fractions collected will then be analyzed using thin-layer chromatography. It should be possible for you to identify most of the pigments discussed above on your thin-layer plate after development.

REQUIRED READING

Review: Introduction to Microscale Laboratory (pp 35–48)
Techniques 1 and 2
Technique 3 Reaction Methods, Section 3.9
Technique 7 Extractions, Sections 7.5 and 7.8
Technique 13 Thin-Layer Chromatography

New: Technique 12 Column Chromatography
    Essay   The Chemistry of Vision

SPECIAL INSTRUCTIONS

Hexane and acetone are both highly flammable. Avoid the use of flames while working with these solvents. The thin-layer chromatography should be performed in the hood. The procedure calls for a centrifuge tube with a tight-fitting cap. If this is not available, a vortex mixer can be used for mixing the liquids. Another alternative is to use a cork to stopper the tube; however, some liquid will be absorbed by the cork.

Fresh spinach is preferable to frozen spinach. Because of handling, frozen spinach contains additional pigments which are difficult to identify. Since the pigments are light-sensitive and can undergo air oxidation, you should work quickly. Samples should be stored in closed containers and kept in the dark when possible. The column chromatography procedure will take less than 15 minutes to perform and cannot be stopped until it is completed. It is very important, therefore, that you have all the materials needed for this part of the experiment prepared in advance and that you are thoroughly familiar with the procedure before running the column. If you need to prepare the 70% hexane-30% acetone solvent mixture, be sure to mix it thoroughly before using.

> **NOTE TO INSTRUCTOR:** The column chromatography should be performed with activated alumina from EM Science (No. AX0612–1). The particle sizes are 80–200 mesh and the material is Type F-20. For thin-layer chromatography, use flexible silica gel plates from Whatman with a fluorescent indicator (No. 4410 222). If you use different alumina or different thin-layer plates, try out the experiment before using it with a class. Other materials than those specified here may give different results than indicated in this experiment.

PROCEDURE

EXTRACTION OF THE PIGMENTS

Weigh about 0.5 g of fresh (or 0.25 g of frozen) spinach leaves (avoid using stems or thick veins). Fresh spinach is preferable, if available. If frozen spinach must be used, dry the thawed leaves by pressing them between several layers of paper towels. Cut or tear the spinach leaves into small pieces and place them in a mortar along with 1.0 mL of cold acetone. Grind with a pestle until the spinach leaves have been broken up into particles too small to be clearly seen. If too much acetone has evaporated, you may need to add an additional portion of acetone (0.5–1.0 mL) to perform the following step. Using a Pasteur pipet, transfer the mixture to a centrifuge tube. Rinse the mortar and pestle with 1.0 mL of cold acetone and transfer the remaining mixture to the centrifuge tube. Centrifuge the mixture (be sure to balance the centrifuge). Using a Pasteur pipet, transfer the liquid to a centrifuge tube with a tight-fitting cap (see Special Instructions, if one is not available).

    Add 2.0 mL of hexane to the tube, cap the tube, and shake the mixture thoroughly. Then add 2.0 mL of water and shake thoroughly with occasional venting. Centrifuge the mixture to break the emulsion, which usually appears as a cloudy, green layer in the middle of the mixture. Remove the bottom aqueous layer with a Pasteur pipet. Using a Pasteur pipet, prepare a column containing anhydrous sodium sulfate to dry the remaining hexane layer, which contains the dissolved pigments. Place a plug of cotton into a Pasteur pipet (5¾-inch) and tamp it into position using a glass rod. The correct position of the cotton is shown in the figure. Add about 0.5 g of powdered or granular anhydrous sodium sulfate and tap the column with your finger to pack the material.

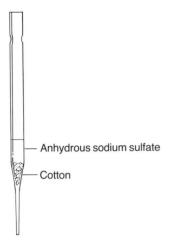

— Anhydrous sodium sulfate

— Cotton

Column for drying extract

Clamp the column in a vertical position and place a dry test tube (13 × 100-mm) under the bottom of the column. Label this test tube with an "E" for "extract" so that you don't confuse it with the test tubes you will be working with later in this experiment. With a Pasteur pipet, transfer the hexane layer to the column. When all the solution has drained, add an additional 0.5 mL of hexane to the column to extract all the pigments from the drying agent. Evaporate the solvent by placing the test tube in a warm sand bath (40–60 °C) and directing a stream of nitrogen gas (or dry air) into the vial. Dissolve the residue in 0.50 mL of hexane. Stopper the test tube and place it in your drawer until you are ready to run the alumina chromatography column.

COLUMN CHROMATOGRAPHY

The pigments are separated on a column packed with alumina. Although there are many different components in your sample, they separate into two main bands on the column. The first band to pass through the column is yellow and consists of the carotenes. This band may be less than 1 mm wide and it passes through the column very rapidly. It is easy to miss seeing the band as it passes through the alumina. The second band consists of all the other pigments discussed in the introduction to this experiment. Although it consists of both green and yellow pigments, it appears as a green band on the column. The green band spreads out on the column more than the yellow band and it moves more slowly. As the samples elute from the column, you should collect the yellow band in one test tube and the green band in another test tube.

Before running the column as described below, assemble the following glass-ware and liquids. Obtain five dry test tubes (16 × 100-mm) and number them one through five. Prepare two dry Pasteur pipets with bulbs attached. It is helpful to calibrate one of them to deliver a volume of about 0.25 mL (see essay, p 44). Place 4.0 mL of a mixture of 70% hexane-30% acetone (by volume), 6.0 mL of acetone, and about 5 mL of hexane into three separate containers, such as 10-mL Erlenmeyer flasks. Clearly label each container.

Prepare a chromatography column packed with alumina (see Technique 12, Section 12.6, Part A, p 707, and Part B, p 709). Dry Pack Method 2 on p 711 will be used.

Place a **loose** plug of cotton in a Pasteur pipet (5¾-inch) and push it **gently** into position using a glass rod (see figure on p 115 for the correct position of the cotton). Add 1.25 g of alumina (EM Science, No. AX0612–1) to the pipet while tapping the column gently with your finger. When all the alumina has been added, tap the column with your finger for several seconds to ensure that the alumina is tightly packed. Clamp the column in a vertical position so that the bottom of the column is just above the height of the test tubes you will be using to collect the fractions. Place test tube #1 under the column.

Read the next two paragraphs carefully before proceeding with the experiment. The chromatography procedure will take less than 15 minutes, and you cannot stop until all the material is eluted from the column. You must have a good understanding of the whole procedure before running the column.

Using a Pasteur pipet, slowly add about 3 mL of hexane to the column. The column must be completely moistened by the solvent. Drain the excess hexane until it just reaches the top level of the alumina. Once hexane has been added to the alumina, the column must not be allowed to run dry. If necessary, add more hexane.

**It is essential that the liquid not be allowed to drain below
the surface of the alumina at any point in this procedure.**

When the surface of the hexane reaches the top of the alumina, add about half (0.25 mL) of the dissolved pigments to the column. Leave the remainder in the test tube for the thin-layer chromatography procedure. (Put a stopper on the tube and place it back in your drawer.) Continue collecting the eluent in test tube #1. Just as the pigment solution penetrates the column, add an additional 1 mL of hexane and drain until the surface of the liquid has reached the alumina. Then add, in several portions, 4.0 mL of the 70% hexane-30% acetone solution to the column. You should be able to observe the yellow band as it passes through the column. Just before the yellow band reaches the bottom of the column, place test tube #2 under the column. When the eluent becomes colorless again (the total volume of the yellow material should be less than 2 mL), place test tube #3 under the column.

When the last portion of the hexane-acetone solution has nearly penetrated the alumina, begin adding pure acetone to the column. The green band should begin to move down the column. Although it will be more diffuse than the yellow band, you should be able to observe its movement down the column. When the eluent begins to appear green or when you observe in the column that the green band is beginning to elute from the column, change to test tube #4. Continue to add acetone to the column. Collect the green band until all the green pigments have been removed from the column and the eluent appears to be almost colorless. The volume of this band may be about 3–4 mL. Collect any remaining eluent in tube #5. All the pigments should have been eluted from the column when 6.0 mL of acetone has passed through the column. The column procedure may be stopped when there is very little or no green color in the liquid draining from the column.

Using a warm sand bath (40–60 °C) and a stream of nitrogen gas, evaporate the solvent from the tube containing the yellow band (tube #2), the tube containing the green band (tube #4), and the tube containing the original pigment solution (tube "E"). As soon as all the solvent has evaporated from each of the tubes, remove them from the sand bath. Do not allow any of the tubes to remain in the sand bath after the solvent has evaporated. Stopper the tubes and place them in your drawer.

THIN-LAYER CHROMATOGRAPHY

Technique 13 describes the procedures used for thin-layer chromatography (TLC). Use a 10-cm × 3.3-cm TLC plate (Whatman Silica Gel Plates No. 4410 222). These plates have a flexible backing but should not be bent excessively. They should be handled carefully, or the adsorbent may flake off them. Also, they should be handled only by the edges; the surface should not be touched. Using a lead pencil (not a pen) **lightly** draw a line across the plate (short dimension) about 1 cm from the bottom (see figure). Using a centimeter ruler, move its index about 0.6 cm in from the edge of the plate and lightly mark off three 1-cm intervals on the line. These are the points at which the samples will be spotted.

Prepare three micropipets using Pasteur pipets. The preparation of these pipets is described and illustrated in Technique 13, Section 13.4, p 728, and Figure 13–3, p 729. The only difference in this experiment is that Pasteur pipets are used rather than capillary tubing. Try to draw out the narrow end of the pipet as close to the tip as possible without burning yourself! Prepare a TLC development chamber with 70% hexane-30% acetone (see Technique 13, Section 13.3, p 730). A beaker covered with aluminum foil or a wide-mouth screw cap bottle is a suitable container to use (see Figure 13–5, p 731). The backing on the TLC plates is very thin, so if they touch the filter paper liner of the development chamber **at any point,** solvent will begin to diffuse onto the absorbent surface at that point. To avoid this, be sure that the filter paper liner does not go completely around the inside of the container. A space about two inches wide must be provided.

Using a Pasteur pipet, add two drops of 70% hexane-30% acetone to each of the three test tubes containing dried pigments. Swirl the tubes so that the drops of solvent dissolve as much of the pigments as possible. The TLC plate should be spotted with three samples: the extract, the yellow band from the column, and the green band. For each of the three samples, use a different micropipet to spot the sample on the plate. The correct method of spotting a TLC plate is described in Technique 13, Section 13.4, p 728. Take up part of the sample in the pipet (don't use a bulb; capillary action will draw up the liquid). For the extract (tube labelled "E") and the green band (tube #4), touch the plate once **lightly** and let the solvent evaporate. The spot should be no larger than 2 mm in diameter and should be a fairly dark green. For the yellow band (tube #2), you will need to repeat the spotting technique 5–10 times, until the spot is a definite yellow color. Allow the solvent to evaporate completely between successive applications, and spot the plate in exactly the same position each time. Save the samples in case you need to repeat the TLC.

Place the TLC plate in the development chamber making sure that the plate does not come in contact with the filter paper liner. Remove the plate when the solvent front is 1–2 cm from the top of the plate. Using a lead pencil, mark the position of the solvent front. As soon as the plates have dried, outline the spots with a pencil and indicate the colors. This is important to do soon after the plates have dried, because some of the pigments will change color when exposed to the air.

ANALYSIS OF RESULTS

In the crude extract you should be able to see the following components (in order of decreasing R_f values):

> Carotenes (1 spot) (yellow-orange),
> Pheophytin a (grey, may be nearly as intense as chlorophyll b),
> Pheophytin b (grey, may not be visible),
> Chlorophyll a (blue-green, more intense than chlorophyll b),
> Chlorophyll b (green),
> Xanthophylls (possibly 3 spots: yellow).

Depending on the spinach sample, the conditions of the experiment, and how much sample was spotted on the TLC plate other pigments may be observed. These additional components can result from air oxidation, hydrolysis, or other chemical reactions involving the pigments discussed in this experiment. It is very common to observe other pigments in samples of frozen spinach. It is also common to observe components in the green band which were not present in the extract.

Identify as many of the spots in your samples as possible. Determine which pigments were present in the yellow band and in the green band. Draw a picture of the TLC plate in your notebook. Label each spot with its color and its identity, where possible. Calculate the R_f values for each spot produced by chromatography of the extract (see Technique 13, Section 13.9, p 733). At the instructor's option, submit the TLC plate with your report.

REFERENCE

Anwar, M. H. "Separation of Plant Pigments by TLC." *Journal of Chemical Education, 40* (1963): 29.

QUESTIONS

1. Why is it possible to achieve better separation on the TLC plates than with column chromatography?

2. Why are the chlorophylls less mobile on column chromatography and why do they have lower R_f values than the carotenes?

3. Propose structural formulas for pheophytin *a* and pheophytin *b*.

4. What would happen to the R_f values of the pigments if you were to increase the relative concentration of acetone in the developing solvent?

5. What would happen with the column chromatography procedure in this experiment if only one of the following solvents were used?

  (a) Acetone.

  (b) Hexane.

  (c) 50% hexane-50% acetone.

Part Two

Preparations and Reactions of Organic Compounds

Experiment 10
Reactivities of Some Alkyl Halides

S_N1/S_N2 reactions
Relative rates
Reactivities

The reactivities of alkyl halides in nucleophilic substitution reactions depend on two important factors: reaction conditions and substrate structure. The reactivities of several different substrate types will be examined under both S_N1 and S_N2 reaction conditions in this experiment.

SODIUM IODIDE OR POTASSIUM IODIDE IN ACETONE

A reagent composed of sodium iodide or potassium iodide dissolved in acetone is useful in classifying alkyl halides according to their reactivity in an S_N2 reaction. Iodide ion is an excellent nucleophile, and acetone is a nonpolar solvent. The tendency to form a precipitate increases the completeness of the reaction. Sodium iodide and potassium iodide are soluble in acetone, but the corresponding bromides and chlorides are not soluble. Consequently, as bromide ion or chloride ion is produced, it is precipitated from the solution. According to LeChâtelier's principle, the precipitation of a product from the reaction solution drives the equilibrium toward the right; such is the case in the reaction described here:

$$R\!-\!Cl + Na^+I^- \longrightarrow RI + NaCl\!\downarrow$$
$$R\!-\!Br + Na^+I^- \longrightarrow RI + NaBr\!\downarrow$$

SILVER NITRATE IN ETHANOL

A reagent composed of silver nitrate dissolved in ethanol is useful in classifying alkyl halides according to their reactivity in an S_N1 reaction. Nitrate ion is a poor nucleophile, and ethanol is a moderately powerful ionizing solvent. The silver ion, because of its ability to coordinate the leaving halide ion to form a silver halide precipitate, greatly assists the ionization of the alkyl halide. Again, a precipitate as one of the reaction products also enhances the reaction.

123

$$R-Cl \longrightarrow \begin{array}{c} R^+ \xrightarrow{C_2H_5OH} R-OC_2H_5 \\ + \\ Cl^- \xrightarrow{Ag^+} AgCl\downarrow \end{array}$$

$$R-Br \longrightarrow \begin{array}{c} R^+ \xrightarrow{C_2H_5OH} R-OC_2H_5 \\ + \\ Br^- \xrightarrow{Ag^+} AgBr\downarrow \end{array}$$

REQUIRED READING

Before beginning this experiment, review the chapters dealing with nucleophilic substitution in your lecture textbook.

SPECIAL INSTRUCTIONS

Some compounds used in this experiment, particularly crotyl chloride, chloroacetone, and benzyl chloride, are powerful lachrymators. **Lachrymators** cause eye irritation and the formation of tears.

> **Because of lachrymator compounds, perform these tests in a hood, and be careful to dispose of the test solutions in an appropriate waste container in the hood.**

The experiment requires very little time and may be performed along with another longer experiment.

PROCEDURE

SODIUM IODIDE IN ACETONE

Label a series of 10 clean dry test tubes (10 × 75-mm test tubes may be used) from 1 to 10. In each test tube place 0.2 mL of one of the following halides: (1) 2-chlorobutane; (2) 2-bromobutane; (3) 2-chloro-2-methylpropane (*t*-butyl chloride); (4) 1-chlorobutane;

(5) crotyl chloride, (CH$_3$CH=CHCH$_2$Cl) (Note: see Special Instructions); (6) chloro-acetone, (ClCH$_2$COCH$_3$) (Note: see Special Instructions); (7) benzyl chloride (α-chlorotoluene) (Note: see Special Instructions); (8) bromobenzene; (9) bromocyclohexane; and (10) bromocyclopentane.

Add to the material in each test tube 2 mL of a 15% NaI-in-acetone solution, noting the time of each addition. After the addition, shake the test tube well to ensure adequate mixing of the alkyl halide and the solvent. Record the times needed for any precipitates to form. After about five minutes, place any test tubes that do not contain a precipitate in a 50 °C water bath. Be careful not to allow the temperature of the water bath to exceed 50 °C since the acetone will evaporate or boil out of the test tube or both. At the end of six minutes, cool the test tubes to room temperature and note whether a reaction has occurred. Record the results and explain why each compound has the reactivity that you observed. Explain the reactivities in terms of structure.

Generally, reactive halides give a precipitate within three minutes, moderately reactive halides give a precipitate when heated, and unreactive halides do not give a precipitate even after being heated.

SILVER NITRATE IN ETHANOL

Label a series of 10 clean dry test tubes from 1 to 10, as described in the previous section. Place 0.2 mL of the appropriate halide in each test tube, as described for the sodium iodide test.

Add 2 mL of a 1% ethanolic silver nitrate solution to the material in each test tube, noting the time of each addition. After the addition, shake the test tubes well to ensure adequate mixing of the alkyl halide and the solvent. Record the times required for any precipitates to form.

After about five minutes, heat each solution that has not yielded any precipitate to boiling on the steam bath. Note whether a precipitate forms.

Record the results and explain why each compound has the reactivity observed. Explain the reactivities in terms of structure, as before.

Again, reactive halides will give a precipitate within three minutes, moderately reactive halides will give a precipitate when heated, and unreactive halides do not yield a precipitate even after being heated.

QUESTIONS

1. In the tests with sodium iodide in acetone and silver nitrate in ethanol, why should 2-bromobutane react faster than 2-chlorobutane?

2. In the test with silver nitrate in ethanol, why should the cyclopentyl compound react faster than the cyclohexyl compound?

3. When benzyl chloride is treated with sodium iodide in acetone, it reacts much faster than 1-chlorobutane, even though both compounds are primary alkyl chlorides. Explain this rate difference.

4. How do you expect the four compounds shown on p 126 to compare in behavior in the two tests?

1   2   3   4

5. How do you predict that chlorocyclopropane will behave in each of these tests?

Experiment 11
Nucleophilic Substitution Reactions:
Competing Nucleophiles

Nucleophilic substitution	Refractometry
Heating under reflux	Gas chromatography
Extraction	NMR spectroscopy

The purpose of this experiment is to compare the relative nucleophilicities of chloride ions and bromide ions toward each of the following alcohols: 1-butanol (*n*-butyl alcohol), 2-butanol (*sec*-butyl alcohol), and 2-methyl-2-propanol (*t*-butyl alcohol). The two nucleophiles will be present at the same time in each reaction, in equimolar concentrations, and they will be competing with each other for substrate.

In general, alcohols do not react readily in simple nucleophilic displacement reactions. If they are attacked by nucleophiles directly, hydroxide ion, a strong base, must be displaced. Such a displacement is not energetically favorable, and it cannot occur to any reasonable extent:

$$X^- + ROH \;\;\not\!\!\longrightarrow\;\; R{-}X + OH^-$$

To avoid this problem, one must carry out nucleophilic displacement reactions on alcohols in acidic media. In a rapid initial step, the alcohol is protonated; then water, a very stable molecule, is displaced. This displacement is energetically very favorable, and the reaction proceeds in high yield:

$$ROH + H^+ \rightleftharpoons R{-}\overset{+}{O}{\Big\langle}\!\!\begin{array}{l} H \\ H \end{array}$$

$$X^- + R{-}\overset{+}{O}{\Big\langle}\!\!\begin{array}{l} H \\ H \end{array} \longrightarrow R{-}X + H_2O$$

Once the alcohol is protonated, it reacts by either the S_N1 or the S_N2 mechanism, depending on the structure of the alkyl group of the alcohol. For a brief review of these mechanisms, you should consult the chapters on nucleophilic substitution in your lecture textbook.

You will analyze the products of the three reactions studied in this experiment by a variety of techniques to determine the relative amounts of alkyl chloride and alkyl bromide formed in each reaction. That is, using equimolar concentrations of chloride ions and bromide ions reacting with 1-butanol, 2-butanol, and 2-methyl-2-propanol, you will try to determine which ion is the better nucleophile. In addition, you will try to determine for which of the three substrates (reactions) this difference is important and whether an S_N1 or S_N2 mechanism predominates in each case.

REQUIRED READING

Review: Techniques 1 and 2

Technique 3	Reaction Methods, Sections 3.2, 3.4, 3.5, and 3.7
Technique 7	Extraction, Sections 7.5, 7.8, and 7.10
Technique 14	Gas Chromatography
Technique 17	Refractometry
Technique 18	Preparation of Samples for Spectroscopy, Part B, Nuclear Magnetic Resonance
Appendix 4	Nuclear Magnetic Resonance Spectroscopy

Before beginning this experiment, review the appropriate chapters on nucleophilic substitution in your lecture textbook.

SPECIAL INSTRUCTIONS

Each student will carry out the reaction with 2-methyl-2-propanol. Your instructor will also assign you either 1-butanol or 2-butanol. By sharing your results with other students, you will be able to collect data for all three alcohols. The solvent-nucleophile medium contains a high concentration of sulfuric acid. Sulfuric acid is very corrosive; be careful when handling it.

NOTE TO THE INSTRUCTOR: The solvent-nucleophile medium must be prepared in advance for the entire class (see p 128).

Be certain that the *t*-butyl alcohol has been melted before the beginning of the laboratory period.

The gas chromatograph should be prepared as follows: column temperature, 100 °C; injection and detector temperature, 130 °C; carrier gas flow rate, 50 mL/min. The recommended column is 8 feet long with a stationary phase such as Carbowax 20M.

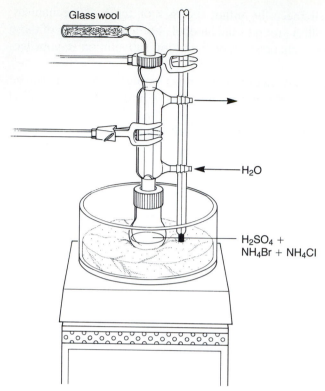

Apparatus for reflux

PROCEDURE

Assemble an apparatus for reflux using a 10-mL round-bottom flask, a reflux condenser, and drying tube as shown in the figure above. Loosely insert dry, glass wool into the drying tube and then add water dropwise onto the glass wool until it is partially moistened. The moistened glass wool will trap the hydrogen chloride and hydrogen bromide gases produced during the reaction. As an alternative, you can use an external gas trap as described in Technique 3, Section 3.7, Part B, p 556. Do not place the round-bottom flask in the sand bath until the reaction mixture has been added to the flask. Five Pasteur pipets and one filter tip pipet, two 3-mL conical vials with Teflon cap liners, and a 5-mL conical vial with a Teflon liner should also be assembled for use. All pipets and vials should be clean and dry. The sand bath should be adjusted to about 140 °C for the procedure with 1-butanol and about 120 °C if you are using 2-butanol.

You should now begin Procedure 11A. During the reflux period in Procedure 11A, you should do Procedure 11B.

THE SOLVENT-NUCLEOPHILE MEDIUM
(Procedure for instructor)

This procedure will provide enough solvent-nucleophile medium for about 20 students (assuming no spillage or other types of waste). Place 100 g of ice in a 500-mL Erlenmeyer flask and carefully add 76 mL of concentrated sulfuric acid. Carefully weigh 19.0 g of ammonium chloride and 35.0 g of ammonium bromide into a beaker. Crush any lumps of these reagents to powder, and then, using a powder funnel, transfer these halides to an Erlenmeyer flask. Exercising caution, add the sulfuric acid mixture to the ammonium salts a little at a time. Swirl the mixture vigorously to induce the salts to dissolve. It will probably be necessary to heat the mixture on a steam bath or a hot plate to achieve total solution. If necessary, you may add as much as 10 mL of water at this stage. Do not worry if just a few small granules do not dissolve. When solution has been achieved, pour the solution into a container which can be kept warm until all students have taken their portions. The temperature of the mixture must be maintained at about 45 °C to prevent precipitation of the salts.

Procedure 11A
Competitive Nucleophiles with
1-Butanol or 2-Butanol

The solvent-nucleophile medium contains a high concentration of sulfuric acid. This liquid will cause severe burns if it comes in contact with your skin.

Using a warm 10-mL graduated cylinder, obtain 6.5 mL of the solvent-nucleophile medium. The graduated cylinder must be warm in order to prevent precipitation of the salts. It can be heated by running hot water over the outside of the cylinder or by putting it in the oven for a few minutes. Immediately pour the mixture into the round-bottom flask. A small portion of the salts in the flask may precipitate as the solution cools. Do not worry about this; the salts will redissolve during the reaction.

Assemble the apparatus shown in the figure. Using the following procedure, add 0.50 mL of 1-butanol (*n*-butyl alcohol) or 0.50 mL of 2-butanol (*sec*-butyl alcohol), depending on which alcohol you were assigned, to the solvent-nucleophile mixture contained in the reflux apparatus. Remove the drying tube, and with a 9-inch Pasteur pipet dispense the alcohol directly into the round-bottom flask by inserting the Pasteur pipet into the opening of the condenser. Also add an inert boiling stone.[1] Replace the drying tube and start circulating the cooling water. Lower the reflux apparatus so that the round-bottom flask is in the sand bath, as shown in the figure. Adjust the heat so that this mixture maintains a **gentle** boiling action. For 1-butanol the sand bath temperature should be about 140 °C, and with 2-butanol the temperature should be about 120 °C. Be very careful to adjust the reflux ring, if one is visible, so that it remains in the lower fourth of the condenser. Violent boiling will cause loss of product. Continue heating the reaction mixture containing 1-butanol for 75 minutes. The mixture containing 2-butanol

---

[1] Calcium carbonate based stones or Boileezers should not be used, because they will partially dissolve in the highly acidic reaction mixture.

should be heated for 60 minutes. During this heating period, go on to Procedure 11B and complete as much of it as possible before returning to this procedure.

REFLUX PERIOD

When the period of reflux has been completed, discontinue heating, remove the sand bath, and allow the reaction mixture to cool. Do not remove the condenser until the flask is cool. Be careful not to shake the hot solution as you remove the sand bath, or a violent boiling and bubbling action will result; this could allow material to be lost out the top of the condenser. After the mixture has cooled for about five minutes, immerse the round-bottom flask in a beaker of cold (not ice) water and wait for this mixture to cool down to room temperature.

There should be an organic layer present at the top of the reaction mixture. Add 0.5 mL of hexane to the mixture and gently swirl the flask. The purpose of the hexane is to increase the volume of the organic layer so that the following operations are easier to accomplish. Using a Pasteur pipet, transfer most (about 5 mL) of the bottom (aqueous) layer to another container. Be very careful that all of the top organic layer remains in the boiling flask. Transfer the remaining aqueous layer and the organic layer to a 3-mL conical vial, taking care to leave behind any solids that may have precipitated. Allow the phases to separate and remove the bottom (aqueous) layer using a Pasteur pipet. Add 1.0 mL of water to the vial and gently shake this mixture. Allow the layers to separate and remove the aqueous layer, which is still on the bottom. Extract the organic layer with 1.0 mL of saturated sodium bicarbonate solution, and remove the bottom aqueous layer.

Using a clean dry Pasteur pipet transfer the remaining organic layer into a small test tube (10 × 75 mm) containing three to four microspatulafuls (using the V-grooved end) of anhydrous granular sodium sulfate. Stir the mixture with a microspatula, put a stopper on the tube, and set it aside for 10–15 minutes or until the solution is clear. If the mixture does not turn clear, add more anhydrous sodium sulfate. Transfer the halide solution with a clean, dry Pasteur pipet to a small, dry, screw cap vial, taking care not to transfer any solid. The screw cap vial should have a cap with a Teflon liner or the cap should be lined with aluminum foil. **Be sure the cap is screwed on tightly.** Do not store the liquid in a container with a cork or a rubber stopper, because these will absorb the halides. This sample can now be analyzed by as many of the methods given below as your instructor indicates. However, it can not be analyzed by refractometry because of the presence of hexane.

Procedure 11B
Competitive Nucleophiles with 2-Methyl-2-propanol

Place 3.0 mL of the solvent-nucleophile medium into a 5-mL conical vial using the same procedure described at the beginning of Procedure 11A. Allow the solution to cool to room temperature. Using an automatic pipet (or a warm graduated pipet and pipet pump), transfer 0.5 mL of 2-methyl-2-propanol (*t*-butyl alcohol, mp 25 °C) to the 5-mL conical vial. Replace the cap, and, with occasional venting, shake the vial vigorously for two minutes. Any solids that were originally present in the vial should dissolve during this period. If all the solids do not dissolve, heat the vial (with the cap on) **gently** in the sand bath. After shaking, allow the layer of alkyl halides to separate (one to two minutes at most). A fairly distinct top layer containing the products should have formed by this time.

_t_-Butyl halides are very volatile and should not be left in
an open container for any longer than necessary.

Slowly remove most of the bottom aqueous layer with a Pasteur pipet and transfer it to a beaker. After a wait of 10 to 15 seconds, remove the remaining lower layer in the vial, including a small amount of the upper organic layer, so as to be certain that the organic layer is not contaminated by any water. Using a dry Pasteur pipet, transfer the remainder of the alkyl halide layer into a small test tube (10 × 75 mm) containing about 0.05 g of solid sodium bicarbonate. As soon as the bubbling stops and a clear liquid is obtained, transfer it with a filter tip pipet into a small, dry, screw cap vial, taking care not to transfer any solid. This screw cap should also have a Teflon liner or the cap should be lined with aluminum foil. Do not store the liquid in a container with a cork or rubber stopper since these will absorb the halides. This sample can now be analyzed by as many of the methods given below as your instructor indicates. When you have finished this procedure, return to Procedure 11A.

ANALYSIS PROCEDURES

The ratio of 1-chlorobutane to 1-bromobutane, 2-chlorobutane to 2-bromobutane, or *t*-butyl chloride to *t*-butyl bromide must be determined. At your instructor's option, you may do this by one of three methods: gas chromatography, refractive index, or NMR spectroscopy. The products obtained from the reactions of 1-butanol and 2-butanol, however, cannot be analyzed by the refractive index method (they contain hexane).

GAS CHROMATOGRAPHY

The instructor or laboratory assistant may either make the sample injections or allow the students in the class to make them. In the latter case, it is essential that instruction beforehand is adequate. A reasonable sample size is 2.5 μL. The sample is injected into the gas chromatograph and the gas chromatogram is recorded. The alkyl chloride, because of its greater volatility, has a shorter retention time than the alkyl bromide.

Once the gas chromatogram has been obtained, determine the relative areas of the two peaks (Technique 14, Section 14.11, p 750). While the peaks may be cut out and weighed on an analytical balance as a method of determining areas, triangulation is the preferred method. Record the percentages of alkyl chloride and alkyl bromide in the reaction mixture.

REFRACTIVE INDEX

Measure the refractive index of the product mixture (Technique 17). To determine the composition of the mixture, assume a linear relation between the refractive index and the molar composition of the mixture. At 20 °C the refractive indices of the alkyl halides are

t-butyl chloride	1.3877
t-butyl bromide	1.4280

If the temperature of the laboratory room is not 20 °C, the refractive index must be corrected. Add 0.0004 refractive index unit to the observed reading for each degree above 20 °C, and subtract the same amount for each degree below this temperature. Record the percentages of alkyl chloride and alkyl bromide in the reaction mixture.

NUCLEAR MAGNETIC RESONANCE

The instructor or a laboratory assistant will record the NMR spectrum of the reaction mixture.[2] Submit a sample vial containing the mixture for this spectral determination. The spectrum will also contain integration of the important peaks (Appendix 4, Nuclear Magnetic Resonance).

If the substrate alcohol was 1-butanol, the resulting halide and hexane mixture will give rise to a complicated spectrum. Each alkyl halide will show a downfield triplet caused by the CH_2 group nearest the halogen. This triplet will appear further downfield for the alkyl chloride than for the alkyl bromide. These triplets will overlap, but one branch of each triplet will be available for comparison. Compare the integral of the **downfield** branch of the triplet for 1-chlorobutane with the **upfield** branch of the triplet for 1-bromobutane. The upper spectrum on p 133 provides an example. The relative heights of these integrals correspond to the relative amounts of each halide in the mixture.

If the substrate alcohol was 2-methyl-2-propanol, the resulting halide mixture will show two peaks in the NMR spectrum. Each halide will show a singlet, since all the CH_3 groups are equivalent and are not coupled. In the reaction mixture, the upfield peak is due to *t*-butyl chloride, while the downfield peak is caused by *t*-butyl bromide. Compare the integrals of these peaks. The lower spectrum on p 133 provides an example. The relative heights of these integrals correspond to the relative amounts of each halide in the mixture.

REPORT

Record the percentages of alkyl chloride and alkyl bromide in the reaction mixture for each of the three alcohols. You will need to share your data from the reaction with 1-butanol or 2-butanol with other students in order to do this. The report must include the percentages of each alkyl halide determined by each method used in this experiment for the two alcohols you studied. On the basis of product distribution, develop an argument

---

[2]It is difficult to determine the ratio of 2-chlorobutane to 2-bromobutane using nuclear magnetic resonance. This method requires at least a 90 MHz instrument.

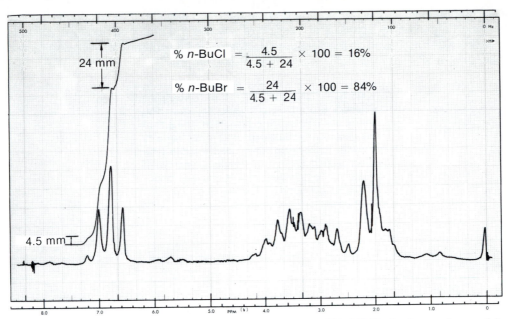

$$\% \; n\text{-BuCl} \; = \frac{4.5}{4.5 + 24} \times 100 = 16\%$$

$$\% \; n\text{-BuBr} \; = \frac{24}{4.5 + 24} \times 100 = 84\%$$

24 mm

4.5 mm

NMR spectrum of 1-chlorobutane and 1-bromobutane, sweep width 250 Hz (no hexane in sample)

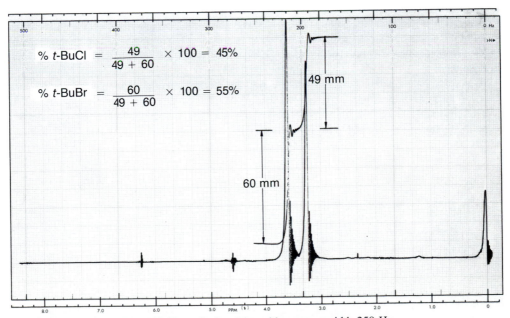

$$\% \; t\text{-BuCl} \; = \frac{49}{49 + 60} \times 100 = 45\%$$

$$\% \; t\text{-BuBr} \; = \frac{60}{49 + 60} \times 100 = 55\%$$

49 mm

60 mm

NMR spectrum of *t*-butyl chloride and *t*-butyl bromide, sweep width 250 Hz

for which mechanism (S_N1 or S_N2) predominated for each of the three alcohols studied. The report should also include a discussion of which is the better nucleophile, chloride ion or bromide ion, based on the experimental results. All gas chromatograms, refractive index data, and spectra should be attached to the report.

> **NOTE TO THE INSTRUCTOR:** If pure samples of each product are available, one can check the assumption (inherent here) that the gas chromatograph responds equally to each substance. Response factors (relative sensitivities) are easily determined by injecting an equimolar mixture of products and comparing the peak areas.

QUESTIONS

1. Which is the better nucleophile, chloride ion or bromide ion? Try to explain this in terms of the nature of the chloride ion and the bromide ion.

2. What is the principal organic by-product of these reactions?

3. A student left some alkyl halides (RCl and RBr) in an open container for several hours. What happened to the composition of the halide mixture during that time?

4. What would happen if all the solids in the nucleophile medium were not dissolved? How might this affect the outcome of the experiment?

5. Why does the alkyl chloride show NMR peaks further downfield than the corresponding peaks in the alkyl bromide for the primary products when the reverse is true for the tertiary products?

6. Draw a complete mechanism that shows why the resultant product distribution observed for the reaction of *t*-butyl alcohol is obtained.

7. What might have been the product ratios observed in this experiment if an aprotic solvent like dimethyl sulfoxide had been used instead of water?

8. Explain the order of elution you observed in doing the gas chromatography for this experiment. What property of the product molecules seems to be the most important in determining relative retention times?

9. Does it seem reasonable to you that the refractive index should be a temperature-dependent parameter? Try to explain.

10. When you calculate the percentage composition of the product mixture, exactly what kind of "percentage" are you dealing with?

Experiment 12
Hydrolysis of Some Alkyl Chlorides

Synthesis of an alkyl halide
Use of a separatory funnel
Titration
Kinetics

Two chemical reactions are of interest in this experiment. The first is the preparation of the alkyl chlorides whose hydrolysis rates are to be measured. The chloride formation is a simple nucleophilic substitution reaction carried out in a separatory funnel. Because the concentration of the initial alkyl chloride does not need to be determined for the kinetic experiment, isolation and purification of the alkyl chloride are not required.

$$ROH + HCl \longrightarrow R{-}Cl + H_2O$$

The second reaction is the actual hydrolysis, and the rate of this reaction will be measured. Under the conditions of this experiment, the reaction proceeds by an S_N1 pathway. The reaction rate is monitored by measuring the rate of appearance of hydrochloric acid. The concentration of hydrochloric acid is determined by titration with aqueous sodium hydroxide.

$$R{-}Cl + H_2O \xrightarrow{\text{solvent}} R{-}OH + HCl$$

The rate equation for the S_N1 hydrolysis of an alkyl chloride is

$$+\frac{d[\text{HCl}]}{dt} = k[\text{RCl}]$$

Let c equal the initial concentration of RCl. At some time, t, x moles per liter of alkyl chloride will have decomposed and x moles per liter of HCl will have been produced. The remaining concentration of alkyl chloride at that value of time equals $c - x$. The rate equation becomes

$$+\frac{dx}{dt} = k(c - x)$$

On integration, this becomes

$$\ln\left(\frac{c}{c-x}\right) = kt$$

which, converted to base 10 logarithms, is

$$2.303 \log\left(\frac{c}{c-x}\right) = kt$$

This equation is of the form appropriate for a straight line $y = mx + b$ with slope m and with intercept b equal to zero. If the reaction is indeed first-order, a plot of $\log (c/c - x)$ versus t will provide a straight line whose slope is $k/2.303$.

Evaluation of the term $c/c - x$ remains a problem, since it is experimentally difficult to determine the concentration of alkyl chloride. We can, however, determine the concentration of hydrochloric acid produced by titrating it with base. Because the stoichiometry of the reaction indicates that the number of moles of alkyl chloride consumed equals the number of moles of hydrochloric acid produced, c must also equal the number of moles of HCl produced when the reaction has gone to completion (the so-called infinity concentration of HCl), and x equals the number of moles of HCl produced at some particular value at time t. From these equalities, we can rewrite the integrated rate expression in terms of volume of base used in the titration. At the end-point of the titration,

$$\text{Number of moles HCl} = \text{number of moles of NaOH}$$

or

$$x = \text{number of moles NaOH at time } t$$

and

$$c = \text{number of moles NaOH at time } \infty$$

$$\text{Number of moles NaOH} = [\text{NaOH}]V$$

where V is the volume. Substituting and cancelling gives

$$\left(\frac{c}{c - x}\right) = \frac{\cancel{[\text{NaOH}]}V_\infty}{\cancel{[\text{NaOH}]}(V_\infty - V_t)}$$

where V_∞ is the volume of NaOH used when the reaction is complete, and V_t is the volume of NaOH used at time t. This integrated rate equation becomes

$$2.303 \log \left(\frac{V_\infty}{V_\infty - V_t}\right) = kt$$

The concentration of the base used in the titration cancels out of this equation, so that it is necessary to know neither the concentration of base nor the amount of alkyl chloride used in the experiment.

A plot of $\log (V_\infty/V_\infty - V_t)$ versus t will provide a straight line whose slope equals $k/2.303$. The slope is determined according to the figure on p 137. If the time is measured in minutes, the units of k are min^{-1}. The experimental points plotted on the graph may contain a certain amount of scatter, but the line drawn is the best **straight** line. The line should pass through the origin of the graph. With some reactions, competing processes may cause the line to contain a certain amount of curvature. In these cases, the slope of the initial portion of the line is used before the curvature becomes too important.

One other value often cited in kinetic studies is the **half-life** of the reaction τ. The half-life is the time required for one half of the reactant to undergo conversion to

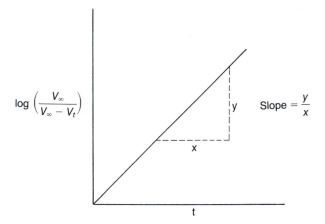

products. During the first half-life, 50% of the available reactant is consumed. At the end of the second half-life, 75% of the reactant has been consumed. For a first-order reaction, the half-life is calculated by

$$\tau = \frac{\ln 2}{k} = \frac{0.69315}{k}$$

Two alkyl chlorides will be studied by the class in a variety of solvents. The class data will be compared, so that relative reactivities of the alkyl chlorides can be determined.

REQUIRED READING

Review: Before beginning this experiment, you should read the material dealing with the methods of kinetics in your lecture textbook.

New: Technique 7   Extractions, Separations, and Drying Agents, especially Section 7.7, p 627

SPECIAL INSTRUCTIONS

Students must work in pairs in this experiment in order to make the measurements rapidly. Students should alternate jobs on each run, one student doing the titrations and the other reading the stopwatch and recording the data.

   Concentrated hydrochloric acid is corrosive; avoid any direct contact, and avoid breathing the vapors.

PROCEDURES

PREPARATION OF THE ALKYL CHLORIDES

The student should select an alcohol. The choices include *t*-butyl alcohol (2-methyl-2-propanol) and α-phenylethyl alcohol (1-phenylethanol). Place the alcohol (11 mL) in a 125-mL separatory funnel along with 25 mL of cold, concentrated hydrochloric acid (specific gravity 1.18; 37.3% hydrogen chloride). Shake the separatory funnel vigorously, with frequent venting to relieve any excess pressure, over 30 minutes. Remove the aqueous layer. Wash the organic phase quickly with three 5-mL portions of cold water, followed by a washing with 5 mL of 5% sodium bicarbonate solution. Place the organic product in a small Erlenmeyer flask over 3–4 g of anhydrous calcium chloride. Shake the flask occasionally over five minutes. Carefully decant the alkyl chloride from the drying agent into a small Erlenmeyer flask, which can then be stoppered tightly. The alkyl chloride is used in this experiment without prior distillation. Because the true concentration of alkyl chloride introduced into the hydrolysis reaction is determined by titration, it is not necessary to purify the product prepared in this part of the experiment.

KINETIC STUDY OF THE HYDROLYSIS OF AN ALKYL CHLORIDE

Because the chlorides hydrolyze rapidly under the conditions used in this experiment, students must perform the kinetic studies working in pairs. One student will perform the titrations, while the other measures the time and records the data.

Prepare a stock solution of alkyl chloride by dissolving about 0.6 g of alkyl chloride in 50 mL of dry, reagent-grade acetone. Store this solution in a stoppered container to protect it from moisture. Use a 125-mL Erlenmeyer flask to carry out the hydrolysis. The flask should contain a magnetic stirring bar, 50 mL of solvent (see the table for the appropriate solvent), and two to three drops of bromthymol blue indicator. Use absolute ethanol in preparing the aqueous ethanol solvent. Do not use denatured ethanol, as the denaturing agents may interfere with the reactions being studied. Bromthymol blue has a yellow color in acid solution and a blue color in alkaline solution.

Place a 50-mL buret filled with approximately 0.01*N* sodium hydroxide above the flask. The exact concentration of sodium hydroxide does not need to be known. Record the initial volume of sodium hydroxide at time *t* equal to 0.0 minutes. Add about 2 mL of sodium hydroxide from the buret to the Erlenmeyer flask and precisely record the new volume in the buret. Start the stirrer. At time 0.0 minutes, **rapidly** add 1.0 mL of the acetone solution of the alkyl chloride from a pipet. Start the timer when the pipet is about half empty. The indicator will undergo a color change, passing from blue through green to yellow when enough hydrogen chloride has been formed in the reaction to neutralize the sodium hydroxide in the flask. Record the time at which the color changed. This color change may not be rapid. One should try to use the same color as the end point each time. Add another 2 mL of sodium hydroxide from the buret, precisely record the volume, and also record the time at which this second volume of sodium hydroxide is consumed. Repeat the sodium hydroxide addition twice more (four total). Finally, allow the reaction to go to completion for an hour without excess sodium hydroxide present. Stopper the Erlenmeyer flask during this period.

After the reaction has gone to completion, **accurately** titrate the amount of hydrogen chloride in solution to the end point. The end point is reached when the color of the solution remains constant for at least 30 seconds. The time corresponding to this final volume is infinity ($t = \infty$). Repeat this process in the other two solvent mixtures indicated in the table below. These experiments can be carried out while you are waiting for the infinity titration of the previous experiments, provided that a separate buret is used for each run, so that the infinity concentrations of hydrogen chloride produced can be accurately determined.

The data should be plotted according to the method described in the introductory section of this experiment. The rate constant k and the half-life τ must be reported. The report to the instructor should include the plot of the data as well as a table of data. A sample table of data is shown. Explain your results, especially the effect of changing the water content of the solvent on the rate of the reaction. If the instructor so desires, the results from the entire class may be compared.

Experimental Conditions

COMPOUND	SOLVENT MIXTURES (volume percentage of organic phase in water)
t-Butyl chloride	40% Ethanol 25% Acetone 10% Acetone
α-Phenylethyl chloride	50% Ethanol 40% Ethanol 35% Ethanol

Hydrolysis of α-Phenylethyl Chloride in 50% Ethanol

TIME (min)	VOL. NaOH RECORDED	VOL. NaOH USED	$V_\infty - V_t$	$\dfrac{V_\infty}{V_\infty - V_t}$	$\ln\left(\dfrac{V_\infty}{V_\infty - V_t}\right)$
0.00	0.2	0.0	6.9	1.00	0.000
8.46	2.2	2.0	4.9	1.41	0.343
18.25	4.2	4.0	2.9	2.37	0.863
31.80	5.9	5.7	1.2	5.75	1.750
47.72	6.8	6.6	0.3	23.00	3.136
100 (∞)	7.1	6.9	0.0	. . .	. . .

QUESTIONS

1. Plot the data given in the table above. Determine the rate constant and the half-life for this example.

2. What are the principal by-products of these reactions? Give the rate equations for these competing reactions? Should the production of these by-products go on at the same rate as the hydrolysis reactions? Explain.

3. Compare the energy diagrams for an S_N1 reaction in solvents with two different percentages of water. Explain any differences in the diagrams and their effect on the reaction rate.

4. Compare the expected rates of hydrolysis of *t*-cumyl chloride (2-chloro-2-phenylpropane) and α-phenylethyl chloride (1-chloro-1-phenylethane) in the same solvent. Explain any differences that might be expected.

Experiment 13
Synthesis of *n*-Butyl Bromide and *t*-Pentyl Chloride

Synthesis of alkyl halides
Extraction
Simple distillation

The synthesis of two alkyl halides from alcohols is the basis for this experiment. In the first procedure, a primary alkyl halide, *n*-butyl bromide is prepared as shown in Equation 1.

$$CH_3CH_2CH_2CH_2OH + NaBr + H_2SO_4 \longrightarrow$$

    n-Butyl alcohol

$$CH_3CH_2CH_2CH_2Br + NaHSO_4 + H_2O \quad (1)$$

    n-Butyl Bromide

In the second procedure, a tertiary alkyl halide, *t*-pentyl chloride (*t*-amyl chloride (*t*-amyl chloride), is prepared as shown in Equation 2.

$$CH_3CH_2-\underset{\underset{OH}{|}}{\overset{\overset{CH_3}{|}}{C}}-CH_3 + HCl \longrightarrow CH_3CH_2-\underset{\underset{Cl}{|}}{\overset{\overset{CH_3}{|}}{C}}-CH_3 + H_2O \quad (2)$$

    t-Pentyl alcohol   *t*-Pentyl chloride

These reactions are an interesting contrast in mechanisms. The *n*-butyl bromide synthesis proceeds by an S_N2 mechanism, while *t*-pentyl chloride is prepared by an S_N1 reaction.

n-BUTYL BROMIDE

The primary alkyl halide, *n*-butyl bromide, can be prepared easily by allowing *n*-butyl alcohol to react with sodium bromide and sulfuric acid by Equation 1. The sodium bromide reacts with sulfuric acid to produce hydrobromic acid.

$$2 NaBr + H_2SO_4 \longrightarrow 2 HBr + Na_2SO_4$$

Excess sulfuric acid serves to shift the equilibrium, and thus to speed the reaction by producing a higher concentration of hydrobromic acid. The sulfuric acid also protonates the hydroxyl group of *n*-butyl alcohol so that water is displaced rather than the hydroxide ion, OH^-. The acid also protonates the water as it is produced in the reaction and deactivates it as a nucleophile. Deactivation of water keeps the alkyl halide from being converted back to the alcohol by nucleophilic attack of water. The reaction of the primary substrate proceeds via an S_N2 mechanism.

$$CH_3CH_2CH_2CH_2-O-H + H^+ \xrightarrow{fast} CH_3CH_2CH_2CH_2-\overset{+}{O}-H$$
$$\qquad\qquad\qquad\qquad\qquad\qquad\qquad\qquad\qquad\qquad\;\; | $$
$$\qquad\qquad\qquad\qquad\qquad\qquad\qquad\qquad\qquad\qquad\;\; H$$

$$CH_3CH_2CH_2CH_2-\overset{+}{O}-H + Br^- \xrightarrow[S_N2]{slow} CH_3CH_2CH_2CH_2-Br + H_2O$$
$$\qquad\qquad\qquad\qquad\quad | $$
$$\qquad\qquad\qquad\qquad\quad H$$

During the isolation of the *n*-butyl bromide, the crude product is washed with sulfuric acid, water, and sodium bicarbonate to remove any remaining acid or *n*-butyl alcohol.

t-PENTYL CHLORIDE

The tertiary alkyl halide can be prepared by allowing *t*-pentyl alcohol to react with concentrated hydrochloric acid according to Equation 2. The reaction is accomplished simply by shaking the two reagents in a sealed conical vial. As the reaction proceeds, the insoluble alkyl halide product forms an upper phase. The reaction of the tertiary substrate occurs via an S_N1 mechanism.

$$CH_3CH_2-\overset{\overset{\displaystyle CH_3}{|}}{\underset{\underset{\displaystyle OH}{|}}{C}}-CH_3 + H^+ \xrightarrow{fast} CH_3CH_2-\overset{\overset{\displaystyle CH_3}{|}}{\underset{\underset{\displaystyle \overset{+}{O}}{|}}{C}}-CH_3$$

$$CH_3CH_2-\overset{\overset{\displaystyle CH_3}{|}}{\underset{\underset{\displaystyle \overset{+}{O}}{|}}{C}}-CH_3 \xrightarrow{slow} CH_3CH_2-\overset{\overset{\displaystyle CH_3}{|}}{\underset{+}{C}}-CH_3 + H_2O$$

$$CH_3CH_2-\overset{\overset{\displaystyle CH_3}{|}}{\underset{+}{C}}-CH_3 + Cl^- \xrightarrow{fast} CH_3CH_2-\overset{\overset{\displaystyle CH_3}{|}}{\underset{\underset{\displaystyle Cl}{|}}{C}}-CH_3$$

A small amount of alkene, 2-methyl-2-butene, is produced as a by-product in this reaction. If sulfuric acid had been used as it was for *n*-butyl bromide, a much larger amount of this alkene would have been produced.

REQUIRED READING

Review: Techniques 1–3
           Technique 6   Physical Constants, Part B, Boiling Points
           Technique 7   Extractions, Separations and Drying Agents
           Technique 8   Simple Distillation

SPECIAL INSTRUCTIONS

Take special care with concentrated sulfuric acid; it causes severe burns.

As your instructor indicates, perform either the *n*-butyl bromide or the *t*-pentyl chloride procedure, or both.

Procedure 13A
n-Butyl Bromide

Using an automatic pipet or a graduated pipet and pipet pump, place 1.4 mL of *n*-butyl alcohol (1-butanol, MW = 74.1) in a preweighed 10-mL round-bottom flask. Reweigh the flask to determine the exact weight of alcohol. Add 2.4 g of sodium bromide and 2.0 mL of water. Cool the mixture in an ice bath and slowly add 1.6 mL of concentrated sulfuric acid dropwise using a Pasteur pipet. Add a small magnetic stirring bar and assemble the reflux apparatus and trap shown in the figure. The trap absorbs the hydrogen bromide gas evolved during the reaction period. With stirring, heat the mixture to a gentle boil (sand bath temperature about 140 °C) for 60 minutes.

Allow the apparatus to cool to room temperature. The *n*-butyl bromide layer should be on top. If the reaction was not yet complete, the remaining *n*-butyl alcohol will sometimes form a **second organic layer** on **top** of the *n*-butyl bromide layer. Treat both these organic layers as if they were one. Remove and discard as much of the aqueous (bottom) layer as possible using a Pasteur pipet, but do not remove any of the organic layer (or layers). Ignore the salts during this separation. If they are drawn into the pipet, treat them as part of the aqueous layer. Transfer the remaining liquid to a 5-mL conical vial. Remove and discard any aqueous layer remaining in the conical vial.

Add 2 mL of 9 M H_2SO_4 to the conical vial. Cap the vial and shake it gently with occasional venting. Allow the layers to separate. Since any remaining *n*-butyl alcohol is extracted by the H_2SO_4 solution, there should now be only one organic layer. The organic layer should be the top layer. Remove and discard the aqueous (bottom) layer.

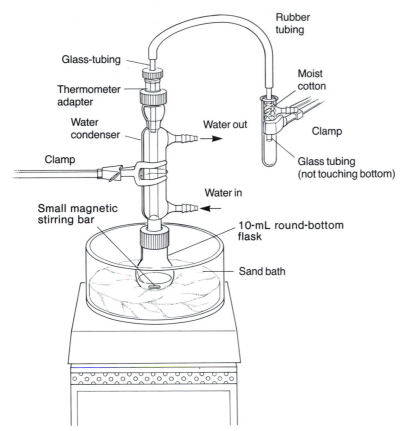

Apparatus for Procedure 13A, *n*-butyl bromide

Add 2 mL of water to the vial. Cap the vial and shake it gently, with occasional venting. Allow the layers to separate. This time the organic layer should be the bottom layer. The bottom layer may form into a globule (ball) instead of separating cleanly. Use a microspatula to prod the ball gently into the bottom of the vial. Using a Pasteur pipet, transfer the bottom layer (or globule) to a clean 5-mL conical vial. Add 2 mL of saturated aqueous sodium bicarbonate solution, a little at a time, with stirring. Cap the vial and shake it vigorously for one minute, venting frequently to relieve any pressure that is produced.

Allow the layers to separate, then carefully transfer the lower alkyl halide layer to a 3-mL conical vial using a Pasteur pipet. Add two microspatulafuls (use the V-grooved end) of granular anhydrous sodium sulfate to dry the solution. Cap the vial and allow it to stand until the product is clear and dry. When it is dry, transfer it to a clean, dry 3-mL vial using a filter tip pipet and distill it (sand bath about 140 °C) using a clean, dry Hickman still (Figure 8–5, p 643). Each time the Hickman head becomes full, transfer the distillate to a preweighed conical vial using a Pasteur pipet.

When the distillation is complete (one or two drops remaining), weigh the vial, calculate the percentage yield and determine a microscale boiling point (Technique 6,

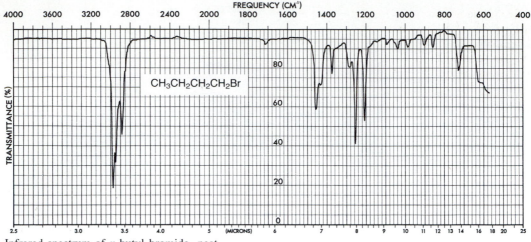

Infrared spectrum of *n*-butyl bromide, neat

Section 6.10, p 607). Determine the infrared spectrum of the product using salt plates (Technique 18, Section 18.2, p 771). Submit the remainder of your sample in a properly labeled vial, along with the infrared spectrum, when you submit your report to the instructor.

Procedure 13B
t-Pentyl Chloride

Note in the following procedures that it may be difficult to see the interfaces between layers since the refractive index of the product will be similar to the refractive indices of the extraction solvents.

Using an automatic pipet or a graduated pipet and pipet pump, place 1.0 mL of *t*-pentyl alcohol (2-methyl-2-butanol, MW = 88.2) in a preweighed 5-mL conical vial. Reweigh the vial to determine the exact weight of alcohol delivered. Add 2.5 mL of concentrated hydrochloric acid, cap the vial, and shake it vigorously for one minute. After shaking the vial, loosen the cap and vent the vial. Recap the vial and shake it for three more minutes with occasional venting. Allow the mixture to stand in the vial until the layer of alkyl halide product separates. The *t*-pentyl chloride (d = 0.865 g/mL) should be the top layer, but be sure to verify this carefully by observation as you add a few drops of hydrochloric acid.

With a Pasteur pipet, separate the layers by placing the tip of the pipet squarely into the bottom of the vial (Figure 7–1, p 618) and transferring the lower layer to a 3-mL conical vial (Technique 7, Section 7.4, p 622). Discard the aqueous layer (Are you sure which one it is?).

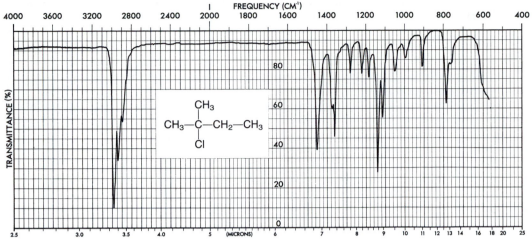

Infrared spectrum of *t*-pentyl chloride, neat

The operations in this paragraph should be carried out as rapidly as possible since the *t*-pentyl chloride is unstable in water and aqueous bicarbonate solution. It is easily hydrolyzed back to the alcohol. Be sure everything you need is at hand. Wash the organic layer by adding 1 mL of water to the conical vial which contains it. Shake the mixtures for a few seconds and then allow the layers to re-form. Once again separate the layers using a Pasteur pipet and discard the aqueous layer after making certain that the proper layer has been saved. Use a few drops of water to test the layers. Add a 1-mL portion of 5% aqueous sodium bicarbonate to the organic layer. Gently mix the two phases in the vial with a stirring rod until they are thoroughly mixed. Now cap the vial and shake it gently for one minute with occasional venting. Following this, vigorously shake the vial for another 30 seconds with occasional venting. Separate the layers and discard the aqueous layer.

Dry the crude *t*-pentyl chloride over two microspatulafuls (use the V-grooved end) of granular anhydrous sodium sulfate. Swirl or stir the alkyl halide with the drying agent to speed the drying process. When the solution is dry (it should be clear), carefully separate the alkyl halide from the drying agent with a Pasteur pipet and transfer it to a clean, dry 3-mL conical vial. Add a microporous boiling stone and distill the crude *t*-pentyl chloride (Figure 8–5, p 643 or, if possible, Figure 8–7B, p 645).

Using a Pasteur pipet, transfer the product to a dry, preweighed conical vial, weigh it, and calculate the percentage yield. Determine a boiling point for the product using a microscale boiling point determination (Technique 6, Section 6.10, p 607). Determine the infrared spectrum of the alkyl halide using salt plates (Technique 18, Section 18.2, p 771). Submit the remainder of your sample in a properly labeled vial, along with the infrared spectrum, when you submit your report to the instructor.

QUESTIONS

n-Butyl Bromide

1. What are the formulas of the salts that precipitate when the reaction mixture is cooled?

2. Why does the alkyl halide layer switch from the top layer to the bottom layer at the point where water is used to extract the organic layer?

3. An ether and an alkene are formed as by-products in this reaction. Draw the structures of these by-products and give mechanisms for their formation.

4. Aqueous sodium bicarbonate was used to wash the crude *n*-butyl bromide.
  (a) What was the purpose of this wash? Give equations.
  (b) Why would it be undesirable to wash the crude halide with aqueous sodium hydroxide?

5. Look up the density of *n*-butyl chloride (1-chlorobutane). Assume that this alkyl halide was prepared instead of the bromide. Decide whether the alkyl chloride would appear as the upper or the lower phase at each stage of the separation procedure: after the reflux, after the addition of water, and after the addition of sodium bicarbonate.

6. Why must the alkyl halide product be dried carefully with anhydrous sodium sulfate before the distillation? (*Hint:* see Technique 10, Section 10.7.)

t-Pentyl Chloride

1. Aqueous sodium bicarbonate was used to wash the crude *t*-pentyl chloride.
  (a) What was the purpose of this wash? Give equations.
  (b) Why would it be undesirable to wash the crude halide with aqueous sodium hydroxide?

2. Some 2-methyl-2-butene may be produced in the reaction as a by-product. Give a mechanism for its production.

3. How is unreacted *t*-pentyl alcohol removed in this experiment? Look up the solubility of the alcohol and the alkyl halide in water.

4. Why must the alkyl halide product be dried carefully with anhydrous sodium sulfate before the distillation? (*Hint:* see Technique 10, Section 10.7.)

5. Will *t*-pentyl chloride (2-chloro-2-methylbutane) float on the surface of water? Look up its density in a handbook.

Experiment 14
Elimination Reactions: 2-Butanol and 2-Bromobutane

Dehydration   Gas chromatography
Dehydrobromination   Regiochemistry
Collection of gaseous products   Saytzeff's rule

Alkenes can be prepared by elimination reactions such as the dehydration of alcohols and the dehydrobromination of alkyl halides. In this experiment, you will study both of these methods for preparing alkenes. The two reactions which will be performed are

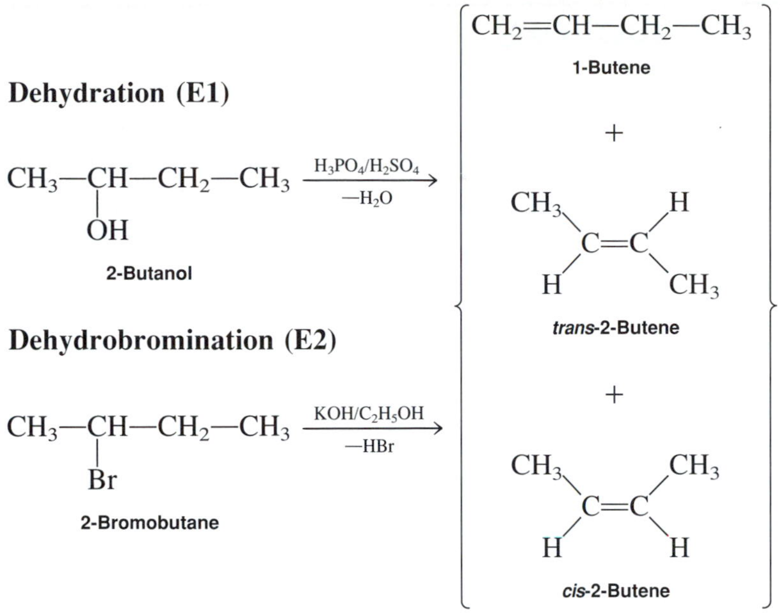

Dehydration (E1)

$$CH_3—CH—CH_2—CH_3 \xrightarrow[-H_2O]{H_3PO_4/H_2SO_4}$$

$$\underset{OH}{|}$$

2-Butanol

Dehydrobromination (E2)

$$CH_3—CH—CH_2—CH_3 \xrightarrow[-HBr]{KOH/C_2H_5OH}$$

$$\underset{Br}{|}$$

2-Bromobutane

$$CH_2=CH—CH_2—CH_3$$

1-Butene

+

$$\underset{H}{\overset{CH_3}{\diagdown}}C=C\underset{CH_3}{\overset{H}{\diagup}}$$

trans-2-Butene

+

$$\underset{H}{\overset{CH_3}{\diagdown}}C=C\underset{H}{\overset{CH_3}{\diagup}}$$

cis-2-Butene

In the first reaction, 2-butanol undergoes dehydration in the presence of strong acid to form the three alkenes shown in the equation. The same three alkenes are also formed in the second reaction by the dehydrobromination of 2-bromobutane in the presence of strong base.

The products of these reactions, which are gases at room temperature, can be analyzed by gas chromatography. For each reaction, the relative percentages of the three alkenes can then be calculated. By applying Saytzeff's rule, you should be able to explain the regiochemistry of these two reactions. That is, there should be a correlation between the relative stabilities of the three alkenes and the relative percentages you obtain.

REQUIRED READING

Review: Techniques 1 and 2
           Technique 14   Gas Chromatography

New: Technique 3   Reaction Methods, Section 3.8

Before beginning this experiment, review the appropriate chapters on elimination reactions in your lecture textbook. Pay special attention to dehydration of alcohols, dehydrohalogenation of alkyl halides, E1 and E2 reactions, and Saytzeff's rule.

SPECIAL INSTRUCTIONS

The two procedures in this experiment can be conveniently scheduled with another experiment since the time required for each procedure is 30–45 minutes. By scheduling this experiment over two laboratory periods, the waiting time for the gas chromatograph will also be minimized.

You may be given a choice of doing one of the two elimination reactions in this experiment; or, alternatively, your instructor may assign one of the reactions to you. In either case, you will need to share your results with a student doing the other reaction so that you can write up the laboratory report.

> **NOTE TO THE INSTRUCTOR:** This experiment has been designed to utilize a specific apparatus for collecting the products (see figure, p 149). Depending on the type of glassware used by your students, it may be necessary to modify the apparatus described here. See Technique 3, Section 3.8, p 558, for possible modifications.
>
> The gas chromatograph should be prepared as follows: column, injector, detector, and outlet should be at room temperature; carrier gas flow rate, 20 mL/min. An 8-foot column containing 20% DC-710 gives good separation.
>
> The acid mixture for Procedure 14A and the ethanolic potassium hydroxide solution for procedure 14B should be prepared ahead of time.

PROCEDURE

Assemble the apparatus shown in the figure, except do not connect the 3-mL conical vial to the thermometer adapter. It is first necessary to make a mark on both the test tube and the glass tube corresponding to a volume of 4 mL. This can be done by inserting a rubber septum into the glass tube, and then filling each tube with 4 mL of water. Using a water resistant marking pen, make a mark on each tube at the level of the water. To place the test tube into the beaker, fill it completely with water; and, while holding your thumb over the opening, invert the test tube and place the open end into the beaker filled with water. Once the tube is in the water, you can release your thumb and allow the test tube to rest on the bottom of the beaker. Place the flexible tubing into the open end of the test tube without allowing air to enter the tube. Repeat this procedure with the glass tube sealed by the rubber septum, except do not insert the flexible tubing into it. The test tube will be used to collect the first 4 mL of gas, which will consist mainly of air. The flexible tubing will then be inserted into the open end of the glass tube to collect the gaseous products.

The section of the Pasteur pipet that fits into the thermometer adapter is prepared from a 5¾-inch Pasteur pipet. Break off the wide end of the pipet about one inch from

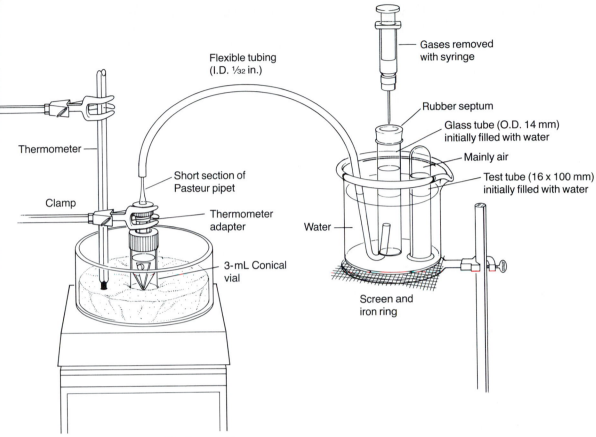

Apparatus for collecting products

where the constriction begins. Connect the narrow end of the Pasteur pipet to the flexible tubing and insert the wide end of the pipet into the thermometer adapter. Be sure that the connection between the pipet and the thermometer adapter is air-tight. Adjust the temperature of the sand bath to about 80 °C and proceed to either Procedure 14A or Procedure 14B.

Procedure 14A
Dehydration of 2-Butanol

To the 3-mL conical vial add 0.20 mL of 2-butanol and a magnetic spin vane. Using the graduated pipet provided, add 0.30 mL of the mixture of concentrated phosphoric acid and concentrated sulfuric acid[1] to the vial. Stir the mixture for a few seconds. Connect

[1]NOTE TO THE INSTRUCTOR: The acid mixture should be prepared for the entire class by mixing 6.0 mL of concentrated phosphoric acid and 3.0 mL of concentrated sulfuric acid. This will provide enough acid for 20 students, assuming little spillage or other types of waste. A graduated pipet and pipet pump should be provided for dispensing this mixture.

the thermometer adapter to the vial and place the vial in the sand bath as shown in the figure. (If possible, use a clamp to secure the assembly in the sand bath.) Stir the mixture and increase the heat slowly until 4 mL of gas (mainly air) are collected in the inverted test tube. Carefully withdraw the flexible tubing from the test tube and insert it into the open end of the glass tube with the rubber septum. Continue heating the reaction mixture (increasing the temperature if necessary) until 4–5 mL of the gaseous products are collected. It should not be necessary to heat the reaction mixture much above 100 °C.

CAUTION: Before turning the heat down or removing the vial from the sand bath, you must first remove the flexible tubing from the gas collection tube or disconnect the thermometer adapter from the conical vial. Otherwise, water may be sucked back into the vial as the reaction mixture cools. Be sure to leave the tube containing the product in the water when you perform this operation.

Using a 1-mL syringe, remove about 0.5 mL of gaseous product by injecting the needle through the rubber septum. With assistance from the instructor or lab assistant, analyze this sample on the gas chromatograph. The order of elution will be 1-butene, *trans*-2-butene, and *cis*-2-butene.

Once the gas chromatogram has been obtained, determine the relative amounts of the products (see Technique 14, Section 14.11, p 750). While the peaks may be cut out and weighed on an analytical balance as a method of determining areas, triangulation is the preferred method. Record the percentages of the three alkenes in the product.

Procedure 14B
Dehydrobromination of 2-Bromobutane

Using the graduated pipet provided, add 2.0 mL of an ethanolic potassium hydroxide solution[2] to the 3-mL conical vial. Add 0.16 mL of 2-bromobutane and a spin vane to the vial. Attach the thermometer adapter to the vial and place this assembly in the sand bath as shown in the figure. (If possible, clamp the apparatus to hold it more securely in the sand bath.)

With stirring, slowly increase the temperature of the sand bath until gas begins to collect in the test tube. Continue heating the reaction mixture at this temperature or slightly higher until 4 mL of gas (mainly air) are collected in the tube. Carefully remove the flexible tubing from the test tube and insert it into the open end of the glass tube with the rubber septum. Continue to heat the reaction mixture until 4–5 mL of gaseous product are collected in the tube.

---

[2] NOTE TO THE INSTRUCTOR: To prepare enough solution for 20 students (assuming little spillage or other types of waste), add 15.0 g of potassium hydroxide to 50.0 mL of 95% ethanol. Stir the mixture until the potassium hydroxide is completely dissolved. A graduated pipet and pipet pump should be provided to dispense this solution.

        CAUTION: Before turning the heat down or taking the sand bath away, you must first remove the flexible tubing from the gas collection tube or disconnect the thermometer adapter from the conical vial. Otherwise, water may be sucked back into the vial. Be sure to leave the tube containing the product in the water when you perform this operation.

        Analyze the products on the gas chromatograph, as described in Procedure 14A. Determine the percentages of the three alkenes produced by the dehydrobromination of 2-bromobutane.

REPORT

It will be necessary to exchange your results with a student who completed the other procedure in order to complete this report. Record the percentages of the isomeric butenes produced in each elimination reaction. Rationalize the relative product distribution for each reaction by relating the percentages to Saytzeff's rule. More specifically, compare the amount of *trans*-2-butene to *cis*-2-butene and compare the total amount of the 2-butenes to 1-butene. Are these relative amounts consistent with Saytzeff's rule? Make this comparison for both reactions; and then, in a general way, compare the results obtained from the dehydration of 2-butanol with the results obtained from the dehydrohalogenation of 2-bromobutane.

QUESTIONS

1. Give the mechanisms for the dehydration of 2-butanol and the dehydrohalogenation of 2-bromobutane.

2. Explain what it means for a reaction to be regioselective.

3. Were the reactions in this experiment regioselective? Explain.

4. Explain the order of elution you observed in doing the gas chromatography for this experiment. What property of the product molecules seems to be the most important in determining the relative retention times?

5. What alkenes (including *cis* and *trans* isomers) would be produced by the dehydration of the following alcohols? Where possible, predict the relative amounts of each product according to Saytzeff's rule.

   (a) 3-Pentanol.
   (b) 2-Methyl-2-butanol.
   (c) 1-Butanol.
   (d) 2-Methyl-1-butanol.

6. What alkenes (including *cis* and *trans* isomers) would be produced by the dehydrobromination of the following alkyl bromides? Where possible, predict the relative amounts of each product according to Saytzeff's rule.

   (a) 2-Bromo-2-methylbutane.
   (b) 1-Bromobutane.
   (c) 2-Bromo-2,3-dimethylbutane.
   (d) 3-Bromopentane.

Experiment 15
Cyclohexene

Preparation of an alkene
Dehydration of an alcohol
Distillation
Bromine and permanganate tests for unsaturation

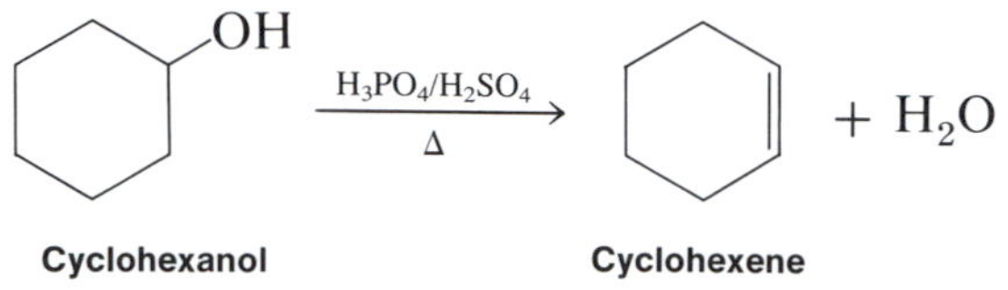

Alcohol dehydration is an acid-catalyzed reaction performed by strong, concentrated mineral acids such as sulfuric and phosphoric acids. The acid protonates the alcoholic hydroxyl group, permitting it to dissociate as water. Loss of a proton from the intermediate (elimination) brings about an alkene. Since sulfuric acid often causes extensive charring in this reaction, phosphoric acid, which is comparatively free of this problem, is a better choice. In order to make the reaction proceed faster, however, a minimal amount of sulfuric acid will also be used.

The equilibrium that attends this reaction will be shifted in favor of the product, by distilling it from the reaction mixture as it is formed. The cyclohexene (bp 83 °C) will co-distill with the water that is also formed. By continuously removing the products, one can obtain a high yield of cyclohexene. Since the starting material, cyclohexanol, is also rather low-boiling (bp 161 °C), the distillation must be done carefully so that cyclohexanol does not distill.

Unavoidably, a small amount of phosphoric acid co-distills with the product. It is removed by washing the distillate mixture with a saturated sodium chloride solution. This step also partially removes the water from the cyclohexene layer, which will be completed by drying the product over anhydrous sodium sulfate.

Compounds containing double bonds react with a bromine solution (red) to decolorize it. Similarly, they react with a solution of potassium permanganate (purple) to discharge its color and produce a brown precipitate (MnO_2). These reactions are often used as qualitative tests to determine the presence of a double bond in an organic molecule (see Procedure 56C). Both tests will be performed on the cyclohexene formed in this experiment.

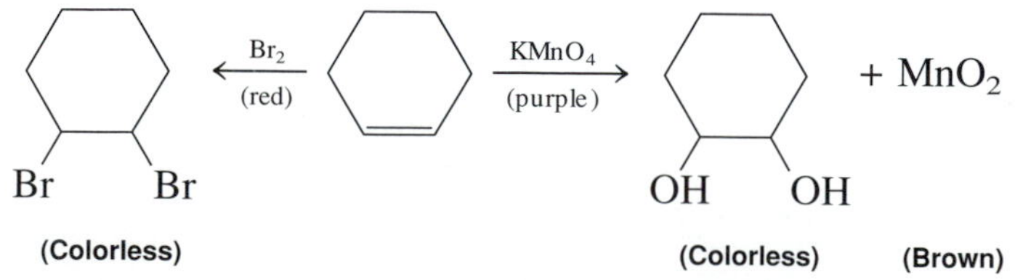

152

REQUIRED READING

Review: Techniques 1 and 2
          Technique 7   Extractions, Sections 7.5 and 7.8

New: Technique 8   Simple Distillation

          If performing the optional boiling point or infrared spectroscopy, also read:

          Technique 6   Physical Constants, Part B, Boiling Points
          Technique 18   Preparation of Samples for Spectroscopy, Part A, Infra-
                                 red Spectroscopy
          Appendix 3   Infrared Spectroscopy

SPECIAL INSTRUCTIONS

Phosphoric acid and sulfuric acid are very corrosive. Do not allow either acid to touch your skin.

PROCEDURE

Place 1.0 mL of cyclohexanol (MW = 100.2) in a tared 3-mL conical vial and reweigh to determine an accurate weight for the alcohol. Add 0.25 mL of 85% phosphoric acid and four drops of concentrated sulfuric acid to the vial. Mix the liquids thoroughly using a glass stirring rod and add a boiling stone or a magnetic spin vane. Assemble a distillation apparatus as shown in Figure 8–5, p 643. A water-cooled condenser should be used, and the drying tube is not necessary. Start circulating the cooling water in the condenser and heat the mixture until the product begins to distill (sand bath about 150–160 °C). Stir the mixture if you are using a spin vane instead of a boiling stone. The heating should be regulated so that the distillation requires about 30–45 minutes.

During the distillation, use a Pasteur pipet to remove the distillate from the well of the Hickman head when necessary. It will be necessary to remove the condenser when performing this procedure. Transfer the distillate to a clean, dry 3-mL conical vial, which should be capped except when liquid is being added. Continue the distillation until no more liquid is being distilled. This can be best determined by observing when boiling in the conical vial has ceased. Also, the volume of liquid remaining in the vial should be about 0.5 mL when distillation is complete.

When distillation is complete, remove as much distillate as possible from the Hickman head and transfer it to the 3-mL conical vial. Then, using a Pasteur pipet with the tip slightly bent, rinse the sides of the inside wall of the Hickman head with 1.0 mL of saturated sodium chloride. Do this thoroughly so that as much liquid as possible is washed down into the well of the Hickman head. Transfer this liquid to the 3-mL conical vial.

Shake the mixture in the capped conical vial. Remove the bottom aqueous layer. Using a dry Pasteur pipet, transfer the organic layer to a small test tube containing two microspatulafuls of granular anhydrous sodium sulfate. Place a stopper in the tube and set it aside for 10–15 minutes to remove the last traces of water. During this time, wash and dry the Hickman head and the 3-mL conical vial.

Transfer as much of the dried liquid as possible to the clean, dry 3-mL conical vial, being careful to leave the solid behind. Add a boiling stone to the vial and assemble the distillation apparatus shown in Figure 8–7A or 8–7B, p 645. Because cyclohexene is so volatile, you will recover more product if the assembly in Figure 8–7B can be used. Distill the cyclohexene, collecting the material that boils over the range 80–84 °C. There will be little or no forerun, and very little liquid will remain in the distilling flask at the end of the distillation. Carefully transfer as much distillate as possible to a small tared vial with a screw cap (lined with aluminum foil). Weigh the product (MW = 82.1) and calculate the percentage yield.

At the instructor's option, determine a more accurate boiling point on your sample using the micro boiling point method (Technique 6, Section 6.10, p 607), and obtain the infrared spectrum of cyclohexene (Technique 18, Section 18.2, p 771 or Section 18.3, p 773). Because cyclohexene is so volatile, you must work quickly to obtain a good spectrum using sodium chloride plates. A better method is to use silver chloride plates with the orientation shown in Figure 18–3B, p 774. Compare the spectrum with the one shown in this experiment. After performing the tests below, submit your sample, along with the report, to the instructor.

UNSATURATION TESTS

Place four to five drops of cyclohexanol in each of two small test tubes. In each of another pair of small test tubes, place four to five drops of the cyclohexene you prepared. Do not confuse the test tubes. Take one test tube from each group and add to the contents of each a solution of bromine in carbon tetrachloride or methylene chloride, drop by drop, until the red color is no longer discharged. Record the result in each case. Test the remaining two test tubes in a similar fashion with a solution of potassium permanganate. Since aqueous potassium permanganate is not miscible with organic compounds, you will have to add about 0.3 mL of 1,2-dimethoxyethane to each test tube before making the test. Record your results and explain them.

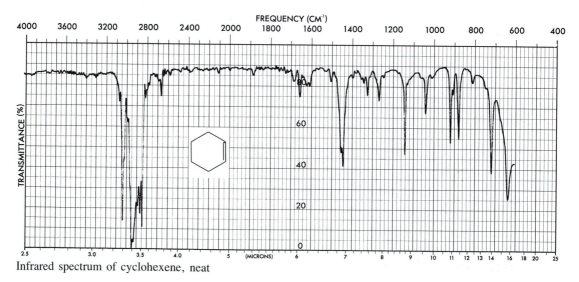

Infrared spectrum of cyclohexene, neat

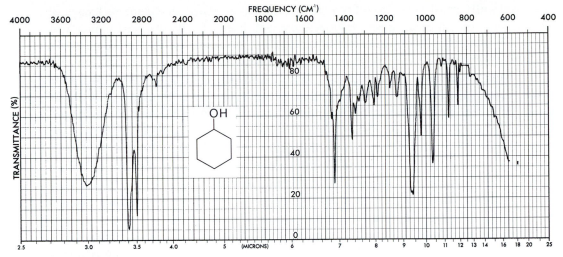

Infrared spectrum of cyclohexanol, neat

QUESTIONS

1. Draw a mechanism for the dehydration of cyclohexanol catalyzed by phosphoric acid.

2. What major alkene product is produced by dehydration of the following alcohols?
   (a) 1-Methylcyclohexanol.
   (b) 2-Methylcyclohexanol.
   (c) 4-Methylcyclohexanol.
   (d) 2,2-Dimethylcyclohexanol.
   (e) 1,2-Cyclohexanediol.

3. Compare and interpret the infrared spectra of cyclohexene and cyclohexanol.

4. In this experiment, 1.0 mL of saturated sodium chloride is used to rinse the Hickman head after the initial distillation. Why is saturated sodium chloride rather than water used for this procedure and the subsequent washing of the organic layer?

Essay
PETROLEUM AND FOSSIL FUELS

Crude petroleum is a liquid that consists of hydrocarbons as well as some related sulfur, oxygen, and nitrogen compounds. Other elements, including metals, may be present in trace amounts. Crude oil is formed by the decay of marine animal and plant organisms that lived millions of years ago. Over many millions of years, under the influence of temperature, pressure, catalysts, radioactivity, and bacteria, the decayed matter was converted into what we now know as crude oil. The crude oil is trapped in pools beneath the ground by various geological formations.

Most crude oils have a specific gravity between 0.78 and 1.00 g/mL. As a liquid, crude oil may be as thick and black as melted tar or as thin and colorless as water. Its characteristics depend on the particular oil field from which it comes. Pennsylvania crude oils are high in straight-chain alkane compounds (called **paraffins** in the petroleum industry); those crude oils are therefore useful in the manufacture of lubricating oils. Oil fields in California and Texas produce crude oil with a higher percentage of cycloalkanes (also called **naphthenes** by the petroleum industry). Some Middle East fields produce crude oil containing up to 90% cyclic hydrocarbons. Petroleum contains molecules in which the number of carbons ranges from 1 to 60.

When petroleum is refined to convert it into a variety of usable products, it is initially subjected to a fractional distillation. Table 1 lists the various fractions obtained from fractional distillation. Each of these fractions has its own particular uses. Each fraction may be subjected to further purification, depending on the desired application.

TABLE 1. Fractions Obtained from the Distillation of Crude Oil

PETROLEUM FRACTION	COMPOSITION	COMMERCIAL USE
Natural gas	C_1 to C_4	Fuel for heating
Gasoline	C_5 to C_{10}	Motor fuel
Kerosene	C_{11} to C_{12}	Jet fuel and heating
Light gas oil	C_{13} to C_{17}	Furnaces, diesel engines
Heavy gas oil	C_{18} to C_{25}	Motor oil, paraffin wax, petroleum jelly
Residuum	C_{26} to C_{60}	Asphalt, residual oils, waxes

The gasoline fraction obtained directly from the distillation of crude oil is called **straight-run gasoline.** An average barrel of crude oil will yield about 19% straight-run gasoline. This yield presents two immediate problems. First, there is not enough gasoline contained in crude oil to satisfy current needs for fuel to power automobile engines. Second, the straight-run gasoline obtained from crude oil is a poor fuel for modern engines. It must be ''refined'' at a chemical refinery.

The initial problem of the small quantity of gasoline available from crude oil can be solved by **cracking** and **polymerization.** Cracking is a refinery process by which large hydrocarbon molecules are broken down into smaller molecules. Heat and pressure are required for cracking, and a catalyst must also be used. Silica-alumina and silica-magnesia are among the most effective cracking catalysts. A mixture of saturated and unsaturated hydrocarbons is produced in the cracking process. If gaseous hydrogen is also present during the cracking, only saturated hydrocarbons are produced. The hydrocarbon mixtures produced by these cracking processes tend to have a fairly high proportion of branched-chain isomers. These branched isomers improve the quality of the fuel.

$$C_{16}H_{34} + H_2 \xrightarrow[\text{heat}]{\text{catalyst}} 2C_8H_{18} \qquad \text{Cracking}$$

In the polymerization process, also carried out at a refinery, small molecules of alkenes are caused to react with one another to form larger molecules, which are also alkenes.

$$2CH_2{=}C\begin{smallmatrix}CH_3\\CH_3\end{smallmatrix} \xrightarrow[\text{heat}]{\text{catalyst}} CH_3{-}\underset{CH_3}{\overset{CH_3}{C}}{-}CH{=}C\begin{smallmatrix}CH_3\\CH_3\end{smallmatrix} \qquad \text{Polymerization}$$

**2-Methylpropene
(isobutylene)**   **2,4,4-Trimethyl-2-pentene**

The newly formed alkenes may be hydrogenated to form alkanes. The reaction sequence shown here is a very common and important one in petroleum refining since the product, 2,2,4-trimethylpentane (or "iso-octane"), forms the basis for determining the quality of gasoline. By these refining methods, the percentage of gasoline that can be obtained from a barrel of crude oil may rise to as much as 45% or 50%.

$$CH_3{-}\underset{CH_3}{\overset{CH_3}{C}}{-}CH{=}C\begin{smallmatrix}CH_3\\CH_3\end{smallmatrix} + H_2 \xrightarrow{\text{catalyst}} CH_3{-}\underset{CH_3}{\overset{CH_3}{C}}{-}CH_2{-}\underset{}{\overset{CH_3}{CH}}{-}CH_3$$

**2,2,4-Trimethylpentene
("iso-octane")**

The internal combustion engine, as it is found in most automobiles, operates in four cycles or **strokes.** They are illustrated in the figure. The power stroke is of greatest

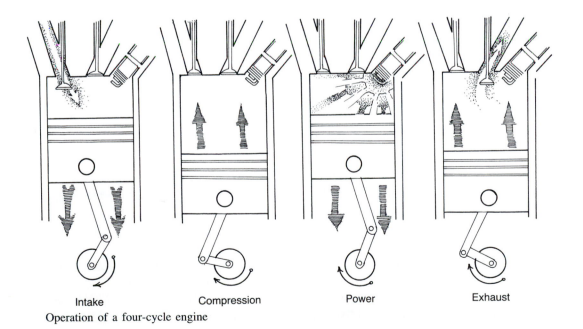

Intake   Compression   Power   Exhaust

Operation of a four-cycle engine

interest to us from the chemical point of view since combustion occurs during this stroke.

When the fuel-air mixture is ignited, it does not explode. Rather, it burns at a controlled, uniform rate. The gases closest to the spark are ignited first; then they in turn ignite the molecules farther from the spark; and so on. The combustion proceeds in a wave of flame or the **flame front,** which starts at the spark plug and proceeds uniformly outward from that point until all the gases in the cylinder have been ignited. Because a certain time is required for this burning, the initial spark is timed to ignite just before the piston has reached the top of its travel. In this way, the piston will be at the very top of its travel at the precise instant that the flame front and the increased pressure that accompanies it reach the piston. The result is a smoothly applied force to the piston, driving it downward.

If heat and compression should cause some of the fuel-air mixture to ignite before the flame front has reached it or to burn faster than expected, the timing of the combustion sequence is disturbed. The flame front arrives at the piston before the piston has reached the very top of its travel. When the combustion is not perfectly coordinated with the motion of the piston, we observe **knocking** or **detonation** (sometimes called ''pinging''). The transfer of power to the piston under these conditions is much less effective than in normal combustion. The wasted energy is merely transferred to the engine block in the form of additional heat. The opposing forces that occur in knocking may eventually damage the engine.

It has been found that the tendency of a fuel to knock is a function of the structures of the molecules composing the fuel. Normal hydrocarbons, those with straight carbon chains, have a greater tendency to lead to knocking than do those alkanes with highly branched chains. The quality of a gasoline, then, is a measure of its resistance to knocking, and this quality is improved by increasing the proportion of branched-chain alkanes in the mixture. Such chemical refining processes as **reforming** and **isomerization** are used to convert normal alkanes to branched-chain alkanes, thus improving the knock-resistance of gasoline.

$$CH_3(CH_2)_6CH_3 \xrightarrow{\text{catalyst}} CH_3{-}\overset{\displaystyle CH_3}{\underset{\displaystyle CH_3}{C}}{-}CH_2{-}\overset{\displaystyle CH_3}{CH}{-}CH_3 \qquad \textbf{Reforming}$$

$$CH_3(CH_2)_5CH_3 \xrightarrow{\text{catalyst}} \qquad \textbf{Reforming}$$

$$CH_3{-}CH_2{-}CH_2{-}CH_2{-}CH_3 \xrightarrow{\text{AlBr}_3} CH_3{-}\overset{\displaystyle CH_3}{CH}{-}CH_2{-}CH_3 \qquad \textbf{Isomerization}$$

None of these processes converts all the normal hydrocarbons into branched-chain isomers; consequently, additives are also put into gasoline to improve the knock-resistance of the fuel. Aromatic hydrocarbons can be considered additives that are effective in improving the knock-resistance of gasoline, and they are used extensively in unleaded as well as leaded gasolines. The most common additive used to reduce knocking has been **tetraethyllead.** Gasoline which contains tetraethyllead is called **leaded** gasoline, while gasoline produced without tetraethyllead is called **unleaded** gasoline. In recent years, because of concern over the possible health hazard associated with emission of lead into the atmosphere and because the presence of lead will inactivate the catalytic converters found on new cars, the Environmental Protection Agency has issued regulations that will lead to the gradual elimination of tetraethyllead in gasoline. As a consequence, oil companies are testing various other additives that will improve the antiknock properties of gasoline without producing harmful emissions.

$$CH_3{-}CH_2{-}\underset{\underset{CH_2{-}CH_3}{|}}{\overset{\overset{CH_2{-}CH_3}{|}}{Pb}}{-}CH_2{-}CH_3$$

Tetraethyllead

New cars are designed to operate on unleaded gasoline, which contains no lead compounds. The quality of the gasoline is maintained by adding increased quantities of hydrocarbons that have very high antiknock properties themselves. Typical are the aromatic hydrocarbons, including benzene, toluene, and xylene.

Benzene   Toluene   Xylene
(1,3-dimethylbenzene)

More expensive refining processes such as **hydrocracking** (cracking in the presence of hydrogen gas) and **reforming** produce mixtures of hydrocarbons that are more knock-resistant than typical gasoline components. Adding the products of hydrocracking and reforming to gasoline improves its performance. Increasing the proportion of aromatic hydrocarbons brings with it certain hazards, however. These substances are toxic, and benzene is considered a serious carcinogenic hazard. The risk that illness will be contracted by workers in refineries, and especially by persons who work in service stations, is increased. Research is also being directed toward development of non-hydrocarbon compounds that can improve the quality of unleaded gasoline. To this end, compounds such as *tert*-butyl alcohol and methyl *tert*-butyl ether have been considered as potential antiknock additives.

$$CH_3{-}\underset{\underset{CH_3}{|}}{\overset{\overset{CH_3}{|}}{C}}{-}OH \qquad CH_3{-}O{-}\underset{\underset{CH_3}{|}}{\overset{\overset{CH_3}{|}}{C}}{-}CH_3$$

tert-Butyl alcohol   Methyl *tert*-butyl ether

A much publicized additive for gasoline is ethanol, to make the mixture known as **gasohol.** The ethanol is added to ordinary unleaded gasoline to yield a solution that is about 10% ethanol. Adding ethanol improves the antiknock property of the fuel, and it extends the amount of fuel that can be produced from a barrel of crude oil. It is not clear, however, that the energy needed to produce the ethanol is significantly smaller than the amount of energy that could be produced when the ethanol is burned in an engine. It seems that more development is needed before gasohol can become a very important motor fuel nationwide.

$$CH_3—CH_2—OH$$

Ethanol

A fuel can be classified according to its antiknock characteristics. The most important rating system is the **octane rating** of gasoline. In this method of classification, the antiknock properties of a fuel are compared in a test engine with the antiknock properties of a standard mixture of heptane and 2,2,4-trimethylpentane. This latter compound is called "iso-octane," hence the name *octane rating*. A fuel that has the same antiknock properties as a given mixture of heptane and "iso-octane" has an octane rating numerically equal to the percentage of "iso-octane" in that reference mixture. Today's 87 octane unleaded gasoline is a mixture of compounds that have, taken together, the same antiknock characteristics as a test fuel composed of 13% heptane and 87% "iso-octane." Other substances besides hydrocarbons may also have high resistance to knocking. Table 2 presents a list of organic compounds with their octane ratings.

TABLE 2   Octane Ratings of Organic Compounds*

COMPOUND	OCTANE NUMBER	COMPOUND	OCTANE NUMBER
Octane	−19	1-Butene	97
Heptane	0	2,2,4-Trimethylpentane	100
Hexane	25	Cyclopentane	101
Pentane	62	Ethanol	105
Cyclohexane	83	Benzene	106
1-Pentene	91	Methanol	106
2-Hexene	93	*m*-Xylene	118
Butane	94	Toluene	120
Propane	97		

*The octane values in this table are determined by the **research method.**

The number of grams of air required for the complete combustion of one mole of gasoline (assuming the formula C_8H_{18}) is 1735 g. This gives rise to a theoretical air-fuel ratio of 15.1:1 for complete combustion. For several reasons, however, it is neither easy nor advisable to supply each cylinder with a theoretically correct air-fuel mixture. The power and performance of an engine improve with a slightly richer mixture (lower air-fuel ratio). Maximum power is obtained from an engine when the air-fuel ratio is near 12.5:1, while maximum economy is obtained when the air-fuel ratio is near 16:1. Under conditions of idling or full load (that is, acceleration), the air-fuel ratio is lower than what would be theoretically correct. As a result, complete combustion does not take place in an internal combustion engine, and carbon monoxide, CO, is produced in the exhaust gases. Other types of nonideal combustion behavior give rise to the presence of unburned hydrocarbons in the exhaust. The high combustion temperatures cause the nitrogen and oxygen of the air to react, forming a variety of nitrogen oxides in the exhaust. Each of these materials contributes to air pollution. Under the influence of sunlight, which has enough energy to break covalent bonds, these materials may react with each other and with air to produce **smog.** Smog consists of **ozone,** which deteriorates rubber and damages plant life; **particulate matter,** which produces haze; **oxides of nitrogen,** which produce a brownish color in the atmosphere; and a variety of eye irritants, such as **peroxyacetyl nitrate** (PAN). Lead particles from tetraethyllead may also cause problems since they are toxic. Sulfur compounds in the gasoline may lead to the production of noxious gases in the exhaust.

$$CH_3-\overset{\overset{\displaystyle O}{\|}}{C}-O-O-NO_2$$

Peroxyacetyl nitrate
(PAN)

Current efforts to reverse the trend of deteriorating air quality caused by automotive exhaust have taken many forms. Initial efforts at modifying the air-fuel mixture of engines produced some improvements in emissions of carbon monoxide, but at the cost of increased nitrogen oxide emissions and poor engine performance. With the more stringent air-quality standards imposed by the Environmental Protection Agency, attention has been turned to alternative sources of power. Recently there has been much interest in the **diesel engine** as a power plant for passenger cars. The diesel engine has the advantage of producing only very small quantities of carbon monoxide and unburned hydrocarbons. It does, however, produce significant amounts of nitrogen oxides, soot (containing polynuclear aromatic hydrocarbons), and odor-causing compounds. At present, there are no legally established standards for the emission of soot or odor by motor vehicles. This does not mean that these substances are harmless; it means merely that there is no reliable method of analyzing exhaust gases quantitatively for the presence of these materials. It may well be that soot and odor may prove to be harmful, but today the emission of these substances remains unregulated. An additional advantage of diesel engines, important in these times of high crude oil prices, is that they tend to yield higher fuel mileage than gasoline engines of a similar size. Research has also been directed at developing internal combustion engines which operate using propane, methane, or even hydrogen as fuels. These engines are not likely to appear in commercial use in the near future, however, because significant technical problems remain to be solved.

In the meantime, since the standard gasoline engine remains the most attractive power plant because of its great flexibility and reliability, efforts at improving its emissions continue. The advent of **catalytic converters,** which are muffler-like devices containing catalysts that can convert carbon monoxide, unburned hydrocarbons, and nitrogen oxides into harmless gases has resulted from such efforts. Unfortunately, the catalysts are rendered inactive by the lead additives in leaded gasoline. Unleaded gasoline must be used, but it takes more crude oil to make a gallon of unleaded gasoline than it does to make leaded gasoline. Other hydrocarbons must be added as antiknock agents to replace tetraethyllead. The active metals in the catalytic converters, principally platinum, palladium, and rhodium, are scarce and extremely expensive. Also, there has been concern that traces of other harmful substances may be produced in the exhaust gases by reactions catalyzed by these metals.

Some success in reducing exhaust emissions has been attained by modifying the design of combustion chambers of internal-combustion engines. Additionally, the use of computerized control of ignition systems shows promise. Efforts have also been directed at developing alternative fuels that would give greater mileage, lower emissions, better performance, and a lower demand on crude oil supplies. Methanol has been proposed as an alternative to gasoline as a fuel. Some preliminary tests have indicated that the amount of the principal air pollutants in automobile exhaust is greatly lowered when methanol is used instead of gasoline in a typical automobile. Experiments with methane have also been promising. Methane has a very high octane number, and the proportion of carbon monoxide and unburned hydrocarbons in the exhaust of a methane-powered engine is very small. The production of methane does not require the expensive and inefficient refining processes that are needed to produce gasoline. The mixture of gasoline and ethanol, **gasohol,** is also being examined carefully. Experiments are even under way toward developing hydrogen gas as a future fuel. Although the technology for using these alternative fuels remains to be developed fully, the future should bring some interesting advances in engine design in an effort to solve our transportation needs while improving the quality of our air.

Experiment 16
Gas Chromatographic Analysis of Gasolines

Gasoline
Gas chromatography

It is possible to analyze samples of gasoline by gas chromatography. From your analysis, you should learn something about the composition of these fuels. Although all gasolines are compounded from the same basic hydrocarbon components, different companies blend these components in different proportions in order to obtain a gasoline with properties similar to those of other brands.

Sometimes the composition of the gasoline may vary depending on the composition of the crude petroleum from which the gasoline was derived. Frequently, refineries vary the composition of gasoline in response to differences in climate or seasonal changes. In the winter or in cold climates, the relative proportion of butane and pentane isomers is increased to increase the volatility of the fuel. This increased volatility permits easier starting. In the summer or in warm climates, the relative proportion of these volatile hydrocarbons is reduced. The decreased volatility thus achieved reduces the possibility of vapor-lock formation. Occasionally, differences in composition can be detected by examining the gas chromatograms of a particular gasoline over several months. In this experiment, we do not try to detect such differences.

There are different octane rating requirements for "regular" and "premium" gasolines. You will be able to observe differences in the composition of these two types of fuels. You should pay particular attention to increases in the proportions of those hydrocarbons that raise octane ratings in the premium fuels. If you analyze an unleaded gasoline, you should be able to observe differences in the composition of this type of gasoline compared with leaded fuels of a similar octane rating.

You will be asked to analyze a sample of regular leaded and one of regular unleaded gasoline, preferably from the same company. If a premium unleaded gasoline from that company is also available, you may be required to analyze it also. Other students in the class may be analyzing gasolines from other companies. If different brands are analyzed, equivalent grades from the different companies should be compared. If there is a supplier of gasohol in the region, a sample of that fuel should be examined as well.

Discount service stations usually buy their gasoline from one of the large petroleum-refining companies. If you analyze gasoline from a discount service station, you may find it interesting to compare that gasoline with an equivalent grade from a major supplier, noting particularly the similarities.

REQUIRED READING

New: Technique 14   Gas Chromatography
      Essay   Petroleum and Fossil Fuels

SPECIAL INSTRUCTIONS

Your instructor may want each student in the class to collect samples of gasoline from different service stations. A list should be compiled of all the different gasoline companies represented in the nearby area, and each student should be assigned to collect a sample from a different company. You should collect the gasoline sample in a labeled screw-cap jar. It may be handy to take a funnel along with you. An easy way to collect a gasoline sample for this experiment is to drain the excess gasoline from the nozzle and hose of the pump into the jar after the gasoline tank of a car has been filled. The collection of gasoline in this manner must be done **immediately after** the gas pump has been used. If not, the volatile components of the gasoline may evaporate, thus changing the composition of the gasoline. Only a very small sample (a few **milli**liters) of gasoline is required, since the gas chromatographic analysis requires no more than a few **micro**liters (μL) of material. Be certain to close the cap of the sample jar tightly to prevent the selective evaporation of the most volatile components. The label on the jar should list the brand of gasoline and the grade.

If you live in a state in which collecting gasoline in glass containers is illegal, your instructor will supply you with a fireproof metal container in which to collect the gasoline sample. Alternatively, the instructor may provide you with gasoline samples.

Always remember that gasoline contains many highly volatile and flammable components. Do not breathe the vapors and do not use open flames near gasoline. Also recall that gasoline may contain tetraethyllead and is therefore toxic.

This experiment may be assigned along with another short one, since it requires only a few minutes of each student's time to carry out the actual gas chromatography. To work as efficiently as possible, it may be convenient to arrange an appointment schedule for using the gas chromatograph.

> **NOTE TO THE INSTRUCTOR:** The gas chromatograph should be prepared as follows: column temperature, 110–115 °C; injection port temperature, 110–115 °C; carrier gas flow rate, 40–50 mL/min. The column should be approximately 12 ft long and should contain a nonpolar stationary phase similar to silicone oil (SE-30) on Chromosorb W or some other stationary phase that separates components principally according to boiling point.

PROCEDURE

The instructor may require you to analyze a series of standard materials. Reference substances that should be used include pentane, hexane, cyclohexane, heptane, toluene, and *m*-xylene, although others may be included appropriately (ethanol, for example). If the samples are each analyzed individually, a sample size of about 0.5 microliters (μL) will be adequate for each reference substance. A better alternative is to analyze a reference mixture that contains all these standard substances and compare it with a similar gas chromatogram previously recorded by the instructor. The instructor may either post a copy of this chromatogram (with the peaks identified) or provide each student with a copy. An example of such a determination is provided at the end of this experiment. If a reference mixture is to be analyzed, a reasonable sample size is 1 μL. The sample is injected into the gas chromatograph, and the retention times of each of the components are measured and recorded (Technique 14, Section 14.7, p 747).

The instructor may prefer to perform the sample injections or have a laboratory assistant perform them. The sample injection procedure requires careful technique, and the special microliter syringes that are required are very delicate and expensive. If students are to perform the sample injections themselves, they must have adequate instruction beforehand.

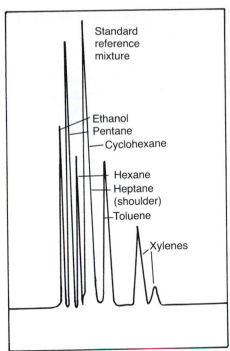
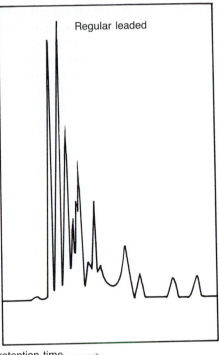

Standard
reference
mixture

Ethanol
Pentane
Cyclohexane

Hexane
Heptane
(shoulder)
Toluene

Xylenes

Regular leaded

Increasing retention time ⟶

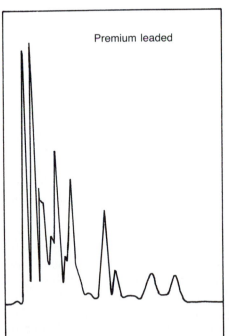
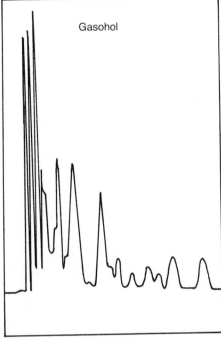

Premium leaded

Gasohol

Increasing retention time ⟶

Gas chromatograms of a standard reference mixture, a regular leaded gasoline, a premium leaded gasoline, and gasohol

Inject the sample of regular leaded gasoline onto the gas chromatography column and wait for the gas chromatogram to be recorded. Next inject the sample of unleaded gasoline and obtain its gas chromatogram. Compare these chromatograms with each other and identify as many of the components as possible. For comparison, sample gas chromatograms of a regular leaded gasoline, a premium leaded gasoline, and gasohol are provided in the figure on p 165. Determine the difference between the two grades of gasoline. If a sample of premium unleaded gasoline was also obtained, record its gas chromatogram and compare it with the chromatograms of the regular and the premium gasoline. Do the same for a sample of gasohol if it is available. Be certain to compare very carefully the retention times of the components in each fuel sample with the standards in the reference mixture. Retention times of compounds vary with the conditions under which they are determined. It is best to analyze the reference mixture and each of the gasoline samples in succession to reduce the variations in retention times that may occur over time. If there are unidentified peaks in the gas chromatograms of the fuel samples, try to guess their probable identity. Compare the gas chromatograms with those of students who have analyzed gasolines from other dealers.

The report to the instructor should include the actual gas chromatograms as well as an identification of as many of the components in each grade of fuel as possible.

QUESTIONS

1. How do regular and premium grades of gasoline differ in this analysis?
2. Assuming one had a mixture of benzene, toluene, and *m*-xylene, what would be the expected order of retention times? Explain.
3. What do you expect the analysis of an unleaded gasoline to reveal in this experiment?
4. Is it possible to detect tetraethyllead in this experiment?
5. If you were a forensic chemist working for the police department, and the fire marshal brought you a sample of gasoline found at the scene of an arson attempt, could you identify the service station at which the arsonist purchased the gasoline? Explain.

Essay
FATS AND OILS

In the normal human diet, about 25% to 50% of the caloric intake consists of fats and oils. These substances are the most concentrated form of food energy in our diet. When metabolized, fats produce about 9.5 kcal of energy per gram. Carbohydrates and proteins produce less than half this amount. For this reason, animals tend to build up fat deposits as a reserve source of energy. They do this, of course, only when their food intake exceeds their energy requirements. In times of starvation, the body metabolizes these stored fats. Even so, some fats are required by animals for bodily insulation and as a protective sheath around some vital organs.

The constitution of fats and oils was first investigated by the French chemist Chevreul during the years 1810 to 1820. He found that when fats and oils were hydrolyzed, they gave rise to several "fatty acids" and the trihydroxylic alcohol, glycerol. Thus, fats and oils are **esters** of glycerol, called **glycerides** or **acylglycerols.** Since glycerol has three hydroxyl groups, it is possible to have mono-, di-, and triglycerides. Fats and oils are predominantly triglycerides (triacylglycerols), constituted as follows:

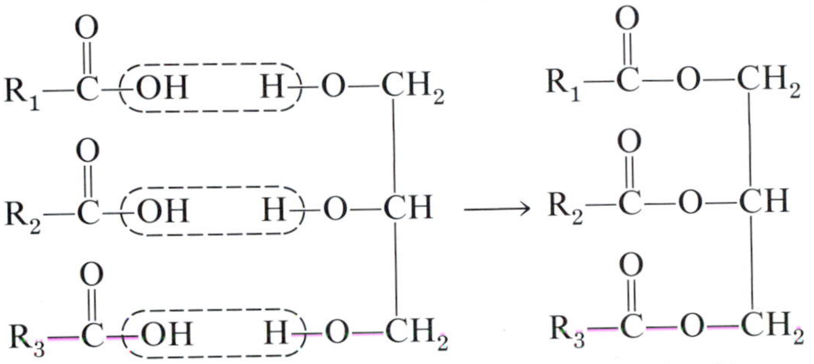

3 Fatty acids + glycerol = **A triglyceride**

Thus, most fats and oils are esters of glycerol, and their differences result from the differences in the fatty acids with which glycerol may be combined. The most common fatty acids have 12, 14, 16, or 18 carbons, although acids with both lesser and greater numbers of carbons are found in several fats and oils. These common fatty acids are listed in the following table along with their structures. As you see, these acids are both saturated and unsaturated. The saturated acids tend to be solids, while the unsaturated acids are usually liquids. This circumstance also extends to fats and oils. Fats are made up of fatty acids that are mostly saturated, while oils are primarily composed of fatty acid portions that have greater numbers of double bonds. In other words, unsaturation lowers the melting point. Fats (solids) are usually obtained from animal sources, while oils (liquids) are commonly obtained from vegetable sources. Therefore vegetable oils usually have a higher degree of unsaturation.

Common Fatty Acids

C_{12} Acids	Lauric	$CH_3(CH_2)_{10}COOH$
C_{14} Acids	Myristic	$CH_3(CH_2)_{12}COOH$
C_{16} Acids	Palmitic	$CH_3(CH_2)_{14}COOH$
	Palmitoleic	$CH_3(CH_2)_5CH=CH-CH_2(CH_2)_6COOH$
C_{18} Acids	Stearic	$CH_3(CH_2)_{16}COOH$
	Oleic	$CH_3(CH_2)_7CH=CH-CH_2(CH_2)_6COOH$
	Linoleic	$CH_3(CH_2)_4(CH=CH-CH_2)_2(CH_2)_6COOH$
	Linolenic	$CH_3CH_2(CH=CH-CH_2)_3(CH_2)_6COOH$
	Ricinoleic	$CH_3(CH_2)_5CH(OH)CH_2CH=CH(CH_2)_7COOH$

Average Fatty Acid Composition (By Percentage) of Selected Fats and Oils

	SATURATED FATTY ACIDS (NO DOUBLE BONDS)						UNSATURATED (1 DOUBLE BOND)			UNSATURATED (>1 DOUBLE BOND)			UNSATURATED
	C_{10} C_8 C_6 C_4	C_{12} Lauric	C_{14} Myristic	C_{16} Palmitic	C_{18} Stearic	C_{20} C_{22} C_{24}	C_{16} Palmitoleic	C_{18} Oleic	C_{18} Ricinoleic	C_{18} Linoleic (2)	C_{18} Linolenic (3)	C_{18} Eleostearic (3)	C_{20} C_{22} C_{24}
Animal fats													
Tallow			2–3	24–32	14–32		1–3	35–48		2–4			
Butter	7–10	2–3	7–9	23–26	10–13		5	30–40		4–5			2
Lard			1–2	28–30	12–18		1–3	41–48		6–7			2
Animal oils													
Neat's foot			4–5	17–18	2–3		13–18	74–77					
Whale			6–8	11–18	2–4		6–15	33–38		↕ 24–30			17–31
Sardine				10–16	1–2								12–19
Vegetable oils													
Corn				7–11	3–4		0–2	43–49		34–42			
Olive			0–2	5–15	1–4		0–1	69–84		4–12			
Peanut				6–9	2–6	3–10	0–1	50–70		13–26			
Soybean			0–1	6–10	2–6			21–29		50–59	4–8		
Safflower				6–10	1–4			8–18		70–80	2–4		
Castor bean				0–1				0–9	80–92	3–7			
Cottonseed			0–2	19–24	1–2		0–2	23–33		40–48			
Linseed				4–7	2–5			9–38		3–43	25–58		
Coconut	10–22	45–51	17–20	4–10	1–5	←2–6→		2–10		0–2			
Palm			1–3	34–43	3–6			38–40		5–11			
Tung								4–16		0–1		74–91	

About 20 to 30 different fatty acids are found in fats and oils, and it is not uncommon for a given fat or oil to be composed of as many as 10 to 12 (or more) different fatty acids. Typically, these fatty acids are randomly distributed among the triglyceride molecules, and the chemist cannot identify anything more than an average composition for a given fat or oil. The average fatty acid composition of some selected fats and oils is given in the table on p 168. As indicated, all the values in the table may vary in percentage, depending, for instance, on the locale in which the plant was grown or on the particular diet on which the animal subsisted. Thus, perhaps there is a basis for the claims that corn-fed hogs or cattle taste better than animals maintained on other diets.

Vegetable fats and oils are usually found in fruits and seeds and are recovered by three principal methods. In the first method, **cold pressing,** the appropriate part of the dried plant is pressed in a hydraulic press to squeeze out the oil. The second method is **hot pressing,** which is the same as the first method but done at a higher temperature. Of the two methods, cold pressing usually gives a better grade of product (more bland); the hot pressing method gives a higher yield, but with more undesirable constituents (stronger odor and flavor). The third method is **solvent extraction.** Solvent extraction gives the highest recovery of all and can now be regulated to give bland, high-grade food oils.

Animal fats are usually recovered by **rendering,** which involves cooking the fat out of the tissue by heating it to a high temperature. An alternative method involves placing the fatty tissue in boiling water. The fat floats to the surface and is easily recovered. The most common animal fats, lard (from hogs) and tallow (from cattle), can be prepared in either way.

Many triglyceride fats and oils are used for cooking. We use them to fry meats and other foods and to make sandwich spreads. Almost all commercial cooking fats and oils, except lard, are prepared from vegetable sources. Vegetable oils are liquids at room temperature. If the double bonds in a vegetable oil are hydrogenated, the resultant product becomes solid. Manufacturers, in making commercial cooking fats (Crisco, Spry, Fluffo, etc.), hydrogenate a liquid vegetable oil until the desired degree of consistency is achieved. This makes a product that still has a high degree of unsaturation (double bonds) left. The same technique is used for margarine. ''Polyunsaturated'' oleomargarine is produced by the partial hydrogenation of oils from corn, cottonseed, peanut, and soybean sources. The final product has a yellow dye (β-carotene) added to make it look like butter; milk, about 15% by volume, is mixed into it to form the final emulsion. Vitamins A and D are also commonly added. Since the final product is tasteless (try Crisco), salt, acetoin, and biacetyl are often added. The latter two additives mimic the characteristic flavor of butter.

$$CH_3-\overset{\overset{\displaystyle HO}{|}}{\underset{\underset{\displaystyle H}{|}}{C}}-\overset{\overset{\displaystyle O}{\|}}{C}-CH_3 \qquad CH_3-\overset{\overset{\displaystyle O}{\|}}{C}-\overset{\overset{\displaystyle O}{\|}}{C}-CH_3$$

Acetoin   **Biacetyl**

Many producers of margarine claim it to be more beneficial to health because it is "high in polyunsaturates." Animal fats are low in unsaturated fatty acid content and are generally excluded from the diets of persons who have a high cholesterol level in the blood. Such people have difficulty in metabolizing saturated fats correctly and should avoid them since they encourage cholesterol deposits to form in the arteries. This ultimately leads to high blood pressure and heart trouble. Persons with normal metabolism, however, have no real need to avoid saturated fats.

Butter, when unrefrigerated and left exposed to air, turns rancid, giving an unpleasant odor and taste. This is due to the hydrolysis of the triglycerides by the moisture in air and to oxidation of the double bonds in the fatty acid components. Oxygen adds to allylic positions by an abstraction-addition reaction to give hydroperoxides.

$$R-CH{=}CH-CH_2-CH{=}CH-(CH_2)_7COOH + O_2 \longrightarrow$$

$$\overset{\displaystyle OOH}{\underset{\displaystyle |}{R-CH{=}CH-CH-CH{=}CH-(CH_2)_7COOH}}$$

Rancid butter smells bad compared with a partially hydrolyzed margarine or cooking fat, because it contains, along with the fatty acids listed in the table on page 167, triglycerides, which are composed of butyric (C_4), caproic (C_6), caprylic (C_8), and capric (C_{10}) acid moieties as well. These low-molecular-weight carboxylic acids are the source of the well-known objectionable odor. The hydroperoxides also decompose to low-molecular-weight aldehydes, which also have objectionable odors and tastes. The same reaction takes place when fats are burned, as in an oven fire. On combustion, an unsaturated fat produces large amounts of acrolein, a potent **lachrymator** (tear inducer), and other aldehydes that also irritate the eyes.

$$\overset{\displaystyle OOH}{\underset{\displaystyle |}{R-CH{=}CH-CH-R'}} \longrightarrow \overset{\displaystyle O}{\underset{\displaystyle \|}{R-CH{=}CH-C-H}}$$

Acrolein if R = H

Some oils, mainly those that are highly unsaturated (for example, linseed oil), thicken on exposure to air and eventually harden to give a smooth, clear resin. Such oils are called **drying oils,** and they are widely used in the manufacture of shellac, varnish, and paint. Apparently the double bonds in these compounds undergo both partial oxidation and polymerization on exposure to light and air.

REFERENCES

Dawkins, M. J. R., and Hull, D. "The Production of Heat by Fat." *Scientific American, 213* (August 1965): 62.

Eckey, E. W., and Miller, L. P. *Vegetable Fats and Oils.* ACS Monograph No. 123. New York: Reinhold, 1954.

Green, D. E. "The Metabolism of Fats." *Scientific American, 190* (January 1954): 32.

Green, D. E. "The Synthesis of Fats." *Scientific American, 202* (February 1960): 46.

Nawar, W. W. "Chemical Changes in Lipids Produced by Thermal Processing." *Journal of Chemical Education, 61* (April 1984): 299.

Shreve, R. N., and Brink, J. "Oils, Fats, and Waxes." *The Chemical Process Industries.* 4th ed. New York: McGraw-Hill, 1977.

Experiment 17
Methyl Stearate from Methyl Oleate

Catalytic hydrogenation
Filtration (Pasteur pipet)
Recrystallization
Unsaturation tests

In this experiment you will convert the liquid methyl oleate, an "unsaturated" fatty acid ester, to solid methyl stearate, a "saturated" fatty acid ester, by catalytic hydrogenation.

$$CH_3(CH_2)_7-CH=CH-(CH_2)_7-\overset{\overset{\displaystyle O}{\|}}{C}-OCH_3 \xrightarrow[H_2]{Pd/C}$$

Methyl oleate
(methyl *cis*-9-octadecenoate)

$$CH_3(CH_2)_7-\underset{\underset{\displaystyle H}{|}}{CH}-\underset{\underset{\displaystyle H}{|}}{CH}-(CH_2)_7-\overset{\overset{\displaystyle O}{\|}}{C}-OCH_3$$

Methyl stearate
(methyl octadecanoate)

By commercial methods like those described in this experiment, the unsaturated fatty acids of vegetable oils are converted to margarine (see the essay, "Fats and Oils" that precedes this experiment). However, rather than using the mixture of triglycerides that would be present in a cooking oil such as Mazola (corn oil), we use as a model the pure chemical methyl oleate.

For this procedure, a chemist would usually use a cylinder of hydrogen gas. Since many students will be following the procedure simultaneously, however, we shall use the simpler expedient of causing zinc metal to react with dilute sulfuric acid:

$$Zn + H_2SO_4 \xrightarrow{H_2O} H_2(g) + ZnSO_4$$

The hydrogen so generated will be passed into a solution containing methyl oleate and the palladium on carbon catalyst (10% Pd/C).

REQUIRED READING

Review: Techniques 1 and 2

New: Technique 4   Filtration, Sections 4.3, 4.4, and 4.5
          Technique 6   Physical Constants, Part A, Melting Points
          Essay   Fats and Oils

You should also read those sections in your lecture textbook that deal with catalytic hydrogenation. If the instructor indicates that you should perform the optional unsaturation tests on your starting material and product, you should read the descriptions of the Br_2/CH_2Cl_2 test on p 175 and in Experiment 15 on p 154.

SPECIAL INSTRUCTIONS

Since this experiment calls for generating hydrogen gas, no flames will be allowed in the laboratory.

No flames allowed

Since a build-up of hydrogen is possible within the apparatus, it is especially important to remember to wear your safety goggles; you can thus protect yourself against the possibility of minor "explosions" from joints popping open, from fires, or from any glassware accidentally cracking under pressure.

Wear safety goggles

When you operate the hydrogen generator, be sure to add sulfuric acid at a rate that does not cause hydrogen gas to evolve too rapidly. The hydrogen pressure in the vial should not rise much above atmospheric pressure. Neither should the hydrogen evolution be allowed to stop. If this happens, your reaction mixture may be "sucked back" into your hydrogen generator.

Finally, it is often the case, depending on the supplier, that commercial methyl oleate (MW 296) is only about 70% pure, containing several impurities. Consult your instructor about your sample, and whether you should base your percentage yield calculation on 70% of theoretical rather than the usual 100%.

APPARATUS

Assemble the apparatus as illustrated in the figure on page 173. This apparatus consists of basically three parts:

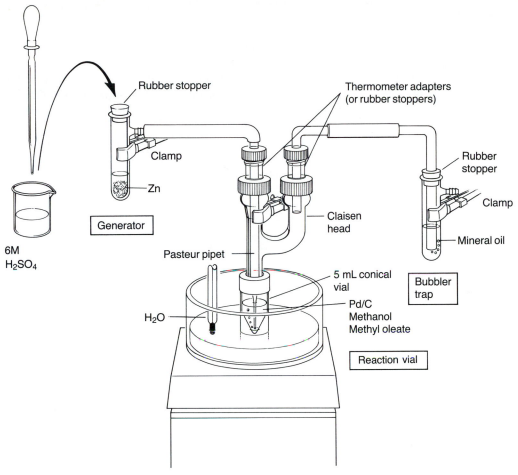

Hydrogenation Apparatus

1. Hydrogen generator,
2. Reaction vial,
3. Mineral-oil bubbler trap.

The mineral-oil bubbler trap has two functions. First, it allows one to keep a pressure of hydrogen within the system that is slightly above atmospheric. Second, it prevents back-diffusion of air into the system. The functions of the other two units are self-explanatory.

So that hydrogen leakage is prevented, the tubing used to connect the various subunits of the apparatus should be either relatively new rubber tubing, without cracks or breaks, or Tygon tubing. The tubing can be checked for cracks or breaks simply by stretching and bending it before use. It should be of such a size that it will fit onto all connections tightly. Similarly if any rubber stoppers are used, they should be fitted with a size of glass tubing that fits firmly through the holes in their centers. If the seal is tight, it will not be easy to slide the glass tubing up and down in the hole. The inlet tube (Pasteur pipet) in the reaction vial should reach almost to the bottom of the conical vial. Hydrogen must bubble through the solution.

PROCEDURE

Fill the bubbler trap (second side-arm test tube) about one-third full with mineral oil. The end of the glass tube should be submerged below the surface of the oil.

    To charge the hydrogen generator, weigh out about 1 g of mossy zinc and place it in the side arm test tube. Seal the large opening at its top using a rubber stopper. Obtain about 5 mL of 6M sulfuric acid and place it in a small Erlenmeyer flask or beaker, **but do not add it yet.**

    Weigh a 5-mL conical vial and then place about 0.5 mL of methyl oleate into it. Reweigh the vial in order to obtain the exact amount of methyl oleate used. Following this, add about 3.0 mL of methanol solvent to the vial. Using smooth weighing paper, weigh about 0.015 g (15 mg) of 10% Pd/C and place it into the vial.

    Be careful when adding the catalyst; sometimes it will cause a flash. Have a watch glass handy to cover the opening and smother the flame should this occur.

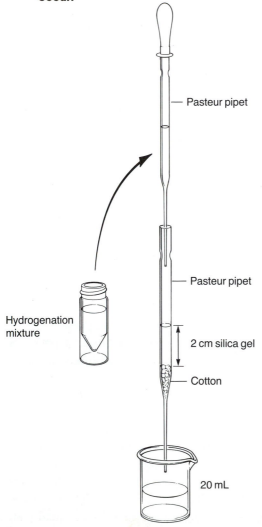

Filtration of catalyst

Complete the assembly of the apparatus making sure that all of the seals are gas tight. Place the reaction vial in a warm water bath maintained at 40 °C. This will help to keep the product dissolved in the solution throughout the course of the reaction. If the temperature rises above 40 °C, you will lose a significant amount of the methanol solvent. If this occurs, do not hesitate to add more methanol to the reaction vial through the side arm of the Claisen head, using a Pasteur pipet. Start the evolution of hydrogen by removing the rubber stopper and adding a portion of the 6*M* sulfuric acid solution (about 2–4 mL) to the hydrogen generator. Replace the rubber stopper. A good rate of bubbling in the reaction vial is about three to four bubbles every second. Continue the evolution of hydrogen for about 45–60 minutes. If necessary, open the generator, empty it, and refresh the zinc and sulfuric acid. (Keep in mind that the acid is used up and becomes more dilute as the zinc reacts.)

After the reaction is complete, stop the reaction by disconnecting the reaction vial. Dilute the acid in the generator with water and decant the solution down the sink with lots of water (do not dump the zinc). Rinse the test tube several times with water, and then place any unreacted zinc in a waste container provided for this purpose.

If you maintained your reaction vial at 40 °C, there should not be any white solid (product) in the hydrogenation vial. If there is a white solid, heat the mixture gently and add more methanol until it dissolves and remains dissolved on cooling.

Prepare a Pasteur pipet for filtering (see figure on p 174) by placing a small, loose plug of cotton at the constriction in the bottom. Add about **2 cm** of silica gel and tap the pipet with your finger to pack the material. Using a second Pasteur pipet and bulb, transfer the hydrogenation solution from the conical vial to the filter pipet, collecting the filtered solution in a small beaker (20 mL). This operation should remove the black catalyst from the solution. If it does not, repeat the filtering operation a second time through the same column.

Place the filtered solution (beaker) in an ice bath to induce crystallization. If crystals do not form, it may be necessary to reduce the volume of solvent by about one-half. Do this by heating in a sand bath and directing a slow stream of air into the beaker using a Pasteur pipet for a nozzle (Figure 3–12A, p 561). Allow the solution to cool and then place it in an ice bath.

Collect the crystals by vacuum filtration using a small Hirsch funnel (Technique 4, Section 4.3, p 571). Save both the crystals and the filtrate for the tests below. After the crystals are dry, weigh them and determine their melting point (literature, 39 °C). Calculate the percentage yield. Submit your remaining sample to your instructor in a properly labeled container along with your report.

UNSATURATION TESTS (optional)

Using a solution of bromine in methylene chloride, test for the number of drops of this solution decolorized by:

1. About 0.1 mL of methyl oleate dissolved in a small amount of methylene chloride.
2. A small spatulaful of your methyl stearate product dissolved in a small amount of methylene chloride.
3. About 0.1 mL of the filtrate that you saved as directed under "Procedure."

Use small test tubes and disposable pipets to make these tests. Include the results of the tests and your conclusions in your report.

QUESTIONS

1. Using the information in the essay on fats and oils, draw the structure of the triglyceride formed from oleic acid, linoleic acid, and stearic acid. Give a balanced equation showing how much hydrogen would be needed to reduce the triglyceride completely; show the product.

2. A 0.150-g sample of a pure compound subjected to catalytic hydrogenation takes up 250 mL of H_2 at 25 °C and 1 atm pressure. Calculate the molecular weight of the compound, assuming that it has only one double bond.

3. A compound with the formula C_5H_6 takes up two moles of H_2 on catalytic hydrogenation. Give one possible structure that would fit the information given.

4. A compound of formula C_6H_{10} takes up one mole of H_2 on reduction. Give one possible structure that would fit the information.

Experiment 18
Markovnikov and Anti-Markovnikov Hydration of an Alkene

Addition to double bonds
Hydroboration-oxidation
Oxymercuration
Gas chromatography
Nuclear magnetic resonance
Regiospecificity-regioselectivity

The most characteristic reaction of alkenes is the **addition reaction.** This reaction can be described by the general equation

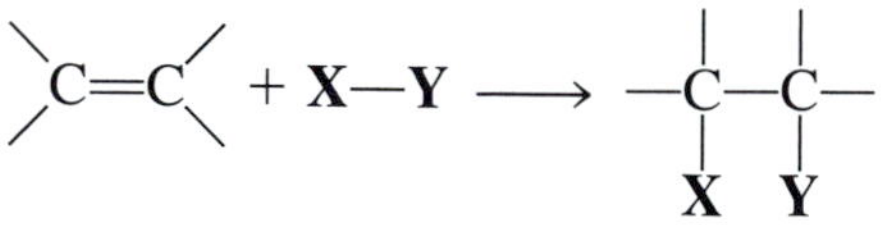

$$\text{C=C} + \text{X—Y} \longrightarrow \underset{\text{X Y}}{-\text{C}-\text{C}-}$$

In this reaction, the π-bond of the alkene is broken, and two σ-bonds are formed.

Because the double bond is an electron-rich center, it behaves like a Lewis base or a nucleophile, donating a pair of electrons to form a bond to a sufficiently electrophilic reagent. Some part of the attacking reagent **XY** must be electrophilic. As an example, in the reaction of an alkene with an acid, **HA,** the first step of the mechanism involves the reaction of the double bond with the electrophilic species, H^+. This step produces a cation:

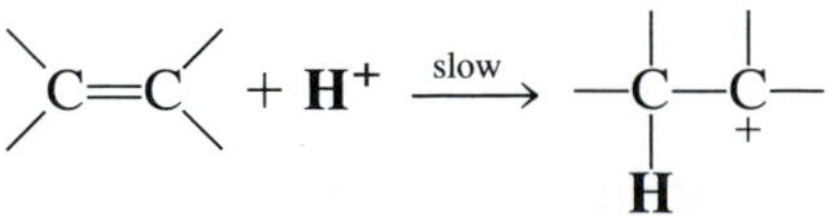

$$\text{C=C} + H^+ \xrightarrow{\text{slow}} \underset{H}{-\overset{}{\text{C}}-\overset{+}{\text{C}}-}$$

This intermediate cation can react rapidly with the anion, A^-, to yield the final product:

$$-\overset{\scriptstyle|}{\underset{\scriptstyle|}{C}}-\overset{\scriptstyle|}{\underset{\scriptstyle|}{\underset{\scriptstyle+}{C}}}- \;+\; A^- \;\xrightarrow{\text{fast}}\; -\overset{\scriptstyle|}{\underset{\scriptstyle|}{\underset{\textbf{H}}{C}}}-\overset{\scriptstyle|}{\underset{\scriptstyle|}{\underset{\textbf{A}}{C}}}-$$

A complication arises when an unsymmetrical reagent, **HA,** adds to an unsymmetrically substituted double bond. In this case, there are two possible products:

$$R\text{—}CH{=}CH_2 + \textbf{HA} \longrightarrow R\text{—}\underset{\textbf{H}}{\overset{}{CH}}\text{—}\underset{\textbf{A}}{\overset{}{CH_2}} \;\; \text{or} \;\; R\text{—}\underset{\textbf{A}}{\overset{}{CH}}\text{—}\underset{\textbf{H}}{\overset{}{CH_2}}$$

An empirical rule, known as **Markovnikov's rule,** was developed in 1868 by the Russian chemist V. V. Markovnikov to apply to such cases. Simply stated, the rule says

> **In the ionic addition of an acid to the carbon-carbon double bond of an alkene, the hydrogen of the acid attaches itself to the carbon atom that already holds the *greater* number of hydrogens.**

In recent years, chemists have determined that the reason Markovnikov's rule holds is that addition according to the rule always leads to the more highly substituted, and hence the more stable, of the two possible cationic intermediates.

The addition of water to alkenes under acidic conditions (hydration) follows Markovnikov's rule. The hydrogen of water attaches itself to the carbon atom that already carries the greater number of hydrogens:

$$R\text{—}CH{=}CH_2 + \textbf{H}_2\textbf{O} \longrightarrow R\text{—}\underset{\textbf{OH}}{\overset{}{CH}}\text{—}\underset{\textbf{H}}{\overset{}{CH_2}} \qquad \textbf{Markovnikov addition}$$

If the addition of water to the alkene had proceeded contrary to Markovnikov's rule, the reaction would be said to have gone in an "anti-Markovnikov" fashion, and the product would be designated the anti-Markovnikov product:

$$R\text{—}CH{=}CH_2 + \textbf{H}_2\textbf{O} \longrightarrow R\text{—}\underset{\textbf{H}}{\overset{}{CH}}\text{—}\underset{\textbf{OH}}{\overset{}{CH_2}} \qquad \textbf{Anti-Markovnikov addition}$$

Our current experiment deals with two practical methods for carrying out the hydration of an alkene in both the Markovnikov and the anti-Markovnikov fashion. These methods are the **hydroboration-oxidation** and the **oxymercuration** reactions.

HYDROBORATION-OXIDATION

Diborane can react with alkenes by addition of the boron-hydrogen bond across the carbon-carbon double bond. This reaction is called **hydroboration.** Diborane is prepared by the reaction of sodium borohydride with boron trifluoride:

$$3NaBH_4 + 4BF_3 \xrightarrow{\text{ether}} 2B_2H_6 + 3NaBF_4$$

Diborane, B_2H_6, is an unusual substance. It has two three-center bonds. In these bonds, one pair of electrons bonds three atoms together—two boron atoms and one hydrogen atom. Without getting into the details of this bonding scheme, it seems a reasonable simplification to describe the reactions of diborane as if it were a dimer, composed of two units of borane, BH_3. Some studies even suggest that BH_3 may be in equilibrium with diborane.

Simplified approximation

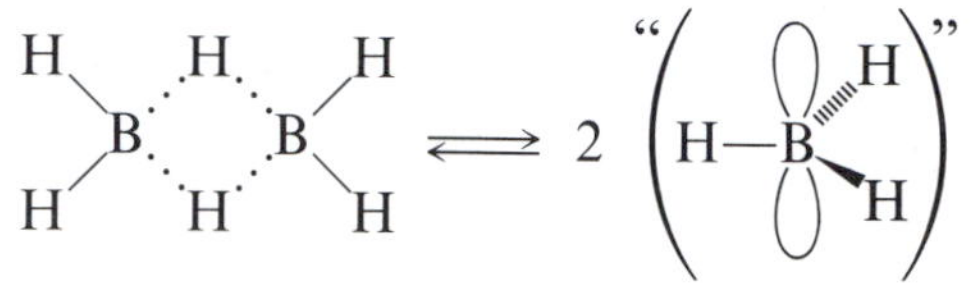

Diborane, because of its **three-center bonds,** is electron-deficient and is a good electrophile. Similarly, BH_3 is also a good electrophile, since it contains an incomplete octet of electrons. With BH_3 as the boron hydride, an alkene will donate electrons to boron as shown in the first step of the reaction below. This creates a boron atom that is electron-rich and a carbon atom that is electron-deficient. This situation is remedied in the next step by the transfer of a hydride ion ($H:^-$) to the carbon atom:

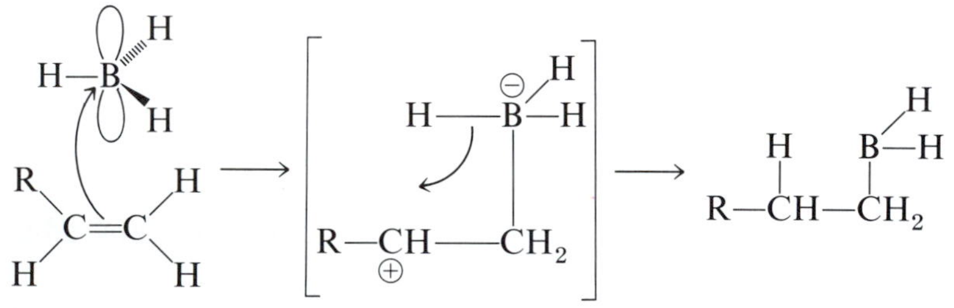

Experimental evidence shows that this entire process is **concerted** without the presence of intermediates and that the addition is stereospecific. The addition of borane to an alkene proceeds exclusively with **syn** stereochemistry. Both the boron and the hydrogen atoms add to the same side of the double bond:

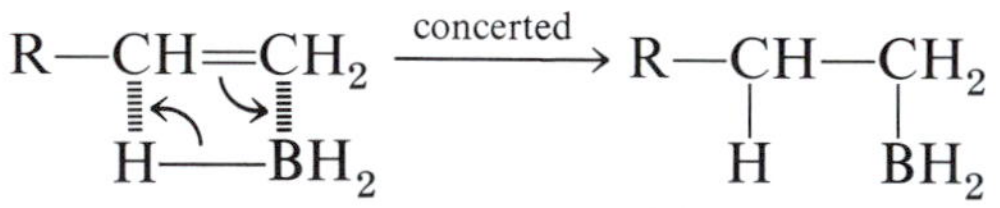

$$R-CH=CH_2 \xrightarrow{\text{concerted}} R-CH-CH_2$$

Notice that this addition is an **anti-Markovnikov** addition. In contrast to the situation that applies for the addition of an acid, **HA,** the hydrogen from diborane does not become attached to the carbon atom with the greater number of hydrogens. However, the electrophilic species that was added to the double bond in this case was not H^+ but the electron-deficient **boron** atom. The lowest-energy cationic intermediate would still predict the course of the reaction (even though it does not exist in hydroboration), and the hydrogen is transferred as $H:^-$ rather than as H^+. This addition occurs three times because there are three **B—H** bonds and causes a trialkylborane to form.

$$R-CH=CH_2 + \text{"}BH_3\text{"} \longrightarrow R-\underset{\underset{H}{|}}{C}H-CH_2-BH_2 \xrightarrow{RCH=CH_2} \left(R-\underset{\underset{H}{|}}{C}H-CH_2 \right)_2 BH$$

$$\downarrow RCH=CH_2$$

$$\left(R-\underset{\underset{H}{|}}{C}H-CH_2 \right)_3 B$$

A trialkylborane

If the trialkylborane compound is treated with alkaline hydrogen peroxide, it is cleaved with oxidation to form three moles of an alcohol. As you can see from the mechanism shown, the replacement of a boron atom by an oxygen atom is an intramolecular process. As a result, this oxidation step is also stereospecific, proceeding entirely with **retention of configuration.** This oxidation proceeds via a migration of the alkyl group from boron to oxygen, generating a trialkoxyborane as an important intermediate.

$$(RCH_2CH_2)_3B + {}^-OOH \rightleftharpoons RCH_2CH_2-\underset{\underset{RCH_2CH_2}{|}}{\overset{\overset{RCH_2CH_2}{|}}{\overset{\ominus}{B}}}-O-O-H \longrightarrow RCH_2CH_2-\underset{\underset{RCH_2CH_2}{|}}{\overset{\overset{RCH_2CH_2}{|}}{B}}-O + OH^-$$

$$\Updownarrow OOH^-$$

$$(RCH_2CH_2O)_3B \longleftarrow \overset{{}^-OOH}{\rightleftharpoons} RCH_2CH_2-\underset{\underset{RCH_2CH_2-O}{|}}{\overset{\overset{RCH_2CH_2}{|}}{B}}-O \longleftarrow RCH_2CH_2-\underset{\underset{RCH_2CH_2-O}{|}}{\overset{\overset{RCH_2CH_2}{|}}{\overset{\ominus}{B}}}-O-O-H$$

Trialkoxyborane $+ OH^-$

$$\downarrow H_2O/OH^-$$

$$3RCH_2CH_2OH + B(OH)_3$$

Hydroboration-oxidation of an alkene therefore leads to an alcohol that corresponds to the anti-Markovnikov addition of water across the double bond of an alkene. This seems to violate Markovnikov's rule. However, analysis of the mechanism of the hydroboration–oxidation reactions shows that this is not the case:

$$3RCH=CH_2 \xrightarrow{B_2H_6} (RCH_2CH_2)_3B \xrightarrow[OH^-]{H_2O_2} 3RCH_2CH_2OH$$

Finally, the anti-Markovnikov addition of water across the double bond proceeds with **syn** stereochemistry; this can best be illustrated using a cyclic alkene as an example. The stereospecificity results from the intramolecular nature of the hydroboration and the oxidation steps, as already described.

$$\xrightarrow[\text{(2) } H_2O_2/OH^-]{\text{(1) } B_2H_6}$$

1-Methylcyclopentene   syn Addition   *trans*-2-Methylcyclopentanol

OXYMERCURATION

The second means of hydrating a double bond we consider is the **oxymercuration** reaction. In this reaction, mercuric acetate is added to an alkene to form an organomercury derivative:

$$RCH{=}CH_2 + Hg(OCOCH_3)_2 \xrightarrow{H_2O} \underset{\underset{OH}{|}}{RCH}{-}CH_2{-}HgOCOCH_3$$

The mechanism of this reaction involves initial ionization of mercuric acetate, followed by addition of the mercury atom across the double bond of the alkene to form a bridged cationic intermediate. This intermediate is then opened by the water, which is a nucleophile. This reaction is also stereospecific, proceeding with **anti** stereochemistry.

$$Hg(OCOCH_3)_2 \rightleftharpoons {}^{+}HgOCOCH_3 + CH_3COO^{-}$$

$$R{-}CH{=}CH_2 + {}^{+}HgOCOCH_3 \xrightarrow{\text{slow}} \underset{Hg{-}OCOCH_3}{R{-}CH{-}CH_2}$$

$$\downarrow$$

$$\underset{\underset{HgOCOCH_3}{|}}{\overset{\overset{OH}{|}}{R{-}CH{-}CH_2}} \qquad + H^{+}$$

The oxymercuration reaction is different from many addition reactions of alkenes since it involves no rearrangements.

    The organomercury intermediate is reduced with sodium borohydride, with the result that the mercury atom is replaced by hydrogen. The reduction of the organomercury intermediate is not stereospecific.

$$\underset{\underset{OH}{|}}{RCH}{-}CH_2{-}HgOCOCH_3 \xrightarrow[\text{NaOH}]{\text{NaBH}_4} \underset{\underset{OH}{|}}{RCH}{-}CH_3$$

    The product obtained from the oxymercuration of an alkene is the same as what would be expected for the hydration of an alkene according to Markovnikov's rule. In this particular reaction, the mercury atom is the original electrophile, so it adds to the end of the double bond that bears the greater number of hydrogens. When the mercury atom is replaced by hydrogen in the reduction step of the reaction, hydrogen is then attached to the carbon atom in the manner the rule predicts.

    In these two reactions, we see contrasting behavior as to the direction of addition of water. Although each of the reactions has been described as though it produced exclusively the product shown, in truth it must be mentioned that in some cases there may be a minor product with the opposite orientation. That is to say, while hydration of an alkene may produce the product predicted by Markovnikov's rule as the principal product, the anti-Markovnikov alcohol may also be produced as a minor product. Reactions that produce a product with only one of several possible orientations are called **regiospecific.** Reactions that produce one substance as the predominant product and small amounts of isomers with other orientations are called **regioselective.** One of the objects of this experiment is to contrast the hydroboration-oxidation and the oxymercuration of an alkene to determine whether the reactions are regiospecific or regioselective.

Procedure 18A
Hydroboration–Oxidation of an Alkene

The balanced equation for the hydroboration of an alkene, using a borane-tetrahydrofuran solution, is

$$3RCH{=}CH_2 + BH_3 \cdot C_4H_8O \longrightarrow (RCH_2CH_2)_3B + C_4H_8O$$

Once the organoborane intermediate is formed, it is oxidatively hydrolyzed with a basic solution of hydrogen peroxide:

$$(RCH_2CH_2)_3B + 3H_2O_2 \xrightarrow{\text{OH}^-} 3RCH_2CH_2{-}OH + B(OH)_3$$

In the method used in this experiment, a solution of diborane in tetrahydrofuran is added to the alkene. The diborane is consumed as it is added. The solution is a commercially available product. The addition method avoids the need for handling the potentially hazardous (toxic and flammable) diborane.

REQUIRED READING

Review: Technique 3   Section 3.9
          Technique 7   Sections 7.5, 7.8, and 7.10
          Technique 14
          Technique 18   Part B
          Appendix 4

New: You should consult your organic chemistry lecture textbook for detailed information about the hydroboration–oxidation reaction.

SPECIAL INSTRUCTIONS

The student may choose either 1-octene or styrene as the alkene starting material. If 1-octene is chosen, the analysis by nuclear magnetic resonance spectroscopy is not recommended.

Because this reaction involves a somewhat lengthy period of preparation of apparatus and addition of reagents, it is important that it be started at the beginning of the laboratory period. The best place to stop is when the ether extracts are stored over sodium sulfate, although the reaction can be stopped at any time after water has been added to the reaction, if necessary.

This reaction involves the use of tetrahydrofuran, which is a potentially toxic and flammable solvent. Do not conduct this reaction in the presence of open flames. Avoid contact with 30% hydrogen peroxide, as it is a strong oxidant.

This experiment requires the use of syringes to measure and transfer reagents. It is advisable to use glass-bodied syringes with metal or glass plungers, as 1-octene and tetrahydrofuran may attack some plastic syringes. Glass plungers, in particular, slide very easily. They can expel reagents from the syringe faster than may be intended. In addition, they may slide out of the syringe barrel, fall, and break. Exercise care when using the syringes.

PROCEDURE

> **NOTE TO THE INSTRUCTOR: This procedure may be used with either 1-octene or styrene as the starting alkene. If 1-octene is used, the analysis of the products by NMR spectroscopy is not recommended. If styrene is analyzed by gas chromatography, over time a build-up of polymerized styrene residue may cause a degradation of the gas chromatography column. Analyze the styrene products by NMR as an alternative.**

Thoroughly dry a 5-mL conical vial by placing it in an oven at 110 °C for at least 15 minutes. Also dry the glass parts of a hypodermic syringe. If droplets of water are visible on the glassware before drying, rinse the glassware with acetone before placing it in the oven. When the vial has cooled, weigh the vial and add 0.30 mL of 1-octene or 0.22 mL of styrene. Reweigh the vial to determine an accurate weight of the alkene. Place a magnetic spin vane in the vial and seal it with a rubber septum cap, as shown in Figure 1.

Insert a hypodermic syringe needle through the rubber septum, being careful to allow enough room for a second syringe needle which will be inserted later. Place the vial in an ice-water bath on top of a magnetic stirrer. Using the dried syringe, draw 0.75 mL of a commercial borane-tetrahydrofuran solution from the stock bottle. Carefully insert this second syringe through the rubber septum (it may be necessary to hold the

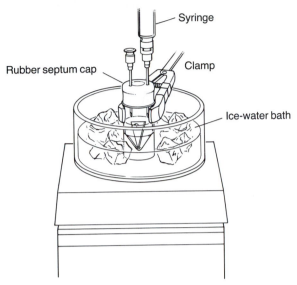

FIGURE 1.   Apparatus for the addition of borane · THF to a reaction

syringe upright with a clamp attached to a ring stand). Begin stirring the solution. Carefully add the borane-tetrahydrofuran solution to the reaction vial at a rate of about one drop per second. The addition will require approximately 25 minutes. An effective method of addition is to twist the plunger carefully down into the body of the syringe; this technique affords the fine control necessary to establish the correct addition rate. After the borane-tetrahydrofuran solution has been added, allow the reaction mixture to stand in the ice bath, with stirring, for an hour.

At the end of this period, remove the septum cap from the reaction vial. While maintaining the reaction vial in the ice bath with stirring, slowly add 0.21 mL of 30% hydrogen peroxide, using a Pasteur pipet to add this solution. During this addition, maintain the pH of the solution near 8, by adding $3M$ sodium hydroxide, dropwise, as needed. Dip a microspatula or glass stirring rod into the solution and touch it to pH paper in order to determine the pH. Add the sodium hydroxide carefully, as the reaction mixture may bubble very vigorously during this addition. It is very important to maintain the pH as near to 8 as possible to prevent premature precipitation of solid materials.

Slowly add 1 mL of water to the reaction mixture, followed by 1 mL of diethyl ether. Allow the mixture in the vial to stir in order to mix the liquid and solid materials adequately. Remove the lower aqueous layer and transfer it to a 3-mL conical vial. Extract the aqueous layer with three successive 1-mL portions of ether. Combine the ether extracts in a centrifuge tube, wash them with 2 mL of saturated sodium bicarbonate solution, and dry them over two or three microspatulafuls of granular anhydrous sodium sulfate.

Remove the ether solution from the drying agent, and evaporate the ether under a stream of air while heating the container on a warm sand bath. When the ether has been evaporated, remove any residual tetrahydrofuran that may remain in the product by evaporation under reduced pressure, according to the following procedure (see Figure 2). Transfer the solution to a 3-mL conical vial and add a spin vane to the vial. Insert an air condenser into the vial containing the product and residual tetrahydrofuran. Attach a thermometer adapter or a glass tube inserted into a stopper to the top of the air

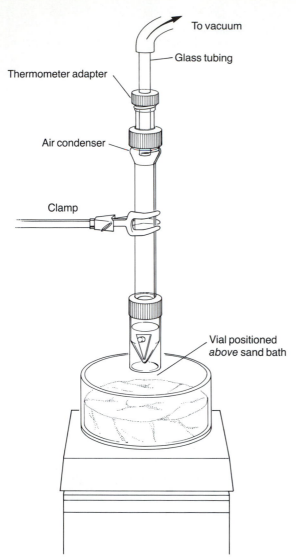

To vacuum

Glass tubing

Thermometer adapter

Air condenser

Clamp

Vial positioned
above sand bath

FIGURE 2.   Assembly for removal of solvent under reduced pressure

condenser (Figure 9–2, p 653). Position the apparatus above the sand bath (temperature about 50 °C). **Do not heat the mixture until after the vacuum has been applied.** Place a trap assembly between your apparatus and the aspirator (as shown in Figure 4–6, p 571). Apply a **gentle** vacuum to the apparatus by using **reduced water flow** through your aspirator while stirring the mixture. Once the vacuum has been applied, **gently** heat the mixture so that the tetrahydrofuran slowly evaporates under reduced pressure. Do not overheat the solution, because some of the residual alkene may polymerize.

When the bubbling in the vial ceases, remove the vacuum and weigh the residual alcohols to determine the percentage yield obtained from the reaction.[1] Analyze the

---

[1] Some unreacted alkene may be present in the mixture of alcohols. If an appreciable amount of alkene is present, the percentage of alkene determined from the gas chromatogram can be used to subtract the contribution of the alkene from the total weight of products. This will give the actual weight of the isomeric alcohols.

mixture of alcohols using gas chromatography to determine the relative percentages of Markovnikov and anti-Markovnikov products (Technique 14, Section 14.11, p 750).[2]

If you used styrene as the alkene starting material, at the option of the instructor, you may use NMR spectroscopy to analyze the composition of the product mixture. The NMR spectra of 2-phenylethanol and 1-phenylethanol are included for reference. Compare the integrals for the four hydrogens of the side chain in each of the two products to determine the relative percentages of the two isomers.

Compare your results with the results obtained by students who performed the oxymercuration of the same alkene that you used. In your laboratory report, submit the gas chromatograms and NMR spectra you obtained. Calculate the percentage yield of alcohol isomers and the ratios of the two isomers from your chromatographic, and, if applicable, the NMR spectral data. Comment on whether or not the reaction proceeded according to Markovnikov's rule and whether the reaction was regiospecific or regioselective.

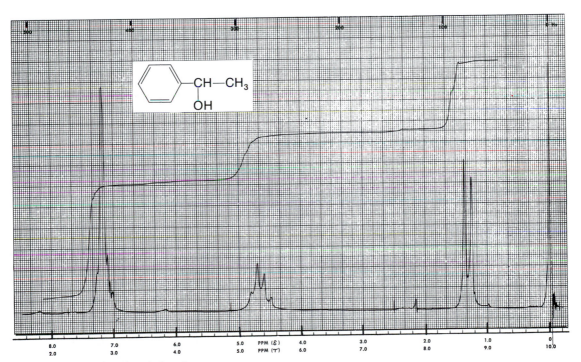

NMR spectrum of 1-phenylethanol, neat

[2] The correct conditions for the gas chromatographic analysis are

        Column temperature: 130 °C
        Injection port temperature: 200 °C
        Column packing: SE-30 silicone oil on Chromosorb W
        Relative order of retention times: 1-octene, 2-octanol, 1-octanol; styrene, 1-phenylethanol, 2-phenylethanol

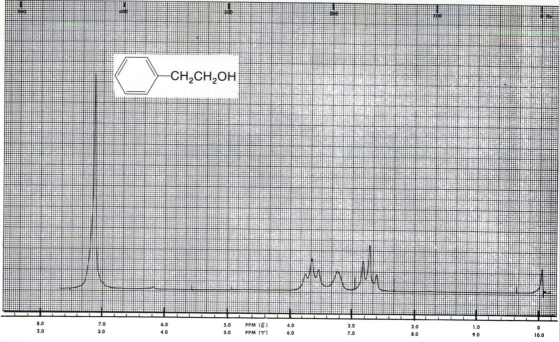

NMR spectrum of 2-phenylethanol, neat

Procedure 18B
Oxymercuration of an Alkene

The balanced equation for the oxymercuration of an alkene is

$$RCH{=}CH_2 + Hg(OCOCH_3)_2 + H_2O \longrightarrow$$

$$R{-}\underset{\underset{OH}{|}}{CH}{-}CH_2{-}Hg{-}OCOCH_3 + CH_3{-}COOH$$

Reduction of the organomercury derivative with sodium borohydride yields the alcohol:

$$4R{-}\underset{\underset{OH}{|}}{CH}{-}CH_2{-}Hg{-}OCOCH_3 + NaBH_4 + 3NaOH \longrightarrow$$

$$4R{-}\underset{\underset{OH}{|}}{CH}{-}CH_3 + 4Hg^0 + B(OH)_3 + 4CH_3COONa$$

REQUIRED READING

Review: Technique 3   Section 3.9
              Technique 7   Sections 7.5, 7.6, and 7.8
              Technique 14
              Technique 18   Part B
              Appendix 4

New: You should consult your organic chemistry lecture textbook for detailed information about the oxymercuration reaction.

SPECIAL INSTRUCTIONS

The student may choose either 1-octene or styrene as the alkene starting material. If 1-octene is chosen, the analysis by nuclear magnetic resonance spectroscopy is not recommended.

The experiment requires that the reaction mixture be allowed to stand overnight. As two lengthy periods of stirring are necessary, this experiment must be started at the beginning of the laboratory period.

This experiment involves the use of mercuric acetate. Like all mercury compounds, mercuric acetate is very toxic. Wash your hands thoroughly after handling this substance. Do not get it on your hands or face.

PROCEDURE

> **NOTE TO THE INSTRUCTOR:** This procedure may be used with either 1-octene or styrene as the starting alkene. If 1-octene is used, the analysis of the products by NMR spectroscopy is not recommended. If styrene is analyzed by gas chromatography, over time a build-up of polymerized styrene residue may cause a degradation of the gas chromatography column. Analyze the styrene products by NMR as an alternative.

Set up an apparatus consisting of a 10-mL round-bottom flask equipped with a magnetic stirring bar, and fitted with a water-cooled condenser. Place the flask in a water bath and assemble all components on a magnetic stirring motor. Place 0.600 g of mercuric acetate and 2 mL of water in the flask. After the mercuric acetate has dissolved, add 2 mL of ether. While stirring the mixture vigorously, add about 0.30 mL of 1-octene or 0.22 mL of styrene (measure the volume of alkene accurately and calculate the weight of alkene using the density; d = 0.711 g/mL for 1-octene, d = 0.906 g/mL for styrene), dropwise down the reflux condenser and continue stirring at room temperature for one hour. After one hour, add 1 mL of 6M sodium hydroxide solution followed by a solution of 0.040 g of sodium borohydride in 2 mL of 3M sodium hydroxide.

> **CAUTION: Mercuric acetate is a highly poisonous substance.**

Stir this mixture for 30 minutes, after which time most of the mercury should have settled to the bottom of the flask. Allow this mixture to settle until the next laboratory period. After the mixture has settled, carefully withdraw the supernatant liquid from the mercury that has been deposited on the bottom of the flask using a Pasteur pipet. Transfer the supernatant liquid to a 15-mL centrifuge tube. Discard the mercury in a suitable waste container. NEVER DISCARD THE MERCURY DOWN THE SINK DRAIN. Using a Pasteur pipet, transfer the aqueous layer to another 15-mL centrifuge tube and save the ether layer in the original centrifuge tube. Extract the aqueous layer with three successive 1-mL portions of ether. After each extraction, combine the ether layer with the ether solution in the first centrifuge tube. Using a dry Pasteur pipet, transfer the ether solution to a dry test tube. Add three microspatulafuls (measured with the V-grooved end) of anhydrous granular sodium sulfate to dry the liquid. Allow the solution to stand for 10 to 15 minutes, with occasional stirring.

Remove the ether from the drying agent using a filter tip pipet. It is important to draw the liquid into the pipet very carefully in order to avoid transferring particles of drying agent with the product. Transfer the ether solution to a clean, pre-weighed 5-mL conical vial and evaporate the ether solution under a stream of air in the hood, while heating the vial in a warm sand bath. If the resulting product has a suspended fine precipitate, centrifuge the liquid before analyzing it by gas chromatography. Otherwise, the fine particles will likely plug the syringe used in the gas chromatographic analysis. Weigh the vial to determine the yield of product. Calculate the percentage yield of alcohol isomers obtained. Analyze the resulting liquid, using gas chromatography, to determine the relative percentages of Markovnikov and anti-Markovnikov products[3] (See Technique 14, Section 14.11, p 750).

If you used styrene as the alkene starting material, at the option of the instructor, you may use NMR spectroscopy to analyze the composition of the product mixture. The NMR spectra of 2-phenylethanol and 1-phenylethanol are given in Procedure 18A for reference. Compare the integrals for the four hydrogens of the side chain in each of the two products to determine the relative percentages of the two isomers.

In your report to the instructor, submit the gas chromatograms and NMR spectra you obtained. Calculate the ratios of the two alcohols, using the gas chromatograms, and, if applicable, the NMR spectra. Report the percentage yield of alcohol isomers. Comment on whether or not the reaction followed Markovnikov's rule and whether the reaction was regiospecific or regioselective. Compare your results with the results obtained by students who performed Procedure 18A, using the same alkene as yours.

---

[3]Some unreacted alkene may be obtained in the products. The conditions for the gas chromatography are identical to those given in Procedure 18A, p 185.

REFERENCES

Brown, H. C. "Hydride Reductions: A 40-Year Revolution in Organic Chemistry." *Chemical and Engineering News, 57* (March 5, 1979): 24.

Brown, H. C. *Hydroboration.* Reading, MA: Benjamin/Cummings, 1980.

Brown, H. C., and Zweifel, G. "Hydroboration. VII. Directive Effects in the Hydroboration of Olefins." *Journal of the American Chemical Society, 82* (1960): 4708.

Jerkunica, J. M., and Traylor, T. G. "Oxymercuration-Reduction: Alcohols from Olefins: 1-Methylcyclohexanol." *Organic Syntheses, 53* (1973): 94.

Zweifel, G., and Brown, H. C. "Hydroboration of Olefins: (+)-Isopinocampheol." *Organic Syntheses, 52* (1972): 59.

QUESTIONS

1. Compare the relative percentages of 1-phenylethanol and 2-phenylethanol formed from styrene or of 2-octanol and 1-octanol formed from 1-octene in each of the two reactions. Classify each reaction as regiospecific or regioselective.

2. Predict the products of the hydroboration-oxidation and the oxymercuration reactions for each of the following compounds. Include the correct stereochemistry, when appropriate.

(a)

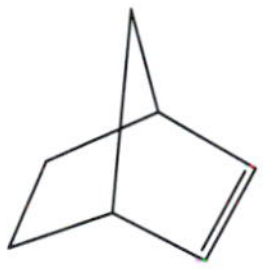

(b)

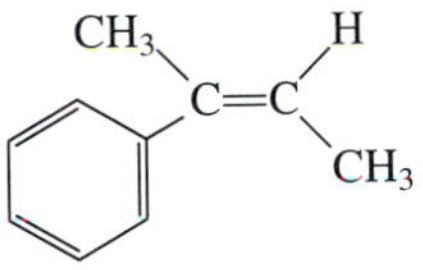

(c)

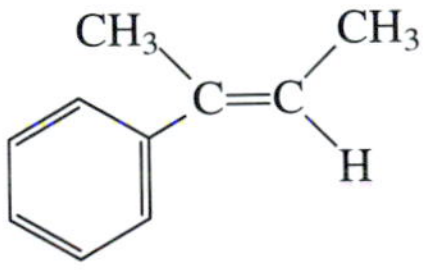

(d)

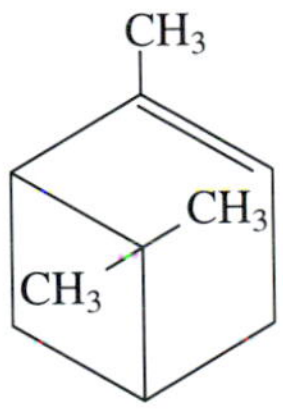

Experiment 19
Phase-Transfer Catalysis: Addition of Dichlorocarbene to Cyclohexene

Carbene formation
Phase-transfer catalysis

It has long been known that a haloform, CHX_3, will react with a strong base to give a highly reactive carbene species, CX_2, by Reactions 1 and 2. In the presence of an alkene, this carbene adds to the double bond to produce a cyclopropane ring (Reaction 3).

$$X-\underset{X}{\overset{X}{C}}-H + {}^-\!\!:\overset{..}{\underset{..}{O}}-H \rightleftharpoons X-\underset{X}{\overset{X}{C}}:{}^- + H\overset{..}{O}H \qquad (1)$$

$$X-\underset{X}{\overset{X}{C}}:{}^- \xrightarrow{\text{slow}} X-\underset{X}{C}: + :\overset{..}{\underset{..}{X}}:{}^- \qquad (2)$$

$$X-\underset{X}{C}: + \ \overset{}{C}{=}\overset{}{C} \longrightarrow -\overset{|}{C}\underset{\underset{X}{\overset{}{C}}\overset{}{X}}{-}\overset{|}{C}- \qquad (3)$$

Traditionally, the reaction has been carried out in **one homogeneous phase** in anhydrous *t*-butyl alcohol solvent, using *t*-butoxide ion as the base [*t*-Bu = $C(CH_3)_3$].

$$HCX_3 + {}^-Ot\text{-Bu} + \ \overset{}{C}{=}\overset{}{C} \xrightarrow[\text{(solvent)}]{t\text{-BuOH}} -\overset{|}{C}\underset{\underset{X}{\overset{}{C}}\overset{}{X}}{-}\overset{|}{C}- + HOt\text{-Bu} + X^-$$

Unfortunately, this technique requires time and effort to give good results. In addition, water must be avoided to prevent conversion of the haloform and carbene to formate ion and carbon monoxide by the undesirable base-catalyzed Reactions 4 and 5.

$$H-\underset{X}{\overset{X}{C}}-X + 2H_2O \longrightarrow H-\underset{\overset{||}{O}}{C}-OH + 3HX \qquad (4)$$

$$H\!-\!\underset{\underset{O}{\|}}{C}\!-\!OH + {}^-OH \longrightarrow H\!-\!\underset{\underset{O}{\|}}{C}\!-\!O^- + H_2O$$

$$:CX_2 + H_2O \xrightarrow{\,{}^-OH\,} \; :C\!\equiv\!O + 2HX \tag{5}$$

QUATERNARY AMMONIUM SALT CATALYSIS

As an alternative to a homogeneous reaction, a *two-phase* reaction can be considered when the organic phase contains the alkene and a haloform, CHX_3, and the aqueous phase contains the base, OH^-. Unfortunately, under these conditions the reaction will be very slow, since the two primary reactants, CHX_3 and OH^-, are in different phases. The reaction rate can be substantially increased, however, by adding a quaternary ammonium salt such as benzyltriethylammonium chloride as a **phase-transfer catalyst.**

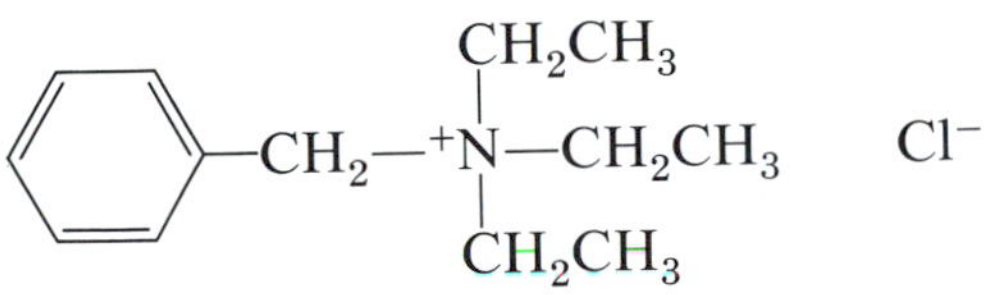

A phase-transfer catalyst: Benzyltriethylammonium chloride

Other common catalysts are tetrabutylammonium bisulfate, trioctylmethylammonium chloride and cetyltrimethylammonium chloride. All these catalysts, including benzyltriethylammonium chloride, have at least 13 carbon atoms. The numerous carbon atoms give the catalyst organic character (hydrophobic) and allow it to be soluble in the organic phase. At the same time, the catalyst also has ionic character (hydrophilic) and can therefore be soluble in the aqueous phase.

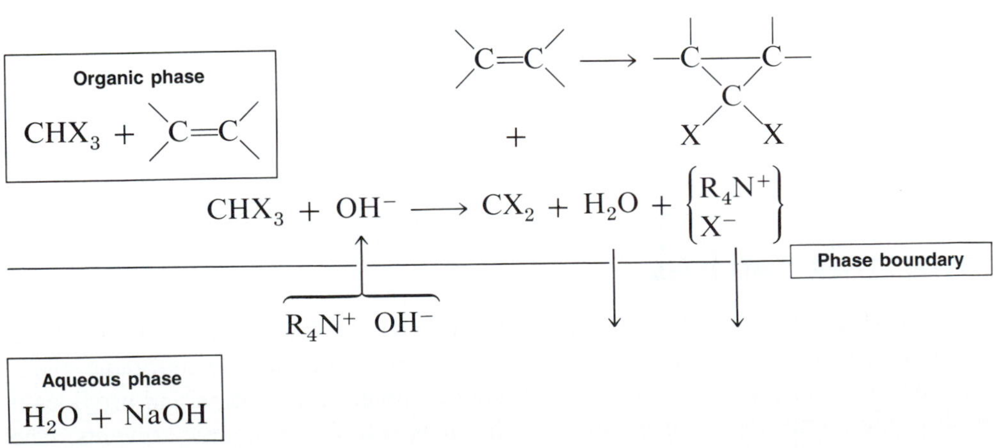

Because of this *dual* nature, the large cation can cross the phase boundary efficiently and transport a hydroxide ion from the aqueous phase to the organic phase (see figure below). Once in the organic phase, the hydroxide ion will react with the haloform to give dihalocarbene by Reactions 1 and 2. Water, a product of the reaction, will move from the organic phase to the aqueous phase, thus keeping the water concentration in the organic phase at a very low level. Because the water content in the organic phase is low, it will not interfere with the desirable reaction of the carbene with an alkene by Reaction 3. Thus, the undesirable side Reactions 4 and 5 are minimized. Finally, the halide ion, which is also produced in Reactions 1 and 2, is transported to the aqueous phase by the tetraalkylammonium cation. In this way, electrical neutrality is maintained and the phase-transfer catalyst, R_4N^+, is returned to the aqueous phase, to repeat the whole procedure. The figure on page 191 summarizes the overall process. This process probably goes on at the interface rather than in the bulk, organic phase.

There are numerous examples of other reactions that might be effectively accelerated by a quaternary ammonium salt or other phase-transfer catalyst (see references). These reactions often involve simple experimental techniques, give shorter reaction times than noncatalyzed reactions, and avoid relatively expensive aprotic solvents that have been widely used to give one phase. Examples of reactions are shown.

Ether synthesis

$$ROH + R'Cl \xrightarrow[\text{aq. NaOH}]{\text{catalyst}} ROR'$$

Nitrile synthesis

$$R{-}Cl + CN^- \xrightarrow[\text{H}_2\text{O}]{\text{catalyst}} RCN$$

Oxidation reactions

$$\underset{H}{\overset{R}{>}}C{=}C\underset{H}{\overset{R}{<}} + K^+MnO_4^- \xrightarrow{\text{catalyst}} \underset{H\ \ OH}{\overset{R}{>}}C{-}C\underset{OH\ H}{\overset{R}{<}}$$

Alkylation reactions

$$PhCH_2\underset{O}{\overset{\|}{C}}CH_3 + RBr \xrightarrow[\text{aq. NaOH}]{\text{catalyst,}} PhCH\underset{R\ \ O}{\overset{\|}{C}}CH_3$$

Wittig reactions

$$\underset{R}{\overset{R}{>}}C{=}O + Ph_3{}^+P{-}CH_2{-}R \xrightarrow[\text{aq. NaOH}]{\text{catalyst,}} \underset{R}{\overset{R}{>}}C{=}C\underset{R}{\overset{H}{<}}$$

Phosphonium salts act as catalysts.

Increased nucleophilicity   Anions are heavily solvated in an aqueous solvent and are therefore poor nucleophiles in some S_N2 reactions. When they are transported into the organic phase with the catalyst, R_4N^+ X^-, the anion, X^-, is no longer solvated with water and may have increased reactivity.

CROWN ETHER CATALYSIS

Another important class of phase-transfer catalysts includes the crown ethers (not used in this experiment). Crown ethers are used to dissolve organic and inorganic alkali metal salts in organic solvents. The crown ether complexes the cation and provides it with an organic exterior (hydrophobic) so that it is soluble in organic solvents. The anion is carried along into solution as the counterion. One example of a crown ether is dicyclohexyl-18-crown-6. Potassium permanganate, $KMnO_4$, complexed to the crown ether is soluble in benzene and is known as "purple benzene." It is useful in various oxidation reactions.

$$+ \ K^{+}MnO_4{}^{-} \ \longrightarrow$$

$$MnO_4{}^{-}$$

   The crown ethers catalyze many of the same types of reactions listed in the preceding section on quaternary ammonium salt catalysis. Crown ethers are very expensive relative to the ammonium salts and are not used as widely for large-scale reactions. In some cases, however, these ethers may be necessary to obtain an efficient and high-yield reaction.

THE EXPERIMENT

We shall prepare 7,7-dichlorobicyclo[4.1.0]heptane, also known as 7,7-dichloronorcarane, by the reaction

$$+ \ CHCl_3 \ + \ OH^{-} \ \longrightarrow \qquad + \ H_2O \ + \ Cl^{-}$$

Chloroform, $CHCl_3$, and base are used in excess in this reaction. Although most of the chloroform reacts to give the 7,7-dichloronorcarane via the carbene intermediate, a significant portion is hydrolyzed by the base to formate ion and carbon monoxide (Equations 4 and 5, pp 190–191). Bromoform, $CHBr_3$, can be used to prepare the corresponding 7,7-dibromonorcarane via the dibromocarbene.

REQUIRED READING

Review: Technique 7   Sections 7.4, 7.8, and 7.9
          Technique 18   Sections 18.2, 18.9, and 18.10

New: Appendix 4   NMR Spectroscopy
          Appendix 5   Carbon-13 NMR Spectroscopy

SPECIAL INSTRUCTIONS

Chloroform is a suspected carcinogen; therefore do not let it touch your skin, and avoid breathing the vapor. Chloroform and cyclohexene must be transferred to the conical vial in the hood.

> **CAUTION: Chloroform is a suspected carcinogen. Work in a hood during the measuring and transferring of this toxic substance. Do not breathe the vapor.**

Once these reagents have been added and the cap attached, the remaining laboratory operations may be conducted at your laboratory bench. The separation procedure may also be done at the bench since most of the chloroform will have been consumed in the reaction and little of the toxic substance will remain. Care must be taken to avoid contact with the caustic 50% aqueous sodium hydroxide.

PROCEDURE

Weigh 0.020 g of the phase-transfer catalyst, benzyltriethylammonium chloride, on a smooth piece of paper and **reclose the bottle** immediately (It is hygroscopic!).[1] Save it for addition later in the procedure. Preweigh a 5-mL conical vial with cap and transfer 0.20 mL of cyclohexene (MW = 82.2) to the vial in a hood. Cap the vial and reweigh it to determine the amount of cyclohexene in the vial. Add 0.50 mL of 50% aqueous sodium hydroxide[2] to the vial being careful to avoid getting any solution on the glass joint. In a hood, add 0.50 mL of chloroform (MW = 119.4, d = 1.49 g/mL) to the conical vial. Add a magnetic spin vane and the benzyltriethylammonium chloride to the vial.

> **CAUTION: Chloroform and cyclohexene should be kept in a hood. Do your measuring and transferring operations in the hood. Avoid contact with these substances. Do not breathe the vapors.**

---

[1]Note to the instructor: The activity of benzyltriethylammonium chloride varies depending upon the source of the catalyst. The reaction should be tried in advance of the laboratory to make sure it works properly. We use Aldrich Chemical Co., #14,655–2.

[2]This reagent should be prepared by the instructor: Dissolve 15 g of sodium hydroxide in 15 mL of water. Cool the solution to room temperature and store it in a plastic bottle.

Cap the vial to help control the loss of vapor and place the vial in a small beaker. Stir the mixture **rapidly** on a magnetic stirrer for 1.5 hours.[3] Vent the cap occasionally to release any pressure that may develop. An emulsion forms during this time.

Following this reaction time, add 1.5 mL of water and 0.5 mL of methylene chloride to the mixture. Stir this mixture for five minutes. Stop the stirring and allow the layers to separate. Swirl the vial gently to help break up the emulsion. Remove the lower methylene chloride layer with a **filter tip** pipet and transfer it to another storage vial (you will not need to remove the spin vane). The small amount of emulsion that forms at the interface should be left behind with the aqueous layer. Add another 0.5 mL portion of methylene chloride, stir the mixture for five minutes, remove the lower organic phase with the filter tip pipet, and transfer the organic layer to the same storage vial. Repeat this extraction with another 0.5 mL portion of methylene chloride and transfer the organic phase to the storage vial. Remove the spin vane from the conical vial containing the remaining aqueous layer, clean the spin vane, and discard the aqueous layer. Avoid contact with the basic solution.

Place the clean spin vane in the vial containing the combined organic phases, add 0.5 mL of saturated sodium chloride to the vial, and stir the mixture for five minutes. Remove the lower organic layer with a filter tip pipet. Transfer this phase to a **dry** conical vial and add two microspatulafuls of granular anhydrous sodium sulfate. Cap the vial and swirl it occasionally for at least 10 minutes to dry the organic phase.

Remove the dried organic phase with a dry Pasteur pipet and transfer the liquid to a dry **preweighed** conical vial. Evaporate the methylene chloride together with any remaining cyclohexene and chloroform in a hood in a warm sand bath. Use a stream of dry air or nitrogen to aid the evaporation process. If there are droplets of water in the residue, methylene chloride should be added and the drying procedure repeated with fresh anhydrous sodium sulfate. After removing the drying agent, evaporate the solvent.

Following removal of methylene chloride, you are left with 7,7-dichloronorcarane of sufficient purity for spectroscopy. Weigh the conical vial, determine the weight of product, and calculate the percentage yield. Obtain the infrared spectrum (Technique 18, Section 18.2, p 771). At the option of the instructor, obtain the proton NMR spectrum. It may also be of interest to obtain the decoupled and coupled carbon-13 spectrum of your product. Submit any remaining product in a labeled vial with your laboratory report.

REFERENCES

Dehmlow, E. V. ''Phase-Transfer Catalyzed Two-Phase Reactions in Preparative Organic Chemistry.'' *Angewandte Chemie, International Edition in English, 13* (1974): 170.

Dehmlow, E. V. ''Advances in Phase-Transfer Catalysis.'' Ibid., *16* (1977): 493.

Gokel, G. W., and Weber, W. P. ''Phase Transfer Catalysis.'' *Journal of Chemical Education, 55* (1978): 350, 429.

Makosza, M., and Wawrzyniewicz, M. ''Catalytic Method for Preparation of Dichlorocyclopropane Derivatives in Aqueous Medium.'' *Tetrahedron Letters* (1969): 4659.

Starks, C. M. ''Phase-Transfer Catalysis.'' *Journal of the American Chemical Society, 93* (1971): 195.

Weber, W. P., and Gokel, G. W. *Phase Transfer Catalysis in Organic Synthesis.* Berlin: Springer-Verlag, 1977.

---

[3] A 30 minute reaction period produces a very low yield (12%); 1.5 hour yields about 45%; 2.0 hour yields about 55%. Longer reaction times produce higher yields.

QUESTIONS

1. Why did you need to stir the mixture vigorously during reaction?

2. Why did you wash the organic phase with saturated sodium chloride solution?

3. What short chemical test could you make on the product to indicate whether cyclohexene is present or absent?

4. Would you expect 7,7-dichloronorcarane to give a positive sodium iodide in acetone test?

5. Assign the C—H stretch for the cyclopropane ring hydrogens in the infrared spectrum.

6. Suggest why it may be necessary to use a large excess of chloroform in this reaction.

7. A student obtained an NMR spectrum of the product isolated in this experiment. The spectrum shows peaks at about 7.3 and 5.6 ppm. What do you think these peaks indicate? Are they part of the 7,7-dichloronorcarane spectrum?

8. Draw the structures of the products that you would expect from the reactions of *cis*- and *trans*-2-butene with dichlorocarbene.

9. Draw the structure of the expected dichlorocarbene adduct of methyl methacrylate (methyl 2-methylpropenoate). With compounds of this type, another product could have been obtained. It is the chloroform adduct to the double bond (Michael-type reaction). What would this structure look like?

10. Provide mechanisms for the following abnormal dichlorocarbene addition reactions. In both cases, the usual adduct is first obtained, and then a subsequent reaction occurs.

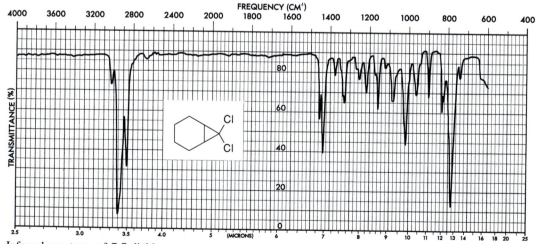

Infrared spectrum of 7,7-dichloronorcarane, neat

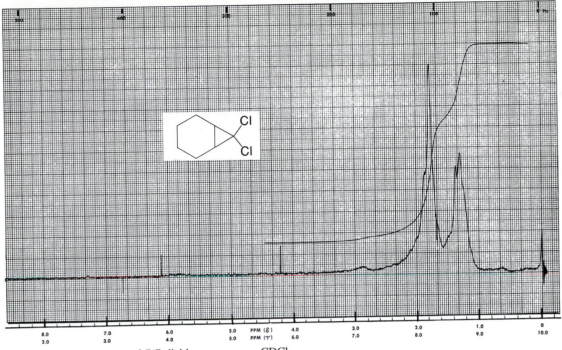

NMR spectrum of 7,7-dichloronorcarane, CDCl$_3$

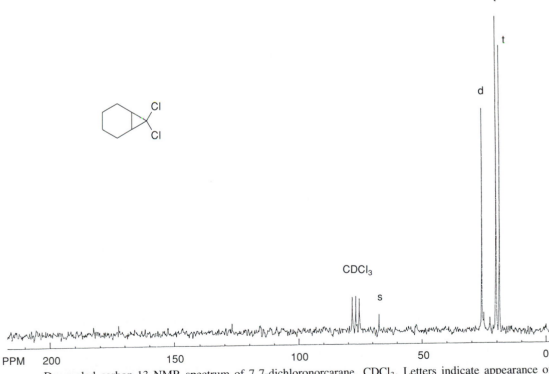

Decoupled carbon-13 NMR spectrum of 7,7-dichloronorcarane, CDCl$_3$. Letters indicate appearance of spectrum when carbons are coupled to protons (s = singlet, d = doublet, t = triplet).

Essay
ETHANOL AND FERMENTATION CHEMISTRY

The fermentation processes involved in making bread, making wine, and brewing are among the oldest chemical arts. Even though fermentation had been known as an art for centuries, not until the nineteenth century did chemists begin to understand this process from the point of view of science. In 1810, Gay-Lussac discovered the general chemical equation for the breakdown of sugar into ethanol and carbon dioxide. The manner in which the process took place was the subject of much conjecture until Louis Pasteur began his thorough examination of fermentation. Pasteur demonstrated that yeast was required in the fermentation. He was also able to identify other factors that controlled the action of the yeast cells. His results were published in 1857 and 1866.

For many years, scientists believed that the transformation of sugar into ethanol and carbon dioxide by yeasts was inseparably connected with the life process of the yeast cell. This view was abandoned in 1897, when Büchner demonstrated that yeast extract will bring about alcoholic fermentation in the absence of any yeast cells. The fermenting activity of yeast is due to a remarkably active catalyst of biochemical origin, the enzyme zymase. It is now recognized that most of the chemical transformations that go on in living cells of plants and animals are brought about by enzymes. The enzymes are organic compounds, generally proteins, and establishment of structures and reaction mechanisms of these compounds is an active field of present-day research. Zymase is now known to be a complex of at least 22 separate enzymes, each of which catalyzes a specific step in the fermentation reaction sequence.

Enzymes show an extraordinary specificity—a given enzyme acts on a specific compound or a closely related group of compounds. Thus, zymase acts on only a few select sugars and not on all carbohydrates; the digestive enzymes of the alimentary tract are equally specific in their activity.

The chief sources of sugars for fermentation are the various starches and the molasses residue obtained from refining sugar. Corn (maize) is the chief source of starch in the United States, and ethyl alcohol made from corn is known commonly as **grain alcohol.** In preparing alcohol from corn, the grain, with or without the germ, is ground and cooked to give the **mash.** The enzyme diastase is added in the form of **malt** (sprouted barley that has been dried in air at 40 °C and ground to a powder) or of a mold such as *Aspergillus oryzae*. The mixture is kept at 40 °C until all the starch has been converted to the sugar **maltose** by hydrolysis of ether and acetal bonds. This solution is known as the **wort.**

The wort is cooled to 20 °C and diluted with water to 10% maltose, and a pure yeast culture is added. The yeast culture is usually a strain of *Saccharomyces cerevisiae* (or *ellipsoidus*). The yeast cells secrete two enzyme systems: maltase, which converts

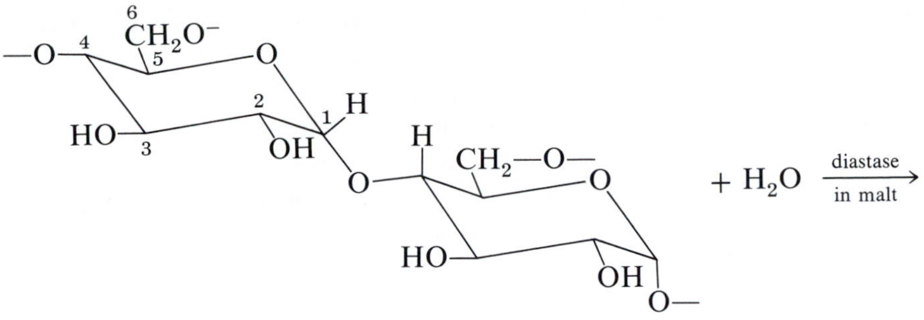

Starch

This is a glucose polymer with 1,4- and
1,6- glycosidic linkages. The linkages
at C-1 are α.

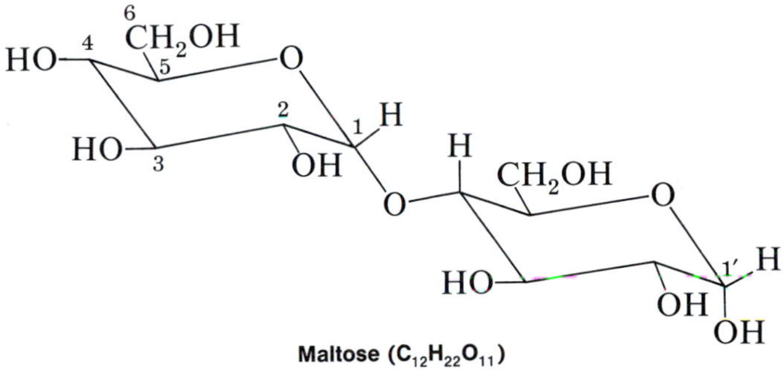

Maltose ($C_{12}H_{22}O_{11}$)

The α linkage still exists at C-1.
The —OH is shown α at the 1' position
(axial), but it can also be β (equatorial).

the maltose into glucose, and zymase, which converts the glucose into carbon dioxide
and alcohol. Heat is liberated, and the temperature must be kept below 35 °C by
cooling to prevent destruction of the enzymes. Oxygen in large amounts is initially
necessary for the optimum reproduction of yeast cells, but the actual production of
alcohol is anaerobic. During fermentation, the evolution of carbon dioxide soon estab-
lishes anaerobic conditions. If oxygen were freely available only carbon dioxide and
water would be produced.

$$\text{Maltose} + H_2O \xrightarrow{\text{maltase}} 2$$

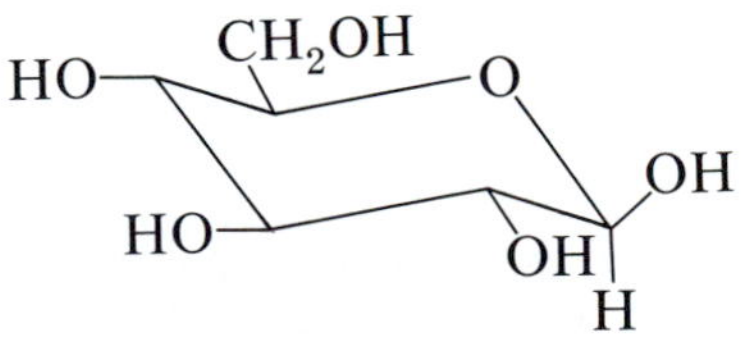

β-D-(+)-Glucose

(α-D-(+)-Glucose, with an axial
—OH, is also produced.)

$$\text{Glucose} \xrightarrow{\text{zymase}} 2CO_2 + 2CH_3CH_2OH + 26 \text{ kcal}$$

$C_6H_{12}O_6$

After 40 to 60 hours, fermentation is complete, and the product is distilled to remove the alcohol from solid matter. The distillate is fractionated by means of an efficient column. A small amount of acetaldehyde, bp 21 °C, distills first and is followed by 95% alcohol. Fusel oil is contained in the higher-boiling fractions. The fusel oil consists of a mixture of higher alcohols, chiefly 1-propanol, 2-methyl-1-propanol, 3-methyl-1-butanol, and 2-methyl-1-butanol. The exact composition of fusel oil varies considerably; it particularly depends on the type of raw material that is fermented. These higher alcohols are not formed by fermentation of glucose. They arise from certain amino acids derived from the proteins present in the raw material and in the yeast. These fusel oils cause the headaches familiarly associated with drinking alcoholic beverages.

Industrial alcohol is ethyl alcohol used for nonbeverage purposes. Most of the commercial alcohol is denatured, to avoid payment of taxes, the biggest cost in the price of liquor. The denaturants render the alcohol unfit for drinking. Methanol, aviation fuel, and other substances are used for this purpose. The difference in price between taxed and nontaxed alcohol is more than $20 a gallon. Before efficient synthetic processes were developed, the chief source of industrial alcohol was fermented blackstrap molasses, the noncrystallizable residue from refining cane sugar (sucrose). Most industrial ethanol in the United States is now manufactured from ethylene, a product of the "cracking" of petroleum hydrocarbons. By reaction with concentrated sulfuric acid, ethylene becomes ethyl hydrogen sulfate, which is hydrolyzed to ethanol by dilution with water. The alcohols 2-propanol, 2-butanol, 2-methyl-2-propanol, and higher secondary and tertiary alcohols also are produced on a large scale from alkenes derived from cracking.

Yeasts, molds, and bacteria are used commercially for the large-scale production of various organic compounds. An important example, in addition to ethanol production, is the anaerobic fermentation of starch by certain bacteria to yield 1-butanol, acetone, ethanol, carbon dioxide, and hydrogen.

REFERENCES

Amerine, M. A. "Wine." *Scientific American, 211* (August 1964): 46.
"Chemical Technology: Key to Better Wines." *Chemical and Engineering News, 51* (July 2, 1973): 14.
"Chemistry Concentrates on the Grape." *Chemical and Engineering News, 51* (June 25, 1973): 16.
Church, L. B. "The Chemistry of Winemaking." *Journal of Chemical Education, 49* (1972): 174.
Ough, C. S. "Chemicals Used in Making Wine." *Chemical and Engineering News, 65* (January 5, 1987): 19.
Ray, O. S. "Alcohol." *Drugs, Society, and Human Behavior*. St. Louis: C. V. Mosby, 1972.

Students wanting to investigate alcoholism and possible chemical explanations for alcohol addiction may consult the following references:

Cohen, G., and Collins, M. "Alkaloids from Catecholamines in Adrenal Tissue: Possible Role in Alcoholism." *Science, 167* (1970): 1749.
Davis, V. E., and Walsh, M. J. "Alcohol Addiction and Tetrahydropapaveroline." *Science, 169* (1970): 1105.

Davis, V. E., and Walsh, M. J. "Alcohols, Amines, and Alkaloids: A Possible Biochemical Basis for Alcohol Addiction." *Science, 167* (1970): 1005.

Labianca, D. A. "Acetaldehyde Syndrome and Alcoholism." *Chemistry, 47* (October 1974): 21.

Levitt, A. E. "Alcoholism: Costly Health Problem for U.S. Industry." *Chemical and Engineering News, 49* (September 13, 1971): 25.

Seevers, M. H., Davis, V. E., and Walsh, M. J. "Morphine and Ethanol Physical Dependence: A Critique of a Hypothesis." *Science, 170* (1970): 1113.

Yamanaka, Y., Walsh, M. J., and Davis, V. E. "Salsolinol, An Alkaloid Derivative of Dopamine Formed in Vitro during Alcohol Metabolism." *Nature, 227* (1970): 1143.

Experiment 20
Ethanol from Sucrose

Fermentation
Fractional distillation
Azeotropes

Sucrose as well as maltose can be used as the starting material for making ethanol. Sucrose is a disaccharide with the formula $C_{12}H_{22}O_{11}$. It has one glucose molecule combined with fructose, whereas maltose consists of two glucose molecules. The enzyme **invertase** is used to catalyze the hydrolysis of sucrose. **Maltase** is more effective in catalyzing the hydrolysis of maltose. The hydrolysis of maltose is discussed in the essay on ethanol and fermentation. **Zymase** is used to convert the sugars to alcohol and carbon dioxide. Pasteur observed that growth and fermentation were promoted by adding small amounts of mineral salts to the nutrient medium. Later it was found that before fermentation actually begins, the hexose sugars combine with phosphoric acid, and the resulting hexose–phosphoric acid combination is then degraded into carbon dioxide and ethanol. The carbon dioxide is not wasted in the commercial process but is converted to dry ice.

The fermentation is inhibited by ethanol; it is not possible to prepare solutions containing more than 10% to 15% ethanol by this method. More concentrated ethanol can be isolated by fractional distillation. Ethanol and water form an azeotropic mixture consisting of 95% ethanol and 5% water by weight, which is the most concentrated ethanol that can be obtained by fractionation of dilute ethanol-water mixtures.

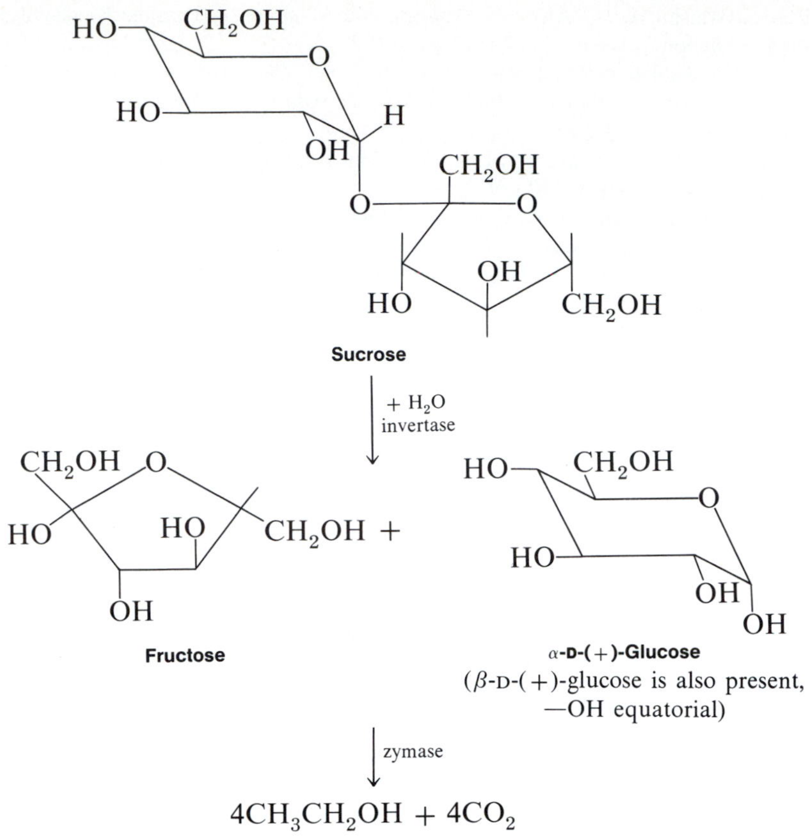

Sucrose

$+ H_2O$
invertase

Fructose   **α-D-(+)-Glucose**
(β-D-(+)-glucose is also present,
—OH equatorial)

zymase

$$4CH_3CH_2OH + 4CO_2$$

REQUIRED READING

Review: Technique 4   Filtration, Sections 4.3 and 4.4
          Technique 6   Physical Constants, Part B, Boiling Points

New: Technique 6   Physical Constants, Part D, Density
          Technique 10   Fractional Distillation, Azeotropes
          Essay   Ethanol and Fermentation Chemistry

SPECIAL INSTRUCTIONS

The fermentation must be started at least one week before the period in which the ethanol will be isolated. When the aqueous ethanol solution is to be separated from the yeast cells, it is important to transfer carefully as much of the clear, supernatant liquid as possible, without agitating the mixture.

> **NOTE TO THE INSTRUCTOR:** Because the volume of the fermentation mixture is only about 10 mL, it is necessary to use an external heat source to maintain a temperature of 30–35 °C. If a barium hydroxide trap is used, it is essential that the temperature remain constant throughout the fermentation; otherwise, pressure changes within the flask may cause the barium hydroxide solution and mineral oil to be sucked into the fermentation flask. An incubator will provide the necessary temperature control. A less reliable method is to place a cardboard box over a light bulb that is turned on during the fermentation. Aluminum foil can be used to seal any openings and to help reflect the heat inward. If this method is used, it is advisable to use a balloon to protect the fermentation flask from the air.

PROCEDURE

Place 1.00 g of sucrose in a 25-mL Erlenmeyer flask. Add 9.0 mL of water warmed to 25–30 °C; 1.0 mL of Pasteur's salts[1]; and 0.1 g of **dried** baker's yeast. Shake the contents vigorously to mix them and fit the flask with a one-hole rubber stopper with a glass tube (O.D. 4 mm) leading to a test tube (16 × 100-mm) containing about 2 mL of a saturated solution of barium hydroxide, as shown in the figure on page 204. The end of the glass tube should be positioned so that it is about 2 mm below the surface of the barium hydroxide solution. Protect the barium hydroxide from air by adding 0.5 mL of mineral oil to form a layer above the barium hydroxide. A precipitate of barium carbonate will form, indicating that carbon dioxide is being evolved. Alternatively, a balloon may be substituted for the barium hydroxide trap assembly. Attach the balloon directly to the Erlenmeyer flask. The gas will cause the balloon to expand as the fermentation continues. Oxygen from the atmosphere is excluded from the chemical reaction by these techniques. If oxygen were allowed to continue in contact with the fermenting solution, the ethanol could be further oxidized to acetic acid or even all the way to carbon dioxide and water. As long as carbon dioxide continues to be liberated, ethanol is being formed.

---

[1]A solution of Pasteur's salts consists of potassium dihydrogen phosphate, 1.0 g; calcium phosphate (monobasic), 0.10 g; magnesium sulfate, 0.10 g; and ammonium tartrate (diammonium salt), 5.0 g, dissolved in 430 mL water.

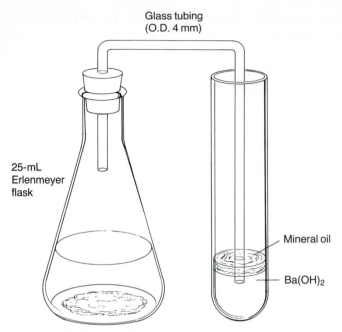

Apparatus for fermentation

Allow the mixture to stand at about 30–35 °C until fermentation is complete, as indicated by the cessation of gas evolution. Usually about one week is required. After this time, **carefully** move the flask away from the heat source and remove the stopper. Without disturbing the sediment, transfer the clear, supernatant liquid solution to another container with a Pasteur pipet. Try to avoid drawing any of the sediment into the pipet.

If it is not possible to remove the solution completely without drawing up sediment, filter the solution in the following manner. Place about 0.5 g of Celite (Filter Aid) in a beaker with about 5 mL of water. Stir the mixture vigorously and then pour the contents into a Hirsch funnel (with filter paper) or a small Büchner funnel while applying a **gentle** vacuum, as in a vacuum filtration (Technique 4, Section 4.3, and Figure 4–6, p 571). Be careful not to let the Filter Aid dry completely. This procedure will cause a thin layer of Celite to be deposited on the filter paper (Technique 4, Section 4.3, p 571). Discard the water that passes through this filter. Pass the fermentation solution through this filter, using **very gentle** suction. The extremely tiny yeast particles are trapped in the pores of the Celite. The liquid contains ethanol in water, plus smaller amounts of dissolved metabolites (fusel oils) from the yeast.

Add about 4.6 g of anhydrous potassium carbonate to the filtered solution for each 10 mL of liquid. The solution, after becoming saturated with potassium carbonate, will be subjected to fractional distillation.

Assemble the apparatus shown in Figure 10–2, p 670. Pack the air condenser with about 1 g of stainless steel cleaning pad material (No soap!).

You should wear gloves when handling the stainless steel cleaning pad. The edges are very sharp and can easily cut into the skin.

Wrap the glass section of the air condenser between the two plastic caps with several layers of aluminum foil. Place a boiling stone and 9.0 mL of the fermentation mixture in the 10-mL round-bottom flask.[2] The apparatus should be clamped so that the bottom half of the flask is buried in the sand. Use a thermometer in the Hickman head to monitor the temperature of vapors rising from the liquid. Also use a thermometer to monitor the temperature of the sand bath. Cover the top of the sand bath with a square of aluminum foil with a tear from the center of one edge to the middle.

The temperature of the sand bath should be adjusted to about 165–170 °C. Once distillation begins, the temperature in the Hickman head will remain at about 65–78 °C until the ethanol fraction is distilled. As distillate condenses in the Hickman head, transfer the liquid from the reservoir to a preweighed 3-mL conical vial, using a 9-inch Pasteur pipet. In order to withdraw the distillate without removing the thermometer, it is helpful to bend the tip of the pipet slightly by heating it in a flame. Be sure to cap the conical vial used for storage each time after you transfer the distillate. Continue to distill the mixture and tranfer the distillate to the vial until the temperature in the Hickman head increases above 78 °C or until you can no longer observe any distillate collecting in the reservoir of the Hickman head. You should have collected about 0.2–0.4 mL of distillate. The distillation should then be interrupted by removing the apparatus from the hot sand bath.

Determine the weight of the distillate and its density using a disposable micropipet (see Technique 6, Section 6.14, p 614). The density can also be determined by transferring a known volume of the liquid with an automatic pipet or graduated pipet to a tared vial. Using the following table, determine the percentage composition by weight of ethanol in your distillate from the density of your sample. The extent of purification of the ethanol is limited since ethanol and water form a constant-boiling mixture or an azeotrope, with a composition of 95% ethanol and 5% water. No amount of distillation will remove the last 5% of water.

PERCENTAGE ETHANOL BY WEIGHT	DENSITY AT 20 °C (g/mL)
75	0.856
80	0.843
85	0.831
90	0.818
95	0.804
100	0.789

Calculate the percentage yield of alcohol. At the option of the instructor, determine the boiling point of the distillate using a micro boiling point method (Technique 6,

---

[2] If you have less than 9.0 mL, use all of your solution. If you have more than 9.0 mL, use only 9.0 mL and record the total volume. When calculating the percentage yield of alcohol, you will need to take into account the volume of liquid not distilled.

Section 6.10, p 607). The boiling point of the azeotrope is 78.1 °C. Submit the ethanol to the instructor in a labeled vial.[3]

[3]A careful analysis by flame-ionization gas chromatography on a typical student-prepared ethanol sample provided the following results:

Acetaldehyde	0.060%
Diethylacetal of acetaldehyde	0.005
Ethanol	88.3 (by hydrometer)
1-Propanol	0.031
2-Methyl-1-propanol	0.092
5-Carbon and higher alcohols	0.140
Methanol	0.040
Water	11.3 (by difference)

QUESTIONS

1. By doing some library research, see whether you can find out the method or methods commercially used to produce **absolute** ethanol.

2. Why is the air trap necessary in the late stages of fermentation?

3. How does acetaldehyde impurity arise in the fermentation?

4. Diethylacetal can be detected by gas chromatography. How does this impurity arise in fermentation?

5. Calculate how many milliliters of carbon dioxide would be produced theoretically from 1.0 g of sucrose at 25 °C and 1-atmosphere pressure.

Experiment 21
Thin-Layer Chromatography for Monitoring the Oxidation of Borneol to Camphor

Chromic acid oxidation
Monitoring reactions
Thin-layer chromatography

A typical reaction of secondary alcohols is their oxidation to ketones by such reagents as chromic acid, sodium dichromate, potassium permanganate, or sodium hypochlorite. In this experiment we use chromic acid to oxidize borneol to camphor. In Experiment 22, the oxidation of borneol to camphor is performed using sodium hypochlorite.

The half-reactions and the overall reaction used in the oxidation of borneol are complicated by the fact that chromium trioxide, chromate ion (chromic acid), and dichromate ion are all species which are in equilibrium in aqueous acid solution.

$$2CrO_3 + 2H_2O \rightleftharpoons 4H^+ + 2CrO_4^{2-} \rightleftharpoons Cr_2O_7^{2-} + H_2O + 2H^+$$

The mechanism generally accepted by chemists assumes that borneol (an alcohol) reacts with chromate ion to form an ester. The ester is decomposed by loss of an α-hydrogen to yield camphor, a ketone.

$$\text{borneol} + CrO_4^{2-} + 2H^+ \underset{\text{fast}}{\rightleftharpoons} \text{ester}$$

$$\text{ester} \xrightarrow{\text{slow}} \text{camphor} + H_3O^+ + {}^-CrO_3H$$

It is interesting to compare this mechanism to the sodium hypochlorite oxidation of borneol presented in Experiment 22. In this second mechanism, the loss of an α-hydrogen from a hypochlorite ester yields the ketone.

The purpose of this experiment is to show how thin-layer chromatography (TLC) may be used to follow the progress of a reaction, and to determine when the reaction is complete. As we shall see, the reaction actually proceeds quite slowly. Although both the product (camphor) and the starting material (borneol) are colorless, they can be visualized by iodine vapor, and we will use this method to view the TLC plates.

While we have not presented the exact method in this experiment, it is quite easy to use the same technique presented here to follow the progress of the oxidation in Experiment 22.

REQUIRED READING

New: Technique 13   Thin-Layer Chromatography

Also read the introductory material presented in Experiment 22.

SPECIAL INSTRUCTIONS

This experiment uses diethyl ether which is a very flammable solvent, and it uses chromic acid which is a suspected carcinogen.

Avoid the use of flames when working with diethyl ether. Also avoid any direct contact with chromic acid residues which are corrosive and possibly carcinogenic.

The quantities of these materials used are quite small and should present no great danger, however, safety precautions should be exercised. Dispose of all chromium wastes in an appropriate waste container.

PROCEDURE

Using a slurry of silica gel G in a methylene chloride-methanol (2:1) solvent, prepare about 10 hand-dipped microscope-slide TLC plates by the method described in Technique 13, Section 13.3A, p 725. Prepare a developing chamber from a 4-oz wide-mouthed screw-cap jar, as described in Technique 13, Section 13.5, p 730. Add some of the methylene chloride-methanol solvent to the developing chamber. Finally, prepare several capillary micropipets as described in Technique 13, Section 13.4, p 728.

For this experiment, you use three solutions: Solution A is a 2% solution of borneol in diethyl ether; Solution B is a solution of 10% chromium trioxide and 5% sulfuric acid in water; and Solution C is a 2% solution of camphor in diethyl ether. All percentages are by mass. Solutions A and C provide reference spots on each TLC plate, and Solution B provides the oxidizing medium to oxidize borneol to camphor.

Mix about 1 mL of Solution A with 1 mL of Solution B in a small test tube. Write down the time at which this mixture is prepared. Shake the mixture briefly. Spot a TLC plate with Solution A, Solution C, and the upper (ether) layer of the reaction mixture from the test tube. Place the plate in the developing chamber and develop it as described in Technique 13, Section 13.5, p 730. After the reaction has been allowed to take place for five minutes, spot a second TLC plate with Solution A, Solution C, and the upper layer of the reaction mixture. Develop this plate as before. Continue spotting and developing plates at five-minute intervals for a total reaction time of 40 minutes. During this reaction period, shake the test tube periodically.

When the TLC plates have all been developed, place each plate in a jar containing a few iodine crystals, cap the jar, and warm it gently on a steam bath (or in a sand bath) until spots begin to appear (Technique 13, Section 13.7, p 732). You will notice that the R_f values for borneol and camphor differ, with camphor more mobile than borneol. By comparing these R_f values with the R_f values for the spots formed in the reaction mixture, you should be able to identify the presence of borneol and camphor in the mixture. Compare the intensities of the borneol and camphor spots in the reaction mixture on each of the TLC plates you have developed. Using this information, determine the time required for the oxidation to reach completion. Report the reaction time and the R_f values for borneol and camphor to your instructor in your report and provide sketches of a typical TLC plate.

REFERENCE

Davis, M. ''Using TLC to Follow the Oxidation of a Secondary Alcohol to a Ketone.'' *Journal of Chemical Education, 45* (March 1968): 192.

QUESTIONS

1. What physical property differences between borneol and camphor make the separation of the two compounds on thin-layer chromatography plates possible?

2. Why is it necessary to shake the mixture periodically during the reaction?

3. Describe how you would use TLC for monitoring the progress of a reaction to follow the esterification of benzoic acid with methanol to form methyl benzoate. Indicate the relative order of R_f values expected for benzoic acid and methyl benzoate.

4. A student spotted the TLC plates with the borneol-camphor reaction mixture as stated in the experiment and placed them in a storage cabinet for a few days. When the student returned and placed the plates in the iodine chamber, no spots were observed. What happened?

5. Are the structures given for borneol and camphor in this experiment different from those in Experiment 22? Explain the differences, if any.

Experiment 22
An Oxidation-Reduction Scheme: Borneol, Camphor, Isoborneol

Hypochlorite (bleach) oxidation
Sodium borohydride reduction
Stereochemistry
Spectroscopy (IR, NMR, carbon-13 NMR)

Borneol   Camphor   Isoborneol

This experiment illustrates the use of an oxidizing agent (hypochlorous acid) for converting a secondary alcohol (borneol) to a ketone (camphor). The camphor is then reduced by sodium borohydride to give the **isomeric** alcohol isoborneol. The spectra of borneol, camphor, and isoborneol are compared to detect structural differences and to determine the extent to which the final step produces a pure alcohol isomeric with the starting material.

OXIDATION OF BORNEOL WITH HYPOCHLORITE

Sodium hypochlorite, bleach, can be used to oxidize secondary alcohols to ketones. Because this reaction occurs more rapidly in an acidic environment, it is likely that the actual oxidizing agent is hypochlorous acid, HOCl. This acid is generated by the reaction between sodium hypochlorite and acetic acid:

$$NaOCl + CH_3COOH \longrightarrow HOCl + CH_3COONa$$

Although the mechanism is not fully understood, there is evidence that an alkyl hypochlorite intermediate is produced, which then gives the product via an E2 elimination:

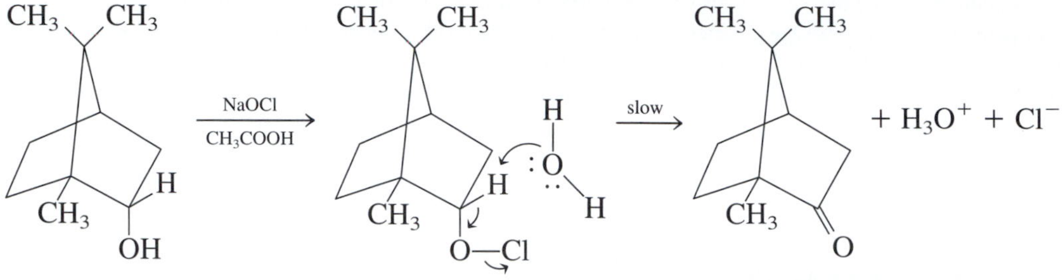

To ensure that complete oxidation of camphor occurs, it is necessary that an excess of sodium hypochlorite be present. Since the concentration of sodium hypochlorite in bleach can vary, a test for excess hypochlorite is performed by using starch-iodide indicator paper. In an acidic solution, iodide is oxidized by hypochlorite to iodine:

$$OCl^- + 2I^- + 2H^+ \longrightarrow Cl^- + I_2 + H_2O$$

The iodine then reacts with the starch in the test paper to produce the characteristic blue-black color of this complex.

REDUCTION OF CAMPHOR WITH SODIUM BOROHYDRIDE

Metal hydrides (sources of H:$^-$) of the Group III elements such as lithium aluminum hydride, LiAlH$_4$, and sodium borohydride, NaBH$_4$, are widely used in reducing carbonyl groups. Lithium aluminum hydride, for example, reduces many compounds containing carbonyl groups, such as aldehydes, ketones, carboxylic acids, esters, or amides, whereas sodium borohydride reduces only aldehydes and ketones. The reduced reactivity of borohydride allows it to be used even in alcohol and water solvents, whereas lithium aluminum hydride reacts violently with these solvents to produce hydrogen gas and thus must be used in nonhydroxylic solvents. In the present experiment, sodium borohydride is used because it is easily handled, and the results of reductions using either of the two reagents are essentially the same. The same care need not be taken in keeping sodium borohydride away from water as is required with lithium aluminum hydride.

The mechanism of action of sodium borohydride in reducing a ketone is as follows:

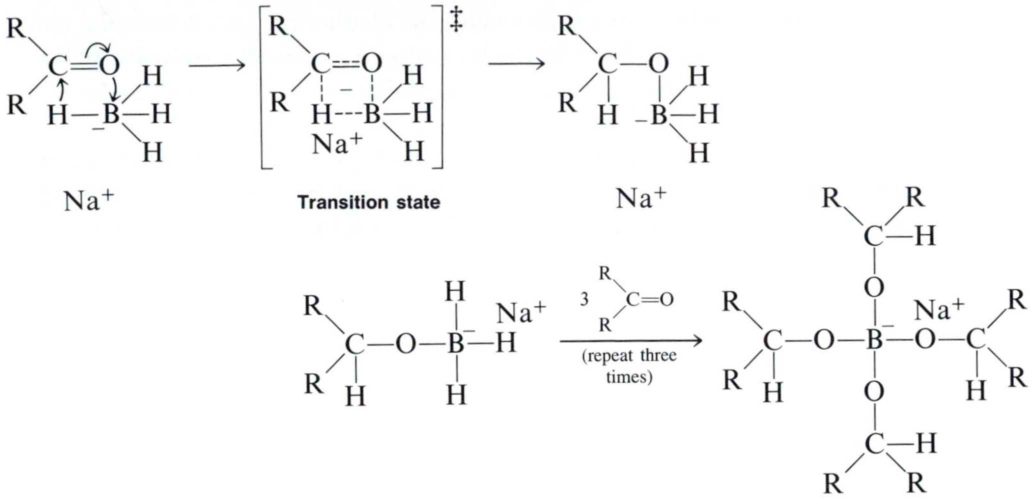

(I)

Note in this mechanism that all four hydrogen atoms are available as hydrides (H:$^-$), and thus one mole of borohydride can reduce four moles of ketones. All the steps are irreversible. Usually excess borohydride is used because there is uncertainty regarding the purity of the material.

Once the final tetraalkoxyboron compound (I) is produced, it can be decomposed (along with excess borohydride) at elevated temperatures as shown:

$$(R_2CH{-}O)_4B^-Na^+ + 4R'OH \longrightarrow 4R_2CHOH + (R'O)_4B^-Na^+$$

(I)

The stereochemistry of the reduction is very interesting. The hydride can approach the camphor molecule more easily from the bottom side (**endo** approach) than from the top side (**exo** approach). If attack occurs at the top, a large steric repulsion is created by one of the two **geminal** methyl groups. Attack at the bottom avoids this steric interaction.

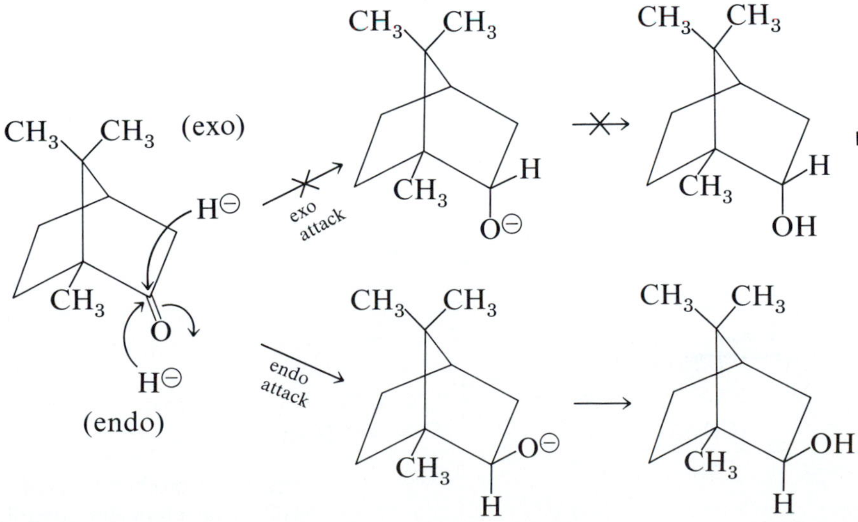

Borneol

Isoborneol

It is expected, therefore, that **isoborneol,** the alcohol produced from the attack at the **least-**hindered position, will **predominate but will not be the exclusive product** in the final reaction mixture. The percentage composition of the mixture can be determined by spectroscopy.

It is interesting to note that when the methyl groups are removed (as in 2-norbornanone), the top side (**exo** approach) is favored, and the opposite stereochemical result is obtained. Again, the reaction does not give exclusively one product.

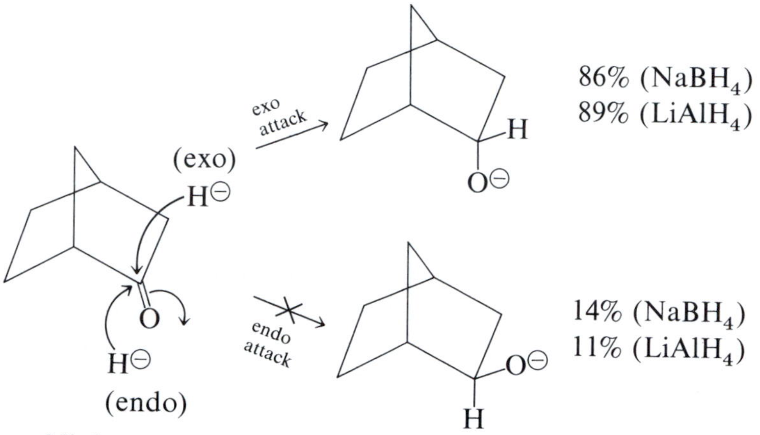

86% ($NaBH_4$)
89% ($LiAlH_4$)

14% ($NaBH_4$)
11% ($LiAlH_4$)

2-Norbornanone

Bicyclic systems such as camphor and 2-norbornanone react predictably according to steric influences. This effect has been termed **steric approach control.** In the reduction of simple acyclic and monocyclic ketones, however, the reaction seems to be influenced primarily by thermodynamic factors. This effect has been termed **product development control.** In the reduction of 4-*t*-butylcyclohexanone, the thermodynamically more stable product is produced by product development control.

10%

4-*t*-Butyl-cyclohexanone

90%

equatorial product favored;
''product development control''

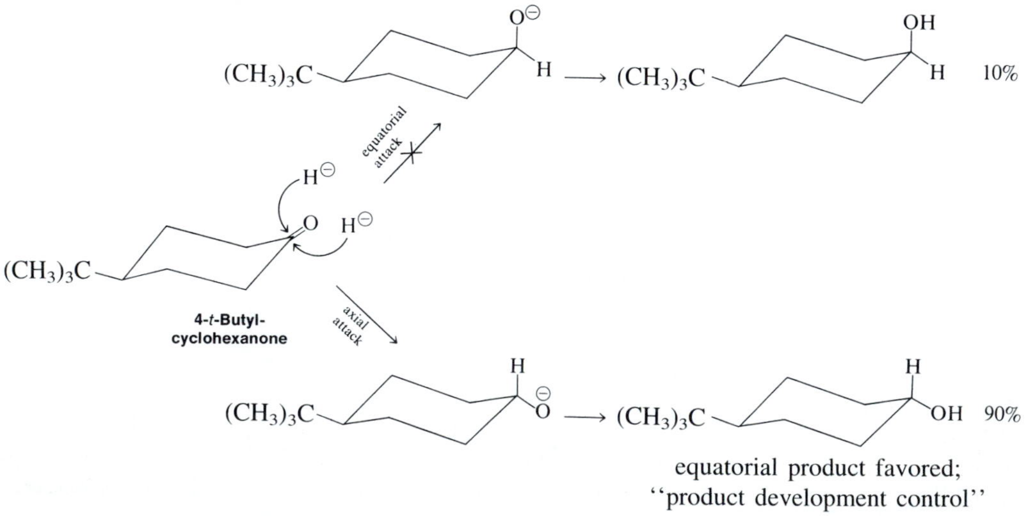

REQUIRED READING

Review: Technique 3   Reaction Methods, Section 3.9
           Technique 4   Filtration, Section 4.1
           Technique 7   Extractions, Separations, and Drying Agents
           Technique 18   Preparation of Samples for Spectroscopy, Sections 18.2
                                and 18.9
              Appendices 3, 4, and 5

SPECIAL INSTRUCTIONS

The reactants and products are all highly volatile and must be stored in tightly closed containers. The reaction should be carried out in a well ventilated room since a small amount of chlorine gas will be emitted from the reaction mixture. The reduction of camphor to isoborneol involves diethyl ether, which is extremely flammable. Be certain that no open flames of any sort are in your vicinity when you are using ether. If the instructor chooses, you may be asked to follow the progress of the oxidation of borneol to camphor by TLC (Experiment 21).

> **NOTE TO THE INSTRUCTOR:** Since the concentration of bleach decreases with time, it is best to use a new, unopened bottle of bleach.

PROCEDURE

OXIDATION OF BORNEOL TO CAMPHOR

To a 5-mL conical vial add 0.180 g of racemic borneol, 0.50 mL of acetone, and 0.15 mL of glacial acetic acid. After adding a spin vane to the vial, attach an air condenser and place the conical vial in a water bath at about 45 °C, as shown in Figure 2–5, p 544. It is important that the temperature of the water bath remain between 40–50 °C during the entire reaction period. Stir the mixture until the borneol is dissolved.

   While continuing to stir the reaction mixture, add dropwise 2.0 mL of a bleach solution (5.25% sodium hypochlorite) through the top of the air condenser over a period of about 30 minutes. When the addition is complete, stop stirring the mixture and remove a few drops of the bottom aqueous layer with a Pasteur pipet. Transfer this liquid onto a wet piece of starch-iodide indicator paper to determine if a sufficient amount of bleach has been added. A blue-black color due to the formation of the starch-iodine complex indicates that an excess of hypochlorite is present. If there is no color change, add an additional 0.2 mL of bleach to the reaction mixture, stir for several minutes, and repeat the starch-iodide test. Continue this process until the paper turns blue. Stir the mixture for 10 minutes after the last addition of bleach and repeat the starch-iodide test. If it is negative (absence of blue-black color), add an additional 0.2 mL of bleach. Whether additional bleach was added or not, allow the reaction to continue for 10 minutes more.

When the reaction time is complete, allow the mixture to cool to room temperature. Remove the air condenser and add 1.0 mL of methylene chloride to extract the camphor (Technique 7, Section 7.4, p 622). Cap the vial and shake well with venting. Remove the spin vane with forceps and rinse the spin vane and forceps with a few drops of methylene chloride. Using a filter tip pipet, transfer the lower methylene chloride layer into another 5-mL conical vial. Extract the aqueous layer with a second 1.0-mL portion of methylene chloride and combine it with the first methylene chloride solution. Wash the combined methylene chloride layers with 1.0 mL of saturated sodium bicarbonate solution. Stir the liquid with a stirring rod or spatula until bubbling produced by the formation of carbon dioxide ceases. Cap the vial and shake with frequent venting to release any pressure produced. Transfer the lower methylene chloride layer to another container and remove the aqueous layer. Return the methylene chloride layer to the vial and wash this solution successively with 1.0 mL of saturated sodium bisulfite and 1.0 mL of water. Using a dry filter tip pipet, transfer the methylene chloride layer to a dry test tube or conical vial. Add three to four microspatulafuls of granular anhydrous sodium sulfate and let dry for 10–15 minutes with occasional shaking. After taring a 10-mL Erlenmeyer flask, transfer the methylene chloride solution to the flask. Evaporate the solvent in the hood with a gentle stream of dry air or nitrogen gas while heating the Erlenmeyer flask in a sand bath at 40–50 °C (see Figure 3–12A, p 561). As an alternative, leave the flask in the hood until the methylene chloride has evaporated.

When all the liquid has evaporated and a solid has appeared, remove the flask from the heat source immediately; otherwise, the product may sublime and be lost. Weigh the flask to determine the weight of your product and calculate the percentage yield. Determine the melting point in a sealed capillary tube to prevent sublimation. The melting point of pure racemic camphor is 174 °C.[1] Save a small amount of the camphor for an infrared spectrum determination. The remainder of the camphor is reduced in the next step to isoborneol, which will be carried out in the same flask. Store the camphor with the flask tightly sealed until needed. For the infrared spectrum, dissolve the sample in carbon tetrachloride, place the solution between the salt plates, and mount the plates in a holder (see Technique 18, Section 18.2, p 771).

REDUCTION OF CAMPHOR TO ISOBORNEOL

Reweigh the 10-mL Erlenmeyer flask to determine the weight of camphor remaining. If the amount is less than 0.100 g, obtain some camphor from the supply shelf to supplement your yield. If it is more than 0.100 g, scale up the reagents appropriately from the amounts given below.

Add 0.5 mL of methanol to the camphor in the 10-mL Erlenmeyer flask. Stir with a glass stirring rod until the camphor has dissolved. In portions, cautiously and intermit-

---

[1] The observed melting point of camphor is often low. A small amount of impurity drastically reduces the melting point and increases the range. (See Question 5.)

tently add 0.060 g of sodium borohydride to the solution.[2] If necessary, cool the flask in an ice bath to keep the reaction mixture at room temperature. When all the borohydride is added, boil the contents of the flask on a steam bath or a sand bath at 70 °C for two minutes.

Allow the reaction mixture to cool for a couple of minutes and carefully add 3.5 mL of ice water. Collect the white solid by filtering on a Hirsch funnel and allow to dry for several minutes with suction. Transfer the solid to a 10-mL Erlenmeyer flask. Add 4.5 mL of ether to dissolve the product and three to four microspatulafuls of anhydrous magnesium sulfate to dry the solution. To remove the drying agent, filter the mixture with a filtering pipet (see Figure 4–4, p 569) into a 25-mL Erlenmeyer flask that has been tared. Use another 1.0 mL of ether to rinse the 10-mL Erlenmeyer flask and also filter this. Evaporate the solvent in a hood, as described above.

Determine the weight of the product and calculate the percentage yield. Determine the melting point (sealed tube); pure racemic isoborneol melts at 212 °C. Determine the infrared spectrum of the product by the method given above. Compare it with the spectra for borneol and isoborneol shown.

PERCENTAGES OF ISOBORNEOL AND BORNEOL OBTAINED FROM THE REDUCTION OF CAMPHOR

The percentage of each of the isomeric alcohols in the borohydride reduction mixture can be determined from the NMR spectrum.[3] (See Technique 18, Section 18.9, p 782, and Appendix 4.) The NMR spectra of the pure alcohols are shown on p 218. The hydrogen on the carbon bearing the hydroxyl group appears at 4.0 ppm for borneol and 3.6 ppm for isoborneol. One can obtain the product ratio by integrating these peaks (using an expanded presentation) in the NMR spectrum of "isoborneol" obtained after the borohydride reduction. In the spectrum shown on p 220, the isoborneol-borneol ratio 5:1 was obtained. The percentages obtained are 83% isoborneol and 17% borneol.

REFERENCES

Brown, H. C., and Muzzio, J. "Rates of Reaction of Sodium Borohydride with Bicyclic Ketones." *Journal of the American Chemical Society, 88* (1966): 2811.

Dauben, W. G., Fonken, G. J., and Noyce, D. S. "Stereochemistry of Hydride Reductions." *Journal of the American Chemical Society, 78* (1956): 2579.

Flautt, T. J., and Erman, W. F. "The Nuclear Magnetic Resonance Spectra and Stereochemistry of Substituted Boranes." *Journal of the American Chemical Society, 85* (1963): 3212.

[2] NOTE TO THE INSTRUCTOR: The sodium borohydride should be checked to see whether it is active. Place a small amount of powdered material in some methanol and heat it gently. The solution should bubble vigorously if the hydride is active.

[3] Approximate percentages can also be obtained by gas chromatography, using a Gow-Mac 69–360 instrument. Use an 8-foot column of 10% Carbowax 20M and operate the device at 180 °C with a 40 mL/min flow rate. The compounds are dissolved in methylene chloride for analysis. The retention times for camphor, isoborneol, and borneol are 8, 10, and 11 minutes, respectively.

Markgraf, J. H. "Stereochemical Correlations in the Camphor Series." *Journal of Chemical Education,* *44* (1967): 36.
Mohrig, J. R., Nienhuis, C. F., Van Zoeren, C., Fox, B. G., and Mahaffy, P. G. "The Design of Laboratory Experiments in the 1980's." *Journal of Chemical Education, 62* (1985): 519.

QUESTIONS

1. Interpret the major absorption bands in the infrared spectra of camphor, borneol, and isoborneol.

2. Explain why the **gem**-dimethyl groups appear as separate peaks in the proton NMR spectrum of isoborneol although they are not resolved in borneol.

3. A sample of isoborneol prepared by reduction of camphor was analyzed by infrared spectroscopy and showed a band at 1760 cm^{-1}. This result was unexpected. Why?

4. Are the structures for borneol and camphor given in this experiment different from the structures given in Experiment 21? Explain.

5. The observed melting point of camphor is often low. Look up the molal freezing-point-depression constant K for camphor and calculate the expected depression of the melting point of a quantity of camphor that contains 0.5 molal impurity. Hint: Look in a general chemistry book under freezing point depression or colligative properties of solutions.

6. Why was the methylene chloride layer washed with sodium bicarbonate in the procedure for the preparation of camphor?

7. The peak assignments are shown on the carbon-13 NMR spectrum of camphor. Using these assignments as a guide, assign as many peaks as possible in the carbon-13 spectra of borneol and isoborneol.

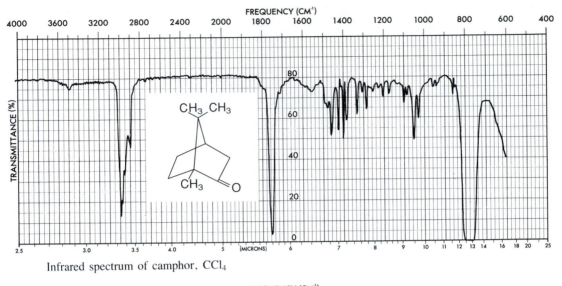

Infrared spectrum of camphor, CCl₄

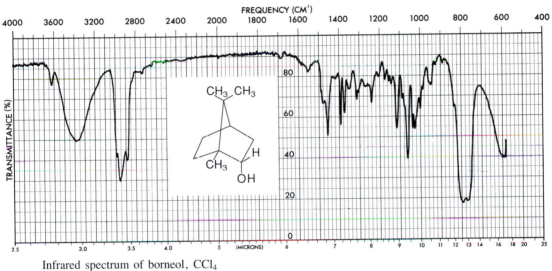

Infrared spectrum of borneol, CCl₄

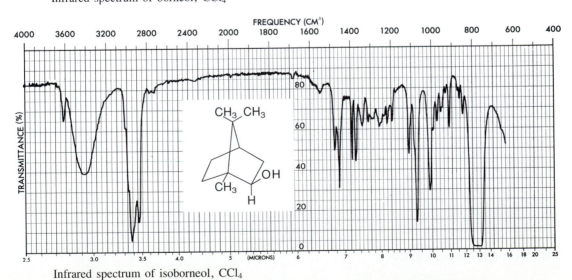

Infrared spectrum of isoborneol, CCl₄

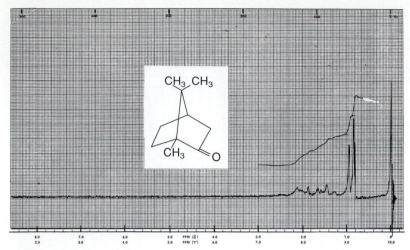

NMR spectrum of camphor, CCl_4

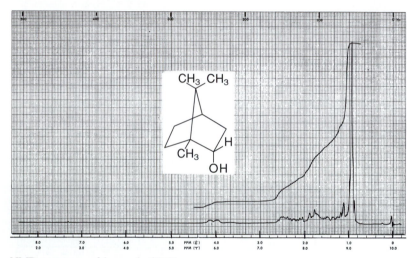

NMR spectrum of borneol, $CDCl_3$

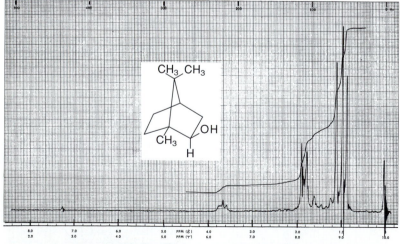

NMR spectrum of isoborneol, $CDCl_3$

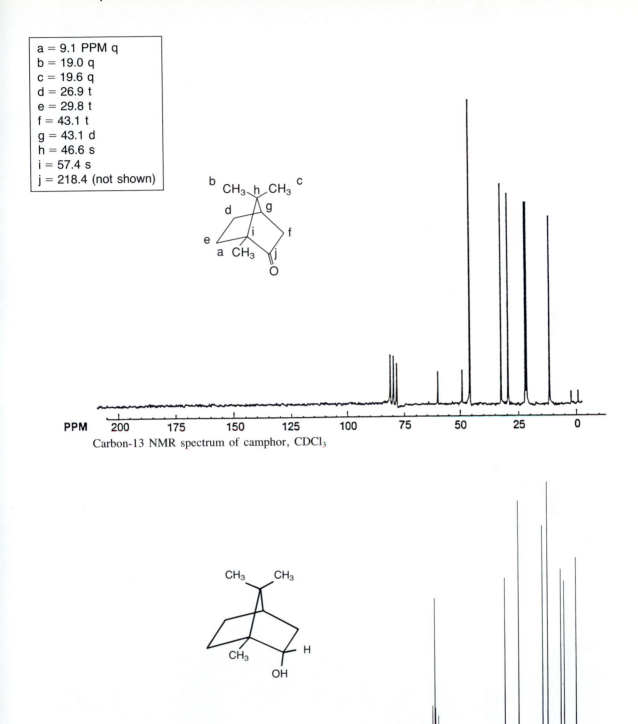

a = 9.1 PPM q
b = 19.0 q
c = 19.6 q
d = 26.9 t
e = 29.8 t
f = 43.1 t
g = 43.1 d
h = 46.6 s
i = 57.4 s
j = 218.4 (not shown)

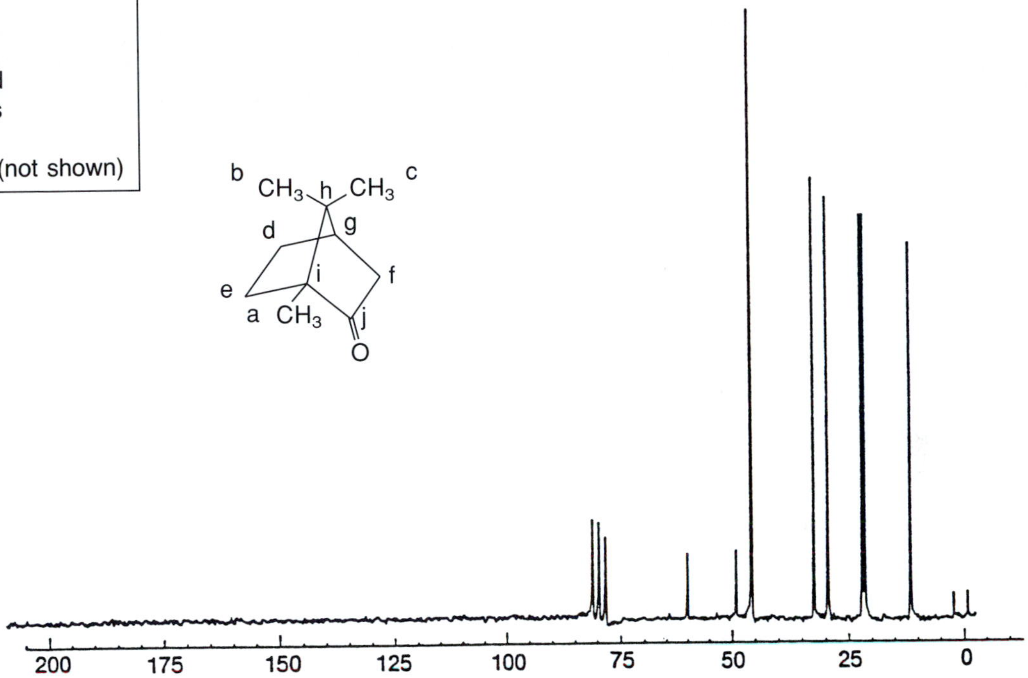

Carbon-13 NMR spectrum of camphor, $CDCl_3$

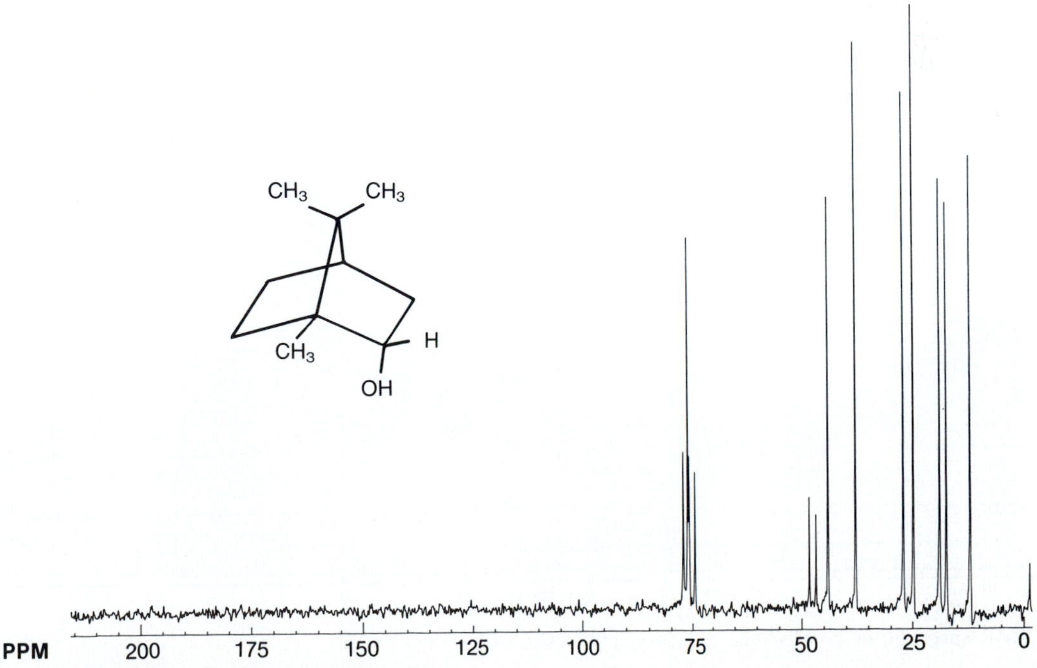

Carbon-13 NMR spectrum of borneol, $CDCl_3$

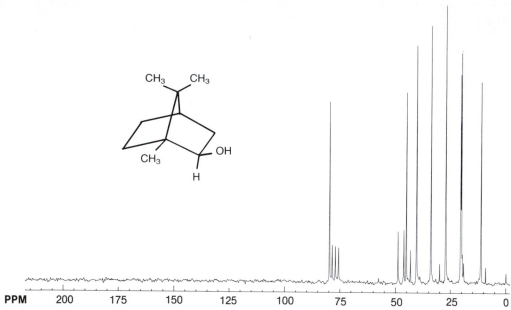

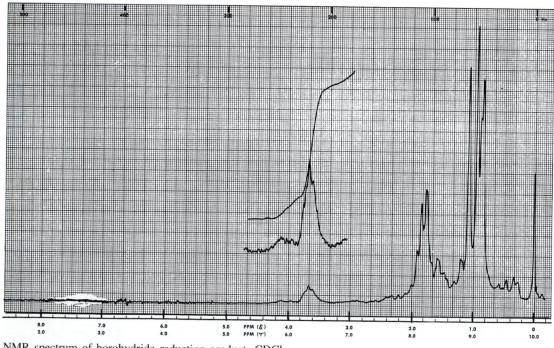

Carbon-13 NMR spectrum of isoborneol, CDCl$_3$. (Small peaks at 9, 19, 30, and 43 PPM are due to impurities.)

NMR spectrum of borohydride reduction product, CDCl$_3$

Experiment 23
Chiral Reduction of Ethyl Acetoacetate

Fermentation
Stereochemistry
Reduction with yeast
Polarimetry
Use of a separatory funnel

This experiment uses common baker's yeast as a chiral reducing agent to transform an achiral starting material, ethyl acetoacetate, into a chiral product, (S)-(+)-ethyl 3-hydroxybutanoate. The chiral product is used as an important building block in the laboratory syntheses of complex natural products.

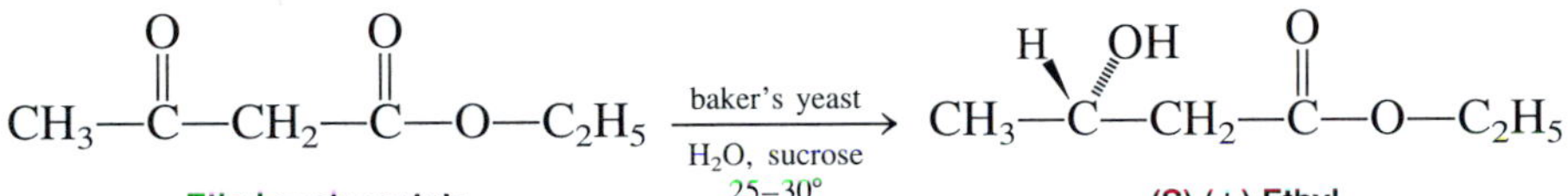

$$CH_3-\overset{\overset{O}{\|}}{C}-CH_2-\overset{\overset{O}{\|}}{C}-O-C_2H_5 \xrightarrow[\substack{H_2O,\ \text{sucrose} \\ 25-30°}]{\text{baker's yeast}} CH_3-\overset{\overset{H}{\diagdown}\overset{OH}{\diagup}}{C}-CH_2-\overset{\overset{O}{\|}}{C}-O-C_2H_5$$

Ethyl acetoacetate   **(S)-(+)-Ethyl
3-hydroxybutanoate**

The product, ethyl 3-hydroxybutanoate, is formed principally as the enantiomer with the (S) configuration. The reaction does produce a small amount (generally less than 10%) of the opposite enantiomer, (R)-(−)-ethyl 3-hydroxybutanoate.

REQUIRED READING

Review: Technique 3   Section 3.8
                Technique 4   Section 4.4
                Technique 7   Section 7.7
                Technique 12

New: Technique 16   Polarimetry

SPECIAL INSTRUCTIONS

The fermentation requires at least three days; the experiment should be begun in advance of the time set aside for product isolation and polarimetry. If the fermentation is allowed to continue for longer than three days, the optical purity of the product will increase. Under these conditions, however, it may become difficult to identify the minor enantiomer in Experiment 24. The observed rotation for a sample isolated by a single student may be only a few degrees, which limits the precision of the optical purity determination. Better results can be obtained if several students combine their products for the polarimetric analysis.

PROCEDURE

Equip a 500-mL Erlenmeyer flask with a magnetic stirring bar and a one-hole rubber stopper with a glass tube leading to a beaker or a test tube containing a solution of barium hydroxide. Protect the barium hydroxide from air by adding some mineral oil or xylene to form a layer above the barium hydroxide. The figure on page 223 depicts the apparatus for this experiment. A precipitate of barium carbonate will form, indicating that carbon dioxide is being evolved during the course of the reaction. Oxygen from the atmosphere is excluded through the use of the trap.

Add 150 mL of tap water, 30 g of sucrose, and about 3.5 g (one package) of dry baker's yeast to the flask. Add these materials, with stirring, in the order indicated. Attach the trap to the fermentation flask. Stir this mixture for about an hour, preferably in a warm location. Add 4.0 grams of ethyl acetoacetate and allow the fermenting mixture to stand for 18 hours, with stirring, at room temperature.

After this time, prepare a second warm (about 40 °C) solution of 30 g sucrose in 100 mL of tap water. Add this solution, along with 3.5 g (one package) of dry baker's yeast, to the fermenting mixture, and allow it to stir for 48 hours (with the trap attached) at room temperature.

Place about 8 g of Filter Aid (Johns-Manville Celite) in a beaker with about 20 mL of water. Stir the mixture vigorously and then pour the contents into a small Büchner funnel (with filter paper) while applying a **gentle** vacuum, as in a vacuum filtration. Be careful not to let the Filter Aid dry completely. This procedure will cause a thin layer of Filter Aid to be deposited on the filter paper. Discard the water that passes through this filter. Decant as much of the clear supernatant liquid as possible and pass it through this filter, using **very gentle** suction. Filter the remaining residue through the same filter. The extremely tiny yeast particles are trapped in the pores of the Filter Aid (Technique 4, Section 4.4, p 572). Wash the residue with 20 mL of water, allowing the water to pass into the flask containing the filtered reaction mixture. Saturate this liquid with 30 g of sodium chloride and stir the mixture vigorously for five minutes. Extract the aqueous solution with three 50-mL portions of diethyl ether using a separatory funnel (Technique 7, Section 7.7, p 627). Evaporate the ether using a sand bath in the hood and a stream of air or nitrogen to recover the liquid ester.

Prepare a small chromatography column in the following manner. Place a small plug of cotton in a 5¾-inch Pasteur pipet. Tamp the cotton to form a loose plug. Add alumina on top of the cotton plug to form a column 1 cm high. Tap the pipet with your finger to pack the alumina. Using a second Pasteur pipet, add the crude hydroxyester to the column. Rinse the remaining crude product onto the column using 1–2 mL of dichloromethane. Collect the eluted product in a 10-mL Erlenmeyer flask. Use a dropper bulb to force the liquid material through the chromatography column. Dry the organic layer over anhydrous magnesium sulfate for about 10 minutes. Decant the dried solution into a preweighed 10-mL beaker. Evaporate the solvent in a sand bath that has been heated to approximately 60 °C using a gentle stream of air or nitrogen. Weigh the beaker again in order to determine the weight of pure hydroxyester obtained.

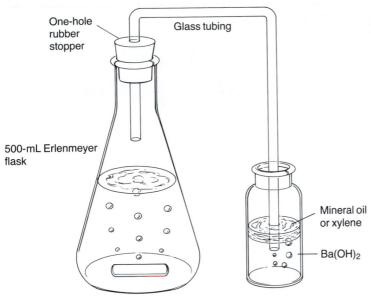

One-hole rubber stopper

Glass tubing

500-mL Erlenmeyer flask

Mineral oil or xylene

Ba(OH)$_2$

Apparatus for the fermentation of ethyl acetoacetate

Combine your product with the products of three other students in order to proceed with the polarimetry part of this experiment. Using a Pasteur pipet, transfer the hydroxyester to a pre-weighed 10-mL volumetric flask. Transfer each student's product carefully to the volumetric flask. Weigh the volumetric flask again in order to determine the concentration of the sample. Fill the volumetric flask to the mark with methylene chloride. Stopper the volumetric flask and invert it several times to mix the solution thoroughly. The concentration in grams/milliliter of this solution can be determined. Transfer the solution to a 0.5-dm polarimeter tube and determine its observed rotation. The published value for the **specific** rotation of (+)-ethyl 3-hydroxybutanoate is $[\alpha]_D^{25} = +43.5°$. Report the value of the specific rotation and the optical purity to the instructor. Calculate the percentage of **each** of the enantiomers in the sample (Technique 16, Section 16.5, p 763).

REFERENCES

Seebach, D., Sutter, M. A., Weber, R. H., and Züger, M. F. "Yeast Reduction of Ethyl Acetoacetate: (S)-(+)-Ethyl 3-Hydroxybutanoate." *Organic Syntheses, 63* (1984): 1.

Experiment 24
NMR Determination of the Optical Purity of (S)-(+)-Ethyl 3-Hydroxybutanoate

Nuclear magnetic resonance
Chemical shift reagents
Optical purity

In Experiment 23, a method for the chiral reduction of ethyl acetoacetate was given. This reduction produces a product that is predominantly the (S)-(+) enantiomer of ethyl 3-hydroxybutanoate. In this experiment, we will use NMR to determine the actual optical purity of the product. An NMR spectrum of racemic ethyl-3-hydroxybutanoate is shown on p 226. In this spectrum there is no discernible difference between the two enantiomers. The methyl hydrogens on carbons 2' and 4 appear together at about 1.25 ppm, the methylene hydrogens on carbon 2 appear at 2.4 ppm, the hydroxyl proton appears at 3.6 ppm, and the methylene hydrogens on carbon 1' and the methine hydrogen on carbon 3 appear together at about 4.2 ppm.

$$CH_3 \underset{4}{} - \underset{3}{\overset{\overset{\displaystyle OH}{|}}{CH}} - \underset{2}{CH_2} - \underset{1}{\overset{\overset{\displaystyle O}{\|}}{C}} - O - \underset{1'}{CH_2} - \underset{2'}{CH_3}$$

Although the normal spectrum shows no visible difference for the two enantiomers, there is a method that will allow the spectra of the two enantiomers to be distinguished. This method uses a chiral shift reagent. A general discussion of chemical shift reagents is found in Appendix 4, Section NMR.13. These reagents "spread out" the resonances of the compound with which they are used, increasing the chemical shifts of the protons that are nearest the center of the metal complex by the largest amount. Since the spectra of both (+)- and (−)-ethyl 3-hydroxybutanoate are identical, the usual chemical shift reagent would not help our analysis. However, if one uses a chemical shift reagent that is itself chiral, one can begin to distinguish the two separate enantiomers by their NMR spectra. The two enantiomers, which are chiral, will interact differently with the chiral shift reagent. The complexes formed from the (R) and (S) isomers and the (+)-camphor-containing shift reagent will be diastereomers. Diastereomers usually have different physical properties, and the NMR spectra are no exception. The two complexes will be formed with slightly differing geometries. Although the effect is small, it is large enough to begin to see differences in the NMR spectra of the two enantiomers. In particular, the originally superimposed methylene and methine multiplets will begin to be resolved (see upper spectrum on p 227).

The chiral shift reagent used in this experiment is tris[3-(heptafluoropropylhydroxymethylene)-(+)-camphorato]europium(III), or Eu(hfc)$_3$. In this complex, the europium is in a chiral environment because it is complexed to camphor, which is a chiral molecule.

Eu(hfc)$_3$ has the following structure:

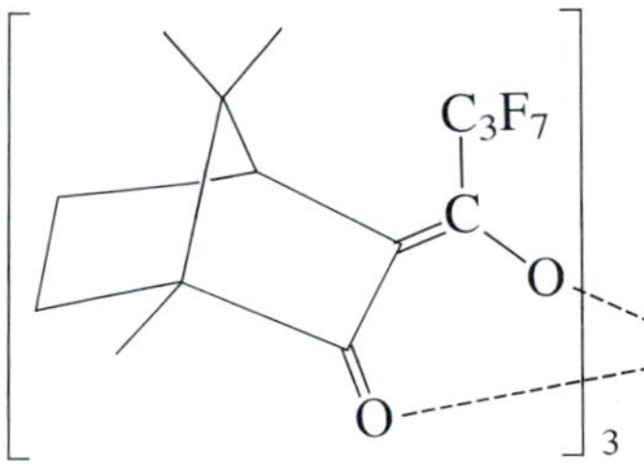

REQUIRED READING

New: Appendix 4   Nuclear Magnetic Resonance Spectroscopy

SPECIAL INSTRUCTIONS

This experiment requires the use of an NMR spectrometer. It is a short experiment which can be done in conjunction with Experiment 23.

PROCEDURE

Place approximately 0.030 g of ethyl 3-hydroxybutanoate (prepared in Experiment 23) in an NMR tube. Use a Pasteur pipet and an analytical balance to perform this operation. It is not important to weigh an exact quantity of the ester; any amount from 0.025 to 0.050 g will suffice, but you must know the exact weight.

Divide the quantity of ester that you weighed by 1.35 to determine the amount of shift reagent you will need. Using smooth weighing paper, use the analytical balance to weigh out this quantity of shift reagent. Again, it is not necessary to be perfectly exact, but you must record the amount. Carefully add this shift reagent to the NMR sample. Add a small quantity of CDCl$_3$ solvent, which contains tetramethylsilane (TMS), but do not add more than double the initial volume of the sample of ester. Allow the sample to stand for 20 minutes.

Determine the NMR spectrum of the sample. The peaks of interest are the methyl hydrogens on carbon 4 and the methylene hydrogens on carbon 1′. You should be able to see two sets of overlapping multiplets. If you do not see this pattern, you may not have added enough shift reagent, the amount of one of the enantiomers may be too small, or it may just appear as shoulders on the base of the peaks from the larger multiplet. If you wish, add a second portion of shift reagent, similar to the portion added originally.

It is a good idea to test the ability of your instrumentation by preparing a reference sample containing equal quantities of racemic ethyl 3-hydroxybutanoate and your sample obtained from Experiment 23. This sample should contain about 75% (S)-(+)-isomer and 25% (R)-(−)-isomer. As before, use about 0.030 g of this sample. In this way you can tell how well your method is working and also assign the upfield and downfield peaks to the correct enantiomer.

Determine the percentage of each isomer in both of your samples in the following manner. Compare the heights of the two inner peaks of the quartets which correspond to the methylene hydrogens on carbon 1′. Determine the ratios of the heights of these peaks. Repeat the comparison, using the two peaks of the doublets which correspond to the methyl hydrogens on carbon 4. Refer to the sample determination illustrated on the NMR spectrum shown on the bottom of page 227. Average all of the ratios that you have determined and calculate the percentages of each enantiomer from this average ratio. On a typical NMR spectrometer, this comparison-of-peak-heights method gives results that are accurate to within 2% to 3% when measuring accurately-prepared reference samples.

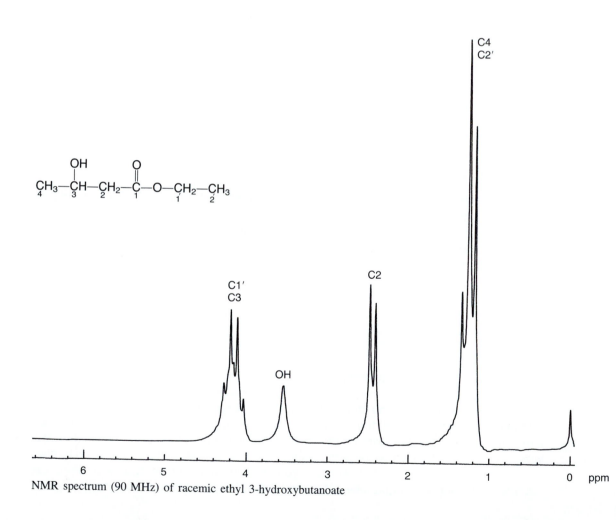

NMR spectrum (90 MHz) of racemic ethyl 3-hydroxybutanoate

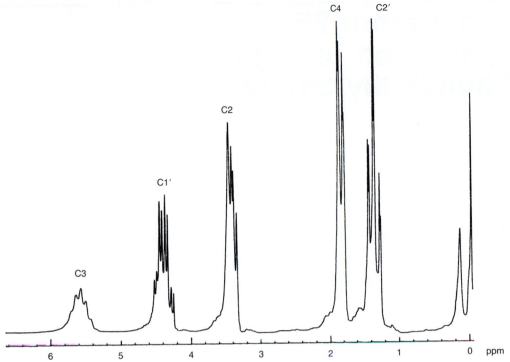

NMR spectrum (90 MHz), with chiral shift reagent, of racemic ethyl 3-hydroxybutanoate. (Note: The OH resonance is off scale.)

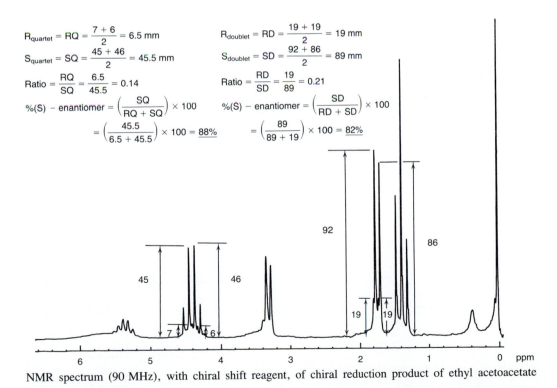

$$R_{quartet} = RQ = \frac{7+6}{2} = 6.5 \text{ mm}$$

$$S_{quartet} = SQ = \frac{45+46}{2} = 45.5 \text{ mm}$$

$$\text{Ratio} = \frac{RQ}{SQ} = \frac{6.5}{45.5} = 0.14$$

$$\%(S) - \text{enantiomer} = \left(\frac{SQ}{RQ+SQ}\right) \times 100$$

$$= \left(\frac{45.5}{6.5+45.5}\right) \times 100 = \underline{88\%}$$

$$R_{doublet} = RD = \frac{19+19}{2} = 19 \text{ mm}$$

$$S_{doublet} = SD = \frac{92+86}{2} = 89 \text{ mm}$$

$$\text{Ratio} = \frac{RD}{SD} = \frac{19}{89} = 0.21$$

$$\%(S) - \text{enantiomer} = \left(\frac{SD}{RD+SD}\right) \times 100$$

$$= \left(\frac{89}{89+19}\right) \times 100 = \underline{82\%}$$

NMR spectrum (90 MHz), with chiral shift reagent, of chiral reduction product of ethyl acetoacetate

Experiment 25
Resolution of (±)-α-Phenylethylamine

Resolution
Polarimetry
Use of a separatory funnel

Although racemic α-phenylethylamine is readily obtained commercially, it is much more difficult to obtain one of the enantiomers in optically pure form. We obtain the pure form through **resolution,** separation, of the enantiomers. In this experiment, you isolate only one of the enantiomers, the levorotatory one, since it can be more easily isolated than the dextrorotatory enantiomer. The resolving agent to be used is (+)-tartaric acid, which forms diastereomeric salts with racemic α-phenylethylamine. The important reactions for this experiment are indicated below.

Optically pure (+)-tartaric acid is abundant in nature. It is readily obtained as a by-product of wine-making. The (−)-amine-(+)-tartrate diastereomeric salt has a lower solubility than the (+)-amine-(+)-tartrate salt, and it separates from the solution as crystals. The crystals are removed by filtration and purified. The salt is then treated with dilute base to regenerate the free (−)-amine. In principle, the mother liquor, which contains mostly the (+)-amine-(+)-tartrate salt, can also be purified to yield the other diastereomeric salt. Hydrolysis of this salt would yield the (+)-amine.

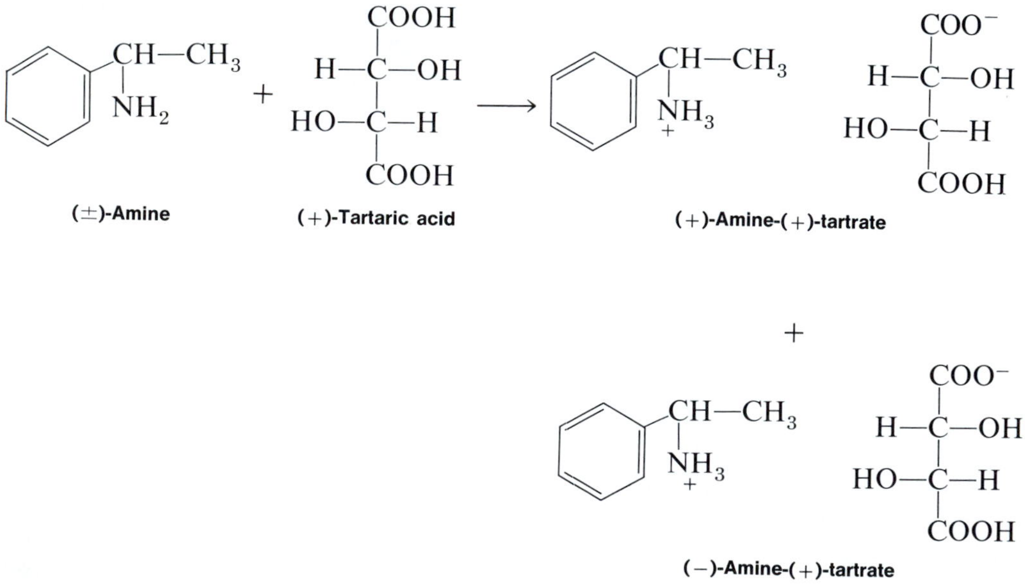

(±)-Amine   (+)-Tartaric acid   (+)-Amine-(+)-tartrate

+

(−)-Amine-(+)-tartrate

REQUIRED READING

Review: Technique 4   Section 4.3
          Technique 7   Sections 7.7 and 7.8

New: Technique 16   Polarimetry

SPECIAL INSTRUCTIONS

The observed rotation for a sample isolated by a single student may be only a few degrees, which limits the precision of the optical purity determination. Better results can be obtained if four students combine their resolved amine products for the polarimetric analysis.

PROCEDURE

> **NOTE TO THE INSTRUCTOR: This experiment is designed for students to work individually, but for four students to combine their products for polarimetry.**

Place 7.8 g of L-(+)-tartaric acid and 125 mL of methanol in a 250-mL Erlenmeyer flask. Heat this mixture on a steam bath until the solution is nearly boiling. Slowly add 6.25 g of racemic α-phenylethylamine (α-methylbenzylamine) to this hot solution.

> **NOTE: Caution should be exercised at this step, because the mixture is very likely to froth and boil over.**

Next, stopper the flask and allow it to stand overnight. The crystals that form should be **prismatic.** If needles are obtained, they should be dissolved and recrystallized; you can do this by seeding the mixture with a prismatic crystal, if one is available. Alternatively, the mixture may be heated until **most** of the solid has dissolved. The needle crystals dissolve easily and usually a small amount of the prismatic crystals remain to seed the solution. Allow the solution to cool slowly to form prismatic crystals. When needles are formed, they are not optically pure enough to give a complete resolution of the enantiomers, and they must be dissolved and the material crystallized again. Filter the crystals, using a Büchner funnel (Technique 4, Section 4.3, and Figure 4–6, p 571), and wash them with a few portions of cold methanol.

Partially dissolve the amine-tartrate salt in 25 mL of water, add 4 mL of 50% sodium hydroxide, and extract this mixture with three 10-mL portions of diethyl ether using a separatory funnel (Technique 7, Section 7.7, p 627). Dry the organic layer over about one gram of anhydrous magnesium sulfate for about 10 minutes. Decant the dried solution into a 50-mL beaker and evaporate the ether on a warm sand bath in the hood. A gentle stream of nitrogen or air can be directed into the beaker to increase the rate of evaporation. Using a small amount of ether, rinse the residue remaining in the beaker into a 10-mL beaker that has been previously weighed to within ± 0.003 g. Evaporate the solvent in a sand bath that has been heated to approximately 60 °C. Weigh the beaker again in order to determine the weight of pure amine obtained.

Combine your product with the products of three other students in order to proceed with the polarimetry part of this experiment. Using a Pasteur pipet, transfer the amine to a preweighed 10 mL volumetric flask. Transfer each student's product carefully to the volumetric flask. Weigh the volumetric flask again in order to determine the concentration of the sample. Fill the volumetric flask to the mark with absolute methanol and mix thoroughly. The concentration in grams/milliliter of this solution can be determined. Transfer the solution to a 0.5-dm polarimeter tube and determine its observed rotation. The published value for the **specific** rotation is $[\alpha]_D^{22} = -40.3°$. Be sure to keep the pure amine tightly stoppered. Report the value of the specific rotation and the optical purity to the instructor. Calculate the percentage of **each** of the enantiomers in the sample (Technique 16, Section 16.5, p 763).

REFERENCES

Ault, A. "Resolution of D,L-α-Phenylethylamine." *Journal of Chemical Education, 42* (1965): 269.
Jacobus, J., and Raban, M. "An NMR Determination of Optical Purity." *Journal of Chemical Education, 46* (1969): 351.

QUESTIONS

1. Using a reference textbook, find examples of reagents used in performing chemical resolutions of acidic, basic, and neutral racemic compounds.

2. Propose methods of resolving each of the following racemic compounds:

(a)

$$CH_3-CH-\overset{\overset{\displaystyle O}{\|}}{C}-OH$$
$$\underset{Br}{|}$$

(b)

(c)

Essay
DETECTION OF ALCOHOL: THE BREATHALYZER

If one places organic compounds on a scale ranking their extent of oxidation, a general order such as

$$R—CH_3 < R—CH_2OH < R—CHO \text{ (or } R_2CO) < R—COOH < CO_2$$

is obtained. According to this scale, you can see that alcohols represent a relatively reduced form of organic compound, while carbonyl compounds and carboxylic acid derivatives represent highly oxidized structures. Using appropriate oxidizing agents, it should be possible to oxidize an alcohol to an aldehyde, a ketone, or a carboxylic acid depending on the substrate and the oxidation conditions.

Primary alcohols can be oxidized to aldehydes by various oxidizing agents, including potassium permanganate, potassium dichromate, and nitric acid:

$$R—CH_2OH \xrightarrow{[O]} \left[R—\overset{\overset{\displaystyle O}{\|}}{C}—H \right] \xrightarrow{[O]} R—\overset{\overset{\displaystyle O}{\|}}{C}—OH$$

The aldehyde formed in this oxidation is unstable relative to further oxidation, and consequently the aldehyde is usually oxidized further to the corresponding carboxylic acid. The aldehyde is seldom isolated from such an oxidation, unless the oxidizing agent is relatively mild.

Chromium(VI) is a very useful oxidizing agent. It appears in various chemical forms, including chromium trioxide, CrO_3, chromate ion, CrO_4^{2-}, and dichromate ion, $Cr_2O_7^{2-}$. The chromium(VI) compounds are typically red to yellow. During the oxidation, they are reduced to Cr^{3+}, which is green. As a result, an oxidation reaction can be monitored by the color change. A typical chromium(VI) oxidation to illustrate the role of both the oxidizing and the reducing species is the dichromate oxidation of ethanol to acetaldehyde:

$$3CH_3CH_2OH + Cr_2O_7^{2-} + 8H^+ \longrightarrow 3CH_3—\overset{\overset{\displaystyle O}{\|}}{C}—H + 2Cr^{3+} + 7H_2O$$

Because the aldehyde is also susceptible to oxidation, a second oxidation step of acetaldehyde to acetic acid can also take place:

$$3CH_3—\overset{\overset{\displaystyle O}{\|}}{C}—H + Cr_2O_7^{2-} + 8H^+ \longrightarrow 3CH_3—\overset{\overset{\displaystyle O}{\|}}{C}—OH + 2Cr^{3+} + 4H_2O$$

231

This oxidation reaction of alcohols by dichromate ion leads to a standard method of analysis for alcohols. The material to be tested is treated with acidic potassium dichromate solution, and the green chromic ion formed in the oxidation of the alcohol is measured spectrophotometrically by measuring the amount of light absorbed at 600 nm. By this method, it is possible indirectly to determine from 1 to 10 mg of ethanol per liter of blood with an accuracy of 5%. The alcohol content of beer can be determined within 1.4% accuracy.

THE BREATHALYZER

An interesting application of the oxidation of alcohols is in a quantitative method of determining the amount of ethanol in the blood of a person who has been drinking. The ethanol contained in alcoholic beverages may be oxidized by dichromate according to the equation shown on page 231. During this oxidation, the color of the chromium-containing reagent changes from reddish orange ($Cr_2O_7^{2-}$) to green (Cr^{3+}). Law-enforcement officials use the color change in this reaction to estimate the alcohol content of the breath of suspected drunken drivers. This value can be converted to an alcohol content of the blood.

In most states, the usual legal definition of being under the influence of alcohol is based on a 0.10% alcohol content in the blood. Because the air deep within the lungs is in equilibrium with the blood passing through the pulmonary arteries, the amount of alcohol in the blood can be determined by measuring the alcohol content of the breath. The proper breath-blood ratio can be determined by simultaneous blood and breath tests. As a result of this equilibration, police officers do not need to be trained to administer blood tests. Instead, a simple instrument, a breath analyzer, which does not require any particular sophistication for its operation, can be used in the field.

In the simplest form, a breath analyzer contains a potassium dichromate–sulfuric acid reagent impregnated on particles of silica gel in a sealed glass ampoule. Before the instrument is to be used, the ends of the ampoule are broken off, and one end is fitted with a mouthpiece while the other is attached to the neck of an empty plastic bag. The person being tested blows into the tube to inflate the plastic bag. As air containing ethanol passes through the tube, a chemical reaction takes place, and the reddish-orange dichromate reagent is reduced to the green chromium sulfate, Cr^{3+}. When the green color extends beyond a certain point along the tube (the halfway point), it is determined that the motorist has a relatively high alcohol concentration in his breath, and he is usually taken to the police station for more precise tests. The device described here is simple, and its precision is not high. It is used primarily as a **screening** device for suspected drunken drivers. An example of this simple device is shown in Figure 1.

A more precise instrument, the "Breathalyzer," is shown in Figure 2. Air is blown into a cylinder A, whereupon a piston is raised. When the cylinder is full, the piston is allowed to fall and pump the measured volume of breath through a reaction ampoule B containing the potassium dichromate solution in sulfuric acid. As the alcohol-laden air is bubbled through this solution, the alcohol is oxidized to acetaldehyde

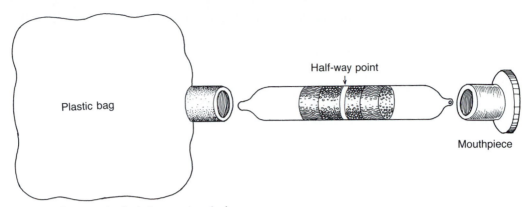

FIGURE 1. Breath alcohol screening device

and further to acetic acid, while the dichromate ion is reduced to Cr^{3+}. The instrument contains a light source C. Filters are used to select light in the blue region of the spectrum. This blue light passes through the reaction ampoule and is detected by a photocell D. The light also passes through a sealed standard reference ampoule E, which contains exactly the same concentration of potassium dichromate in sulfuric acid as the reaction ampoule B had originally. No alcohol is allowed to enter this reference ampoule. The light passing through the reference ampoule is detected by another photocell. A meter F, calibrated in milligrams of ethanol per 100 mL of blood, or in percentage of blood alcohol, registers the difference in voltages between the two photocells. Before the test, both ampoules transmit blue light to the same extent, so the meter

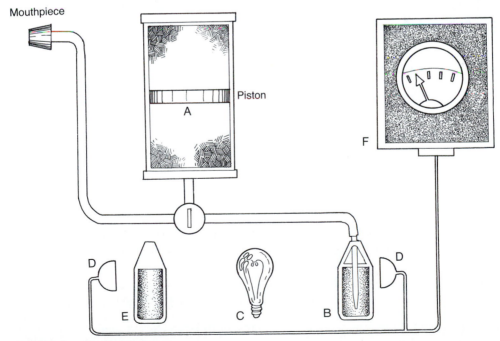

FIGURE 2. The Breathalyzer

reads zero. After the test, the reaction ampoule transmits more blue light than the reference ampoule, and a voltage is registered on the meter.

Such an instrument, while more complicated and more delicate than the simple device shown in Figure 1, can be used in the field without a support laboratory. The instrument is portable, permitting it to be easily transported in the trunk of a patrol car.

A similar method is used in Experiment 26 to follow the rate of oxidation of several alcohols by dichromate ion. The color change that accompanies the oxidation is monitored by a spectrophotometer.

REFERENCES

Denney, R. C. "Analysing for Alcohol." *Chemistry in Britain, 6* (1970): 533.

Treptow, R. S. "Determination of Alcohol in Breath for Law Enforcement." *Journal of Chemical Education, 51* (1974): 651.

Lovell, W. S. "Breath Tests for Determining Alcohol in the Blood." *Science, 178* (1972): 264.

Symposium on Breath Alcohol Tests. *Journal of Forensic Science, 5* (1960): 395.

Timmer, W. C. "An Experiment in Forensic Chemistry—The Breathalyzer." *Journal of Chemical Education, 63* (October 1986): 897.

Experiment 26
Chromic Acid Oxidation
of Alcohols

Chromic acid oxidation of an alcohol
Kinetics
Ultraviolet-visible spectrophotometry

The chemical reaction of interest in this experiment is the oxidation of an alcohol to the corresponding aldehyde by an acidic solution of potassium dichromate:

$$3RCH_2OH + Cr_2O_7^{2-} + 8H^+ \longrightarrow 3R\overset{\overset{\displaystyle O}{\|}}{C}-H + 2Cr^{3+} + 7H_2O$$

Normally, the aldehyde formed is also susceptible to oxidation by the dichromate ion, yielding the corresponding acid:

$$3R\overset{\overset{\displaystyle O}{\|}}{C}-H + Cr_2O_7^{2-} + 8H^+ \longrightarrow 3R\overset{\overset{\displaystyle O}{\|}}{C}-OH + 2Cr^{3+} + 4H_2O$$

In this experiment, however, the alcohol is present in large excess, and the likelihood that the second reaction will take place is thereby greatly reduced. A secondary alcohol is oxidized to a ketone by a similar process. A ketone is not easily oxidized further by the dichromate reagent.

$$3R-\underset{\underset{R'}{|}}{C}-OH + Cr_2O_7^{2-} + 8H^+ \longrightarrow 3R-\overset{\overset{O}{\|}}{C}-R' + 2Cr^{3+} + 7H_2O$$

Although various mechanisms have been proposed to explain how dichromate ion oxidizes alcohols, the most commonly accepted mechanism is the one F. H. Westheimer first proposed in 1949. In acid solution, dichromate ion forms two molecules of chromic acid, H_2CrO_4:

$$2H_3O^+ + Cr_2O_7^{2-} \overset{fast}{\rightleftharpoons} 2H_2CrO_4 + H_2O \tag{1}$$

The chromic acid, in a rapid, reversible step, forms a chromate ester with the alcohol:

$$RCH_2OH + H_2CrO_4 \overset{fast}{\rightleftharpoons} RCH_2-O-\overset{\overset{O}{\|}}{\underset{\underset{O}{\|}}{Cr}}-OH + H_2O \tag{2}$$

The chromate ester undergoes a rate-determining decomposition by a two-electron transfer with cleavage of the α-carbon-hydrogen bond, as seen in Step 3.

$$R-\underset{\underset{H}{|}}{\overset{\overset{H}{|}}{C}}-O-\underset{\underset{O}{\|}}{\overset{\overset{O}{\|}}{Cr}}-OH \overset{slow}{\longrightarrow} R-\underset{\underset{H}{|}}{C}=O + H_2CrO_3 \tag{3}$$

Cr in +6   Cr in +4
oxidation state   oxidation state

The H_2CrO_3 is further reduced to Cr^{3+} by interaction with chromium in various oxidation states and by further interaction with the alcohol. All these subsequent steps are very rapid relative to Step 3. Consequently they are not involved in the rate-determining step of the mechanism and need not be considered further.

The rate-determining step, Step 3, involves only one molecule of the chromate ester, which in turn arises from a prior equilibrium involving the combination of one molecule of alcohol and one molecule of chromic acid (Step 2). As a result, this reaction, which is first-order in chromate ester, turns out to be *second-order* for the reacting alcohol and the dichromate reagent. The kinetic equation therefore is

$$-\frac{d[Cr_2O_7^{2-}]}{dt} = k[RCH_2OH][Cr_2O_7^{2-}]$$

The presence of the chromium atom strongly affects the distribution of electrons in the remainder of the chromate ester molecule. Electrons need to be transferred to the chromium atom during the cleavage step. If the R group includes an electron-withdrawing group, it diminishes the necessary electron density needed for reaction. Consequently, the reaction proceeds more slowly. An electron-releasing group would be expected to have the opposite effect.

THE EXPERIMENTAL METHOD

The rate of the reaction is measured by following the rate of disappearance of the dichromate ion as a function of time. The dichromate ion, $Cr_2O_7^{2-}$, is yellow orange, absorbing light at 350 and 440 nm. The chromium is reduced to the green Cr^{3+} during the reaction. The ion Cr^{3+} does not absorb light significantly at 350 or 440 nm, but rather at 406, 574, and 666 nm. Therefore, if we measure the amount of light absorbed at a single wavelength, such as 440 nm, we can follow the rate of disappearance of dichromate ion without any interfering absorption of light by the ion Cr^{3+}.

The instrument used to measure the amount of light absorbed at a particular wavelength, when that light lies within the visible region of the electromagnetic spectrum, is a **colorimeter.** The absorption can also be measured using an ultra-violet spectrophotometer. These instruments can be described simply. Ordinary visible light is passed through the sample and then through a prism, where the light of the particular wavelength being studied is selected. This selected light is directed against a photocell, where its intensity is measured. A meter provides a visible display of the intensity of the light of the desired wavelength.

The true rate equation for this reaction is second-order. However, because a large excess of alcohol will be used, its concentration will change imperceptibly during the reaction. The rate equation, under these conditions, will simplify to that of a pseudo first-order reaction. The mathematics involved will become much simpler as a result.

The rate equation for a first-order (or a pseudo first-order) reaction is

$$-\frac{d[A]}{dt} = k[A]$$

In this experiment, the rate equation becomes

$$-\frac{d[Cr_2O_7^{2-}]}{dt} = k[Cr_2O_7^{2-}]$$

Let a equal the initial concentration of dichromate ion. At some time t, an amount x moles/L of dichromate will have undergone reaction, and x moles/L of aldehyde will have been produced. The remaining concentration of dichromate at that value of time equals $a - x$. The rate equation becomes

$$+\frac{dx}{dt} = k(a - x)$$

Integration provides

$$\ln\left(\frac{a}{a - x}\right) = kt$$

Converting to base 10 logarithms gives

$$2.303 \log\left(\frac{a}{a - x}\right) = kt$$

This equation is of the form appropriate for a straight line with intercept equal to zero. If the reaction is indeed first-order, a plot of $\log (a/a - x)$ versus t will provide a straight line whose slope is $k/2.303$.

Since it is experimentally difficult to measure directly how much dichromate ion is consumed during this reaction, evaluating the term $a/a - x$ requires an indirect approach. What is needed is some measurable quantity from which the concentration of dichromate can be derived. Such a quantity is the amount of light absorbed by the solution at wavelength 440 nm.

The Beer-Lambert law relates the amount of light absorbed by a molecule or an ion to its concentration, according to the equation

$$A = \epsilon c l$$

where A is the absorbance of the solution, ϵ is the molar absorptivity (a measure of the efficiency with which the sample absorbs the light), c is the concentration of the solution, and l is the path length of the cell in which the solution is contained. The absorbance is read by the spectrophotometer.

At the initial concentration a of dichromate ion, we may write

$$A_0 = \epsilon a l \qquad \text{or} \qquad a = \frac{A_0}{\epsilon l}$$

The amount of dichromate ion remaining unreacted at any particular time t, which equals $a - x$, becomes

$$A_t = \epsilon(a - x)l \qquad \text{or} \qquad a - x = \frac{A_t}{\epsilon l}$$

Substituting absorbance values for concentrations and cancelling provides

$$\left(\frac{a}{a - x}\right) = \frac{A_0}{A_t}\frac{\epsilon l}{\epsilon l} \qquad \text{or} \qquad \left(\frac{a}{a - x}\right) = \frac{A_0}{A_t}$$

At this point a correction must be introduced. When the reaction reaches completion, at "infinite" time, a certain degree of absorption of 440-nm light remains. In other words, at time $t = \infty$, the value A_∞ does not equal zero. Therefore this residual absorbance must be subtracted from each of the absorbance terms written above. The difference $A_0 - A_\infty$ gives the actual amount of dichromate ion present initially, and the difference $A_t - A_\infty$ gives the actual amount of dichromate ion remaining unreacted at a value t of time. Introducing these corrections, we have

$$\left(\frac{a}{a - x}\right) = \frac{A_0 - A_\infty}{A_t - A_\infty}$$

The integrated rate equation becomes

$$2.303 \log \left(\frac{A_0 - A_\infty}{A_t - A_\infty} \right) = kt$$

Since the dimensions of the cell and the molar absorptivity cancel out of this equation, it is not necessary to have any particular knowledge of these parameters. A plot of $\log \left(\frac{A_0 - A_\infty}{A_t - A_\infty} \right)$ versus time will provide a straight line whose slope equals $k/2.303$. The slope is determined as shown on the graph. If time is measured in minutes, the units of k are $\min^{-1}$. The experimental points plotted on the graph may contain a certain amount of scatter, but the line drawn is the best **straight** line (use some mathematical method such as the method of averages or of least squares).

One other value often cited in kinetic studies is the **half-life, τ,** of the reaction. The half-life is the time required for one-half the reactant to undergo conversion to products. During the first half-life, 50% of the available reactant is consumed. At the end of the second half-life, 75% of the reactant has been consumed. For a first-order reaction, the half-life is calculated by

$$\tau = \frac{\ln 2}{k} = \frac{0.69315}{k}$$

The class will study several alcohols in this experiment. The class data will be compared in determining the relative reactivities of the alcohols. Two particular alcohols, 2-methoxyethanol and 2-chloroethanol, react more slowly than the other alcohols used in this experiment. In spite of this lower reactivity, the reactions will not be followed for more than a few minutes, because in these compounds, the second reaction—the oxidation of the aldehyde product to the corresponding carboxylic acid—becomes more important over longer periods. As a result of this second reaction, dichromate ion becomes consumed more rapidly than the calculations would suggest,

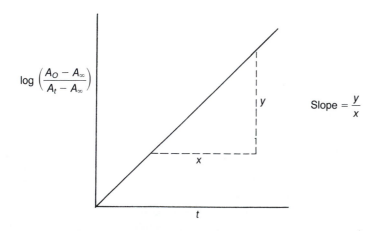

and the graph of $\log \left(\dfrac{A_0 - A_\infty}{A_t - A_\infty} \right)$ versus time becomes curved. So that this complication is avoided, only the first few minutes of the reaction are used to calculate an initial reaction rate, which corresponds to the reaction being studied in this experiment. The other alcohols are sufficiently reactive that the second reaction does not introduce any significant error.

REQUIRED READING

Review: Read the sections on kinetics in your lecture textbook.

New: Essay   Detection of Alcohol: the Breathalyzer

SPECIAL INSTRUCTIONS

Primary and secondary alcohols are oxidized in this experiment to aldehydes and ketones, respectively. The experimental procedure is identical for both types of alcohols. The procedure described in this experiment is based on the controls found on a typical ultraviolet-visible spectrophotometer. Your instructor will need to show you how the specific controls must be adjusted on your instrument. This experiment can also be conducted using a colorimeter. Each kinetic run, including the temperature equilibration, requires from one and one-half to two hours, although most of that time is not involved in actually using the spectrophotometer.

This experiment involves using an acidic solution of potassium dichromate. Potassium dichromate solutions have been determined to be potential carcinogens. In this experiment, the dichromate solution will be prepared as a stock solution for the entire class to use. This stock solution should be stored in a hood. Students should wear gloves and use pipet bulbs when using this stock solution. Dispose of all chromium waste in an appropriate waste container.

PROCEDURE

Select an alcohol from one of the following: ethanol, 1-propanol, 2-propanol, 2-methoxyethanol, 2-chloroethanol, ethylene glycol, and 1-phenylethanol (methylbenzyl alcohol). A stock solution of $3.9M$ sulfuric acid and a carefully prepared solution of $0.0196M$ potassium dichromate solution (prepared using distilled water in a volumetric flask) should be available for the entire class to use.

Turn on the instrument and allow it to warm up. Select the tungsten lamp as the light source. Select an operating mode that allows the instrument to operate at a fixed wavelength of **440 nm** and to record data as **absorbance.**

Using a small flask, prepare the test solution by transferring 1 mL of the stock dichromate solution and 10 mL of the stock sulfuric acid solution by pipet (**USE A PIPET**

BULB) into the flask. Shake the solution well. Rinse a sample cuvette three times with this acidic dichromate solution and then fill the cell. Wipe the cuvette clean and dry. Place the cuvette into the sample cell compartment, and place a cuvette filled with distilled water into the reference cell compartment. Close the cell compartment lid and allow the chromic acid solution to reach temperature equilibrium by allowing it to remain in the instrument (with the instrument running) for 20 minutes. This preheating minimizes the problem of the solution being slowly heated during the experiment by the tungsten lamp, which is the light source in the spectrophotometer. Such heating would tend to accelerate the reaction as time passed.

At the end of the preheating, record the absorbance A_0 of the chromic acid solution, along with a time value of 0.0 minutes. Withdraw a 10.0-μL sample of the alcohol being studied into a hypodermic syringe and transfer it rapidly to the chromic acid solution. As the transfer is made, start the timer. Withdraw the sample cuvette from the cell compartment, shake it vigorously for 20 to 30 seconds, and return it to the cell compartment. Be sure to wipe off the cell again. Close the compartment lid. The measurements can now be started.

Take readings of the absorbance A_t and the corresponding time at one-minute intervals over a six-minute period (eight minutes for 2-propanol). At the end of this time, remove the cuvette from the cell compartment of the spectrophotometer and allow the solution in the cuvette to stand undisturbed for at least an hour. After this period, turn on the instrument as before and allow it to warm up. Return the cuvette to the cell compartment and read the absorbance value. This final value corresponds to "infinite" time (A_∞).

The instructor may require each student to perform a duplicate run. If so, repeat the experiment under precisely the same conditions used for the first run.

Plot the data according to the method described in the introductory section of the experiment. Report the value of each rate constant determined in this experiment (and the average of the rate constants, if duplicate determinations were made). Also, report the value of the half-life, τ. Include all data and graphs in the report. A table of sample data is shown. The results from the entire class may be compared, at the option of the instructor.

Oxidation of Ethanol

TIME (min)	ABSORBANCE (440 nm)	$A_t - A_\infty$	$\dfrac{A_0 - A_\infty}{A_t - A_\infty}$	$\log\left(\dfrac{A_0 - A_\infty}{A_t - A_\infty}\right)$
0.0	0.630	0.578	1.000	0.000
1.0	0.535	0.483	1.197	0.078
2.0	0.440	0.388	1.490	0.173
3.0	0.365	0.313	1.847	0.266
4.0	0.298	0.246	2.350	0.371
5.0	0.247	0.195	2.964	0.472
6.0	0.202	0.150	3.853	0.586
66.0 (∞)	0.052	0.000	. . .	. . .

REFERENCES

Lanes, R. M., and Lee, D. G. "Chromic Acid Oxidation of Alcohols." *Journal of Chemical Education,* 45 (1968): 269.

Westheimer, F. H. "The Mechanisms of Chromic Acid Oxidations." *Chemical Reviews, 45* (1949): 419.

Westheimer, F. H., and Nicolaides, N. "Kinetics of the Oxidation of 2-Deuterio-2-propanol by Chromic Acid." *Journal of the American Chemical Society, 71* (1949): 25.

Pavia, D. L., Lampman, G. M., and Kriz, G. S., Jr. *Introduction to Spectroscopy: A Guide for Students of Organic Chemistry*. Philadelphia: W. B. Saunders, 1979. Chap. 5.

QUESTIONS

1. Plot the data given in the table. Determine the rate constant and the half-life for this example.

2. Using data collected by the class, compare the relative rates of ethanol, 1-propanol, and 2-methoxyethanol. Explain the observed order of reactivities in terms of the mechanism of the oxidation reaction.

3. Using data collected by the class, compare the relative rates of 1-propanol and 2-propanol. Account for any differences that might be observed.

4. Balance the following oxidation-reduction reactions:

(a) $HO{-}CH_2CH_2{-}OH + K_2Cr_2O_7 \xrightarrow{H_2SO_4} H{-}\overset{O}{\overset{\|}{C}}{-}\overset{O}{\overset{\|}{C}}{-}H + Cr^{3+}$

(b) 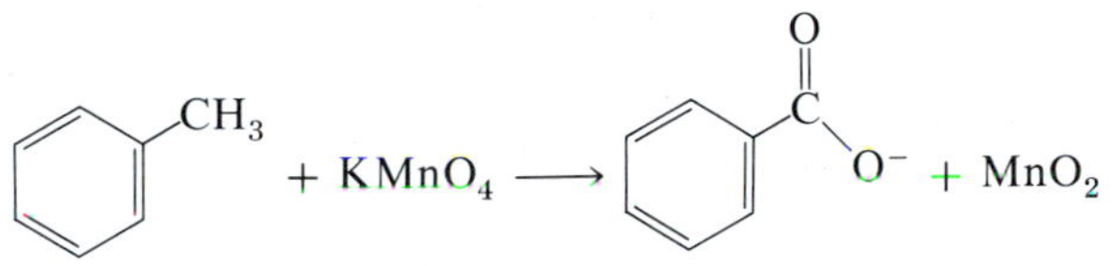 $+ KMnO_4 \longrightarrow$ $+ MnO_2$

(c) $HO{-}CH_2CH_2CH_2CH_2{-}OH + K_2Cr_2O_7 \xrightarrow{H_2SO_4} HO{-}\overset{O}{\overset{\|}{C}}{-}CH_2CH_2{-}\overset{O}{\overset{\|}{C}}{-}H + Cr^{3+}$

(d) $CH_3CH_2CH_2CH{=}CH_2 + KMnO_4 \xrightarrow{KOH} CH_3CH_2CH_2\overset{O}{\overset{\|}{C}}{-}O^- + CO_2 + MnO_2$

(e) 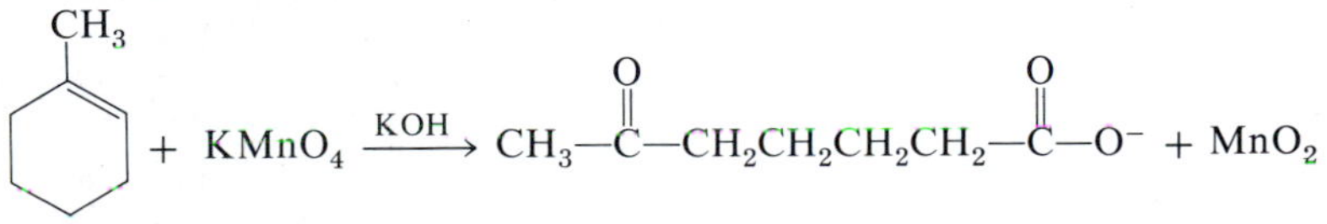 $+ KMnO_4 \xrightarrow{KOH} CH_3{-}\overset{O}{\overset{\|}{C}}{-}CH_2CH_2CH_2CH_2{-}\overset{O}{\overset{\|}{C}}{-}O^- + MnO_2$

Experiment 27
Triphenylmethanol and Benzoic Acid

Grignard reactions
Extraction
Crystallization

In this experiment a Grignard reagent, organomagnesium reagent, is prepared. The reagent is phenylmagnesium bromide.

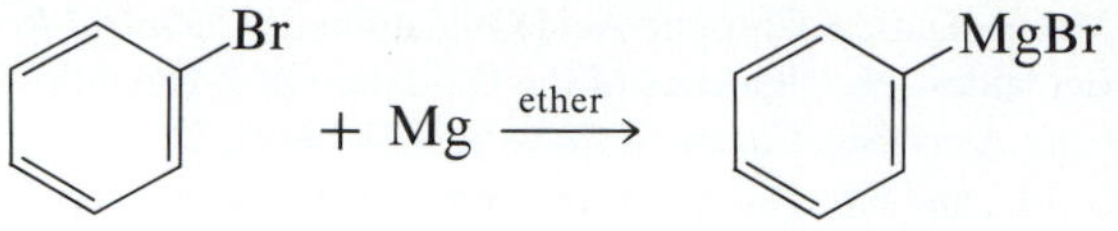

Bromobenzene   **Phenylmagnesium bromide**

This reagent will be converted to a tertiary alcohol or a carboxylic acid, depending on the procedure selected.

PROCEDURE 27A

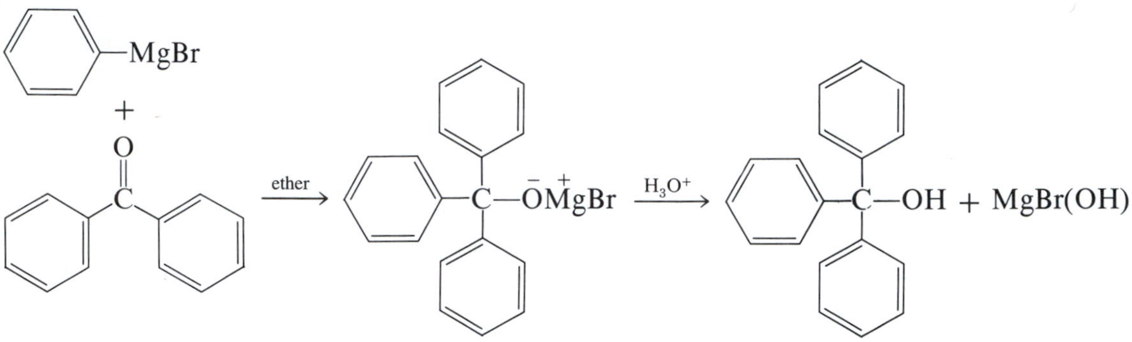

Benzophenone   **Triphenylmethanol**

PROCEDURE 27B

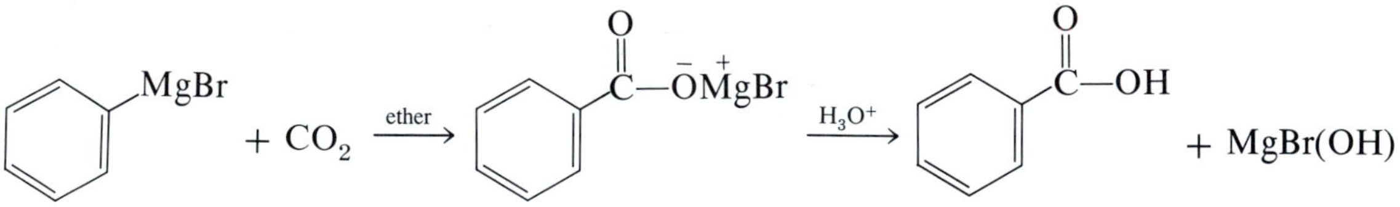

Benzoic acid

The alkyl portion of the Grignard reagent behaves as if it had the characteristics of a **carbanion.** We may write the structure of the reagent as a partially ionic compound: $^{\delta-}R\cdots^{\delta+}MgX$. This partially bonded carbanion is a Lewis base. It reacts with acids, as you would expect, to give an alkane:

$$\overset{\delta-}{R}\cdots\overset{\delta+}{Mg}X + HX \longrightarrow R\!-\!H + MgX_2$$

Any compound with a suitably acidic hydrogen will donate a proton to destroy the reagent. Water, alcohols, terminal acetylenes, phenols, and carboxylic acids are all acidic enough to bring about this reaction.

The Grignard reagent also functions as a good nucleophile in nucleophilic addition reactions of the carbonyl group. The carbonyl group has electrophilic character at its carbon atom (due to resonance), and a good nucleophile seeks out this center for addition.

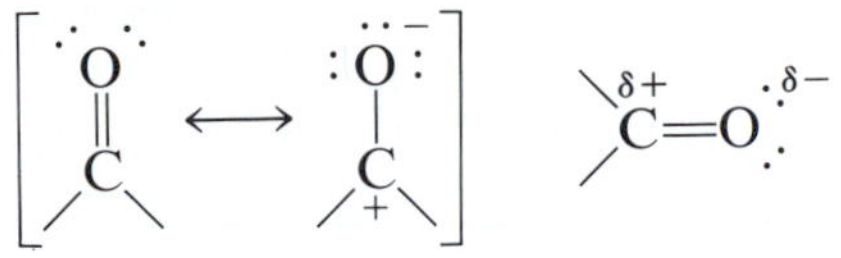

The magnesium salts produced form a complex with the addition product, an alkoxide salt, and in a second step of the reaction, these must be protonated by addition of dilute aqueous acid:

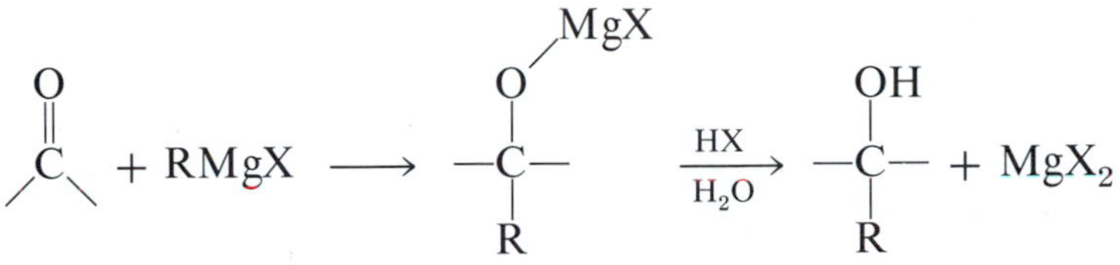

Step 1   **Step 2**

The Grignard reaction is used synthetically to prepare secondary alcohols from aldehydes and tertiary alcohols from ketones. The Grignard reagent will react with esters twice to give tertiary alcohols. Synthetically, it also can be allowed to react with carbon dioxide to give carboxylic acids and with oxygen to give hydroperoxides:

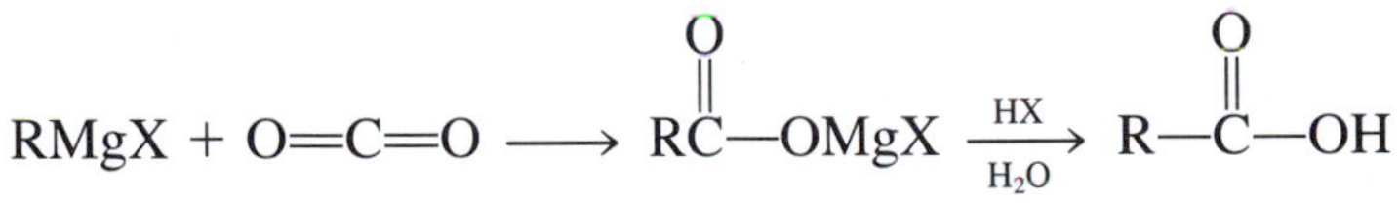

$$\text{RMgX} + \text{O}{=}\text{C}{=}\text{O} \longrightarrow \text{RC—OMgX} \xrightarrow[\text{H}_2\text{O}]{\text{HX}} \text{R—C—OH}$$

Carboxylic acid

$$\text{RMgX} + \text{O}_2 \longrightarrow \text{ROOMgX} \xrightarrow[\text{H}_2\text{O}]{\text{HX}} \text{ROOH}$$

Hydroperoxide

Because the Grignard reagent reacts with water, carbon dioxide, and oxygen, it must be protected from air and moisture when it is used. The apparatus in which the reaction is to be conducted must be scrupulously dry (recall that 18 mL of H_2O is one mole), and the solvent must be free of water, or anhydrous. During the reaction, the flask must be protected by a calcium chloride drying tube. Oxygen should also be excluded. In practice this can be done by allowing the solvent ether to reflux. This blanket of solvent vapor keeps air from the surface of the reaction mixture.

In the experiment described here, the principal impurity is **biphenyl,** which is formed by a heat- or light-catalyzed coupling reaction of the Grignard reagent and unreacted bromobenzene. A high reaction temperature favors the formation of this product. Biphenyl is highly soluble in petroleum ether, and it is easily separated from triphenylmethanol. Biphenyl can be separated from benzoic acid by extraction.

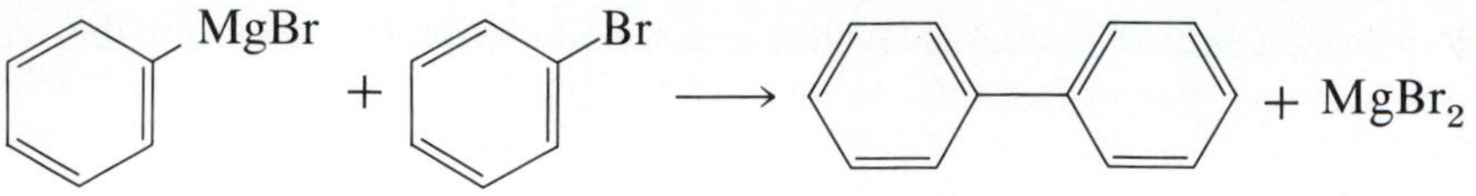

$$\text{C}_6\text{H}_5\text{MgBr} + \text{C}_6\text{H}_5\text{Br} \longrightarrow \text{biphenyl} + \text{MgBr}_2$$

Biphenyl

REQUIRED READING

Review: Technique 4   Sections 4.3 and 4.7
         Technique 5   Section 5.4
         Technique 7   Sections 7.5, 7.8, and 7.10
         Technique 18   Section 18.4

SPECIAL INSTRUCTIONS

This experiment must be conducted in one laboratory period either to the point after which benzophenone is added (Procedure 27A) or the point after which the Grignard reagent is poured over dry ice (Procedure 27B). The Grignard reagent cannot be stored. This reaction involves the use of diethyl ether, which is extremely flammable. Be certain that no open flames of any sort are in your vicinity when you are using ether.

During this experiment you will need to use **anhydrous** diethyl ether which is usually contained in metal cans with a screw cap. You are instructed in the experiment to transfer a small portion of this solvent to a conical vial. Be certain to minimize exposure to atmospheric water. Always recap the container after use. Solvent grade ether must not be used because it contains water.

All students will prepare the Grignard reagent, phenylmagnesium bromide. At the option of the instructor, you should proceed to either Procedure 27A (triphenylmethanol) or Procedure 27B (benzoic acid).

PROCEDURE

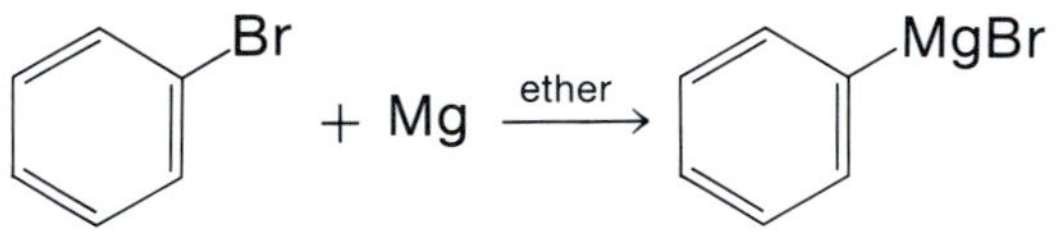

$$\text{C}_6\text{H}_5\text{Br} + \text{Mg} \xrightarrow{\text{ether}} \text{C}_6\text{H}_5\text{MgBr}$$

PREPARATION OF THE GRIGNARD REAGENT: PHENYLMAGNESIUM BROMIDE

All glassware used in a Grignard reaction must be **scrupulously** dried. Surprisingly large amounts of water adhere to the walls of glassware, even glassware that is apparently dry. Dry all of the pieces of **glassware** shown in the figure in an oven at 110 °C for at least 30 minutes. Prepare a drying tube with anhydrous calcium chloride and place it in the oven. In addition, dry the following: two 3-mL and one 5-mL conical vials that will be needed for solutions and solvents, and a calibrated Pasteur pipet (0.5-mL and 1.0-mL calibration marks) for use in dispensing ether.

> **Do not place any plasticware or plastic connectors in the oven as they may melt or soften. This also applies to a plastic syringe. If the syringe is glass, you may dry it in the oven. Check with your instructor if in doubt.**

If there are visible signs of water in the apparatus you must first dry the glassware and any plastic connectors by rinsing them with acetone and air drying them before placing them in the oven.

Assemble the apparatus as shown in the figure as soon as the glassware has cooled somewhat. Obtain a 4.5-cm length of magnesium ribbon and scrape both sides of the ribbon with the side of a spatula to remove any oxide coating that may be present.

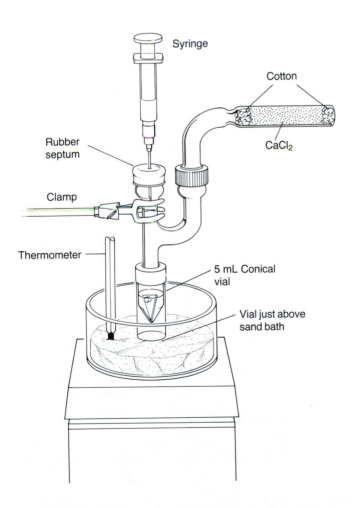

Syringe

Cotton

Rubber septum

CaCl$_2$

Clamp

Thermometer

5 mL Conical vial

Vial just above sand bath

Weigh the strip of magnesium and trim both ends until you have a weight of about 0.037 g (record the actual weight in your notebook). While holding the ribbon with tweezers, cut the strip into about 2-mm sections with sharp scissors and allow the pieces to fall directly into a **dry** beaker. Remove the 5-mL conical vial from the apparatus and transfer the magnesium to the vial. Place a dry magnetic spin vane into the vial and reassemble the apparatus.

Transfer about 4 mL of **anhydrous diethyl ether** into a **dry** 5-mL conical vial and cap the vial. Use this vial to store your dry ether during the course of this experiment. During the experiment, remove the ether from this vial with a dry calibrated Pasteur pipet.

Place 0.17 mL of bromobenzene (MW = 157.0) into a preweighed 3-mL conical vial and determine the weight of the material transferred. Add 1.0 mL of anhydrous ether to the vial. After the bromobenzene dissolves, draw this solution into the syringe and cap the vial that you used for the bromobenzene solution for later use. After inserting the syringe needle through the rubber septum, add 0.2 mL of the bromobenzene solution to the magnesium in the vial. Position the apparatus just above the sand bath and stir the mixture gently to avoid throwing the magnesium onto the side of the vial. You should begin to notice the evolution of bubbles from the surface of the metal that signals that the reaction is starting. It will probably be necessary to heat the mixture using a warm sand bath (about 60 °C) to start the reaction. Since ether has a low boiling point (35 °C), it may be sufficient to heat the vial by placing it just above the warm sand bath. Check to see if the bubbling action continues after the apparatus is removed from the heat. The reaction should start, but if you experience difficulty, proceed to the next paragraph.

You may need to employ one or more of the following procedures if heating fails to start the reaction. If you are experiencing difficulty, remove the syringe and rubber septum. Place a **dry** glass stirring rod into the vial and gently twist the stirring rod so as to crush the magnesium against the glass surface. Reattach the rubber septum and again heat the mixture. Repeat the crushing procedure several times, if necessary, to start the reaction. If the crushing procedure fails to start the reaction, then add one small crystal of iodine to the vial. Again, heat the mixture gently. The most drastic action, other than starting the experiment over again, is to prepare a small sample of the Grignard reagent in a test tube. When this reaction is started, it is added to the main reaction mixture in the vial.

When the reaction has started, you should observe the formation of a brownish-gray, cloudy solution. Add the remaining solution of bromobenzene slowly over a period of five minutes. It may be necessary to heat the mixture occasionally with the sand bath during the addition, but if the reaction becomes too vigorous, slow the addition of the bromobenzene solution and remove the vial from the sand bath. Ideally the mixture will boil without the application of external heat. **It is important that you heat the mixture if the reflux slows or stops.** As the reaction proceeds, you should observe the gradual disintegration of the magnesium metal. When all of the bromobenzene has been added, place 0.50 mL of **anhydrous** ether in the vial that originally contained the bromobenzene solution, draw it into the syringe, and add the ether to the reaction mixture. You should add additional anhydrous ether to replace any that is lost during the reflux period. After a period of about 45 minutes from the beginning of the addition of bromobenzene, most or all of the magnesium metal should have reacted. Cool the mixture to room temperature. As your instructor designates, go on to either Procedure 27A or Procedure 27B.

Procedure 27A
Triphenylmethanol

Adduct

While the phenylmagnesium bromide solution is being heated and stirred under reflux, make a solution of 0.273 g of benzophenone in 0.50 mL of **anhydrous** ether in a 3-mL conical vial. Cap the vial until the reflux period is over. Once the Grignard reagent is cooled to room temperature, draw the benzophenone solution into the syringe. Add this solution as rapidly as possible to the stirred Grignard reagent, but not so rapidly that the solution boils. Once the addition has been completed, cool the mixture to room temperature. The solution turns red and then gradually solidifies as the adduct is formed. When stirring is no longer effective, remove the syringe and septum, and stir the mixture with a spatula. Rinse the vial that contained the benzophenone solution with about 0.2 mL of anhydrous ether and add it to the mixture. Remove the reaction vial from the apparatus and cap it. Occasionally stir the contents of this vial. Recap the vial when it is standing to avoid contact with water vapor. The adduct should be fully formed after about 15 minutes. You may stop here.

Add 1.5 mL of 6M hydrochloric acid **(dropwise at first)** to neutralize the reaction mixture. The acid converts the adduct to triphenylmethanol and inorganic compounds (MgX_2). Eventually, you should obtain two distinct phases: the upper ether layer will contain triphenylmethanol; the lower aqueous hydrochloric acid layer will contain the inorganic compounds. Use a spatula to break up the solid during the addition of hydrochloric acid. You may need to cap the vial and shake it vigorously to dissolve the solid. Since the neutralization procedure evolves heat, some ether will be lost due to evaporation. You should add enough additional ether to maintain a 3-mL volume in the upper organic phase (use graduations on the vial for measurement). Make sure that you have two distinct liquid phases before separating the layers. More ether or hydrochloric acid may be added, if necessary, to dissolve any remaining solid.

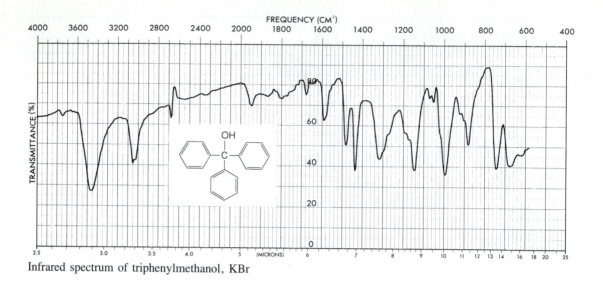

Infrared spectrum of triphenylmethanol, KBr

If a small amount of unreacted magnesium metal is present, you will observe bubbles of hydrogen being formed. You may remove the aqueous layer even though the magnesium is still producing hydrogen. Remove the magnetic spin vane with tweezers and rinse it with a little ether. Draw off the lower aqueous phase with a Pasteur pipet and place it in another conical vial. **Remember to save the ether layer** in the first vial because it contains the triphenylmethanol product. Reextract the aqueous phase in the second vial with 0.5 mL of ether. Remove the lower aqueous phase and discard it. Combine the remaining ether phase with the first ether extract. Transfer the combined ether layers to a dry 5-mL conical vial with a dry Pasteur pipet. Dry the ether solution with granular anhydrous sodium sulfate (two microspatulafuls measured in the V-grooved end).

Remove the dried ether solution from the drying agent with a filter tip pipet, transfer it to a small Erlenmeyer flask, and rinse the drying agent with more diethyl ether. Evaporate the solvent in a hood by heating the flask in a warm sand bath. You should be left with a mixture that varies from a brown oil to a colored solid mixed with an oil. This crude mixture contains the desired triphenylmethanol and the byproduct biphenyl. Most of the biphenyl can be removed by adding 1 mL of **petroleum ether (bp 30 to 60 °C).** Petroleum ether is a mixture of hydrocarbons that easily dissolves the hydrocarbon biphenyl and leaves behind the alcohol triphenylmethanol. Do not confuse this solvent with diethyl ether ("ether"). Heat the mixture slightly, stir it, and then cool the mixture to room temperature. Collect the triphenylmethanol by vacuum filtration on a Hirsch funnel and rinse it with small portions of petroleum ether (Technique 4, Section 4.3, and Figure 4–6, p 571). Air dry the solid, weigh it, and calculate the percentage yield of the crude triphenylmethanol (MW = 260.3).

Crystallize all of your product from hot isopropyl alcohol using a Craig tube (Technique 5, Section 5.4, and Figure 5–5, p 586). Step 2 in Figure 5–5 (removal of insoluble impurities) should not be required in this crystallization. Set the crystals aside to air-dry. Report the melting point of the purified triphenylmethanol (literature value, 162 °C) and recovered yield in grams. Submit the sample to the instructor.

At the option of the instructor, determine the infrared spectrum of the purified material in a KBr pellet (Technique 18, Section 18.4, p 775). Your instructor may assign certain tests on the product you prepared. These test are described in the instructor's manual.

Procedure 27B
Benzoic Acid

When the phenylmagnesium bromide has cooled to room temperature, use a Pasteur pipet to transfer this reagent as quickly as possible to 1 g of crushed dry ice contained in a small beaker. The dry ice should be weighed as quickly as possible to avoid contact with atmospheric moisture. It need not be weighed precisely. Rinse the conical vial with 1 mL of anhydrous ether and add it to the beaker.

> **CAUTION:** Exercise caution in handling dry ice. Contact with the skin can cause severe frostbite. Always use gloves or tongs. The dry ice is best crushed by wrapping large pieces in a clean, dry towel and striking it with a hammer or with a wooden block. It should be used as soon as possible after crushing it to avoid contact with atmospheric water.

Cover the reaction mixture with a watch glass and allow it to stand until the excess dry ice has completely sublimed. The Grignard addition compound will appear as a viscous glassy mass.

Hydrolyze the Grignard adduct by slowly adding 2.5 mL of 6M hydrochloric acid, with stirring, to the beaker. Any remaining magnesium chips will react with acid to evolve hydrogen. At this point you should have two distinct liquid phases in the beaker. If you have solid present (other than magnesium), try adding a little more ether. If the solid is insoluble in ether, try adding a little 6M hydrochloric acid solution. Benzoic acid is soluble in ether, while inorganic compounds (MgX_2) are soluble in the acid solution. Transfer the liquid phases to a 5-mL conical vial with a Pasteur pipet leaving behind any residual magnesium. Add more ether to the beaker to rinse the beaker. Again, transfer the ether solution to the conical vial. You may stop here and continue with the experiment during the next laboratory period. Cap the vial.

Remove the lower aqueous layer with a Pasteur pipet and keep the upper ether layer in the vial. The aqueous phase contains inorganic salts and may be discarded. The ether layer contains the product, benzoic acid, and the by-product, biphenyl. Add 1.0 mL

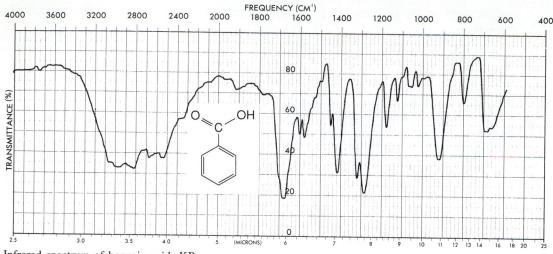

Infrared spectrum of benzoic acid, KBr

of 5% sodium hydroxide solution, cap the vial, and shake it. Allow the layers to separate, **remove the lower aqueous layer with a Pasteur pipet, and save this layer in a beaker.** This extraction removes benzoic acid from the ether layer by converting it to the water soluble sodium benzoate. The by-product, biphenyl, stays in the ether layer along with some remaining benzoic acid. Again, shake the remaining ether phase in the conical vial with a second 1.0-mL portion of 5% sodium hydroxide and transfer the lower aqueous layer into the beaker with the first extract. Repeat the extraction process with a third portion (1 mL) of 5% sodium hydroxide and save the aqueous layer, as before. Discard the ether layer that contains the biphenyl impurity.

Heat the combined basic extracts with stirring in a sand bath (100–120 °C) for about five minutes to remove any ether which may be dissolved in this aqueous phase. Ether is soluble in water to the extent of 7%. During this heating period you may observe slight bubbling, but the volume of liquid **will not decrease** substantially. Unless the ether is removed before the benzoic acid is precipitated, the product may appear as a waxy solid instead of crystals.

Cool the alkaline solution and precipitate the benzoic acid by adding 1.0 mL of 6M hydrochloric acid with stirring. Cool the mixture in an ice bath. Collect the solid by vacuum filtration on a Hirsch funnel (Technique 4, Section 4.3, and Figure 4–6, p 571). The transfer may be aided and the solid washed with several small portions of cold water (total volume, 2 mL). Allow the crystals to dry thoroughly at room temperature at least overnight. Weigh the solid and calculate the percentage yield of benzoic acid (MW = 122.1).

Crystallize 0.08 g of your product (or all of the material if your yield is less than this amount) from hot water using a Craig tube (Technique 5, Section 5.4, and Figure 5–5, p 586). Step 2 in Figure 5–5 (removal of insoluble impurities) should not be required in this crystallization. Set the crystals aside to air-dry at room temperature before determining the melting point of the purified benzoic acid (literature value, 122 °C) and recovered yield in grams. Submit your product to your instructor in a properly labeled vial.

At the option of the instructor, determine the infrared spectrum of the purified material in a KBr pellet (Technique 18, Section 18.4, p 775). Your instructor may assign certain tests on the product you prepared. These tests are described in the instructor's manual.

QUESTIONS

1. Benzene is often produced as a side product during Grignard reactions using phenylmagnesium bromide. How can its formation be explained? Give a balanced equation for its formation.

2. Write a balanced equation for the reaction of benzoic acid with hydroxide ion. Why is it necessary to extract the ether layer with sodium hydroxide?

3. Interpret the principal peaks in the infrared spectrum of either triphenylmethanol or benzoic acid, depending on the procedure used in this experiment.

4. Outline a separation scheme for isolating either triphenylmethanol or benzoic acid from the reaction mixture, depending on the procedure used in this experiment.

5. Provide methods for preparing the following compounds by the Grignard method:

(a) $CH_3CH_2CHCH_2CH_3$
             |
             OH

(c) $CH_3CH_2CH_2CH_2CH_2-\overset{\overset{\displaystyle O}{\|}}{C}-OH$

(b) $CH_3CH_2-\overset{\overset{\displaystyle CH_3}{|}}{\underset{\underset{\displaystyle OH}{|}}{C}}-CH_2CH_3$

(d) 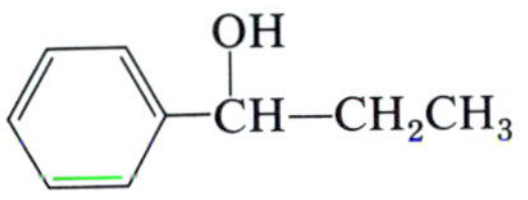 a phenyl group bonded to $\overset{\overset{\displaystyle OH}{|}}{CH}-CH_2CH_3$

Essay
CYANOHYDRINS IN NATURE

The addition of hydrogen cyanide to aldehydes is a typical nucleophilic addition reaction of the aldehyde functional group. The adduct is called a **cyanohydrin.** Because cyanide ion is required to be the attacking nucleophile, the reaction is catalyzed by base. The reaction is usually carried out with sodium or potassium cyanide, by adding only that amount of acid needed to keep the pH near 7 or 8. In the presence of excess

$$R-CHO + HCN \rightleftharpoons R-CH\overset{\diagup OH}{\diagdown CN}$$

A cyanohydrin

acid, the cyanide ion, which is a fairly strong base, is protonated and its effectiveness as a nucleophile is diminished.

In the cyanohydrin, the hydrogen derived from the original aldehyde is made acidic due to its new position alpha to the cyano group. The cyano group provides resonance stabilization.

$$R\text{—}CH\begin{array}{c}OH\\C\equiv N:\end{array} \xrightarrow{B:^-} \left[R\text{—}C\begin{array}{c}OH\\:^-\\C\equiv N:\end{array} \longleftrightarrow R\text{—}C\begin{array}{c}OH\\C\\N:^-\end{array} \right]$$

In a slightly basic KCN solution, the removal of this proton promotes the formation of the cyanohydrin by immediately converting it to its resonance-stabilized conjugate anion (LeChâtelier's principle). This anion can also behave as a nucleophile toward an unreacted molecule of benzaldehyde. An example of this type of behavior can be found in the benzoin condensation as described in Experiment 28.

The cyanohydrin formed from benzaldehyde is called mandelonitrile since it yields mandelic acid when hydrolyzed in strongly **basic** solutions:

$$C_6H_5\text{—}\underset{\underset{C\equiv N}{|}}{\overset{\overset{OH}{|}}{C}}\text{—}H \xrightarrow[\text{(2) } H_3O^+]{\text{(1) } OH^-/H_2O} C_6H_5\text{—}\underset{\underset{COOH}{|}}{\overset{\overset{OH}{|}}{C}}\text{—}H \quad \textbf{Mandelic acid}$$

$$\xrightarrow{H_3O^+} C_6H_5\text{—}\overset{\overset{O}{\|}}{C}\text{—}H + HCN$$

Mandelonitrile

In contrast, when mandelonitrile is hydrolyzed under **acidic** conditions, it decomposes to regenerate benzaldehyde and hydrogen cyanide.

Surprisingly, mandelonitrile is found in many forms in nature. Perhaps the most interesting example is in the millipede *Apheloria corrugata*. This millipede uses the compound as a part of its protective apparatus. It synthesizes and stores the cyanohydrin in a series of 22 glands, which are arranged in pairs on several of the body sections just above the legs. Each gland has two compartments. The inner compartment is a large saclike reservoir lined with cells that secrete mandelonitrile as an aqueous emulsion. The inner compartment is separated from the outer compartment by a muscular valve. The second compartment, the vestibule, is lined with cells that secrete an enzyme that decomposes mandelonitrile into benzaldehyde and hydrogen cyanide, a poisonous gas. When the millipede is alarmed by a predator, it opens the valve separating the two compartments, and, by contraction of the storage reservoir, forces mandelonitrile into the vestibule, where it is mixed with the enzyme and forced outside. The products of the dissociation either kill or repel the predator. A single *Apheloria* can

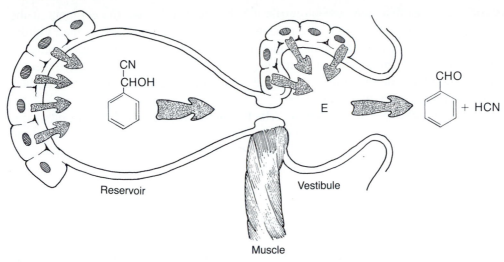

Reactor gland of *Apheloria corrugata*

secrete enough hydrogen cyanide to kill a small bird or a small mouse. The benzaldehyde is also an effective repellent against many of the millipede's predators.

Mandelonitrile is also found in several plants of the Rosaceae family. It is found not in the free form but as a **glycoside,** that is, with its hydroxyl group attached through an acetal linkage to a sugar. The most common glycoside is **amygdalin,** which is found

Amygdalin

in the seeds (pits), leaves, and bark of many plants such as bitter almond (*Prunus amygdalus*), apricot, wild cherry, peach, and plum.

Two similar glycosides, **prunlaurasin** and **sambunigrin,** are found in cherry laurel and *Sambucus nigra,* respectively. These glycosides differ from amygdalin only in the stereochemistry. On hydrolysis, amygdalin yields two molecules of glucose and D-mandelonitrile. Sambunigrin yields L-mandelonitrile, and prunlaurasin yields racemic (D,L)-mandelonitrile.

The seeds of most of these plants also contain two enzymes, **emulsin** and **prunase.** Emulsin hydrolyzes one glucose molecule from amygdalin to produce **prunasin,** or mandelonitrile monoglucoside. Prunasin is often found along with amygdalin. The second enzyme, prunase, cleaves the second glucose molecule from amygdalin (prunasin) to give mandelonitrile. The presence of the two enzymes, along with the glycoside, ensures that free mandelonitrile will be produced when the contents of the seed are digested in the stomach of a predator. The free mandelonitrile is then rapidly hydrolyzed to benzaldehyde and hydrogen cyanide in the acidic stomach medium. Thus, amygdalin, prunlaurasin, and sambunigrin probably constitute protective mechanisms for the plants in which they are found. Many of these plants produce an otherwise luscious and attractive fruit.

The controversial drug **Laetrile,** which is made from crushed apricot pits, consists mainly of amygdalin. Because it contains cyanide, it was originally considered too toxic for human use by its "discoverer," Dr. Ernst Krebs, Sr., a California physician. His son, Ernst Krebs, Jr., however, found an apparent way to purify it. Since that time they, along with others, have advocated its use as an effective treatment for cancer. They also developed an explanation for its action. Laetrile, it is maintained, goes directly to cancer cells, where an abundant enzyme releases cyanide, which in turn kills those cells. Normal cells are held to be low in this enzyme and therefore not affected. In addition, normal cells are held to contain a "protective" enzyme that detoxifies the Laetrile, whereas cancer cells lack the enzyme. Thus, normal cells live while cancer cells die, these workers conclude.

Unfortunately, other workers have shown that releasing enzymes are more abundant in normal cells than in cancer cells and that protective enzymes are distributed about equally. Further, the cyanide does not migrate preferentially to the cancerous cells but diffuses rapidly to all parts of the body.

In 1953, the Cancer Commission of the California Medical Association investigated Laetrile and found it ineffective. In one case study involving 44 patients treated with Laetrile, a follow-up showed that all but one of the patients either had died or still had cancer. Other studies gave similar results. Animal studies, some completed as recently as 1976, have corroborated clinical records. On the strength of this evidence, most states have banned Laetrile, and the Food and Drug Administration has banned its use in interstate commerce.

The proponents of Laetrile, however, vigorously maintain that it is effective, and that cancer patients should have free choice in deciding their treatment. In an apparent attempt to circumvent drug laws, Laetrile was promoted in 1970 as "Vitamin B-17," a vitamin needed to **prevent** rather than **cure** cancer. Vitamins are over-the-counter preparations and are exempt from the usual drug regulations. In 1974, the FDA won action in court stopping production of vitamin B-17. The claims for it as a vitamin could not be substantiated and were deemed fraudulent advertising.

Most scientific and medical authorities deem Laetrile ineffective, and, in fact, dangerous since it may keep some persons from receiving proper treatment until it is too late. Additionally, numerous deaths have resulted from overdoses, accidental ingestion by children, and misuse.

REFERENCES

Claus, E. P., Tyler, V. E., and Brady, L. R. *Pharmacognosy.* Philadelphia: Lea & Febiger, 1970. Pp 114–118, "Cyanophore Glycosides."

Consumer's Union. "Laetrile: The Political Success of a Scientific Failure." *Consumer Reports* (August 1977): 444.

Culliton, B. J. "Sloan-Kettering: The Trials of an Apricot Pit." *Science, 182* (December 7, 1973): 1000.

Holden, C. "Laetrile: Quack Cancer Remedy Still Brings Hope to Sufferers." *Science, 193* (September 10, 1976): 982.

"Laetrile Poisoning." *Newsweek* (August 22, 1977): 80.

Sondheimer, E., and Simeone, J. B., eds. *Chemical Ecology.* New York: Academic Press, 1970. Chap. 8, "Chemical Defense against Predation in Arthropods."

Experiment 28
Benzoin Condensation

Condensation reaction

Benzaldehyde does not possess alpha hydrogens and therefore does not undergo a aldol condensation. In fact, in strongly basic solution, benzaldehyde, like other aldehydes that lack alpha hydrogens, undergoes the Cannizzaro reaction, yielding benzyl alcohol and sodium benzoate:

$$2C_6H_5CHO + NaOH \longrightarrow C_6H_5CH_2OH + C_6H_5COO^-Na^+$$

In the presence of cyanide ion, however, benzaldehyde undergoes a unique self-condensation reaction called the **benzoin condensation,** yielding an α-hydroxy ketone called **benzoin.** In this experiment, we use the benzoin condensation to synthesize benzoin:

Benzaldehyde   Benzoin

The complete mechanism for this reaction is shown on page 256. The first step is the formation of the cyanohydrin **1**, which in the basic reaction medium immediately forms its conjugate anion **2**. The conjugate anion, **2**, is stabilized by resonance that involves both the cyano group and the aromatic ring. In a second step of the reaction, the cyanohydrin anion makes a nucleophilic addition to a second molecule of benzaldehyde, to give the adduct **3**. After a proton transfer to form the anion, **4**, cyanide is expelled, forming benzoin, an α-hydroxy ketone **5**.

The reaction is carried out in an aqueous ethanol solution, and the product, which is sparingly soluble, crystallizes from the reaction mixture on cooling. The product is collected by vacuum filtration and recrystallized from 95% ethanol.

REQUIRED READING

Review: Technique 5   Crystallization, Sections 5.3, 5.4, and 5.7

New: Essay   Cyanohydrins in Nature

SPECIAL INSTRUCTIONS

Be sure to take all the cautionary measures mentioned in the note and to dispose of all cyanide residues in the appropriate waste container.[1]

---

[1] NOTE TO THE INSTRUCTOR: Excess cyanide solution and cyanide residues should be treated by the following procedure. Adjust the pH of the mixture to 11. Slowly add a 50% excess of commercial laundry bleach, controlling the rate of addition so that the mixture does not get too hot. Adjust the pH if necessary. When all the bleach has been added, set the mixture aside overnight. Cautiously adjust the pH of the mixture to 7 in a hood since vigorous evolution of gas may occur. Filter the solid for burial in a chemical landfill. Adapted from *The Sigma-Aldrich Library of Chemical Safety Data*, edited by R. E. Lenga.

The benzaldehyde used for this experiment **must** be pure. The best results are obtained with a fresh bottle of benzaldehyde that has no benzoic acid, a white precipitate, evident in the bottom of the bottle. If there is any question about the purity of the benzaldehyde, it can be purified using the following procedure. Wash the benzaldehyde in a separatory funnel with an equal volume of 5% sodium carbonate until the evolution of carbon dioxide ceases. Remove the sodium carbonate layer and wash the organic layer with an equal volume of water. After drying the organic layer over calcium chloride, vacuum distill the benzaldehyde. Prevent the hot liquid from coming in contact with air by passing nitrogen through the ebulliator.

> **CAUTION: Sodium cyanide is extremely hazardous and toxic. When you are working with the solution, be careful not to spill any of it and do not allow it to come into contact with your skin. Any spilled material should be cleaned up immediately and disposed of in a waste container. Any area of the skin that has come into contact with cyanide should be washed thoroughly with water. Do not allow sodium cyanide to come into contact with acid, since hydrogen cyanide, a toxic gas, will be generated.**

PROCEDURE

Using a 5-mL conical vial and a water-cooled condenser, assemble an apparatus for heating under reflux as shown in Figure 3–2A, p 550. The temperature of the sand bath should be adjusted to about 115 °C. Place 0.80 mL of benzaldehyde into the tared 5-mL conical vial and reweigh the vial to determine an accurate weight for the benzaldehyde. Add 1.6 mL of 95% ethanol, 0.80 mL of cyanide solution[2], and a spin vane to the conical vial. Attach the condenser, and, with stirring, increase the temperature of the sand bath until the mixture is gently boiling. If the mixture boils too vigorously, lower the temperature of the sand bath or raise the conical vial.

After the mixture has boiled gently for 30 minutes, remove the sand bath and allow the vial to cool to room temperature. Scratch the inside of the vial with a stirring rod as the solution cools to induce crystallization. Then place the vial in an ice bath to complete crystallization. Collect the crude benzoin by vacuum filtration using a Hirsch funnel. Wash the solid with three 2.0-mL portions of cold water to remove all the sodium cyanide (see the caution above). Set the benzoin aside to dry. The crystals obtained should be colorless or pale yellow. Weigh the crude material and record the weight.

[2]NOTE TO THE INSTRUCTOR: To prepare this solution, add 2.0 g of sodium cyanide to 20.0 mL of water and stir until the solid dissolves. This will provide enough solution for 20 students, assuming little material is wasted.

Recrystallize the crude benzoin from 95% ethanol (about 8 mL/g of crude crystals) using an Erlenmeyer flask for the crystallization container and a Hirsch funnel to collect the crystals (see Technique 5, Section 5.3, and Figure 5–3, p 581). Weigh the purified material, calculate the yield, and determine the melting point (mp 134–135 °C). At your instructor's option, determine the infrared spectrum of the benzoin as a KBr mull (Technique 18, Section 18.6, p 779). If the sample is to be used to prepare benzil (Experiment 30), go on to that experiment. If not, place the sample in a labeled vial and submit it with your report.

QUESTIONS

1. Give the structure of the product that would be formed by the action of cyanide ion on acetaldehyde. Will the reaction be the same as benzaldehyde?

2. Interpret the principal peaks in the infrared spectra of benzoin and benzaldehyde.

3. Give all the possible resonance structures of the conjugate base of benzaldehyde cyanohydrin (**2**).

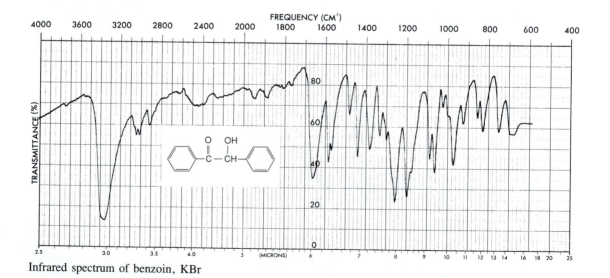

Infrared spectrum of benzoin, KBr

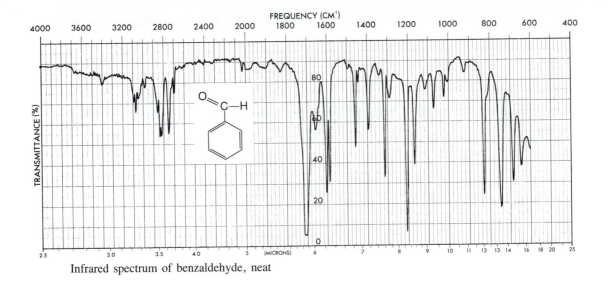

Infrared spectrum of benzaldehyde, neat

Essay
THIAMINE AS A COENZYME

Vitamin B_1, thiamine, as its pyrophosphate derivative, thiamine pyrophosphate, is a coenzyme universally present in all living systems. It was originally discovered as a required nutritional factor (vitamin) in humans by its link with the disease beriberi. **Beriberi** is a disease of the peripheral nervous system caused by a deficiency of Vitamin B_1 in the diet. Symptoms include pain and paralysis of the extremities, emaciation, or swelling of the body. The disease is most common in the Far East.

Thiamine Pyrophosphate

Thiamine serves as a coenzyme (defined later) for three important types of enzymatic reactions:

1. Nonoxidative decarboxylations of α-keto acids

$$R-\underset{\underset{O}{\|}}{C}-COOH \xrightarrow{B_1} R-\underset{\underset{O}{\|}}{C}-H + CO_2$$

2. Oxidative decarboxylations of α-keto acids

$$R-\underset{\underset{O}{\|}}{C}-COOH \xrightarrow{B_1,O_2} R-\underset{\underset{O}{\|}}{C}-OH + CO_2$$

3. Formation of acyloins (α-hydroxy ketones)

$$R-\underset{\underset{O}{\|}}{C}-COOH + R-\underset{\underset{O}{\|}}{C}-H \xrightarrow{B_1} R-\underset{\underset{O}{\|}}{C}-\underset{\underset{OH}{|}}{C}H-R + CO_2$$

or

$$2R-\underset{\underset{O}{\|}}{C}-COOH \xrightarrow{B_1} R-\underset{\underset{O}{\|}}{C}-\underset{\underset{OH}{|}}{C}H-R + 2CO_2$$

or

$$2R-\underset{\underset{O}{\|}}{C}-H \xrightarrow{B_1} R-\underset{\underset{O}{\|}}{C}-\underset{\underset{OH}{|}}{C}H-R$$

Most biochemical processes are no more than organic chemical reactions carried out under special conditions. It is easy to lose sight of this fact. Most of the steps of the ubiquitous metabolic pathways can, if they have been studied well enough, be explained mechanistically. Some simple organic reaction is a model for almost every biological process. Such reactions, however, are modified ingeniously through the intervention of a protein molecule (''enzyme'') to make them more efficient (have greater yield), more selective in choice of substrate (molecule being acted on), and more stereospecific in their result and to enable them to occur under milder conditions (pH) than would normally be possible.

Experiment 29 is designed to illustrate the last circumstance. As a biological reagent, the coenzyme thiamine is used to carry out an organic reaction **without** resorting to an enzyme. The reaction is an acyloin condensation (see above) of benzaldehyde:

$$2C_6H_5-CHO \longrightarrow C_6H_5-\underset{\underset{O}{\|}}{C}-\underset{\underset{OH}{|}}{C}H-C_6H_5$$

In Experiment 28, a similar condensation is described, in which sodium cyanide is used as the catalyst. Nature would clearly prefer to use a reactant that is milder and less toxic than cyanide ion to carry out the acyloin condensations necessary to everyday metabolism. Thiamine constitutes just such a reagent.

In the chemical view, the most important part of the entire thiamine molecule is the central ring—the thiazole ring—which contains nitrogen and sulfur. This ring constitutes the **reagent** portion of the coenzyme. The other portions of the molecule, although important in a biological sense, are not necessary to the chemistry that thiamine initiates. Undoubtedly the pyrimidine ring and the pyrophosphate group have important ancillary functions, such as enabling the coenzyme to make the correct attachment to its associated protein molecule (enzyme) or enabling it to achieve the correct degree of polarity and the correct solubility properties necessary to allow free passage of the coenzyme across the cell membrane boundary (that is, to allow it to get to its site of action). These properties of thiamine are no less important to its biological functioning than to its chemical reagent abilities; only the latter is our concern here, however.

Experiments with the model compound 3,4-dimethylthiazolium bromide have explained how thiamine-catalyzed reactions work. It was found that this model thiazolium compound rapidly exchanged the C-2 proton for deuterium in D_2O solution. At a pD of 7 (No pH here!), this proton was completely exchanged in seconds!

3,4-Dimethylthiazolium bromide   ylide

This indicates that the C-2 proton is more acidic than one would have expected. It is apparently easily removed because the conjugate base is a highly stabilized **ylide.** An ylide is a compound or intermediate with positive and negative formal charges on adjacent atoms.

The sulfur atom plays an important role in stabilizing this ylide. This was shown by comparing the rate of exchange of 1,3-dimethylimidazolium ion with the rate for the thiazolium ion shown. The dinitrogen compound exchanged its C-2 proton more slowly than the sulfur-containing ion. Sulfur, being in the third row of the periodic chart, has *d* orbitals available for bonding to adjacent atoms. Thus, it has fewer geometrical restrictions than carbon and nitrogen atoms do and can form carbon-sulfur multiple bonds in situations in which carbon and nitrogen normally would not.

1,3-Dimethylimidazolium bromide

DECARBOXYLATION OF α-KETO ACIDS

From the knowledge described above, it is now thought that the active form of thiamine is its ylide. The system is interestingly constructed, as is seen in the decarboxylation of pyruvic acid by thiamine (see below). Notice especially how the positively charged nitrogen provides a site to accommodate the electron pair that is released on decarboxylation. Thiamine is regenerated by use of this same pair of electrons that become protonated in vinylogous fashion on carbon. The other product is the protonated form of acetaldehyde, the decarboxylation product of pyruvic acid.

OXIDATIVE DECARBOXYLATION OF α-KETO ACIDS

In oxidative decarboxylations, two additional coenzymes—lipoic acid and coenzyme A—are involved. An example of this type of process, which characterizes all living organisms, is found in the metabolic process **glycolysis.** It is found in the steps that convert pyruvic acid to acetyl coenzyme A, which then enters the citric acid cycle (Krebs cycle, tricarboxylic acid cycle) to provide an energy source for the organism. In this process, the enamine intermediate, **3** (see page 263), is first oxidized by lipoic acid and then transesterified by coenzyme A.

Following this sequence of events, the dihydrolipoic acid is oxidized (through a chain of events involving molecular oxygen) back to lipoic acid, and the acetyl coenzyme A is condensed with oxaloacetic acid to form citric acid. The formation of citric acid begins the citric acid cycle. Notice that acetyl coenzyme A is a thioester of acetic acid and could be hydrolyzed to give acetic acid, not an aldehyde. Thus, an oxidation has taken place in this sequence of events.

ACYLOIN CONDENSATIONS

The enamine intermediate **3** can also function much like the enolate partner in an acid-catalyzed aldol condensation. It can condense with a suitable carbonyl-containing acceptor to form a new carbon-carbon bond. Decomposition of the adduct **9** to regenerate the thiamine ylide yields the protonated acyloin **10.**

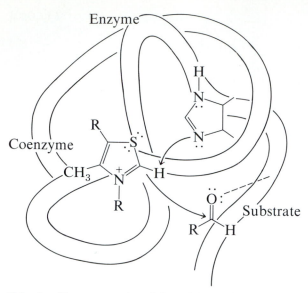

Thiamine (the coenzyme) and the substrate aldehyde are bound to the protein molecule, here called an enzyme. A possible catalytic group (imidazole) is also shown.

FUNCTION OF A COENZYME

In biological terminology, thiamine is a **coenzyme.** It must bind to an enzyme before the enzyme is activated. The enzyme also binds the substrate. The coenzyme reacts with the substrate while they are both bound to the enzyme (a large protein). Without the coenzyme thiamine, no chemical reaction would occur. The coenzyme is the **chemical reagent.** The protein molecule (the enzyme) helps and mediates the reaction by controlling stereochemical, energetic, and entropic factors, but, in this case, it is non-essential to the overall result (see Experiment 29). A special name is given to coenzymes that are essential to the nutrition of an organism. They are called **vitamins.** Many biological reactions are of this type, in which a chemical reagent (coenzyme) and a substrate are bound to an enzyme for reaction and, after the reaction, are again released into the medium.

REFERENCES

Bernhard, S. *The Structure and Function of Enzymes.* New York: W. A. Benjamin, 1968. Chap. 7, "Coenzymes and Cofactors."

Bruice, T. C., and Benkovic, S. *Bioorganic Mechanisms.* Vol. 2. New York: W. A. Benjamin, 1966. Chap. 8, "Thiamine Pyrophosphate and Pyridoxal-5′-Phosphate."

Lowe, J. N., and Ingraham, L. L. *An Introduction to Biochemical Reaction Mechanisms.* Englewood Cliffs, N.J.: Prentice-Hall, 1974. Chap. 5, "Coenzyme Function and Design."

Experiment 29
Coenzyme Synthesis of Benzoin

Coenzyme chemistry
Benzoin condensation

In this experiment, a benzoin condensation of benzaldehyde is carried out with a biological coenzyme, thiamine hydrochloride, as the catalyst:

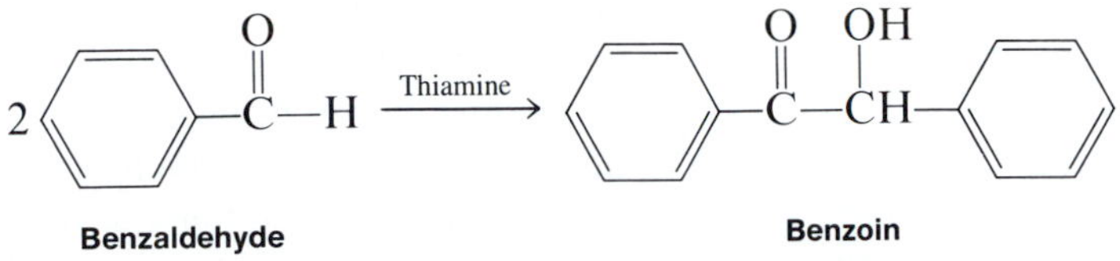

The same reaction can be accomplished with cyanide ion, an inorganic reagent, as the catalyst. A mechanism for the cyanide-catalyzed condensation is given in Experiment 28. The mechanistic information needed for understanding how thiamine accomplishes this same reaction is given in the essay that precedes this experiment.

REQUIRED READING

Review: Techniques 4 and 5

New: Essay   Thiamine as a Coenzyme

SPECIAL INSTRUCTIONS

> **NOTE TO THE INSTRUCTOR:** It is essential that the benzaldehyde used in this experiment is pure. A newly opened bottle which has been recently purchased should be used. Benzaldehyde is easily oxidized in air, and crystals of benzoic acid are often visible in the bottom of the reagent bottle. If solid appears in the bottle of reagent or if the bottle is old, the benzaldehyde must be purified, as described on p 257. Purification by this method, however, does not usually produce as good a material as that found in a fresh bottle of benzaldehyde. It is also advisable to use a fresh bottle of thiamine hydrochloride, which should be stored in the refrigerator.

PROCEDURE

Add 0.15 g of thiamine hydrochloride to a dry 5-mL conical vial. Dissolve the solid in 0.45 mL of water by swirling the vial. Add 1.5 mL of 95% ethanol and cool the solution for a few minutes in an ice bath. Place a spin vane in the vial, and, with stirring, add 0.45 mL of 2*M* NaOH. Weigh the conical vial and solution, add 0.90 mL of benzaldehyde,[1] and reweigh the vial to determine an accurate weight of benzaldehyde used. Attach an air condenser and heat the reaction mixture in a water bath at 60 °C for about 90 minutes (see Figure 2–5, p 544). The liquid should be stirred for a few seconds until the reaction mixture is homogeneous.

At the end of the reaction time remove the spin vane with your forceps. Allow the mixture to cool to room temperature, and then induce crystallization of the benzoin (it may already have begun) by cooling the mixture in an ice-water bath. If the product separates as an oil, reheat the mixture until it is once again homogeneous and allow it to cool more slowly than before. It may be helpful to scratch the vial with a glass rod or seed the mixture by allowing a small amount of solution to dry on the end of a glass rod and then placing this into the mixture. When the crystallization at room temperature is complete, cool the mixture in an ice-water bath.

Collect the product by vacuum filtration on a Hirsch funnel (see Technique 4, Section 4.3, and Figure 4–6, p 571) and wash the crystals with two 1.0 mL portions of ice water. Weigh the crude product, determine the melting point (mp 134–135 °C), and calculate the percentage yield. The benzoin should be of sufficient purity to be converted to benzil (Experiment 30).

At your instructor's option, recrystallize the product from 95% ethanol (about 8 mL/g) using a 10-mL Erlenmeyer flask for the crystallization and a Hirsch funnel to collect the crystals (see Technique 5, Section 5.3, and Figure 5–3, p 581). Determine the infrared spectrum of the benzoin as a KBr pellet (see Technique 18, Section 18.4, p 775). A spectrum is shown here for comparison.

The benzoin may be converted to benzil (Experiment 30). However, if you are not scheduled to do this experiment, submit the sample of benzoin, along with your report, to the instructor.

QUESTIONS

1. The infrared spectrum of benzoin is given in this experiment. Interpret the principal peaks in the spectrum.

2. Why is sodium hydroxide added to the solution of thiamine hydrochloride?

3. Using the information given in the essay that precedes this experiment, formulate a complete mechanism for the thiamine-catalyzed conversion of benzaldehyde to benzoin.

4. How do you think the appropriate enzyme would have affected the reaction (degree of completion, yield, stereochemistry)?

5. What modifications of conditions would be appropriate if the enzyme were to be used?

6. Refer to the essay that precedes this experiment. It gives a structure for thiamine pyrophosphate. Using this structure as a guide, draw a structure for thiamine hydrochloride. The pyrophosphate group is absent in this compound.

---

[1] NOTE TO THE INSTRUCTOR: See box in Special Instructions.

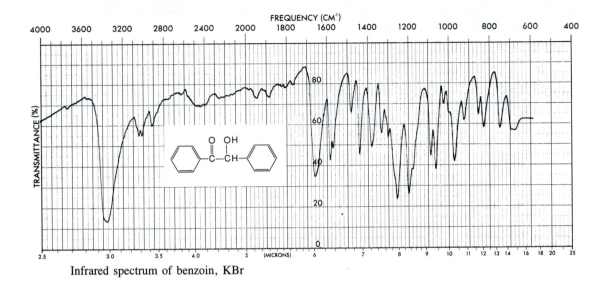

Infrared spectrum of benzoin, KBr

Experiment 30
Benzil

Oxidation
Crystallization

In this experiment an α-diketone, benzil, is prepared by the oxidation of an α-hydroxyketone, benzoin (Experiment 28 or 29).

Benzoin $\xrightarrow{\text{HNO}_3}$ **Benzil**

This oxidation can easily be done with mild oxidizing agents such as Fehling's solution (alkaline cupric tartrate complex) or copper(II) acetate in the presence of ammonium nitrate. In this experiment, the oxidation is performed with nitric acid.

REQUIRED READING

Review: Technique 3   Reaction Methods, Sections 3.2 and 3.7
         Technique 5   Crystallization

SPECIAL INSTRUCTIONS

Concentrated nitric acid is highly corrosive and will cause burns if it is spilled on your skin. The nitrogen oxide fumes are highly toxic and may cause inflammation of the lungs and edema. Do not breathe the nitrogen oxide fumes.

PROCEDURE

Place 0.25 g of benzoin (Experiment 28 or 29) in a 3-mL conical vial with 1.0 mL of concentrated nitric acid. Add a spin vane and attach an air condenser. Set up the apparatus for heating under reflux as shown in Figure 3–2A (inset), p 550. A 9-inch Pasteur pipet connected to an aspirator must be used to remove the nitrogen oxide gases (see Figure 3–9, p 558). Be sure that there is an air space between the Pasteur pipet and the inside of the air condenser. The water flow through the aspirator may be set at a moderate rate, and it is not necessary to use pressure tubing to connect the Pasteur pipet to the aspirator. The temperature of the sand bath should be set at about 110 °C.

With stirring, heat the reaction mixture. Begin timing the reaction when nitrogen oxide gases (red-brown) are visible above the reaction mixture and gas bubbles are present on the spin vane. The reaction mixture is then heated for at least 30 minutes, or until nitrogen oxide gases are not being produced if a longer reaction time is required. Remove the heat source and allow the reaction mixture to cool for one to two minutes. Detach the air condenser and transfer the reaction mixture with a Pasteur pipet to a small beaker containing 3.0 mL of cool tap water. Rinse the conical vial and spin vane with an additional 1.0 mL of water and add this to the beaker. Stir the mixture vigorously with a glass stirring rod as the temperature drops to room temperature. The oil should crystallize as a yellow solid. It is sometimes necessary to add seed crystals of benzil to induce crystallization. Crush the crystals of crude benzil in the beaker with a stirring rod. Vacuum-filter the crude benzil on a Hirsch funnel (see Technique 4, Section 4.3, and Figure 4–6, p 571). Wash it well with cold water (about 5.0 mL) to remove the nitric acid. Continue drawing air through the crystals on the Hirsch funnel by suction for about five minutes.

Weigh the crude benzil and transfer it to a Craig tube. Add 95% ethanol (5 mL/g) to the Craig tube and heat the mixture in a sand bath until the solid dissolves.[1] Place the Craig tube in a 10-mL Erlenmeyer flask containing about 8 mL of hot water (50–60 °C). As the solution cools, seed it with solid benzil that forms on a spatula after the spatula is dipped into the solution. The solution will become supersaturated unless this is done, and crystallization will occur too rapidly. Yellow, needle-like crystals are formed. Cool the mixture in an ice bath to complete the crystallization and isolate the crystals by centrifugation (Technique 4, Section 4.7, p 575, and Figure 4–11, p 576). Allow the crystals to air-dry.

Weigh the benzil and calculate the percentage yield, then determine the melting point. The melting point of pure benzil is 95 °C. The value obtained is often lower than

---

[1] If this mixture does not fit into a Craig tube, it will be necessary to perform the recrystallization in an Erlenmeyer flask and collect the crystals on a Hirsch funnel (see Technique 5, Section 5.3, and Figure 5–3, p 581).

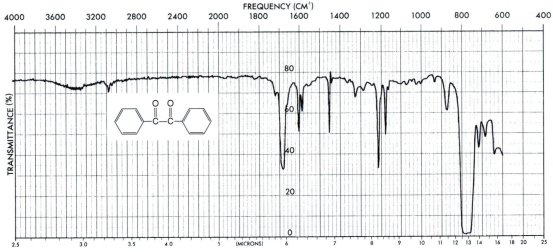

Infrared spectrum of benzil, CCl_4

this, ranging from a low value of 84 °C to 92 °C. Material in this range is pure enough for conversion to benzilic acid (Experiment 31) or tetraphenylcyclopentadienone (Experiment 34). Submit the benzil to the instructor unless it is to be used to prepare benzilic acid or tetraphenylcyclopentadienone. At the instructor's option, obtain the infrared spectrum of benzil in carbon tetrachloride or chloroform. Compare it with the infrared spectrum of benzoin shown in Experiment 28 or 29.

Experiment 31
Benzilic Acid

Anionic rearrangement

In this experiment, benzilic acid is prepared by causing the rearrangement of the α-diketone benzil. Preparation of benzil is described in Experiment 30. The rearrangement of benzil proceeds in the following way:

Benzil
(An α-diketone)

Potassium
benzilate

Benzilic acid
(An α-hydroxyacid)

The driving force for the reaction is provided by the formation of a stable carboxylate salt (potassium benzilate). Once this salt is produced, acidification yields benzilic acid. The reaction can generally be used to convert aromatic α-diketones to aromatic α-hydroxyacids. Other compounds, however, also will undergo benzilic acid type of rearrangement (see questions).

REQUIRED READING

Review: Technique 3   Reaction Methods, Section 3.2
           Technique 5   Crystallization

PROCEDURE

Add 0.100 g of benzil and 0.30 mL of 95% ethanol to a 3-mL conical vial. Place a spin vane in the vial and attach an air condenser. Heat the mixture in a sand bath (90–100 °C) with stirring until the benzil has dissolved (see inset in Figure 3–2A, p 550). Using a 9-inch Pasteur pipet, add dropwise 0.25 mL of an aqueous potassium hydroxide solution[1] downward through the condenser into the vial. Gently boil the mixture in the sand bath (about 110 °C) for 15 minutes. The mixture should be stirred during this period. After the solid has dissolved or after several minutes of heating, the mixture will become blue-black in color. As the reaction proceeds, the color will turn to brown and the solid may or may not be completely dissolved. At the end of the heating period, remove the assembly from the sand bath and allow it to cool for one to two minutes.

Detach the air condenser when the apparatus is cool enough to handle. Transfer the reaction mixture, which may contain some solid, with a Pasteur pipet into a 10-mL beaker. Allow the mixture to cool to room temperature and then cool in an ice-water bath for about 15 minutes until crystallization is complete. Crystallization is complete when virtually the entire mixture has solidified. If this does not occur in 15 minutes, allow the mixture to set overnight, or until crystallization has occurred. Collect the crystals on a Hirsch funnel by vacuum filtration (Technique 4, Section 4.3 and Figure 4–6, p 571) and wash the crystals thoroughly with three 1-mL portions of ice-cold 95% ethanol. The solvent should remove most of the color from the crystals.

Transfer the solid, which is mainly potassium benzilate, to a 10-mL Erlenmeyer flask containing 3 mL of hot water. Stir the mixture until all the solid has dissolved or until it appears that the remaining solid will not dissolve. **If solid still remains in the flask,** transfer the mixture with a Pasteur pipet to a centrifuge tube. Centrifuge the mixture for several minutes (be sure the centrifuge is balanced). Being careful not to disturb the solid on the bottom of the centrifuge tube, transfer the liquid with a filter tip pipet (Figure 4–9, p 575) into a clean 10-mL Erlenmeyer flask. **If no solid remains in the flask,** the centrifugation step may be omitted. In either case, proceed to the next step.

---

[1] The aqueous potassium hydroxide solution should be prepared for the class by dissolving 2.75 g of potassium hydroxide in 6.0 mL of water. This will provide enough solution for 20 students, assuming little solution is wasted.

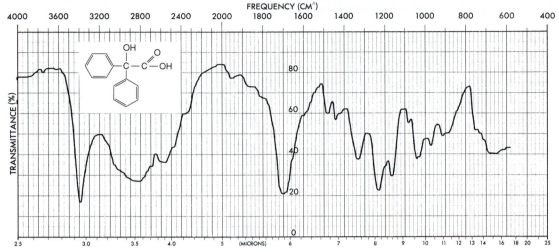

Infrared spectrum of benzilic acid, KBr

With stirring, add dropwise 0.5 mL of 1 M hydrochloric acid to the solution of potassium benzilate. The pH should be about 2; if it is higher than this, add a few more drops of acid and check the pH again. Allow the mixture to cool to room temperature and then complete the cooling in an ice bath. Collect the benzilic acid by vacuum filtration, using a Hirsch funnel. Wash the crystals thoroughly with 3–4 mL of water to remove salts, and remove the wash water by drawing air through the filter. Dry the product thoroughly by allowing it to stand until the next laboratory period.

Determine the melting point of the product. Pure benzilic acid melts at 150 °C. If necessary, recrystallize the product from hot water using a Craig tube (Technique 5, Section 5.4, and Figure 5–5, p 586). If some impurities remain undissolved, transfer the mixture to a test tube with a Pasteur pipet. Clean the Craig tube and filter the mixture by transferring it back to the Craig tube with a filter tip pipet. It will be necessary to keep the mixture hot during this filtration step. Cool the solution and induce crystallization, if necessary. Allow the mixture to stand at room temperature until crystallization is complete (about 15 minutes). Cool the mixture in an ice bath and collect the crystals by centrifugation. Determine the melting point of the crystallized product.

At the instructor's option, determine the infrared spectrum of the benzilic acid in potassium bromide (Technique 18, Section 18.4, p 775). Calculate the percentage yield. Submit the sample to your laboratory instructor in a labeled vial.

QUESTIONS

1. Show how to prepare the following compounds, starting from the appropriate aldehyde (see Experiments 28 and 30).

2. Give the mechanisms for the following transformations:

(a)
$$\xrightarrow[\text{(2) } H^+]{\begin{array}{c}\text{(1) KOH,}\\ \text{alcohol}\end{array}}$$

(b)
$$\text{HO}-\overset{\displaystyle O}{\overset{\|}{\text{C}}}-\text{CH}_2-\overset{\displaystyle O}{\overset{\|}{\text{C}}}-\overset{\displaystyle O}{\overset{\|}{\text{C}}}-\text{CH}_2\overset{\displaystyle O}{\overset{\|}{\text{C}}}-\text{OH} \xrightarrow[\text{(2) } H^+]{\begin{array}{c}\text{(1) KOH,}\\ H_2O\end{array}} \text{HO}-\overset{\displaystyle O}{\overset{\|}{\text{C}}}\text{CH}_2-\underset{\underset{\displaystyle CO_2H}{|}}{\overset{\overset{\displaystyle OH}{|}}{\text{C}}}-\text{CH}_2\overset{\displaystyle O}{\overset{\|}{\text{C}}}-\text{OH}$$

Citric acid

(c)
$$\text{Ph}-\overset{\displaystyle O}{\overset{\|}{\text{C}}}-\overset{\displaystyle O}{\overset{\|}{\text{C}}}-\text{Ph} \xrightarrow[\text{CH}_3\text{OH}]{^-\text{OCH}_3} \text{Ph}-\underset{\underset{\displaystyle Ph}{|}}{\overset{\overset{\displaystyle OH}{|}}{\text{C}}}-\underset{\displaystyle O}{\overset{\displaystyle }{\text{COCH}_3}}$$

3. Interpret the infrared spectrum of benzilic acid.

Experiment 32
The Aldol Condensation Reaction: Preparation of Benzalacetones and Benzalacetophenones

Aldol condensation
Crystallization

Benzaldehyde reacts with a ketone in the presence of base to give α,β-unsaturated ketones. This reaction is an example of a crossed aldol condensation where the intermediate dehydrates to produce the resonance-stabilized unsaturated ketone.

$$\text{C}_6\text{H}_5-\overset{\displaystyle H}{\overset{\diagup}{\underset{\diagdown}{\text{C}}}}\underset{\displaystyle O}{} + \text{CH}_3-\overset{\displaystyle O}{\overset{\|}{\text{C}}}-\text{R} \xrightarrow{\text{OH}^-} \text{C}_6\text{H}_5-\underset{\underset{\displaystyle OH}{|}}{\overset{\overset{\displaystyle H}{|}}{\text{C}}}-\text{CH}_2-\overset{\displaystyle O}{\overset{\|}{\text{C}}}-\text{R} \xrightarrow{-H_2O} \underset{\text{C}_6\text{H}_5}{\overset{H}{\diagdown}}\text{C}=\text{C}\underset{H}{\overset{\overset{\displaystyle O}{\overset{\|}{\text{C}}}-\text{R}}{\diagup}}$$

Intermediate

Crossed aldol condensations of this type proceed in high yield since benzaldehyde cannot react with itself by an aldol condensation reaction because it has no α-hydrogen. Likewise, ketones do not react easily with themselves in aqueous base. Therefore, the only possibility is for a ketone to react with benzaldehyde.

In this experiment, procedures are given for the preparation of benzalacetones and benzalacetophenones (chalcones). You should choose one of the substituted benzaldehydes and react it with one of the ketones: acetone or acetophenone. All of the products are solids that can be recrystallized easily.

BENZALACETONES

Benzalacetones are prepared by the reaction of a substituted benzaldehyde with excess acetone in aqueous base. Piperonaldehyde and *p*-anisaldehyde are used.

A benzaldehyde   Acetone   A benzalacetone
                                            (*trans*)

Piperonaldehyde   *p*-Anisaldehyde

BENZALACETOPHENONES

Benzalacetophenones (chalcones) are prepared by the reaction of a substituted benzaldehyde with acetophenone in aqueous base. Piperonaldehyde, *p*-anisaldehyde, and 3-nitrobenzaldehyde are used.

A benzaldehyde   Acetophenone   A benzalacetophenone
                                            (*trans*)

Piperonaldehyde   p-Anisaldehyde   3-Nitrobenzaldehyde

REQUIRED READING

Review: Technique 4   Sections 4.3 and 4.7
          Technique 5   Section 5.4

SPECIAL INSTRUCTIONS

Before beginning this experiment, you should select one of the procedures and a substituted benzaldehyde. Alternatively, your instructor may assign a particular compound to you.

PROCEDURES

BENZALACETONES

Choose one of two aldehydes for this experiment: piperonaldehyde (solid) or p-anisaldehyde (liquid). Place 0.150 g of piperonaldehyde (3,4-methylenedioxybenzaldehyde, MW = 150.1) into a 5-mL conical vial. Alternatively, transfer 0.13 mL of p-anisaldehyde (4-methoxybenzaldehyde, MW = 136.2) to a **tared** 5-mL conical vial and reweigh the vial to determine the weight of material transferred. With either choice, dissolve the aldehyde in 1.00 mL of acetone. While stirring the mixture with a magnetic spin vane, add 0.10 mL of sodium hydroxide solution[1] to the aldehyde/acetone mixture. Cap the vial and stir the mixture for 45 minutes at room temperature.

Place 20 mL of water (room temperature) in a small beaker and pour a portion of the water into the conical vial containing the reaction mixture. Transfer the contents of the vial into the beaker. Repeat the transfer operation several times, until all of the reaction mixture has been transferred into the beaker.

You should obtain a cloudy mixture that has some yellow oil in the beaker. As quickly as possible, transfer approximately one-half of the mixture into each of two

---

[1] This reagent should be prepared in advance by the instructor in the ratio of 0.60 g of sodium hydroxide to 1 mL of water.

centrifuge tubes. After the solutions are centrifuged, the oil should have collected at the bottom. Pour the liquid on top into an Erlenmeyer flask, leaving the oil behind. Discard the oil.

Stopper the Erlenmeyer flask and store it until the next laboratory period. The product slowly crystallizes from the aqueous solution. Vacuum filter the mixture on a Hirsch funnel, wash it with a small amount of water, and allow the solid to dry thoroughly. Weigh the product and calculate the percentage yield. Determine the melting point of the crystals. The literature melting points for the benzalacetones formed from piperonal-dehyde and *p*-anisaldehyde are 108 °C and 74 °C, respectively. Include a balanced equation for the reaction in your report. At the instructor's option, obtain the proton and/or carbon-13 NMR spectrum. Submit the sample to the instructor in a labeled vial.

BENZALACETOPHENONES

Choose one of three aldehydes for this experiment: piperonaldehyde (solid), 3-nitroben-zaldehyde (solid), or *p*-anisaldehyde (liquid). Place 0.150 g of piperonaldehyde (3,4-methylenedioxybenzaldehyde, MW = 150.1) or 0.151 g of 3-nitrobenzaldehyde (MW = 151.1) into a 5-mL conical vial. Alternatively, transfer 0.13 mL of *p*-anisaldehyde (4-methoxybenzaldehyde, MW = 136.2) to a **tared** conical vial and reweigh the vial to determine the weight of material transferred.

Add 0.12 mL of acetophenone (MW = 120.2, d = 1.03 g/mL) and 0.80 mL of 95% ethanol to the vial containing your choice of aldehyde. While stirring the mixture with a magnetic spin vane, add 0.10 mL of sodium hydroxide solution[2] to the aldehyde/ acetophenone mixture. Cap the vial and stir the mixture at room temperature until it solidifies (approximately four minutes).

Add 2 mL of ice water to the vial, stir the solid with a spatula, and transfer the mixture to a small beaker with 3 mL of ice water. Stir the precipitate to break it up and then collect the solid on a Hirsch funnel. Wash the product with cold water. Allow the solid to air dry for about 30 minutes. Weigh the solid and determine the percentage yield. Crystallize part of the chalcone using a Craig tube as follows:

> **3,4-methylenedioxychalcone (from piperonaldehyde).** Crystallize a 0.040 g sample from about 0.5 mL of hot 95% ethanol; literature melting point is 122 °C.
>
> **4-methoxychalcone (from *p*-anisaldehyde).** Crystallize a 0.075 g sample from about 0.3 mL of hot 95% ethanol. Scratch the tube to induce crystallization while cooling; literature melting point is 74 °C.
>
> **3-nitrochalcone (from 3-nitrobenzaldehyde).** Crystallize a 0.010 g sample from about 1 mL of hot 95% ethanol (ignore the cloudiness of the solution). Scratch the tube to induce crystallization while cooling; literature melting point is 146 °C.

Determine the melting point of your purified product. At the option of the instruc-tor, obtain the proton and/or carbon-13 NMR spectrum. Include a balanced equation for the reaction in your report. Submit the crude and purified samples to the instructor in labeled vials.

---

[2] See Footnote 1.

QUESTIONS

1. Give a mechanism for the preparation of the appropriate benzalacetone or benzalacetophenone using the aldehyde and ketone that you selected in this experiment.

2. Draw the structure of the *cis* and *trans* isomers of the compound that you prepared. Why did you obtain the *trans* isomer?

3. Using proton NMR, how could you experimentally determine that you have the *trans* isomer rather than the *cis* one? (*Hint:* consider the use of coupling constants for the vinyl hydrogens.)

4. When the amount of acetone is decreased significantly, the benzalacetone becomes contaminated with a side-product that has consumed two moles of aromatic aldehyde. What is its structure and why is it produced when the amount of acetone is decreased? Would you expect that the benzalacetophenone preparation should have this problem? Why?

5. Provide the starting materials needed to prepare the following compounds:

(a) $CH_3CH_2CH{=}C{-}\overset{\overset{\displaystyle O}{\|}}{C}{-}H$
     $\underset{\displaystyle CH_3}{|}$

(b) $\overset{\displaystyle CH_3}{\underset{\displaystyle CH_3}{>}}C{=}CH\underset{\underset{\displaystyle O}{\|}}{C}{-}CH_3$

(c) $\overset{\displaystyle Ph}{\underset{\displaystyle CH_3}{>}}C{=}CH{-}\overset{\overset{\displaystyle O}{\|}}{C}{-}Ph$

(d) $CH_3O{-}\langle\!\!\bigcirc\!\!\rangle{-}CH{=}CH{-}\overset{\overset{\displaystyle O}{\|}}{C}{-}CH{=}CH{-}\langle\!\!\bigcirc\!\!\rangle{-}OCH_3$

(e) $O_2N{-}\langle\!\!\bigcirc\!\!\rangle{-}CH{=}CH{-}\overset{\overset{\displaystyle O}{\|}}{C}{-}\langle\!\!\bigcirc\!\!\rangle{-}Br$

(f) $Cl{-}\langle\!\!\bigcirc\!\!\rangle{-}CH{=}CH{-}\overset{\overset{\displaystyle O}{\|}}{C}{-}\underset{\underset{\displaystyle NO_2}{|}}{\langle\!\!\bigcirc\!\!\rangle}$

6. Prepare the following compounds starting from benzaldehyde and the appropriate ketone. Provide reactions for preparing the ketones starting from aromatic hydrocarbon compounds (see Experiment 39).

$\langle\!\!\bigcirc\!\!\rangle{-}CH{=}CH{-}\underset{\underset{\displaystyle O}{\|}}{C}{-}\langle\!\!\bigcirc\!\!\rangle{-}CH_2CH_3$

$\langle\!\!\bigcirc\!\!\rangle{-}CH{=}\underset{\underset{\displaystyle CH_3}{|}}{C}{-}\overset{\overset{\displaystyle O}{\|}}{C}{-}\overset{\displaystyle CH_3}{\underset{\displaystyle}{\langle\!\!\bigcirc\!\!\rangle}}{-}CH_3$

Experiment 33
Enamine Reactions:
2-Acetylcyclohexanone

Enamine reaction
Azeotropic distillation
Column chromatography
Keto-enol tautomerism
Infrared and NMR spectroscopy

Hydrogens on the α-carbon of ketones, aldehydes, and other carbonyl compounds are weakly acidic and are removed in a basic solution (Equation 1). Although resonance stabilizes the conjugate base (**A**) in such a reaction, the equilibrium is still unfavorable because of the high pK_a (about 20) of a carbonyl compound.

$$R-CH_2\overset{\overset{\displaystyle O}{\|}}{C}CH_2R + {}^-OH \rightleftharpoons$$

$$\left[RCH_2\overset{\overset{\displaystyle O}{\|}}{C}-\overset{..}{C}HR \longleftrightarrow RCH_2\overset{\overset{\displaystyle O^-}{|}}{C}=CHR \right] + H_2O \quad (1)$$

(A)

Typically, carbonyl compounds are alkylated (Equation 2) or acylated (Equation 4) only with difficulty in the presence of aqueous sodium hydroxide because of more important secondary side reactions (Equations 3, 5, and 6). In effect, the concentration of the nucleophilic conjugate base species (**A** in Equation 1) is low because of the unfavorable equilibrium (Equation 1), while the concentration of the competing nucleophile (OH$^-$) is very high. A significant side reaction occurs when hydroxide ion reacts with an alkyl halide by Equation 3 or acyl halide by Equation 5. In addition, the conjugate base can react with unreacted carbonyl compound by an aldol condensation reaction (Equation 6). Enamine reactions, described below, avoid many of the problems described here.

Alkylation

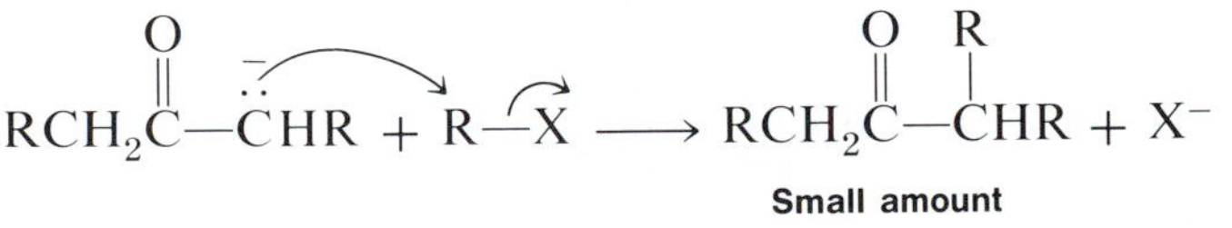

$$RCH_2\overset{\overset{\displaystyle O}{\|}}{C}-\overset{..}{C}HR + R-X \longrightarrow RCH_2\overset{\overset{\displaystyle O}{\|}}{C}-\overset{\overset{\displaystyle R}{|}}{C}HR + X^- \quad (2)$$

Small amount

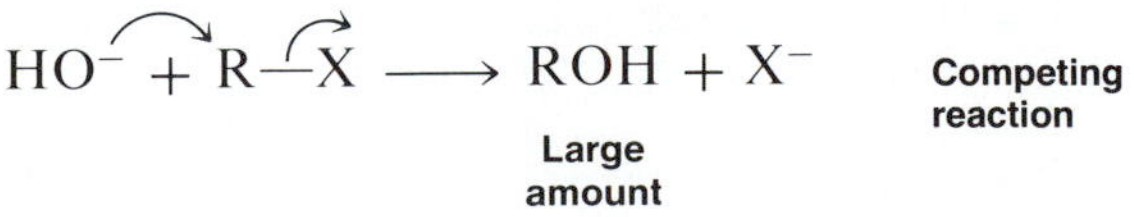

$$HO^- + R-X \longrightarrow ROH + X^- \quad \text{Competing reaction} \quad (3)$$

Large amount

Acylation

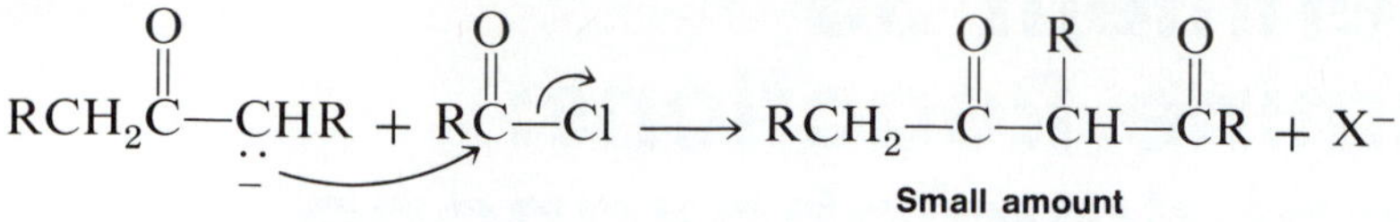

$$RCH_2\overset{O}{\overset{\|}{C}}-\underset{\underset{-}{\cdot\cdot}}{CHR} + R\overset{O}{\overset{\|}{C}}-Cl \longrightarrow RCH_2-\overset{O}{\overset{\|}{C}}-\overset{R}{\underset{|}{CH}}-\overset{O}{\overset{\|}{C}}R + X^- \qquad (4)$$

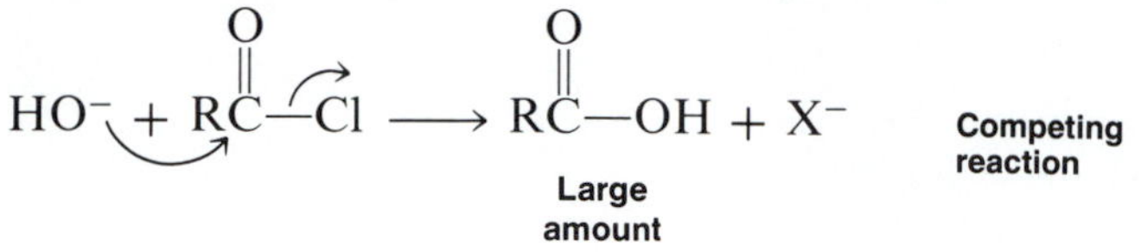

Small amount

$$HO^- + R\overset{O}{\overset{\|}{C}}-Cl \longrightarrow R\overset{O}{\overset{\|}{C}}-OH + X^- \qquad \text{Competing} \atop \text{reaction} \qquad (5)$$

**Large
amount**

Aldol condensation

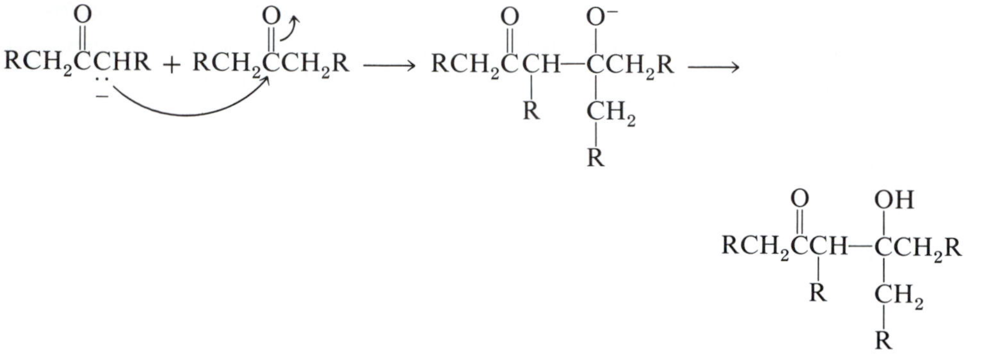

$$RCH_2\overset{O}{\overset{\|}{C}}\underset{\underset{-}{\cdot\cdot}}{CHR} + RCH_2\overset{O}{\overset{\|}{C}}CH_2R \longrightarrow RCH_2\overset{O}{\overset{\|}{C}}\overset{}{\underset{\underset{R}{|}}{CH}}-\overset{O^-}{\underset{\underset{\underset{R}{|}}{CH_2}}{\overset{|}{C}}}CH_2R \longrightarrow$$

$$RCH_2\overset{O}{\overset{\|}{C}}\overset{}{\underset{\underset{R}{|}}{CH}}-\overset{OH}{\underset{\underset{\underset{R}{|}}{CH_2}}{\overset{|}{C}}}CH_2R \qquad (6)$$

FORMATION AND REACTIVITY OF ENAMINES

Enamines are prepared easily from carbonyl compounds (for example, cyclohexanone) and a secondary amine (for example, pyrrolidine) by an acid-catalyzed addition-elimination reaction. Water, the other product of the reaction, is removed by azeotropic distillation with toluene which drives the equilibrium to the right:

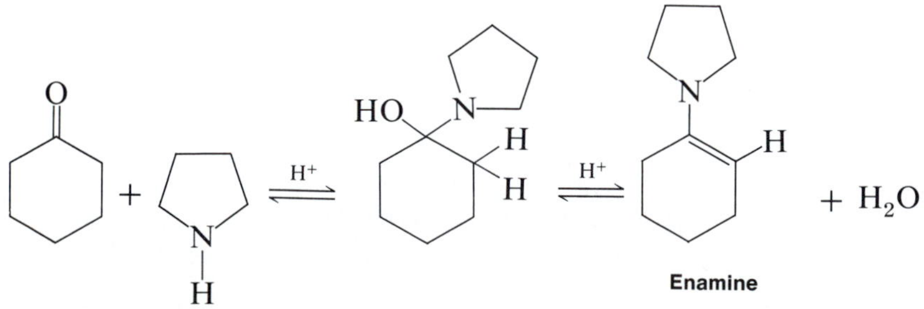

If the water were not removed, the equilibrium would be unfavorable, and only a small amount of enamine would be produced. Azeotropic distillation of water is an important "trick" used in organic chemistry to produce desired products in spite of an unfavorable equilibrium.

An enamine has the desirable property of being nucleophilic (carbon alkylation

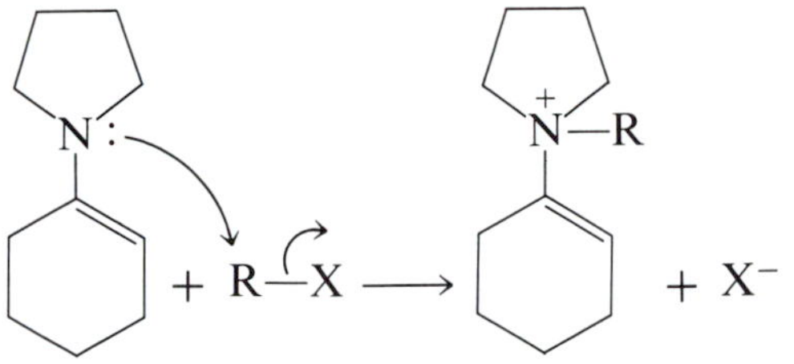

N Alkylation
(minor)

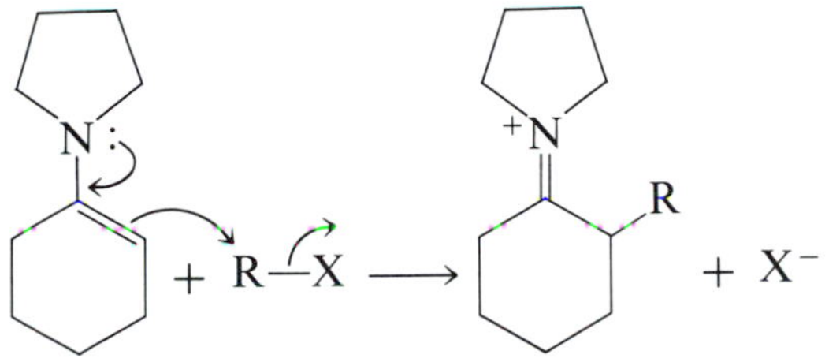

C Alkylation
(major)

is more important than nitrogen alkylation) and is easily alkylated. The **key point** is that the resonance hybrid (**B**) is like the resonance hybrid (**A**) shown in Equation 1. However, **B** has been produced under nearly neutral conditions so that it is the *only* nucleophile present.

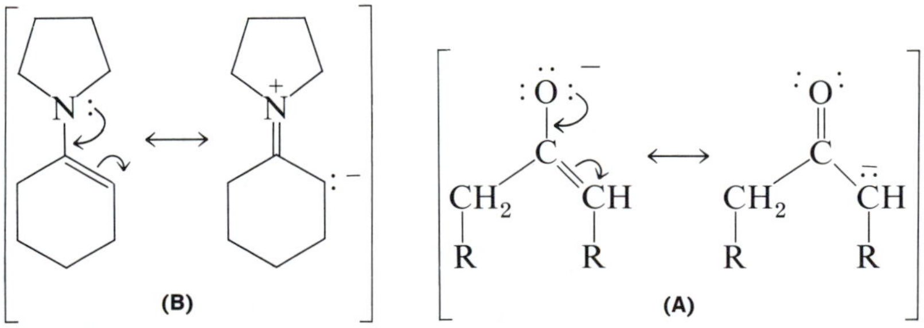

Contrast this situation to the one in Equation 1 where hydroxide ion, present in large amount, produces undesirable side reactions (Equations 3 and 5).

The alkylation step is followed by removal of the secondary amine by an acid-catalyzed hydrolysis:

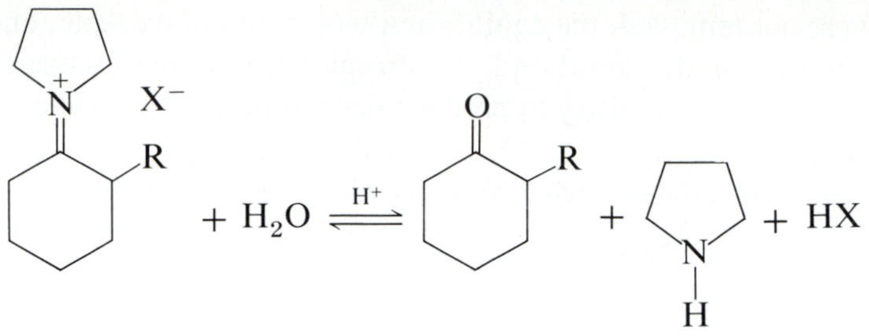

EXAMPLES OF ENAMINE REACTIONS

Alkylation

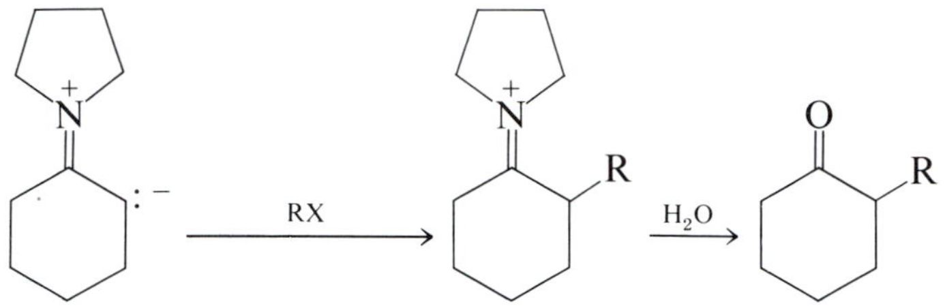

Acylation

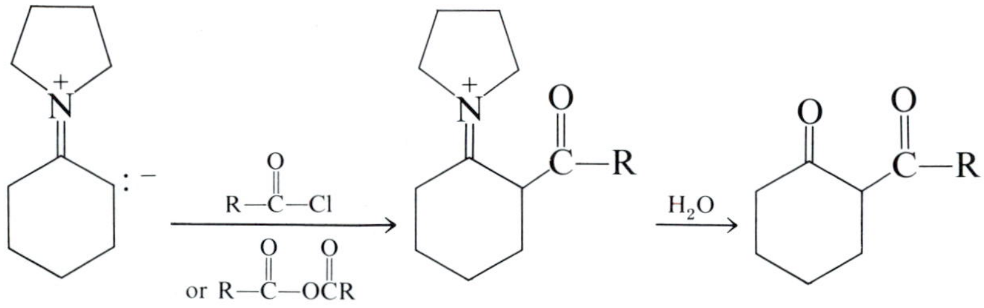

Conjugate addition (Michael reaction)

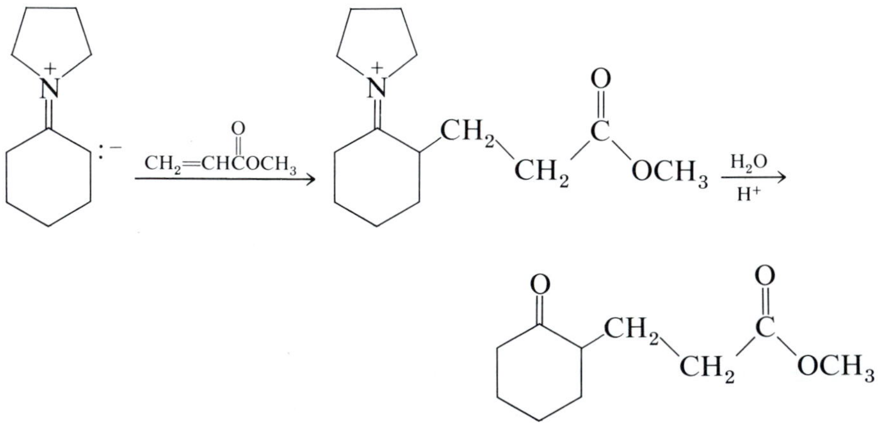

ROBINSON ANNELATION (RING-FORMATION) REACTION

Reactions that combine the Michael addition reaction and aldol condensation to form a six-membered ring fused on another ring are well known in the steroid field. These reactions are known as **Robinson annelation reactions.** An example is the formation of $\Delta^{1,9}$-2-octalone.

**Michael addition
(conjugate addition)**

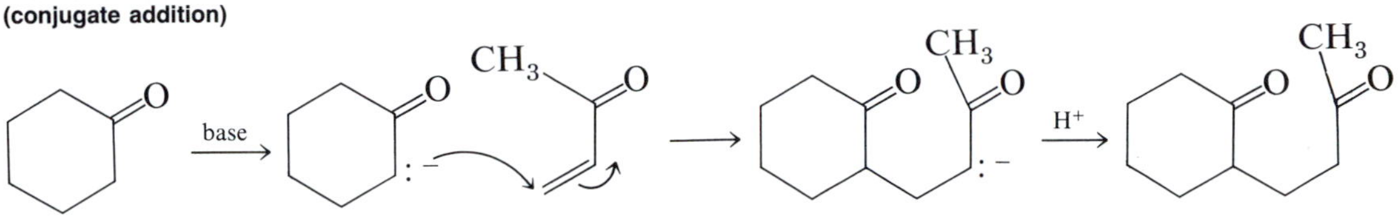

Aldol condensation

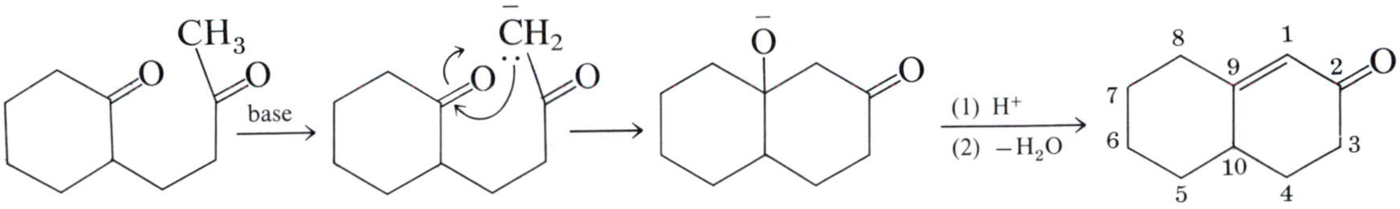

$\Delta^{1,9}$**-2-Octalone**

Robinson annelation reactions can also be conducted by enamine chemistry. One advantage of enamines is that the unsaturated ketones are not easily polymerized under the mild conditions of this reaction. Base-catalyzed reactions often give large amounts of polymer.

THE EXPERIMENT

In this experiment, pyrrolidine reacts with cyclohexanone to give the enamine. This enamine is used to prepare 2-acetylcyclohexanone.

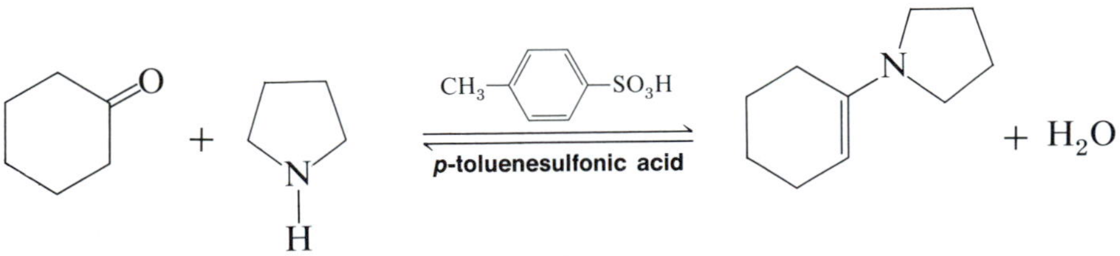

Cyclohexanone   Pyrrolidine   Enamine

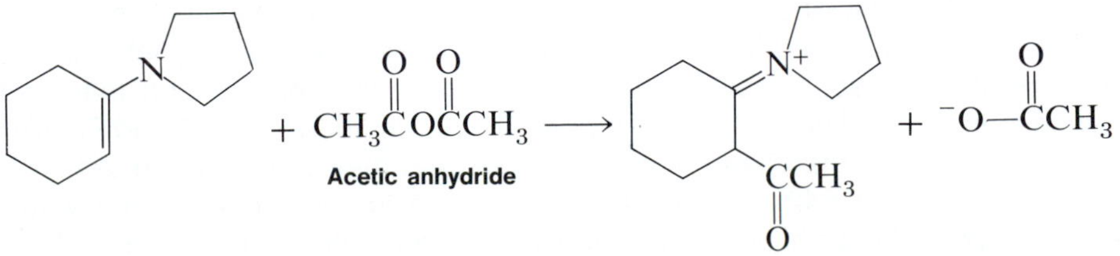

Acetic anhydride

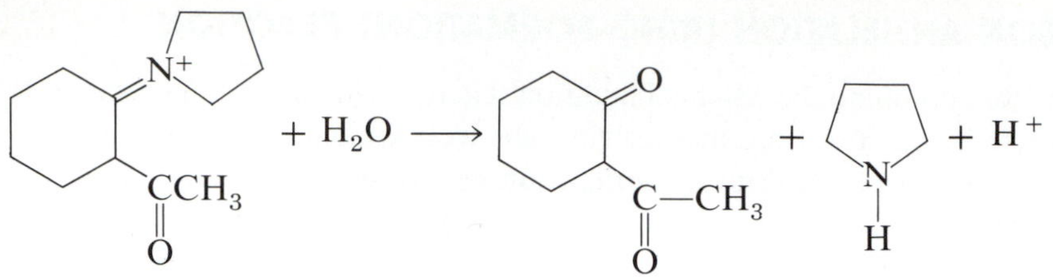

2-Acetylcyclohexanone

REQUIRED READING

Review: Technique 7   Sections 7.5, 7.8, and 7.10
         Technique 8   Section 8.3
         Technique 9   Section 9.1
         Technique 12   Sections 12.6, 12.7, and 12.8

New: Technique 10   Part B, Azeotropes, (Sections 10.7, and 10.8)

SPECIAL INSTRUCTIONS

Pyrrolidine and acetic anhydride are toxic and noxious. You must measure and transfer these substances in a hood. If you are not careful, the entire room will be filled with vapors of pyrrolidine, and it will not be pleasant to work in the laboratory.

The enamine should be made during the first part of the laboratory period and used as soon as possible. Once the acetic anhydride has been added, the reaction mixture must be allowed to stand in your drawer for at least 48 hours to complete the reaction. The second period is used for the work-up and column chromatography. The yields in these reactions are low (less than 20%), partly due to reduced reaction periods necessary to fit the experiment into convenient 3-hour laboratory periods.

PROCEDURE

PREPARATION OF ENAMINE

Place 0.32 mL of cyclohexanone (MW = 98.1) into a preweighed 5-mL conical vial and determine the weight of the material transferred. Add 2.0 mL of toluene to the vial. Place two crystals (about 10 mg) of *p*-toluenesulfonic acid monohydrate in the mixture. In a hood, transfer 0.27 mL of pyrrolidine (MW = 71.1, d = 0.85 g/mL) to this vial from a bottle that has been cooled in ice to reduce its volatility. Add a magnetic spin vane. Attach a Hickman head, a water-cooled condenser, and a drying tube that contains moistened glass wool. The apparatus is shown in Figure 10–15, p 686 and Figure 8–5, p 643. If a ported Hickman head is not available, you may use an unported Hickman head (Figure 8–4A, p 642). In the procedure which follows, toluene and water are collected in the Hickman head, while the enamine remains in the conical vial.

Distill the mixture with a sand bath at about 190 °C and collect the distillate in the Hickman head (Technique 8, Section 8.3, p 641 and Figure 8–5, p 643). You will need to remove 1 mL of distillate. Since some Hickman stills have capacities of less than 1 mL, you should remove the distillate each time the reservoir is filled. Detach the condenser (or open the port), remove the distillate, and transfer it to a conical vial for storage. Continue to remove the distillate until 1 mL of liquid has been removed from the Hickman head and transferred to the conical vial (use the graduations on the vial for measurement).

Water is formed in the reaction as the enamine is produced, and it azeotropes with the solvent, toluene, and collects in the Hickman head. Only a small amount of water is produced in this reaction, and it is soluble at elevated temperatures in the Hickman head (no cloudiness). When you remove the distillate from the Hickman head, the liquid cools rapidly, and the mixture will become cloudy as water separates from the toluene. Discard the azeotrope that you collected in the conical vial.

After the distillation has been completed, allow the reaction mixture to cool to room temperature. Remove the vial and prepare 2-acetylcyclohexanone as described in the next section. Proceed to the next step during this laboratory period.

2-ACETYLCYCLOHEXANONE

In a hood, dissolve 0.32 mL of acetic anhydride (MW = 102.1, d = 1.08 g/mL) in 0.5 mL of toluene in another conical vial. Using a Pasteur pipet, add this solution to the enamine that you have prepared above. Cap the vial, stir it for a few minutes at room temperature, and allow the mixture to stand for at least 48 hours.

Following this period, add 0.5 mL of water. Attach a water-cooled condenser (without the Hickman head) and boil the mixture for 30 minutes in a sand bath at about 150 °C. Cool the vial to room temperature and remove the magnetic spin vane with forceps. Add another 0.5 mL of water, cap the vial, shake it, and allow the layers to separate. The 2-acetylcyclohexane is contained in the upper toluene layer. Remove the lower aqueous layer and discard it (Technique 7, Section 7.5, p 624).

Add 1 mL of 6M hydrochloric acid to the toluene layer remaining in the vial, and shake the mixture to extract any nitrogen-containing contaminants from the organic phase. After allowing the layers to separate, remove the lower aqueous layer and discard it. Finally, shake the organic phase with 0.5 mL of water, and remove the lower aqueous layer (discard). Using a dry Pasteur pipet, transfer the organic layer to a dry conical vial, and add granular anhydrous sodium sulfate (2 microspatulafuls measured in the V-grooved end) to dry the organic layer. Using a **dry** filter tip pipet, transfer the dried organic phase from the drying agent and place it in a **dry** 5-mL conical vial. Rinse the drying agent with a minimum amount of fresh toluene and add this to the vial.

Evaporate the toluene in a sand bath at about 80 °C, using a stream of dry air or nitrogen. When the toluene has all been removed, the volume of liquid will remain constant (0.1 to 0.3 mL). Save the yellow liquid residue for purification by column chromatography.

Preweigh a 5-mL conical vial for use in collecting the material eluted from the column. Prepare a column for column chromatography using a 5¾ inch Pasteur pipet as a column (Technique 12, Section 12.6, Part A, p 707, and Part B, p 709). Use the Dry

Pack Method 2 on p 711 with alumina as the adsorbent and methylene chloride as the eluent. Place a small piece of cotton in the pipet, and gently push it into the constriction. Pour 1.0 g of alumina[1] into the column. Obtain 4 mL of methylene chloride in a graduated cylinder. The methylene chloride will be used to prepare the column, dissolve the crude product, and elute the purified product as described below.

Dissolve the crude product in 10 drops of methylene chloride. Clamp the column above the preweighed 5-mL conical vial. Then add about 1 mL of the methylene chloride to the column and allow it to percolate through the alumina. Allow the solvent to drain until the solvent surface just begins to enter the alumina. Add the crude product to the top of the column, and allow the mixture to pass onto the column. Use about 0.5 mL of methylene chloride to rinse the vial that contained the crude product. When the first batch of crude product has drained so that the surface of the liquid just begins to enter the alumina, place the methylene chloride rinse on the column.

When the solvent level has again reached the top of the alumina, add more methylene chloride with a Pasteur pipet to elute the product into the conical vial. You should place 2 mL of methylene chloride, in portions, on the column to elute the product. Collect all of the liquid that passes through the column as one fraction. Place the conical vial in a warm sand bath (about 70 °C) and evaporate the methylene chloride with a light stream of air or nitrogen in a hood to give the 2-acetylcyclohexanone as a yellow liquid. If necessary use a few drops of methylene chloride to rinse the product from the side of the vial into the bottom. Evaporate this solvent. When the solvent has been removed, reweigh the vial to determine the weight of product. Calculate the percentage yield (MW = 140.2).

At the option of the instructor, determine the infrared spectrum and/or the NMR spectrum. The NMR spectrum may be used to determine the percentage enol content for 2-acetylcyclohexanone. This compound is highly enolic, giving a calculated value in excess of 70%.[2] Submit the remaining sample to the instructor in a labeled vial with your laboratory report.

REFERENCES

Augustine, R. L., and Caputo, J. A. "$\Delta^{1,9}$-2-Octalone." *Organic Syntheses*, Coll. Vol. 5 (1973): 869.
Cook, A. G., ed. *Enamines: Synthesis, Structure, and Reactions.* New York: Marcel Dekker, 1969.
Dyke, S. F. *The Chemistry of Enamines.* London: Cambridge Univ. Press, 1973.

---

[1]EM Science (No. AX0612-1). The particle sizes are 80-200 mesh and the material is Type F-20.

[2]The percentage enol content can be calculated using the NMR spectrum reproduced in this experiment. The offset peak is assigned to the enolic hydrogen (integral height, 10 mm). The remaining absorptions at 1.5 to 2.85 δ (integral height, 155 mm) are assigned to the 11 protons remaining in the enol structure and the 12 protons in the keto structure. Thus, 110 mm (10 × 11) of the 155 mm integral height is assigned to the enol hydrogens. Enol % = 110/155 = 71; keto % = 45/155 = 29.

Mundy, B. P. "The Synthesis of Fused Cycloalkenones via Annelation Methods." *Journal of Chemical Education, 50* (1973): 110.

Stork, G., Brizzolara, A., Landesman, H., Szmuszkovicz, J., and Terrell, R. "The Enamine Alkylation and Acylation of Carbonyl Compounds." *Journal of the American Chemical Society, 85* (1963): 207.

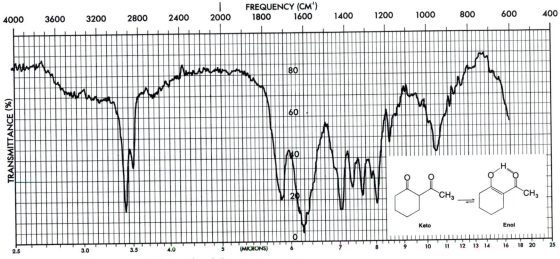

Infrared spectrum of 2-acetylcyclohexanone, neat

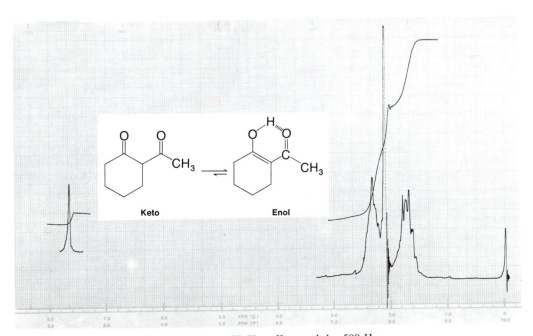

NMR spectrum of 2-acetylcyclohexanone, CDCl$_3$, offset peak by 500 Hz

QUESTIONS

1. Draw a mechanism for the enamine synthesis of $\Delta^{1,9}$-2-octalone. Why is this octalone rather than the $\Delta^{9,10}$-2-octalone the main product in the reaction? On the other hand, why is there a relatively large amount of the $\Delta^{9,10}$-2-octalone produced in the reaction?

2. (a) The enamine formed from pyrrolidine and 2-methylcyclohexanone has the **A** structure. What reason can you give for the less substituted enamine being formed instead of the more substituted enamine, **B**? (*Hint:* Consider steric effects.)

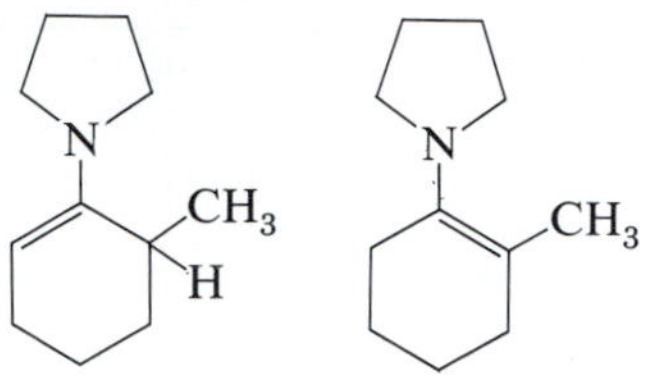

(A)   (B)

(b) Draw the structure of the product that would result from the reaction of enamine **A** with methyl vinyl ketone. Compare its structure with the product obtained in Question 3.

3. (a) The enolate formed from 2-methylcyclohexanone has the structure shown below. What is the structure of the other possible enolate, and why is it not as stable as the one shown here?

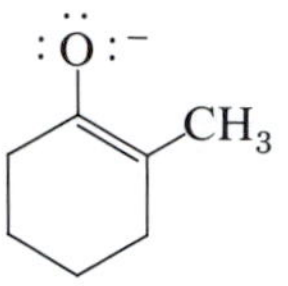

(b) Draw the structure of the product that would result from the reaction with methyl vinyl ketone. Compare its structure with the product obtained in Question 2.

4. Draw the structures of the Robinson annelation products that would result from the following reactions.

(a) 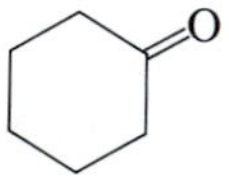 $+ \ HC\equiv CCCH_3 \longrightarrow$

(b) 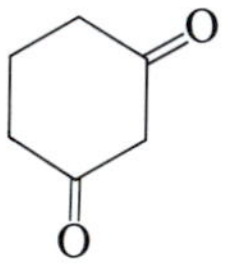 $+ \ CH_3CH_2CCH=CH_2 \longrightarrow$

5. Draw the structures of the products that would result from the following enamine reactions. Use pyrrolidine as the amine and write equations for the reaction sequence.

(a) 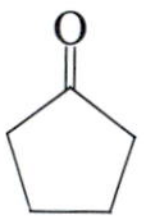 $+ \ Cl-C-OCH_3 \longrightarrow$

(b) 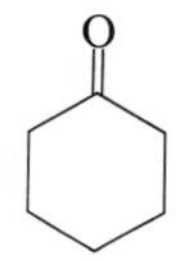 $+ \ CH_2=C-CO_2CH_3 \longrightarrow$

(c) + $CH_3I \longrightarrow$ (see Question 2)

(d) + $CH_2{=}CH{-}CH_2{-}Br \longrightarrow$

6. Interpret the infrared spectrum of 2-acetylcyclohexanone, especially in the O—H and C=O stretch regions of the spectrum.

7. Calculate the amount of water produced during the formation of the enamine in this experiment.

8. Write equations showing how one could carry out the following multistep transformation, starting from the indicated materials. One need not use an enamine synthesis.

$+ CH_3\overset{O}{\overset{\|}{C}}CH_2Cl \longrightarrow$

Experiment 34
Tetraphenylcyclopentadienone

Aldol condensation

In this experiment tetraphenylcyclopentadienone is prepared by the reaction of dibenzyl ketone (1,3-diphenyl-2-propanone) with benzil (Experiment 30) in the presence of base.

| Dibenzyl ketone | Benzil | Tetraphenylcyclopentadienone |

This reaction proceeds via an aldol condensation reaction, with dehydration giving the purple unsaturated cyclic ketone. A stepwise mechanism for the reaction may proceed as follows:

The aldol intermediate, **A**, readily loses water to give the highly conjugated system, **B**, which reacts further by intramolecular aldol condensation to give the dienone product following dehydration.

REQUIRED READING

Review: Technique 4   Filtration, Section 4.3
          Technique 5   Crystallization, Section 5.4

SPECIAL INSTRUCTIONS

This reaction can be completed in about one hour. The product can then be converted to 1,2,3,4-tetraphenylnaphthalene (Experiment 50).

The ethanolic potassium hydroxide solution should be prepared in advance by the instructor.

PROCEDURE

Add 0.100 g of benzil (Experiment 30), 0.100 g of dibenzyl ketone (1,3-diphenyl-2-propanone, 1,3-diphenylacetone), and 0.80 mL of absolute ethanol to a 3-mL conical vial. Place a spin vane in the vial and attach a water-cooled condenser. Heat the mixture with stirring in a sand bath at about 80 °C until the solids dissolve.

Raise the temperature of the sand bath until the mixture is just below its boiling point. Continue to stir the mixture. Using a 9-inch Pasteur pipet, carefully add dropwise 0.15 mL of ethanolic potassium hydroxide solution[1] downward through the condenser into the vial.

> **CAUTION: Foaming may occur.**

The mixture will immediately turn deep purple. Once the potassium hydroxide has been added, increase the temperature of the sand bath until the mixture is gently boiling. Heat the mixture with stirring at a gentle boil for 15 minutes.

At the end of the heating period, remove the vial from the sand bath and allow the mixture to cool to room temperature. Then place the vial in an ice-water bath for five minutes to complete crystallization of the product. Collect the deep purple crystals on a Hirsch funnel. Hold the spin vane with forceps and scrape off as much solid as possible. Wash the crystals with three 0.5-mL portions of cold 95% ethanol. The rinse solvent can also be used to aid in the transfer of crystals from the conical vial to the Hirsch funnel and in removing the remainder of the crystals from the spin vane. Dry the tetraphenyl-cyclopentadienone in an oven for 30 minutes or in air overnight.

The crude product is pure enough (mp 218–220 °C) for the preparation of 1,2,3,4-tetraphenylnaphthalene (Experiment 50). Weigh the product and calculate the percentage yield. Determine the melting point. A small portion may be recrystallized, if desired, from a 1:1 mixture of 95% ethanol and toluene (12 mL/0.5 g; mp 219–220 °C). A Craig tube (Technique 5, Section 5.4, and Figure 5–5, p 586) should be used for the crystallization. At the instructor's option, determine the infrared spectrum of tetraphenyl-cyclopentadienone in potassium bromide (Technique 18, Section 18.4, p 775). Submit the product to the instructor in a labeled vial or save it for Experiment 50.

QUESTIONS

1. Interpret the infrared spectrum of tetraphenylcyclopentadienone.
2. Draw the structure of the product you would expect from the reaction of benzaldehyde and acetophenone with base.
3. Suggest several possible by-products of this reaction.

[1] NOTE TO THE INSTRUCTOR: This solution is prepared by dissolving 0.40 g of potassium hydroxide in 4.0 mL of absolute ethanol. It will take about 30 minutes for the solid to dissolve with vigorous stirring. As the solid dissolves, crush the pieces with a spatula to aid in the solution process. This will provide enough solution for 20 students, assuming little material is wasted.

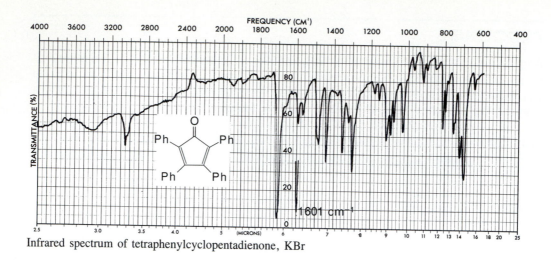

Infrared spectrum of tetraphenylcyclopentadienone, KBr

Experiment 35
5,5-Dimethyl-1,3-cyclohexanedione (Dimedone)

Working with sodium ethoxide
Michael reaction (conjugate addition)
Claisen condensation (ketone + ester)
Decarboxylation reaction
Keto-enol tautomerism
Derivative formation

In this experiment, you prepare 5,5-dimethyl-1,3-cyclohexanedione (dimedone) by a series of important organic reactions: a Michael addition reaction, a Claisen-type reaction of a ketone with an ester, and a decarboxylation. In addition, you will gain experience in handling sodium ethoxide and in conducting a reaction under anhydrous conditions.

The first step involves the use of a strong base, sodium ethoxide, to deprotonate diethyl malonate to give the sodium salt that is used in the Michael addition reaction.

$$CH_3CH_2-O-\overset{\displaystyle O}{\overset{\displaystyle \|}{C}} \diagdown CH_2 + Na^+ \ ^- :\overset{..}{\underset{..}{O}}-CH_2CH_3 \longrightarrow$$
$$CH_3CH_2-O-\underset{\displaystyle O}{\underset{\displaystyle \|}{C}} \diagup$$

Diethyl malonate   **Sodium ethoxide**

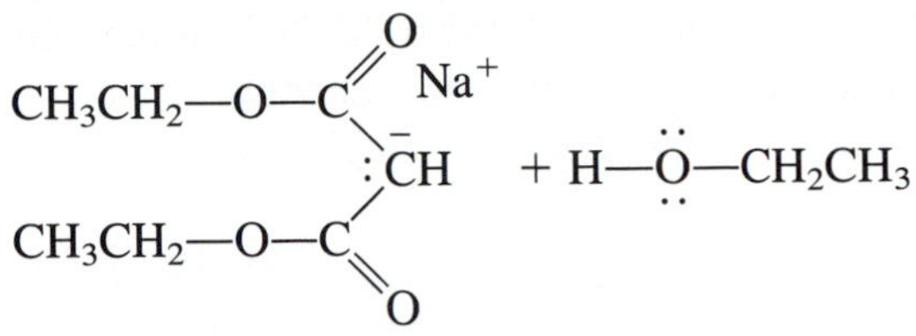

$$CH_3CH_2-O-C \overset{O}{\underset{}{}} \quad Na^+$$
$$\overset{-}{:}CH \quad + \quad H-\overset{..}{\underset{..}{O}}-CH_2CH_3$$
$$CH_3CH_2-O-C \overset{}{\underset{O}{}}$$

Sodium salt

The sodium salt of diethyl malonate, acting as a nucleophile, attacks the β position of the α, β-unsaturated ketone, 4-methyl-3-penten-2-one (mesityl oxide). This reaction is an example of a Michael reaction or a conjugate addition reaction. The carbonyl group in the unsaturated ketone stabilizes the intermediate conjugate base by resonance. Nucleophilic addition to a C=C double bond cannot proceed without the carbonyl group.

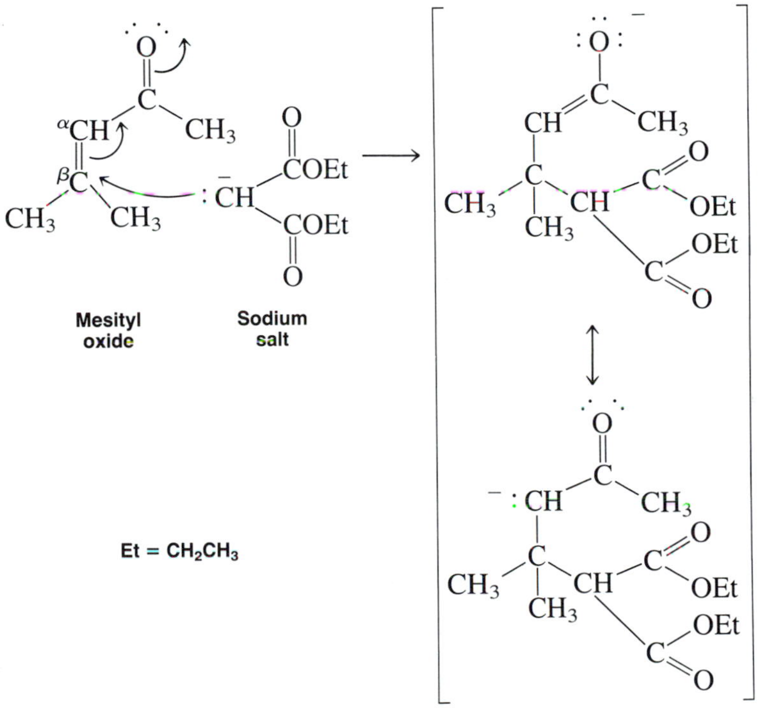

Mesityl oxide   **Sodium salt**

Et = CH₂CH₃

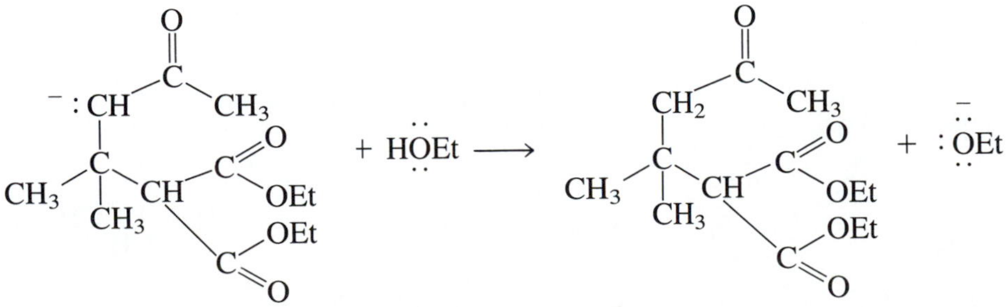

Intermediate (I)

The intermediate (I) is now cyclized by a Claisen-type condensation reaction. Sodium ethoxide is again used as a base to generate a nucleophile. Intramolecular attack by this nucleophile on the ester carbonyl group gives a six-membered ring compound.

At this point in the synthetic scheme, the ester functional group is hydrolyzed with aqueous potassium hydroxide. Once the mixture is neutralized with acid, the resulting β-keto carboxylic acid (II) decarboxylates (loses CO_2) when heated to yield the desired product, 5,5-dimethyl-1,3-cyclohexanedione (dimedone).

Intermediate (II)

5,5-Dimethyl-1,3-cyclo-
hexanedione (dimedone)

Ordinarily, carboxylic acids do not decarboxylate easily. However, when the compound has a keto carbonyl group in the β position (a β-keto acid), it activates the loss of a carboxyl group by way of a cyclic mechanism. The enol tautomer is isomerized to the more stable keto tautomer.

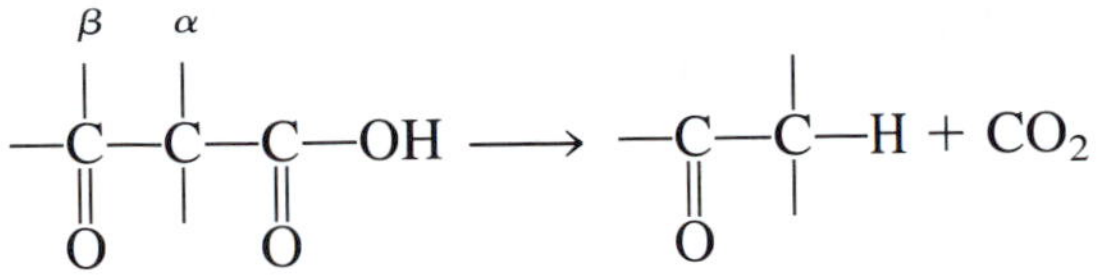

A β-keto acid   A ketone

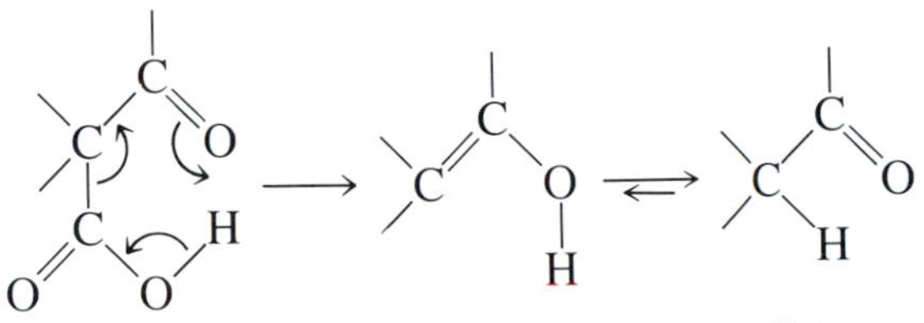

Enol   Keto

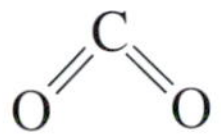

Carbon dioxide

Dimedone, like most 1,3-diketones, exists partially in the enolic form, and both the enol and keto tautomers can be observed by means of proton NMR and infrared spectroscopy.

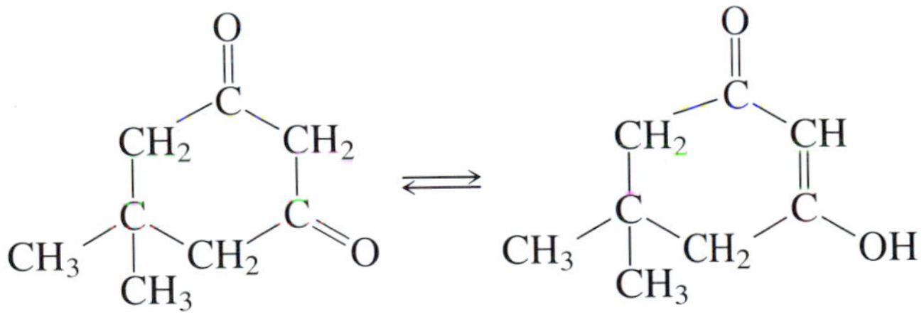

Keto tautomer   Enol tautomer

Dimedone can be used to form derivatives of aldehydes. For example, dimedone reacts readily with formaldehyde in aqueous ethanol to form a solid product. Ketones do not form this derivative.

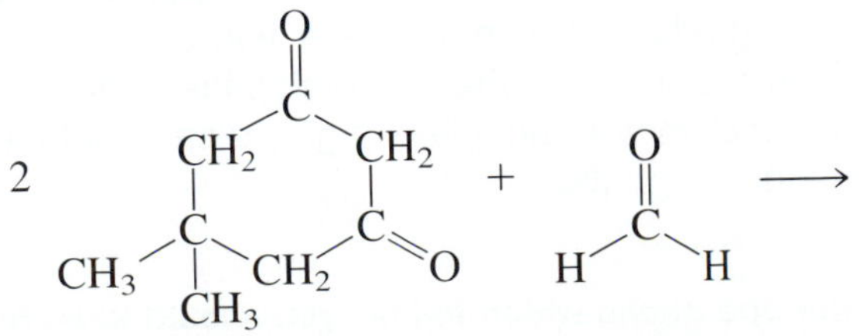

Dimedone   Formaldehyde

**Dimedone derivative
mp 189 °C**

REQUIRED READING

Review: Technique 3   Section 3.2
          Technique 4   Sections 4.3 and 4.7
          Technique 5   Section 5.4

SPECIAL INSTRUCTIONS

The reaction must be carried out under strictly anhydrous conditions; the apparatus must be dry. If possible, dry the necessary glassware prior to starting this experiment and store the equipment in a desiccator. There are two lengthy reflux periods used in this experiment (0.5 and 1.5 hours). You should plan your time carefully so that you can complete the experiment through the 1.5 hour reflux period.

PROCEDURE

It is important to maintain anhydrous conditions while following the procedures below. Dry a 5-mL conical vial, a water-cooled condenser, and a drying tube packed with calcium chloride in an oven at 110 °C for at least 30 minutes. In addition, you should dry a 1-mL and two 0.5-mL graduated pipets, unless you are using an automatic pipet. When the glassware has cooled somewhat, assemble the apparatus for reflux. You can save time by setting the dial on your hot plate to give a sand bath temperature of about 150 °C prior to measuring reagents.

**In the operations which follow, you should keep the
conical vial capped between additions of reagents.**

Ask the instructor or assistant for help in dispensing the sodium ethoxide in etha-nol reagent (restopper it after each use). Transfer 0.80 mL of sodium ethoxide/ethanol[1] solution with a **dry** graduated pipet directly into the **dry** 5-mL conical vial. Place a magnetic spin vane into the vial. Weigh the vial, add 0.30 mL of diethyl malonate (MW = 160.2), and reweigh the vial to give the weight of diethyl malonate transferred. Then add 0.22 mL of 4-methyl-3-penten-2-one (mesityl oxide, MW = 98.1)[2], and reweigh the vial to give the weight of mesityl oxide transferred.

Attach the water-cooled condenser and drying tube. While stirring, **gently** boil the mixture using a sand bath at 150–160 °C for 30 minutes (Figure 3–2A, p 550, with drying tube). During that period, a solid forms and the stirrer stops turning. It may be necessary to raise the position of the vial in the sand bath to prevent bumping.

Remove the apparatus from the sand bath so that it cools for a few minutes and remove the drying tube. Using a graduated pipet, transfer 1.80 mL of an aqueous potas-sium hydroxide solution[3] into a test tube. Use a Pasteur pipet to transfer the aqueous base through the condenser into the conical vial. Most of the solid should dissolve. Boil the mixture gently for 1.5 hours in a sand bath maintained at 145 to 150 °C, while stirring the solution. If you cannot finish the reflux during this laboratory period, remove the condenser and cap the vial until the next period. You may finish the reflux during the next period.

After the reflux period is over, allow the mixture to cool to the touch and remove the vial. Cool the vial to room temperature in an ice bath. Add concentrated hydrochloric acid dropwise (about 25 drops from a Pasteur pipet) until the pH is about 3. Stir the mixture with a spatula. Note the volume of liquid contained in the vial using the gradua-tions, and boil the mixture (open vial, no condenser) in a sand bath at 145–150 °C for 30 minutes. During this time evaporate 1 mL of liquid. You should notice the evolution of bubbles (CO_2), along with normal boiling action, indicating that decarboxylation is occur-ring. Later, during the heating process, the gas evolution ceases, and the solution changes from a colored homogeneous solution to a cloudy colored mixture. Once this change occurs, the temperature of the sand bath may be increased somewhat to speed the removal of the 1 mL volume specified. An oily layer will form during this 30-minute heating period.

Cool the mixture to the touch and then place the vial in an ice bath. Swirl the mixture. The product will either precipitate after 10 or 15 minutes, or a layer of brownish oil will form. In either case, cap the vial and allow the vial to stand until the next labora-tory period to complete the crystallization process. During that period, crystals should form and the oil should solidify.

Collect all of the solid, even the dark material, on a Hirsch funnel under vacuum (Technique 4, Section 4.3, and Figure 4–6, p 571). Use about 1 mL of **cold water** to

---

[1] This reagent is prepared in advance by the instructor. Carefully dry a 250-mL Erlenmeyer flask and insert a drying tube filled with calcium chloride into a one-hole rubber stopper. Obtain a large piece of sodium, clean it by cutting off the oxidized surface, weigh out a 2.30 g piece, cut it into 20 smaller pieces, and store it under xylene. Using tweezers remove each piece, wipe off the xylene, and add the sodium slowly over a period of about 30 minutes to 40 mL of absolute (**anhydrous**) ethanol in the 250-mL Erlenmeyer flask. After the addition of each piece, replace the stopper. The ethanol will warm as the sodium reacts, but do not cool the flask. After the sodium has been added, warm the solution and shake it gently until the sodium all reacts. Cool the sodium ethoxide solution to room temperature. This reagent may be prepared in advance of the laboratory period, but it **must be stored in a refrigerator between laboratory periods.** When it is stored in a refrigerator, it may be kept for about one week. It should be brought to room temperature, and swirled gently in order to redissolve any precipitated sodium ethoxide. Keep the flask stoppered between each use.

[2] The mesityl oxide must be distilled before use. The instructor should distill a large batch prior to its use in class; bp 128 to 131 °C.

[3] This solution should be prepared by the instructor, dissolve 12 g of potassium hydroxide in 60 mL of water.

wash the product and to aid in the transfer. Remove the product from the magnetic spin vane with a spatula and add the solid to the filter. Weigh the crude dimedone after it has dried for about 10 minutes on the filter.

Transfer all of the solid to a Craig tube. Add the calculated amount of acetone to the tube (1 mL acetone/0.14 g of crude product) and heat the mixture in a sand bath to dissolve the product. Use a microspatula to stir the solution during heating to prevent the solution from bumping. Once the solid dissolves, remove the Craig tube and allow the solution to cool to room temperature. Crystals should separate. After cooling to room temperature, place the mixture in an ice-water bath to complete the crystallization process. Remove the crystals from the filtrate in the usual way (Technique 4, Section 4.7, p 575, and Figure 4–11, p 576). Remove the purified dimedone from the Craig tube. Allow the product to air-dry thoroughly and weigh the product. Calculate the percentage yield and obtain the melting point of the dimedone (literature, 148 °C). At the option of the instructor, obtain the infrared spectrum (KBr) and/or the NMR spectrum (CDCl$_3$). Assign the peaks in each to the enol and keto tautomers. Submit any remaining sample in a labeled vial with your laboratory report.

DIMEDONE DERIVATIVE

Place a small amount (0.010 g) of dimedone in a small test tube (10 × 75 mm). Using a Pasteur pipet, add 25 drops of 95% ethanol and 20 drops of water to dissolve the solid. Add four drops of 37% aqueous formaldehyde solution to the solution. Gently tap the test tube to mix the components. Allow the mixture to stand for at least 30 minutes with occasional agitation of the mixture. You will observe the formation of fine needle crystals. Add 1 mL of water to the tube, collect the crystalline solid by vacuum filtration on a Hirsch funnel, wash them with about 5 mL of water, and allow them to dry until the next laboratory period. Determine the melting point of the dimedone derivative (literature, 189 °C) and submit the sample in a labeled vial with your laboratory report.

QUESTIONS

1. Write products that you would expect from the following Michael addition reactions.

(a)
$$CH_3-C(=O) \atop CH_3CH_2O-C(=O)} \overset{|}{CH} + CH_2{=}CH-\overset{O}{\overset{\|}{C}}-O-CH_3 \xrightarrow{CH_3CH_2OH}$$

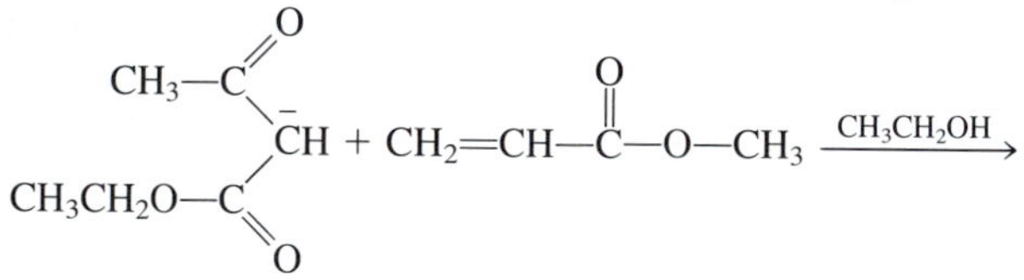

(b) $CH_3CH_2O-\overset{O}{\overset{\|}{C}}$
$\overset{|}{CH} + CH_2{=}CH-C{\equiv}N \xrightarrow{CH_3CH_2OH}$
$N{\equiv}C$

(c) $\overset{O}{\underset{^-O}{{\diagdown}{\overset{+}{N}}{\diagup}}}{-}\overset{-}{CH_2} + CH_2{=}CH-\overset{O}{\overset{\|}{C}}CH_2CH_3 \xrightarrow{CH_3CH_2OH}$

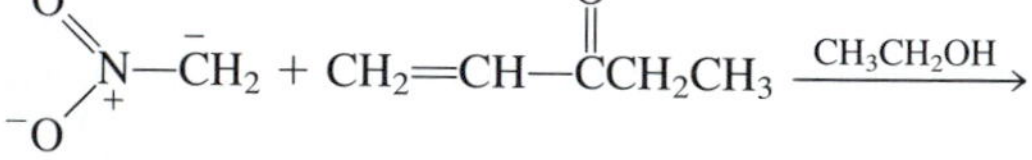

2. Give the structure of the cyclized products expected from the following reactions.

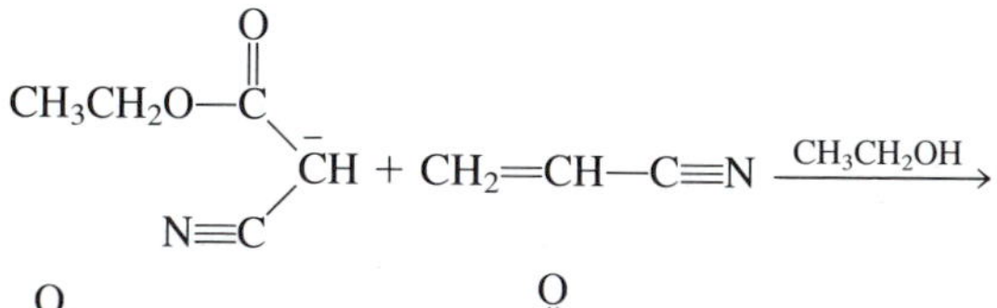

(a) $CH_3-\overset{O}{\overset{\|}{C}}-CH_2-CH_2-CH_2-\overset{O}{\overset{\|}{C}}-OCH_3 \xrightarrow[CH_3CH_2OH]{CH_3CH_2O^-}$

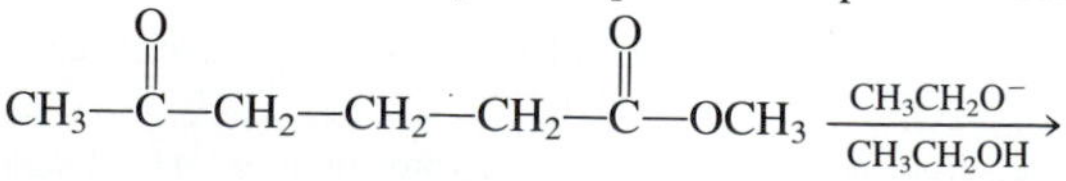

(b) $CH_3CH_2\overset{\displaystyle O}{\overset{\|}{C}}-CH_2-CH_2-CH_2-\overset{\displaystyle O}{\overset{\|}{C}}-OCH_3 \xrightarrow[CH_3CH_2OH]{CH_3CH_2O^-}$

3. What would happen to the sodium ethoxide if water were present in the conical vial or reagents used in the experiment?

4. Draw a mechanism for the preparation of the dimedone derivative of formaldehyde.

5. It is possible to make a dimedone derivative of aldehydes other than formaldehyde. Show the structure for the derivative formed from propanal.

Experiment 36
1,4-Diphenyl-1,3-Butadiene

Wittig reaction
Working with sodium ethoxide
Thin-layer chromatography
UV/NMR spectroscopy

The Wittig reaction is often used to form alkenes from carbonyl compounds. In this experiment the isomeric dienes *cis-, trans-,* and *trans, trans*-1,4-diphenyl-1,3-butadiene will be formed from cinnamaldehyde and benzyltriphenylphosphonium chloride Wittig reagent. Only the *trans, trans* isomer will be isolated.

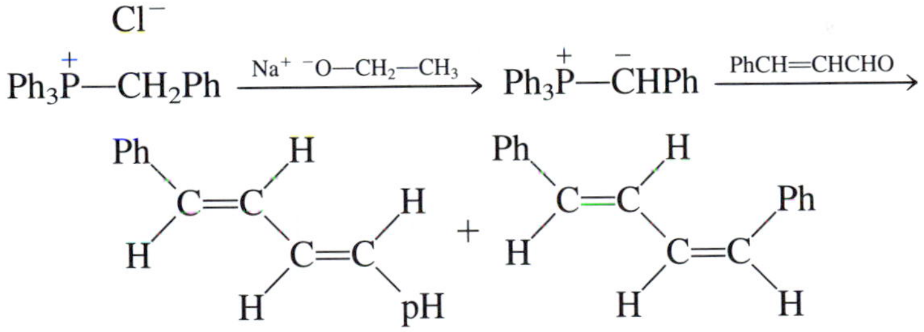

trans,trans   cis,trans

The reaction is carried out in two steps. First, the phosphonium salt is formed by the reaction of triphenylphosphine with benzyl chloride. The reaction is a simple nucleophilic displacement of chloride ion by triphenylphosphine. The salt that is formed is called the "Wittig reagent" or "Wittig salt."

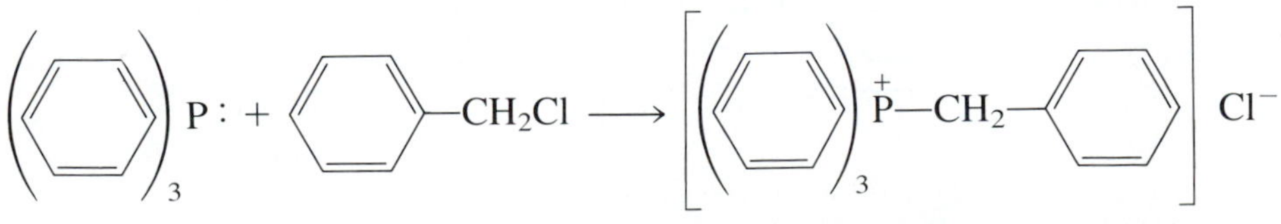

Benzyltriphenylphosphonium chloride

"Wittig salt"

When treated with base, the Wittig salt forms an **ylide.** An ylide is a species having adjacent atoms oppositely charged. The ylide is stabilized due to the ability of phosphorus to accept more than eight electrons in its valence shell. Phosphorus uses its 3d orbitals to form the overlap with the 2p orbital of carbon that is necessary for resonance stabilization. Resonance stabilizes the carbanion.

An ylide

The ylide is a carbanion that acts as a nucleophile, and it adds to the carbonyl group in the first step of the mechanism. Following the initial nucleophilic addition, a remarkable sequence of events occurs, as outlined in the following mechanism:

Triphenylphosphine oxide   **An alkene**

The addition intermediate, formed from the ylide and the carbonyl compound, cyclizes to form a four-membered-ring intermediate. This new intermediate is unstable and fragments into an alkene and triphenylphosphine oxide. Notice that the ring breaks open in a different way than it was formed. The driving force for this ring opening process is the formation of a very stable substance, triphenylphosphine oxide. A large decrease in potential energy is achieved upon the formation of this thermodynamically stable compound.

In this experiment, cinnamaldehyde is used as the carbonyl compound and yields mainly the *trans, trans*-1,4-diphenyl-1,3-butadiene which is obtained as a solid. The *cis, trans* isomer is formed in smaller amounts, but it is an oil which is not isolated in this experiment. The *trans, trans* isomer is the more stable isomer and is formed preferentially.

Cinnamaldehyde

***trans,trans*-1,4-Diphenyl-1,3-butadiene**   **Triphenylphosphine oxide**

REQUIRED READING

Review: Technique 4   Section 4.3
          Technique 13

SPECIAL INSTRUCTIONS

A 1.5-hour reflux period is employed in this experiment. Considerable time may be saved if this reflux is conducted concurrently with another experiment. Triphenylphosphine is rather toxic. Be careful not to inhale the dust. Benzyl chloride is a skin irritant and a lachrymator. It should be handled in the hood with care. The prepared sodium ethoxide solution must be kept tightly stoppered when not in use as it reacts readily with atmospheric water. Cinnamaldehyde is easily oxidized in air. If the bottle is old, the cinnamaldehyde must be purified, as described on p 257 for benzaldehyde.

PROCEDURE

BENZYLTRIPHENYLPHOSPHONIUM CHLORIDE (WITTIG SALT)

Place 0.550 g of triphenylphosphine (MW = 262.3) into a 5-mL conical vial. In a hood, transfer 0.36 mL of benzyl chloride (MW = 126.6, d = 1.10 g/mL) to the vial and add 2.0 mL of xylenes (mixture of o-, m-, and p-isomers).

> **CAUTION: Benzyl chloride is a lachrymator: a tear producing substance.**

Add a magnetic spin vane to the conical vial and attach a water-cooled condenser. Boil the mixture in a sand bath at 195–200 °C for at least 1.5 hours. An increased yield may be expected when the mixture is heated for longer periods. In fact, you may begin heating the mixture before the temperature has reached the values given, but do not include this time in the 1.5-hour reaction period. The solution will be homogeneous at first, and then the Wittig salt will begin to precipitate. Maintain the stirring during the entire heating period or bumping may occur. Remove the apparatus from the sand bath and allow it to cool for a few minutes. Remove the vial and cool it thoroughly in an ice bath for about five minutes.

Collect the Wittig salt by vacuum filtration using a Hirsch funnel. Use three 1-mL portions of cold petroleum ether (bp 60–90 °C) to aid the transfer and to wash the crystals free of the xylene solvent. Dry the crystals, weigh them, and calculate the percentage yield of the Wittig salt. At the option of the instructor, obtain the proton NMR spectrum of the salt in $CDCl_3$. The methylene group appears as a doublet (J = 14 Hz) at 5.5δ because of ^1H-^{31}P coupling.

1,4-DIPHENYL-1,3-BUTADIENE

In the following operations, cap the 5-mL conical vial whenever possible to avoid contact with moisture from the atmosphere. Place 0.320 g of benzyltriphenylphosphonium chloride (MW = 388.9) in a **dry** 5 mL-conical vial. Add a magnetic spin vane. Transfer 1.5 mL of absolute (anhydrous) ethanol to the vial and stir the mixture to dissolve the phosphonium salt (Wittig salt). Add 0.50 mL of sodium ethoxide solution[1] to the vial using a **dry** pipet, while maintaining continuous stirring. Cap the vial and stir this mixture for 15 minutes. During this period, the cloudy solution acquires the characteristic yellow color of the ylide.

Measure 0.10 mL of cinnamaldehyde (MW = 132.2, d = 1.11 g/mL) and place it in another small conical vial. To the cinnamaldehyde, add 0.50 mL of absolute ethanol. Cap the vial until it is needed. After the 15 minute period, use a Pasteur pipet to mix the cinnamaldehyde with the ethanol and add this solution to the ylide in the reaction vial. A color change should be observed as the ylide reacts with the aldehyde and the product precipitates. Stir the mixture for 10 minutes.

Cool the vial thoroughly in an ice-water bath (10 min), stir the mixture with a spatula, and transfer the material from the vial to a Hirsch funnel under vacuum. Use two 1-mL portions of ice cold absolute ethanol to aid the transfer and to rinse the product. Dry the crystalline *trans, trans*-1,4-diphenyl-1,3-butadiene by drawing air through the solid. The product has a small amount of sodium chloride that is removed as described below. The cloudy material in the filter flask contains triphenylphosphine oxide, the *cis, trans*- isomer, and some *trans, trans* product. Pour the filtrate into a beaker and save it for the thin-layer chromatography experiment described in the next paragraph. Remove the solid product from the filter paper, place the solid in a 10-mL beaker, and add 3 mL of water. Stir the mixture and filter it on a Hirsch funnel, under vacuum, to collect the nearly colorless crystalline *trans, trans*-1,4-diphenyl-1,3-butadiene. Use about 1 mL of water to aid the transfer. Allow the solid to dry thoroughly.

Analyze the filtrate that you saved in the beaker by thin-layer chromatography during the same laboratory period so that the *cis, trans* isomer will not be photochemically converted to the *trans, trans* compound. Use a 2 × 8 cm silica gel TLC plate that has a fluorescent indicator (Eastman Chromatogram Sheet, No. 13181). At one position on the TLC plate, spot the filtrate, as is, without dilution. Dissolve a few crystals of the *trans, trans*-1,4-diphenyl-1,3-butadiene in a few drops of acetone and spot it at another position on the plate. Use petroleum ether (bp 60–90°) as a solvent to develop (run) the plate, and visualize the spots with a UV lamp using both the long and short wavelength settings. A typical set of R_f values is:

Ph_3P^+—O^-	0
trans, trans diene	0.36 (fluoresces brilliantly)
cis, trans diene	0.47

The filtrate contains all of the above compounds, and you should observe spots for each of these substances. The solid that you spotted on the plate should be nearly pure *trans, trans*-isomer. Report the results that you obtain including R_f values and the appearance of the spots under illumination. Discard the filtrate.

---

[1]The sodium ethoxide solution is prepared by dissolving 2.30 g of sodium in 40 mL of absolute (anhydrous) ethanol as described in the instructor's manual and in Footnote 1 of Experiment 35.

When the *trans, trans*-1,4-diphenyl-1,3-butadiene is dry, determine the melting point (literature, 152 °C). Weigh the solid and determine the percentage yield. If the melting point is below 145 °C, recrystallize a portion of the compound from hot 95% ethanol (20 mg/1.3 mL ethanol) in a Craig tube. Redetermine the melting point.

At the option of the instructor, obtain the proton NMR in $CDCl_3$ or the UV spectrum in hexane. For the UV spectrum of the product, dissolve a 10-mg sample in 100 mL of hexane in a volumetric flask. Remove 10 mL of this solution and dilute it to 100 mL in another volumetric flask. This concentration should be adequate for analysis. The *trans, trans* isomer absorbs at 328 nm and possesses fine structure, while the *cis,trans* isomer absorbs at 313 nm and has a smooth curve.[2] See if your spectrum is consistent with the above observations. Submit the spectral data with your laboratory report.

QUESTIONS

1. There is an additional isomer of 1,4-diphenyl-1,3-butadiene, mp 70 °C, which has not been shown in this experiment. Draw the structure and name it. Why is it not produced in this experiment? (*Hint:* the cinnamaldehyde has *trans* stereochemistry.)

2. Why should the *trans, trans* isomer be the thermodynamically most stable one?

3. A lower yield of phosphonium salt is obtained in refluxing benzene than in xylene. Look up the boiling points for these solvents and explain why the difference in boiling points might influence the yield.

4. Outline a synthesis for *cis* and *trans* stilbene (the 1,2-diphenylethenes) using the Wittig reaction.

5. The sex attractant of the female housefly (*Musca domestica*) is called **muscalure,** and its structure is shown below. Outline a synthesis of muscalure, using the Wittig reaction. Will your synthesis lead to the required *cis* isomer?

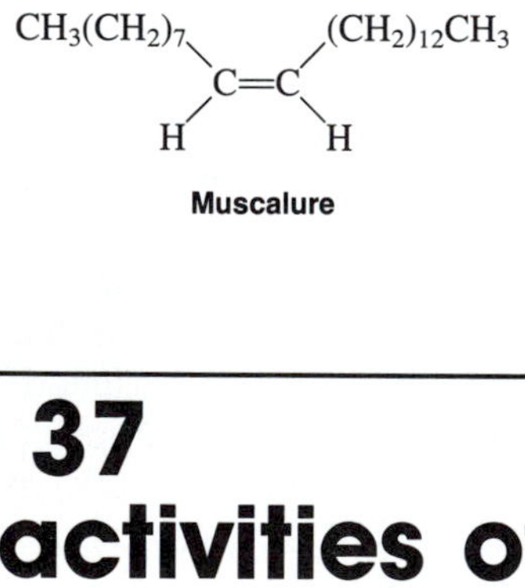

$$CH_3(CH_2)_7 \diagdown \qquad \diagup (CH_2)_{12}CH_3$$
$$C = C$$
$$H \diagup \qquad \diagdown H$$

Muscalure

Experiment 37
Relative Reactivities of Several Aromatic Compounds

Aromatic substitution
Relative activating ability of aromatic substituents
Crystallization

When substituted benzenes undergo electrophilic aromatic substitution reactions, both the reactivity and the orientation of the electrophilic attack are affected by the nature of

---

[2]The comparative study of the stereoisomeric 1,4-diphenyl-1,3-butadienes has been published: J. H. Pinckard, B. Wille and L. Zechmeister, *Journal of the American Chemical Society, 70* (1948): 1938.

the original group attached to the benzene ring. Substituent groups that make the ring more reactive than benzene are called **activators.** Such groups are also said to be **ortho, para** directors because the products formed are those in which substitution occurs either ortho or para to the activating group. Various products may be formed depending on whether substitution occurs at the ortho or para position and the number of times substitution occurs on the same molecule. Some groups may activate the benzene ring so strongly that multiple substitution consistently occurs, while other groups may be moderate activators and benzene rings containing such groups may undergo only a single substitution. The purpose of this experiment is to determine the relative activating effect of several substituent groups.

In this experiment you will study the bromination of acetanilide, aniline, and anisole:

Acetanilide   Aniline   Anisole

The acetamido group, $-NHCOCH_3$; the amino group, $-NH_2$; and the methoxy group, $-OCH_3$ are all activators and ortho, para directors. Each student will carry out the bromination of one of these compounds and determine its melting point. By sharing your data, you will have information on the melting points of the brominated products for acetanilide, aniline, and anisole. Using the following table it will then be possible for you to rank the three substituents in order of activating strength.

Melting Points of Relevant Compounds

COMPOUND	MELTING POINTS (°C)
o-Bromoacetanilide	99
p-Bromoacetanilide	168
2,4-Dibromoacetanilide	145
2,6-Dibromoacetanilide	208
2,4,6-Tribromoacetanilide	232
o-Bromoaniline	32
p-Bromoaniline	66
2,4-Dibromoaniline	80
2,6-Dibromoaniline	87
2,4,6-Tribromoaniline	122
o-Bromoanisole	3
p-Bromoanisole	13
2,4-Dibromoanisole	60
2,6-Dibromoanisole	13
2,4,6-Tribromoanisole	87

The classical method of brominating an aromatic compound is to use Br_2 and a catalyst such as $FeBr_3$, which acts as a Lewis acid. The first step is the reaction between bromine and the Lewis acid:

$$Br_2 + FeBr_3 \longrightarrow [FeBr_4{}^- \ Br^+]$$

The positive bromine ion then reacts with the benzene ring in an aromatic electrophilic substitution reaction:

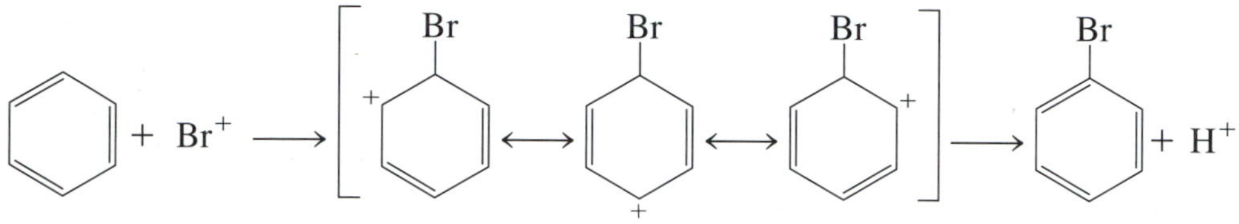

Aromatic compounds that contain activating groups can be brominated without the use of the Lewis acid catalyst since the π electrons in the benzene ring are more available and polarize the bromine molecule sufficiently to produce the required electrophile, Br^+. This is illustrated by the first step in the reaction between anisole and bromine:

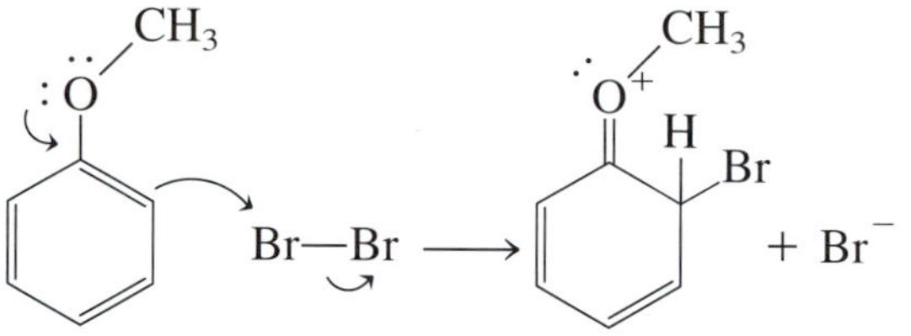

In this experiment, the brominating mixture consists of bromine, hydrobromic acid (HBr), and acetic acid. The presence of bromide ion from the hydrobromic acid helps to solubilize the bromine and increase the concentration of the electrophile.

REQUIRED READING

Review: Technique 5

You should review the chapters in your lecture textbook that deal with electrophilic aromatic substitution. Pay special attention to halogenation reactions and the effect of activating groups.

SPECIAL INSTRUCTIONS

Bromine is a skin irritant and its vapors cause severe irritation to the respiratory tract. It will also oxidize many pieces of jewelry. Hydrobromic acid may cause skin or eye irritation. Aniline is highly toxic and a suspected teratogen. All bromoanilines are toxic.

Each person will carry out the bromination of only one of the aromatic compounds according to your instructor's directions. The procedures are identical except for the initial compound used and the final recrystallization step.

> **NOTE TO THE INSTRUCTOR:** The brominating mixture should be prepared in advance.

PROCEDURE

To a tared 5-mL conical vial, add the given amount of **one** of the following compounds: 0.090 g of acetanilide, 0.060 mL of aniline, or 0.070 mL of anisole. Reweigh the conical vial to determine the actual weight of the aromatic compound. Add 0.5 mL of glacial acetic acid along with a spin vane to the conical vial. Attach an air condenser and place the conical vial in a water bath at 23–27 °C, as shown in Figure 2–5, p 544. Stir the mixture until the aromatic compound is completely dissolved. While the compound is dissolving, pack a drying tube loosely with glass wool. Add about 0.5 mL of 1M sodium bisulfite dropwise to the glass wool until it is moistened but not soaked. This apparatus will capture any bromine given off during the following reaction.

Under the hood, obtain 1.0 mL of the bromine/hydrobromic acid mixture[1] in a 3-mL conical vial. Place the cap on the vial before returning to your lab bench. With stirring, add all of the bromine/hydrobromic acid mixture through the top of the air condenser using a Pasteur pipet. **Be careful not to spill any of this mixture.** Attach the drying tube prepared above. Continue stirring the reaction mixture for 20 minutes.

When the reaction is complete, transfer the mixture to a 10-mL Erlenmeyer flask containing 5 mL of water and 0.5 mL of saturated sodium bisulfite solution. Stir this mixture with a glass stirring rod until the red color of bromine disappears. If an oil has formed, it may be necessary to stir the mixture for several minutes. Place the Erlenmeyer flask in an ice bath for 10 minutes. If the product doesn't solidify, scratch the bottom of the flask with a glass stirring rod to induce crystallization. It may take 10–15 minutes to induce crystallization of the brominated anisole product. Filter the product on a Hirsch funnel with suction and rinse with several 1-mL portions of cold water. Air dry the product on the funnel for about five minutes with the aspirator on.

If you started with **aniline,** transfer the solid to a 10-mL Erlenmeyer flask and recrystallize the product from 95% ethanol (see Technique 5, Section 5.3 and Figure 5–3, p 581). Filter the crystals on a Hirsch funnel and dry them for several minutes with suction. The brominated products from either **acetanilide** or **anisole** should be crystallized using a Craig tube (Technique 5, Section 5.4, and Figure 5–5, p 586). Use 95% ethanol to crystallize the acetanilide product and hexane to crystallize the brominated anisole compound. Allow the crystals to air-dry and determine the weight and melting point.

---

[1] NOTE TO THE INSTRUCTOR: The brominating mixture is prepared by adding 2.6 mL of bromine to 17.4 mL of 48% hydrobromic acid. This will provide enough solution for 20 students, assuming no waste of any type. This solution should be stored in the hood.

Based on the melting point and the preceding table, you should be able to identify your product. Calculate the percentage yield and submit your product, along with your report, to your instructor.

REPORT

By collecting data from other students, you should be able to determine which product was obtained from the bromination of each of the three aromatic compounds. Using this information, arrange the three substituent groups (acetamido, amino, and methoxy) in order of decreasing ability to activate the benzene ring.

REFERENCE

Zaczek, N. M., and Tyszklewicz, R. B. "Relative Activating Ability of Various Ortho, Para-Directors." *Journal of Chemical Education, 63* (1986): 510.

QUESTIONS

1. Using resonance structures, show why the amino group is activating. Consider an attack by the electrophile, E^+, at the *para* position.
2. For the substituent in this experiment that was found to be least activating, explain why bromination took place at the position on the ring indicated by the experimental results.
3. What other experimental techniques (including spectroscopy) might be used to identify the products in this experiment?

Experiment 38
Nitration of Methyl Benzoate

Aromatic substitution
Crystallization

The nitration of methyl benzoate to prepare methyl *m*-nitrobenzoate is an example of an electrophilic aromatic substitution reaction, in which a proton of the aromatic ring is replaced by a nitro group:

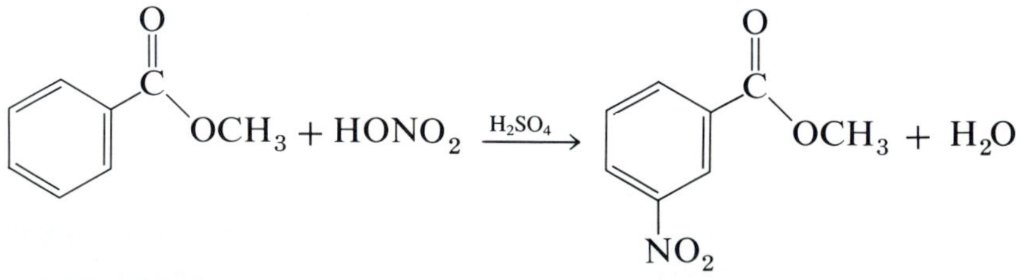

Methyl benzoate   **Methyl *m*-nitrobenzoate**

Many such aromatic substitution reactions are known to occur when an aromatic substrate is allowed to react with a suitable electrophilic reagent, and many other groups besides nitro may be introduced into the ring.

You may recall that alkenes (which are electron-rich due to an excess of electrons in the π system) can react with an electrophilic reagent. The intermediate formed is electron-deficient. It reacts with the nucleophile to complete the reaction. The overall sequence is called **electrophilic addition.** Addition of HX to cyclohexene is an example.

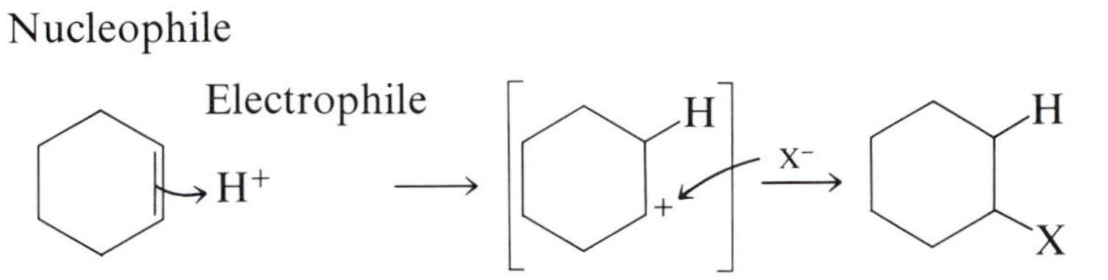

Cyclohexene

ATTACK OF ALKENE ON ELECTROPHILE (H^+) | CARBOCATION INTERMEDIATE | NET ADDITION OF HX

Aromatic compounds are not fundamentally different from cyclohexene. They can also react with electrophiles. However, due to resonance in the ring, the electrons of the π system are generally less available for addition reactions since an addition would mean the loss of the stabilization that resonance provides. In practice this means that aromatic compounds react only with **powerfully electrophilic reagents,** usually at somewhat elevated temperatures.

Benzene, for example, can be nitrated at 50 °C with a mixture of concentrated nitric and sulfuric acids; the electrophile is NO_2^+ (nitronium ion), whose formation is promoted by action of the concentrated sulfuric acid on nitric acid:

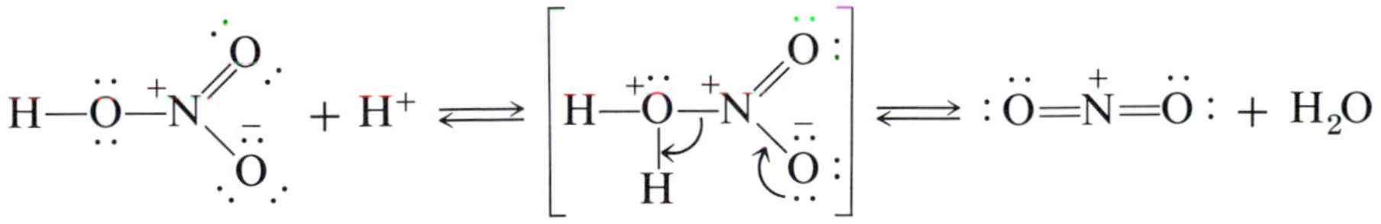

Nitric acid | Nitronium ion

The nitronium ion thus formed is sufficiently electrophilic to add to the benzene ring, **temporarily** interrupting ring resonance:

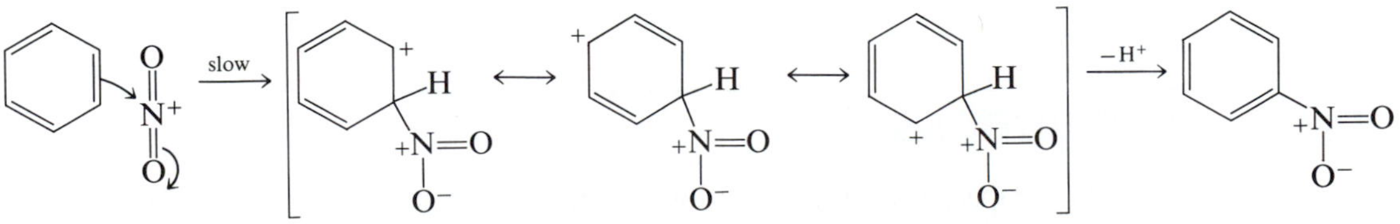

The intermediate first formed is somewhat stabilized by resonance and does not rapidly undergo reaction with a nucleophile; in this behavior, it is different from the unstabilized carbocation formed from cyclohexene plus an electrophile. In fact, aromaticity

can be restored to the ring if **elimination** occurs instead. (Recall that elimination is often a reaction of carbocations.) Removal of a proton, probably by HSO_4^-, from the sp^3-ring carbon **restores the aromatic system** and yields a net **substitution** wherein a hydrogen has been replaced by a nitro group. Many similar reactions are known, and they are called **electrophilic aromatic substitution reactions.**

The substitution of a nitro group for a ring hydrogen occurs with methyl benzoate in the same way it does with benzene. In principle, one might expect that any hydrogen on the ring could be replaced by a nitro group. However, for reasons beyond our scope here (see your lecture textbook), the carbomethoxy group directs the aromatic substitution preferentially to those positions that are *meta* to it. As a result, methyl *m*-nitrobenzoate is the principal product formed. In addition, one might expect the nitration to occur more than once on the ring. However, both the carbomethoxy group and the nitro group that has just been attached to the ring **deactivate** the ring against further substitution. Consequently, the formation of a methyl dinitrobenzoate product is much less favorable than the formation of the mononitration product.

While the products described above are the principal ones formed in the reaction, it is possible to obtain as impurities in the reaction small amounts of the ortho and para isomers of methyl *m*-nitrobenzoate and of the dinitration products. These side products are removed when the desired product is washed with methanol and purified by crystallization.

Water has a retarding effect on the nitration since it interferes with the nitric acid-sulfuric acid equilibria that form the nitronium ions. The smaller the amount of water present, the more active the nitrating mixture. Also, the reactivity of the nitrating mixture can be controlled by varying the amount of sulfuric acid used. This acid must protonate nitric acid, which is a **weak** base, and the larger the amount of acid available, the more numerous the protonated species (and hence NO_2^+) in the solution. Water interferes since it is a stronger base than H_2SO_4 or HNO_3. Temperature is also a factor in determining the extent of nitration. The higher the temperature, the greater will be the amounts of dinitration products formed in the reaction.

REQUIRED READING

Review: Technique 5

SPECIAL INSTRUCTIONS

It is important that the temperature of the reaction mixture be maintained below 15 °C. Nitric acid and sulfuric acid, especially when mixed, are very corrosive substances. Be careful not to get these acids on your skin. If you do get some of these acids on your skin, flush the affected area liberally with water.

PROCEDURE

Add 0.210 mL of methyl benzoate to a tared 3-mL conical vial and determine the actual weight of methyl benzoate. Add 0.45 mL of concentrated sulfuric acid to the methyl benzoate along with a magnetic spin vane. Attach an air condenser to the conical vial. The purpose of the air condenser is to make it easier to hold the conical vial in place. Prepare an ice bath in a 250-mL beaker using both ice and water. Clamp the air condenser so that the conical vial is immersed in the ice bath as shown in Figure 2–5, p 544. (Note that in Figure 2–5 a water bath is shown rather than an ice bath.) While stirring, **very slowly** add a cool mixture of 0.15 mL of concentrated sulfuric acid and 0.15 mL of concentrated nitric acid over a period of about 15 minutes. The acid mixture should be added with a 9-inch Pasteur pipet through the top of the air condenser. If the addition is too fast, the formation of by-product increases rapidly, bringing about a decrease in the yield of the desired product.

After all the acid has been added, warm the mixture to room temperature by replacing the ice water in the 250-mL beaker with water at room temperature. Allow the reaction mixture to set for 15 additional minutes without stirring. Then, using a Pasteur pipet, transfer the reaction mixture to a 20-mL beaker containing 2.0 g of crushed ice. After the ice has melted, isolate the product by vacuum filtration using a Hirsch funnel and wash it with two 1.0-mL portions of cold water and then with two 0.3-mL portions of ice-cold methanol. Weigh the crude, dry product and recrystallize it from methanol using a Craig tube (see Technique 5, Section 5.4, p 586).

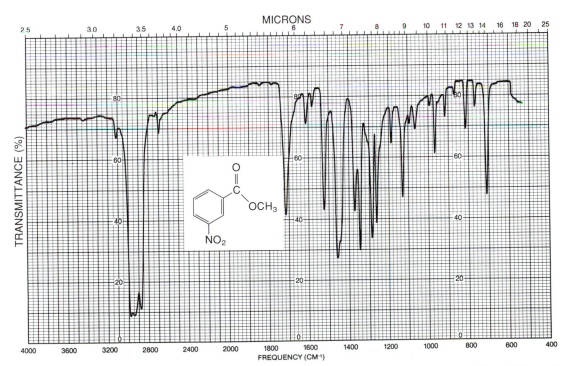

Infrared spectrum of methyl *m*-nitrobenzoate, Nujol mull (Nujol peaks: 2850–3000, 1460, and 1380 cm^{-1})

Determine the melting point of the product. The melting point of the recrystallized product should be 78 °C. Obtain the infrared spectrum of the product as a Nujol mull (Technique 18, Section 18.6, p 779). Submit the product to your instructor in a labeled vial.

QUESTIONS

1. Why is methyl *m*-nitrobenzoate formed in this reaction instead of the ortho or para isomers?
2. Why does the amount of the dinitration increase at high temperatures?
3. Why is it important to add the nitric acid-sulfuric acid mixture slowly over a 15 minute period?
4. Interpret the infrared spectrum of methyl *m*-nitrobenzoate.
5. Indicate the product formed on nitration of each of the following compounds: benzene, toluene, chlorobenzene, and benzoic acid.

Experiment 39
Friedel-Crafts Acylation

Aromatic substitution
Directive groups
Vacuum distillation
Infrared spectroscopy
NMR spectroscopy
  proton/carbon-13
Structure proof

In this experiment a Friedel-Crafts acylation of an aromatic compound is undertaken, using acetyl chloride:

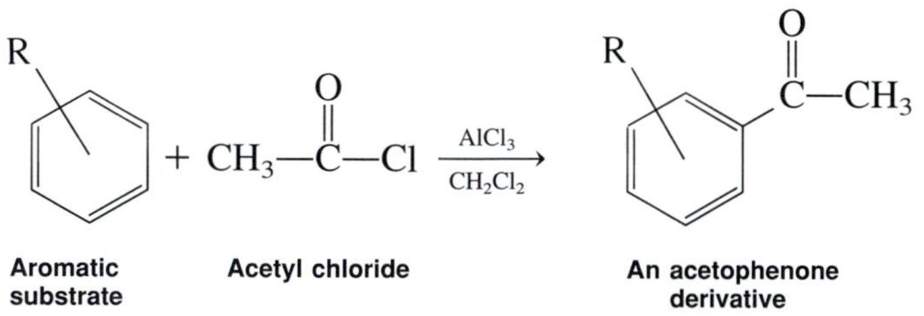

|Aromatic substrate | Acetyl chloride | An acetophenone derivative |

If benzene (R = H) were used as the substrate, the product would be a ketone, acetophenone. Instead of using benzene, however, you will perform the acylation on one of the following compounds:

Toluene	Ethylbenzene
o-Xylene ⎫	Mesitylene (1,3,5-trimethylbenzene)
m-Xylene ⎬ Dimethylbenzenes	Cumene (isopropylbenzene)
p-Xylene ⎭	Anisole (methoxybenzene)

Each of these products will give a single product, a **substituted** acetophenone. You are to isolate this product by vacuum distillation and to determine its structure by IR and NMR spectroscopy. That is, you are to determine at which position of the original compound the new acetyl group becomes attached.

This experiment is much the same kind that a professional chemist performs every day. A standard procedure, Friedel-Crafts acylation, is applied to a new compound for which the results are not known (at least not to you). A chemist who knows reaction theory well should be able to predict the result in each case. However, once the reaction is completed, it must be proved that the expected product has actually been obtained. If it has not, and sometimes surprises do occur, then the structure of the unexpected product must be determined.

To determine the position of substitution, several features of the product's spectra should be examined closely. These include the following.

INFRARED SPECTRUM

- The C–H out-of-plane bending modes found between 900 and 690 cm^{-1}.

    The C–H out-of-plane absorptions (Figure IR–6A, p 822) often allow one to determine the type of ring substitution by their numbers, intensities, and positions.

- The weak combination and overtone absorptions that occur between 2000 and 1667 cm^{-1} (5–6 μ).

    These combination bands (Figure IR–6B, p 822), may not be as useful as those mentioned above since the spectral sample must be very concentrated for them to be visible. They are often weak. In addition, a broad carbonyl absorption may overlap and obscure this region, rendering it useless.

PROTON NMR SPECTRUM

- The **integral ratio** of the downfield peaks in the aromatic ring resonances found between 6 ppm and 8 ppm.

    The acetyl group has a significant anisotropic effect, and those protons found *ortho* to this group on an aromatic ring usually have a greater chemical shift than the other ring protons (see Appendix 4, Section NMR.6, p 838, and Section NMR.10, p 845).

• A splitting analysis of the patterns found in the 6–8 ppm region of the NMR spectrum.

The coupling constants for protons in an aromatic ring differ according to their positional relations:

ortho   J = 6–10 Hz
meta   J = 1–4 Hz
para   J = 0–2 Hz

If complex second-order splitting interaction does not occur, a simple splitting diagram will often suffice to determine the positions of substitution for the protons on the ring. For several of these products, however, such an analysis will be difficult. In other cases, an easily interpretable pattern like those described in Section NMR.10 (p 845) will be found.

CARBON-13 NMR SPECTRUM

• In **completely decoupled** carbon-13 spectra, the number of resonances for the aromatic ring carbons (at about 120–130 ppm) will give some help in deciding the substitution patterns of the ring.

Ring carbons that are equivalent by symmetry will give rise to a single peak thereby causing the number of aromatic carbon peaks to fall below the maximum of six. A *p*-disubstituted ring, for instance, will only show four resonances. Carbons that bear a hydrogen usually will have a larger intensity than ''quaternary'' carbons. (See Appendix 5, Carbon-13 Nuclear Magnetic Resonance Spectroscopy, p 851.)

• In **coupled** carbon-13 spectra, the ring carbons that bear hydrogen atoms will be split into doublets, allowing them to be easily recognized.

> **NOTE TO THE INSTRUCTOR:** For those not equipped to perform carbon-13 NMR spectroscopy, carbon-13 NMR spectra of all the products can be found reproduced in the instructor's manual.

As a final note, you should not eschew using the library. Technique 19 (p 787) outlines how to find several important types of information. Once you think you know the identity of your compound, you might well try to find whether it has been reported previously in the literature, and, if so, whether or not the reported data match your own findings. You may also wish to consult some spectroscopy books, such as Pavia, Lampman, and Kriz, *Introduction to Spectroscopy,* or one of the other textbooks listed at the end of either Appendix 3 (Infrared Spectroscopy) or Appendix 4 (NMR Spectroscopy) for additional help in interpreting your spectra.

REQUIRED READING

Review: Techniques   1, 2, 7, and 18
               Technique 3   Reaction Methods, Sections 3.5 and 3.7
               Technique 6   Physical Constants, Part B, Boiling Points
               Appendices   3,4,5, Infrared, NMR, and Carbon-13 Spectroscopy

New: Technique 9   Vacuum Distillation, Manometers, Sections 9.1, 9.2, 9.4, 9.8, and 9.9

Before you begin this experiment, you should review the chapters in your lecture text that deal with electrophilic aromatic substitution. Pay special attention to Friedel-Crafts acylation and to the explanations of directing groups. You should also review what you have learned about the infrared and NMR spectra of aromatic compounds.

SPECIAL INSTRUCTIONS

Both acetyl chloride and aluminum chloride are corrosive reagents. You should not allow them to come in contact with your skin, nor should you breathe them since they generate HCl on hydrolysis. They may even react explosively on contact with water. Weighing and dispensing operations should be carried out in a hood. The work-up procedure wherein excess aluminum chloride is decomposed with ice water should also be done in the hood.

Your instructor will either assign you a compound or have you choose one yourself from the list given on page 311. While you will acetylate only one of the above compounds, you should learn much more from this experiment by comparing results with other students.

Notice that the details of the vacuum distillation are left for you to figure out on your own. However, here are two hints. First, all the products boil between 100–150 °C at 20 mm pressure. Second, if your chosen substrate is anisole, the product will be a low-melting solid and will solidify soon after the vacuum distillation is completed. In this case it might be worthwhile to pre-weigh the Hickman head itself. It will be difficult to transfer all of the solidified product to another container to determine a yield.

PROCEDURE

Assemble the reaction apparatus shown in the figure. It consists of a 10-mL round-bottom flask and a Claisen head with one opening fitted with a rubber septum and the other attached to an inverted-funnel trap for acidic gases. Secure the Claisen head and the gas trap funnel with clamps. The funnel should be about 2 mm **above** the water. Remove the Claisen head and add 2 mL of methylene chloride, 0.8 g of AlCl$_3$, and a magnetic stirring bar to the 10-mL round-bottom flask. Replace the Claisen head and begin stirring.

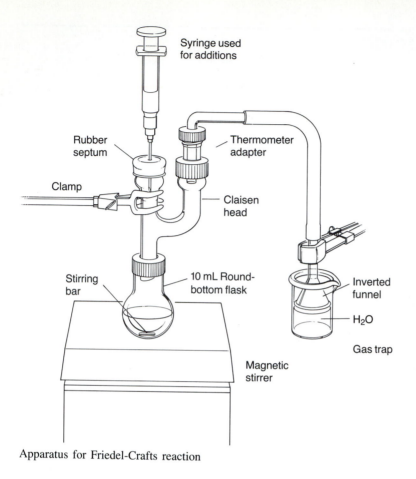

Apparatus for Friedel-Crafts reaction

CAUTION: Both aluminum chloride and acetyl chloride are corrosive and noxious. Avoid contact and conduct all weighings in a hood. On contact with water, either compound may react violently.

Fill your 1-mL syringe (needle attached) with no less than 0.5 mL of fresh acetyl chloride. Insert the syringe through the rubber septum cap, (see figure) and add the acetyl chloride slowly over a 2 minute period. (Rapid addition of the acetyl chloride may cause foaming.) Using a graduated pipet and pipet pump, transfer exactly 0.5 mL of your chosen aromatic compound to a preweighed 3-mL conical vial. Determine the weight of material delivered by weighing on a balance. Take up the aromatic compound with your syringe and slowly add it through the rubber septum over a 5 minute period. (This should not be done hastily because the reaction is very exothermic; the mixture may boil up into the Claisen head.) When the aromatic compound has been added, rinse the vial with 1 mL of methylene chloride and, using the syringe, add this rinse to the reaction flask. Continue stirring at least 5 minutes after the final addition has been made.

ISOLATION OF PRODUCT

Remove the gas trap from the Claisen head and take the remaining apparatus, including the stirrer, to the hood. With your syringe, slowly add 4 mL of **ice cold water** to the reaction mixture over a 5 minute period while stirring slowly. Next add 4 mL of concentrated HCl with a Pasteur pipet, and then stir the mixture vigorously with the magnetic stirrer until all of the aluminum salts dissolve. At this point, discontinue stirring and allow the organic layer to separate. If the organic layer does not separate cleanly, add 0.5 mL of methylene chloride, stir again, and then allow the organic layer to separate. You may have to add up to 1.5 mL of methylene chloride in order to induce the organic layer to separate cleanly.

Decant the entire mixture into a 15-mL centrifuge tube, leaving the stirring bar behind. Transfer the lower organic layer to a 5-mL conical vial with a filter tip pipet. Avoid transferring any of the aqueous layer. If necessary, add a small amount of water and reseparate the layers that have been transferred to the conical vial. If a significant amount of the original highly-acidic aqueous layer is present, violent foaming will occur in the next step. Add about 1 mL of 5% sodium bicarbonate to the conical vial containing the organic layer. Cap the vial and shake it gently. Carefully vent the vial by loosening the cap and resealing it after a few moments. Repeat this mixing several times until the evolution of CO_2 is no longer apparent.

Transfer the organic layer to a dry 3-mL conical vial (5-mL if necessary) and add three to four microspatulafuls of anhydrous sodium sulfate (use the V-grooved end). Cap the vial and set it aside for 10–15 minutes while the liquid is dried. If the liquid appears cloudy, shake the vial several times during the drying period, or add more sodium sulfate. The final product should be clear, but might be colored a light brown, green, or yellow depending on which starting material you used. Transfer the organic layer to a clean, dry 3-mL conical vial using a filter tip pipet. In a hood, place the vial in a sand bath regulated to a temperature of 40 °C and direct a stream of air into the vial to evaporate the methylene chloride (Figure 3–12A, p 561). Do not rush this process. Allow the methylene chloride to be driven off completely or it will cause foaming during the vacuum distillation. Monitor the evaporation by checking the volume markings on the side of the vial. When the volume is constant, the methylene chloride has been removed.

VACUUM DISTILLATION

If you are using a sand bath to heat, you should preheat it to about 165 °C while assembling the apparatus described below. Assemble the apparatus **above** the sand bath; do not lower it into the sand bath until you are ready to distill. If you are using an aluminum block, preheating will not be necessary.

Review Technique 9, Sections 9.1, 9.2, and 9.4 before proceeding.

Assemble an apparatus for vacuum distillation using an aspirator as shown in Figure 9–5, p 656. A manometer should be attached as shown in Figure 9–14, p 667. A piece of stainless steel sponge should be placed in the bottom portion of the neck of the Hickman still to protect the distilled product from any bumping action. Do not pack the stainless steel sponge too tightly. You may wish to preweigh the Hickman head (without the packing) to avoid having to transfer the product in order to determine the yield. This will be especially convenient if anisole was used as the substrate in the reaction. Using an **empty** conical vial, evacuate the system and check for any leaks. When there are no significant leaks, add a spin vane to the 3-mL conical vial containing the product (methylene chloride removed). Attach the vial to the distillation apparatus and reestablish the vacuum.

If using a sand bath, lower the apparatus to begin the distillation and cover the sand bath with aluminum foil. If using an aluminum block, begin heating after lowering the apparatus. Adjust the spin vane to its maximum rate of spin. If boiling, bumping, or refluxing has not occurred after three minutes of heating, the heat may be increased. A sand bath or aluminum block temperature in the range of 165–200 °C will be required, depending on your compound. Once the distillate begins to appear on the walls of the Hickman still, the distillation proceeds very rapidly. When no liquid remains in the 3-mL vial or when liquid is no longer distilling, raise the apparatus immediately to discontinue the distillation. If you overheat the vial, it may crack. Turn the hot plate off. Allow the apparatus to cool to room temperature and then vent the system.

Transfer the product to a preweighed storage container and determine its weight. (If you preweighed your Hickman still, remove the stainless steel sponge and transfer the still to a beaker for weighing.) Calculate the percentage yield. Determine the boiling point of your product using the micro boiling point method (Technique 6, Section 6.10, p 607). Determine both the Infrared and the NMR spectra (proton and carbon-13). The infrared spectra may be determined neat, using salt plates (Technique 18, Section 18.2, p 771), except for the product from anisole, which is a solid. For this product, one of the solution spectrum techniques (Technique 18, Section 18.5, p 777) should be used. Again, except for anisole, the proton NMR spectra can be determined neat as described in Technique 18, Section 18.9, p 782. If the samples are viscous, add a little carbon tetrachloride. The solid product from anisole will have to be dissolved in carbon tetrachloride or deuteriochloroform. Deuteriochloroform is also an excellent solvent for all of the carbon-13 samples as described in Technique 18, Section 18.10, p 785.

THE REPORT

In the usual fashion, you should report the boiling point (or melting point) of your product, calculate the percentage yield, and construct a separation scheme diagram. You should also give the actual structure of your product. Include the Infrared and NMR spectra and discuss carefully what you learned from each spectrum. If they did not help your structure determination, explain why not. As many peaks as possible should be assigned on each spectrum and all important features explained, including the NMR splitting patterns, if possible. Consult a handbook for the boiling point (or melting point) of the possible products. Discuss any literature you consulted and compare the reported results with your own.

You should explain in terms of aromatic substitution theory why the substitution occurred at the position observed, and why a single substitution product was obtained. Could you have predicted the result in advance?

REFERENCE

Schatz, Paul F. ''Friedel-Crafts Acylation.'' *Journal of Chemical Education, 56* (July 1979): 480.

QUESTIONS

1. The following are all relatively inexpensive aromatic compounds that could have been used as substrates in this reaction. Predict the product or products, if any, that would be obtained on acylation of each of them using acetyl chloride.

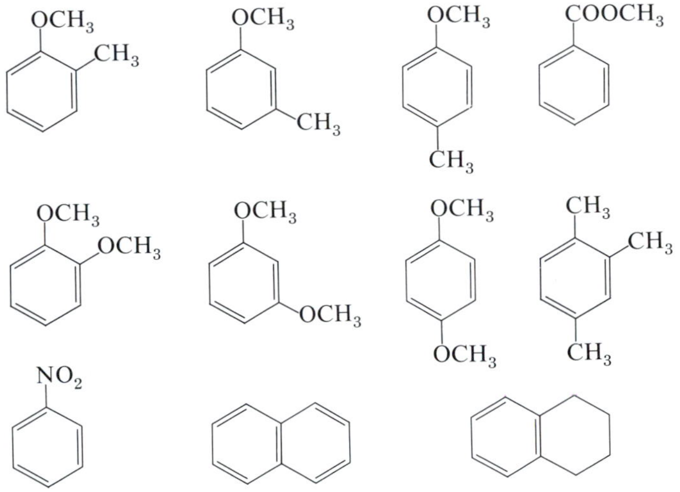

2. Why is it that only monosubstitution products are obtained in the acylation of the substrate compounds chosen for this experiment?

3. Draw a full mechanism for the acylation of the compound you chose for this experiment. Include attention to any relevant directive effects.

4. Why do none of the substrates given as choices for this experiment include any with meta-directing groups?

5. Acylation of *n*-propylbenzene gives an unexpected (?) side-product. Explain this occurrence and give a mechanism.

6. Write equations for what happens when aluminum chloride is hydrolyzed in water and do the same for acetyl chloride.

7. Explain carefully, with a drawing, why the protons substituted ortho to an acetyl group normally have a greater chemical shift than the other protons on the ring.

8. The compounds shown are possible acylation products from 1,2,4-trimethylbenzene (pseudocumene). Explain the only way you could distinguish these two products by NMR spectroscopy.

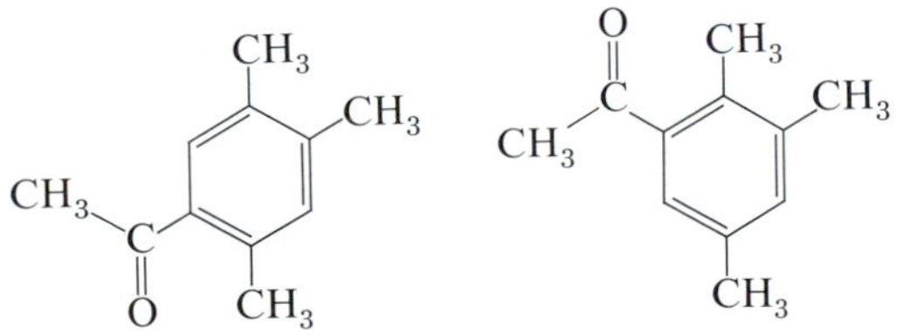

Essay
SYNTHETIC DYES

The practice of using dyes is an ancient art. Substantial evidence exists that plant dyestuffs were known long before humans began to keep written history. Before this century, practically all dyes were obtained from natural plant or animal sources. Dyeing was a complicated and secret art passed from one generation to the next. Dyes were extracted from plants mainly by macerating the roots, leaves, or berries in water. The extract was often boiled and then strained before use. In some cases, it was necessary to make the extraction mixture acidic or basic before the dye could be liberated from the plant tissues. Applying the dyes to cloth was also a complicated process. **Mordants** were used to fix the dye to the cloth or even to modify its color.

Madder is one of the oldest known dyes. Alexander the Great was reputed to have used the dye to trick the Persians into overconfidence during a critical battle. Using madder, a root bearing a brilliant red dye, he simulated bloodstains on the tunics of his soldiers. The Persians, seeing the apparently incapacitated Greek army, became overconfident and much to their surprise were overwhelmingly defeated. Through modern chemical analysis, we now know the structure of the dye found in madder root. It is called **alizarin** (see structures) and is very similar in structure to another ancient dye, **henna,** which has been responsible for a long line of synthetic redheads. Madder is obtained from the plant *Rubia tinctorum*. Henna is a dye prepared from the leaves of the Indian henna plant (*Lawsonia alba*) and an extract of *Acacia catechu.*

Indigo is another plant dyestuff with a long history. This dye, obtained from the plant *Indigofera tinctoria,* has been known in Asia for more than 4000 years. By the ancient process for producing indigo, the leaves of the indigo plant are cut and allowed to ferment in water. During the fermentation, **indican** (see chart) is extracted into the solution, and the attached glucose molecule is split off to produce **indoxyl.** The fermented mixture is transferred to large open vats, in which the liquid is beaten with bamboo sticks. During this process, the indoxyl is air-oxidized to indigo. Indigo, a strong blue dye, is insoluble in water, and it precipitates. Today, indigo is made synthetically, and its principal use is in dyeing denim to produce "blue jeans" material.

Many plants yield dyestuffs that will dye wool or silk, but there are few of these that dye cotton well. Most do not dye synthetic fibers like polyester or rayon. In addition, the natural dyes, with a few exceptions, do not cover a wide range of colors, nor do they yield "brilliant" colors. Even though some people prefer the softness of the "homespun" colors from natural dyes, the **synthetic dyes,** which give rise to deep, brilliant colors, are much in demand today. Also, synthetic dyes that will dye the popular synthetic fibers can now be manufactured. Thus today we have available an almost infinite variety of colors as well as dyes to dye any type of fabric.

Before 1856, all dyes in use came from natural sources. However, an accidental discovery by W. H. Perkin, an English chemist, started the development of a huge

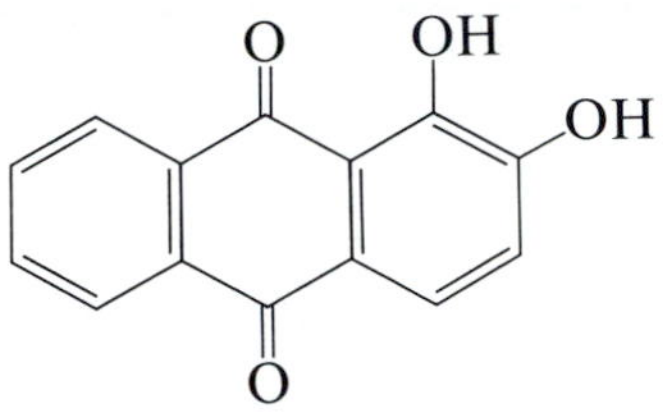

Alizarin

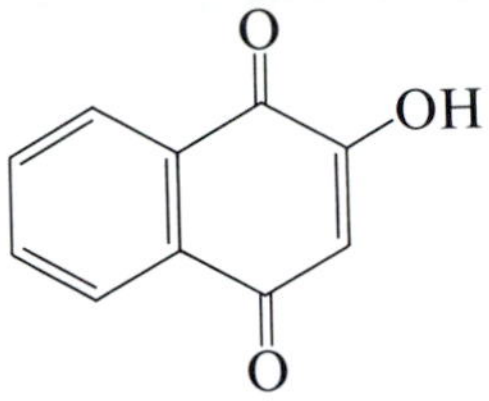

Henna

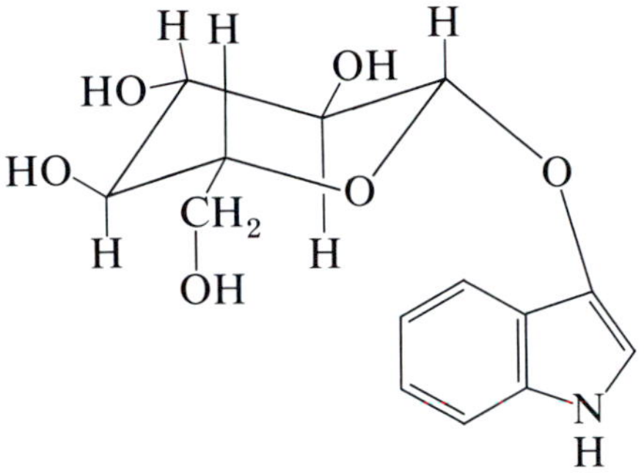

Indican

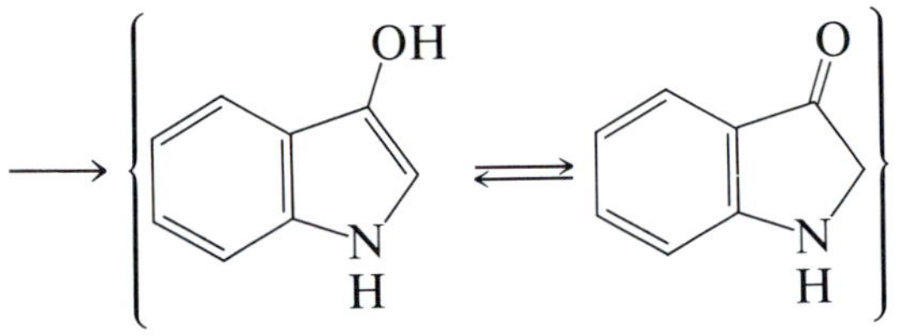

Indoxyl

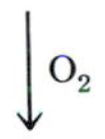

O_2

Tyrian purple

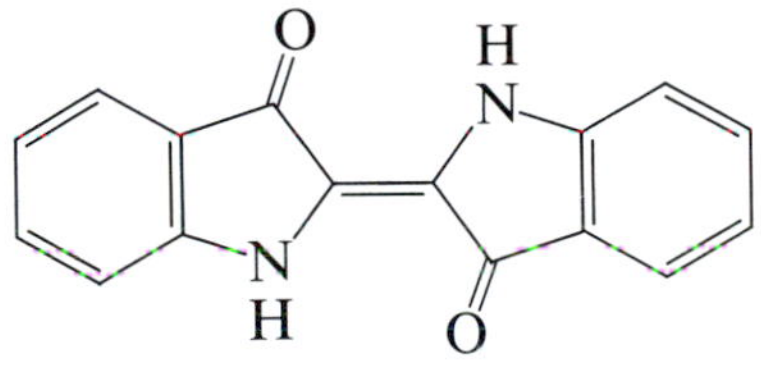

Indigo

synthetic dye industry, mostly in England and Germany. Perkin, then only aged 18, was trying to synthesize quinine. Structural organic chemistry was not very well developed at that time, and the chief guide to the structure of a compound was its molecular formula. Perkin thought, judging from the formulas, that it might be possible to synthesize quinine by the oxidation of allyltoluidine:

$$2C_{10}H_{13}N + 3[O] \longrightarrow C_{20}H_{24}N_2O_2 + H_2O$$

Allyltoluidine   **Quinine**

He made allyltoluidine and oxidized it with potassium dichromate. The reaction was unsuccessful, because allyltoluidine bore no structural relation to quinine. He obtained no quinine, but he did recover a reddish-brown precipitate with properties that interested him. He decided to try the reaction with a simpler base, aniline. On treating aniline sulfate with potassium dichromate, he obtained a black precipitate, which could be extracted with ethanol to give a beautiful purple solution. This purple solution subsequently proved to be a good dye for fabrics. After receiving favorable comments from dyers, Perkin resigned his post at the Royal College and went on to found the

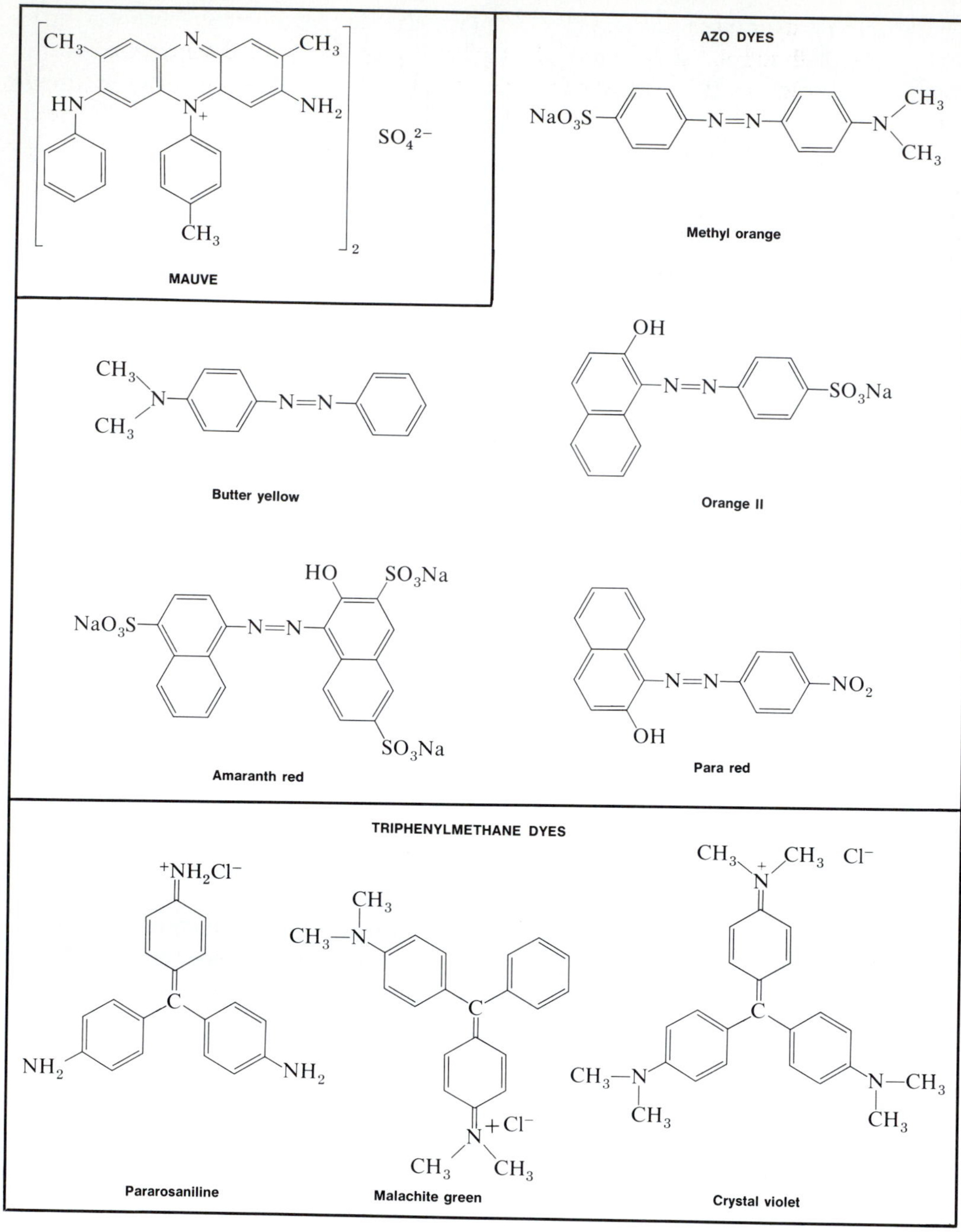

MAUVE

AZO DYES

Methyl orange

Butter yellow

Orange II

Amaranth red

Para red

TRIPHENYLMETHANE DYES

Pararosaniline

Malachite green

Crystal violet

British coal tar dye industry. He became a very successful industrialist and retired at age 36 (!) to devote full time to research. The dye he synthesized became known as **mauve.** The structure of mauve was not proved until much later. From the structure (see above) it is clear that the aniline Perkin used was not pure and that it contained the *o*-, *m*-, and *p*-toluidines also.

Mauve was the first synthetic dye, but soon (1859) the triphenylmethyl dyes pararosaniline, malachite green, and crystal violet (see figures) were discovered in

France. These dyes were produced by treating mixtures of aniline or of the toluidines, or of both, with nitrobenzene, an oxidizing agent, and in a second step, with concentrated hydrochloric acid. The triphenylmethyl dyes were soon joined by **synthetic** alizarin (Lieberman, 1868), **synthetic** indigo (Baeyer, 1879), and the azo dyes (Griess, 1862). The azo dyes, also manufactured from aromatic amines, revolutionized the dye industry.

The azo dyes are one of the most common types of dye still in use today. They are used as dyes for clothing, as food dyes (see essay preceding Experiment 41), and as pigments in paints. In addition, they are used in printing inks and in certain color printing processes. Azo dyes have the basic structure

$$Ar—N{=}N—Ar'$$

Several of these dyes are illustrated on p 320. The unit containing the nitrogen-nitrogen bond is called an **azo** group, a strong chromophore that imparts a brilliant color to these compounds.

Producing an azo dye involves treating an aromatic amine with nitrous acid to give a **diazonium ion** intermediate. This process is called **diazotization:**

$$Ar—NH_2 + HNO_2 + HCl \longrightarrow Ar—\overset{+}{N}{\equiv}N{:} + Cl^- + 2H_2O$$

The diazonium ion is an electron-deficient (electrophilic) intermediate. A nucleophilic aromatic compound will react with the diazonium ion. The most common nucleophilic species are aromatic amines and phenols. Both these types of compounds are usually more nucleophilic at a ring carbon than at either nitrogen or oxygen. This is due to resonance of the following types:

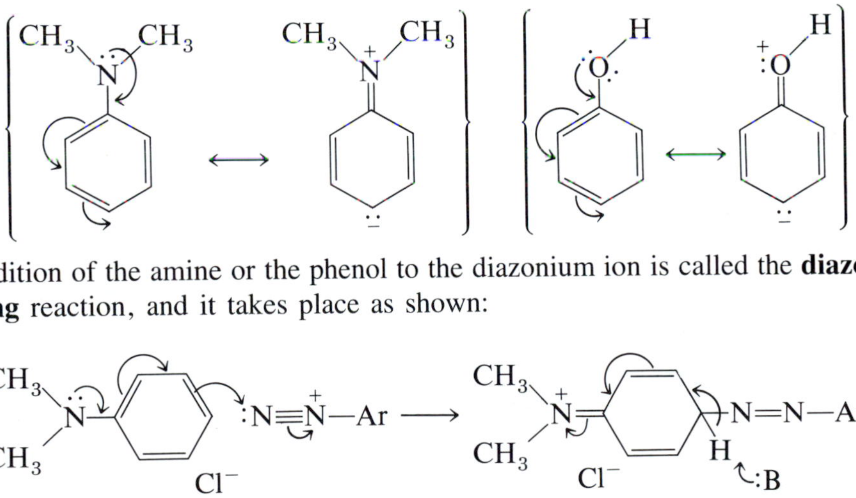

The addition of the amine or the phenol to the diazonium ion is called the **diazonium coupling** reaction, and it takes place as shown:

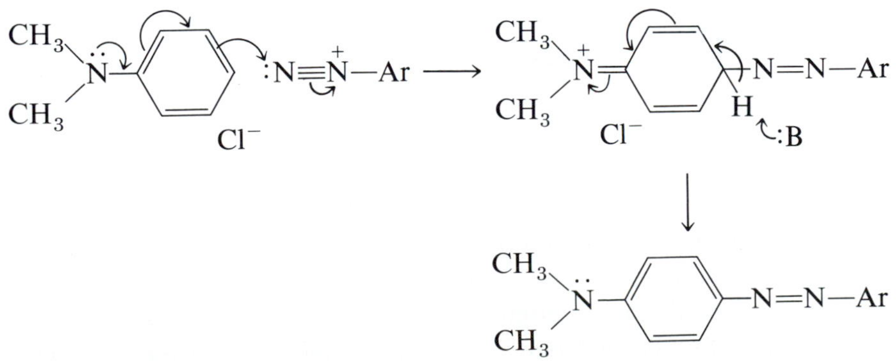

An azo dye

Azo dyes are both the largest and the most important group of synthetic dyes. In the formation of the azo linkage, many combinations of $ArNH_2$ and $Ar'NH_2$ (or $Ar'OH$) are possible. These combinations give rise to dyes with a broad range of colors, encompassing yellows, oranges, reds, browns, and blues. The preparation of an azo dye is given in Experiment 40.

The azo dyes, the triphenylmethyl dyes, and mauve are all synthesized from the anilines (aniline, o-, m-, and p-toluidine) and aromatic substances (benzene, naphthalene, anthracene). All these substances can be found in **coal tar,** a crude material that is obtained by distilling coal. Perkin's discovery led to a multimillion-dollar industry based on coal tar, a material that was once widely regarded as a foul-smelling nuisance. Today these same materials can be recovered from crude oil or from petroleum as by-products in the refining of gasoline. Although we no longer use coal tar, many of the dyes are still widely used.

REFERENCES

Abrahart, E. N. *Dyes and Their Intermediates*. London: Pergamon Press, 1968.
Allen, R. L. M. *Colour Chemistry*. New York: Appleton-Century-Crofts, 1971.
Decelles, C. "The Story of Dyes and Dyeing." *Journal of Chemical Education, 26* (1949): 583.
Jaffee, H. E., and Marshall, W. J. "The Origin of Color in Organic Compounds." *Chemistry, 37* (December 1964): 6.
Juster, N. J. "Color and Chemical Constitution." *Journal of Chemical Education, 39* (1962): 596.
Robinson, R. "Sir William Henry Perkin: Pioneer of Chemical Industry." *Journal of Chemical Education, 34* (1957): 54.
Sementsov, A. "The Medical Heritage from Dyes." *Chemistry, 39* (November 1966): 20.
Shaar, B. E. "Chance Favors the Prepared Mind. Part 1: Aniline Dyes." *Chemistry, 39* (January 1966): 12.

Experiment 40
Methyl Orange

Diazotization
Diazonium coupling
Azo dyes

In this experiment, the azo dye **methyl orange** is prepared by the diazo coupling reaction. It is prepared from sulfanilic acid and *N,N*-dimethylaniline. The first product obtained from the coupling is the bright red acid form of methyl orange, called **helianthin.** In base, helianthin is converted to the orange sodium salt, called methyl orange.

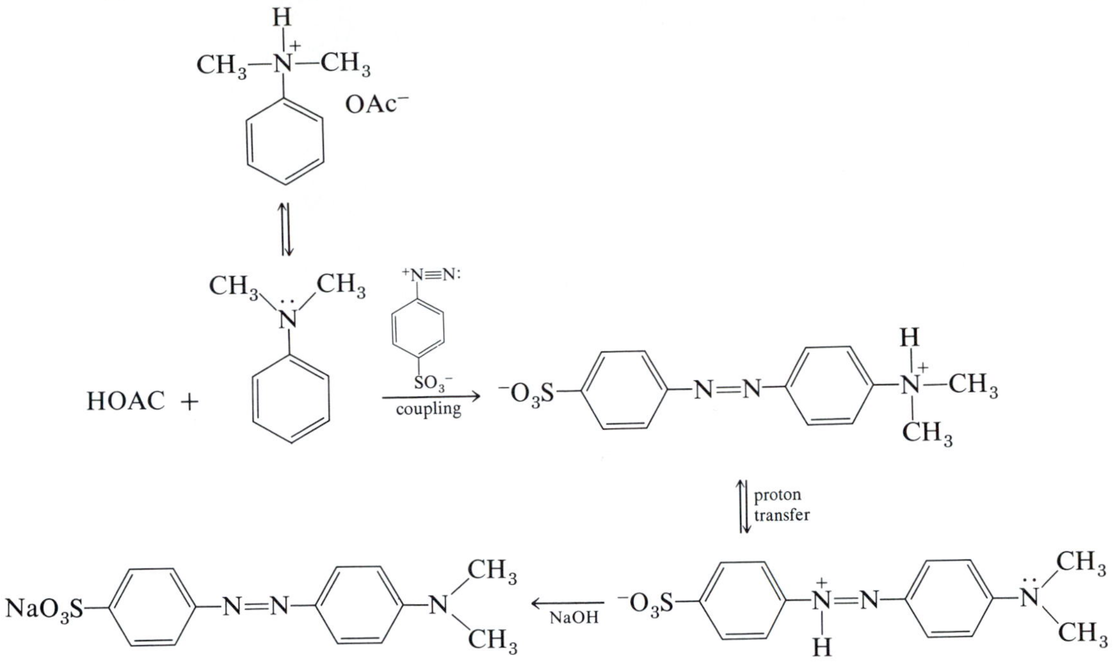

Methyl orange

Helianthin (red)

Although sulfanilic acid is insoluble in acid solutions, it is nevertheless necessary to carry out the diazotization reaction in an acid (HNO_2) solution. This problem can be avoided by precipitating sulfanilic acid from a solution in which it is initially soluble. The precipitate is a fine suspension and reacts instantly with nitrous acid. The first step is to dissolve sulfanilic acid in basic solution.

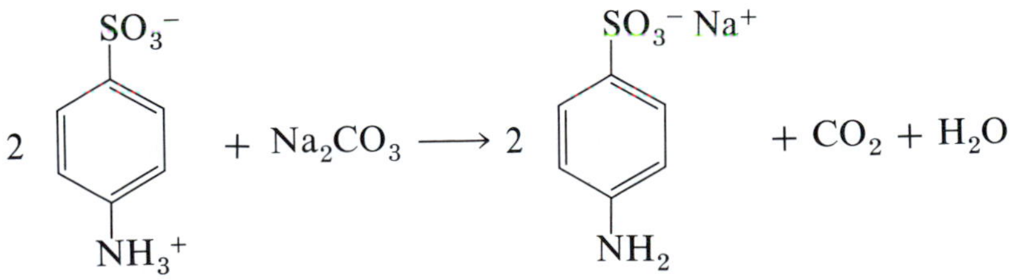

Sulfanilic acid

When the solution is acidified during the diazotization to form nitrous acid,

$$NaNO_2 + HCl \longrightarrow NaCl + HNO_2$$

the sulfanilic acid is precipitated out of solution as a finely divided solid, which is immediately diazotized. The finely-divided diazonium salt is allowed to react immediately with dimethylaniline in the solution in which it was precipitated.

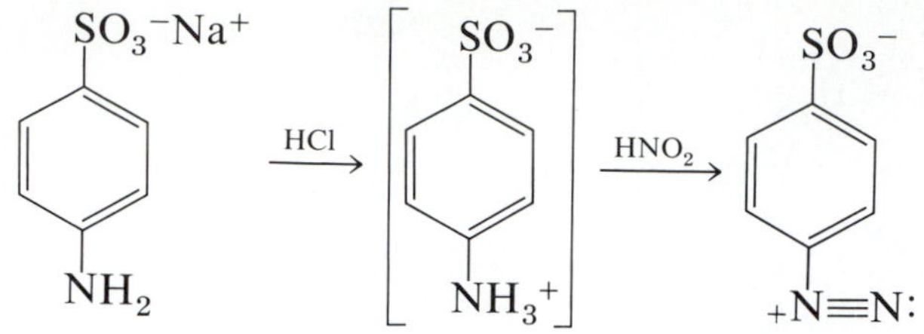

$$\text{(structure: } SO_3^-Na^+ \text{ benzene ring with } NH_2) \xrightarrow{HCl} [\text{(structure: } SO_3^- \text{ benzene ring with } NH_3^+)] \xrightarrow{HNO_2} \text{(structure: } SO_3^- \text{ benzene ring with } +N\!\equiv\!N\!:)$$

Methyl orange is often used as an acid-base indicator. In solutions that are more basic than pH 4.4, methyl orange exists almost entirely as the **yellow** negative ion. In solutions that are more acidic than pH 3.2, it is protonated to form a **red** dipolar ion.

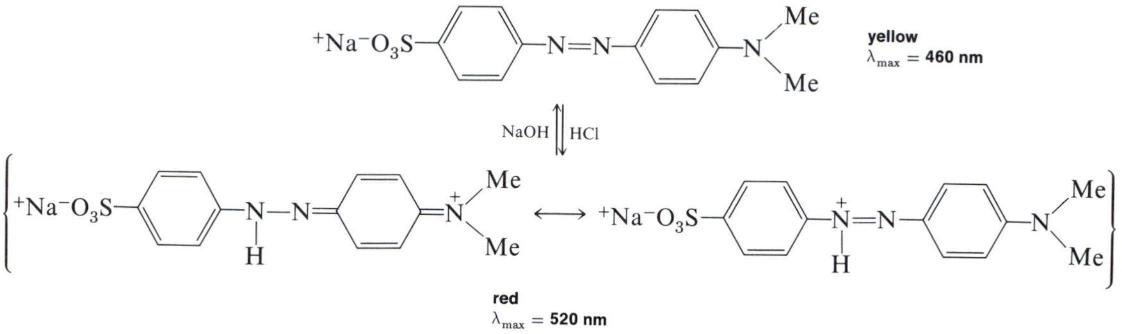

$$^+Na^-O_3S\!-\!\!\bigcirc\!\!-\!N\!=\!N\!-\!\!\bigcirc\!\!-\!N\!\!\begin{array}{c}Me\\Me\end{array} \quad \begin{array}{l}\textbf{yellow}\\\lambda_{max}=460\text{ nm}\end{array}$$

$$\text{NaOH} \updownarrow \text{HCl}$$

$$\left\{ ^+Na^-O_3S\!-\!\!\bigcirc\!\!-\!\underset{H}{N}\!-\!N\!=\!\!\bigcirc\!\!=\!\overset{+}{N}\!\!\begin{array}{c}Me\\Me\end{array} \longleftrightarrow \;\;^+Na^-O_3S\!-\!\!\bigcirc\!\!-\!\underset{H}{\overset{+}{N}}\!=\!N\!-\!\!\bigcirc\!\!-\!N\!\!\begin{array}{c}Me\\Me\end{array} \right\}$$

$$\begin{array}{l}\textbf{red}\\\lambda_{max}=520\text{ nm}\end{array}$$

Thus, methyl orange can be used as an indicator for titrations that have their end points in the pH 3.2 to pH 4.4 region. The indicator is usually prepared as a 0.01% solution in water. In higher concentrations in basic solution, of course, methyl orange appears **orange.**

Azo compounds are easily reduced at the nitrogen-nitrogen double bond by reducing agents. Sodium hydrosulfite, $Na_2S_2O_4$, is often used to bleach azo compounds.

$$R'\!-\!N\!=\!N\!-\!R \xrightarrow{Na_2S_2O_4} R'\!-\!NH_2 + H_2N\!-\!R$$

Other good reducing agents, such as stannous chloride in concentrated hydrochloric acid, will also work.

REQUIRED READING

Review: Technique 4   Filtration, Section 4.3

New: Essay   Synthetic Dyes

SPECIAL INSTRUCTIONS

This experiment will present a real challenge to your laboratory technique. See if you can complete it **without dyeing your fingers orange!**

N,N-Dimethylaniline is quite toxic; avoid any direct contact, and avoid breathing its vapors.

PROCEDURE

DIAZOTIZED SULFANILIC ACID

Dissolve 0.06 g of anhydrous sodium carbonate in 5 mL of water in a 25-mL Erlenmeyer flask. Add 0.2 g of sulfanilic acid monohydrate to the solution and heat it in a sand bath (about 100 °C) until it dissolves. Cool the solution to room temperature, add 0.08 g of sodium nitrite, and stir the mixture until it is completely dissolved. Cool the solution in an ice-water bath for 5–10 minutes with frequent swirling action, or until the temperature is **below** 10 °C. When the solution is completely cooled, add 0.25 mL of concentrated hydrochloric acid to the flask. The diazonium salt of sulfanilic acid should separate as a finely-divided white precipitate. Keep this suspension in an ice bath until it is to be used.

METHYL ORANGE

Combine 0.14 mL of *N,N*-dimethylaniline and 0.10 mL of glacial acetic acid in a 3-mL conical vial. Using a Pasteur pipet, add this solution to the cooled suspension of diazotized sulfanilic acid in the 25-mL Erlenmeyer flask. Stir the mixture vigorously with a stirring rod or microspatula. Within a few minutes, a red precipitate of helianthin should form. Keep this mixture cooled for about 10 minutes in an ice bath. Next, add 1.5 mL of 10% aqueous sodium hydroxide. Do this slowly, with stirring, while you continue to cool the mixture in the ice bath. Using a glass rod, transfer a drop of the supernatant liquid to a strip of litmus or pH paper to determine if the solution is basic. If the solution is not basic, add more sodium hydroxide solution and check it again.

Heat the basic solution to boiling for 10–15 minutes to dissolve most of the newly formed methyl orange. It will not be possible to dissolve it all. When most of the methyl orange has dissolved, add 0.5 g of sodium chloride, and allow the mixture to cool. When the solution reaches room temperature, place the flask in an ice bath to complete the crystallization. Collect the product by vacuum filtration (Technique 4, Section 4.3, p 571) using a small Büchner funnel. Rinse the Erlenmeyer flask with two **ice-cold** portions of saturated aqueous sodium chloride solution, and wash the filter cake of material you just collected with these rinse solutions.

To further purify the product, transfer the filter cake and paper to a 50-mL beaker containing about 15 mL of boiling water. Maintain the solution at a gentle boil for a few minutes, stirring it constantly with a glass stirring rod or microspatula. Not all the dye will dissolve, but the salts with which it is contaminated will dissolve. Remove the filter paper and allow the solution to cool to room temperature. Cool the mixture in an ice bath. When it is cold, collect the product by vacuum filtration, using a small Büchner or a Hirsch funnel. Allow the product to dry until the next lab period, weigh it, and calculate the percentage yield. Since salts do not generally have well-defined melting points, the melting point determination should not be attempted.

TESTS (OPTIONAL)

Obtain a square of Multifiber Fabric 10A from your instructor. This cloth contains alternate bands of acetate rayon, cotton, nylon, polyester, acrylic, and wool woven in sequence[1]. Prepare a dye bath by dissolving 0.050 g of methyl orange (your crude material will suffice) in 30 mL of water to which 1 mL of a 15% aqueous sodium sulfate solution and one drop of concentrated sulfuric acid have been added. Heat the solution to just below its boiling point. Immerse the fabric in the bath for 5–10 minutes. Remove the fabric, rinse it well, and note the results.

Make the dye bath basic by adding sodium carbonate. Then add a solution of sodium dithionite (sodium hydrosulfite) until the color of the bath is discharged. Add a slight excess. Now place the very end portion of the dyed fabric in the bath for a few minutes. Note the result.

INDICATOR ACTION (OPTIONAL)

Dissolve a few crystals of methyl orange in a small amount of water in a test tube. Alternately add a few drops of dilute hydrochloric acid and a few drops of dilute sodium hydroxide solution until the color changes are apparent in each case.

QUESTIONS

1. Why does the dimethylaniline couple with the diazonium salt at the *para* position of the ring?

2. Give a mechanism for producing a phenol from the diazonium salt that was prepared from sulfanilic acid.

3. What would be the result if cuprous chloride were added to the diazonium salt prepared in this reaction?

4. The diazonium coupling reaction is an electrophilic aromatic substitution reaction. Give a mechanism that clearly indicates this fact.

5. In the essay on food colors that precedes Experiment 41, the structures of several azo food colors are given. Indicate how you would synthesize each of these dyes by the diazo coupling reaction.

6. Immediately after the coupling reaction in this experiment, a proton transfer occurs. The proton is transferred from the dimethylamino group to the azo linkage. Why is the latter protonated form lower in energy than the product formed initially?

[1] Available from Testfabrics, Inc., P.O. Box 118, 200 Blackford Ave., Middlesex, NJ 08846.

Essay
FOOD COLORS

Before 1850, most of the colors added to foods were derived from natural biological sources. Some of these natural colors are listed below.

Red	Alkanet root	Yellow	Annato seed (bixin)
	Beets (betanin)		Carrots (β-carotene)
	Cochineal insects		Crocus stigmas (saffron)
	(carminic acid)		Turmeric (rhizome)
	Sandalwood	Green	Chlorophyll
Orange	Brazilwood	Blue	Purple grape skins (oenin)
Brown	Caramel (charred sugar)		

A wide variety of colors can be obtained from these natural sources, many of which are still used today, but they have been largely supplanted by synthetic dyes.

After 1856, when Perkin succeeded in synthesizing mauve, the first coal tar dye (see the essay on synthetic dyes preceding Experiment 40), and when chemists began to discover other new synthetic dyes, artificial colors began to find their way into foodstuffs with increasing regularity. Today, more than 90% of the coloring agents added to foods are synthetic.

The synthetic dyes have certain advantages over the natural coloring agents. Many natural dyes are sensitive to degradation by light and oxygen or by bacterial action; therefore, they are not stable or long-lasting. Synthetic colors can be devised that have a much longer shelf life. The synthetic dyes are also stronger and give more intense colors, and they can be used in smaller quantity to achieve a given color. Often the artificial coloring materials are cheaper than the natural colors. This fact of economics is often especially true when the smaller amounts that are required are taken into account.

Why should artificial colors be added to foods at all? It is easier to answer this question from the point of view of the manufacturer rather than of the consumer. The manufacturer knows that, to a certain extent, the eye appeal of a product will affect its sales. For example, a consumer is more likely to buy an orange that has a bright orange skin than to buy one with a mottled green and yellow skin. This is true, even though the flavor and nutritive value of the orange may not be affected at all by the color of the skin. Sometimes more than eye appeal is involved. The consumer is a creature of habit and is accustomed to having certain foodstuffs a particular color. How would you react to green margarine or blue steak? For obvious reasons, these products would not sell very well. Both butter and margarine are artificially colored yellow. Natural butter has a yellow color only in the summer; in the winter it is colorless, and manufacturers customarily add yellow coloring. Margarine must always be artificially colored yellow.

327

Thus, the colors that are added to foodstuffs are added for a different reason from what prompts the use of other types of food additives. Other additives may be added to foods for either nutritional or technological reasons. Some of these additives can be justified by good arguments. For instance, during the processing of many foods, valuable vitamins and minerals are lost. Many manufacturers replace these lost nutrients by "enriching" their product. In another instance, preservatives are sometimes added to food to forestall spoilage from oxidation or the growth of bacteria, yeasts, and molds. With modern marketing practices, which involve the shipping and warehousing of products over long distances and periods, preservatives are often a virtual necessity. Other additives, such as thickeners and emulsifiers, are often added for technological reasons, for example, to improve the texture of the foodstuff.

There is no nutritional or technological necessity for the use of food colors, however. In fact, in some cases, dyes have been used to deceive customers. For instance, yellow dyes have been used in both cake mixes and egg noodles to suggest a higher egg content than what is actually present. On the grounds that synthetic food dyes are unnecessary and perhaps dangerous, many persons have advocated that their use be abandoned.

Of all the food additives, dyes have come under the heaviest attack. As early as 1906, the government took steps to protect the consumer. At the turn of the century, more than 90 dyes were used in foods. There were no governmental regulations, and the same dyes that were used for dyeing clothes could be used to color foodstuffs. The first legislation governing dyes was passed in 1906, when food colors known to be harmful were removed from the market. At that time only seven colors were approved for use in food. In 1938 the law was extended, and any batch of dye destined for use in food had to be **certified** for chemical purity; previously, certification had been voluntary for the manufacturer. At that time there were 15 food colors in general use and each was given a color and a Food, Drug, and Cosmetic (F,D&C) number designation rather than a chemical name. In 1950, when the number of dyes in use had expanded to 19, an unfortunate incident led to the discontinuation of three of the dyes: two were F,D&C Oranges Number 1 and Number 2, and the other was F,D&C Red Number 32. These dyes were removed when several children became seriously ill after eating popcorn colored by them.

Since that time, research has revealed that many of these dyes are toxic, that they can cause birth defects, that they can cause heart trouble, or that they are carcinogenic (cancer-inducing). Because of experimental evidence, mainly with chick embryos, rats, and dogs, Reds Numbers 1 and 4 and Yellows Numbers 1, 2, 3, and 4 were also removed from the approved list in 1960. Subsequently, Reds Numbers 4 and 32 were reinstated but restricted to particular uses. In 1965, the ban on Red Number 4 was partly lifted to allow it to be used to color maraschino cherries. This use was allowed because there was no substitute dye available that would dye cherries, and it was thought that since maraschino cherries are mainly decorative, they are not properly a foodstuff. This use of Red Number 4 was considered to be a minor use. Similarly, Red Number 32, which may not be used to color food to be eaten, is now called Citrus Red Number 2, and is allowed only for dyeing the skins of oranges.

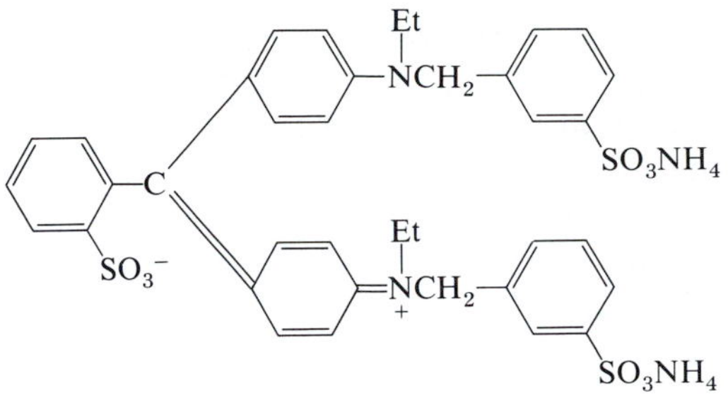

F,D&C Blue No. 1
(Brilliant blue FCF)

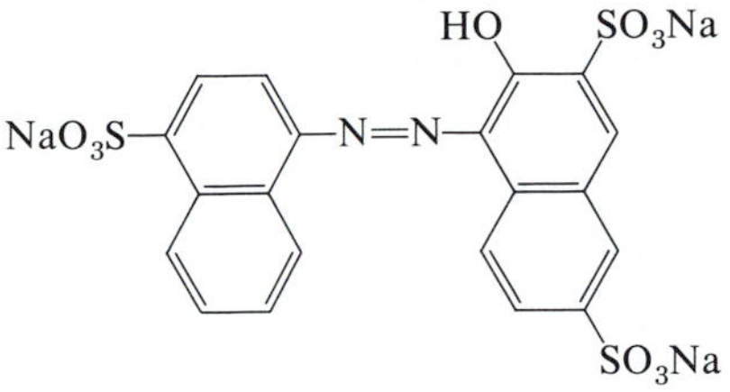

F,D&C Red No. 2
(Amaranth)

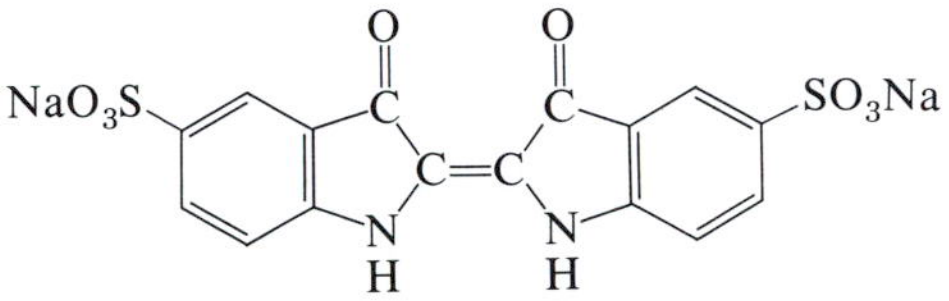

F,D&C Blue No. 2
(Indigo carmine)

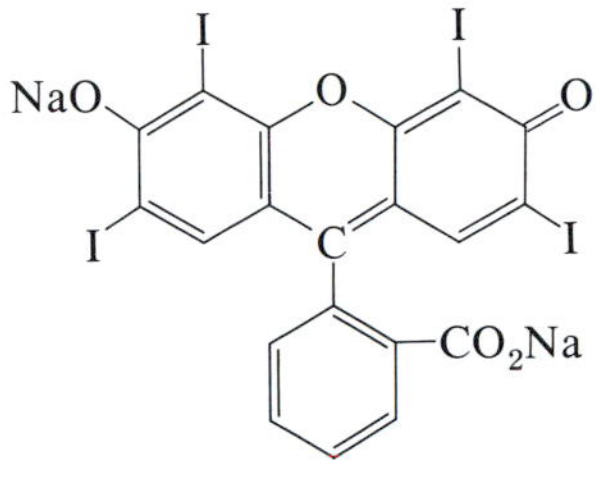

F,D&C Red No. 3
(Erythrosine)

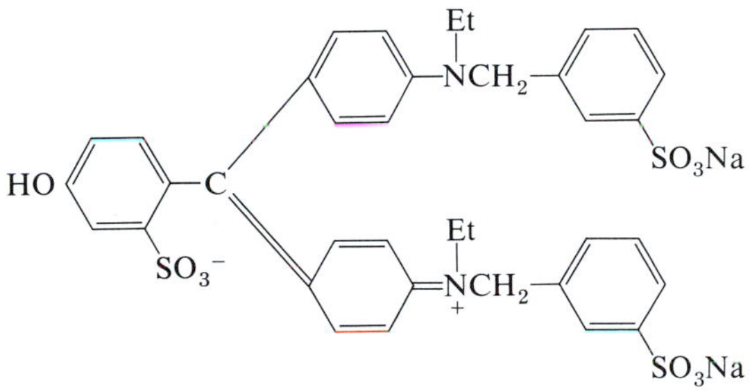

F,D&C Green No. 3
(Fast green FCF)

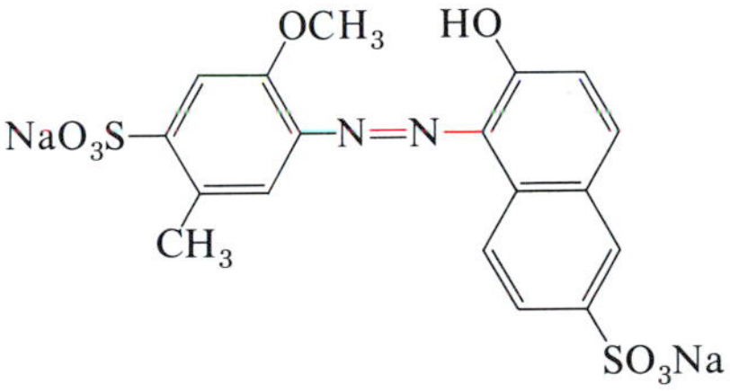

F,D&C Red No. 40
(Allura red)

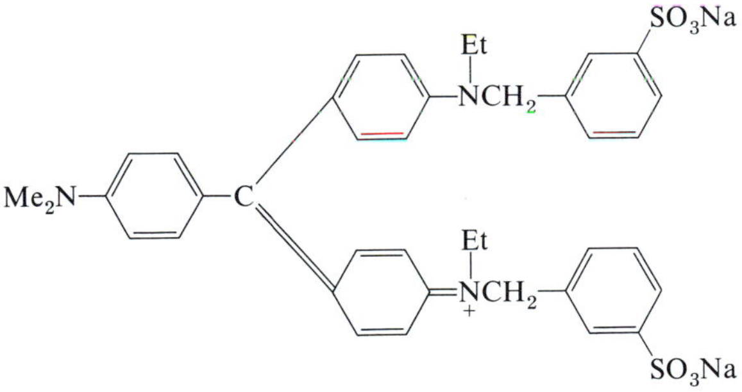

F,D&C Violet No. 1
(Benzyl violet)

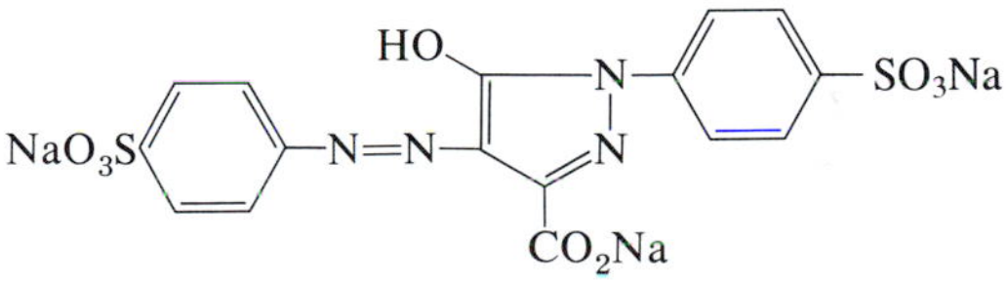

F,D&C Yellow No. 5
(Tartrazine)

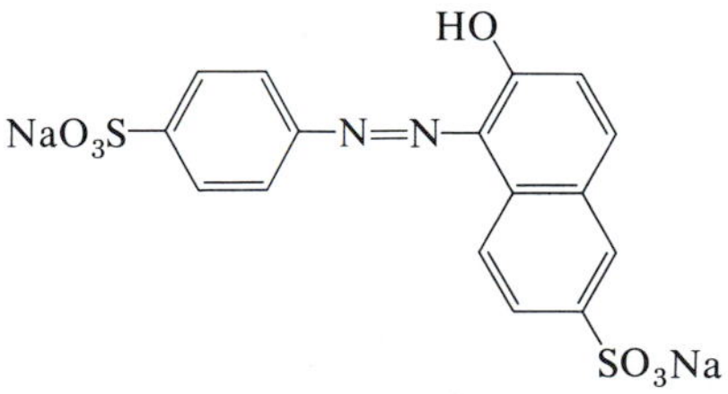

F,D&C Yellow No. 6
(Sunset yellow)

Nine food colors approved by the Food and Drug Administration in 1975. All are still in use except for Red No. 2 (Amaranth), which was banned in 1976.

The structures of the main food dyes are shown in the figure on page 329. Note that a good many of them are azo dyes. Since many of the dyes with the azo linkage have been shown to be carcinogens, many persons suspect all such dyes. In 1960, the law was amended to require that any new dyes submitted for approval should undergo extensive scientific testing before they could be approved. They must be shown to be free from causing birth defects, organic dysfunction, and cancer. Old dyes may be subject to reconsideration if experimental evidence suggests that this is necessary.

Several recent studies have suggested that synthetic food dyes may be responsible, at least in part, for hyperkinetic activity in certain young children. It was shown that when these children were maintained on diets that excluded synthetic food dyes, many of them reverted to more normal behavior patterns. On the contrary, when they were administered a synthetic mixture of food dyes as a capsule along with this diet, the hyperkinetic syndrome would often manifest itself once again. Currently several groups of workers are involved in studying this apparent relationship.

Red Number 2 is the dye that has most recently been involved in a controversy concerning its safety. In many tests, some even performed by Food and Drug Administration (FDA) chemists, mounting evidence was found that this dye might be harmful, causing birth defects, spontaneous abortion of fetuses, and possibly cancer. However, the results of other workers were found to contradict these findings. Much controversy, involving the FDA, the opponents, the proponents, and the courts, ensued. Finally, in February 1976, this dye was banned for food use after the FDA and the courts decided that the bulk of the evidence argued for its discontinuation. More of this interesting story may be found in the references.

While Red Number 2 is proscribed in the United States, it is still approved for use in Canada and within the European Economic Community, and it may be found in products originating in those countries. Before the ban in the United States, Red Number 2 was the most widely used food dye in the industry, appearing in everything from ice cream to cherry soda. Fortunately, proscription of Red Number 2 has not been disastrous for the industry, since for most uses, either Red Number 3 or Red Number 40 are ready substitutes. This knowledge probably had much to do with the court decision finally to ban the dye.

Red Number 40, the most recently accepted food dye, was approved in 1971. Before gaining approval, the Allied Chemical Corporation, which holds exclusive patent rights to the dye, carried out the most thorough and expensive testing program ever given to a food dye. These tests even included a study of possible birth defects. Red Number 40, called Allura red, seems destined to replace Red Number 2, since it has an extremely wide variety of applications, including the dyeing of maraschino cherries; in this it can replace the provisionally listed Red Number 4, which was also banned in 1976.

There are currently eight allowed dyes for food use. The structures of these eight approved food dyes are given in the accompanying chart. The use of these eight dyes is unrestricted. In addition, two other dyes are approved for restricted uses. Citrus Red Number 2 (old Red Number 32) may be used to color the skins of oranges, and Orange B may be used to color the skins of sausages.

REFERENCES

Augustine, G. J., Jr. and Levitan, H. "Neurotransmitter Release from a Vertebrate Neuromuscular Synapse Affected by a Food Dye." *Science, 207* (March 28, 1980): 1489.

Boffey, P. M. "Color Additives: Botched Experiment Leads to Banning of Red Dye No. 2." *Science, 191* (February 6, 1976): 450.

Boffey, P. M. "Color Additives: Is Successor to Red Dye No. 2 Any Safer?" *Science, 191* (February 22, 1976): 832.

Feingold, B. F. *Why Your Child Is Hyperactive.* New York: Random House, 1975.

Hopkins, H. "Countdown on Color Additives." *FDA Consumer* (November 1976): 5.

Jacobson, M. F. *Eater's Digest: The Consumer's Factbook of Food Additives.* New York: Doubleday and Co. (Anchor No. A 861), 1972.

Mebane, R. C., and Rybolt, T. R. "Chemistry in the Dyeing of Eggs." *Journal of Chemical Education, 64* (April 1987): 291.

National Academy of Sciences, National Research Council. *Food Colors.* Printing and Publishing Office of the National Academy of Sciences, 2101 Constitution Avenue, NW, Washington, DC 20418 (1971).

"Red Food Coloring: How Safe Is It?" *Consumer Reports* (February 1973): 130.

Sanders, H. J. "Food Additives." *Chemical and Engineering News,* Part 1 (October 10, 1966): 100; Part 2 (October 17, 1966): 108.

Swanson, J. M., and Kinsbourne, M. "Food Dyes Impair Performance of Hyperactive Children on a Laboratory Learning Test." *Science, 207* (March 28, 1980): 1485.

Turner, J. *The Chemical Feast.* New York: Grossman, 1970.

Weiss, B., *et. al.* "Behavioral Responses to Artificial Food Colors." *Science, 207* (March 28, 1980): 1487.

Winter, R. *A Consumer's Dictionary of Food Additives.* New York: Crown Publishers, 1972.

Experiment 41 Chromatography of Some Dye Mixtures

Thin-layer chromatography
Prepared plates
Hand-dipped plates
Paper chromatography
Column chromatography

In this experiment several different types of chromatography are used to separate mixtures of dyes. Three types of dye mixtures are involved. The first type of mixture will be represented by the commercial food colors that can be bought in any grocery store (Parts I and II). These are usually available in small packages containing bottles of red, yellow, blue, and green food dye mixtures. As the experiment will show, each of these colors is rarely compounded of only a single dye. For instance, the blue dye usually has a small admixture of a red dye to make it more brilliant in color. A red dye is often added to the yellow food dye for a similar reason. The green dye will normally be a mixture of blue and yellow dyes.

The second type of dye mixture to be used is one compounded by the instructor from three F,D&C dyes approved for food use. This artificial mixture contains a red dye, a blue dye, and a yellow dye. The identities of these dyes will not be given; rather, you will be asked to identify them by thin-layer chromatography (Parts IV and V) using standard solutions of the individual pure dyes. Your instructor may also ask you to separate this mixture using column chromatography (Part VI).

The third type of mixture consists of the dyes obtained from a commercial powdered drink mix, such as Kool-Aid (Part II). In this case, you will be asked to try to identify the particular dyes used in its preparation.

For those students who are interested, the references listed at the end of this experiment give methods of extracting food dyes from various other types of foods. Also given is information on how to differentiate the various food dyes by their visible absorption spectra.

REQUIRED READING

New: Technique 12   Column Chromatography, Sections 12.1–12.4
       Technique 13   Thin-Layer Chromatography
       Essay   Food Colors

If you are instructed to complete Part VI (Separation of a Dye Mixture by column chromatography), it will be necessary to read **all** of Technique 12.

SPECIAL INSTRUCTIONS

The instructor may choose to do only selected portions of this experiment or perhaps all of the parts. Several experiments may be done at one time since much of the time is spent waiting for the solvent to ascend the chromatograms. To aid in your planning, an estimate of the amount of time required for development or separation is given at the beginning of each section. This time does not include preparation time for the solvents, development chambers, columns, or spotting procedures.

PROCEDURES

I. Paper Chromatography of Food Colors
(Development Time: 40 Minutes)

At least 12 capillary micropipets will be required for the experiment. Prepare these according to the method described and illustrated in Technique 13, Section 13.4, p 728.

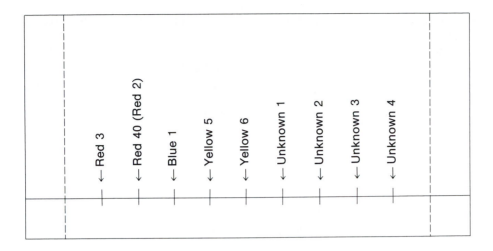

Prepare about 90 mL of a development solvent consisting of

30 mL 2*N* NH$_4$OH (4 mL conc. NH$_4$OH + 26 mL H$_2$O)
30 mL 1-Pentanol (*n*-amyl or *n*-pentyl alcohol)
30 mL Absolute ethanol

The entire mixture may be prepared in a 100-mL graduated cylinder. Mix the solvent well and pour it into the development chamber for storage. A 32-oz wide-mouthed screw-cap jar (or a Mason jar) is an appropriate development chamber. Cap the jar tightly to prevent losses of solvent from evaporation.

Next, obtain a 12-cm × 24-cm sheet of Whatman No. 1 paper. Using a pencil (not a pen), lightly draw a 24-cm-long line about 2 cm up from the long edge of the sheet. Using a centimeter ruler and the pencil, measure and mark off two dashed lines, each about 2 cm from each short end of the paper. Then make nine small marks at 2-cm intervals along the line on the long axis of the paper. These are the positions at which the samples will be spotted (see the illustration).

If they are available, starting from left to right, spot the reference dyes F,D&C Red No. 3 (Erythrosine), F,D&C Red No. 40 (Allura red),[1] F,D&C Blue No. 1 (Erioglaucine), F,D&C Yellow No. 5 (Tartrazine), and F,D&C Yellow No. 6 (Sunset yellow). These should be available in 2% aqueous solutions. It may be wise to practice the spotting technique on a small piece of Whatman No. 1 filter paper before trying to spot the actual chromatogram. The correct method of spotting is described in Technique 13, Section 13.4, p 728. It is important that the spots be made as small as possible and that the paper not be overloaded. If either of these conditions is not met, the spots will tail and overlap after development. The applied spot should be 1–2 mm (¹⁄₁₆ in.) in diameter.

On the remaining four positions (nine if standards are not used) you may spot any dyes of your choice. Use of red, blue, green, and yellow dyes from a single manufacturer is suggested. If the dyes are supplied in screw-cap bottles, the pipets can be filled simply by dipping them into the bottle. If the dyes are supplied in squeeze bottles, however, it will be easiest to place a drop of the dye on a microscope slide and to insert the pipet into the drop. One microscope slide should suffice for all the samples.

---

[1]In Canada or the United Kingdom, substitute F,D&C Red No. 2 (Amaranth) for Red No. 40. Even in the United States, if the food color samples are old (predating 1977), they may contain Amaranth.

When the samples have been spotted, hold the paper upright with the spots at the bottom and coil it into a cylinder. Overlap the areas indicated by the dashed lines and fasten the cylinder together (spots inside) with a paper clip or a staple. When the spots have dried, place the cylinder, spotted edge down, in the development chamber. The solvent level should be below the spots, or they will dissolve in the solvent. Cap the jar and wait until the solvent ascends to the top of the paper. This will take about 40 minutes, and the remaining parts of the experiment (if required) can be done while waiting.

When the solvent has ascended to within 1 cm from the top of the paper, remove the cylinder, open it quickly, and mark the level of the solvent with a pencil. This upper-most level is the solvent front. Allow the chromatogram to dry. Then, using a ruler marked in millimeters, measure the distance that each spot has traveled relative to the solvent front; calculate its R_f value (see Technique 13, Section 13.9, p 773). Using the list of approved food dyes in the essay, "Food Colors" and the reference dyes (if used), try to determine which particular dyes were used to formulate the food colors you tested. Be sure to examine the dye package (or the bottles) to see whether the desired information is given. What conclusions can you draw? Include your chromatogram along with your report.

II. Paper Chromatography of the Dyes from a Powdered Drink Mix or a Gelatin Dessert (Development Time: 40 Minutes)

Place a quantity of the powdered drink mix or gelatin dessert in a small test tube and add warm water dropwise until the sample just dissolves. Use this concentrated solution to spot the paper as described in the section above. Four drink mixes can be spotted on the same piece of Whatman No. 1 paper along with the five standards. Use a **pencil** to label each spot, and then develop the chromatogram in the solvent containing equal parts of $2N$ NH_4OH, pentanol, and ethanol as previously described. Try to identify which dyes are used in samples of several drink mixes (for example, black cherry, cherry, grape, lemon-lime, lime, orange, punch, raspberry, or strawberry). Calculate and compare the R_f values of the standards as well as those of the dyes from the drink mixes. Methods of treating other types of foods to extract and identify the dyes that have been added are described in the references listed at the end of this experiment.

III. Separation of Food Colors Using Prepared TLC Plates (Development Time: 90 Minutes)

Obtain from the instructor a 5-cm $\times$ 10-cm sheet of a prepared silica gel TLC plate (Eastman Chromagram Sheet No. 13180 or No. 13181). These plates have a flexible backing, but they should not be bent excessively. They should be handled carefully, or the adsorbent may flake off of them. In addition, they should be handled only by the edges. The surface should not be touched.

Using a lead pencil (not a pen), **lightly** draw a line across the short dimension of the plate about 1 cm from the bottom. Using a centimeter ruler, mark off four 1-cm intervals on the line (see figure). These are the points at which the samples will be spotted.

Prepare at least four capillary micropipets as described and illustrated in Technique 13, Section 13.4, p 728. Starting from left to right, spot first a red food dye, then a blue dye, a green dye, and a yellow dye. The correct method of spotting a TLC plate is described in Technique 13, Section 13.4. It is important that the spots be made as small as possible and that the plate not be overloaded. If either of these cautions is disregarded, the spots will tail and will overlap after development. The applied spot should be about 1–2 mm ($\frac{1}{16}$ in.) in diameter. If small scrap pieces of the TLC plates are available, it would be a good idea to practice spotting on these before using the actual sample plate.

Prepare a development chamber from an 8-oz wide-mouthed screw-cap jar. It **should not** have the filter paper liner described in Technique 13, Section 13.5, p 730. These plates are very thin, and if they touch a liner at any point, solvent will begin to diffuse onto the plate from that point. The development solvent, which can be prepared in a 10-mL graduated cylinder, should be a 4:1 mixture of isopropyl alcohol (2-propanol) and concentrated ammonium hydroxide.[2] Mix the solvent well and pour enough into the development chamber to give a solvent depth of about 0.5 cm (or less). If the solvent level is too high, it will cover the spotted substances, and they will dissolve into the solvent reservoir.

Place the spotted TLC plate in the development chamber, cap the jar tightly, and wait for the solvent to rise almost to the top of the plate. When the solvent is close to the top edge, remove the plate, and using a pencil (not a pen), quickly mark the position of the solvent front. Allow the plate to dry. Using a ruler marked in millimeters, measure the distance that each spot has travelled relative to the solvent front and calculate its R_f value (see Technique 13, Section 13.9, p 733).

At your instructor's option, and if the dyes are available, you may be asked to spot a second plate with a set of reference dyes. The reference dyes will include F,D&C Red No. 40 (Allura red),[3] F,D&C Blue No. 1 (Erioglaucine), F,D&C Yellow No. 5 (Tartrazine), and F,D&C Yellow No. 6 (Sunset yellow). If this second set of dyes is analyzed, it should be possible (using the list of approved dyes in the Essay Food Colors) to determine the identity of the dyes used to formulate the food colors tested on the first plate. Be sure to examine the package (or bottles) of the food dyes to determine if the desired information is given.

---

[2]An alternative solvent mixture, suggested by McKone and Nelson (see references), is a 50:25:25:10 mixture of 1-butanol, ethanol, water, and concentrated ammonia.

[3]See Footnote 1.

Since the plates are fragile, a sketch, rather than the actual plates, should be included along with your report.

IV. Separation of a Dye Mixture Using Hand-dipped TLC Slides (Development Time: 40 Minutes)

Using a silica gel slurry, prepare three hand-dipped microscope-slide TLC plates by the method described in Technique 13, Section 13.3A, p 725. Prepare developing chambers from 4-oz wide-mouthed screw-cap jars as described in Technique 13, Section 13.5, p 730. Finally, prepare several capillary micropipets as described in Technique 13, Section 13.4, p 728.

Next, obtain from the reagent shelf a bottle containing a mixture of three unknown dyes.[4] Using a capillary micropipet, spot the dye mixture twice on each plate. The correct method of spotting a TLC plate is described in Technique 13, Section 13.4. It is important that the spots be made as small as possible and that the plates not be over-loaded. If either of these errors is made, the spots will tail and will overlap after develop-ment. The applied spot should be 1–2 mm ($\frac{1}{16}$ in.) in diameter.

Develop the first slide, using isopropyl alcohol (2-propanol) as the development solvent. The second slide should be developed in methanol, and the third in a 4:1 mixture of isopropyl alcohol and concentrated ammonium hydroxide.[5] Time will be saved if all three slides are run simultaneously. Two slides will easily fit in a single development chamber, and at least two students can develop slides simultaneously in the same jar.

When the solvents have risen to within 0.5 cm of the top of the slides, remove them, and with a pencil, quickly mark the position of the solvent fronts. Set the slides aside to dry. When they are dry, using a ruler calibrated in millimeters, measure the distance that each spot traveled relative to its solvent front and calculate an R_f value for that spot (Technique 13, Section 13.9, p 733).

When you have determined which of the three solvents is the best development solvent (separates the dyes best), you can proceed in the same manner as before to try to identify which dyes are contained in the mixture. As an example of how to proceed, using another hand-dipped slide, spot two yellow dye standards and a spot of the dye mixture on the same slide. Three spots will easily fit on the same slide if they are small. If your mixture contains a yellow dye, you should readily be able to identify *which* yellow dye is in the mixture when you develop this slide. Proceed in a similar fashion to identify any red, blue, or green dyes present. Include sketches of your plates in your report, along with R_f values, and explain the results. By consulting the structures given for these dyes in the essay that precedes this experiment, you should be able to explain the relative R_f values of the three dyes. (*Hint:* Consider the substituent groups.) Also explain the function of the ammonia.

V. Separation of a Dye Mixture Using Prepared TLC Plates (Development Time: 90 Minutes)

Obtain from the instructor three 5-cm × 10-cm sheets cut from a large prepared silica gel TLC plate (Eastman Chromagram Sheet No. 13180 or No. 13181). Although these

---

[4]The composition of this three-component dye mixture will be found in the Instructor's Manual. The standard solutions have 0.1 g of each dye dissolved in 20 mL of methanol.

[5]See Footnote 2.

plates have a flexible backing, they should not be bent excessively. They should be handled carefully, or the adsorbent may flake off them. They should be handled only by the edges; the surface should not be touched.

Using a lead pencil (not a pen), **lightly** draw a line across the short dimension of the plate about 1 cm from the bottom. Using a centimeter ruler, mark off four 1-cm intervals on the line (see figure shown in Section III). These are the points at which the samples will be spotted.

Prepare at least twelve capillary micropipets as described and illustrated in Technique 13, Section 13.4, p 728. On each of the three plates, spot the unknown dye mixture in the lower left corner.[6] The remaining three positions on each slide may be used to spot standard solutions of known dyes. When spotting, it is important that the spots be made as small as possible and that the slide not be overloaded. If either of these errors is made, the spots will tail and will overlap one another after development. The applied spot should be about 1–2 mm (¹⁄₁₆ in.) in diameter. If small scrap pieces of the TLC plates are available, it would be a good idea to practice spotting on these prior to preparing the actual sample plate.

Prepare a development chamber from an 8-oz wide-mouth screw-cap jar. It *should not* have the filter paper liner that is described in Technique 13, Section 13.5, p 730. These plates are very thin, and if they touch a liner at any point, solvent will begin to diffuse onto the plate from that point. The development solvent, which can be prepared in a 10-mL graduated cylinder, should be a 4:1 mixture of isopropyl alcohol and concentrated ammonium hydroxide.

Place the spotted TLC slides in the development chamber, cap the jar tightly, and wait for the solvent to rise almost to the top of the slide. When the solvent is close to the top edge, remove the slide, and using a pencil (not a pen), quickly mark the position of the solvent front. Allow the plate to dry. Using a ruler marked in millimeters, measure the distance that each spot has traveled relative to the solvent front and calculate its R_f value (see Technique 13, Section 13.9, p 733).

If you have spotted the correct reference dye solutions along with the unknown mixture, you should now be able to identify which dyes are contained in the mixture. Since the plates are rather fragile, sketches, instead of the actual plates, should be included in your report. The spots should be labeled with their R_f values. Be sure to give your conclusions. By consulting the structures given for these dyes in the essay that precedes this experiment, you should be able to explain the relative R_f values of the dyes in the mixture. (*Hint:* Consider the substituent groups.) Also explain the function of the ammonia.

VI. Separation of a Dye Mixture by Column Chromatography (Elution Time for Column: 2½–3 Hours)[7]

Obtain a dry chromatography column made from 10-mm glass tubing and having a length of about 20 cm.[8] Place a small, loose plug of glass wool in the bottom of the

---

[6]See Footnote 4.

[7]To perform this separation, it is necessary to read **all** of Technique 12.

[8]Two columns can be made from a 42-cm-long piece of 10-mm glass tubing. Heat the tubing at the center while rotating it, and when it is hot, constrict the center by pulling the two ends about 4 cm apart. When the tubing is cool, score the constriction in the middle with a file; then, with light pressure from the thumbs while pulling on the two ends, break the tubing into two equivalent columns. (See an illustration of the equivalent operation using capillary tubing to make micropipets in Figure 13–3, p 729.)

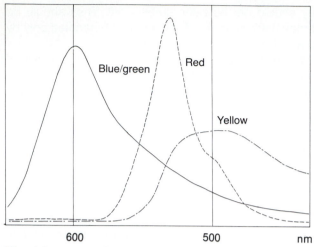

Ultraviolet spectra of unknown dyes, Experiment 41

column, and, with a length of glass rod, gently tamp it level. Cover the glass wool with about 4–5 mm of white sand. With a utility clamp, attach the column vertically to a ring stand. Place a small piece of flexible tubing over the column's exit, loosely attach a screw clamp to the tubing, and close it.

Place a beaker under the exit of the column and fill the column ⅔ full with isopropyl alcohol (2-propanol). Start the column dripping **slowly** and add enough dry absorption alumina (about 5 g) to give a column of adsorbent 6–8 cm high.[9] Rinse the sides of the column with a little solvent. Allow any excess isopropyl alcohol to drain until the top of the column just becomes dry, and then close the screw clamp. Obtain the synthetic dye mixture from the supply shelf, and with a disposable pipet, carefully transfer enough of this mixture to the top of the column to make a layer of liquid 3 mm deep. Open the screw clamp, drain this mixture into the column, and then once again stop the flow. In the same way, place a similar amount of isopropyl alcohol on top of the column and then drain it also into the column, again stopping the flow when the surface of the alumina becomes dry. Finally, carefully add as much of 4:1 isopropyl alcohol–concentrated ammonia elution solvent on top of the column as the unfilled portion allows, and open the screw clamp to allow the chromatography to begin.

From time to time, refill the reservoir with more solvent, never allowing the column to run dry. Continue elution and carefully collect each of the dyes in a separate container. Once the first dye has been collected, you may switch to 4:1 ethanol–concentrated ammonia to elute the second dye. The third dye may be eluted by changing to 4:1 methanol–concentrated ammonia. If the last dye does not elute easily, the ratio of ammonia to methanol may be increased.

Each of the eluted dyes may be spotted on hand-dipped TLC slides (described in Part IV) and compared with standard solutions of the three dyes, using 4:1 isopropyl alcohol–concentrated ammonia for development. Give your conclusions in your report, and give the R_f values you found for each of the dyes when making the TLC compari-

---

[9]We use Fisher No. A-540 Adsorption Alumina. Longer columns should be avoided since they lead to very long elution and separation times.

sons. Finally, consult the structures given in the essay preceding this experiment and try to explain the order of elution (R_f values) observed for the three dyes. (*Hint:* Consider the substituent groups.) Also explain the function of the ammonia.

REFERENCES

McKone, H. T. "Identification of F,D&C Dyes by Visible Spectroscopy." *Journal of Chemical Education, 54* (June 1977): 376.

McKone, H. T., and Nelson, G. J. "Separation and Identification of Some F,D&C Dyes by TLC." *Journal of Chemical Education, 53* (November 1976): 722.

Essay
LOCAL ANESTHETICS

Local anesthetics, or "painkillers," are a well-studied class of compounds with which chemists have shown their ability to study the essential features of a naturally occurring drug and to improve on them by substituting totally new, synthetic surrogates. Often such substitutes are superior in desired medical effects and also in lack of unwanted side effects or hazards.

The coca shrub (*Erythroxylon coca*) grows wild in Peru, specifically in the Andes Mountains, at elevations of 1500 to 6000 ft above sea level. The natives of

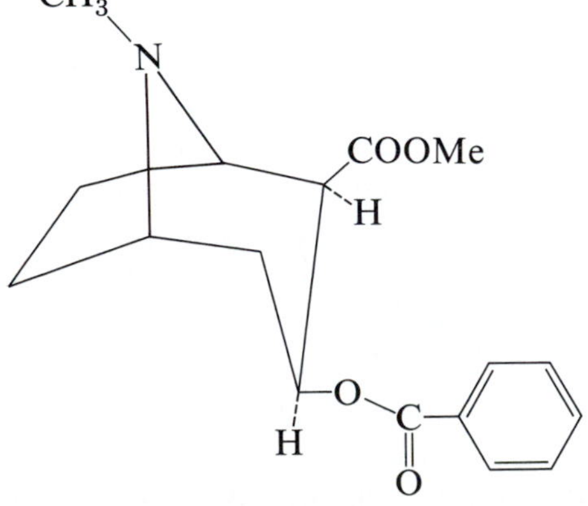

Cocaine

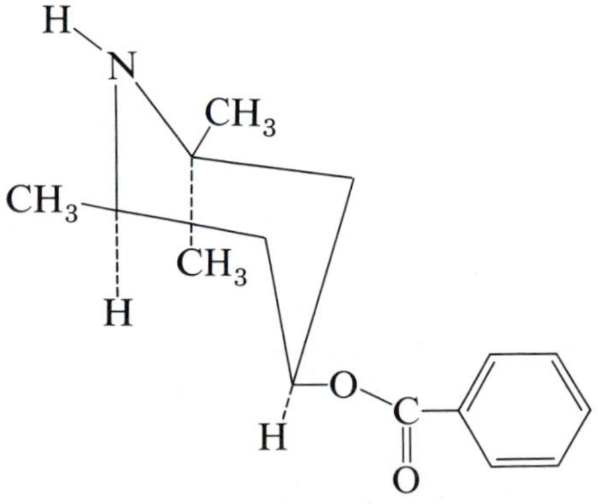

Eucaine

South America have long chewed these leaves for their stimulant effects. Leaves of the coca shrub have even been found in pre-Inca Peruvian burial urns. The leaves bring about a definite sense of mental and physical well-being and have the power to increase endurance. For chewing, the Indians smear the coca leaves with lime and roll them. The lime, $Ca(OH)_2$, apparently releases the free alkaloid components; it is remarkable that the Indians learned this subtlety long ago by some empirical means. The pure alkaloid responsible for the properties of the coca leaves is **cocaine.**

The amounts of cocaine consumed in this way by the Indians are extremely small. Without such a crutch of central-nervous-system stimulation, the natives of the Andes would probably find it more difficult to perform the nearly Herculean tasks of their daily lives, such as carrying heavy loads over the rugged mountainous terrain. Unfortunately, overindulgence can lead to mental and physical deterioration and eventually an unpleasant death.

The pure alkaloid in large quantities is a common drug of addiction. Sigmund Freud first made a detailed study of cocaine in 1884. He was particularly impressed by the ability of the drug to stimulate the central nervous system, and he used it as a replacement drug to wean one of his addicted colleagues from morphine. This attempt was successful, but unhappily, the colleague became the world's first known cocaine addict.

An extract from coca leaves was one of the original ingredients in Coca-Cola. However, early in the present century, government officials, with much legal difficulty, forced the manufacturer to omit coca from its beverage. The company has managed to this day to maintain the *coca* in its trademarked title even though "Coke" contains none!

Our interest in cocaine lies in its anesthetic properties. The pure alkaloid was isolated in 1862 by Niemann, who noted that it had a bitter taste and produced a queer numbing sensation on the tongue, rendering it almost devoid of sensation. (Oh, those brave, but foolish chemists of yore who used to taste everything!) In 1880, Von Anrep found that the skin was made numb and insensitive to the prick of a pin when cocaine was injected subcutaneously. Freud and his assistant, Karl Koller, having failed at attempts to rehabilitate morphine addicts, turned to a study of the anesthetizing properties of cocaine. Eye surgery is made difficult by involuntary reflex movements of the eye in response to even the slightest touch. Koller found that a few drops of a solution of cocaine would overcome this problem. Not only can cocaine serve as a local anesthetic, but it can also be used to produce mydriasis (dilation of the pupil). The ability of cocaine to block signal conduction in nerves (particularly of pain) led to its rapid medical use in spite of its dangers. It soon found use as a "local" in both dentistry (1884) and in surgery (1885). In this type of application, it was injected directly into the particular nerves it was intended to deaden.

Soon after the structure of cocaine was established, chemists began to search for a substitute. Cocaine has several drawbacks for wide medical use as an anesthetic. In eye surgery, it also produces mydriasis. It can also become a drug of addiction. Finally, it has a dangerous effect on the central nervous system.

The first totally synthetic substitute was eucaine. This was synthesized by Harries in 1918 and retains many of the essential skeletal features of the cocaine molecule.

The development of this new anesthetic partly confirmed the portion of the cocaine structure essential for local anesthetic action. The advantage of eucaine over cocaine is that it does not produce mydriasis and is not habit-forming. Unfortunately, it is highly toxic.

A further attempt at simplification led to piperocaine. The molecular portion common to cocaine and eucaine is outlined by dotted lines in the structure shown. Piperocaine is only a third as toxic as cocaine itself.

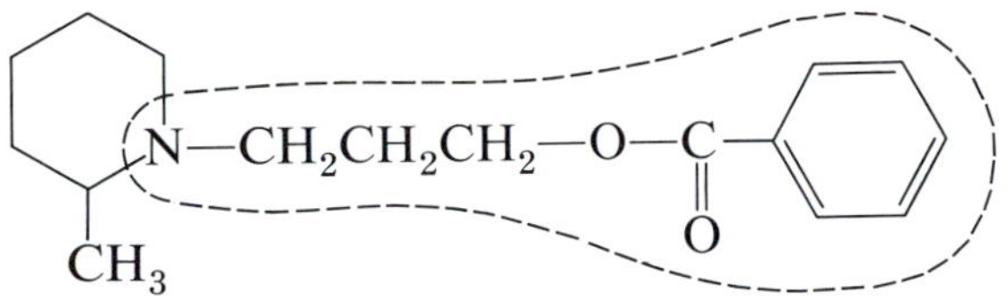

$$\text{N}-\text{CH}_2\text{CH}_2\text{CH}_2-\text{O}-\underset{\text{O}}{\overset{}{\text{C}}}- \quad\quad \textbf{Piperocaine}$$

$$\text{CH}_3$$

The most successful synthetic for many years was the drug procaine, also known more commonly by its trade name Novocain (see table). Novocain is only a fourth as toxic as cocaine, giving a better margin of safety in its use. The toxic dose is almost 10 times the effective amount, and it is not a habit-forming drug.

Over the years, hundreds of new local anesthetics have been synthesized and tested. For one reason or another, most have not come into general use. The search for the perfect local anesthetic is still under way. All the drugs found to be active have certain structural features in common. At one end of the molecule is an aromatic ring. At the other is a secondary or tertiary amine. These two essential features are separated by a central chain of atoms usually one to four units long. The aromatic part is usually an ester of an aromatic acid. The ester group is important to the bodily detoxification of these compounds. The first step in deactivating them is a hydrolysis of this ester linkage, a process that occurs in the bloodstream. Compounds that do not have the ester link are both longer lasting in their effect and generally more toxic. An exception is lidocaine, which is an amide. The tertiary amino group is apparently necessary to enhance the solubility of the compounds in the injection solvent. Most of these compounds are used in their hydrochloride salt forms, which can be dissolved in water for injection. Benzocaine, in contrast, is active as a local anesthetic but is not used for injection. It does not suffuse well into tissue and is not water-soluble. It is used primarily in skin preparations, in which it can be included in an ointment or salve for direct application. It is an ingredient of many sunburn preparations.

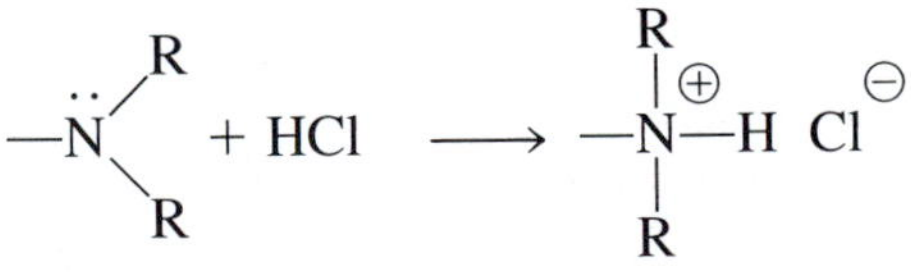

$$-\overset{\text{R}}{\underset{\text{R}}{\text{N}}} + \text{HCl} \longrightarrow -\overset{\text{R}}{\underset{\text{R}}{\overset{\oplus}{\text{N}}}}-\text{H} \quad \text{Cl}^{\ominus}$$

How these drugs act to stop pain conduction is not well understood. Their main site of action is at the nerve membrane. They seem to compete with calcium at some receptor site, altering the permeability of the membrane and keeping the nerve slightly depolarized electrically.

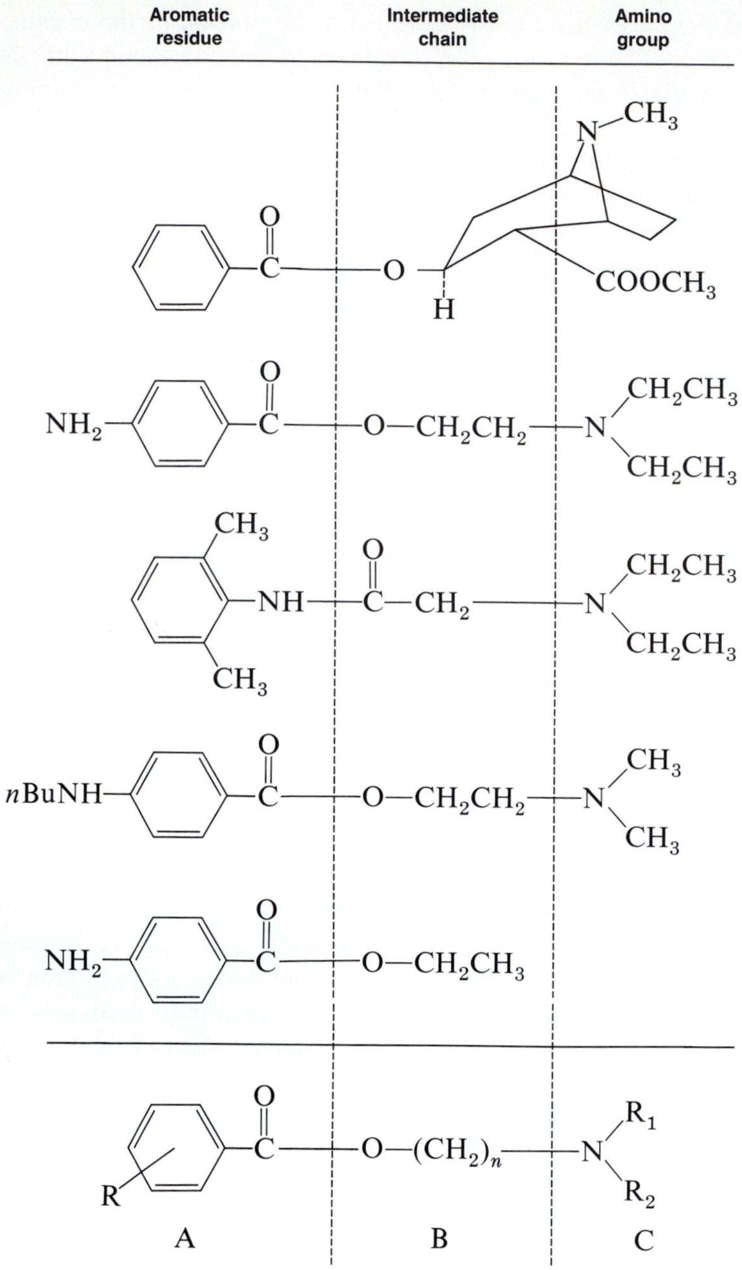

Aromatic residue	Intermediate chain	Amino group	

Cocaine

Procaine

Lidocaine

Tetracaine

Benzocaine

Generalized structure for a local anesthetic

A   B   C

Local anesthetics

REFERENCES

Foye, W. O. *Principles of Medicinal Chemistry*. Philadelphia: Lea & Febiger, 1974. Chap. 14, "Local Anesthetics."

Goodman, L. S., and Gilman, A. *The Pharmacological Basis of Therapeutics*. 7th ed. New York: Macmillan, 1985. Chap. 20, "Cocaine: Procaine and Other Synthetic Local Anesthetics," J. M. Ritchie, et al.

Ray, O. S. *Drugs, Society, and Human Behavior*. 3rd ed. St. Louis: C. V. Mosby, 1983. Chap. 11, "Stimulants and Depressants."

Snyder, S. H. "The Brain's Own Opiates." *Chemical and Engineering News* (November 28, 1977): 26–35.

Taylor, N. *Narcotics: Nature's Dangerous Gifts*. New York: Dell, 1970. Paperbound revision of *Flight from Reality*. Chap. 3, "The Divine Plant of the Incas."

Taylor, N. *Plant Drugs that Changed the World*. New York: Dodd, Mead, 1965. Pp 14–18.

Wilson, C. O., Gisvold, O., and Doerge, R. F. *Textbook of Organic Medicinal and Pharmaceutical Chemistry*. 6th ed. Philadelphia: J. B. Lippincott, 1971. Chap. 22, "Local Anesthetic Agents," R. F. Doerge.

Experiment 42
Benzocaine

Esterification

In this experiment, a procedure is given for the preparation of a local anesthetic, benzocaine, by the direct esterification of *p*-aminobenzoic acid with ethanol. At the instructor's option, you may test the prepared anesthetic on a frog's leg muscle.

p-Aminobenzoic acid + CH_3CH_2OH $\underset{}{\overset{H^+}{\rightleftharpoons}}$ Ethyl *p*-aminobenzoate (Benzocaine) + H_2O

REQUIRED READING

Review:	Filtration	Section 4.3
	Crystallization	Sections 5.4 and 5.9
New:	Essay	Local Anesthetics

SPECIAL INSTRUCTIONS

Sulfuric acid is very corrosive. Do not allow it to come in contact with your skin.

> **NOTE TO THE INSTRUCTOR:** If instruction for testing benzocaine on a frog's leg muscle is needed, please refer to the instructor's manual.

PROCEDURE

Place 0.120 g of p-aminobenzoic acid and 1.20 mL of absolute ethanol into a 3-mL conical vial. Add a magnetic spin vane and stir the mixture until the solid dissolves completely. With stirring, add 0.10 mL of concentrated sulfuric acid dropwise. A large amount of precipitate will form when the sulfuric acid is added, but this solid will slowly dissolve during the reflux that follows. Attach a water-cooled condenser and heat the mixture at a gentle boil for 60–75 minutes with a sand bath at about 115 °C. Stir the mixture during this heating period.

At the end of the reaction time, remove the apparatus from the sand bath and allow the reaction mixture to cool for several minutes. Using a Pasteur pipet, transfer the contents of the vial to a small beaker containing 3.0 mL of water. When the liquid has cooled to room temperature, add a 10% sodium carbonate solution (about 1 mL needed) dropwise to neutralize the mixture. Stir the contents of the beaker with a stirring rod or spatula. After each addition of the sodium carbonate solution, extensive gas evolution (frothing) will be perceptible until the mixture is nearly neutralized. As the pH increases, a white precipitate of benzocaine is produced. When gas no longer evolves as you add a drop of sodium carbonate, check the pH of the solution and add further portions of sodium carbonate until the pH is about 8.

Collect the benzocaine by vacuum filtration using a Hirsch funnel. Use three 1-mL portions of water to aid in the transfer and to wash the product in the funnel. Be sure that the solid is rinsed thoroughly with the water. After the product has dried overnight, weigh it, calculate the percentage yield, and determine its melting point. The melting point of pure benzocaine is 92 °C.

Although the product should be fairly pure, it may be recrystallized by the mixed solvent method using methanol and water (Technique 5, Section 5.9, p 593). Place the product in a Craig tube; add several drops of methanol; and, while heating the Craig tube in a sand bath (70–80 °C) and stirring the mixture with a microspatula, add methanol dropwise until all the solid dissolves. Add two to three additional drops of methanol and then add hot water dropwise until the mixture turns cloudy or a white precipitate forms. Add methanol again until the solid dissolves completely. Insert the inner plug of the Craig tube and allow the solution to cool slowly to room temperature. Complete the crystallization by cooling the mixture in an ice bath, and collect the crystals by centrifugation (Technique 4, Section 4.7, p 575). Weigh the purified benzocaine and determine its melting point.

At the option of the instructor, obtain the Infrared spectrum in chloroform (Technique 18, Section 18.5, p 777) and the NMR spectrum in carbon tetrachloride or CDCl$_3$, (Technique 18, Section 18.9, p 782). Submit the sample in a labeled vial to the instructor.

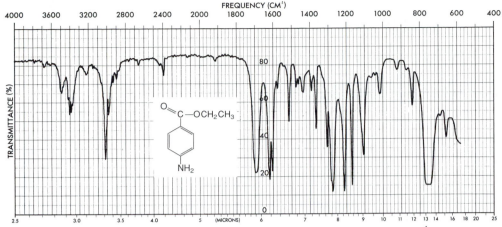

Infrared spectrum of benzocaine, $CHCl_3$. ($CHCl_3$ solvent: 3030, 1220, and 750 cm^{-1})

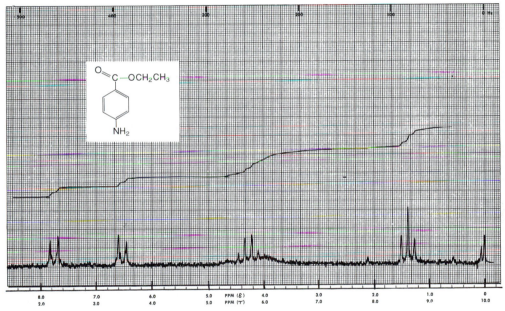

NMR spectrum of benzocaine, CCl_4

QUESTIONS

1. Interpret the Infrared and NMR spectra of benzocaine.

2. What is the structure of the precipitate that forms after the sulfuric acid has been added?

3. When 10% sodium carbonate solution is added, a gas evolves. What is the gas? Give a balanced equation for this reaction.

4. Explain why benzocaine precipitates during the neutralization.

5. Refer to the structure of procaine in the table in the essay, "Local Anesthetics." Using *p*-aminobenzoic acid, give equations showing how procaine and procaine monohydrochloride could be prepared. Which of the two possible amino functional groups in procaine will be protonated first? Defend your choice. (*Hint:* Consider resonance.)

Experiment 43
Methyl Salicylate (Oil of Wintergreen)

Synthesis of an ester   Extraction
Heating under reflux   Vacuum distillation

In this experiment you prepare a familiar-smelling organic ester—oil of wintergreen. Methyl salicylate was first isolated in 1843 by extraction from the wintergreen plant (*Gaultheria*). It was soon found that this compound had analgesic and antipyretic character almost identical to that of salicylic acid (see the essay, "Aspirin") when taken internally. This medicinal character probably derives from the ease with which methyl salicylate is hydrolyzed to salicylic acid under the alkaline conditions found in the intestinal tract. Salicylic acid is known to have such analgesic and antipyretic properties. Methyl salicylate can be taken internally or absorbed through the skin; thus, it finds much use in liniment preparations. Applied to the skin, it produces a mild burning or soothing sensation, which probably comes from the action of its phenolic hydroxyl group. This ester also has a pleasant odor, and it is used to a small extent as a flavoring principle.

Salicylic acid   + CH_3OH $\underset{}{\overset{H^+}{\rightleftharpoons}}$ Methyl salicylate (Oil of wintergreen)   + H_2O

Methyl salicylate will be prepared from salicylic acid, which is esterified at the carboxyl group with methanol. You should recall from your organic chemistry lecture course that esterification is an acid-catalyzed equilibrium reaction. The equilibrium does not lie far enough to the right to favor the formation of the ester in high yield. More product can be formed by increasing the concentrations of one of the reactants. In this experiment, a large excess of methanol will shift the equilibrium to favor a more complete formation of the ester.

This experiment also illustrates the use of distillation under reduced pressure for purifying high-boiling liquids. Distillation of high-boiling liquids at atmospheric pressure is often unsatisfactory. At the high temperatures required, the material being distilled (the ester, in this case) may partially or even completely decompose causing loss of product and contamination of the distillate. When the total pressure inside the distillation apparatus is reduced, however, the boiling point of the substance is lowered. In this way the substance can be distilled without being decomposed.

346

REQUIRED READING

Review: Techniques 1–3 and 7
            Technique 6   Physical Constants, Part B, Boiling Points

New: Technique 9   Vacuum Distillation
            Technique 18   Preparation of Samples for Spectroscopy
            Essay   Esters—Flavors and Fragrances

SPECIAL INSTRUCTIONS

The experiment must be started at the beginning of the laboratory period since a long reflux time is needed to esterify salicylic acid and obtain a respectable yield. A supplementary experiment may be performed during the reaction period, or work that is pending from previous experiments may be completed. Enough time should remain at the end of the period to perform the extractions, to place the product over the drying agent, and to assemble the apparatus and perform the vacuum distillation.

Handle the concentrated sulfuric acid with caution; it can cause severe burns.

When a distillation is conducted under reduced pressure, it is important to guard against the dangers of an implosion. Inspect the glassware for flaws and cracks, and replace any that is defective.

Wear your safety glasses.

Because the amount of methyl salicylate obtained in this experiment is small, your instructor may want two students to combine their products for the final vacuum distillation.

PROCEDURE

Assemble equipment for reflux using a 5-mL conical vial and a water-cooled condenser (Figure 3–2A, p 550). Top the apparatus with a calcium chloride drying tube. Use a hot plate with a sand bath. Place 0.65 g of salicylic acid, 2.0 mL of methanol (density = 0.792 g/mL), and a spin vane in the vial. Stir the mixture until the salicylic acid dissolves. Carefully add 0.75 mL of concentrated sulfuric acid, **in small portions,** to the mixture in the vial while stirring. A white precipitate may form, but it will redissolve during the reflux period. Complete assembly of the apparatus, and, while stirring, gently boil the mixture (sand bath 80–100 °C) for 60–75 minutes.

After the mixture has cooled, extract it with three 1-mL portions of methylene chloride (Technique 7, Section 7.4, p 622). Add the methylene chloride, cap the vial, shake it, and then loosen the cap. When the layers separate, transfer the lower layer with a filter tip pipet to another container. After completing the three extractions, discard the aqueous layer and return the three methylene chloride extracts to the vial. Extract the methylene chloride layers with a 1-mL portion of 5% aqueous sodium bicarbonate. Transfer the lower organic layer to a clean, dry conical vial. Discard the aqueous layer. Add two to three microspatulafuls of anhydrous sodium sulfate to the organic layer and cap the vial. When the solution is dry (about 10 minutes), transfer it to a clean, dry 3-mL conical vial with a filter tip pipet. Evaporate the methylene chloride using a hot plate and a sand bath (40–50 °C) in the hood. A stream of nitrogen or air will accelerate the evaporation (Figure 3–12A, p 561). The product may be stored in the capped vial and saved for the next period, or it may be distilled under vacuum during the same period.

VACUUM DISTILLATION

Using the procedure described in Technique 9, Section 9.4, p 658, distill the product by vacuum distillation using an apparatus fitted with a Hickman still and a water-cooled condenser (Figure 9–5, p 656). Place a small piece of a stainless steel sponge in the lower stem of the Hickman still to prevent bump-over, and stir vigorously with a magnetic spin vane. An aspirator should be used for the vacuum source and a manometer should be attached if one is available (See Figure 9–14, p 667). You may use either a sand bath or an aluminum block to heat the distillation mixture. The sand bath (or aluminum block) temperature will be about 130 °C (with 20 mmHg vacuum). If you have less than 0.75 mL, you should combine your product with that of another student.

When the distillation is complete, transfer the distillate to a tared 3-mL conical vial with a Pasteur pipet and weigh it to determine the percentage yield. Determine a microscale boiling point (Technique 6, Section 6.10, p 607) for your product.

INFRARED AND NMR SPECTROSCOPY (optional)

At your instructor's option, record the infrared spectrum and NMR spectrum of the product. Use salt plates (Technique 18, Section 18.2, p 771) to determine the Infrared spectrum. Compare the spectra with the ones reproduced in this experiment.

QUESTIONS

1. Write a mechanism for the acid-catalyzed esterification of salicylic acid with methanol. You may need to consult the chapter on carboxylic acids in your lecture textbook.

2. What is the function of the sulfuric acid in this reaction? Is it consumed in the reaction?

3. In this experiment, excess methanol was used to shift the equilibrium toward the formation of more ester. Describe other methods for achieving the same result.

4. How are sulfuric acid and the excess methanol removed from the crude ester after the reaction has been completed?

5. Why was 5% $NaHCO_3$ used in the extraction? What would have happened if 5% $NaOH$ had been used?

6. Interpret the principal absorption bands in the Infrared and NMR spectra of methyl salicylate.

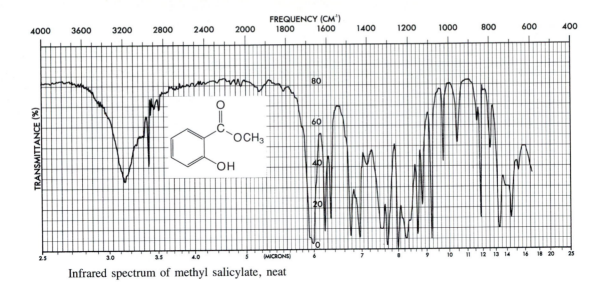

Infrared spectrum of methyl salicylate, neat

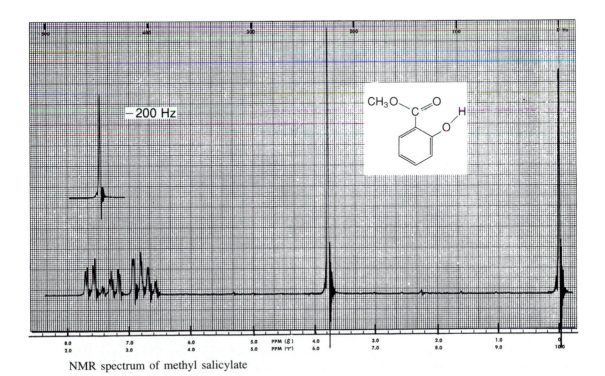

NMR spectrum of methyl salicylate

Essay
SOAPS AND DETERGENTS

Soaps as we know them today were virtually unknown before the first century A.D. Clothes were cleaned primarily by the abrasive action of rubbing them on rocks in water. Somewhat later, it was discovered that certain types of leaves, roots, nuts, berries, and barks formed soapy lathers that solubilized and removed dirt from clothes. We now know these natural materials that lather as **saponins.** Many saponins contain pentacyclic triterpene carboxylic acids, such as oleanolic acid or ursolic acid, chemically combined with a sugar molecule. These acids also appear in the uncombined state. Saponins were probably the first known "soaps." They may have also been an early source of pollution in that they are known to be toxic to fish. The pollution problem associated with the development of soap and detergents has been long and controversial.

Oleanolic acid   **Ursolic acid**

Soap as we know it today has probably evolved over many centuries from experimentation with crude mixtures of alkaline and fatty materials. Pliny the Elder described the manufacture of soap during the first century A.D. A modest soap factory was even built in Pompeii. During the Middle Ages, cleanliness of the body or clothing was not considered important. Those who could afford perfumes used them to hide their body odor. Perfumes, like fancy clothes, were a status symbol for the rich. An interest in cleanliness again emerged during the eighteenth century, when disease-causing microorganisms were discovered.

SOAPS

The process of making soap has remained practically unchanged for 2000 years. The procedure involves the basic hydrolysis (saponification) of a fat. Chemically, fats are usually referred to as **triglycerides.** They contain ester functional groups. Saponifica-

350

tion involves heating fat with an alkaline solution. This alkaline solution was originally obtained by leaching wood ashes or from the evaporation of natural alkaline waters. Today, lye (sodium hydroxide) is used as the source of the alkali. The alkaline solution hydrolyzes the fat to its component parts, the salt of a long-chain carboxylic acid (soap) and an alcohol (glycerol). When common salt is added, the soap precipitates. The soap is washed free of unreacted sodium hydroxide and molded into bars. The carboxylic acid salts of soap usually contain 12–18 carbons arranged in a straight chain. The carboxylic acids containing even numbers of carbon atoms predominate, and the chains may contain unsaturation. The chemical structures of fats, and the related oils, are shown in the essay, "Fats and Oils." The equation below shows how soap is produced

$$
\begin{array}{c}
\underset{\text{Triglyceride (Fat)}}{
\begin{array}{l}
R{-}\overset{\displaystyle O}{\overset{\|}{C}}{-}O{-}CH_2 \\[4pt]
R{-}\overset{\displaystyle O}{\overset{\|}{C}}{-}O{-}CH \\[4pt]
R{-}\overset{\displaystyle O}{\overset{\|}{C}}{-}O{-}CH_2
\end{array}}
\; + \; \underset{\text{Lye}}{3NaOH}
\;\longrightarrow\;
\underset{\substack{\text{Sodium salt}\\\text{of an acid (Soap)}}}{3R{-}\overset{\displaystyle O}{\overset{\|}{C}}{-}O^- Na^+}
\;+\;
\underset{\text{Glycerol}}{
\begin{array}{l}
HO{-}CH_2 \\[4pt]
HO{-}CH \\[4pt]
HO{-}CH_2
\end{array}}
\end{array}
$$

from a fat. It is an idealized reaction, in which the R groups are the same. Usually, the R groups are **not** the same. Therefore, soap is actually a mixture of salts of carboxylic acids.

A disadvantage of soap is that it is an ineffective cleanser in hard water. Hard water contains salts of magnesium, calcium, and iron in solution. When soap is used in hard water, "calcium soap," the insoluble calcium salts of the fatty acids, and other precipitates are deposited as **curds.** This precipitate, or curd, is referred to as bathtub ring. Although soap is a poor cleanser in hard water, it is an excellent cleanser in soft water.

$$
\underset{\text{Soap}}{2R{-}\overset{\displaystyle O}{\overset{\|}{C}}{-}O^- Na^+} \;+\; Ca^{2+}
\;\longrightarrow\;
\underset{\text{Curd}}{\left(R{-}\overset{\displaystyle O}{\overset{\|}{C}}{-}\overset{-}{O} \right)_2 Ca^{2+}}
$$

Water softeners are added to soaps to help remove the troublesome hard-water ions so that the soap will remain effective in hard water. Sodium carbonate or trisodium phosphate will precipitate the ions as the carbonate or phosphate. Unfortunately, the precipitate may become lodged in the fabric of items being laundered, causing a grayish or streaked appearance.

$$Ca^{2+} + CO_3^{2-} \longrightarrow CaCO_3 \quad \downarrow$$

$$3Ca^{2+} + 2PO_4^{3-} \longrightarrow Ca_3(PO_4)_2 \quad \downarrow$$

An important advantage of soap is that it is **biodegradable.** Microorganisms can consume the linear soap molecules and convert them to carbon dioxide and water. The soap is thus eliminated from the environment.

ACTION OF SOAP IN CLEANING

Dirty clothes, skin, or other surfaces have particles of dirt suspended in a layer of oil or grease. Polar water molecules cannot remove the dirt embedded in nonpolar oil or grease. One can remove the dirt with soap, however, because of its dual nature. The soap molecule has a polar, **water-soluble** head (carboxylate salt) and a long, **oil-soluble** tail (the hydrocarbon chain). The hydrocarbon tail of soap dissolves in the oily substance but the ionic end remains outside the oily surface. When enough soap molecules have oriented themselves around an oil droplet with their hydrocarbon ends dissolved in the oil, the oil droplet, together with the suspended dirt particles, is removed from the surface of the cloth or skin. The oil droplet is removed because the heavily negatively charged oil droplet is now strongly attracted to water and solvated by the water. The solvated oil droplet is called a **micelle;** one is shown below.

DETERGENTS

Detergents are synthetic cleaning compounds, often referred to as "syndets." They were developed as an alternative to soaps because they are effective in **both** soft and hard water. No precipitates form when calcium, magnesium, or iron ions are present in

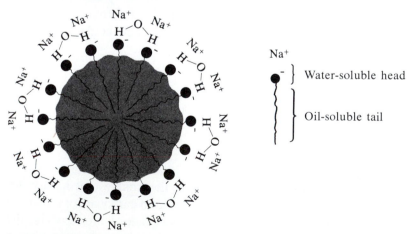

A soap micelle solvating a droplet of oil (from Linstromberg, Walter W., *Organic Chemistry: A Brief Course*, D. C. Heath, 1978).

a detergent solution. One of the earliest detergents developed was sodium lauryl sulfate. It is prepared by the action of sulfuric acid or chlorosulfonic acid on lauryl alcohol (1-dodecanol). This detergent is relatively expensive, however. The following reactions show one industrial method of preparation:

$$CH_3(CH_2)_{10}CH_2OH + H_2SO_4 \longrightarrow CH_3(CH_2)_{10}CH_2OSO_3H + H_2O$$

Lauryl alcohol

$$CH_3(CH_2)_{10}CH_2OSO_3H + Na_2CO_3 \longrightarrow CH_3(CH_2)_{10}CH_2OSO_3{}^-Na^+ + NaHCO_3$$

Sodium lauryl sulfate

The first of the inexpensive detergents appeared about 1950. These detergents, called alkylbenzenesulfonates (ABS), can be prepared from inexpensive petroleum sources by the following set of reactions:

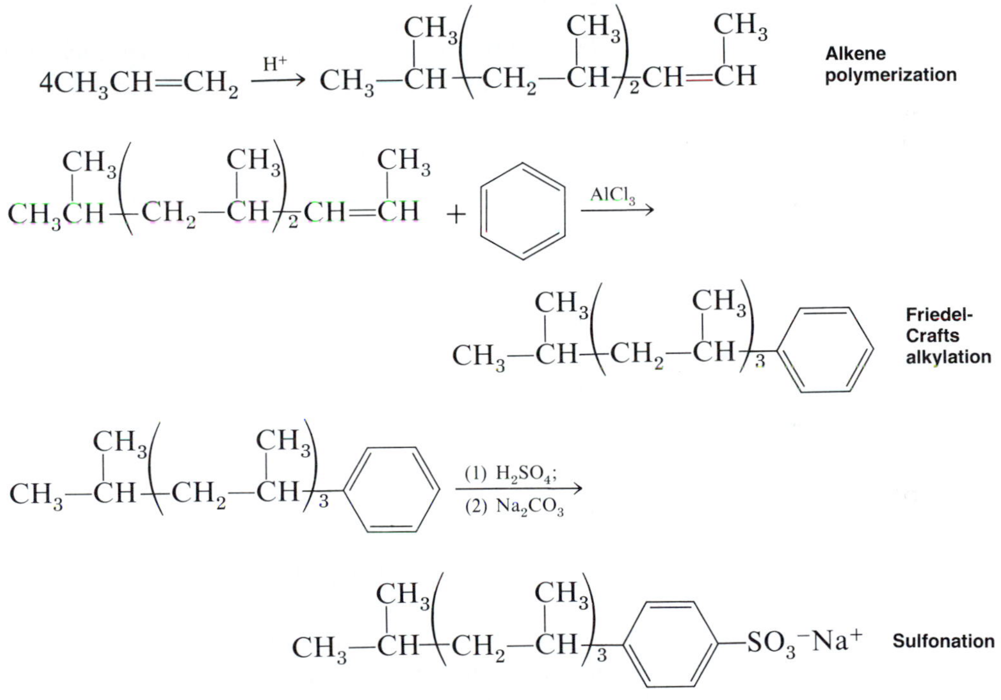

Alkene polymerization

Friedel-Crafts alkylation

Sulfonation

An alkylbenzenesulfonate (ABS)

Detergents became very popular because they could be used effectively in all types of water and were cheap. They rapidly displaced soap as the most popular cleaning agent. A problem with the detergents was that they passed through sewage-treatment plants without being degraded by the microorganisms present, a process necessary for the full sewage treatment. Rivers and streams in many sections of the country became polluted with detergent foam. The detergents even found their way into the drinking water supplies of numerous cities. The reason for the persistence of the detergent was that bacterial enzymes, which could degrade straight-chain soaps and sodium lauryl sulfate, could not destroy the highly branched detergents such as ABS.

It was soon found that the bacterial enzymes could degrade only a chain of carbons that contained, at the most, one branch. As numerous cities and states banned the sale of the nonbiodegradable detergents, by 1966 they were replaced by the new biodegradable detergents called linear alkylsulfonates (LAS). One example of an LAS detergent is shown below. Notice that there is one branch next to the aromatic ring.

$$CH_3(CH_2)_9\!-\!\overset{\displaystyle CH_3}{\underset{\displaystyle |}{CH}}\!-\!\!\left\langle\!\!\bigcirc\!\!\right\rangle\!-\!SO_3^-Na^+$$

A linear alkylsulfonate detergent (LAS)

NEW PROBLEMS WITH DETERGENTS

Detergents (also soaps) are not sold as pure compounds. A typical heavy-duty, controlled "sudser" may contain only 8–20% of the linear alkylsulfonate. A large quantity (30–50%) of a "builder" such as sodium tripolyphosphate, $Na_5P_3O_{10}$, may be present. Other additives include corrosion inhibitors, antideposition agents, and perfumes. Optical brighteners are also added. Brighteners absorb invisible ultraviolet light and re-emit it as visible light, so laundry appears white and thus "clean." The phosphate builder is added to complex the hard-water ions, calcium and magnesium, and keep them in solution. Builders seem to enhance the washing ability of the LAS and also act as a cheap filler.

Unfortunately, phosphates speed the **eutrophication** of lakes and other bodies of water. The phosphates, along with other substances, are nutrients for algae. When algae begin to die and decompose, they consume so much dissolved oxygen from the water that no other life can exist in that water. The lake rapidly "dies." This is eutrophication.

Because phosphates have this undesirable effect, a search was initiated for a replacement for the phosphate builders. Some replacements have been made, but most also have problems associated with them. Two replacement builders, sodium metasilicate and sodium perborate, are highly basic substances, and they have caused injuries to children. In addition, they appear to destroy bacteria in sewage-treatment plants and may have other unknown environmental effects.

Many people have suggested a return to soap. The main problem is that we probably cannot produce enough soap to meet the demand because of the limited amount of animal fat available. Where do we go from here?

REFERENCES

Davidsohn, J., et al. *Soap Manufacture*. New York: Interscience, 1953.
Garner, W. *Detergents*. Amsterdam: Elsevier, 1966.
Kushner, L. M., and Hoffman, J. I. "Synthetic Detergents." *Scientific American, 185* (October 1951): 26.

"LAS Detergents End Stream Foam." *Chemical and Engineering News, 45* (1967): 29.

Levey, M. "The Early History of Detergent Substances." *Journal of Chemical Education, 31* (1954): 521.

"Makers Plan New Detergent Formulas." *Chemical and Engineering News, 48* (1970): 18.

Meloan, C. E., "Detergents—Soaps and Syndets." *Chemistry, 49* (September 1976): 6.

Puplett, P. A. R. *Synthetic Detergents*. London: Sidgwick and Jackson, 1957.

Snell, F. D. "Soap and Glycerol." *Journal of Chemical Education, 19* (1942): 172.

Snell, F. D., and Snell, C. T. "Syndets and Surfactants." *Journal of Chemical Education, 35* (1958): 271.

Stinson, S. C. "Consumer Preferences Spur Innovation in Detergents." *Chemical and Engineering News, 65* (January 26, 1987): 21.

Experiment 44
Preparation of a Soap

Hydrolysis of a fat (ester)
Filtration

In this experiment we prepare soap from animal fat (lard). Animal fats and vegetable oils are esters of carboxylic acids; they have a high molecular weight and contain the alcohol, glycerol. Chemically, these fats and oils are called **triglycerides.** The principal acids in animal fats and vegetable oils can be prepared from the natural triglycerides by alkaline hydrolysis (saponification).

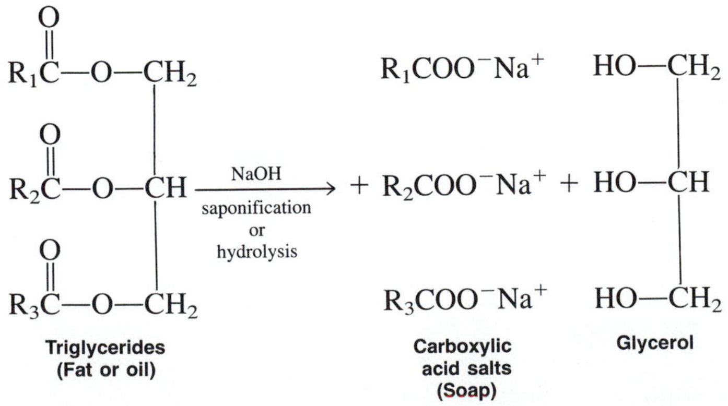

$$R_1\overset{\displaystyle O}{\overset{\|}{C}}\!-\!O\!-\!CH_2$$

$$R_2\overset{\displaystyle O}{\overset{\|}{C}}\!-\!O\!-\!CH \quad \xrightarrow[\substack{\text{saponification}\\ \text{or}\\ \text{hydrolysis}}]{\text{NaOH}} \quad + \quad R_2COO^-Na^+ \quad + \quad HO\!-\!CH$$

$$R_3\overset{\displaystyle O}{\overset{\|}{C}}\!-\!O\!-\!CH_2 \qquad\qquad R_3COO^-Na^+ \qquad HO\!-\!CH_2$$

$R_1COO^-Na^+ \qquad HO\!-\!CH_2$

**Triglycerides
(Fat or oil)** · **Carboxylic
acid salts
(Soap)** · **Glycerol**

The natural acids are rarely of a single type in any given fat or oil. In fact, a single triglyceride molecule in a fat may contain three different acid residues (R_1COOH, R_2COOH, R_3COOH), and not every triglyceride in the substance will be identical. Each fat or oil, however, has a characteristic **statistical distribution** of the various types of acids possible. The composition of the common fats and oils is given in the essay, "Fats and Oils" (p 168).

The fats and oils that are most common in soap preparations are lard and tallow from animal sources, and coconut, palm, and olive oils from vegetable sources. The length of the hydrocarbon chain and the number of double bonds in the carboxylic acid portion of the fat or oil determine the properties of the resulting soap. For example, a salt of a saturated long-chain acid makes a harder, more insoluble soap. Chain length also affects solubility.

Tallow is the principal fatty material used in making soap. The solid fats of cattle are melted with steam, and the tallow layer formed at the top is removed. Soap-makers usually blend tallow with coconut oil and saponify this mixture. The resulting soap contains mainly the salts of palmitic, stearic, and oleic acids from the tallow, and the salts of lauric and myristic acids from the coconut oil. The coconut oil is added to produce a softer, more soluble soap. Lard (from hogs) differs from tallow (from cattle or sheep) in that lard contains more oleic acid.

Tallow $CH_3(CH_2)_{14}COOH$ $CH_3(CH_2)_{16}COOH$

Palmitic acid   **Stearic acid**

$CH_3(CH_2)_7CH{=}CH(CH_2)_7COOH$

Oleic acid

Coconut oil $CH_3(CH_2)_{10}COOH$ $CH_3(CH_2)_{12}COOH$

Lauric acid   **Myristic acid**

Pure coconut oil yields a soap that is very soluble in water. The soap contains essentially the salt of lauric acid with some myristic acid. It is so soft (soluble) that it will lather even in seawater. Palm oil contains mainly two acids, palmitic acid and oleic acid, in about equal amounts. Saponification of this oil yields a soap that is an important constituent of toilet soaps. Olive oil contains mainly oleic acid. It is used to prepare Castile soap, named after the region in Spain in which it was first made.

Toilet soaps generally have been carefully washed free of any alkali remaining from the saponification. As much glycerol as possible is usually left in the soap, and perfumes and medicinal agents are sometimes added. Floating soaps are produced by blowing air into the soap as it solidifies. Soft soaps are made by using potassium hydroxide, yielding potassium salts rather than the sodium salts of the acids. They are used in shaving creams and liquid soaps. Scouring soaps have abrasives added, such as fine sand or pumice.

REQUIRED READING

Review: Technique 3   Sections 3.1–3.3
           Technique 4   Section 4.3

New: Essay   Soaps and Detergents (p 350)
            Essay   Fats and Oils (p 166)

SPECIAL INSTRUCTIONS

This experiment is short and can easily be scheduled with another experiment.

PROCEDURE

Prepare a solution of about 0.25 g of sodium hydroxide (two to three pellets) dissolved in a mixture of 1.0 mL of distilled water and 1.0 mL of 95% ethanol.

> **Do not touch the sodium hydroxide pellets as they are very caustic. Use a scoop. The pellets should be weighed as rapidly as possible as they tend to draw moisture from air and become sticky.**

You may use a calibrated Pasteur pipet to measure and transfer the water and ethanol. Place about 0.25 g of lard in a 10-mL Erlenmeyer flask and add the sodium hydroxide solution to the flask. Heat the mixture in a sand bath at about 120 °C. Place an inverted 20-mL beaker over the neck of the flask to help to reduce evaporation. Swirl the Erlenmeyer flask every few minutes.

The soap often begins to precipitate from the boiling mixture within about 20 minutes. If it appears that some of the alcohol and water is evaporating from the flask, you may add up to 0.4 mL of a 50% water/alcohol mixture to replace the solvent that is lost. Heat the mixture for a total of 25 minutes in the sand bath.

Place 4 mL of salt solution[1] in a 20-mL beaker and transfer the saponified mixture from the Erlenmeyer flask to the beaker. Stir the mixture while cooling the beaker in an ice-water bath. Collect the prepared soap on a Hirsch funnel by vacuum filtration on fast filter paper (Technique 4, Sections 4.2 and 4.3, pp 570–572). Wash the soap with two 3-mL portions of ice cold distilled water to remove any excess sodium hydroxide. Continue to draw air through the filter for a few minutes to partially dry the product. **Test your soap while it is still damp** using the procedure given below. Allow the remaining sample to dry in your locker until the next period. Weigh the product. Submit the sample to your instructor in a labeled vial.

TESTS ON SOAPS AND DETERGENTS

Soap. After rinsing your product, remove a 4-mm piece (about 0.01 g) from the filter paper and place it in a clean 10-mL graduated cylinder. Add 3 mL of distilled water,

---

[1] The instructor should prepare the aqueous sodium chloride solution for the class in the ratio of 40 g of sodium chloride to 150 mL of distilled water.

place your thumb over the opening of the cylinder, and shake the mixture vigorously for about 15 seconds. After about 30 seconds, observe the level of the foam. Add two drops of 4% calcium chloride solution to the soap mixture from a Pasteur pipet. Shake the mixture for 15 seconds and allow it to stand for 30 seconds. Observe the effect of the calcium chloride on the foam. Do you observe anything else? Add 0.5 g of trisodium phosphate and shake the mixture again for 15 seconds. After 30 seconds, again observe the results. Explain the results of these tests in your laboratory notebook.

Detergent.   Place a 2-mm piece (about 0.005 g) of sodium lauryl sulfate (sodium dodecyl sulfate or dodecyl sodium sulfate) in a 10-mL graduated cylinder. Add 3 mL of distilled water to the sample. Place your thumb over the opening of the cylinder and shake the mixture vigorously for about 15 seconds. Allow the mixture to stand for about 30 seconds and observe the level of the foam. Add two drops of 4% calcium chloride solution. Shake the mixture for 15 seconds. After about 30 seconds, observe the effect of the calcium chloride on the foam. Explain the results of these tests in your laboratory notebook.

QUESTIONS

1. Why should the potassium salts of fatty acids yield soft soaps?
2. Why is the soap derived from coconut oil so soluble?
3. Why do you suppose a mixture of ethanol and water instead of simply water itself is used for saponification?
4. Sodium acetate and sodium propanoate are poor soaps. Why?

Experiment 45
Preparation of a Detergent

Preparation of a sulfonate ester
Properties of soaps and detergents

In this experiment, you will prepare a detergent, sodium lauryl sulfate. A detergent is usually defined as a synthetic cleaning agent, whereas a soap is derived from a natural source—a fat or an oil.

$$CH_3(CH_2)_{10}CH_2O-\overset{\displaystyle O}{\underset{\displaystyle O}{\overset{\|}{\underset{\|}{S}}}}-O^-Na^+ \qquad CH_3(CH_2)_{16}-\overset{\displaystyle O}{\overset{\|}{C}}-O^-Na^+$$

Sodium lauryl sulfate   Sodium stearate
(a detergent)   (a soap)

The differences between the two basic types of cleaning agents are discussed in the essay, ''Soaps and Detergents,'' which precedes Experiment 44. Following the preparation of sodium lauryl sulfate, the properties of soap are compared with the properties of the prepared detergent.

In the first step of the synthesis, lauryl alcohol is allowed to react with chlorosulfonic acid to give the lauryl ester of sulfuric acid. In the second step, aqueous sodium carbonate is added to produce the sodium salt (detergent).

$$CH_3(CH_2)_{10}CH_2OH + Cl\!-\!\overset{\displaystyle O}{\underset{\displaystyle O}{\overset{\|}{\underset{\|}{S}}}}\!-\!OH \longrightarrow CH_3(CH_2)_{10}CH_2O\!-\!\overset{\displaystyle O}{\underset{\displaystyle O}{\overset{\|}{\underset{\|}{S}}}}\!-\!OH + HCl \quad (1)$$

Lauryl alcohol **Chlorosulfonic acid** **Lauryl ester of sulfuric acid**

$$2CH_3(CH_2)_{10}CH_2\!-\!O\!-\!\overset{\displaystyle O}{\underset{\displaystyle O}{\overset{\|}{\underset{\|}{S}}}}\!-\!OH + Na_2CO_3 \longrightarrow$$

$$2CH_3(CH_2)_{10}CH_2O\!-\!\overset{\displaystyle O}{\underset{\displaystyle O}{\overset{\|}{\underset{\|}{S}}}}\!-\!O^-Na^+ + H_2O + CO_2 \quad (2)$$

Sodium lauryl sulfate

The aqueous mixture is saturated with solid sodium carbonate and extracted with 1-butanol. Sodium carbonate must be added to give phase separation; otherwise, 1-butanol would be soluble in water. The sodium salt (detergent) is more soluble in 1-butanol than in the aqueous layer because of the long hydrocarbon chain, which gives the salt considerable organic (nonpolar) character.

REQUIRED READING

Review: Technique 3   Section 3.7A
          Technique 7   Section 7.5

New: Essay   Soaps and Detergents (p 350)

SPECIAL INSTRUCTIONS

Chlorosulfonic acid must be handled with care since it is a corrosive liquid and reacts violently with water. Be certain to use dry glassware.

> **NOTE TO THE INSTRUCTOR:** The 1-dodecanol (lauryl alcohol) is best handled as a liquid. If necessary, melt the alcohol (mp 24–27 °C) and pour the liquid into a small container. Keep the alcohol in the liquid state by placing the container in a warm sand bath or on a hot plate. In this way, the alcohol will be available to the class as a liquid.

PROCEDURE

Transfer 0.100 mL of concentrated (glacial) acetic acid into a **dry** 5-mL conical vial. Cap the vial and cool the vial in a small beaker, with ice, for about five minutes. In a hood, remove 0.035 mL of chlorosulfonic acid (d = 1.77 g/mL) using the graduated pipet provided for you and add it **dropwise** to the conical vial containing the acetic acid in the ice bath (Use safety glasses!).

> **CAUTION:** Use chlorosulfonic acid with extreme care. Avoid getting water or ice in the vial. Chlorosulfonic acid reacts violently with water to form hydrochloric acid. Transfer the material directly into your vial without dripping the liquid. Chlorosulfonic acid is an extremely strong acid similar to concentrated sulfuric acid. It will cause immediate burns on the skin.

The following operations may be conducted at your laboratory bench if care is taken. Prepare a gas trap by placing some cotton into a drying tube and adding a few drops of water to moisten the cotton (Technique 3, Section 3.7A, p 555). Avoid excess water or it may accidentally run down into the vial. Remove the vial from the ice bath, place a magnetic spin vane in the vial, and add 0.12 mL of 1-dodecanol (lauryl alcohol, d = 0.831 g/mL) into the conical vial. Place the drying tube (gas tap) onto the vial, clamp the vial securely, and stir the mixture for 15 minutes at room temperature. After this time, **carefully add 20 drops of ice-cold water** to the vial with a Pasteur pipet over a period of two minutes. Continue to stir the mixture while adding the water.

> **CAUTION:** Excess chlorosulfonic acid will react violently with water. Replace the gas trap (drying tube) after each addition of water.

Add 0.30 mL of 1-butanol to the conical vial and stir the mixture with the spin vane for five minutes. While stirring, add slowly 0.15 g of sodium carbonate (anhydrous) to neutralize the acids and to aid in the separation of layers. The sodium carbonate will dissolve. After stirring the mixture, cap the vial, and shake the conical vial so that the

1-butanol will extract the detergent from the aqueous layer. Allow the layers to separate for 5–10 minutes, or until a complete separation has been achieved. The 1-butanol layer will be on top. Remove the magnetic spin vane from the vial with forceps. With care, remove the lower aqueous layer with a Pasteur pipet (Technique 7, Section 7.5, p 624) and place it in a 3-mL conical vial. Save the organic layer (1-butanol) in the original 5-mL vial, as it contains your detergent product.

Reextract the aqueous layer. To do this, add 0.3 mL of 1-butanol to the vial, cap the vial, and shake it. Allow the vial to stand for about 10 minutes or until a complete separation has been achieved. Remove the lower aqueous phase with a Pasteur pipet and discard it.

Combine the **two** 1-butanol organic phases together in one of the vials. Allow these combined phases to stand for a few minutes to see if any further separation of layers occurs. If some further separation is observed, remove the lower aqueous layer and discard it. Otherwise, transfer the 1-butanol extracts into a preweighed 10-mL beaker and store it in your locker until the next period. During this time, the 1-butanol should evaporate to give the detergent. If an odor of 1-butanol still remains, place the beaker in an oven maintained at about 80 °C until the solid is thoroughly dry and odor-free. Use your spatula to break up the solid. Weigh the product and calculate the percentage yield (MW = 288.4). If the detergent is not totally free of 1-butanol, the apparent yield may exceed 100%. If necessary, continue to dry the sample. After doing the tests which follow, submit the remaining detergent to the instructor in a labeled vial.

TESTS ON SOAPS AND DETERGENTS

Soap. Pour 3 mL of soap solution[1] into a 10-mL graduated cylinder. Place your thumb over the opening of the cylinder and shake it vigorously for about 15 seconds. Allow the solution to stand for 30 seconds and observe the level of the foam. Add two drops of 4% calcium chloride solution from a Pasteur pipet. Shake the mixture for 15 seconds and allow it to stand for about 30 seconds. Observe the effect of the calcium chloride on the foam. Do you observe anything else? Add 0.3 g of trisodium phosphate and shake the mixture again for about 15 seconds. Allow the solution to stand for 30 seconds. What do you observe? Explain these tests in your laboratory report.

Detergent. Place a 2 mm piece (about 0.005 g) of your prepared detergent in a 10-mL graduated cylinder and add 3 mL of distilled water. Hold your thumb over the opening and shake the graduated cylinder vigorously for 30 seconds. Allow the solution to stand for about 30 seconds and observe the level of the foam. Add two drops of 4% calcium chloride solution. Shake the mixture for 15 seconds and allow it to stand for 30 seconds. What do you observe? Explain the results of these tests in your laboratory report.

---

[1] A large batch of soap solution should be prepared by the instructor as follows: Add one bar of Ivory soap to 1 L of distilled water. Stir the solution occasionally and allow the mixture to stand overnight. Remove the remainder of the bar. The mixture can be used directly. Alternatively, a wet 0.01 g sample of soap from Experiment 44 can be added to 3 mL of distilled water.

QUESTIONS

1. Draw a mechanism for the reaction of lauryl alcohol with chlorosulfonic acid.

2. Why do you suppose sodium carbonate, instead of some other base, is used for neutralization?

3. Propose a model to explain how a cationic detergent works. A cationic detergent has its polar end positively charged.

4. Sodium methyl sulfate, $CH_3OSO_2^-Na^+$, is a poor detergent. Why?

5. Sodium lauryl sulfate can be prepared by replacing chlorosulfonic acid with another reagent. What could be used? Show the equations.

6. Suggest a method for synthesizing the linear alkyl sulfonate detergent shown on p. 354, starting with lauryl alcohol, benzene, and any needed inorganic compounds.

Essay
PHEROMONES: INSECT ATTRACTANTS AND REPELLENTS

It is difficult for humans, who are accustomed to heavy reliance on visual and verbal forms of communication, to imagine that there are forms of life that depend primarily on the release and perception of **odors** to communicate with one another. Among insects, however, this is perhaps the chief form of communication. Many species of insects have developed a virtual "language" based on the exchange of odors. These insects have well-developed scent glands, often of several different types, which have as their sole purpose the synthesis and release of chemical substances. When these chemical substances, known as **pheromones,** are secreted by insects and detected by other members of the same species, they induce a specific and characteristic response. Pheromones are usually of two distinct types: releaser pheromones and primer pheromones. **Releaser pheromones** produce an immediate **behavioral** response in the recipient insect; **primer pheromones** trigger a series of **physiological** changes in the recipient. Some pheromones, however, combine both releaser and primer effects.

SEX ATTRACTANTS

Among the most important types of releaser pheromones are the sex attractants. **Sex attractants** are pheromones secreted by either the female, or, less commonly, the male of the species to attract the opposite member for the purpose of mating. Since in large concentrations, sex pheromones also induce a physiological response in the recipient (for example, the changes necessary to the mating act), they also have a primer effect and are therefore misnamed.

Anyone who has owned a female cat or dog will know that sex pheromones are not limited to insects. Female cats or dogs widely advertise, by odor, their sexual availability when they are "in heat." This type of pheromone is not uncommon to mammals. Some persons even believe that there are human pheromones that are responsible for attracting certain sensitive males and females to one another. This idea is, of course, responsible for many of the perfumes now widely available. Whether or not the idea is correct cannot yet be established, but there are proven sexual differences in the ability of humans to smell certain substances. For instance, Exaltolide, a synthetic lactone of 14-hydroxytetradecanoic acid, can be perceived only by females, or by males after they have been injected with an estrogen. Exaltolide is very similar in overall structure to civetone (civet cat) and muskone (musk deer), which are two naturally-occurring compounds believed to be mammalian sex pheromones. Whether or not human males emit pheromones has never been established. Curiously, Exaltolide is used in perfumes intended for female as well as male use! But while the odor may lead a woman to believe that she smells pleasant, it cannot possibly have any effect on the male. The "musk oils," civetone and muskone, have also been long used in expensive perfumes.

One of the first identified insect attractants belonged to the gypsy moth, *Porthetria dispar*. This moth is a common agricultural pest, and it was hoped that the sex attractant that females emitted could be used to lure and trap males. Such a method of insect control would be preferable to inundating large areas with DDT and would be species-specific. Nearly 50 years of work were expended in identifying the chemical substance responsible. Early in this period, workers had found that an extract from the tail sections of female gypsy moths would attract males, even from a great distance. In experiments with the isolated gypsy moth pheromone, it was found that the male gypsy moth has an almost unbelievable ability to detect extremely small amounts of the substance. He can detect it in concentrations lower than a few hundred **molecules** per cubic centimeter (about 10^{-19}–10^{-20} g/cc)! When a male moth encounters a small concentration of pheromone, he immediately turns into the wind and flies upward in search of higher concentrations and the female. In only a mild breeze, a continuously emitting female can activate a space 300 ft high, 700 ft wide, and almost 14,000 ft (nearly 3 miles) long!

In subsequent work, 20 mg of a pure chemical substance was isolated from solvent extracts of the two extreme tail segments collected from each of 500,000 female gypsy moths (about 0.1 μg/moth). This emphasizes that pheromones are effective in very minute amounts and that chemists must work with very small amounts to isolate them and prove their structures. It is not unusual to have to process thousands of insects to get even a very small sample of these substances. Very sophisticated analytical and instrumental methods, like spectroscopy, must be used to determine the structure of a pheromone.

In spite of these techniques, the original workers assigned an incorrect structure to the gypsy moth pheromone and proposed for it the name *gyplure*. Because of its great promise as a method of insect control, gyplure was soon synthesized. The synthetic material turned out to be totally inactive. After some controversy about why the synthetic material was incapable of luring male gypsy moths (see the references for the complete story), it was finally shown that the proposed structure for the pheromone (that is, the gyplure structure) was incorrect. The actual pheromone was found to be *cis*-7,8-epoxy-2-methyloctadecane, as shown below. This material was soon synthesized, found to be active, and given the name **disparlure.** In recent years, disparlure traps have been found to be a convenient and economical method for controlling the gypsy moth.

A similar story of mistaken identity can be related for the structure of the pheromone of the pink bollworm, *Pectinophora gossypiella*. The originally proposed structure was called propylure. Synthetic propylure turned out to be inactive. Subsequently the pheromone was shown to be a mixture of two isomers of 7,11-hexadecadien-1-yl acetate, the *cis,cis* (7Z, 11Z) isomer and the *cis,trans* (7Z, 11E) isomer. It turned out to be quite easy to synthesize a 1:1 mixture of these two isomers, and the 1:1 mixture was named **gossyplure.** Curiously, adding as little as 10% of either of the other two possible isomers, either *trans,cis* (7E, 11Z) or *trans,trans* (7E, 11E), to the 1:1 mixture greatly diminishes its activity, apparently masking it. Geometric isomerism can be important! The details of the gossyplure story can also be found in the references.

Both these stories have been partly repeated here to point out the difficulties of research on pheromones. The usual method is to propose a structure determined by work on **very tiny** amounts of the natural material. The margin for error is great. Such proposals are usually not considered "proved" until synthetic material is shown to be as biologically effective as the natural pheromone.

INSECT SEX ATTRACTANTS

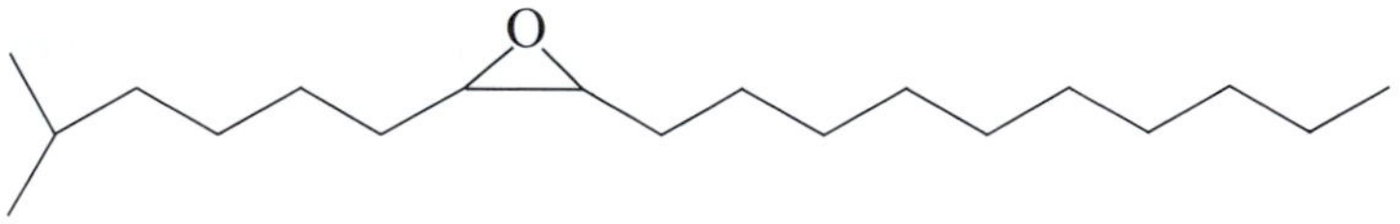

Disparlure
(gypsy moth)

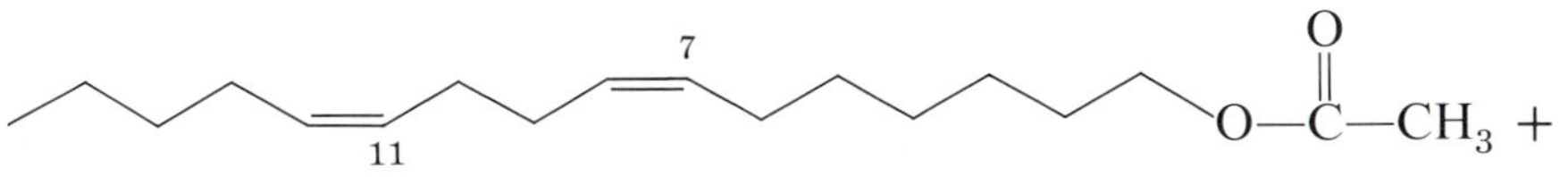

Gossyplure
(pink bollworm)

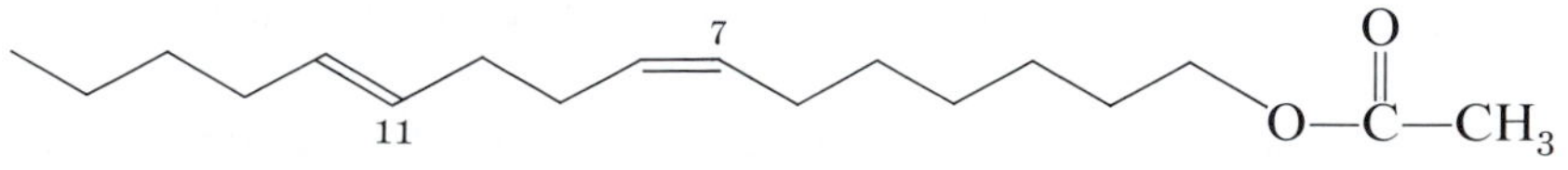

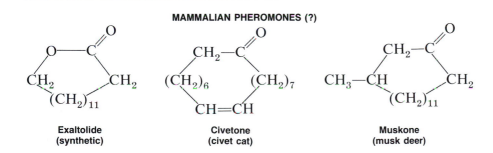

RECRUITING PHEROMONE

Geraniol
(honeybee)

Citral
(honeybee)

PRIMER PHEROMONE

Queen substance
(honeybee)

ALARM PHEROMONES

Isoamyl acetate
(honeybee)

Citral   Citronellal

(ant species)

Periplanone B
(American cockroach)

MAMMALIAN PHEROMONES (?)

Exaltolide
(synthetic)

Civetone
(civet cat)

Muskone
(musk deer)

OTHER PHEROMONES

The most important example of a **primer pheromone** is found in honeybees. A bee
colony consists of one queen bee, several hundred male drones, and thousands of
worker bees, or undeveloped females. It has recently been found that the queen, the
only female that has achieved full development and reproductive capacity, secretes a
primer pheromone called the **queen substance.** The worker females, while tending the
queen bee, continuously ingest quantities of this queen substance. This pheromone
prevents the workers from rearing any competitive queens and also prevents the devel-

opment of ovaries in all other females in the hive. The substance is also active as a sex attractant; it attracts drones to the queen during her "nuptial flight."

Honeybees also produce several other important types of pheromones. It has long been known that bees will swarm after an intruder. It has also been known that isoamyl acetate induces a similar type of behavior in bees. Isoamyl acetate (Experiment 6) is an **alarm pheromone.** When an angry worker bee stings an intruder, she discharges, along with the sting venom, a mixture of pheromones that incites the other bees to swarm upon and attack the intruder. Isoamyl acetate is an important component of the alarm pheromone mixture. Alarm pheromones have also been identified in many other insects. In less aggressive insects than bees or ants, the alarm pheromone may take the form of a **repellent,** which induces the insects to go into hiding or leave the immediate vicinity.

Honeybees also release **recruiting** or **trail pheromones.** These pheromones attract others to a source of food. Honeybees secrete recruiting pheromones when they locate flowers in which large amounts of sugar syrup are available. Although the recruiting pheromone is a complex mixture, both geraniol and citral have been identified as components. In a similar fashion, when ants locate a source of food, they drag their tails along the ground on their way back to the nest, continuously secreting a trail pheromone. Other ants follow this trail to the source of food.

In some species of insects, **recognition pheromones** have been identified. In carpenter ants, a caste-specific secretion has been found in the mandibular glands of the males of five different species. These secretions have several functions, one of which is to allow members of the same species to recognize one another. Insects not having the correct recognition odor are immediately attacked and expelled from the nest. In one species of carpenter ant, the recognition pheromone has been shown to have methyl anthranilate as an important component.

We do not yet know all the types of pheromones that any given species of insect may use, but it seems that as few as 10 or 12 pheromones could constitute a "language" that could adequately regulate the entire life cycle of a colony of social insects.

INSECT REPELLENTS

Currently, the most widely used **insect repellent** is the synthetic substance *N,N*-diethyl-*m*-toluamide (Experiment 46), also called Deet. It is effective against fleas, mosquitoes, chiggers, ticks, deerflies, sandflies, and biting gnats. A specific repellent is known for each of these types of insects, but none has the wide spectrum of activity that this repellent has. Exactly why these substances repel insects is not yet fully understood. The most extensive investigations have been carried out on the mosquito.

Originally, many investigators thought that repellents might simply be compounds that provided unpleasant or distasteful odors to a wide variety of insects. Others thought that they might be alarm pheromones for the species affected, or that they might be the alarm pheromones of a hostile species. Early research with the mosquito indicates that at least for several varieties of mosquitoes, none of these is the correct answer.

Mosquitoes seem to have hairs on their antennae that are receptors enabling them to find a warm-blooded host. These receptors detect the convection currents arising from a warm and moist living animal. When a mosquito encounters a warm and moist convection current, it moves steadily forward. If it passes out of the current into dry air, it turns until it finds the current again. Eventually it finds the host and lands. Repellents cause a mosquito to turn in flight and become confused. Even if it should land, it becomes confused and flies away again.

Researchers have found that the repellent prevents the moisture receptors of the mosquito from responding normally to the raised humidity of the subject. At least two sensors are involved, one responsive to carbon dioxide and the other responsive to water vapor. The carbon dioxide sensor is activated by the repellent, but if exposure to the chemical continues, adaptation occurs and the sensor returns to its usual low output of signal. The moisture sensor, on the other hand, simply seems to be deadened, or turned off, by the repellent. Therefore, mosquitoes have great difficulty in finding and interpreting a host when they are in an environment saturated with repellent. They fly right through warm and humid convection currents as if the currents did not exist. Only time will tell if other biting insects respond likewise.

REFERENCES

Batra, S. W. T. "Polyester-Making Bees and Other Innovative Insect Chemists." *Journal of Chemical Education, 62* (February 1985): 121.

Beroza, M., ed. *Chemicals Controlling Insect Behavior.* New York: Academic Press, 1970.

Beroza, M., "Insect Sex Attractants." *American Scientist, 59* (May–June 1971): 320.

Jacobson, M., and Beroza, M. "Insect Attractants." *Scientific American, 211* (August 1964): 20.

Katzenellenbogen, J. A. "Insect Pheromone Synthesis: New Methodology." *Science, 194* (October 8, 1976): 139.

Prestwick, G. D. "The Chemical Defenses of Termites." *Scientific American, 249* (August 1983): 78.

Silverstein, R. M. "Pheromones: Background and Potential Use for Insect Control." *Science, 213* (September 18, 1981): 1326.

Sondheimer, E., and Simeone, J. B. *Chemical Ecology.* New York: Academic Press, 1970.

Vollmer, J. J., and Gordon, S. A. "Chemical Communication." *Chemistry, 47* (November 1974):6; ibid., *48* (April 1975): 6; ibid., *48* (May 1975): 6.

Wilson, E. O. "Pheromones." *Scientific American, 208* (May 1963): 100.

Wood, W. F. "Chemical Ecology: Chemical Communication in Nature." *Journal of Chemical Education, 60* (July 1983): 531.

Wright, R. H. "Why Mosquito Repellents Repel." *Scientific American, 233* (July 1975): 105.

Gypsy moth

Beroza, M., and Knipling, E. F. "Gypsy Moth Control with the Sex Attractant Pheromone." *Science, 177* (1972): 19.

Bierl, B. A., Beroza, M., and Collier, C. W. "Potent Sex Attractant of the Gypsy Moth: Its Isolation, Identification, and Synthesis." *Science, 170* (1970): 87.

Pink bollworm

Anderson, R. J., and Henrick, C. A. "Preparation of the Pink Bollworm Sex Pheromone Mixture, Gossyplure." *Journal of the American Chemical Society, 97* (1975): 4327.

Hummel, H. E., Gaston, L. K., Shorey, H. H., Kaae, R. S., Byrne, K. J., and Silverstein, R. M.
     "Clarification of the Chemical Status of the Pink Bollworm Sex Pheromone." *Science, 181*
     (1973): 873.

American cockroach

Adams, M. A., Nakanishi, K., Still, W. C., Arnold, E. V., Clardy, J. and Persoon, C. J., *Journal of the*
     American Chemical Society, 101 (1979): 2495.
Still, W. C. *Journal of the American Chemical Society, 101* (1979): 2493.
Stinson, S. C. *Chemical and Engineering News, 57* (April 30, 1979): 24.

Experiment 46
N,N-Diethyl-m-Toluamide:
The Insect Repellent "OFF"

Preparation of an amide
Extraction
Column chromatography

In this experiment, we synthesize the active ingredient of the insect repellent "OFF,"
N,N-diethyl-*m*-toluamide. This substance belongs to the class of compounds called
amides. Amides have the generalized structure

$$R{-}\overset{\overset{\displaystyle O}{\|}}{C}{-}NH_2$$

The amide to be prepared in this experiment is a disubstituted amide. That is, two of the
hydrogens on the amide —NH_2 group have been replaced with ethyl groups. Amides
cannot be prepared directly by mixing a carboxylic acid with an amine. If an acid and
an amine are mixed, an acid-base reaction occurs, giving the conjugate base of the
acid, which will not react further while in solution:

$$RCOOH + R_2NH \longrightarrow [RCOO^- R_2NH_2^+]$$

However, if the amine salt is isolated as a crystalline solid and strongly heated, the
amide can be prepared:

$$[RCOO^- R_2NH_2^+] \xrightarrow{\text{heat}} RCONR_2 + H_2O$$

This is not a convenient laboratory method because of the high temperature required for
this reaction.

     Amines are usually prepared via the acid chloride, as in this experiment. In Step
1, *m*-toluic acid is converted to its acid chloride derivative using thionyl chloride
($SOCl_2$). The acid chloride is not isolated or purified, and it is allowed to react directly
with diethylamine in Step 2. An excess of diethylamine is used in this experiment to
react with the hydrogen chloride produced in Step 2.

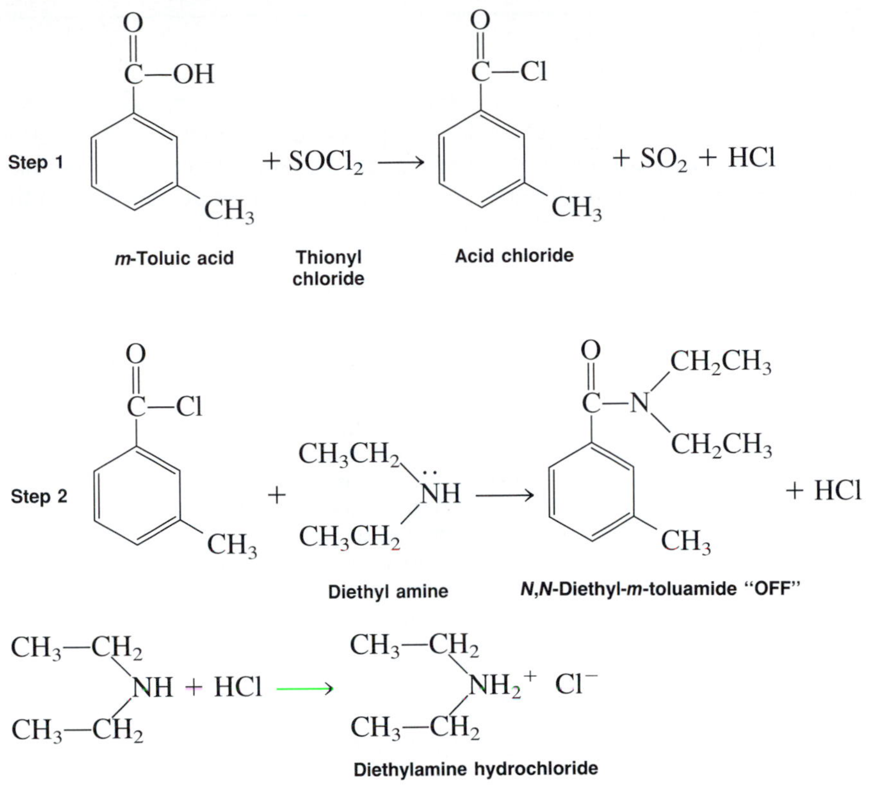

Step 1

m-Toluic acid + SOCl₂ → Acid chloride + SO₂ + HCl

$$\text{m-Toluic acid} + SOCl_2 \longrightarrow \text{Acid chloride} + SO_2 + HCl$$

m-Toluic acid   Thionyl chloride   Acid chloride

Step 2

Acid chloride + Diethyl amine → N,N-Diethyl-m-toluamide "OFF" + HCl

Diethyl amine   *N,N*-Diethyl-*m*-toluamide "OFF"

$$ (CH_3-CH_2)_2NH + HCl \longrightarrow (CH_3-CH_2)_2NH_2^+ \; Cl^- $$

Diethylamine hydrochloride

REQUIRED READING

Review: Technique 3   Sections 3.2, 3.3, 3.5, 3.7A, and 3.9
            Technique 4   Section 4.6
            Technique 7   Sections 7.5, 7.8, and 7.10
            Technique 12   Sections 12.6, 12.7, and 12.8

New: Essay   Pheromones: Insect Attractants and Repellents

SPECIAL INSTRUCTIONS

All equipment used in this experiment should be dry since thionyl chloride reacts with water to liberate HCl and SO_2. Likewise, **anhydrous** ether should be used because water reacts with both thionyl chloride and the intermediate acid chloride.

Thionyl chloride is a noxious and corrosive chemical and should be handled with care. If it is spilled on the skin, serious burns will result. Thionyl chloride and diethylamine must be dispensed **in the hood** from bottles that should be kept tightly closed when not in use. Diethylamine is also noxious and corrosive. In addition, it is quite volatile (bp 56 °C) and must be cooled in a hood prior to use.

PROCEDURE

Assemble the apparatus as shown in the figure, except for the syringe. The drying tube is packed with glass wool and a few drops of water are added to the drying tube. Excess water should be avoided so that water does not get into the conical vial. The moistened glass wool traps the hydrogen chloride and sulfur dioxide that are evolved in the reaction. You can save time by setting the dial on your hot plate to give a sand bath temperature of about 90 °C prior to measuring reagents.

Place 0.136 g of *m*-toluic acid (3-methylbenzoic acid, MW = 136.1) into a **dry** 5-mL conical vial. In a hood, transfer 0.15 mL of thionyl chloride (MW = 118.9, d = 1.64 g/mL) into the vial with the **dry** graduated pipet provided.

> **CAUTION: The thionyl chloride is kept in a hood. Do not breathe the vapors of this noxious and corrosive chemical. Use dry equipment when handling this material as it reacts violently with water. Do not get it on your skin. Once the drying tube containing the moistened glass wool has been attached, the apparatus may be taken to your desk.**

Add a magnetic spin vane, start the circulation of water in the reflux condenser, and heat the mixture with stirring in a sand bath at about 90 °C. Boil the mixture gently for 15 minutes.

Raise the apparatus from the sand bath and allow the vial to **cool to room temperature.** Remove the sand bath from the hot plate. Turn off the heater and allow the unit to cool. You may need to place an insulating pad between the vial and the stirring unit to avoid heating the vial with residual heat from the hot plate. The next part of this reaction sequence is conducted at room temperature.

Inject 2.0 mL of **anhydrous** ether into the reaction vial using a syringe and stir the mixture at room temperature until a homogeneous solution is obtained. In a hood, place 0.33 mL of ice cold diethylamine (MW = 73.1, d = 0.71 g/mL) in a small conical vial and dissolve it in 0.66 mL of **anhydrous** ether. Draw this solution into the syringe and insert the needle through the rubber septum of your apparatus. Add this mixture of diethylamine and ether **dropwise** over a 10–15 minute period to the conical vial. As the solution is added, a voluminous white cloud of diethylamine hydrochloride will form in the vial. Allow the cloud to settle before more diethylamine is added. The mixture may be stirred with the spin vane for few seconds to help break up the cloud as the addition occurs.

After adding the diethylamine, stir the mixture for 10 minutes at room temperature. After this time, inject 1 mL of a 5% aqueous sodium hydroxide solution into the conical vial and stir the mixture for 15 minutes. During this time, the sodium hydroxide converts any remaining acid chloride to the sodium salt of *m*-toluic acid. This salt is soluble in the aqueous layer. Diethylamine hydrochloride is also water soluble. Any remaining thionyl chloride is destroyed by water. The desired amide is soluble in ether.

Remove the drying tube (gas trap), the condenser, and the Claisen head. After the two layers separate, remove the magnetic spin vane with forceps and draw out the lower aqueous layer with a Pasteur pipet so that the desired ether layer remains in the vial (Technique 7, Section 7.5, p 624). Discard the aqueous layer. Add another 1-mL

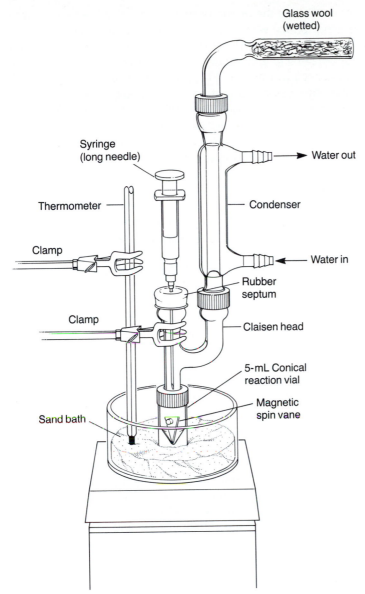

Glass wool
(wetted)

Syringe
(long needle)

Thermometer

Clamp

Clamp

Sand bath

Water out

Condenser

Water in

Rubber
septum

Claisen head

5-mL Conical
reaction vial

Magnetic
spin vane

Note: A long syringe needle is recommended

portion of 5% sodium hydroxide to the remaining ether layer, cap the vial, shake the mixture occasionally over a period of five minutes, allow the layers to separate, and again remove the lower aqueous layer. Discard the aqueous layer. Add additional ether to replace solvent lost by evaporation during these extractions.

Now extract the ether layer with a 1-mL portion of 10% hydrochloric acid to remove any remaining diethylamine as its hydrochloride salt. Finally, extract the ether layer with a 1-mL portion of water. Each time, shake the mixture vigorously, allow time for the phases to separate, and remove the lower aqueous layer with a pipet. Discard all of the aqueous phases and keep the ether layer.

Transfer the ether layer containing the amide product with a **dry** Pasteur pipet to a **dry** conical vial and dry the ether phase with granular anhydrous sodium sulfate (two microspatulafuls measured in the V-grooved end). Remove the solution from the drying agent with a filter tip pipet and transfer it to another dry vial. A small amount of additional ether may be used to aid in a complete transfer. Place the vial in a warm sand bath (about 70 °C) and evaporate the ether using a stream of air or nitrogen in a hood to give the crude dark brown amide (Technique 3, Section 3.9, p 558 and Figure 3–12A, p 561). Column chromatography will be used to remove much of the dark color from the product.

Preweigh a 5-mL conical vial for use in collecting the material eluted from the column. Prepare a column for column chromatography using a 5¾-inch Pasteur pipet as a column (Technique 12, Section 12.6, Part A, p 707 and Part B, p 709). Use the Dry Pack Method 2 on p 710 with alumina as the adsorbent and hexane as the eluent. Place a small piece of cotton in the pipet and gently push it into the constriction. Pour 1.0 g of alumina[1] into the column. Obtain 4 mL of hexane in a graduated cylinder. The hexane will be used to prepare the column, dissolve the crude product, and elute the purified product as described below.

Dissolve the crude product in 10 drops of hexane. Clamp the column above the preweighed 5-mL conical vial. Then add about 1 mL of the hexane to the column and allow it to percolate through the alumina. Allow the solvent to drain until the solvent level just begins to enter the alumina. Add the crude product to the top of the column, and allow the mixture to pass onto the column. Use about 0.5 mL of hexane to rinse the vial that contained the crude product. When the first batch of crude product has drained, so that the liquid just begins to enter the alumina, place the hexane rinse on the column.

When the solvent level has again reached the top of the alumina, add more hexane with a Pasteur pipet to elute the product into the conical vial. You should place 2 mL of hexane, in portions, on the column to elute the product. Collect all of the liquid that passes through the column as one fraction (yellow material). Place the conical vial in a warm sand bath (about 70 °C) and evaporate the hexane with a light stream of air or nitrogen in a hood to give the N,N-diethyl-m-toluamide as a light tan liquid. If necessary use a few drops of hexane to rinse the product from the side of the vial into the bottom. Evaporate this solvent. When the hexane has been removed, reweigh the vial to determine the weight of product. Calculate the percentage yield (MW = 193.1). Determine the infrared spectrum of your product. Submit the remaining sample to the instructor.

REFERENCE

Wang, B. J-S. "An Interesting and Successful Organic Experiment." *Journal of Chemical Education, 51* (October 1974): 631. (The synthesis of N,N-diethyl-m-toluamide.)

QUESTIONS

1. Write an equation that describes the reaction of thionyl chloride with water.
2. What reaction would take place if the acid chloride of m-toluic acid were mixed with water?

---

[1]EM Science (No. AX0612–1). The particle sizes are 80–200 mesh and the material is Type F-20.

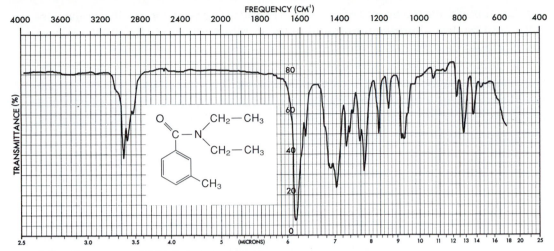

Infrared spectrum of *N,N*-diethyl-*m*-toluamide, neat

3. Why is the reaction mixture extracted with 5% aqueous sodium hydroxide? Write an equation.

4. Write a mechanism for each step in the preparation of *N,N*-diethyl-*m*-toluamide.

5. Interpret each of the principal peaks in the infrared spectrum of *N,N*-diethyl-*m*-toluamide.

6. A student determined the infrared spectrum of the product and found an absorption at 1780 cm^{-1}. The rest of the spectrum resembled the one given in this experiment. Assign this peak and provide an explanation for this unexpected result.

Essay
SULFA DRUGS

The history of chemotherapy extends back as far as 1909 when Paul Ehrlich first used the term. Although Ehrlich's original definition of chemotherapy was limited, he is recognized as one of the giants of medicinal chemistry. **Chemotherapy** might be defined as "the treatment of disease by chemical reagents." It is preferable that these chemical reagents exhibit a toxicity toward only the pathogenic organism, and not toward both the organism and the host. A chemotherapeutic agent would not be useful if it poisoned the patient at the same time that it cured the patient's disease!

In 1932, the German dye manufacturing firm I. G. Farbenindustrie patented a new drug, Prontosil. Prontosil is a red azo dye, and it was first prepared for its dye properties. Remarkably, it was discovered that Prontosil showed antibacterial action when it was used to dye wool. This discovery led to studies of Prontosil as a drug capable of inhibiting the growth of bacteria. The following year, Prontosil was success-fully used against staphylococcal septicemia, a blood infection. In 1935, Gerhard Domagk published the results of his research which indicated that Prontosil was capa-ble of curing streptococcal infections of mice and rabbits. Prontosil was shown to be active against a wide variety of bacteria in later work. This important discovery, which paved the way for a tremendous amount of research on the chemotherapy of bacterial infections, earned for Domagk the 1939 Nobel Prize in Medicine, but an order from Hitler prevented Domagk from accepting this honor.

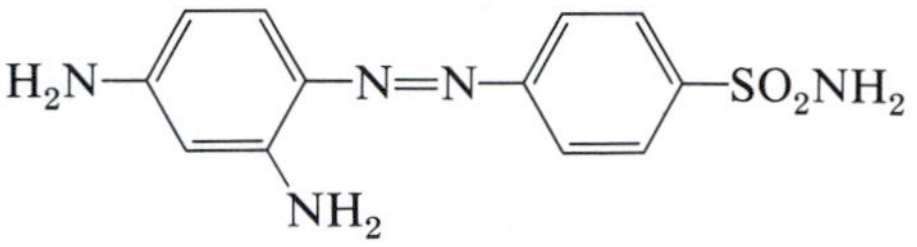

$$H_2N—\overset{}{\underset{NH_2}{\bigcirc}}—N{=}N—\bigcirc—SO_2NH_2 \qquad H_2N—\bigcirc—SO_2NH_2$$

Prontosil 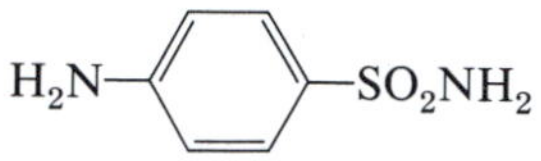　　　　　　　　　　　　　　　　**Sulfanilamide**

Prontosil is an effective antibacterial substance **in vivo,** that is, when injected into a living animal. Prontosil is not medicinally active when the drug is tested **in vitro,** that is, on a bacterial culture grown in the laboratory. In 1935, the research group at the Pasteur Institute in Paris headed by J. Tréfouël learned that Prontosil is metabolized in animals to **sulfanilamide.** Sulfanilamide had been known since 1908. Experiments with sulfanilamide showed that it had the same action as Prontosil in vivo and that it was also active in vitro, where Prontosil was known to be inactive. It was concluded that the active portion of the Prontosil molecule was the sulfanilamide moiety. This discovery led to an explosion of interest in sulfonamide derivatives. Well over a thou-sand sulfonamide substances were prepared within a few years of these discoveries.

Although many sulfonamide compounds were prepared, only a relative few showed useful antibacterial properties. As the first useful antibacterial drugs, these few medicinally active sulfonamides, or **sulfa drugs,** became the wonder drugs of their day. An antibacterial drug may be either **bacteriostatic** or **bactericidal.** A bacterio-static drug suppresses the growth of bacteria; a bactericidal drug kills bacteria. Strictly speaking, the sulfa drugs are bacteriostatic. The structures of some of the most com-mon sulfa drugs are shown below. These more complex sulfa drugs have various important applications. Although they do not have the simple structure characteristic of sulfanilamide, they tend to be less toxic than the simpler compound.

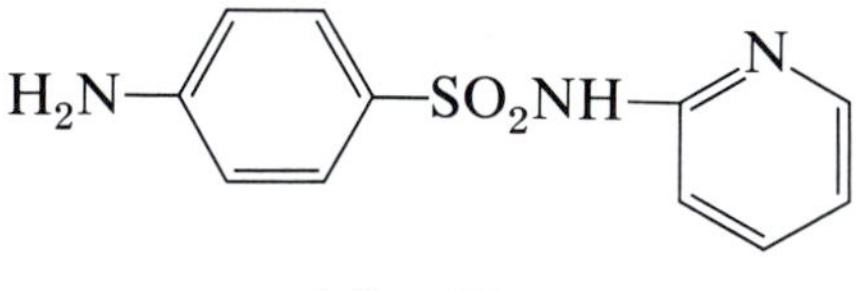

$$H_2N—\bigcirc—SO_2NH—\bigcirc\!\!\!\diagdown_{N} \qquad H_2N—\bigcirc—SO_2NH—\overset{N}{\underset{S}{\bigcirc}}$$

Sulfapyridine 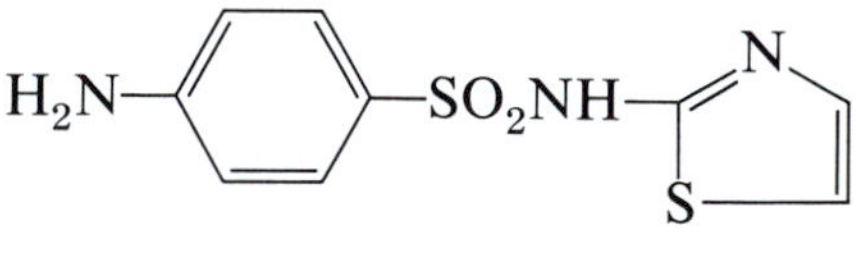　　　　　　　　　　　　　　**Sulfathiazole**

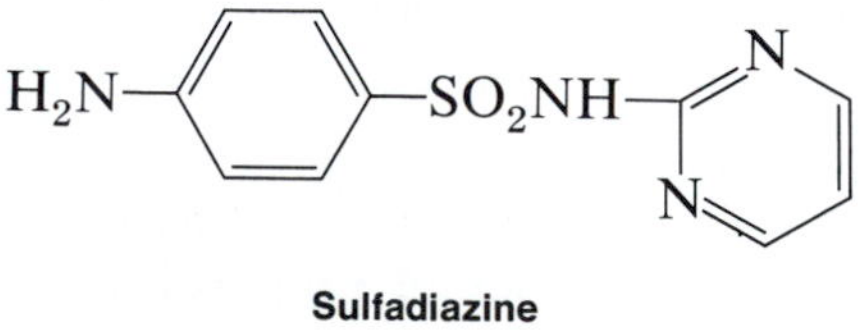

$$H_2N—\bigcirc—SO_2NH—\overset{N}{\underset{N}{\bigcirc}} \qquad H_2N—\bigcirc—SO_2NH—\overset{NH}{\underset{}{\overset{\|}{C}}}—NH_2$$

Sulfadiazine 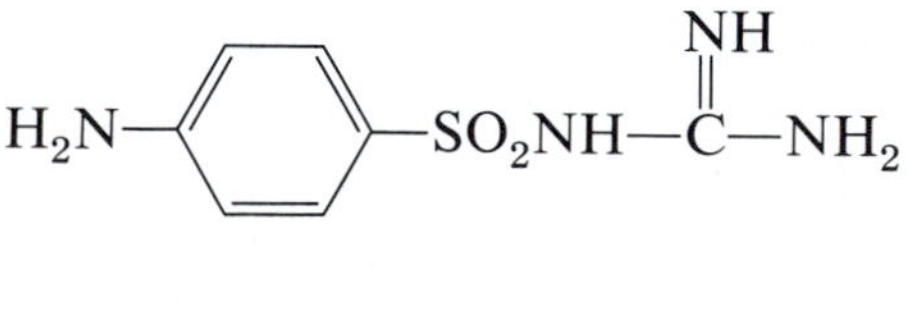　　　　　　　　　　　　　　**Sulfaguanidine**

Sulfisoxazole

Sulfa drugs began to lose their importance as generalized antibacterial agents when production of antibiotics in large quantity began. In 1929, Sir Alexander Fleming made his famous discovery of **penicillin.** In 1941, penicillin was first used successfully on humans. Since that time, the study of antibiotics has spread to molecules that bear little or no structural similarity to the sulfonamides. Besides penicillin derivatives, antibiotics that are derivatives of **tetracycline,** including Aureomycin and Terramycin, were also discovered. These newer antibiotics have high activity against bacteria, and they do not usually have the severe unpleasant side effects of many of the sulfa drugs. Nevertheless, the sulfa drugs are still widely used in treating malaria, tuberculosis, leprosy, meningitis, pneumonia, scarlet fever, plague, respiratory infections, and infections of the intestinal and urinary tracts.

Penicillin G   **Tetracycline**

Even though the importance of sulfa drugs has declined, studies of how these materials act provide a very interesting insight into how chemotherapeutic substances might behave. In 1940, Woods and Fildes discovered that *p*-aminobenzoic acid (PABA) inhibits the action of sulfanilamide. They concluded that sulfanilamide and PABA, because of their structural similarity, must compete with each other within the organism even though they cannot carry out the same chemical function. Further studies indicated that sulfanilamide does not kill bacteria but inhibits their growth. In order to grow, bacteria require an enzyme-catalyzed reaction that uses **folic acid** as a cofactor. Bacteria synthesize folic acid, using PABA as one of the components. When sulfanilamide is introduced into the bacterial cell, it competes with PABA for the active site of the enzyme that carries out the incorporation of PABA into the molecule of folic acid. Because sulfanilamide and PABA compete for an active site due to their structural similarity, and because sulfanilamide cannot carry out the chemical transformations characteristic of PABA once it has formed a complex with the enzyme, sulfanilamide is called a **competitive inhibitor** of the enzyme. The enzyme, once it has formed a complex with sulfanilamide, is incapable of catalyzing the reaction required for the synthesis of folic acid. Without folic acid, the bacteria cannot synthesize the nucleic acids required for growth. As a result, bacterial growth is arrested until the body's immune system can respond and kill the bacteria.

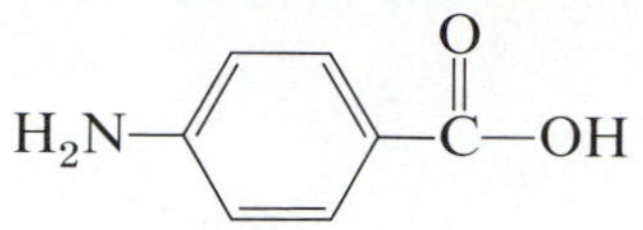

$$H_2N \text{---} \underset{}{\bigodot} \text{---} \overset{\displaystyle O}{\underset{}{C}} \text{---} OH$$

p-**Aminobenzoic acid**
(PABA)

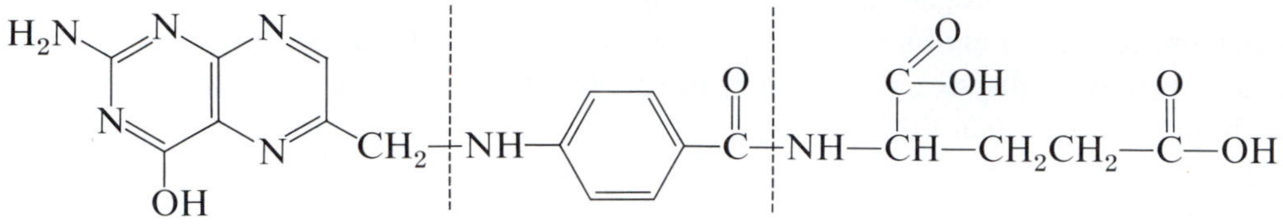

Folic acid structure with PABA residue

PABA residue

Folic acid

One might well ask the question, "Why, when someone takes sulfanilamide as a drug, doesn't it inhibit the growth of **all** cells, bacterial and human alike?" The answer is simple. Animal cells cannot synthesize folic acid. Folic acid must be a part of the diet of animals and is therefore an essential vitamin. Since animal cells receive their fully synthesized folic acid molecules through the diet, only the bacterial cells are affected by the sulfanilamide, and only their growth is inhibited.

For most drugs, a detailed picture of their mechanism of action is unavailable. The sulfa drugs, however, provide a rare example from which we can theorize how other therapeutic agents carry out their medicinal activity.

REFERENCES

Amundsen, L. H. "Sulfanilamide and Related Chemotherapeutic Agents." *Journal of Chemical Education, 19* (1942): 167.

Evans, R. M. *The Chemistry of Antibiotics Used in Medicine*. London: Pergamon Press, 1965.

Fieser, L. F., and Fieser, M. *Topics in Organic Chemistry*. New York: Reinhold, 1963. Chap. 7, "Chemotherapy."

Garrod, L. P., and O'Grady, F. *Antibiotic and Chemotherapy*. Edinburgh: E. and S. Livingstone, Ltd., 1968.

Goodman, L. S., and Gilman, A. *The Pharmacological Basis of Therapeutics*. 7th ed. New York: Macmillan, 1985. Chap. 56, "The Sulfonamides," by L. Weinstein.

Sementsov, A. "The Medical Heritage from Dyes." *Chemistry, 39* (November 1966): 20.

Zahner, H., and Maas, W. K. *Biology of Antibiotics*. Berlin: Springer-Verlag, 1972.

Experiment 47
Sulfa Drugs: Preparation of Sulfanilamide

Crystallization
Protecting groups
Testing the action of drugs on bacteria
Preparation of a sulfonamide
Aromatic substitution

In this experiment you will prepare a sulfa drug, sulfanilamide, by the following synthetic scheme. The synthesis involves converting acetanilide to the intermediate *p*-acetamidobenzenesulfonyl chloride in Step 1. This intermediate is converted to sulfanilamide by way of *p*-acetamidobenzenesulfonamide in Step 2.

$$(1)$$

Acetanilide   **p-Acetamidobenzenesulfonyl chloride**

$$(2)$$

p-Acetamidobenzene sulfonamide   **Sulfanilamide**

Acetanilide, which can easily be prepared from aniline, is allowed to react with chlorosulfonic acid to yield *p*-acetamidobenzenesulfonyl chloride. The acetamido group directs substitution almost totally to the *para* position. The reaction is an example of an electrophilic aromatic substitution reaction. Two problems would result if aniline itself were used in the reaction. First, the amino group in aniline would be protonated in strong acid to become a *meta* director; and, secondly, the chlorosulfonic acid would react with the amino group rather than with the ring, to give

377

C_6H_5—$NHSO_3H$. For these reasons, the amino group has been "protected" by acety-lation. The acetyl group will be removed in the final step, after it is no longer needed, to regenerate the free amino group present in sulfanilamide.

   p-Acetamidobenzenesulfonyl chloride is isolated by adding the reaction mixture to ice water, which decomposes the excess chlorosulfonic acid. This intermediate is fairly stable in water; nevertheless, it is converted slowly to the corresponding sulfonic acid (Ar—SO_3H). Thus, it should be isolated as soon as possible from the aqueous medium by filtration.

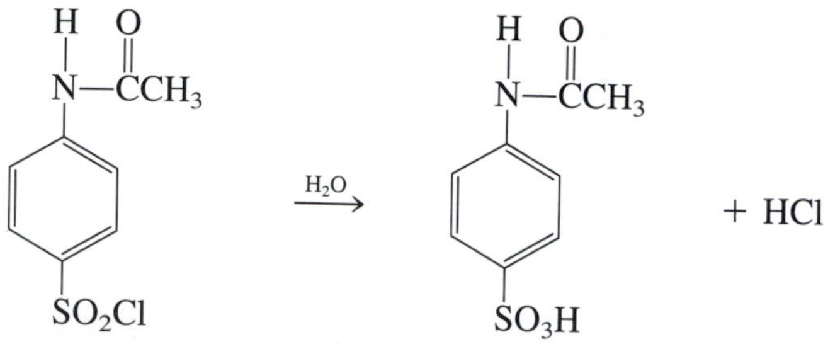

p-Acetamidobenzenesulfonyl
chloride

p-Acetamidobenzenesulfonic
acid

   The intermediate sulfonyl chloride is converted to p-acetamidobenzenesul-fonamide by a reaction with aqueous ammonia (Step 2). Excess ammonia neutralizes the hydrogen chloride produced. The only side reaction is the hydrolysis of the sulfonyl chloride to p-acetamidobenzenesulfonic acid.

   The protecting acetyl group is removed by acid-catalyzed hydrolysis to generate the hydrochloride salt of the product, sulfanilamide. Notice that of the two amide linkages present, only the carboxylic acid amide (acetamido group) was cleaved, not the sulfonic acid amide (sulfonamide). The salt of the sulfa drug is converted to sulfa-nilamide when the base, sodium bicarbonate, is added.

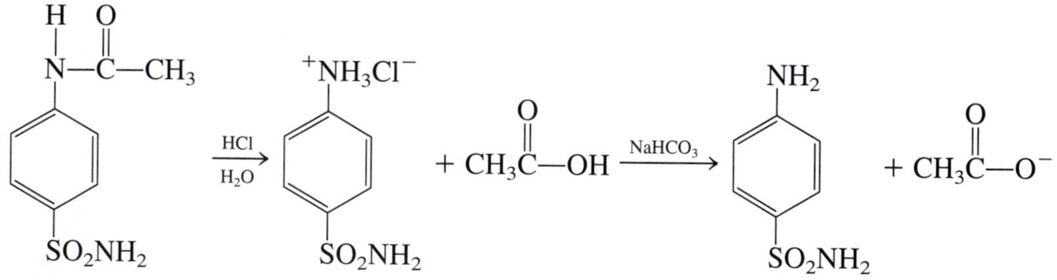

p-Acetamidobenzene-
sulfonamide

Sulfanilamide

REQUIRED READING

Review: Technique 3   Sections 3.2 and 3.7A
            Technique 4   Sections 4.3 and 4.7

Technique 5   Section 5.4
Technique 18   Section 18.4

New: Essay   Sulfa Drugs

SPECIAL INSTRUCTIONS

Chlorosulfonic acid must be handled with care since it is a corrosive liquid and reacts violently with water. The p-acetamidobenzenesulfonyl chloride should be used during the same laboratory period in which it is prepared. It is unstable and will not survive long storage. The sulfa drug may be tested on several kinds of bacteria (Instructor's Manual).

PROCEDURE

p-ACETAMIDOBENZENESULFONYL CHLORIDE

Assemble the apparatus as shown in Figure 3–2A (inset) on p 550 using dry glassware. You will need a 5-mL conical vial, an air condenser, and a drying tube which will be used as a gas trap. Prepare the drying tube for use as a gas trap by packing the tube loosely with dry glass wool (Technique 3, Section 3.7A, p 555). Moisten the wool slightly with several drops of water. The moistened glass wool traps the hydrogen chloride that is evolved in the reaction. Attach the 5-mL conical vial after the acetanilide and chlorosulfonic acid have been added as directed below. You should adjust the temperature of the sand bath to about 120 °C for use later in the experiment.

Place 0.18 g of acetanilide in the dry 5-mL conical vial and connect the air condenser, but not the drying tube. Melt the acetanilide (mp 113 °C) by heating the vial in a community sand bath, at about 160 °C. Remove the vial from the sand bath and swirl the heavy oil while holding the vial at an angle so that it solidifies uniformly on the cone-shaped bottom of the vial. Allow the conical vial to cool to room temperature and then cool it further in an ice-water bath. (Don't place the hot vial directly into the ice-water bath without prior cooling or the vial will crack!)

> **CAUTION:** Chlorosulfonic acid is an extremely noxious and corrosive chemical and should be handled with care. Use only dry glassware with this reagent. Should the chlorosulfonic acid be spilled on your skin, wash it off immediately with water. Wear safety glasses.

Remove the air condenser. In a hood, transfer 0.50 mL of chlorosulfonic acid, $ClSO_2OH$ (MW = 116.5, d = 1.77 g/mL), to the acetanilide in the conical vial using the graduated pipet provided. Reattach the air condenser and drying tube. Allow the mixture to stand for five minutes and then heat the reaction vial in the sand bath at about 120 °C for 10 minutes to complete the reaction. When all of the solid has dissolved, remove the vial from the sand bath. Allow the vial to cool to the touch and then cool it in an ice-water bath.

The operations in this paragraph should be conducted as rapidly as possible since the *p*-acetamidobenzenesulfonyl chloride reacts with water. Add 3 g of crushed ice to a 20-mL beaker. In a hood, transfer the cooled reaction mixture dropwise (it may splatter somewhat) with a Pasteur pipet onto the ice while stirring the mixture with a glass stirring rod. (The remaining operations in this paragraph may be completed at your laboratory bench.) Rinse the conical vial with a few drops of cold water and transfer the contents to the beaker containing the ice. Stir the precipitate to break up the lumps and then filter the *p*-acetamidobenzenesulfonyl chloride on a Hirsch funnel (Technique 4, Section 4.3, and Figure 4–6, p 571). Rinse the conical vial and beaker with two 1-mL portions of ice water. Use the rinse water to wash the crude product on the funnel. Any remaining solid in the conical vial should be left there since this vial will be used again in the next step. Convert the solid into sulfanilamide in the same laboratory period.

SULFANILAMIDE

Prepare a hot water bath at 70 °C. Place the crude *p*-acetamidobenzenesulfonyl chloride into the original 5-mL conical vial and add 1.1 mL of dilute ammonium hydroxide solution.[1] Stir the mixture well with a spatula and reattach the air condenser and drying tube (gas trap) using **fresh** moistened glass wool. Heat the mixture in the hot water bath for 10 minutes. Allow the conical vial to cool to the touch and place it in an ice-water bath for several minutes. Collect the *p*-acetamidobenzenesulfonamide on a Hirsch funnel and rinse the vial and product with a small amount of ice water.

Transfer this solid into the conical vial, and add 0.53 mL of dilute hydrochloric acid solution.[2] Attach the air condenser and heat the mixture in the sand bath at about 140 °C until all of the solid has dissolved. Then, heat the solution for an additional five minutes. Allow the mixture to cool to room temperature. If a solid (unreacted starting material) appears, heat the mixture for several minutes at 140 °C. When the vial has cooled to room temperature, no further solids should appear. With a Pasteur pipet transfer the solution to a 20-mL beaker. While stirring with a glass rod, cautiously add dropwise a slurry of 0.5 g of sodium bicarbonate in about 1 mL of water to the mixture in the beaker. Foaming will occur after each addition of the bicarbonate solution because of carbon dioxide evolution. Allow gas evolution to cease before making the next addition. Eventually, sulfanilamide will begin to precipitate. At this point begin to check the pH of the solution. Add the aqueous sodium bicarbonate until the pH of the solution is between 4 and 6. Cool the mixture thoroughly in an ice-water bath. Collect the product on a Hirsch funnel and rinse the beaker and solid with about 0.5 mL of cold water. Allow the sulfanilamide to air dry on the Hirsch funnel with suction for several minutes.

Weigh the crude product and crystallize it from hot water (1.0 to 1.2 mL water/ 0.1 g) using a Craig tube (Technique 5, Section 5.4, and Figure 5–5, p 586). Step 2 in Figure 5–5 (removal of insoluble impurities) should not be required in this crystallization.

---

[1] Prepared by mixing 11 mL of concentrated ammonium hydroxide with 11 mL of water.

[2] Prepared by mixing 7.0 mL of water with 3.6 mL of concentrated hydrochloric acid.

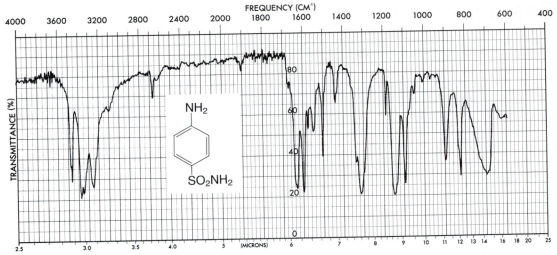

Infrared spectrum of sulfanilamide, KBr

Allow the purified product to dry until the next laboratory period. Weigh the dry sulfanila-mide and calculate the percentage yield (MW = 172.2). Determine the melting point (pure sulfanilamide melts at 163–164 °C) and obtain the infrared spectrum in potassium bromide (Technique 18, Section 18.4, p 775). Submit the sulfanilamide to the instructor in a labeled vial or save it for the tests with bacteria (Instructor's Manual).

QUESTIONS

1. Write an equation showing how excess chlorosulfonic acid is decomposed in water.

2. In the preparation of sulfanilamide, why was aqueous sodium bicarbonate used, rather than aqueous sodium hydroxide, to neutralize the solution in the final step?

3. At first glance, it might seem possible to prepare sulfanilamide from sulfanilic acid by the set of reactions shown below.

When the reaction is conducted in this way, however, a polymeric product is produced after Step 1. What is the structure of the polymer? Why does *p*-acetamidobenzenesulfonyl chloride not produce a polymer?

Essay
POLYMERS AND PLASTICS

Chemically, plastics are composed of chainlike molecules of high molecular weight, called polymers. Polymers have been built up from simpler chemicals, called monomers. A different monomer or combination of monomers is used to manufacture each different type or family of polymers. There are many polymers around us that are familiar. Examples of synthetic polymers are Teflon, nylon, Dacron, polyethylene, polyester, Orlon, epoxy, vinyl, polyurethane, silicones, Lucite, and boat resin. Examples of natural polymers are starch and cellulose (from glucose), rubber (from isoprene), and proteins (from amino acids). Certainly, polymers have had a great influence on our society. They are rapidly replacing many metals in manufacturing. In addition, synthetic polymeric textiles are replacing natural fibers for making cloth. As these materials have been created, a problem has arisen in disposing of them since many are not biodegradable.

CHEMICAL STRUCTURES OF POLYMERS

Basically a polymer is made up of many repeating molecular units formed by sequential addition of monomer molecules to one another. Many monomer molecules of A, say 1000 to 1 million, can be linked to form a gigantic polymeric molecule:

$$\text{Many A} \longrightarrow \text{etc.}\text{—A-A-A-A-A—etc.} \qquad \text{or} \qquad -(A)_n$$

Monomer molecules   **Polymer molecule**

Monomers that are different can also be linked to form a polymer with an alternating structure. This type of polymer is called a copolymer.

$$\text{Many A + many B} \longrightarrow \text{etc.}\text{—A-B-A-B-A-B—etc.} \qquad \text{or} \qquad -(A\text{-}B)_n$$

Monomer molecules   **Polymer molecule**

TYPES OF POLYMERS

For convenience, chemists classify polymers in several main groups, depending on method of synthesis.

1. **Addition polymers** are formed by a reaction in which monomer units simply add to one another to form a long-chain (generally linear or branched) polymer. The monomers usually contain carbon-carbon double bonds. Familiar examples of addition polymers are polyethylene and Teflon. The process can be represented as follows:

382

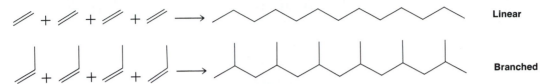

Linear

Branched

2. Condensation polymers are formed by reaction of bifunctional or polyfunctional molecules, with the elimination of some small molecule (such as water, ammonia, or hydrogen chloride) as a by-product. Familiar examples of condensation polymers are nylon, Dacron, and polyurethane. The process can be represented as follows:

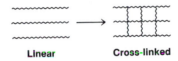

3. Cross-linked polymers are formed when long chains are linked in one gigantic, three-dimensional structure with tremendous rigidity. Addition and condensation polymers can exist with a cross-linked network, depending on the monomers used in the synthesis. Familiar examples of cross-linked polymers are Bakelite, rubber, and casting (boat) resin. The process can be represented as follows:

Linear   Cross-linked

Industrialists and technologists often classify polymers in other categories also:

1. Thermoplastics are materials that can be softened (melted) by heat and re-formed (molded) into another shape. Weaker, noncovalent bonds are broken during the heating. Technically, thermoplastics are the materials we call plastics. Both addition and condensation polymers can be so classified. Familiar examples include polyethylene (addition polymer) and nylon (condensation polymer).

2. Thermoset plastics are materials that melt initially but on further heating become permanently hardened. They cannot be softened and remolded without destruction of the polymer because covalent bonds are broken. Chemically, thermoset plastics are cross-linked polymers. Bakelite is an example of a thermoset plastic.

Polymers can also be classified in other ways; for example, many varieties of rubber are often referred to as elastomers, Dacron is a fiber, and polyvinyl acetate is an adhesive. The addition and condensation classification will be used in this essay.

ADDITION POLYMERS

Most of the polymers made are of the addition type. The monomers generally contain a carbon-carbon double bond. The most important example of an addition polymer is the well-known polyethylene, for which the monomer is ethylene. Countless numbers of ethylene molecules are linked in long-chain polymeric molecules by breaking the pi bond and creating two new single bonds between the monomer units. The number of recurring units may be large or small, depending on the polymerization conditions.

Many $\begin{array}{c} H \\ \diagdown \\ H \end{array} C = C \begin{array}{c} H \\ \diagup \\ H \end{array}$ $\longrightarrow$ etc. $-\overset{\displaystyle H}{\underset{\displaystyle H}{\overset{|}{C}}} - \overset{\displaystyle H}{\underset{\displaystyle H}{\overset{|}{C}}} - \overset{\displaystyle H}{\underset{\displaystyle H}{\overset{|}{C}}} - \overset{\displaystyle H}{\underset{\displaystyle H}{\overset{|}{C}}} -$ etc.   or $\left(\overset{\displaystyle H}{\underset{\displaystyle H}{\overset{|}{C}}} - \overset{\displaystyle H}{\underset{\displaystyle H}{\overset{|}{C}}} \right)_n$

**Ethylene
monomer**   **Polyethylene
polymer**

This reaction can be promoted by heat, pressure, and a chemical catalyst. The molecules produced in a typical reaction vary in the number of carbon atoms in their chains. In other words, a mixture of polymers of varying length is produced, rather than a pure compound.

Polyethylenes with linear structures can pack together easily and are referred to as high-density polyethylenes. They are fairly rigid materials. Low-density polyethylenes consist of branched-chain molecules, with some cross-linking in the chains. They are more flexible than the high-density polyethylenes. The reaction conditions and the catalysts that produce polyethylenes of low and high density are quite different. The monomer, however, is the same in each case.

Another example of an addition polymer is polypropylene. In this case, the monomer is propylene. The polymer that results has a branched methyl on alternate carbon atoms of the chain.

Many $\begin{array}{c} H \\ \diagdown \\ H \end{array} C = C \begin{array}{c} H \\ \diagup \\ CH_3 \end{array}$ $\longrightarrow$ etc. $-\overset{\displaystyle H}{\underset{\displaystyle H}{\overset{|}{C}}} - \overset{\displaystyle H}{\underset{\displaystyle CH_3}{\overset{|}{C}}} - \overset{\displaystyle H}{\underset{\displaystyle H}{\overset{|}{C}}} - \overset{\displaystyle H}{\underset{\displaystyle CH_3}{\overset{|}{C}}} -$ etc.   or $\left(\overset{\displaystyle H}{\underset{\displaystyle H}{\overset{|}{C}}} - \overset{\displaystyle H}{\underset{\displaystyle CH_3}{\overset{|}{C}}} \right)_n$

**Propylene
monomer**   **Polypropylene
polymer**

Several common addition polymers are shown in Table 1. Some of their principal uses are also listed. The last three entries in the table all have a carbon-carbon double bond remaining after the polymer is formed. These bonds activate or participate in a further reaction to form cross-linked polymers called elastomers; this term is almost synonymous with *rubber,* since they designate materials with common characteristics.

CONDENSATION POLYMERS

Condensation polymers, for which the monomers contain more than one type of functional group, are more complex than addition polymers. In addition, most condensation polymers are copolymers made from more than one type of monomer. You will recall that addition polymers, in contrast, are all prepared from substituted ethylene molecules. The single functional group in each case is one or more double bonds, and a single type of monomer is generally used.

TABLE 1. Addition Polymers

EXAMPLE	MONOMER(S)	POLYMER	USE
Polyethylene	$CH_2{=}CH_2$	$-CH_2-CH_2-$	Most common and important polymer; bags, insulation for wires, squeeze bottles
Polypropylene	$CH_2{=}CH$ $\quad\;\; CH_3$	$-CH_2-CH-$ $\qquad\;\; CH_3$	Fibers, indoor-outdoor carpets, bottles
Polystyrene	$CH_2{=}CH$ (phenyl)	$-CH_2-CH-$ (phenyl)	Styrofoam, inexpensive household goods, inexpensive molded objects
Polyvinyl chloride (PVC)	$CH_2{=}CH$ $\quad\;\; Cl$	$-CH_2-CH-$ $\qquad\;\; Cl$	Synthetic leather, clear bottles, floor covering. phonograph records, water pipe
Polytetrafluoroethylene (Teflon)	$CF_2{=}CF_2$	$-CF_2-CF_2-$	Nonstick surfaces, chemically resistant films
Polymethyl methacrylate (Lucite, Plexiglas)	$\qquad\; CO_2CH_3$ $CH_2{=}C$ $\qquad\;\; CH_3$	$\qquad\; CO_2CH_3$ $-CH_2-C-$ $\qquad\;\; CH_3$	Unbreakable ''glass,'' latex paints
Polyacrylonitrile (Orlon, Acrilan, Creslan)	$CH_2{=}CH$ $\quad\;\; CN$	$-CH_2-CH-$ $\qquad\;\; CN$	Fiber used in sweaters, blankets, carpets
Polyvinyl acetate (PVA)	$CH_2{=}CH$ $\quad\;\; OCCH_3$ $\qquad\;\;\; O$	$-CH_2-CH-$ $\qquad\;\; OCCH_3$ $\qquad\qquad O$	Adhesives, latex paints, chewing gum, textile coatings
Natural rubber	$\qquad\; CH_3$ $CH_2{=}CCH{=}CH_2$	$\qquad\;\; CH_3$ $-CH_2-C{=}CH-CH_2-$	The polymer is cross-linked with sulfur (vulcanization).
Polychloroprene (neoprene rubber)	$\qquad\; Cl$ $CH_2{=}CCH{=}CH_2$	$\qquad\;\; Cl$ $-CH_2-C{=}CH-CH_2-$	Cross-linked with ZnO; resistant to oil and gasoline
Styrene butadiene rubber (SBR)	$CH_2{=}CH$ (phenyl) $CH_2{=}CHCH{=}CH_2$	$-CH_2CH-CH_2CH{=}CHCH_2-$ (phenyl)	Cross-linked with peroxides; most common rubber; used for tires; 25% styrene 75% butadiene

Dacron, a polyester, can be prepared by causing a dicarboxylic acid to react with a bifunctional alcohol (a diol):

$$HO-\overset{\overset{O}{\|}}{C}-\underset{\text{Terephthalic acid}}{\bigcirc}-\overset{\overset{O}{\|}}{C}-\boxed{OH \quad H}-OCH_2CH_2OH \quad \longrightarrow$$

Terephthalic
acid

Ethylene
glycol

$$-\overset{\overset{O}{\|}}{C}-\bigcirc-\overset{\overset{O}{\|}}{C}-OCH_2CH_2-O- \; + \; H_2O$$

Dacron

Nylon 6-6, a polyamide, can be prepared by causing a dicarboxylic acid to react with a bifunctional amine:

$$HO-\overset{\overset{O}{\|}}{C}(CH_2)_4\overset{\overset{O}{\|}}{C}-\boxed{OH \quad H}-\overset{\overset{H}{|}}{N}(CH_2)_6\overset{\overset{H}{|}}{N}H \longrightarrow -\overset{\overset{O}{\|}}{C}(CH_2)_4\overset{\overset{O}{\|}}{C}-\underset{\overset{|}{H}}{N}(CH_2)_6\underset{\overset{|}{H}}{N}- \; + \; H_2O$$

Adipic
acid

Hexamethylene-
diamine

Nylon

Notice, in each case, that a small molecule, water, is eliminated as a product of the reaction. Several other condensation polymers are listed in Table 2. Linear (or branched) chain polymers as well as cross-linked polymers are produced in condensation reactions.

The nylon structure contains the amide linkage at regular intervals,

$$-\overset{\overset{O}{\|}}{C}-\overset{\overset{H}{|}}{N}-$$

This type of linkage is extremely important in nature because of its presence in proteins and polypeptides. Proteins are gigantic polymeric substances made up of monomer units of amino acids. They are linked by the peptide (amide) bond.

Other important natural condensation polymers are starch and cellulose. They are polymeric materials made up of the sugar monomer glucose. Another important natural condensation polymer is the DNA molecule. A DNA molecule is made up of the sugar deoxyribose linked with phosphates to form the backbone of the molecule. A portion of a DNA molecule is shown in the essay that precedes Experiment 5.

PROBLEMS WITH PLASTICS

Plastics have certainly become very common in our society. However, they are not without problems. There are disposal problems, health hazards, littering problems, fire hazards, and energy shortages associated with their manufacture and use.

TABLE 2. Condensation Polymers

EXAMPLE	MONOMERS	POLYMER	USE
Polyamides (nylon)	$HOC(CH_2)_nCOH$ $H_2N(CH_2)_nNH_2$	$-C(CH_2)_nC-NH(CH_2)_nNH-$	Fibers, molded objects
Polyesters (Dacron, Mylar, Fortrel)	$HOC-\!\!\!\!\bigcirc\!\!\!\!-COH$ $HO(CH_2)_nOH$	$-C-\!\!\!\!\bigcirc\!\!\!\!-C-O(CH_2)_nO-$	Linear polyesters, fibers, recording tape
Polyesters (Glyptal resin)	(phthalic anhydride) $HOCH_2CHCH_2OH$ OH	$-COCH_2CHCH_2O-$	Cross-linked polyester, paints
Polyesters (casting resin)	$HOCCH\!=\!CHCOH$ $HO(CH_2)_nOH$	$-CCH\!=\!CHC-O(CH_2)_nO-$	Cross-linked with styrene and peroxide, fiberglass boat resin
Phenol-formaldehyde resin (Bakelite)	(phenol) $CH_2\!=\!O$	$-CH_2$ $\cdots$ OH $\cdots$ CH_2- ... CH_2	Mixed with fillers, molded electrical goods, adhesives, laminates, varnishes
Cellulose acetate*	CH_2OH ... OH ... OH CH_3COOH	CH_2OAc ... OAc ... OAc	Photographic film
Silicones	CH_3 $Cl-Si-Cl \quad H_2O$ CH_3	CH_3 $-O-Si-O-$ CH_3	Water-repellent coatings, temperature-resistant fluids and rubbers (CH_3SiCl_3 cross-links in water)
Polyurethanes	CH_3, $N\!=\!C\!=\!O$ $N\!=\!C\!=\!O$ $HO(CH_2)_nOH$	CH_3, $NHC-O(CH_2)_nO-$ $NHC-O(CH_2)_nO-$	Rigid and flexible foams, fibers

*Cellulose, a polymer of glucose, is used as the monomer.

Plasticizers and Health Hazards

Certain types of plastics such as polyvinyl chloride (PVC) are mixed with plasticizers that soften the plastic so that it is more pliable. If plasticizers were not added, the plastic would be hard and brittle. Some of the plasticizers used in vinyl plastics are phthalate esters. The structure of a phthalate ester is shown below. These esters are volatile compounds of low molecular weight. Part of the new car "smell" comes from the odor of these esters as they evaporate from the vinyl upholstery. The vapor often condenses on the windshield as an oily, insoluble film. After some time, the vinyl material may lose enough plasticizer to cause it to crack. Phthalate esters may constitute a health hazard. Sometimes vinyl containers incorporating phthalate plasticizers are used to store blood. The esters are leached from blood bags made of PVC and may be partly responsible for shock lung, a condition that sometimes leads to death during a blood transfusion. The long-term effects of these plasticizers are not known.

Recently, a rare and fatal form of liver cancer (angiosarcoma) was discovered among small numbers of workers in chemical companies making polyvinyl chloride. The monomer used in making PVC is vinyl chloride, a gas. The structure is shown below. Currently, industry is required to eliminate this health hazard by reducing or eliminating vinyl chloride from the atmosphere.

Other types of plasticizers once used were the polychlorinated biphenyls (PCB). These compounds and DDT have similar physiological effects, and they are even more persistent in the environment! The PCBs are actually a mixture of compounds that have had the hydrogens on the basic hydrocarbon structure, biphenyl, replaced with chlorines (from 1 to 10 hydrogens can be replaced). One typical PCB that may be present in a plasticizer mixture is shown. PCBs are no longer being sold except for use in closed systems, where they cannot leak into the environment.

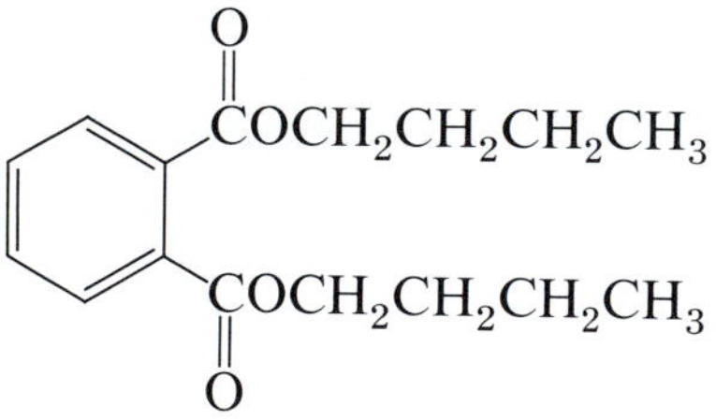

Dibutyl phthalate   Vinyl chloride   A polychlorinated biphenyl (PCB)

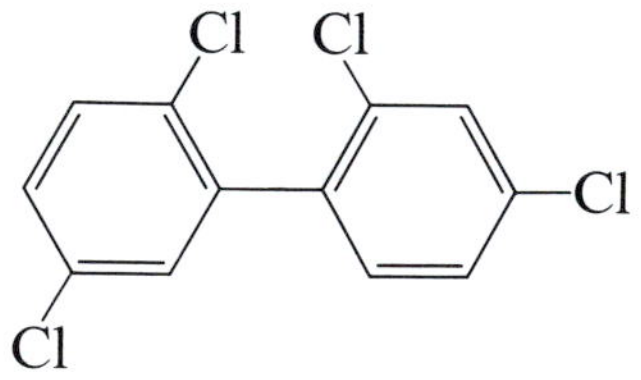

Disposability Problems

What do we do with all our wastes? Currently, the most popular method is to bury our garbage in sanitary landfills. However, as we run out of good places to bury our garbage, incineration appears to be an attractive method for solving the solid waste problem. Plastics, which compose about 2% of our garbage, burn readily. The new high-temperature incinerators are extremely efficient and can be operated with very little air pollution. It should also be possible to burn our garbage and generate electrical power from it.

Ideally, we should either recycle all our wastes or not produce the waste in the first place. Plastic waste consists of about 55% polyethylene and polypropylene, 20% polystyrene, and 11% PVC. All these polymers are thermoplastics and can be recycled. They can be resoftened and remolded into new goods. Unfortunately, thermosetting plastics (cross-linked polymers) cannot be remelted. They decompose on high-temperature heating. Thus, thermosetting plastics should not be used for "disposable" purposes. To recycle plastics effectively, we must sort the materials according to the various types. This requires will power as well as knowledge about the plastics that we are discarding. Neither requirement is easily effected.

Littering Problems

Plastics, if they are well made, will not corrode or rust, and they last almost indefinitely. Unfortunately, these desirable properties also lead to a problem when plastics are buried in a landfill or thrown on the landscape—they do not decompose. Currently, research is being undertaken to discover plastics that are biodegradable or photodegradable, so that either microorganisms or light from the sun can decompose our litter and garbage. Some success has been achieved.

Fire Hazards

Numerous injuries are caused by clothing made of polymers, especially children's clothing. Many of these organic fibers burn readily. To combat this problem, chemists have developed flame-retardant fabrics, especially for children's sleepwear.

Toxic gases are sometimes liberated when plastics burn. For example, hydrogen chloride is generated when PVC is burned, and hydrogen cyanide when polyacrylonitriles are burned. This presents a problem that compounds the fire danger.

Energy Shortage

The demand for energy has increased at an alarming rate, leading to the energy crisis. The production of polymers requires petroleum as a raw material and as a source of energy to conduct manufacturing. Unfortunately, fossil fuels are a nonrenewable resource, and as their availability decreases, we shall have an even greater problem. On the other hand, natural substances, such as cotton, are renewable resources; perhaps for some uses they would actually be better and less costly than the synthesized polymers. There are many plastics, however, that are superior to natural materials. The answer lies in using plastics wisely.

REFERENCES

Anderson, B. C., Bartron, L. R., and Collette, J. W. "Trends in Polymer Development." *Science, 208* (1980): 807.
"Biodegradability: Lofty Goal for Plastics." *Chemical and Engineering News* (September 11, 1972): 37–38.

Carraher, C. E., Jr. "What Are Polymers?" *Chemistry, 51* (June 1978): 6. Other articles in the same issue of *Chemistry* deal with polymers and the quality of life, sizes and shapes of giant molecules, biopolymers, fire and polymers, polymers in medicine and surgery, and polymers for the future.

"Degradable Plastic Lids Are Now Commercial." *Chemical and Engineering News* (June 19, 1972): 32.

Flory, P. J. "Understanding Unruly Molecules." *Chemistry* (May 1964): 6–13.

Heckert, W. W. "Synthetic Fibers." *Journal of Chemical Education, 30* (1953): 166.

Kaufman, M. *Giant Molecules*. New York: Doubleday, 1968.

Keller, E. "Nylon—from Test Tube to Counter." *Chemistry* (September 1964): 8–23.

McGrew, F. C. "Structure of Synthetic High Polymers." *Journal of Chemical Education, 35* (1958): 178.

Mark, H. F. "Giant Molecules." *Scientific American, 197* (1957): 80.

Mark, H. F. "The Nature of Polymeric Materials." *Scientific American, 217* (1967): 149.

"More Trouble Brewing for Vinyl Chloride." *Chemical and Engineering News* (September 2, 1974): 12–13.

Morton, M. "Big Molecules." *Chemistry* (January 1964): 13–17.

Morton, M. "Design and Formation of Long Chain Polymers." *Chemistry* (March 1964): 6–11.

Oster, B. "Polyethylene." *Scientific American, 197* (1957): 139.

Natta, G. "How Giant Molecules Are Made." *Scientific American, 197* (1957): 98.

"Plasticizers: Found in Heart Cells." *Chemical and Engineering News* (November 8, 1971): 7–8.

"Plastics Face Growing Pressure from Ecologists." *Chemical and Engineering News* (March 29, 1971): 12–13.

"PVC Makers Hit with Health Hazard Suit." *Chemical and Engineering News* (September 2, 1974): 9–10.

Sanders, H. J. "Flame Retardants." *Chemical and Engineering News, 56* (April 24, 1978): 22.

Several issues of *Journal of Chemical Education* have been devoted to aspects of polymer chemistry. Articles on this topic may be found in:

Journal of Chemical Education, 58 (November 1981).
Journal of Chemical Education, 58 (July 1981).
Journal of Chemical Education, 61 (March 1984).
Journal of Chemical Education, 63 (May 1986).
Journal of Chemical Education, 64 (January 1987).

Experiment 48
Preparation of Polymers: Polyester, Nylon, Polystyrene, and Polyurethane

Condensation polymers
Addition polymers
Cross-linked polymers

In this experiment, the syntheses of two polyesters (Procedure 48A), nylon (Procedure 48B), polystyrene (Procedure 48C), and polyurethane (Procedure 48D) are described. These polymers represent the most important commercial plastics. They also represent the main classes of polymers: condensation (linear polyester, nylon), addition (polystyrene), and cross-linked (Glyptal polyester, polyurethane).

REQUIRED READING

New: Essay   Polymers and Plastics

SPECIAL INSTRUCTIONS

Toluene diisocyanate (TDI) and styrene are toxic. They will irritate the skin and eyes. Avoid breathing the vapors of these materials. These substances must be dispensed and stored in a good hood. Benzoyl peroxide is flammable and may detonate on impact or on heating.

Procedure 48A
Polyesters

Linear and cross-linked polyesters are prepared in this experiment. Both are examples of condensation polymers.

The linear polyester is prepared as follows:

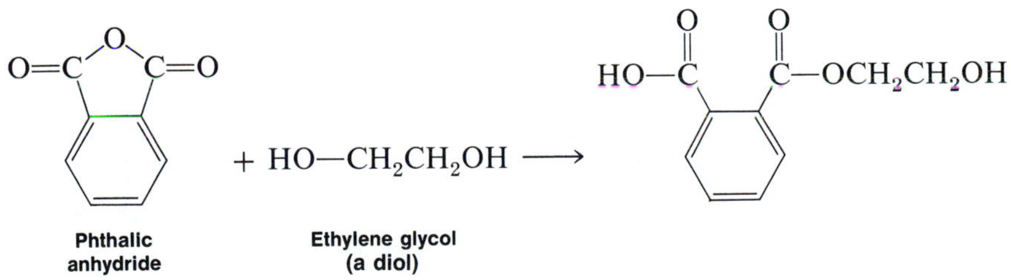

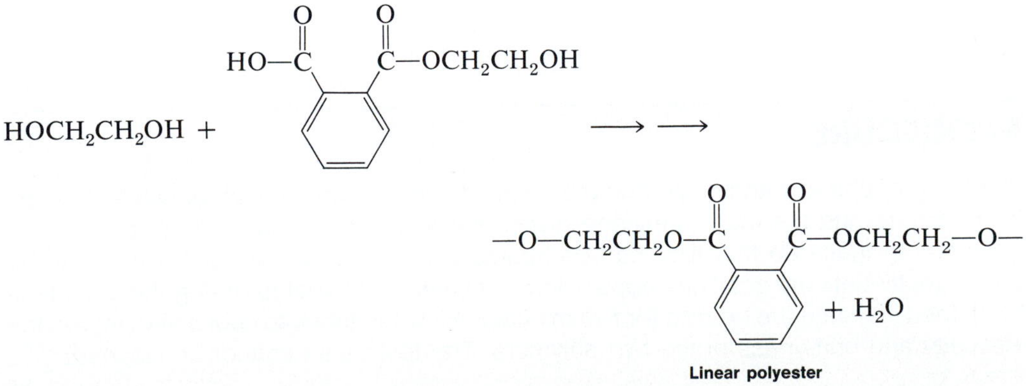

This linear polyester is isomeric with Dacron, which is prepared from terephthalic acid and ethylene glycol (see the preceding essay). Dacron and the linear polyester made in this experiment are both thermoplastics.

If more than two functional groups are present in one of the monomers, the polymer chains can be linked to one another (cross-linked) to form a three-dimensional network. Such structures are usually more rigid than linear structures and are useful in making paints and coatings. They may be classified as thermosetting plastics. The polyester Glyptal is prepared as follows:

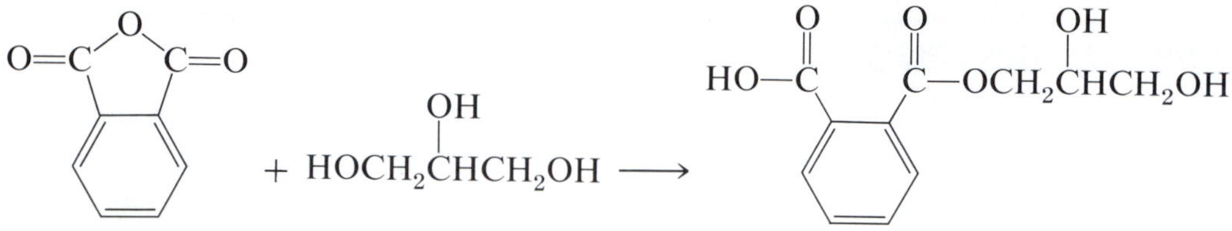

Phthalic anhydride

Glycerol (a triol)

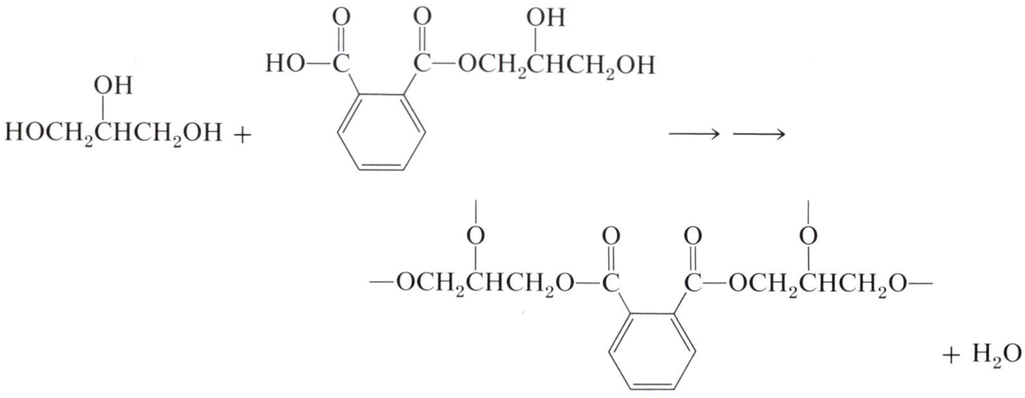

Cross-linked polyester (glyptal resin)

The reaction of phthalic anhydride with a diol (ethylene glycol) is described in the procedure. This linear polyester is compared with the cross-linked polyester (Glyptal) prepared from phthalic anhydride and a triol (glycerol).

PROCEDURE

Place 1 g of phthalic anhydride and 0.05 g of sodium acetate in each of two test tubes. To one tube add 0.4 mL of ethylene glycol and to the other add 0.4 mL of glycerol. Clamp both tubes so that they can be heated simultaneously with a flame. Heat the tubes gently until the solutions appear to boil (water is eliminated during the esterification), then continue the heating for five minutes. Allow the tubes to cool and compare the viscosity and brittleness of the two polymers. The test tubes cannot be cleaned.

Procedure 48B
Polyamide (Nylon)

Reaction of a dicarboxylic acid, or one of its derivatives, with a diamine leads to a linear polyamide through a condensation reaction. Commercially, nylon 6-6 (so called

$$Cl-\overset{\overset{\displaystyle O}{\|}}{C}CH_2CH_2CH_2CH_2\overset{\overset{\displaystyle O}{\|}}{C}-Cl \; + \; H-\overset{\overset{\displaystyle H}{|}}{N}CH_2CH_2CH_2CH_2CH_2CH_2\overset{\overset{\displaystyle H}{|}}{N}-H \longrightarrow$$

Adipoyl chloride   **Hexamethylenediamine**

$$-\overset{\overset{\displaystyle O}{\|}}{C}CH_2CH_2CH_2CH_2\overset{\overset{\displaystyle O}{\|}}{C}-\overset{\overset{\displaystyle H}{|}}{N}CH_2CH_2CH_2CH_2CH_2CH_2\overset{\overset{\displaystyle H}{|}}{N}-$$

Nylon 6-6

because each monomer has six carbons) is made from adipic acid and hexamethylenediamine. In this experiment, you use the acid chloride instead of adipic acid. The acid chloride is dissolved in cyclohexane and this is added **carefully** to hexamethylenediamine dissolved in water. These liquids do not mix and two layers will form. It can then be drawn out continuously to form a long strand nylon. Imagine how many molecules have been linked in this long strand! It is a fantastic number.

PROCEDURE

Pour 10 mL of a 5% aqueous solution of hexamethylenediamine (1,6-hexanediamine) into a 50-mL beaker. Add 10 drops of 20% sodium hydroxide solution. Carefully add 10 mL of a 5% solution of adipoyl chloride in cyclohexane to the solution by pouring it down the wall of the slightly tilted beaker. Two layers will form (see figure on page 394), and there will be an immediate formation of a polymer film at the liquid-liquid interface. Using a copper-wire hook (a 6-in. piece of wire bent at one end), gently free the walls of the beaker from polymer strings. Then hook the mass at the center, and slowly raise the wire so that polyamide forms continuously, producing a rope that can be drawn out for many feet. The strand can be broken by pulling it faster. Rinse the rope several times with water and lay it on a paper towel to dry. With the piece of wire, vigorously stir the remainder of the two-phase system to form additional polymer. Decant the liquid and wash the polymer thoroughly with water. Allow the polymer to dry. Do not discard the nylon in the sink. Use a waste container.

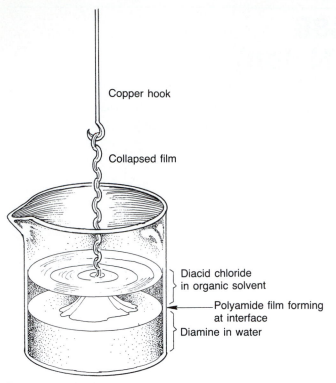

Preparation of nylon

Procedure 48C
Polystyrene

An addition polymer, polystyrene, is prepared in this experiment. Reaction can be brought about by free-radical, cationic, or anionic catalysts, the first of these being most common. In this experiment, polystyrene is prepared by free-radical-catalyzed polymerization.

The reaction is initiated by a free-radical source. The initiator will be benzoyl peroxide, a relatively unstable molecule, which at 80–90 °C decomposes with homolytic cleavage of the oxygen-oxygen bond:

Benzoyl peroxide

$\xrightarrow{\text{heat}}$ 2

Benzoyl radical

If an unsaturated monomer is present, the catalyst radical adds to it, initiating a chain reaction by producing a new free radical. If we let R stand for the catalyst radical, the reaction with styrene can be represented as

$$R\cdot \ + \ CH_2{=}CH \longrightarrow R{-}CH_2{-}CH\cdot$$

The chain continues to grow:

$$R{-}CH_2{-}CH\cdot \ + \ CH_2{=}CH \longrightarrow R{-}CH_2{-}CH{-}CH_2{-}CH\cdot, \ etc.$$

The chain can be terminated by causing two radicals to combine (either both polymer radicals or one polymer radical and one initiator radical) or by causing a hydrogen atom to become abstracted from another molecule.

PROCEDURE

Since it is difficult to clean the glassware, this experiment is best performed by the laboratory instructor. One large batch should be made for the entire class (at least 10 times the amounts given). Perform the experiment in a hood. Place several thicknesses of newspaper in the hood.

> **CAUTION:** Styrene vapor is very irritating to the eyes, mucous membranes, and upper respiratory track. Do not breathe the vapor and do not get it on your skin. Exposure can cause nausea and headaches. All operations with styrene must be conducted in a good hood.
>
> Benzoyl peroxide is flammable and may detonate on impact or on heating (or grinding). It should be weighed on glassine (glazed, not ordinary) paper. Clean up all spills with water. Wash the glassine paper with water before discarding it.

Place 12–15 mL of styrene monomer in a 100-mL beaker and add 0.35 g of benzoyl peroxide. Heat the mixture on a hot plate until the mixture turns yellow. When the color disappears and bubbles begin to appear, immediately take the beaker of styrene off the hot plate since the reaction is exothermic (use tongs or an insulated glove).

After the reaction subsides, put the beaker of styrene back on the hot plate and continue heating it until the liquid becomes very syrupy. With a stirring rod, draw out a long filament of material from the beaker. If this filament can be cleanly snapped after a few seconds of cooling, the polystyrene is ready to be poured. If the filament does not break, continue heating the mixture and repeat the above process until the filament breaks easily. Pour the syrupy liquid on a watch glass. After being cooled, the polystyrene can be lifted from the glass surface by gentle prying with a spatula.

Procedure 48D
Polyurethane Foam

A cross-linked polymer, polyurethane foam, is prepared in this experiment from a diisocyanate and a triol. The main reaction is the addition of the alcohol across the —N=C— bond of an isocyanate:

$$R-N=C=O \longrightarrow R-N-C=O$$

H—OR •   H OR

Isocyanate + alcohol   **Urethane**

With a diisocyanate and a triol, the reaction can proceed in **three** directions, leading to a large molecule that is rigidly held into a three-dimensional structure.

The foaming is caused by the evolution of carbon dioxide, much as in the baking of bread. In baking, carbon dioxide is evolved by the fermentation of sugars with yeast, which causes the bread to rise. In the present preparation, the carbon dioxide is produced by the small amount of water present, which decomposes a small amount of the isocyanate:

$$R-N=C=O + H-O-H \longrightarrow R-N-C-OH \longrightarrow R-N-H + CO_2$$

The evolution of carbon dioxide bubbles creates pores in the viscous mixture as the foam sets into a rigid mass. Thus, the foam has excellent buoyant properties. The cell size and structure of the foams are controlled by adding silicone oil.

The structure of the polymer is as follows:

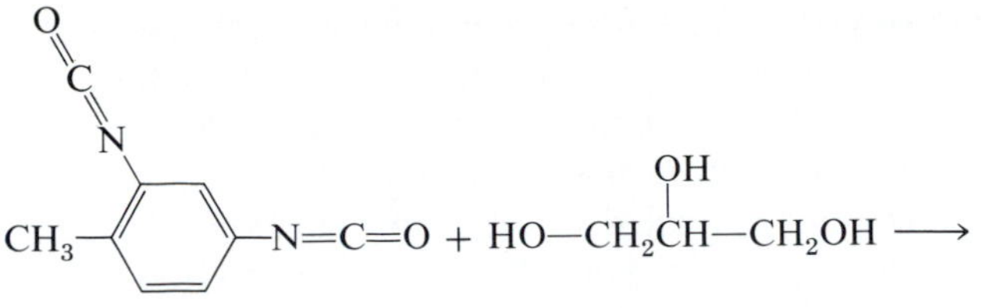

Toluene diisocyanate   **Glycerol**

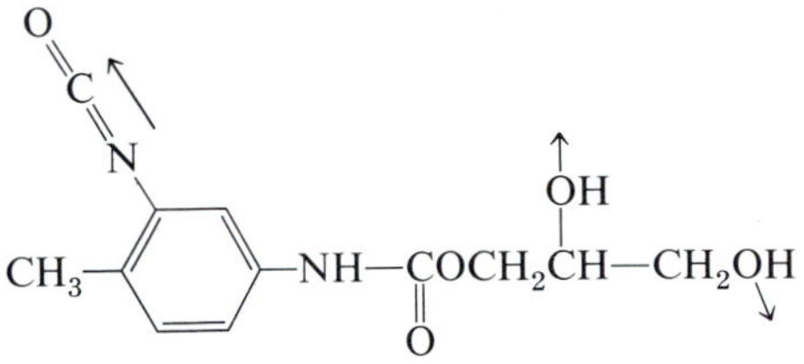

The polymer can grow in all the indicated directions.

Castor oil, a triol, can also react in the same way. It is a triglyceride (fat) of ricinoleic acid (see the essay, ''Fats and Oils,'' p 166):

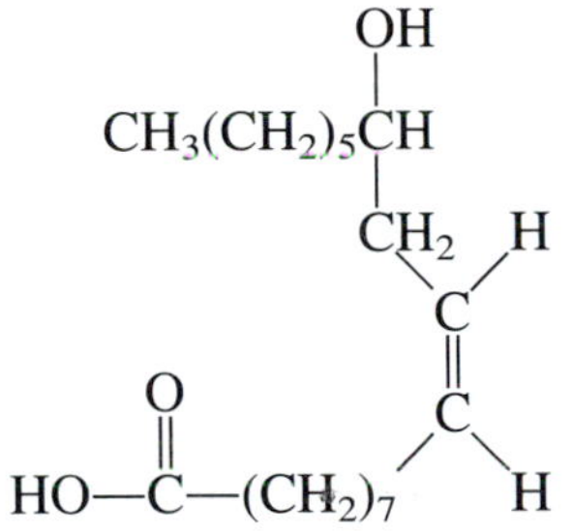

Ricinoleic acid

In the present experiment, glycerol, castor oil, small amounts of water, silicone oil (a foaming agent), and stannous octoate (a catalyst) are mixed together. The diisocyanate (TDI) is added to this mixture. The mixture is stirred and foaming begins. Commercial foams are not usually prepared from these simple materials. A polymeric diol or triol is generally used instead of glycerol and castor oil.

PROCEDURE

> **CAUTION:** Toluene diisocyanate (TDI) is toxic. It will irritate the skin and eyes. Avoid breathing the vapor. It may cause an allergic respiratory response. Work in a hood or in an area with adequate ventilation. Keep the container tightly closed when it is not in use (TDI reacts with moisture in the air). After handling TDI, wash your hands thoroughly.

Pour 8.5 mL of mixture A (shake well before using) into a waxed soft-drink cup.[1] Then add 5 mL of toluene diisocyanate (tolylene-2,4-diisocyanate). Stir the mixture rapidly and thoroughly with a stirring rod until the mixture is smooth and creamy. The mixture should become warm and should begin to evolve bubbles of carbon dioxide after about one minute. When the gas begins to evolve, immediately stop stirring (foaming will be spontaneous). Do not breathe the vapors. Place the mixture in the hood. After the foaming has ceased, allow the material to cool and set thoroughly. The polyurethane is initially sticky, but after several hours it will become firm. The paper container can then be removed. The material will shrink noticeably on standing.

QUESTIONS

1. Ethylene dichloride, $ClCH_2CH_2Cl$, and sodium polysulfide, Na_2S_4, react to form a chemically resistant rubber, Thiokol A. Write the structure of the rubber.

2. Vinylidene chloride, $CH_2{=}CCl_2$, is polymerized with vinyl chloride to make Saran. Write a structure that includes at least two units for the copolymer formed.

3. Isobutylene, $CH_2{=}C(CH_3)_2$, is used to prepare cold-flow rubber. Write a structure for the addition polymer formed from this alkene.

4. Kel-F is an addition polymer with the following partial structure. What is the monomer used to prepare it?

$$\begin{array}{ccccccc}
 & F & F & F & F & F & F \\
 & | & | & | & | & | & | \\
-C & -C & -C & -C & -C & -C- \\
 & | & | & | & | & | & | \\
 & F & Cl & F & Cl & F & Cl
\end{array}$$

5. Maleic anhydride reacts with ethylene glycol to produce an alkyd resin. Write the structure of the condensation polymer produced.

Maleic anhydride

6. Kodel is a condensation polymer made from terephthalic acid and 1,4-cyclohexanedimethanol. Write the structure of the resulting polymer.

Terephthalic acid　　　　　　**1,4-Cyclohexanedimethanol**

[1]Mixture A is prepared as follows: Place 350 g of castor oil, 100 g of glycerol, 50 drops of stannous octoate (stannous 2-ethylhexanoate), 50 drops of Dow-Corning 200 silicone oil (this is estimated since it is difficult to measure), and 150 drops of water in a bottle. Cap the bottle and shake it thoroughly. Allow this mixture to stand no more than 12 hours before use.

Essay
DIELS-ALDER REACTION AND INSECTICIDES

Since the 1930s, it has been known that the addition of an unsaturated molecule across a diene system forms a substituted cyclohexene. The original research dealing with this type of reaction was performed by Otto Diels and Kurt Alder in Germany, and the reaction has become known as the **Diels-Alder reaction.** The Diels-Alder reaction is the reaction of a **diene** with a species capable of reacting with the diene, the **dienophile.**

Diene   Dienophile

The product of the Diels-Alder reaction is usually a structure that contains a cyclohexene ring system. If the substituents as shown are simply alkyl groups or hydrogen atoms, the reaction proceeds only under extreme conditions of temperature and pressure. With more complex substituents, however, the Diels-Alder reaction may go on at low temperatures and under mild conditions. The reaction of cyclopentadiene with maleic anhydride (Experiment 49) and the reaction of tetraphenylcyclopentadiene with benzyne (Experiment 50) are examples of Diels-Alder reactions carried out under reasonably mild conditions.

A commercially important use of the Diels-Alder reaction involves using hexachlorocyclopentadiene as the diene. Depending on the dienophile, a variety of chlorine-containing addition products may be synthesized. Nearly all these products are powerful **insecticides.** Three insecticides synthesized by the Diels-Alder reaction are shown on the following page.

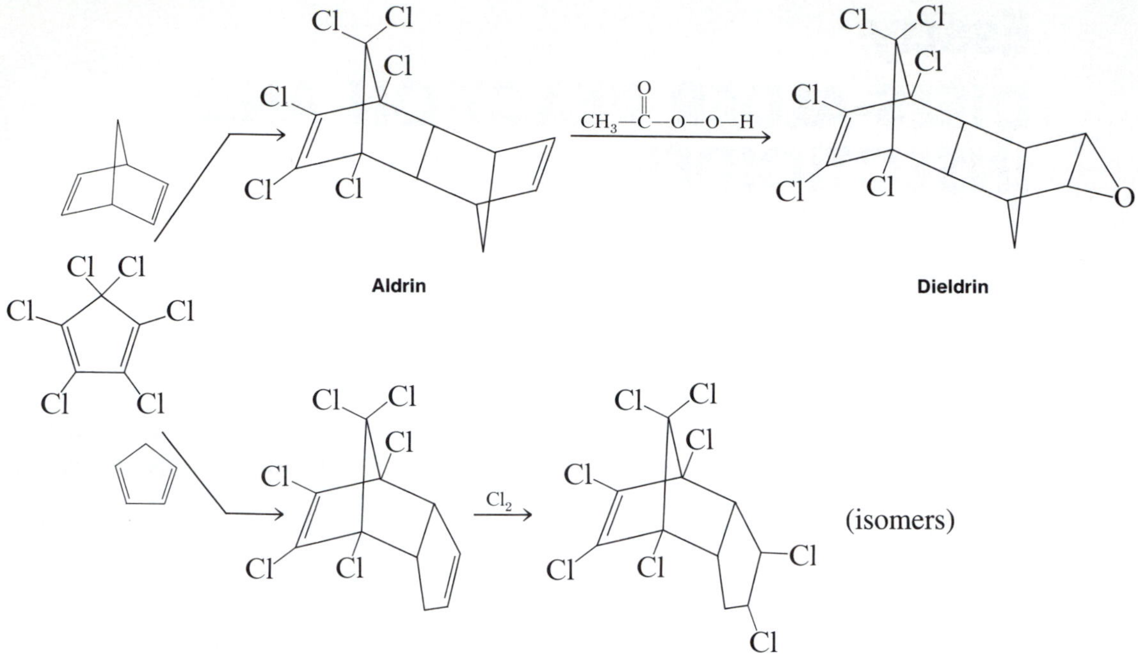

Aldrin

Dieldrin

Chlordane

(isomers)

Dieldrin and Aldrin are named after Diels and Alder. These insecticides have been used against the insect pests of fruits, vegetables, and cotton; against soil insects, termites, and moths; and in treating seeds. Chlordane has been used in veterinary medicine against insect pests of animals, including fleas, ticks, and lice.

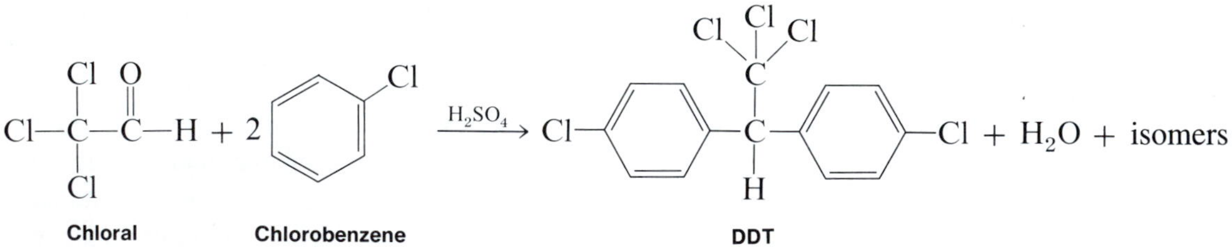

Chloral   **Chlorobenzene**   **DDT**

The best known insecticide, DDT, is not prepared by the Diels-Alder reaction, but is nevertheless the best illustration of difficulties experienced when insecticides are used indiscriminately. DDT was first synthesized in 1874, and its insecticidal properties were first demonstrated in 1939. It is easily synthesized commercially, with inexpensive reagents.

At the time DDT was introduced, it was an important boon to mankind. It was effective in controlling lice, fleas, and malaria-carrying mosquitoes and thus helped to control human and animal disease. The use of DDT rapidly spread to the control of hundreds of insects that damage fruit, vegetable, and grain crops.

Pesticides that persist in the environment for a long time after application are called hard pesticides. Beginning in the 1960s, some of the harmful effects of such "hard" pesticides as DDT and the other chlorocarbon materials became known. DDT is a fat-soluble material and is therefore likely to collect in the fat, nerve, and brain tissues of animals. The concentration of DDT in tissues increases in animals high in the food chain. Thus, birds that eat poisoned insects accumulate large quantities of DDT. Animals that feed on the birds accumulate even more DDT. In birds, at least two undesirable effects of DDT have been recognized. First, birds whose tissues contain large amounts of DDT have been observed to lay eggs having shells too thin to survive until young birds are hatched. Second, large quantities of DDT in the tissues seem to interfere with normal reproductive cycles. The massive destruction of bird populations that sometimes occurs after heavy spraying with DDT has become an issue of great concern. The brown pelican and the bald eagle are in danger of extinction. The use of chlorocarbon insecticides has been identified as the principal reason for the decline in the numbers of these birds.

Because DDT is chemically inert, it persists in the environment without decomposing to harmless materials. It can decompose very slowly, but the decomposition products are every bit as harmful as DDT itself. Consequently, each application of DDT means that still more DDT will pass from species to species, from food source to predator, until it concentrates in the higher animals, possibly endangering their existence. Even humans may be threatened. As a result of evidence of the harmful effects of DDT, in the early 1970s the Environmental Protection Agency banned general use of DDT; it may still be used for certain purposes, although permission of the Environmental Protection Agency is required. In 1974, permission was granted for using DDT against the tussock moth in the forests of Washington and Oregon.

Because the life cycles of insects are short, they are able to evolve an immunity to insecticides within a short period of time. As early as 1948, several strains of DDT-resistant insects were identified. Today, the malaria-bearing mosquitoes are almost completely resistant to DDT, an ironic development. Other chlorocarbon insecticides have been used as alternatives to DDT against resistant insects. Examples of other chlorocarbon materials include Dieldrin, Aldrin, Chlordane, and the substances whose structures are shown below. Heptachlor and Mirex are also prepared using Diels-Alder reactions.

Lindane   Heptachlor   Mirex

In spite of structural similarity, Chlordane and Heptachlor show different behavior. Compared with Heptachlor, Chlordane is short-lived and less toxic to mammals. Nevertheless, all the chlorocarbon insecticides have been the objects of much suspicion. A ban on the use of Dieldrin and Aldrin has also been ordered by the Environmental Protection Agency. In addition, strains of insects resistant to Dieldrin, Aldrin, and other materials have been observed. Some insects become addicted to a chlorocarbon insecticide and thrive on it!

    The problems associated with the chlorocarbon materials have led to the development of the ''soft'' insecticides. These usually are organophosphorus or carbamate derivatives, and they are characterized by a short lifetime before they are decomposed to harmless materials in the environment.

    The organic structures of some organophosphorus insecticides are shown below.

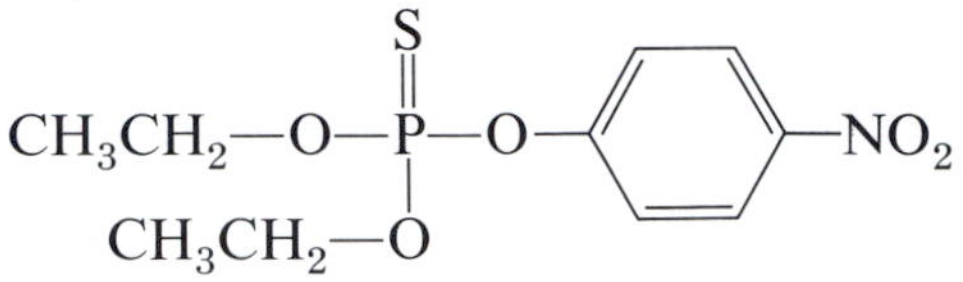

$$CH_3CH_2-O-\overset{\overset{\displaystyle S}{\|}}{\underset{\underset{\displaystyle CH_3CH_2-O}{|}}{P}}-O-\!\!\!\left\langle\!\!\bigcirc\!\!\right\rangle\!\!\!-NO_2$$

Parathion

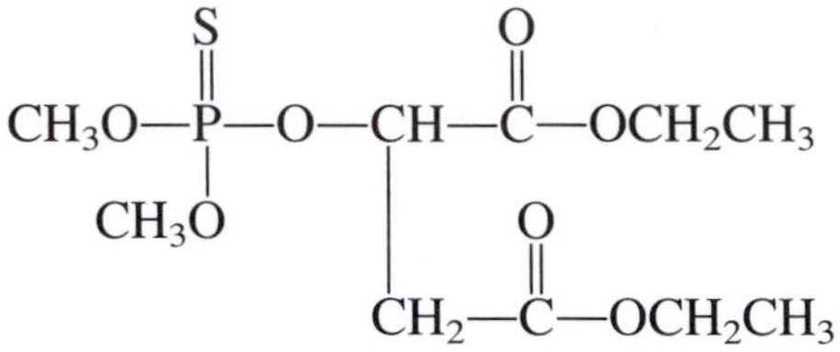

$$CH_3O-\overset{\overset{\displaystyle S}{\|}}{\underset{\underset{\displaystyle CH_3O}{|}}{P}}-O-\overset{}{\underset{\underset{\underset{\displaystyle CH_2-\overset{\overset{\displaystyle O}{\|}}{C}-OCH_2CH_3}{|}}{O}}{CH}}-\overset{\overset{\displaystyle O}{\|}}{C}-OCH_2CH_3$$

Malathion

$$CH_3O-\overset{\overset{\displaystyle O}{\|}}{\underset{\underset{\displaystyle CH_3O}{|}}{P}}-O-CH=C\overset{\displaystyle Cl}{\underset{\displaystyle Cl}{}}$$

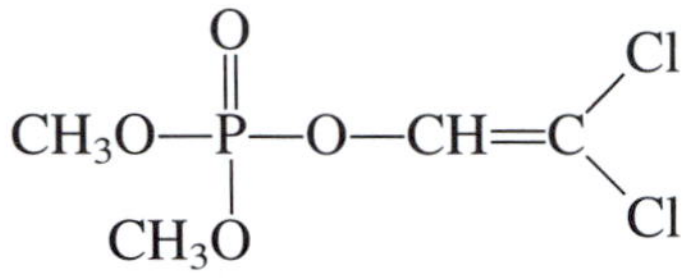

DDVP or Dichlorvos

Parathion and Malathion are used widely for agriculture. DDVP is used in ''pest strips,'' which are used for combating household insect pests. The organophosphorus materials do not persist in the environment, and so they are not passed between species up the food chain, as the chlorocarbon compounds are. However, the organophosphorus compounds are highly toxic to humans. Some loss of life among migrant and other agricultural workers has been caused by accidents involving these materials. Stringent safety precautions must be applied when the organophosphorus insecticides are being used.

    The carbamate derivatives, including Carbaryl, tend to be less toxic than the organophosphorus compounds. They are also readily degraded to harmless materials. Nevertheless, insects resistant to the soft insecticides have also been observed. Furthermore, the organophosphorus and carbamate derivatives destroy many more nontarget pests than the chlorocarbon compounds do. The danger to earthworms, mammals, and birds is very high.

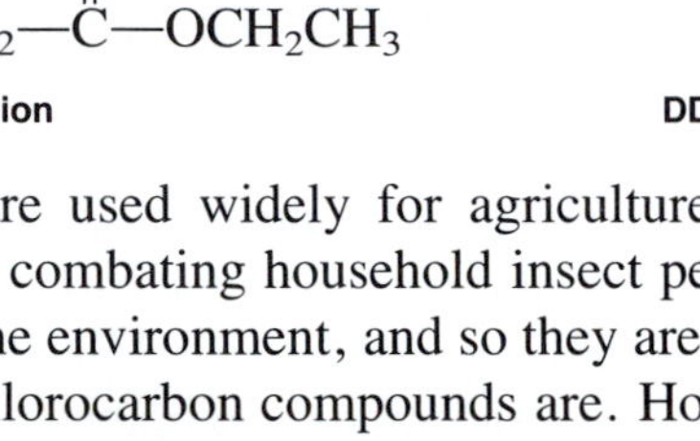

$$CH_3-NH-\overset{\overset{\displaystyle O}{\|}}{C}-O$$

Carbaryl

ALTERNATIVES TO INSECTICIDES

Several alternatives to the massive application of insecticides have recently been explored. Insect attractants, including the pheromones (see the essay preceding Experiment 46), have been used in localized traps. Such methods have been effective against the gypsy moth. A "confusion technique," whereby a pheromone is sprayed into the air in such high concentrations that male insects are no longer able to locate females, has been studied. These methods are specific for the target pest and do not cause repercussions in the general environment.

Recent research has been focused on using an insect's own biochemical processes to control pests. Experiments with **juvenile hormone** have shown promise. Juvenile hormone is one of three internal secretions used by insects to regulate growth and metamorphosis from larva to pupa and thence to the adult. At certain stages in the metamorphosis from larva to pupa, juvenile hormone must be secreted; at other stages it must be absent, or the insect will either develop abnormally or fail to mature. Juvenile hormone is important in maintaining the juvenile, or larval, stage of the growing insect. The male cecropia moth, which is the mature form of the silkworm, has been used as a source of juvenile hormone. The structure of the cecropia juvenile hormone is shown below. This material has been found to prevent the maturation of yellow-fever mosquitoes and human body lice. Since insects are not expected to develop a resistance to their own hormones, it is hoped that insects will not be likely to develop a resistance to juvenile hormone.

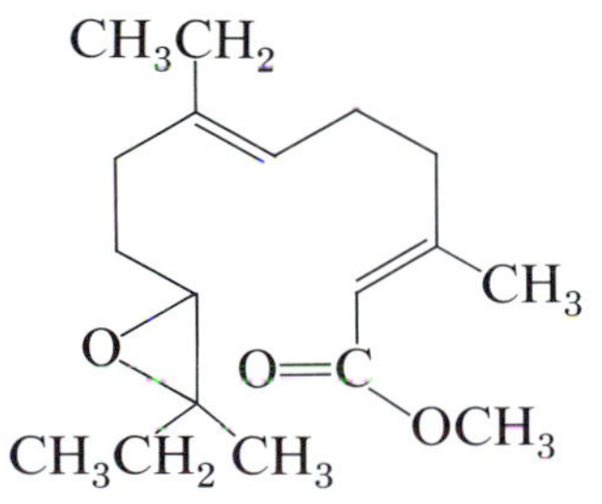

Cecropia juvenile hormone 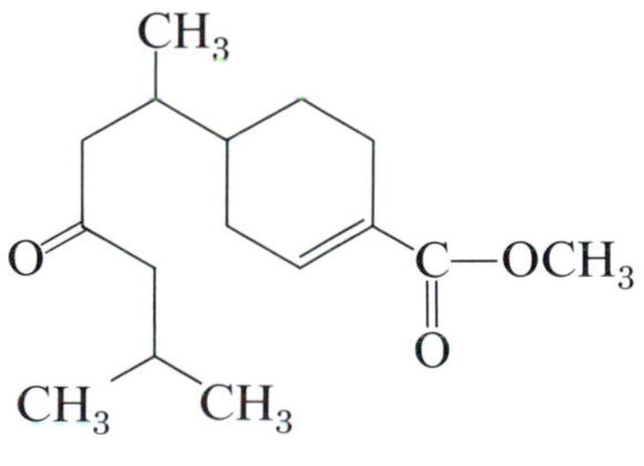   **Paper factor**

While it is very difficult to get enough of the natural substance for use in agriculture, synthetic analogues have been prepared, and they have been shown to be similar in properties and effectiveness to the natural substance. Williams, Sláma, and Bowers have identified and characterized a substance found in the American balsam fir (*Abies balsamea*), known as **paper factor,** which is active against the linden bug, *Pyrrhocoris apterus,* a European cotton pest. This substance is merely one of thousands of terpenoid materials synthesized by the fir tree. Other terpenoid substances are being investigated as potential juvenile hormone analogs.

Certain plants are capable of synthesizing substances that protect them against insects. Included among these natural insecticides are the **pyrethrins** and derivatives of **nicotine.**

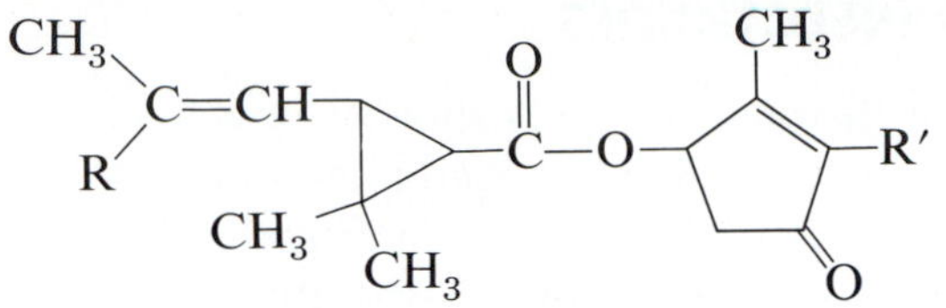

Pyrethrin

R = CH$_3$ or COOCH$_3$
R' = CH$_2$CH=CHCH=CH$_2$ or CH$_2$CH=CHCH$_3$
    or CH$_2$CH=CHCH$_2$CH$_3$

The search for environmentally suitable means of controlling agricultural pests continues with a great sense of urgency. Insects cause billions of dollars of damage to food crops each year. With food becoming increasingly scarce and with the world's population growing at an exponential rate, preventing such losses to food crops becomes absolutely essential.

REFERENCES

Berkoff, C. E. "Insect Hormones and Insect Control." *Journal of Chemical Education, 48* (1971): 577.

Bowers, W. S. and Nishida, R. "Juvocimenes: Potent Juvenile Hormone Mimics from Sweet Basil." *Science, 209* (1980): 1030.

Carson, R. *Silent Spring.* Boston: Houghton Mifflin, 1962.

Keller, E. "The DDT Story." *Chemistry, 43* (February 1970): 8.

O'Brien, R. D. *Insecticides: Action and Metabolism.* New York: Academic Press, 1967.

Peakall, D. B. "Pesticides and the Reproduction of Birds." *Scientific American, 222* (April 1970): 72.

Pryde, L. T. *Environmental Chemistry: An Introduction.* Menlo Park, CA: Cummings, 1973. Chap. 7.

Pyle, J. L. *Chemistry and the Technological Backlash.* Englewood Cliffs, NJ: Prentice-Hall, 1974. Chap. 4.

Rudd, R. L. *Pesticides and the Living Landscape.* Madison: University of Wisconsin Press, 1964.

Saunders, H. J. "New Weapons Against Insects." *Chemical and Engineering News, 53* (July 28, 1975): 18.

Williams, C. M. "Third-Generation Pesticides." *Scientific American, 217* (July, 1967): 13.

Williams, W. G., Kennedy, G. G., Yamamoto, R. T., Thacker, J. D., and Bordner, J. "2-Tridecanone: A Naturally Occurring Insecticide from the Wild Tomato." *Science, 207* (1980): 888.

Experiment 49
The Diels-Alder Reaction of Cyclopentadiene with Maleic Anhydride

Diels-Alder reaction
Fractional distillation

Cyclopentadiene and maleic anhydride react readily in a Diels-Alder reaction to form the adduct, *cis*-norbornene-5,6-*endo*-dicarboxylic anhydride:

Cyclopentadiene   Maleic Anhydride   *cis*-Norbornene-5,6-
endo-dicarboxylic anhydride

Since two molecules of cyclopentadiene can also undergo a Diels-Alder reaction to form dicyclopentadiene, it is not possible to store cyclopentadiene. Therefore, it is necessary to first "crack" dicyclopentadiene to produce cyclopentadiene for use in this experiment. This will be done by heating the dicyclopentadiene to a boil and collecting the cyclopentadiene as it is formed by fractional distillation. The cyclopentadiene must be kept cold and used fairly soon in order to keep it from dimerizing.

Dicyclopentadiene   Cyclopentadiene

REQUIRED READING

Review: Technique 5   Crystallization, Section 5.4

New: Essay   Diels-Alder Reaction and Insecticides

SPECIAL INSTRUCTIONS

The cracking of dicyclopentadiene should be done by the instructor or lab assistant. If a flame is used for this, be sure that there are no leaks in the system since both cyclopentadiene and the dimer are highly flammable. The procedure given below will provide enough cyclopentadiene for about 50 students.

PROCEDURE

PREPARATION OF CYCLOPENTADIENE (Procedure for the instructor)

Working in a hood, assemble a fractional distillation apparatus, as shown in the figure. Although the required temperature control can best be obtained with a micro burner, using a sand bath or heating mantle lessens the possibility of a fire occurring. Place several boiling stones and 15 mL of dicyclopentadiene in the 50-mL distilling flask. Control the heat source so that the cyclopentadiene distills at about 40–43 °C. (If a sand bath is used, the temperature should be 190–200 °C, and it may be necessary to cover the sand bath and distilling flask with aluminum foil.) After 30–45 minutes, 6–7 mL of cyclopentadiene should be collected and the distillation can be stopped. If the cyclopentadiene is cloudy, dry the liquid over granular anhydrous sodium sulfate. Store the product in a sealed container, and keep it cooled in an ice-water bath until all students have taken their portions.

PREPARATION OF THE ADDUCT (Procedure for the student)

To a Craig tube add 0.100 g of maleic anhydride and 0.40 mL of ethyl acetate. Without inserting the plug, shake the tube gently to dissolve the solid (slight heating in a warm sand bath may be necessary). Add 0.40 mL of ligroin (bp 60–90 °C) and shake the tube gently to mix the solvents and reactant thoroughly. Add 0.10 mL of cyclopentadiene and mix thoroughly by shaking until no visible layers of liquid are present. Since this reaction is exothermic, the temperature of the mixture will likely become high enough to keep the product in solution. However, if a solid does form at this point, it will be necessary to heat the mixture gently in a sand bath to dissolve any solids present.

Allow the mixture to cool slowly to room temperature by placing the Craig tube in a 10-mL Erlenmeyer flask that has been filled with about 8 mL of water at 50–60 °C. The inner plug of the Craig tube should be inserted to prevent evaporation of the solvent. Better crystal formation can be achieved by seeding the solution before it cools to room temperature. To seed the solution, dip a spatula or glass stirring rod into the solution after it has cooled for about five minutes. Allow the solvent to evaporate so that a small amount of solid forms on the surface of the spatula or glass rod. Place the spatula or stirring rod back into the solution for a few seconds to induce crystallization. When crystallization is complete at room temperature, cool the mixture in an ice bath for several minutes. Isolate the crystals from the Craig tube by centrifugation (see Technique 4, Section 4.7, p 575, and Figure 4–11, p 576), and allow the crystals to air-dry. Determine the weight and the melting point (164 °C).

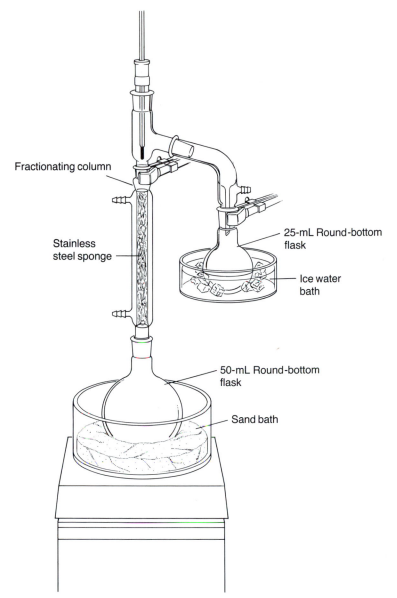

Fractionating column

Stainless
steel sponge

25-mL Round-bottom
flask

Ice water
bath

50-mL Round-bottom
flask

Sand bath

Apparatus for cracking dicyclopentadiene

At the instructor's option, determine the infrared spectrum of the adduct in potassium bromide. Calculate the percentage yield and submit the product to the instructor in a labeled vial.

QUESTIONS

1. Draw a structure for the exo product formed by cyclopentadiene and maleic anhydride.
2. Since the exo form is more stable than the endo form, why is the endo product formed almost exclusively in this reaction?

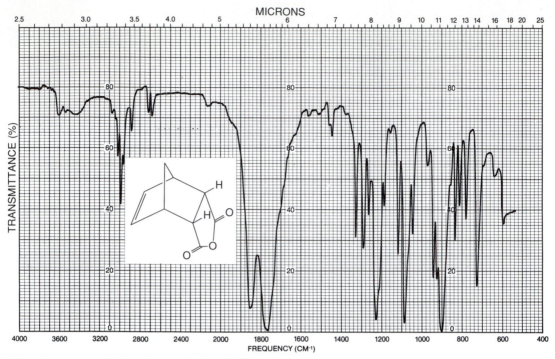

Infrared spectrum of *cis*-norbornene-5,6-*endo*-dicarboxylic anhydride, KBr

3. In addition to the main product, what are two side reactions which could occur in this experiment?
4. The infrared spectrum of the adduct is given in this experiment. Interpret the principal peaks.

Experiment 50
Benzyne Formation and the Diels-Alder Reaction: Preparation of 1,2,3,4-Tetraphenylnaphthalene

Benzyne formation
Diels-Alder reaction

In this experiment, you prepare 1,2,3,4-tetraphenylnaphthalene. In Step 1, benzyne is produced via the unstable diazonium salt. Benzyne is also unstable and cannot be isolated. In Step 2, tetraphenylcyclopentadienone (Experiment 34) "traps" the reactive benzyne as it is formed by a Diels-Alder reaction to give an unstable intermediate. This intermediate readily loses carbon monoxide and yields the fully aromatized naphthalene system. The reaction can be followed easily since the reaction mixture changes from a purple to a yellow-orange solution when the tetraphenylcyclopentadienone is consumed and 1,2,3,4-tetraphenylnaphthalene is produced.

Step 1 BENZYNE FORMATION

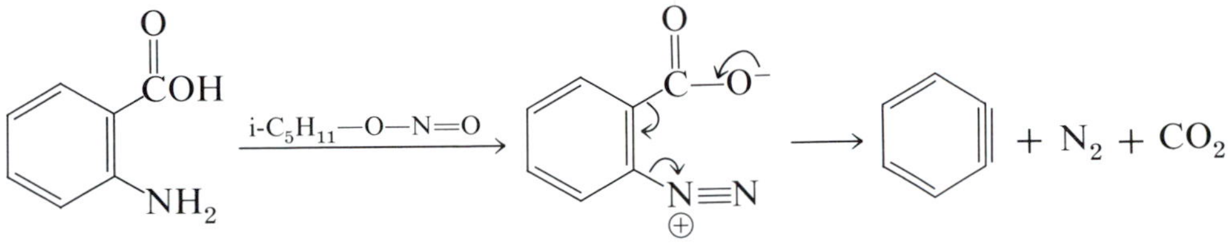

Anthranilic acid   Unstable   Benzyne
                          diazonium
                          salt

Step 2 DIELS-ALDER REACTION

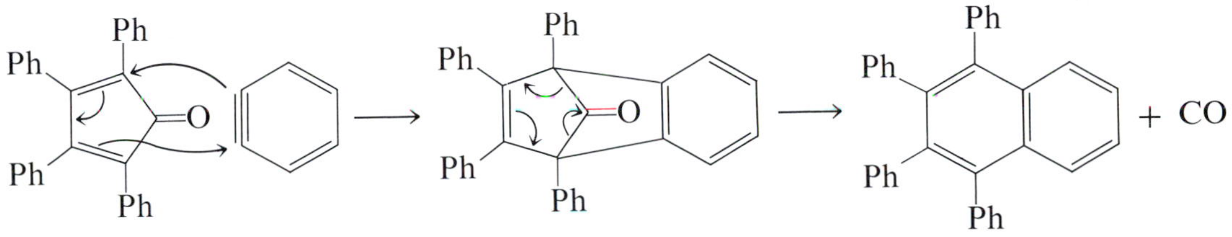

Diels-Alder reaction   Unstable   1,2,3,4-Tetraphenyl-
                              intermediate   naphthalene

REQUIRED READING

Review: Technique 3   Section 3.2
            Technique 4   Section 4.3
            Technique 18   Sections 18.5A and 18.9

New: Essay   Diels-Alder Reaction and Insecticides

SPECIAL INSTRUCTIONS

Special care should be taken to avoid breathing isopentyl nitrite (isoamyl nitrite) as it is a powerful heart stimulant. The isopentyl nitrite must be stored in a refrigerator when not in use. Restopper the bottle after the liquid has been removed to minimize contact with air. A small amount of carbon monoxide is produced in this reaction. Although it may be advisable to run the reaction in a hood, the reaction can be conducted on a bench, if the laboratory has reasonable ventilation.[1]

PROCEDURE

Place 0.100 g of tetraphenylcyclopentadienone (MW = 384.5, Experiment 34), 0.045 g of anthranilic acid (MW = 137.1), and 1.2 mL of 1,2-dimethoxyethane in a 5-mL conical vial. Add a magnetic spin vane and attach a water-jacked condenser. In a hood, transfer 0.06 mL of isopentyl nitrite (isoamyl nitrite, MW = 117.2, d = 0.875 g/mL) to a 3-mL conical vial.

> **CAUTION: Do not breathe the isopentyl nitrite vapor as it is a powerful heart stimulant.**

Cap the vial to prevent loss by evaporation. Replace the lid on the reagent bottle as soon as possible to minimize exposure to air.[2] Dissolve the isopentyl nitrite in 0.50 mL of 1,2-dimethoxyethane.

Heat the mixture containing tetraphenylcyclopentadienone and anthranilic acid with a sand bath at about 155 °C (Figure 3–2A, p 550). When the solution begins to boil, add the solution of isopentyl nitrite through the top of the condenser with a Pasteur pipet over a period of about 30 seconds. (Wear your safety glasses!) Make sure that the pipet is inserted deep into the condenser so that the solution is added directly to the vial. Use a few drops of 1,2-dimethoxyethane to rinse the vial and add this solution to the reaction mixture.

Continue to boil this mixture until the color changes from the deep purple color of the tetraphenylcyclopentadienone to a yellow-orange solution formed after the dienone is consumed (usually less than 10 minutes). If the color has not changed after about 15 minutes, add a drop of pure isopentyl nitrite (no solvent, pipet extended down into the condenser), and continue to boil the solution for an additional 10 minute period. If the color still has not changed to a yellow-orange color after this reflux period, add another drop of isopentyl nitrite and boil the solution for an additional 10 minutes.[3]

After the color changes, cool the mixture to room temperature and use a Pasteur pipet to transfer the solution to a beaker containing 5 mL of water and 2 mL of methanol. Stir the mixture well to break up the precipitate. Collect the solid on a Hirsch funnel

---

[1] The amount of carbon monoxide formed in this reaction is very small (see Question 1). It may be trapped by placing a solution of cuprous chloride and ammonium chloride in aqueous ammonia in a test tube. Gasses which are evolved are led from the condenser to the test tube (Figure 3–7, p 556, omit the moistened glass wool). Immerse the tip of the tubing just below the surface of the liquid so that if the pressure changes, the trapping solution will not be pulled back into the vial. You will need to remove the trap from the top of the condenser when the isopentyl nitrite is added. A large amount of trapping agent is prepared as follows: Dissolve 20 g of cuprous chloride and 25 g of ammonium chloride in 70 mL of water. Add to the solution a third of its volume of concentrated (28%) ammonium hydroxide.

[2] Isopentyl nitrite (isoamyl nitrite) must be stored in a refrigerator. It decomposes in the presence of light and air. This reaction gives the best results if the material has been bought recently. Material from Aldrich Chemical Co. (#15,049–5) works well.

[3] If the color has not changed after the additions of the extra isopentyl nitrite, add a small amount (about 0.010 g) of anthranilic acid. Reflux the mixture for another 15 minutes or until the color changes.

under vacuum (Technique 4, Section 4.3, and Figure 4–6, p 571). In some cases, the filtration process may be slow because the solid plugs the filter paper. If this occurs, add a little ice-cold methanol to the Hirsch funnel while the mixture is being filtered. Use 10 mL of ice-cold methanol to aid the transfer of the solid remaining in the beaker and to wash the solid collected in the Hirsch funnel. Additional product precipitates in the filter flask. Collect this material and add it to the other solid. Weigh the crude product.

Crystallize all of the product from hot isopropyl alcohol in order to remove the remaining colored impurities. Place the crude tetraphenylnaphthalene in a 25-mL Erlenmeyer flask and dissolve it in boiling isopropyl alcohol (2-propanol). For the amount of product which you may expect to obtain, it will require approximately 12 mL of boiling isopropyl alcohol to dissolve your crude product (12 mL of solvent/85 mg of product). The 12 mL of isopropyl alcohol is a rough estimate and you may use more or less than this amount. The solution process is enhanced by breaking up any lumps with a spatula. Once your crude product has dissolved in the boiling solvent, cool the mixture in an ice bath. When it has cooled somewhat, scratch the side of the flask with a stirring rod. The product slowly crystallizes. Allow the mixture to cool for at least 30 minutes in an ice bath. Collect the product on a Hirsch funnel under vacuum and wash it with a small amount of ice-cold isopropyl alcohol. Allow the solid to dry fully, weigh the product, and calculate the percentage yield.

When the solid has dried completely, determine the melting point. Pure 1,2,3,4-tetraphenylnaphthalene melts at 196–199 °C. When the material has melted, remove the capillary tube from the melting-point apparatus and cool the tube until the material solidifies. Redetermine the melting point (literature 203–205 °C).[4] The tetraphenylnaphthalene exists in two crystalline forms, each with a different melting point. At the option of the instructor, obtain the infrared spectrum in CCl_4 (Technique 18, Section 18.5, Method A, p 777) or the proton NMR spectrum in CCl_4 or $CDCl_3$ (Technique 18, Section 18.9, p 782). Submit the sample to the instructor in a labeled vial.

REFERENCES

Dougherty, C. M., Baumgarten, R. L., Sweeney, A., and Concepcion, E. "Phthalimide, Anthranilic Acid and Benzyne." *Journal of Chemical Education, 54* (1977): 643.
Fieser, L. F., and Haddadin, M. J. "Isobenzofurane, A Transient Intermediate." *Canadian Journal of Chemistry, 43* (1965): 1599.

QUESTIONS

1. Calculate the number of moles and milliliters of carbon monoxide gas theoretically produced in this reaction.

---

[4]This compound exhibits a double melting point. The initial melting point varies according to its particle size and is not a reliable index of purity. The remelt melting point is more reproducible and reliable.

(a) [structure: furan] + [structure: benzene] $\longrightarrow$

(c) [structure: tetraphenylcyclopentadienone] + [structure: maleic anhydride] $\longrightarrow$

CO is also produced.

(b) [structure: tetraphenylcyclopentadienone] + Ph—C≡C—Ph $\longrightarrow$

CO is also produced.

(d) [structure: anthracene, positions 9 and 10 labeled] + [structure: benzene] $\longrightarrow$

Benzyne adds to the 9, 10 position on anthracene.

2. Draw the structures of the products that would result from the following reactions.

3. Interpret the principal absorption bands in the infrared spectrum of 1,2,3,4-tetraphenylnaphthalene.

4. Interpret the NMR spectrum of 1,2,3,4-tetraphenylnaphthalene. In interpreting the NMR spectrum, notice that the molecule is symmetrical and that each of the singlets integrates for 10 hydrogens. The multiplet at 7.2 to 7.8 ppm represents four hydrogens.

5. Draw a mechanism for the formation of the diazonium salt from anthranilic acid and isopentyl nitrite.

6. What is the ultimate fate in the reaction of the isopentyl group from the isopentyl nitrite? That is, what compound or compounds are formed?

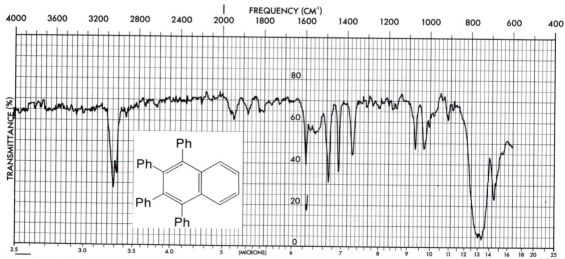

Infrared spectrum of 1,2,3,4-tetraphenylnaphthalene, CCl_4

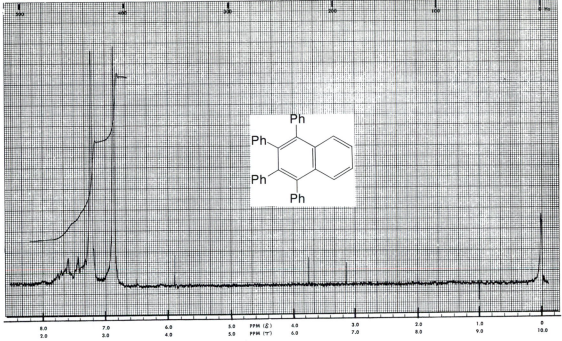

NMR spectrum of 1,2,3,4-tetraphenylnaphthalene, CDCl$_3$

Experiment 51
Photoreduction of Benzophenone

Photochemistry
Photoreduction
Energy transfer

The photoreduction of benzophenone is one of the oldest and most thoroughly studied photochemical reactions. Early in the history of photochemistry, it was discovered that solutions of benzophenone are unstable to light when certain solvents are used. If benzophenone is dissolved in a ''hydrogen-donor'' solvent, such as 2-propanol, and exposed to ultraviolet light, an insoluble dimeric product, benzpinacol, will form.

Benzophenone   **2-Propanol**   **Benzpinacol**

To understand this reaction, one has to review some simple photochemistry as it relates to aromatic ketones. In the typical organic molecule, all the electrons are paired in the occupied orbitals. When such a molecule absorbs ultraviolet light of the appropriate wavelength, an electron from one of the occupied orbitals, usually the one of highest energy, is excited to an unoccupied molecular orbital, usually to the one of lowest energy. During this transition, the electron must retain its spin value, because a change of spin is a quantum-mechanically forbidden process during an electronic transition. Therefore, just as the two electrons in the highest occupied orbital of the molecule originally had their spins paired (opposite), so they will retain paired spins in the first electronically excited state of the molecule. This is true even though the two electrons will be in **different** orbitals after the transition. This first excited state of a molecule is called a **singlet state** (S_1), since its spin multiplicity ($2S + 1$) is one. The original unexcited state of the molecule is also a singlet state since its electrons are paired, and it is called the **ground-state** singlet state (S_0) of the molecule.

The excited state singlet, S_1, may return to the ground state, S_0, by reemission of the absorbed photon of energy. This process is called **fluorescence.** Alternatively, the excited electron may undergo a change of spin to give a state of higher multiplicity, the excited **triplet state,** so called because its spin multiplicity ($2S + 1$) is three. The conversion from the first excited singlet state to the triplet state is called **intersystem crossing.** Because the triplet state has a higher multiplicity, it will inevitably be a lower energy state than the excited singlet state (Hund's rule). Normally this change of spin (intersystem crossing) is a quantum-mechanically forbidden process, just as a direct excitation of the ground state (S_0) to the triplet state (T_1) is forbidden. However, in those molecules in which the singlet and triplet states lie close to one another in energy, the two states inevitably have several overlapping vibrational states, that is, states in common, a situation that allows the "forbidden" transition. In many molecules in which S_1 and T_1 have similar energy ($\Delta E < 10$ kcal/mole), intersystem crossing occurs faster than fluorescence, and the molecule is rapidly converted from its excited singlet state to its triplet state. In benzophenone, S_1 undergoes intersystem crossing to T_1 with a rate of $k_{isc} = 10^{10}$ sec^{-1}, meaning that the lifetime of S_1 is only 10^{-10} second. The rate of fluorescence for benzophenone is $k_f = 10^6$ sec^{-1}, meaning that intersystem crossing occurs at a rate that is 10^4 times faster than fluorescence. Thus, the conversion of S_1 to T_1 in benzophenone is essentially a quantitative process. In

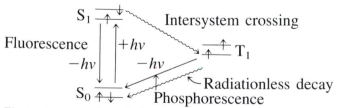

Electronic states of a typical molecule and the possible interconversions. In each state (S_0, S_1, T_1), the lower line represents the highest occupied orbital and the upper line represents the lowest unoccupied orbital of the unexcited molecule. Straight lines represent processes in which a photon is absorbed or emitted. Wavy lines represent radiationless processes—those that occur *without* emission or absorption of a photon.

molecules that have a wide energy gap between S_1 and T_1, this situation would be reversed. As you will see shortly, the naphthalene molecule presents a reversed situation.

Since the excited triplet state is lower in energy than the excited singlet state, the molecule cannot easily return to the excited singlet state. Nor can it easily return to the ground state by returning the excited electron to its original orbital. Once again, the transition $T_1 \longrightarrow S_0$ would require a change of spin for the electron, and this is a forbidden process. Hence, the triplet excited state usually has a long lifetime (relative to other excited states) since it generally has nowhere that it can easily go. Even though the process is forbidden, the triplet, T_1, may eventually return to the ground state, S_0, by a process called a **radiationless transition.** In this process, the excess energy of the triplet is lost to the surrounding solution as heat, thereby "relaxing" the triplet back to the ground state, S_0. This process is the study of much current research and is not well understood. In the second process, in which a triplet state may revert to the ground state, **phosphorescence,** the excited triplet emits a photon to dissipate the excess energy and returns directly to the ground state. Although this process is "forbidden," it nevertheless occurs when there is no other open pathway by which the molecule can dissipate its excess energy. In benzophenone, radiationless decay is the faster process, with rate $k_d = 10^5 \text{ sec}^{-1}$, and phosphorescence, which is not observed, has a lower rate of $k_p = 10^2 \text{ sec}^{-1}$.

Benzophenone is a ketone. Ketones have **two** possible excited singlet states, and, consequently, two excited triplet states as well. This occurs since two relatively low energy transitions are possible in benzophenone. It is possible to excite one of the π electrons in the carbonyl π bond to the lowest-energy unoccupied orbital, a π^* orbital. It is also possible to excite one of the nonbonded or n electrons on oxygen to the same orbital. The first type of transition is called a $\pi-\pi^*$ transition, while the second is called an $n-\pi^*$ transition. These transitions and the states that result are illustrated pictorially.

Spectroscopic studies show that for benzophenone and most other ketones, the $n-\pi^*$ excited states S_1 and T_1 are of lower energy than the $\pi-\pi^*$ excited states. An energy diagram depicting the excited states of benzophenone (along with one that depicts those of naphthalene) is shown.

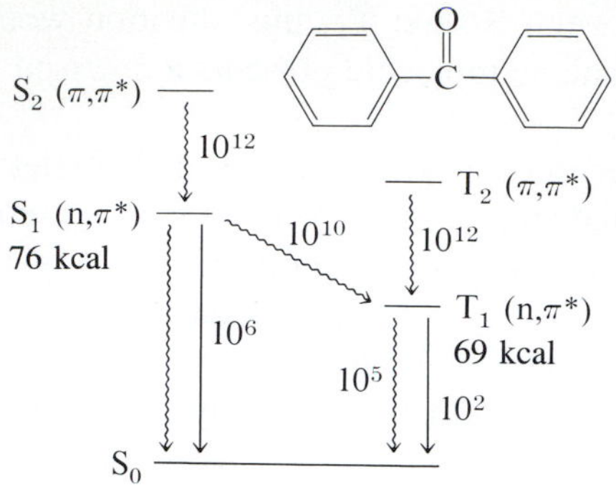

Excited states of benzophenone

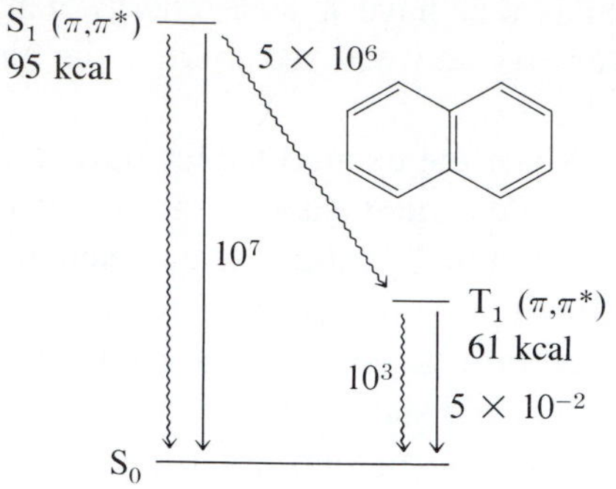

$S_2 (\pi,\pi^*)$ 10^{12}

$S_1 (n,\pi^*)$ 76 kcal 10^{10} $T_2 (\pi,\pi^*)$ 10^{12}

10^6 10^5 $T_1 (n,\pi^*)$ 69 kcal 10^2

S_0

Excited states of naphthalene

$S_1 (\pi,\pi^*)$ 95 kcal 5×10^6

10^7 $T_1 (\pi,\pi^*)$ 61 kcal

10^3 5×10^{-2}

S_0

It is now known that the photoreduction of benzophenone is a reaction of the $n-\pi^*$ triplet state (T_1) of benzophenone. The $n-\pi^*$ excited states have radical character at the carbonyl oxygen atom because of the unpaired electron in the nonbonding orbital. Thus, the radical-like and energetic T_1 excited state species can abstract a hydrogen atom from a suitable donor molecule to form the diphenylhydroxymethyl radical. Two of these radicals, once formed, may couple to form benzpinacol. The complete mechanism for photoreduction is outlined below.

$$Ph_2C{=}O \xrightarrow{hv} P\dot{H_2}C{-}\dot{O} \ (S_1)$$

$$P\dot{H_2}C{-}\dot{O} \ (S_1) \xrightarrow{isc} Ph_2\dot{C}{-}\dot{O} \ (T_1)$$

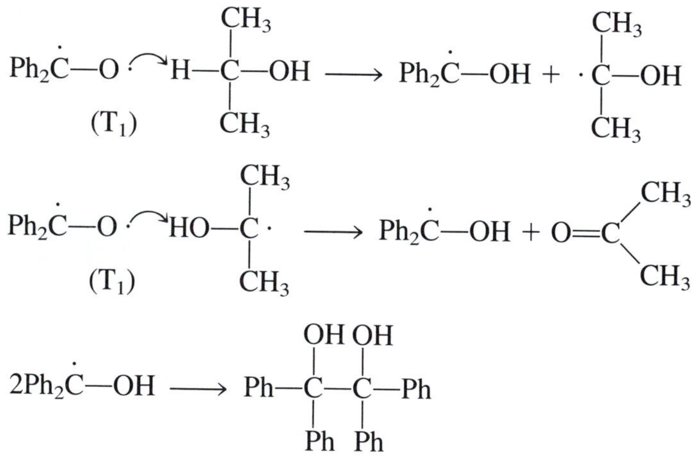

$$Ph_2\dot{C}{-}\dot{O} \ \overset{\frown}{\cdot} H{-}\underset{CH_3}{\overset{CH_3}{\underset{|}{\overset{|}{C}}}}{-}OH \longrightarrow Ph_2\dot{C}{-}OH + \cdot\underset{CH_3}{\overset{CH_3}{\underset{|}{\overset{|}{C}}}}{-}OH$$

$$(T_1)$$

$$Ph_2\dot{C}{-}\dot{O} \ \overset{\frown}{\cdot} HO{-}\underset{CH_3}{\overset{CH_3}{\underset{|}{\overset{|}{C}}}}{\cdot} \longrightarrow Ph_2\dot{C}{-}OH + O{=}C\underset{CH_3}{\overset{CH_3}{\big\langle}}$$

$$(T_1)$$

$$2Ph_2\dot{C}{-}OH \longrightarrow Ph{-}\underset{Ph}{\overset{OH}{\underset{|}{\overset{|}{C}}}}{-}\underset{Ph}{\overset{OH}{\underset{|}{\overset{|}{C}}}}{-}Ph$$

Many photochemical reactions must be carried out in a quartz apparatus because they require ultraviolet radiation of shorter wavelengths (higher energy) than the wavelengths that can pass through Pyrex. Benzophenone, however, requires radiation of approximately 350 nm (350 mμ, or 3500 Å) to become excited to its $n-\pi^*$ singlet state, S_1, a wavelength that readily passes through Pyrex. In the figure on page 417, the ultraviolet absorption spectra of benzophenone and naphthalene are given. Superimposed on their spectra are two curves, which show the wavelengths that can be transmitted by Pyrex and quartz, respectively. Pyrex will not allow any radiation of wavelength shorter than approximately 300 nm to pass, whereas quartz will allow

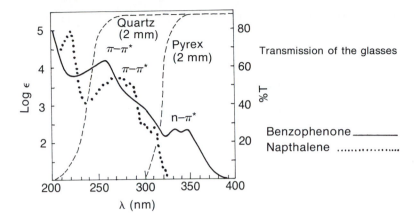

wavelengths as short as 200 nm to pass. Thus, when benzophenone is placed in a Pyrex flask, the only electronic transition possible is the $n-\pi^*$ transition, which occurs at 350 nm.

However, even if it were possible to supply benzophenone with radiation of the appropriate wavelength to produce the second excited singlet state of the molecule, this singlet would rapidly convert to the lowest singlet state, S_1. The state S_2 has a lifetime of less than 10^{-12} second. The conversion process $S_2 \longrightarrow S_1$ is called an **internal conversion.** Internal conversions are processes of conversion between excited states of the same multiplicity (singlet–singlet, or triplet–triplet), and they usually are very rapid. Thus, when a S_2 or T_2 is formed, it readily converts to S_1 or T_1, respectively. As a consequence of their very short lifetimes, very little is known about the properties or the exact energies of S_2 and T_2 of benzophenone.

ENERGY TRANSFER

Using a simple **energy-transfer** experiment, one can show that the photoreduction of benzophenone proceeds via the T_1 excited state of benzophenone, rather than the S_1 excited state. If naphthalene is added to the reaction, the photoreduction is stopped because the excitation energy of the benzophenone triplet is transferred to naphthalene. The naphthalene is said to have **quenched** the reaction. This occurs in the following way.

When the excited states of molecules have long enough lifetimes, they often can transfer their excitation energy to another molecule. The mechanisms of these transfers are complex and cannot be explained here; however, the essential requirements can be outlined. First, for two molecules to exchange their respective states of excitation, the process must occur with an overall decrease in energy. Second, the spin multiplicity of the total system must not change. These two features can be illustrated by the two most common examples of energy transfer—singlet transfer and triplet transfer. In these two examples, the superscript 1 denotes an excited singlet state, the superscript 3 denotes a triplet state, and the subscript 0 denotes a ground state molecule. The designations A and B represent different molecules.

$$A^1 + B_0 \longrightarrow B^1 + A_0 \quad \text{Singlet energy transfer}$$

$$A^3 + B_0 \longrightarrow B^3 + A_0 \quad \text{Triplet energy transfer}$$

In singlet energy transfer, excitation energy is transferred from the excited singlet state of A to a ground-state molecule of B, converting it to its excited singlet state and returning A to its ground state. In triplet energy transfer, there is a similar interconversion of excited state and ground state. Singlet energy is transferred through space by a dipole-dipole coupling mechanism, but triplet energy transfer requires the two molecules involved in the transfer to collide. In the usual organic medium, about 10^9 collisions occur per second. Thus, if a triplet state A^3 has a lifetime longer than 10^{-9} second, and if an acceptor molecule, B_0, which has a lower triplet energy than that of A^3 is available, energy transfer can be expected. If the triplet A^3 undergoes a reaction (like photoreduction) at a rate lower than the rate of collisions in the solution, and if an acceptor molecule is added to the solution, the reaction can be **quenched.** The acceptor molecule, which is called a **quencher,** deactivates, or ''quenches,'' the triplet before it has a chance to react. Naphthalene has the ability to quench benzophenone triplets in this way and to stop the photoreduction.

Naphthalene cannot quench the excited-state singlet S_1 of benzophenone since its own singlet has an energy (95 kcal/mol) that is higher than the energy of benzophenone (76 kcal/mol). In addition, the conversion $S_1 \longrightarrow T_1$ is very rapid (10^{-10} second) in benzophenone. Thus, naphthalene can intercept only the triplet state of benzophenone. The triplet excitation energy of benzophenone (69 kcal/mol) is transferred to naphthalene ($T_1 = 61$ kcal/mol) in an exothermic collision. Finally, the naphthalene molecule does not absorb light of the wavelengths transmitted by Pyrex (see spectra on page 417), and therefore benzophenone is not inhibited from absorbing energy when naphthalene is present in solution. Thus, since naphthalene quenches the photoreduction reaction of benzophenone, we can infer that this reaction proceeds via the triplet state, T_1, of benzophenone. If naphthalene did not quench the reaction, the singlet state of benzophenone would be indicated as the reactive intermediate. In the experiment that follows, the photoreduction of benzophenone is attempted both in the presence and in the absence of added naphthalene.

REQUIRED READING

Review: Technique 4   Filtration, Section 4.3

SPECIAL INSTRUCTIONS

This experiment may be performed concurrently with some other experiment. It requires only 15 minutes during the first laboratory period, and only about 15 minutes in a subsequent laboratory period about one week later (or at the end of the laboratory period if you use a sun lamp.)

Using Direct Sunlight. It is important that the reaction mixture be left where it will receive direct sunlight. If it does not, the reaction will be slow and may need more than a week for completion. It is also important that the room temperature not be too low, or the benzophenone will precipitate. If you are performing this experiment in winter and the laboratory is not heated at night, you will need to shake the solutions every morning to redissolve the benzophenone. Benzpinacol should not redissolve easily.

Using a Sun Lamp. If you wish, a 275 W sun lamp may be used instead of direct sun light. Place the lamp in a hood which has had its window covered with aluminum foil (shiny side in). The lamp (or lamps) should be mounted in a ceramic socket attached to a ring stand with a three-pronged clamp.

> **The purpose of the aluminum foil is to protect the eyes of persons in the laboratory. You should not directly view a sun lamp or damage to the eyes may result. Take all possible viewing precautions.**

Samples should be attached to a ring stand placed at least 18 inches from the sun lamp. Placing them at this distance will avoid their being heated by the lamp. Heating may cause loss of the solvent. It is a good idea to agitate the samples every 30 minutes. With a sun lamp, the reaction will be complete in three to four hours.

PROCEDURE

Label two 13 × 100-mm test tubes near the top of the tubes. The labels should have your name and "No. 1" and "No. 2" written on them. Place 0.50 g of benzophenone in the first tube. Place 0.50 g of benzophenone and 0.05 g of naphthalene in the second tube. Add about 2 mL of 2-propanol (isopropyl alcohol) to each tube and warm them in a sand bath or in a beaker of warm water to dissolve the solids. When the solids have dissolved, add one small drop (Pasteur pipet) of glacial acetic acid to each tube and then fill each tube nearly to the top with more 2-propanol. Stopper the tubes tightly with rubber stoppers, shake them well, and place them in a beaker on a window sill where they will receive direct sunlight.

> **You may be directed by your instructor to use a sun lamp instead of direct sunlight (see Special Instructions).**

The reaction will require about one week for completion (three hours with a sun lamp). If the reaction has occurred during this period, the product will have crystallized from the solution. Observe the result in each test tube. Collect the product by vacuum filtration using a small Büchner or a Hirsch funnel (Technique 4, Section 4.3, p 571) and allow it to dry. Weigh the product and determine its melting point and percentage yield.

At your instructor's option, determine the infrared spectrum of the benzpinacol as a KBr mull (Technique 18, Section 18.4, p 775). Submit the product to your instructor along with the report.

REFERENCE

Vogler, A., and Kunkely, H. "Photochemistry and Beer." *Journal of Chemical Education, 59* (January 1982): 25.

QUESTIONS

1. Can you think of a way to produce the benzophenone $n-\pi^*$ triplet T_1 **without** having benzophenone pass through its first singlet state? Explain.

2. A reaction similar to the one here described occurs when benzophenone is treated with the metal magnesium (pinacol reduction).

$$2Ph_2C{=}O \xrightarrow{\text{Mg}} Ph_2\overset{\displaystyle OH}{\underset{}{C}}{-}\overset{\displaystyle OH}{\underset{}{C}}Ph_2$$

Compare the mechanism of this reaction with the photoreduction mechanism. What are the differences?

3. Which of the following molecules do you expect would be useful in quenching benzophenone photoreduction? Explain.

Oxygen	$(S_1 = 22\ \text{kcal/mol})$	Biphenyl	$(T_1 = 66\ \text{kcal/mol})$
9,10-Diphenylanthracene	$(T_1 = 42\ \text{kcal/mol})$	Toluene	$(T_1 = 83\ \text{kcal/mol})$
trans-1,3-Pentadiene	$(T_1 = 59\ \text{kcal/mol})$	Benzene	$(T_1 = 84\ \text{kcal/mol})$
Naphthalene	$(T_1 = 61\ \text{kcal/mol})$		

Essay
FIREFLIES AND PHOTOCHEMISTRY

The production of light as a result of a chemical reaction is called **chemiluminescence.** A chemiluminescent reaction generally produces one of the product molecules in an electronically excited state. The excited state emits a photon, and light is produced. If a reaction that produces light is biochemical, occurring in a living organism, the phenomenon is called **bioluminescence.**

The light produced by fireflies and other bioluminescent organisms has fascinated observers for many years. Many different organisms have developed the ability to emit light. They include bacteria, fungi, protozoans, hydras, marine worms, sponges, corals, jellyfishes, crustaceans, clams, snails, squids, fishes, and insects. Curiously, among the higher forms of life, only fish are included on the list. Amphibians, reptiles, birds, mammals, and the higher plants are excluded. Among the marine species, none is a freshwater organism. The excellent *Scientific American* article by McElroy and Seliger (see references) delineates the natural history, characteristics, and habits of many bioluminescent organisms.

The first significant studies of a bioluminescent organism were by the French physiologist Raphael Dubois in 1887. He studied the mollusk *Pholas dactylis,* a bioluminescent clam indigenous to the Mediterranean Sea. Dubois found that a cold-water extract of the clam was able to emit light for several minutes following the extraction. When the light emission ceased, it could be restored, he found, by a material extracted from the clam by hot water. A hot-water extract of the clam alone did not produce the luminescence. Reasoning carefully, Dubois concluded that there was an enzyme in the cold-water extract that was destroyed in hot water. The luminescent compound, however, could be extracted without destruction in either hot or cold water. He called the luminescent material **luciferin,** and the enzyme that induced it to emit light, **luciferase;** both names were derived from *Lucifer,* a Latin name meaning ''bearer of light.'' Today the luminescent materials from all organisms are called luciferins, and the associated enzymes are called luciferases.

The most extensively studied bioluminescent organism is the firefly. Fireflies are found in many parts of the world and probably thus represent the most familiar example of bioluminescence. In such areas, on a typical summer evening, fireflies, or ''lightning bugs,'' can frequently be seen to emit flashes of light as they cavort over the lawn or in the garden. It is now universally accepted that the luminescence of fireflies is a mating device. The male firefly flies about 2 ft. above the ground and emits flashes of light at regular intervals. The female, who remains stationary on the ground, waits a characteristic interval and then flashes a response. In return, the male will reorient his direction of flight, toward her, and flash a signal once again. The entire cycle is rarely repeated more than 5 to 10 times before the male reaches the female. Fireflies of different species can recognize one another by their flash patterns, which vary in number, rate, and duration among species.

Although the total structure of the luciferase enzyme of the American firefly, *Photinus pyralis,* is unknown, the structure of the luciferin has been established. In spite of a large amount of experimental work, however, the complete nature of the chemical reactions that produce the light is still subject to some controversy. It is possible, nevertheless, to outline the most salient details of the reaction.

Besides the luciferin and the luciferase, other substances—magnesium(II), ATP (adenosine triphosphate), and molecular oxygen—are needed to produce the luminescence. In the postulated first step of the reaction, the luciferase complexes with an ATP molecule. In the second step, the luciferin binds to the luciferase and reacts with the already bound ATP molecule to become ''primed.'' In this reaction, pyrophosphate ion is expelled, and AMP (adenosine monophosphate) becomes attached to the

$$\text{Luciferase} + \text{ATP} \longrightarrow \text{Luciferase-ATP}$$

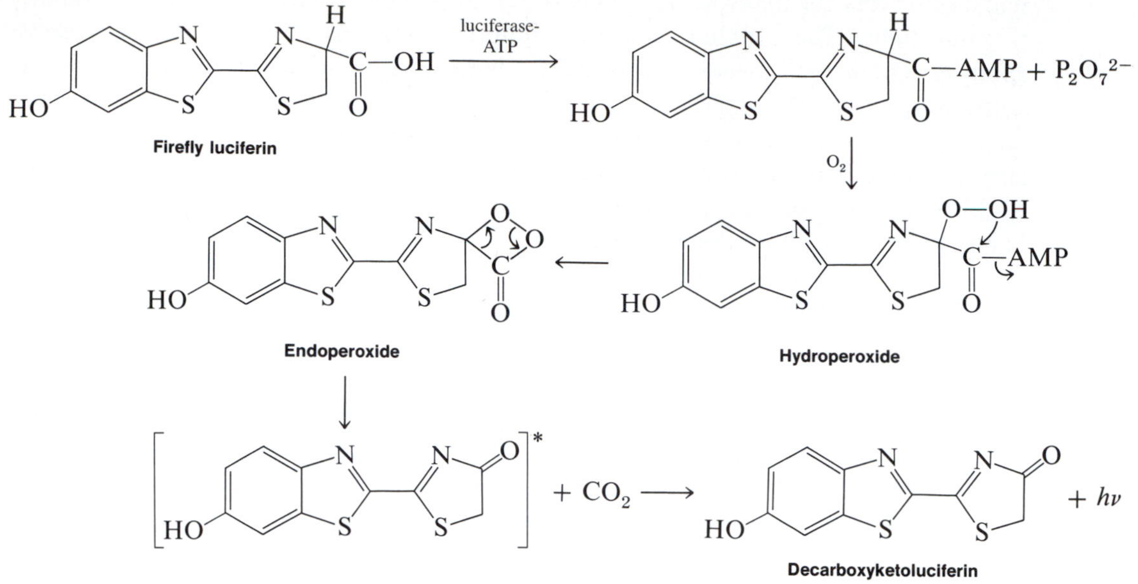

Firefly luciferin

Endoperoxide

Hydroperoxide

$$+ \ CO_2 \longrightarrow$$

$$+ \ h\nu$$

Decarboxyketoluciferin

carboxyl group of the luciferin. In the third step, the luciferin-AMP complex is oxidized by molecular oxygen to form a hydroperoxide; this cyclizes with the carboxyl group, expelling AMP and forming the cyclic endoperoxide. This reaction would be difficult if the carboxyl group of the luciferin had not been primed with ATP. The endoperoxide is unstable and readily decarboxylates, producing decarboxyketoluciferin in an **electronically excited state,** which is deactivated by the emission of a photon (fluorescence). Thus, it is the cleavage of the four-membered-ring endoperoxide that leads to the electronically excited molecule and hence the bioluminescence.

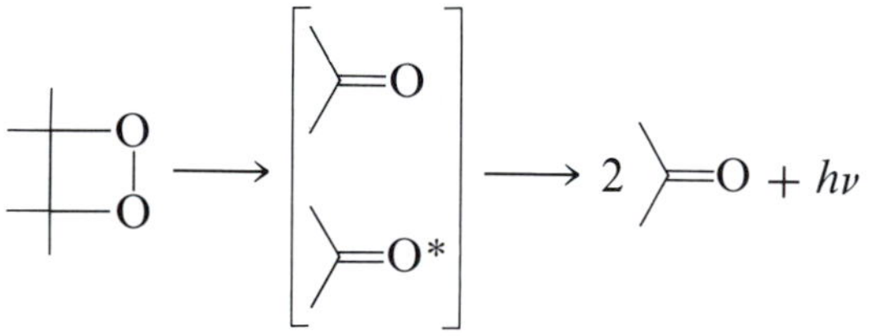

That one of the two carbonyl groups, either that of the decarboxyketoluciferin or that of the carbon dioxide, should be formed in an excited state can be readily predicted from the orbital symmetry conservation principles of Woodward and Hoffmann. This reaction is formally like the decomposition of a cyclobutane ring to yield two ethylene molecules. In analyzing the forward course of that reaction, that is, 2 ethylene $\longrightarrow$ cyclobutane, one can easily show that the reaction, which involves four π electrons, is forbidden for two ground-state ethylenes but allowed for only one ethylene in the ground state and the other in an excited state. This suggests that in the reverse process, one of the ethylene molecules should be formed in an excited state.

Extending these arguments to the endoperoxide also suggests that one of the two carbonyl groups should be formed in its excited state.

The emitting molecule, decarboxyketoluciferin, has been isolated and synthesized. When it is excited photochemically by photon absorption in basic solution (pH $> 7.5-8.0$), it fluoresces, giving a fluorescence emission spectrum that is identical to the emission spectrum produced by the interaction of firefly luciferin and firefly luciferase. The emitting form of decarboxyketoluciferin has thus been identified as the **enol dianion.** In neutral or acidic solution, the emission spectrum of decarboxyketoluciferin does not match the emission spectrum of the bioluminescent system.

The exact function of the enzyme firefly luciferase is not yet known, but it is clear that all these reactions occur while luciferin is bound to the enzyme as a substrate. Also, since the enzyme undoubtedly has several basic groups ($-COO^-$, $-NH_2$, and so on), the buffering action of those groups would easily explain why the enol dianion is also the emitting form of decarboxyketoluciferin in the biological system.

$$\text{Decarboxyketoluciferin} \xrightarrow{-2H^+}$$

Enol dianion

Most chemiluminescent and bioluminescent reactions require oxygen. Likewise, most produce an electronically excited emitting species through the decomposition of a **peroxide** of one sort or another. In the experiment that follows, a **chemiluminescent** reaction that involves the decomposition of a peroxide intermediate is described.

REFERENCES

Clayton, R. K. *Light and Living Matter*. Vol. 2: *The Biological Part*. New York: McGraw-Hill, 1971. Chap. 6, "The Luminescence of Fireflies and Other Living Things."

Fox, J. L. "Theory May Explain Firefly Luminescence." *Chemical and Engineering News* (March 6, 1978): 17.

Harvey, E. N. *Bioluminescence*. New York: Academic Press, 1952.

Hastings, J. W. "Bioluminescence." *Annual Review of Biochemistry*, *37* (1968): 597.

McCapra, F. "Chemical Mechanisms in Bioluminescence." *Accounts of Chemical Research*, *9* (1976): 201.

McElroy, W. D., and Seliger, H. H. "Biological Luminescence." *Scientific American*, *207* (December 1962): 76.

McElroy, W. D., Seliger, H. H., and White, E. H. "Mechanism of Bioluminescence, Chemilumines-cence and Enzyme Function in the Oxidation of Firefly Luciferin." *Photochemistry and Photobiology, 10* (1969): 153.

Seliger, H. H., and McElroy, W. D. *Light: Physical and Biological Action*. New York: Academic Press, 1965.

Experiment 52
Luminol

Chemiluminescence
Energy transfer
Reduction of a nitro group
Amide formation

In this experiment, the chemiluminescent compound **luminol,** or **5-amino-phthalhydrazide,** will be synthesized from 3-nitrophthalic acid.

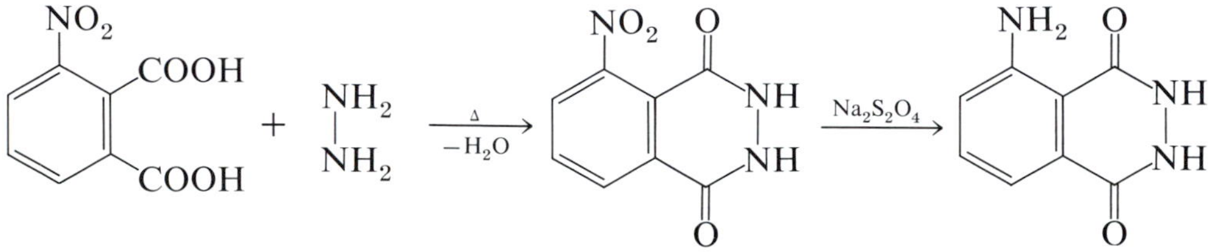

3-Nitrophthalic acid   **Hydrazine**   **5-Nitrophthalhydrazide**   **Luminol**

The first step of the synthesis is the simple formation of a cyclic diamide, 5-nitrophthalhydrazide, by reaction of 3-nitrophthalic acid with hydrazine. Reduction of the nitro group with sodium dithionite affords luminol.

In neutral solution, luminol exists largely as a dipolar anion (zwitterion). This dipolar ion exhibits a weak blue fluorescence after being exposed to light. However, in alkaline solution, luminol is converted to its dianion, which may be oxidized by molecular oxygen to give an intermediate that is chemiluminescent. The reaction is thought to have the following sequence:

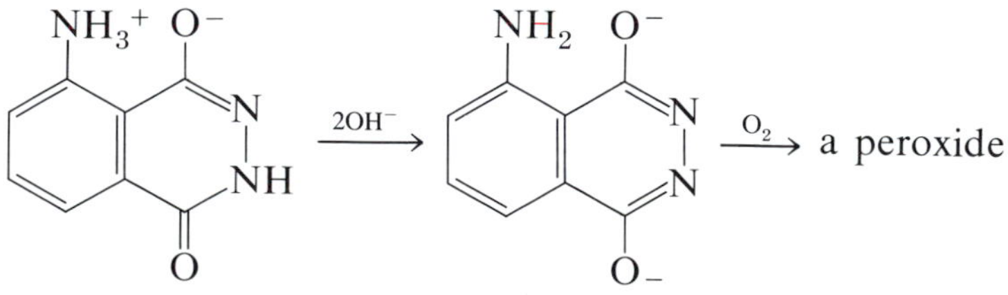

Luminol   **Dianion**

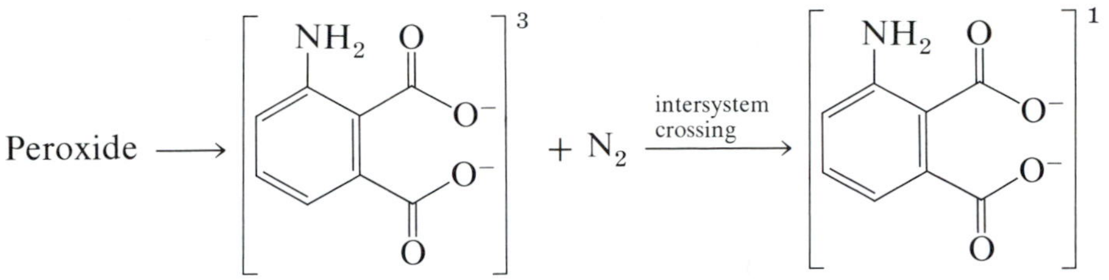

3-Aminophthalate
triplet dianion (T$_1$)

3-Aminophthalate
singlet dianion (S$_1$)

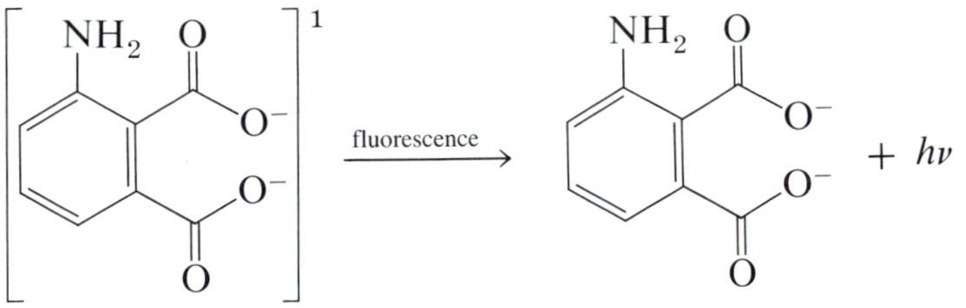

S$_1$

3-Aminophthalate
ground-state dianion, S$_0$

The dianion of luminol undergoes a reaction with molecular oxygen to form a peroxide of unknown structure. This peroxide is unstable and decomposes with the evolution of nitrogen gas, producing the 3-aminophthalate dianion in an electronically excited state. The excited dianion emits a photon that is visible as light. One very attractive hypothesis for the structure of the peroxide postulates a cyclic endoperoxide that decomposes by the following mechanism:

A postulate

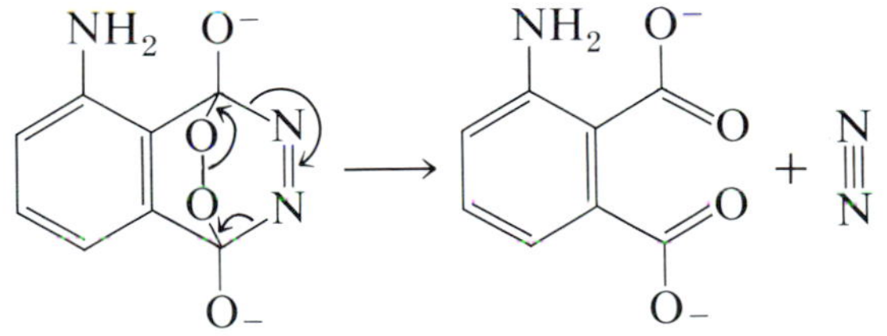

Certain experimental facts argue against this intermediate, however. For instance, certain acyclic hydrazides that cannot form a similar intermediate have also been found to be chemiluminescent.

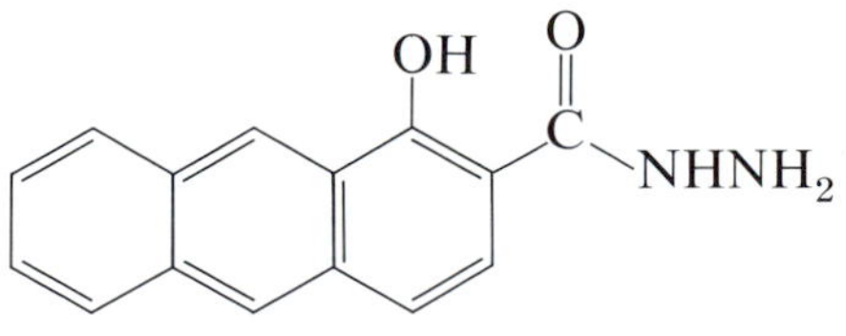

1-Hydroxy-2-anthroic acid
hydrazide (chemiluminescent)

Although the nature of the peroxide is still debatable, the remainder of the reaction is well understood. The chemical products of the reaction have been shown to be the 3-aminophthalate dianion and molecular nitrogen. The intermediate that emits light has been identified definitely as the **excited state singlet** of the 3-aminophthalate dianion.[1] Thus, the fluorescence emission spectrum of the 3-aminophthalate dianion (produced by photon absorption) is identical to the spectrum of the light emitted from the chemiluminescent reaction. However, for numerous complicated reasons, it is believed that the 3-aminophthalate dianion is formed first as a vibrationally excited triplet state molecule, which makes the intersystem crossing to the singlet state before emission of a photon.

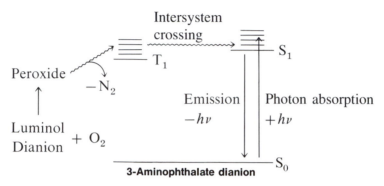

The excited state of the 3-aminophthalate dianion may be quenched by suitable acceptor molecules, or the energy (about 50–80 kcal/mol) may be transferred to give emission from the acceptor molecules. Several such experiments are described in the following procedure.

The system chosen for the chemiluminescence studies of luminol in this experiment uses dimethylsulfoxide, $(CH_3)_2SO$, as the solvent, potassium hydroxide as the base required for the formation of the dianion of luminol, and molecular oxygen. Several alternative systems have been used, substituting hydrogen peroxide and an oxidizing agent for molecular oxygen. An aqueous system using potassium ferricyanide and hydrogen peroxide is an alternative system used frequently.

REFERENCES

Rahaut, M. M. ''Chemiluminescence from Concerted Peroxide Decomposition Reactions.'' *Accounts of Chemical Research, 2* (1969): 80.

White, E. H., and Roswell, D. F. ''The Chemiluminescence of Organic Hydrazides.'' *Accounts of Chemical Research, 3* (1970): 54.

REQUIRED READING

Review: Technique 3   Section 3.9

New: Essay   Fireflies and Photochemistry

---

[1]The terms *singlet, triplet, intersystem crossing, energy transfer,* and *quenching* are explained in Experiment 51.

SPECIAL INSTRUCTIONS

This entire experiment can be completed in less than one hour. When you are working with hydrazine, you should remember that it is toxic and should not be spilled on the skin. It is also a suspected carcinogen. Dimethylsulfoxide may also be toxic; avoid breathing the vapors or spilling it on your skin.

A darkened room is required to observe adequately the chemiluminescence of luminol. A darkened hood that has had its window covered with butcher paper or aluminum foil also works well. Other fluorescent dyes besides those mentioned (for instance, 9,10-diphenylanthracene) can also be used for the energy-transfer experiments. The dyes selected may depend on what is immediately available. The instructor may have each student use one dye for the energy-transfer experiments, with one student making a comparison experiment without a dye.

PROCEDURE

5-NITROPHTHALHYDRAZIDE

Place 0.150 g of 3-nitrophthalic acid and 0.2 mL of a 10% aqueous solution of hydrazine (use gloves) in a small sidearm test tube.[2] At the same time, heat 2 mL of water in a beaker on a steam bath or a sand bath which has been adjusted to about 80 °C. Heat the test tube over a micro burner until the solid dissolves. Add 0.4 mL of triethylene glycol and clamp the test tube in an upright position on a ring stand. Place a thermometer (do not seal the system) and a boiling stone in the test tube and attach a piece of pressure tubing to the sidearm. Connect this tubing to an aspirator (use a trap). Heat the solution with a micro burner until the liquid boils vigorously and the refluxing water vapor is drawn away by the aspirator vacuum (the temperature will rise to about 120 °C). Continue heating and allow the temperature to increase rapidly until it rises above 200 °C. About five minutes will be required for this heating. Remove the burner briefly when this temperature has been achieved, and then resume gentle heating to maintain a fairly constant temperature of 210–220 °C for about two minutes. Allow the test tube to cool to about 100 °C, add the 2.0 mL of hot water that was prepared previously, and cool the test tube to room temperature by allowing tap water to flow over the outside of the test tube. Collect the light yellow crystals of 5-nitrophthalhydrazide by vacuum filtration, using a small Hirsch funnel. It is not necessary to dry the product before you go on with the next reaction step.

LUMINOL (5-AMINOPHTHALHYDRAZIDE)

Transfer the moist 5-nitrophthalhydrazide to a 13 × 100-mm test tube. Add 0.65 mL of a 10% sodium hydroxide solution and agitate the mixture until the hydrazide dissolves.

[2]A 10% aqueous solution of hydrazine can be prepared by diluting 15.6 g of a commercial 64% hydrazine solution to a volume of 100 mL using water.

Add 0.40 g of sodium dithionite dihydrate (sodium hydrosulfite dihydrate, $Na_2S_2O_4 \cdot 2H_2O$). Using a Pasteur pipet, add enough water to wash the solid from the walls of the test tube. Heat the test tube until the solution boils, agitate the solution, and maintain the boiling, with agitation, for five minutes. Add 0.26 mL of glacial acetic acid and cool the test tube to room temperature by allowing tap water to flow over the outside of it. Agitate the mixture during the cooling step. Collect the light yellow crystals of luminol by vacuum filtration, using a small Hirsch funnel. Save a small sample of this product, allow it to dry overnight, and determine its melting point (mp 319–320 °C). The remainder of the luminol may be used without drying for the chemiluminescence experiments.

CHEMILUMINESCENCE EXPERIMENTS

Cover the bottom of a 10-mL Erlenmeyer flask with a layer of potassium hydroxide pellets. Add enough dimethylsulfoxide to cover the pellets. Add about 0.025 g of the moist luminol to the flask, stopper it, and shake it vigorously to mix air into the solution.[3] In a dark room, a faint glow of blue-white light will be visible. The intensity of the glow will increase with continued shaking of the flask and occasional removal of the stopper to admit more air.

To observe energy transfer to a fluorescent dye, dissolve one or two crystals of the indicator dye in about 0.25 mL of water. Add the dye solution to the dimethylsulfoxide solution of luminol, stopper the flask, and shake the mixture vigorously. Observe the intensity and the color of the light produced.

A table of some dyes and the colors produced when they are mixed with luminol is given below. Other dyes not included on this list may also be tested in this experiment.

FLUORESCENT DYE	COLOR
No dye	Faint bluish white
2,6-Dichloroindophenol	Blue
9-Aminoacridine	Blue green
Eosin	Salmon pink
Fluorescein	Yellow green
Dichlorofluorescein	Yellow orange
Rhodamine B	Green
Phenolphthalein	Purple

[3]An alternative method for demonstrating chemiluminescence, using potassium ferricyanide and hydrogen peroxide as oxidizing agents, is described in E. H. Huntress, L. N. Stanley, and A. S. Parker, *Journal of Chemical Education, 11* (1934): 142.

Essay
CHEMISTRY OF MILK

Milk is a food of exceptional interest. Not only is milk an excellent food for the very young, but humans have also adopted milk, specifically cow's milk, as a food substance for persons of all ages. Many specialized milk products like cheese, yogurt, butter, and ice cream are staples of our diet.

Milk is probably the most nutritionally complete food that can be found in nature. This property is important for milk, since it is the only food young mammals consume in the nutritionally significant weeks following birth. Whole milk contains vitamins (principally thiamine, riboflavin, pantothenic acid, and vitamins A, D, and K), minerals (calcium, potassium, sodium, phosphorus, and trace metals), proteins (which include all the essential amino acids), carbohydrates (chiefly lactose), and lipids (fats). The only important elements in which milk is seriously deficient are iron and Vitamin C. Infants are usually born with a storage supply of iron large enough to meet their needs for several weeks. Vitamin C is easily secured through an orange juice supplement. The average composition of the milk of each of several mammals is summarized below.

Average Percentage Composition of Milk from Various Mammals

	COW	HUMAN	GOAT	SHEEP	HORSE
Water	87.1	87.4	87.0	82.6	90.6
Protein	3.4	1.4	3.3	5.5	2.0
Fats	3.9	4.0	4.2	6.5	1.1
Carbohydrates	4.9	7.0	4.8	4.5	5.9
Minerals	0.7	0.2	0.7	0.9	0.4

FATS

Whole milk is an oil-water type of emulsion, containing about 4% fat dispersed as very small (5–10 microns in diameter) globules. The globules are so small that a drop of milk contains about a million of them. Because the fat in milk is so finely dispersed, it is digested more easily than fat from any other source. The fat emulsion is stabilized to some extent by complex phospholipids and proteins that are adsorbed on the surfaces of the globules. The fat globules, which are lighter than water, coalesce on standing and eventually rise to the surface of the milk, forming a layer of **cream.** Since vitamins A and D are fat-soluble vitamins, they are carried to the surface with the cream. Commercially, the cream is often removed by centrifugation and skimming and is either diluted to form coffee cream (''half and half''), sold as **whipping cream,** converted to **butter,** or converted to **ice cream.** The milk that remains is called **skimmed milk.** Skimmed milk, except for lacking the fats and vitamins A and D, has approximately the same composition as whole milk. If milk is **homogenized,** its fatty content will not separate. Milk is homogenized by forcing it through a small hole. This breaks up the fat globules and reduces their size to about one to two microns in diameter.

The structure of fats and oils is discussed in the essay that precedes Experiment 17. The fats in milk are primarily triglycerides. For the saturated fatty acids, the following percentages have been reported:

C_2 (3%)	C_8 (2.7%)	C_{14} (25.3%)	$>C_{18}$ (~5%)
C_4 (1.4%)	C_{10} (3.7%)	C_{16} (9.2%)	
C_6 (1.5%)	C_{12} (12.1%)	C_{18} (1.3%)	

Thus, about two thirds of all the fatty acids in milk are saturated and about one third are unsaturated. Milk is unusual in that about 12% of the fatty acids are **short**-chain fatty acids (C_2–C_{10}) like butyric, caproic, and caprylic acids.

Additional lipids (fats and oils) in milk include small amounts of cholesterol, phospholipids, and lecithins (phospholipids conjugated with choline). The structures of phospholipids and lecithins are shown. The phospholipids help to stabilize the whole milk emulsion; the phosphate groups help to achieve partial water solubility for the fat globules. All the fat can be removed from milk by extraction with petroleum ether or a similar organic solvent.

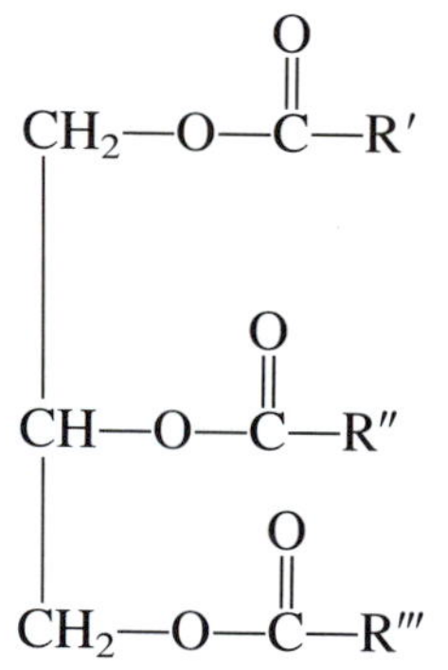

A triglyceride

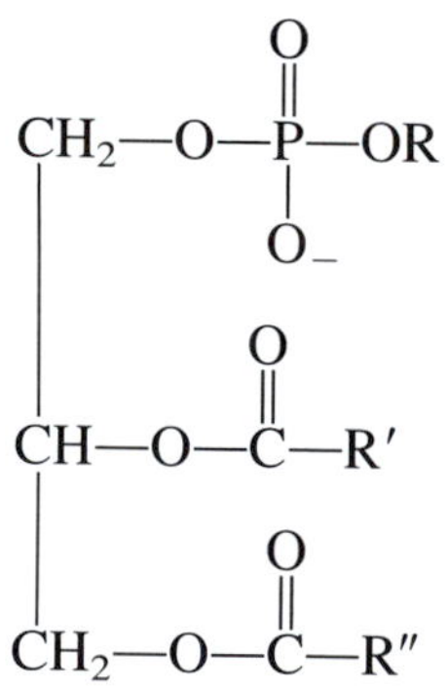

A phospholipid

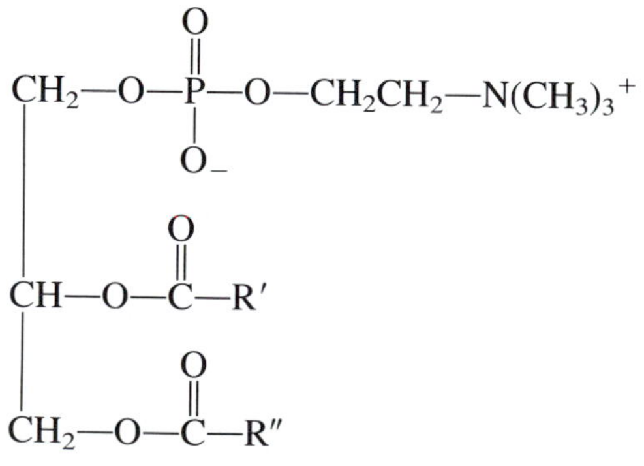

A lecithin

PROTEINS

Proteins may be classified broadly in two general categories: fibrous and globular. Globular proteins are those that tend to fold back on themselves into compact units that approach nearly spheroidal shapes. These types of proteins do not form intermolecular interactions between protein units (hydrogen-bonds, and so on) as fibrous proteins do, and they are more easily solubilized as colloidal suspensions. There are three kinds of proteins in milk: **caseins, lactalbumins,** and **lactoglobulins.** All are globular.

Casein is a phosphoprotein, meaning that phosphate groups are attached to some of the amino acid side-chains. These are attached mainly to the hydroxyl groups of the serine and threonine moieties. Actually, casein is a mixture of at least three similar proteins, principally α, β, and κ caseins. These three proteins differ primarily in molecular weight and amount of phosphorus they contain (number of phosphate groups).

CASEIN	MW	PHOSPHATE GROUPS/MOLECULE
α	27,300	~9
β	24,100	~4–5
κ	~8,000	~1.5

Casein exists in milk as the calcium salt, **calcium caseinate.** This salt has a complex structure. It is composed of α, β, and κ caseins, which form a **micelle,** or a solubilized unit. Neither the α nor the β casein is soluble in milk, and neither is soluble either singly or in combination. If κ casein is added to either one, or to a combination of the two, however, the result is a casein complex that is soluble owing to the formation of the micelle.

A structure proposed for the casein micelle is shown in the figure on page 432. The κ casein is thought to stabilize the micelle. Since both α and β casein are phosphoproteins, they are precipitated by calcium ions. Recall that $Ca_3(PO_4)_2$ is fairly insoluble.

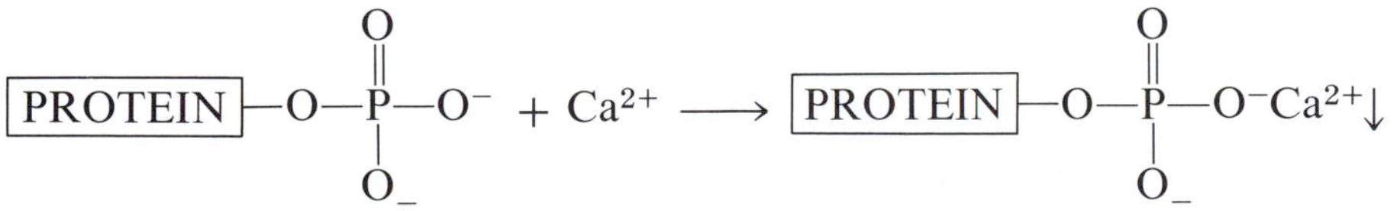

$$\boxed{\text{PROTEIN}}\!-\!\overset{\displaystyle O}{\underset{\displaystyle O_-}{\overset{\|}{\underset{|}{-O-P-O^-}}}}\ +\ Ca^{2+}\ \longrightarrow\ \boxed{\text{PROTEIN}}\!-\!\overset{\displaystyle O}{\underset{\displaystyle O_-}{\overset{\|}{\underset{|}{-O-P-O^-Ca^{2+}\downarrow}}}}$$

Insoluble

The κ casein protein, however, has fewer phosphate groups and a high content of carbohydrate bound to it. It is also thought to have all its serine and threonine residues (which have hydroxyl groups), as well as its bound carbohydrates, on only one side of its outer surfaces. This portion of its outer surface is easily solubilized in water since these polar groups are present. The other portion of its surface binds well to the water-insoluble α and β caseins and solubilizes them by forming a protective colloid or micelle around them. Since the entire outer surface of the micelle can be solubilized in

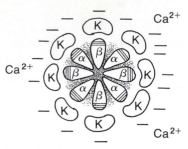

A casein micelle (average diameter, 1200 Å)

water, the unit is solubilized **as a whole,** thus bringing the α and β caseins, as well as κ casein, into solution.

Calcium caseinate has its isoelectric (neutrality) point at pH 4.6. Therefore, it is insoluble in solutions of pH less than 4.6. The pH of milk is about 6.6; therefore, casein has a negative charge at this pH and is solubilized as a salt. If acid is added to milk, the negative charges on the outer surface of the micelle are neutralized (the phosphate groups are protonated) and the neutral protein precipitates:

$$Ca^{2+} \text{ Caseinate} + 2HCl \longrightarrow \text{Casein} \downarrow + CaCl_2$$

The calcium ions remain in solution. When milk sours, lactic acid is produced by bacterial action (see equations on page 434), and the consequent lowering of the pH causes the same **clotting** reaction. The isolation of casein from milk is described in Experiment 53.

The casein in milk can also be clotted by the action of an enzyme called **rennin.** Rennin is found in the fourth stomach of young calves. However, both the nature of the clot and the mechanism of clotting differ when rennin is used. The clot formed using rennin, **calcium paracaseinate,** contains calcium.

$$Ca^{2+} \text{ Caseinate} \xrightarrow{\text{rennin}} Ca^{2+} \text{ Paracaseinate} + \text{a small peptide}$$

Rennin is a hydrolytic enzyme (peptidase) and acts specifically to cleave peptide bonds between phenylalanine and methionine residues. It attacks the κ casein, breaking the peptide chain so as to release a small segment of it. This destroys the water-solubilizing surface of the κ casein, which protects the inner α and β caseins, and causes the entire micelle to precipitate as calcium paracaseinate. Milk can be decalcified by treatment with oxalate ion, which forms an insoluble calcium salt. If the calcium ions are removed from milk, a clot will not be formed when the milk is treated with rennin.

The clot, or **curd,** formed by the action of rennin is sold commercially as **cottage cheese.** The liquid remaining is called the **whey.** The curd can also be used in producing various types of **cheese.** It is washed, pressed to remove any excess whey, and chopped. After this treatment, it is melted, hardened, and ground. The ground curd is then salted, pressed into molds, and set aside to age.

Albumins are globular proteins that are soluble in water and in dilute salt solutions. They are, however, denatured and coagulated by heat. The second most abundant protein types in milk are the **lactalbumins.** Once the caseins have been removed, and the solution has been made acidic, the lactalbumins can be isolated by heating the mixture to precipitate them. The typical albumin has a molecular weight of about 41,000.

A third type of protein in milk is the **lactoglobulins.** They are present in smaller amounts than the albumins and generally denature and precipitate under the same conditions as the albumins. The lactoglobulins carry the immunological properties of milk. They protect the young mammal until its own immune systems have developed.

CARBOHYDRATES

When the fats and the proteins have been removed from milk, the carbohydrates remain, as they are soluble in aqueous solution. The main carbohydrate in milk is lactose.

Lactose, a disaccharide, is the **only** carbohydrate that mammals synthesize. Hydrolyzed, it yields one molecule of D-glucose and one of D-galactose. It is synthesized in the mammary glands. In this process, one molecule of glucose is converted to galactose and joined to another of glucose. The galactose is apparently needed by the developing infant to build developing brain and nervous tissue. Brain cells contain **glycolipids** as a part of their structure. A glycolipid is a triglyceride in which one of the fatty acid groups has been replaced by a sugar, in this case galactose. Galactose is more stable (to metabolic oxidation) than glucose and affords a better material for forming structural units in cells.

Lactose
D-Galactose + D-Glucose

A glycolipid

Although almost all human infants can digest lactose, some adults lose this ability on reaching maturity, since milk is no longer an important part of their diet. An enzyme called **lactase** is necessary to digest lactose. Lactase is secreted by the cells of the small intestine, and it cleaves lactose into its two component sugars, which are easily digested. Persons lacking the enzyme lactase do not digest lactose properly. As it is poorly absorbed by the small intestine, it remains in the digestive tract, where its osmotic potential causes an influx of water. This results in cramps and diarrhea for the affected individual. Persons with a lactase deficiency cannot tolerate more than one glass of milk a day. The deficiency is most common among blacks, but it is also quite common among older whites.

Lactose can be removed from whey by adding ethanol. Lactose is insoluble in ethanol, and when the ethanol is mixed with the aqueous solution, the lactose is forced to crystallize. The isolation of lactose from milk is described in Experiment 53.

When milk is allowed to stand at room temperature, it sours. Many bacteria are present in milk, particularly **lactobacilli.** These bacteria act on the lactose in milk to produce the sour **lactic acid.** These microorganisms actually **hydrolyze** lactose and produce lactic acid only from the galactose portion of the lactose. Since the production of the lactic acid also lowers the pH of the milk, the milk clots when it sours:

$$C_{12}H_{22}O_{11} + H_2O \longrightarrow C_6H_{12}O_6 + C_6H_{12}O_6$$

Lactose   **Galactose**   **Glucose**

$$C_6H_{12}O_6 \xrightarrow{\text{lactobacilli}} CH_3\text{—}\underset{\underset{\text{OH}}{|}}{CH}\text{—}COOH$$

Galactose   **Lactic acid**

Many "cultured" milk products are manufactured by allowing milk to sour before it is processed. For instance, milk or cream is usually allowed to sour somewhat by lactic acid bacteria before it is churned to make butter. The fluid left after the milk is churned is sour and is called **buttermilk.** Other cultured milk products include sour cream, yogurt, and certain types of cheese.

REFERENCES

Fox, B. A., and Cameron, A. G. *Food Science—A Chemical Approach.* New York: Crane, Russak, 1973. Chap. 6, "Oils, Fats, and Colloids."

Kleiner, I. S., and Orten, J. M. *Biochemistry.* 7th ed. St. Louis: C. V. Mosby, 1966. Chap. 7, "Milk."

McKenzie, H. A., ed. *Milk Proteins.* 2 volumes. New York: Academic Press, 1970.

Oberg, C. J., "Curdling Chemistry—Coagulated Milk Products." *Journal of Chemical Education, 63* (September 1986): 770.

Experiment 53
Isolation of Casein and Lactose from Milk

Isolation of a protein
Isolation of a sugar

In this experiment you isolate several of the chemical substances found in milk. You first isolate a phosphorus-containing protein, casein (Procedure 53A). The remaining milk mixture will then be used as a source of a sugar, α-lactose (Procedure 53B). After you have isolated the milk sugar, several chemical tests can be made on this material. Fats, which are present in whole milk, are not isolated in this experiment, since powdered nonfat milk is used.

Here is the procedure you will follow. First, the casein is precipitated by warming the powdered milk and adding dilute acetic acid. It is important that the heating not be excessive or the acid too strong, since these conditions also hydrolyze lactose into its components, glucose and galactose. After the casein has been removed, the excess acetic acid is neutralized with calcium carbonate, and the solution is heated to its boiling point to precipitate the initially soluble protein, albumin. The liquid containing the lactose is poured away from the albumin. Alcohol is added to the solution and any remaining protein is removed by centrifugation. α-Lactose crystallizes on cooling.

Lactose is an example of a disaccharide. It is made up of two sugar molecules: galactose and glucose. In the structures shown below, the galactose portion is on the left and glucose is on the right. Galactose is bonded through an acetal linkage to glucose.

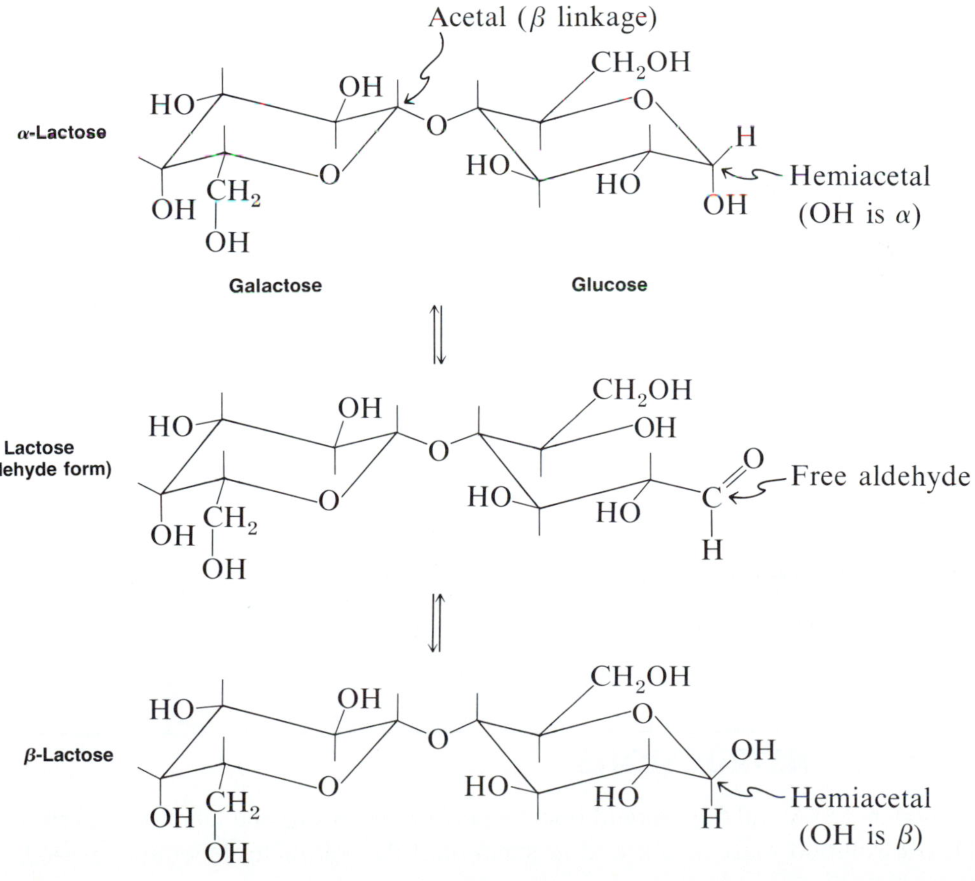

One should notice that the glucose portion can exist in one of two isomeric hemiacetal structures: α-lactose and β-lactose. Glucose can also exist in a free aldehyde form. This aldehyde form (open form) is an intermediate in the equilibration (interconversion) of α- and β-lactose. Very little of this free aldehyde form exists in the equilibrium mixture. The isomeric α- and β-lactose are diastereomers since they differ in the configuration at one carbon atom, called the anomeric carbon atom.

The sugar α-lactose is easily obtainable by crystallization from a water-ethanol mixture at room temperature. On the other hand, β-lactose must be obtained by a more difficult process, which involves crystallization from a concentrated solution of lactose at temperatures above 93.5 °C. In the present experiment, α-lactose is isolated by the simpler experimental procedure indicated above.

It has been found that α-lactose undergoes numerous interesting reactions. First, α-lactose interconverts, via the free aldehyde form, to a large extent, to the β-isomer in aqueous solution. This causes a change in the rotation of polarized light from +92.6° to +52.3° with increasing time. The process that causes change in optical rotation with time is called **mutarotation.** Mutarotation of lactose can be studied in Experiment 54.

A second reaction of lactose is the oxidation of the free aldehyde form by Benedict's reagent. Lactose is referred to as a reducing sugar because it reduces Benedict's reagent (cupric ion to cuprous ion) and produces a red precipitate (Cu_2O). In the process, the aldehyde group is oxidized to a carboxyl group. The reaction that takes place in Benedict's test is

$$R\text{—}CHO + 2Cu^{2+} + 4OH^- \longrightarrow RCOOH + Cu_2O + 2H_2O$$

A third reaction of lactose is the oxidation of the galactose part by the mucic acid test. In this test, the acetal linkage between galactose and glucose units is cleaved by the acidic medium to give free galactose and glucose. Galactose is oxidized with nitric acid to the dicarboxylic acid, galactaric acid (mucic acid). Mucic acid is an insoluble, high-melting solid, which precipitates from the reaction mixture. On the other hand, glucose is oxidized to a diacid (glucaric acid), which is more soluble in the oxidizing medium and does not precipitate, (see page 437).

REQUIRED READING

Review: Technique 4

New: Essay   Chemistry of Milk

SPECIAL INSTRUCTIONS

Procedures 53A and 53B should both be performed during one laboratory period. The lactose solution must be allowed to stand until the following laboratory period.

Lactose $\xrightarrow{\text{H}^+}$

D-Galactose $\xrightarrow{\text{HNO}_3}$ Mucic acid (insoluble)

D-Glucose $\xrightarrow{\text{HNO}_3}$ D-Glucaric acid (soluble)

Procedure 53A
Isolation of Casein from Milk

Place 4.0 g of powdered milk and 10 mL of water into a 50-mL beaker. Heat the mixture in a sand bath to about 40 °C (sand bath at about 50 °C). Check the temperature of the milk solution with a thermometer. Place 1.0 mL of dilute acetic acid solution[1] in a conical vial for temporary storage. When the mixture has reached 40 °C, add the dilute acetic acid dropwise to the warm milk. After every five drops, stir the mixture gently using a glass stirring rod with an attached rubber policeman. Using the rubber police-man, push the casein up onto the side of the beaker so that most of the liquid drains from the solid. Then transfer the congealed casein to a 10-mL beaker in portions. If any liquid separates from the casein in the 10-mL beaker, use a Pasteur pipet to transfer the liquid back into the reaction mixture. Continue to add dropwise the remainder of the 1.0 mL of dilute acetic acid to the milk mixture in the 50-mL beaker to fully precipitate the casein. Remove as much of the casein as possible and transfer it to the 10-mL beaker. Avoid adding an excess of acetic acid to the milk solution, as this will cause the lactose in the milk to hydrolyze into glucose and galactose. When most of the casein has been re-

---

[1]The laboratory instructor should prepare a large batch for the class in the ratio of 2 mL glacial acetic acid to 20 mL of water.

moved from the milk solution, add 0.2 g of calcium carbonate to the milk in the 50-mL beaker. Stir the mixture for a few minutes and save it for use in Procedure 53B. Use this mixture as soon as possible during the laboratory period. This beaker contains lactose and albumins.

Transfer the casein from the 10-mL beaker to a Hirsch funnel (Technique 4, Section 4.3, and Figure 4–6, p 571). Draw a vacuum on the casein to remove as much liquid as possible (about five minutes). Press the casein with a spatula during this time. Transfer the casein to a piece of filter paper (about 7 cm). Using a spatula, move the solid around on the paper so that the remaining liquid is absorbed into the filter paper. When most of the liquid has been removed, transfer the solid to a watch glass to complete the drying operation. Allow the casein to air-dry completely for two to three days before weighing the product. You must remove the casein from the filter paper or it will become "glued" to the paper. (You have nearly prepared white glue!) Submit the casein in a labeled vial to the instructor or save it for Experiment 58. Calculate the weight percent of the casein isolated from the powdered milk.

Procedure 53B
Isolation of Lactose from Milk

Heat the mixture that you saved from Procedure 53A to about 75 °C for about five minutes (heat the mixture directly on the hot plate without a sand bath). This heating operation results in a nearly complete separation of the albumins from the solution. Decant the liquid in the beaker away from the solid into a clean centrifuge tube. You may need to hold the solid with a spatula while transferring the liquid. Press the albumins with a spatula to remove as much liquid as possible and pour the liquid into the centrifuge tube (save the albumins in the original beaker). You should have about 7 mL of liquid. When the liquid has cooled to about room temperature, centrifuge the contents of the tube for two to three minutes. Be sure to place another tube in the centrifuge to balance the unit.

Following centrifugation, decant the liquid away from the solid into a 50-mL beaker. Add 15 mL of 95% ethanol to the beaker. Solids will precipitate. Heat this mixture to about 60 °C on the hot plate (heat the mixture directly on the hot plate) to dissolve some of the solid. Pour the **hot** liquid into a 40-mL centrifuge tube (or two 15-mL tubes) and centrifuge the hot solution as soon as possible before the solution cools appreciably. Centrifuge the mixture for two to three minutes. Be sure to place another tube in the centrifuge to balance the unit. It is important to centrifuge this mixture while it is warm to prevent premature crystallization of the lactose. A considerable quantity of solid forms in the bottom of the centrifuge tube.

Remove the warm supernatant liquid from the tube using a Pasteur pipet and transfer the liquid to a 50-mL Erlenmeyer flask. Discard the solid remaining in the centrifuge tube. Stopper the flask and allow the lactose to crystallize for at least two days. Granular crystals will form during this time.

Collect the lactose by vacuum filtration on a Hirsch funnel. Use about 3 mL of 95% ethanol to aid the transfer and to wash the product. α-Lactose crystallizes with one water of hydration, $C_{12}H_{22}O_{11} \cdot H_2O$. Weigh the product after it is thoroughly dry. Submit the α-lactose in a labeled vial to the instructor or save it for Experiment 54. Calculate the weight percent of the lactose isolated from the powdered milk.

Allow the albumins to dry for two to three days in the original beaker. Break up the solid and weigh it. Calculate the weight percent of albumins isolated from the powdered milk.

BENEDICT'S TEST (optional)

Prepare a hot-water bath (above 90 °C) for this experiment. Dissolve about 0.01 g of your lactose in 1 mL of water in a test tube. Heat the mixture to dissolve most of the lactose (some cloudiness remains). Place about 1 mL each of 1% solutions of glucose and galactose in separate test tubes. Add to each of the three test tubes 2 mL of Benedict's reagent.[2] Place the test tubes in the hot-water bath for two minutes. Remove the tubes and note the results. The formation of a orange to brownish-red precipitate indicates a positive test for a reducing sugar. This test is described on pp 501–503 in Experiment 57.

MUCIC ACID TEST (optional)

Prepare a hot-water bath (above 90 °C) for this experiment or use the one prepared for the Benedict's test. Place 0.1 g of the isolated lactose, 0.05 g of glucose (dextrose), and 0.05 g of galactose in three separate test tubes. Add 1 mL of water to each tube and dissolve the solids, by heating if necessary. The lactose solution may be somewhat cloudy, but will clear when the nitric acid is added. Add 1 mL of concentrated nitric acid to each of the tubes. Heat the tubes in a hot-water bath for one hour in a hood (nitrogen oxide gases are evolved). Remove the tubes and allow them to cool slowly after the heating period. Scratch the test tubes with clean stirring rods to induce crystallization. After the test tubes are cooled to room temperature, place them in an ice bath. A fine precipitate of mucic acid should begin to form in the galactose and lactose tubes about one-half hour after the tubes are removed from the water bath. Allow the test tubes to stand until the next laboratory period to complete the crystallization. Confirm the insolubility of the solid formed by adding about 1 mL of water, then shaking the resulting mixture. If the solid remains, it is mucic acid.

QUESTIONS

1. A student decided to determine the optical rotation of mucic acid. What should be expected as a value? Why?

2. Draw a mechanism for the acid-catalyzed hydrolysis of the acetal bond in lactose.

---

[2]Dissolve 34.6 g of hydrated sodium citrate and 20.0 g of anhydrous sodium carbonate in 160 mL of distilled water by heating. Filter the solution, if necessary. Add to it a solution of 3.46 g of cupric sulfate ($CuSO_4 \cdot 5H_2O$) dissolved in 20 mL of distilled water. Dilute the combined solutions to 200 mL.

3. β-Lactose is present to a larger extent in an aqueous solution when the solution is at equilibrium. Why is this to be expected?

4. Very little of the free aldehyde form is present in an equilibrium mixture of lactose. However, a positive test is obtained with Benedict's reagent. Explain.

5. Outline a separation scheme for isolating casein, albumin, and lactose from milk. Use a flowchart like that shown in the Advance Preparation and Laboratory Records section at the beginning of the book.

Experiment 54
Mutarotation of Lactose

Polarimetry

In this experiment the mutarotation of lactose is studied by polarimetry. The disaccharide α-lactose, made up of galactose and glucose, can be isolated from milk (Experiment 53). As you can see in the structures drawn in Experiment 53, the glucose unit can exist in one of two isomeric hemiacetal structures: α- and β-lactose. These isomers are diastereomers since they differ in configuration at one carbon atom. The glucose part can also exist in a free aldehyde form. This aldehyde form (open form in the equation below) is an intermediate in the equilibration of α- and β-lactose. Very little of this free aldehyde form exists in the equilibrium mixture.

α-Lactose has a specific rotation at 20 °C of $+92.6°$. However, when it is placed in water, the optical rotation **decreases** until it reaches an equilibrium value of $+52.3°$. β-Lactose has a specific rotation of $+34°$. The optical rotation of β-lactose **increases** in water until it reaches the same equilibrium value obtained for α-lactose. At the equilibrium point, both the α and β isomers are present. However, since the equilibrium rotation is closer in value to the initial rotation of β-lactose, the mixture must contain more of this isomer. The process, which results in a change in optical rotation over time to approach an equilibrium value, is called **mutarotation.**

REQUIRED READING

Review:	Essay	The Chemistry of Milk
New:	Technique 16	Polarimetry

SPECIAL INSTRUCTIONS

The procedure for preparing the cells and for operating the instrument are those appropriate for the Zeiss polarimeter. Your instructor will provide instructions for the use of another type of polarimeter if a Zeiss instrument is not available. A 2-dm cell is used for this experiment. If a cell of a different path length is used, adjust the concentrations appropriately. About one hour is required to complete the mutarotation study.

If you are using lactose isolated in Experiment 53, you will need to combine your product with one other student so that you will have an adequate amount of lactose to perform this experiment. One person can measure the rotations while the other student records the data. Student-prepared lactose may yield a cloudy solution when the sample is dissolved in water. This should not be of concern as the cloudiness is caused by a trace of protein that remains in the lactose. Commercial lactose will dissolve completely.

PROCEDURE

Turn on the polarimeter to warm the sodium lamp. After about 10 minutes, adjust the instrument so that the scale reads about $+9°$. This scale reading provides an adjustment of the instrument to the approximate range of rotation that will be observed at the initial reading. Set the timer to zero. Clean and dry a 2-dm cell (Technique 16, Figure 16–6, p 762). Weigh 1.25 g of α-lactose and transfer it completely to a **dry** 25-mL volumetric flask.

The operations described in the next paragraph should be studied carefully **before** starting this part of the experiment. It is essential to complete carefully the described operations in **two minutes** or less. The reason for speedy operation is that the α-lactose immediately begins to mutarotate when it comes in contact with water. The initial rotations obtained are necessary to get a precise value of the rotation at zero time. You should practice with the necessary equipment in a place near the polarimeter before performing the actual operations. Study the scale on the polarimeter so that you can read it rapidly.

Add about half the volume of distilled water to the volumetric flask containing the α-lactose and swirl it to dissolve the solid. When about half the solid is dissolved (a rough estimate), start the timer. As soon as the solid is dissolved (about **20–25** seconds), carefully fill the flask to the mark with distilled water. Use an eye dropper to finish adding the water. Stopper the flask and invert it about five times to mix the contents. Using a funnel, fill the polarimeter cell with the lactose solution. Screw the end piece on the cell and tilt it to transfer any remaining bubbles to the enlarged ring. Place the cell in the polarimeter, close the cover, and adjust the analyzer until the split field is of uniform density (Technique 16, Figure 16–7, p 763). Record the time and rotation in the notebook.

Obtain the optical rotation at one-minute intervals for eight additional minutes (10 minutes total from the time of initial mixing) and record these values, along with the times at which they were determined. After the 10-minute period, obtain readings at two-minute intervals for the next 20 minutes. Record the optical rotations and times.

Remove the cell from the polarimeter and add two drops of concentrated ammonium hydroxide to the lactose solution. The ammonia rapidly catalyzes the mutarotation of lactose to its equilibrium value. If the ammonia is not added, the equilibrium value will not be obtained until after about 22 hours. Shake the tube and replace it in the polarimeter. Follow the decrease in rotation until there is no longer a change with time. This final value, which is the equilibrium optical rotation, should remain constant for about five minutes. Place a thermometer in the polarimeter and determine the temperature in the cell compartment.

Plot the data on a piece of graph paper ruled in millimeters with the optical rotation plotted on the vertical axis and time plotted (up to 30 minutes) on the horizontal axis. Draw the best possible curved line through the points and extrapolate the line to $t = 0$. Remember that there may be some scattering of points about the line, especially at the values for the longer times. The extrapolated value at $t = 0$ corresponds to the optical rotation of α-lactose at the time of initial mixing.

Using the equation in Technique 16, Section 16.2, p 760, calculate the specific rotation, $[\alpha]_D$, of α-lactose at $t = 0$. Likewise, calculate the specific rotation of the equilibrium mixture of α- and β-lactose.

Calculate the percentage of each of the diastereomers at equilibrium, using the experimentally determined specific rotation values for α-lactose and the equilibrium mixture and the literature value for the specific rotation of β-lactose ($+34°$). Assume a linear relation between the specific rotations and the concentrations of the species.

QUESTIONS

1. Explain why β-lactose predominates in the equilibrium mixture of α- and β-lactose.
2. The following rotation data have been obtained for D-glucose at 20 °C:

α-D-glucose	$+112.2°$
β-D-glucose	$+18.7°$
Equilibrium mixture	$+52.7°$

Using these values, calculate the percentage composition of the α and β isomers at equilibrium. Inspect the structures of α- and β-D-glucose in Experiment 57 and rationalize the values obtained in the calculations.

Experiment 55
"Pet Molecule" Project

Literature search

In this project, you will gain experience in using the chemical literature by writing a paper on an organic compound, your "pet molecule." Since a variety of chemical information will be included in this paper, you will need to use many different types of references, ranging from scientific journals and advanced chemistry textbooks to popular science magazines. By selecting a compound which is of particular interest to you, this project will also provide an opportunity for you to relate organic chemistry to a special interest you have.

REQUIRED READING

New: Technique 19   Guide to the Chemical Literature

SPECIAL INSTRUCTIONS

You will either select a "pet molecule" from a list provided by your instructor, or you will make a selection without a list to choose from. If you are allowed to select a compound on your own, your choice should be approved by your instructor.

A number of ideas for the content of this paper are given below. Your instructor may require that you delete some of these items or that you include information not described here. Specific instructions will also be provided by your instructor on how to write the paper, including guidelines such as organization, length, style, and how to handle footnotes and the bibliography. A complete reference on formal writing in the field of chemistry can be found in *The ACS Style Guide—A Manual for Authors and Editors* edited by Janet S. Dodd. It may also be helpful to consult a current issue of *Journal of the American Chemical Society* for guidelines on how footnotes and bibliographical information can be written. In addition, some of the optional activities may be required.

> **NOTE TO THE INSTRUCTOR:** A list of compounds that have been used by the authors is included in the "Instructor's Manual" accompanying this textbook.
>
> If you require that students write a procedure for synthesizing their compound, it is convenient to provide copies of catalogs from chemical companies such as Aldrich, Sigma, or Alfa.
>
> Some instructors may want to require that their students actually attempt a laboratory synthesis. Considerable care must be exercised in selecting a list of possible compounds or in advising students whether their choice is realistic. It is necessary to provide students with a list of available chemicals, if they will be performing a synthesis.

CONTENT OF THE PAPER

(Your instructor will indicate which of the following to include in your paper and how the paper should be written.)

1. General Information

      (A) IUPAC and common names,
      (B) Physical properties,
      (C) Molecular and structural formulas, and
      (D) Where found in nature.

2. History

      (A) Who discovered the compound? When? How?
      (B) Who determined the molecular structure? When? How?

3. Information of Special Interest

This part of the paper is where you have the opportunity to discuss what you find most interesting about the compound. In some cases, this may include information concerning the chemistry of your "pet molecule." You may also discuss why you are personally interested in the compound or how it relates to the "real world." This is your chance to relate organic chemistry to your own particular interests.

4. Chemical Reactions

Give chemical equations for the major reactions your "pet molecule" can undergo. In some cases, this may include reactions that occur within a living organism or other reactions that are of particular importance for your compound. These reactions may also include reactions you have studied in your organic chemistry course. Where possible, you should identify the type of each reaction. For example, a reaction might be identified as an electrophilic aromatic substitution reaction or a Diels-Alder reaction.

5. Synthesis

Give a laboratory procedure for synthesizing the compound using chemicals that would likely be found in an organic chemistry laboratory. This should be a several step procedure which could be carried out in your laboratory. Be as specific as possible about amounts, reaction conditions, glassware and equipment required, length of time required, and special concerns or cautions. You should also discuss each step, including what type of reaction it is.

SOURCES OF INFORMATION

Most of the information required to write your paper can be found in sources discussed in Technique 19, Guide to the Chemical Literature, p 787. In addition to specific sources listed in this chapter, you should consult card catalogs in the library and various guides to periodicals such as the *Readers Guide to Periodical Literature*. There are many books on organic syntheses other than those listed in Technique 19. These can be located by using the card catalog or by browsing in the chemistry section of your library. *Beilstein* and *Chemical Abstracts* (see Technique 19, Section 19.10, p 793) may be helpful in finding specific synthetic methods for your compound. The *Encyclopedia of Chemical Technology* and the *McGraw-Hill Encyclopedia of Science and Technology* are very useful resources for this project.

In order to determine if a specific synthetic method could be done with readily available chemicals, you should consult chemical catalogs from companies such as Aldrich, Sigma, or Alfa. Your instructor will make these sources available.

OPTIONAL ACTIVITIES

Molecular Model

Build a model of your compound. Although this can be done with a molecular model kit, it is more creative (and more fun) to construct a model using materials of your own selection.

Oral Presentation

In relatively small classes, it is valuable for each student to make a short presentation on their "pet molecule." Not only is this a good experience for the student making the presentation, but it is also informative for the other students.

Part Three

Identification of Organic Substances

Experiment 56
Identification of Unknowns

Qualitative organic analysis, the identification and characterization of unknown compounds, is an important part of organic chemistry. Every chemist must learn the appropriate methods for establishing the identity of a compound. In this experiment you will be issued an unknown compound and asked to identify it through chemical and spectroscopic methods. Your instructor may give you a general unknown or a specific unknown. With a **general unknown,** you must first determine the class of compound to which the unknown belongs, that is, identify its main functional group; and then you must determine the specific compound in that class that corresponds to the unknown. With a **specific unknown,** the class of compound (ketone, alcohol, amine, and so on) will be known in advance, and it will only be necessary to determine whatever specific member of that class was issued to you as an unknown. This experiment is designed so that the instructor can issue several general unknowns or as many as six successive specific unknowns, each having a different main functional group.

Although there are well over a million organic compounds that an organic chemist might be called upon to identify, the scope of this experiment is necessarily limited. In this textbook, just over 300 compounds are included in the tables of possible unknowns given for the experiment (see Appendix 1). Your instructor may wish to expand the list of possible unknowns, however. In such a case you will have to consult more extensive tables, such as those found in the work compiled by Rappoport (see references). In addition, the experiment is restricted to include only seven important functional groups:

Aldehydes	Carboxylic acids	Amines	Esters
Ketones	Phenols	Alcohols	

Even though this list of functional groups omits some of the important types of compounds (alkyl halides, alkenes, alkynes, aromatics, ethers, amides, mercaptans, nitriles, acid chlorides, acid anhydrides, nitro compounds, and so on), the methods introduced here can be applied equally well to these other classes of compounds. The list is sufficiently broad to illustrate all the principles involved in identifying an unknown compound.

In addition, although many of the functional groups listed as being excluded will not appear as the major functional group in a compound, several of them will frequently appear as secondary, or subsidiary, functional groups. Three examples of this are presented below.

MAJOR:	**KETONE**	**PHENOL**	**ALDEHYDE**
SUBSIDIARY:	Halide	Nitro	Alkene Aromatic
	Aromatic	Aromatic	Ether

449

The groups included that have subsidiary status are

—Cl	Chloro	—NO$_2$	Nitro	C=C	Double Bond
—Br	Bromo	—C≡N	Cyano	C≡C	Triple Bond
—I	Iodo	—OR	Alkoxy	 ⬡	Aromatic

The experiment presents all of the chief chemical and spectroscopic methods of determining the main functional groups, and it includes methods for verifying the presence of the subsidiary functional groups as well. It will usually not be necessary to determine the presence of the subsidiary functional groups to identify the unknown compound correctly. **Every** piece of information helps the identification, however, and if these groups can be detected easily, one should not hesitate to determine them. Finally, complex bifunctional compounds are generally avoided in this experiment; only a few are included.

HOW TO PROCEED

Fortunately, one can detail a fairly straightforward procedure for determining all the necessary pieces of information. This procedure consists of the following steps:

PART ONE

1. Preliminary classification by physical state, color, and odor.
2. Melting-point or boiling-point determination; other physical data.
3. Purification, if necessary.
4. Determination of solubility behavior in water and in acids and bases.
5. Simple preliminary tests: Beilstein, ignition (combustion).
6. Application of relevant chemical classification tests.

PART TWO

7. Determination of Infrared and NMR spectra.
8. Elemental analysis, if necessary.
9. Preparation of derivatives.
10. Confirmations of identity.

Each of these steps is discussed briefly in the sections below.

PRELIMINARY CLASSIFICATION

One should note the physical characteristics of the unknown. These include its color, its odor, and its physical state (liquid, solid, crystalline form). Many compounds have characteristic colors or odors, or they crystallize with a specific crystal structure. This information can often be found in a handbook and can be checked later. Compounds with a high degree of conjugation are frequently yellow to red. Amines often have a fishlike odor. Esters have a pleasant fruity or floral odor. Acids have a sharp and pungent odor. A part of the training of every good chemist includes a cultivation of the ability to recognize familiar or typical odors. As a note of caution, many compounds have distinctly unpleasant or nauseating odors. Some have corrosive vapors. Any unknown substance should be sniffed with the greatest caution. As a first step, open the container, hold it away from you, and using your hand, carefully waft the vapors toward your nose. If you get past this stage, a closer inspection will be possible.

MELTING-POINT OR BOILING-POINT DETERMINATION

The single most useful piece of information to have for an unknown compound is its melting point or boiling point. Either piece of data will drastically limit the compounds that are possible. The electric melting-point apparatus gives a rapid and accurate measurement (see Technique 6, Sections 6.7 and 6.8). To save time, you can often determine two separate melting points. The first determination can be made rapidly to get an approximate value. Then, you can determine the second melting point more carefully.

The boiling point is easily obtained by a simple distillation of the unknown (Technique 8, Section 8.4) or by a micro boiling-point determination (Technique 6, Section 6.10). The simple distillation has the advantage that it also purifies the compound. A Hickman head should be used if a simple distillation is performed, and you should be sure that the thermometer bulb is fully immersed in the vapor of the distilling liquid. For an accurate boiling-point value, the liquid should be distilled rapidly and you must distill more than about 0.75 mL of liquid.

If the solid is high-melting (>200 °C), or the liquid high-boiling (>200 °C), a thermometer correction may be needed (Technique 6, Sections 6.12 and 6.13). In any event, allowance should be made for errors of as large as ±5 °C in these values.

PURIFICATION

If the melting point of a solid has a wide range (>4–5 °C), it should be recrystallized and the melting point redetermined.

If a liquid was highly colored before distillation, if it yielded a wide boiling-point range, or if the temperature did not hold constant during the distillation, it should be redistilled to determine a new temperature range. A reduced-pressure distillation is in order for high-boiling liquids or for those that show any sign of decomposition on heating.

Occasionally, column chromatography may be necessary to purify solids that have large amounts of impurities and do not yield satisfactory results on crystallization.

Acidic or basic impurities that contaminate a neutral compound may often be removed by dissolving the compound in a low-boiling solvent, such as CH_2Cl_2 or ether, and extracting with 5% $NaHCO_3$ or 5% HCl, respectively. Conversely, acidic or basic compounds can be purified by dissolving them in 5% NaHCO$_3$ or 5% HCl, respectively, and extracting them with a low-boiling organic solvent to remove impurities. After neutralization of the aqueous solution, the desired compound can be recovered by extraction.

SOLUBILITY BEHAVIOR

Tests on solubility are described fully in Procedure 56A. They are extremely important. The solubility of small amounts of the unknown in water, 5% HCl, 5% NaHCO$_3$, 5% NaOH, concentrated H$_2$SO$_4$, and organic solvents is determined. This information reveals whether a compound is an acid, a base, or a neutral substance. The sulfuric acid test reveals whether a neutral compound has a functional group that contains an oxygen, a nitrogen, or a sulfur atom that can be protonated. This information allows one to eliminate or to choose various functional-group possibilities. The solubility tests must be made on **all** unknowns.

PRELIMINARY TESTS

The two combustion tests, the Beilstein test (Procedure 56B) and the ignition test (Procedure 56C), can be performed easily and quickly, and they often give valuable information. It is recommended that they be performed on all unknowns.

CHEMICAL CLASSIFICATION TESTS

The solubility tests usually suggest or eliminate several possible functional groups. The chemical classification tests listed in Procedures 56D to 56I allow one to distinguish among the possible choices. Choose only those tests the solubility tests suggest might be meaningful. Time will be wasted doing unnecessary tests. There is no substitute for a firsthand thorough knowledge of these tests. Each of the sections should be studied carefully until the significance of each test is understood. Also, it will be helpful to actually try the tests on **known** substances. In this way, it will be easier to recognize a positive test. Appropriate test compounds are listed for many of the tests. When you are performing a test that is new to you, it is always good practice to run the test separately on both a known substance and the unknown **at the same time.** This lets you compare results directly.

Once the melting or boiling point, the solubilities, and the main chemical tests have been made, it will be possible to identify the class of compound. At this stage, with the melting point or boiling point as a guide, it will be possible to compile a list of possible compounds. Inspection of this list will suggest additional tests that must be performed to distinguish among the possibilities. For instance, one compound may be a methyl ketone and the other may not. The iodoform test is called for to distinguish the two possibilities. The tests for the subsidiary functional groups may also be required.

These are described in Procedures 56B and 56C. These tests should also be studied carefully; there is no substitute for firsthand knowledge about these either.

One should not perform the chemical tests either willynilly or in a methodical, comprehensive sequence. One should use the tests selectively. Solubility tests automatically eliminate the need for some of the chemical tests. Each successive test will either eliminate the need for another test or dictate its use. One should also examine the tables of unknowns carefully. The boiling point or the melting point of the unknown may eliminate the need for many of the tests. For instance, the possible compounds may simply not include one with a double bond. Efficiency is the key word here. You should not waste time in doing nonsensical or unnecessary tests. Many possibilities may be eliminated on the basis of logic alone.

How you proceed with the following steps may be limited by your instructor's wishes. Many instructors may restrict your access to infrared and NMR spectra until you have narrowed your choices to a few compounds **all within the same class.** Others may have you determine these data routinely. Some instructors may want students to perform elemental analysis on all unknowns; others may restrict it to only the most essential situations. Most unknowns can be identified without either spectroscopy or elemental analysis. Again, some instructors may require derivatives as a final confirmation of the compound's identity; others may not wish to use them at all.

SPECTROSCOPY

Spectroscopy is probably the most powerful and modern tool available to the chemist for determining the structure of an unknown compound. It is often possible to determine structure through spectroscopy alone. On the other hand, there are also situations for which spectroscopy is not of much help and the traditional methods must be relied on. For this reason, you should use spectroscopy not to the exclusion of the more traditional tests but rather as a confirmation of those results. Nevertheless, the main functional groups and their immediate environmental features can be determined quickly and accurately with spectroscopy.

ELEMENTAL ANALYSIS

Elemental analysis, which allows one to determine the presence of nitrogen, sulfur, or a specific halogen atom (Cl, Br, I) in a compound is often useful; however, other information often renders these tests unnecessary. A compound identified as an amine by solubility tests obviously contains nitrogen. Many nitrogen-containing groups (for instance, nitro groups) can be identified by infrared spectroscopy. Finally, it is not usually necessary to identify a specific halogen. The simple information that the compound contains a halogen (any halogen) may be enough information to distinguish between two compounds. A simple Beilstein test provides this information.

DERIVATIVES

One of the principal tests of the correct identification of an unknown compound comes in trying to convert the compound by a chemical reaction to another known compound. This second compound is called a **derivative.** The best derivatives are solid compounds, since the melting point of a solid provides an accurate and reliable identification of most compounds. Solids are also easily purified through crystallization. The derivative is a way of distinguishing two otherwise very similar compounds. Usually they will have derivatives (both prepared by the same reaction) that have different melting points. Tables of unknowns and derivatives are listed in Appendix 1 at the end of the book. Procedures for preparing derivatives are given in Appendix 2.

CONFIRMATION OF IDENTITY

A rigid and final test for identifying an unknown can be made if an ''authentic'' sample of the compound is available for comparison. One can compare infrared and NMR spectra of the unknown compound with the spectra of the known compound. If the spectra match, peak for peak, then the identity is probably certain. Other physical and chemical properties can also be compared. If the compound is a solid, a convenient test is the mixed melting point (Technique 6, Section 6.4). Thin-layer or gas chromatographic comparisons may also be useful. For thin-layer analysis, however, it may be necessary to experiment with several different development solvents to reach a satisfactory conclusion about the identity of the substance in question.

   While we cannot be complete in this experiment in terms of the functional groups covered, or the tests described, the experiment should give a good introduction to the methods and the techniques chemists use to identify unknown compounds. Textbooks that cover the subject more thoroughly are listed in the references. You are encouraged to consult these for more information, including specific methods and classification tests.

REFERENCES

Comprehensive Textbooks

Cheronis, N. D., and Entrikin, J. B. *Identification of Organic Compounds*. New York: Wiley-Interscience, 1963.

Pasto, D. J., and Johnson, C. R. *Laboratory Text for Organic Chemistry*. Englewood Cliffs, NJ: Prentice-Hall, 1979.

Shriner, R. L., Fuson, R. C., Curtin, D. Y., and Morrill, T. C. *The Systematic Identification of Organic Compounds*. 6th ed. New York: Wiley, 1980.

Spectroscopy

Bellamy, L. J. *The Infra-red Spectra of Complex Molecules*. 3rd ed. New York: Methuen, 1975.

Dyer, J. R. *Applications of Absorption Spectroscopy of Organic Compounds*. Englewood Cliffs, NJ: Prentice-Hall, 1965.

Nakanishi, K. *Infrared Absorption Spectroscopy*. San Francisco: Holden-Day, 1962.

Pavia, D. L., Lampman, G. M., and Kriz, G. S., Jr. *Introduction to Spectroscopy: A Guide for Students of Organic Chemistry*. Philadelphia: W. B. Saunders, 1979.

Silverstein, R. M., Bassler, G. C., and Morrill, T. C. *Spectrometric Identification of Organic Compounds*. 4th ed. New York: Wiley, 1981.

Extensive Tables of Compounds and Derivatives

Rappoport, Z., ed. *Handbook of Tables for Organic Compound Identification*. Cleveland: Chemical Rubber Co., 1967.

Procedure 56A
Solubility Tests

Solubility tests should be performed on **every unknown.** They are extremely important in determining the nature of the main functional group of the unknown compound. The tests are very simple and require only small amounts of the unknown. In addition, solubility tests reveal whether the compound is a strong base (amine), a weak acid (phenol), a strong acid (carboxylic acid), or a neutral substance (aldehyde, ketone, alcohol, ester). The common solvents used to determine solubility types are

5% HCl	Concentrated H_2SO_4
5% $NaHCO_3$	Water
5% NaOH	Organic solvents

The solubility chart on p 456 indicates solvents in which compounds containing the various functional groups are likely to dissolve. The summary charts in sections 56D through 56I repeat this information for each functional group included in this experiment. In this section, the correct procedure for determining whether a compound is soluble in a test solvent is given. Also given is a series of explanations detailing the reasons that compounds having specific functional groups are soluble in only specific solvents. This is accomplished by indicating the type of chemistry or the type of chemical interaction that is possible in each solvent.

SOLUBILITY TESTS

Procedure. Place about 1 mL of the solvent in a small test tube. Add one drop of an unknown liquid from a Pasteur pipet or a few crystals of an unknown solid from the end of a spatula, directly into the solvent. Gently tap the test tube with your finger to ensure mixing, and then observe whether any mixing lines appear in the solution. The disappearance of the liquid or solid, or the appearance of the mixing lines, indicates that solution is taking place. Add several more drops of the liquid, or a few more crystals of the solid, to determine the extent of the compound's solubility. A common mistake in determining the solubility of a compound is testing with a quantity of the unknown too large to dissolve in the chosen solvent. Use small amounts. It may take several minutes

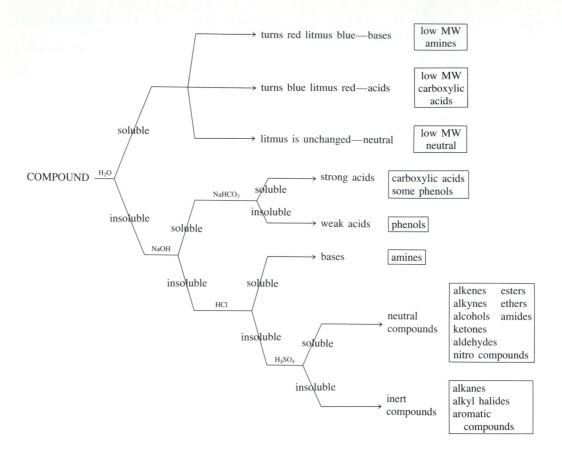

to dissolve solids. Compounds in the form of large crystals will need more time to dissolve than powders or very small crystals. In some cases it is helpful to pulverize a compound with large crystals using a mortar and pestle. Sometimes gentle heating helps, but strong heating is discouraged, as it often leads to reaction. When colored compounds dissolve, the solution often assumes the color.

By the above procedure, the solubility of the unknown should be determined in each of the following solvents: water, 5% NaOH, 5% NaHCO$_3$, 5% HCl, and concentrated H$_2$SO$_4$. With sulfuric acid, a color change may be observed rather than solution. A color change should be regarded as a positive solubility test. Solid unknowns that do not dissolve in any of the test solvents may be inorganic substances. To eliminate this possibility, one must determine the solubility of the unknown in several organic solvents, like ether. If the compound is organic, a solvent that will dissolve it can usually be found.

If a compound is found to dissolve in water, the pH of the aqueous solution should be estimated with pH paper or litmus. Compounds soluble in water are usually soluble in **all** the aqueous solvents. If a compound is only slightly soluble in water, it may be **more** soluble in another aqueous solvent. For instance, a carboxylic acid may be only slightly soluble in water but very soluble in dilute base. It will often not be necessary to determine the solubility of the unknown in every solvent.

Test Compounds.   Five solubility unknowns will be found on the supply shelf. The five unknowns include a base, a weak acid, a strong acid, a neutral substance with an oxygen-containing functional group, and a neutral substance that is inert. Using solubility tests, distinguish these unknowns by type. Verify your answer with the instructor.

SOLUBILITY IN WATER

Compounds that contain four or fewer carbons and also contain oxygen, nitrogen, or sulfur are often soluble in water. Almost any functional group containing these elements will lead to water solubility for low-molecular-weight (C_4) compounds. Compounds having five or six carbons and any of those elements are often insoluble in water or have borderline solubility. Branching of the alkyl chain in a compound lowers the intermolecular forces between its molecules. This is usually reflected in a lowered boiling point or melting point and a greater solubility in water for the branched compound than for the corresponding straight-chain compound. This occurs simply because the molecules of the branched compound are more easily separated from one another. Thus, *t*-butyl alcohol would be expected to be more soluble in water than *n*-butyl alcohol.

When the ratio of the oxygen, nitrogen, or sulfur atoms in a compound to the carbon atoms is increased, the solubility of that compound in water often increases. This is due to the increased number of polar functional groups. Thus, 1,5-pentanediol would be expected to be more soluble in water than 1-pentanol.

As the size of the alkyl chain of a compound is increased beyond about four carbons, the influence of a polar functional group is diminished and the water solubility begins to decrease. A few examples of these generalizations are given below.

Soluble	Borderline	Insoluble

SOLUBILITY IN 5% HCl

The possibility of an amine should be considered immediately if a compound is soluble in dilute acid (5% HCl). Aliphatic amines (RNH_2, R_2NH, R_3N) are basic compounds that readily dissolve in acid because they form hydrochloride salts that are soluble in the aqueous medium:

$$R—NH_2 + HCl \longrightarrow R—NH_3^+ + Cl^-$$

The substitution of an aromatic ring, Ar, for an alkyl group, R, reduces the basicity of an amine somewhat, but the amine will still protonate, and it will still generally be soluble in dilute acid. The reduction in basicity in an aromatic amine is due to the resonance delocalization of the unshared electrons on the amino nitrogen of the free base. The delocalization is lost on protonation, a problem that does not exist for aliphatic amines. The substitution of two or three aromatic rings on an amine nitrogen reduces the basicity of the amine even further. Diaryl and triaryl amines do not dissolve in dilute HCl since they do not protonate easily. Thus, Ar_2NH and Ar_3N are insoluble in dilute acid. Some amines of very high molecular weight, like tribromoaniline (molecular weight [MW] 330), may also be insoluble in dilute acid.

SOLUBILITY IN 5% NaHCO₃ AND 5% NaOH

Compounds that dissolve in sodium bicarbonate, a weak base, are strong acids. Compounds that dissolve in sodium hydroxide, a strong base, may be either strong or weak acids. Thus, one can distinguish weak and strong acids by determining their solubility in both strong (NaOH) and weak ($NaHCO_3$) base. The classification of some functional groups as either weak or strong acids is given in the table on page 459.

In this experiment, carboxylic acids ($pK_a \sim 5$) are generally indicated when a compound is soluble in both bases, while phenols ($pK_a \sim 10$) are indicated when it is soluble in NaOH only.

Compounds dissolve in base because they form sodium salts that are soluble in the aqueous medium. The salts of some high-molecular-weight compounds are not soluble, however, and precipitate. The salts of the long-chain carboxylic acids, such as myristic (C_{14}), palmitic (C_{16}), and stearic (C_{18}) acids, which form soaps, are in this category. Some phenols also produce insoluble sodium salts, and often these are colored due to resonance in the anion.

STRONG ACIDS (Soluble in both NaOH and NaHCO$_3$)	**WEAK ACIDS** (Soluble in NaOH but not NaHCO$_3$)

Sulfonic acids   RSO$_3$H
Carboxylic acids RCOOH
Ortho- and *para*-substituted
  di- and trinitrophenols
	Phenols ArOH
	Nitroalkanes RCH$_2$NO$_2$
	             R$_2$CHNO$_2$
	
	β-Diketones
	
	β-Diesters
	
	Imides
	
	Sulfonamides ArSO$_2$NH$_2$
	                ArSO$_2$NHR

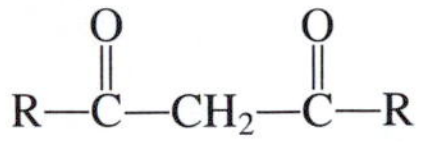

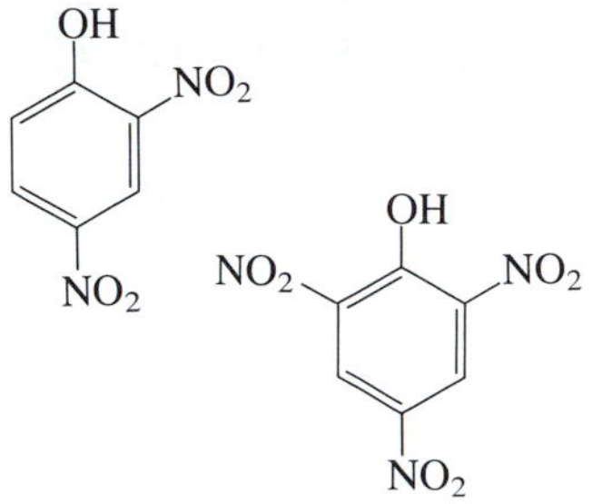

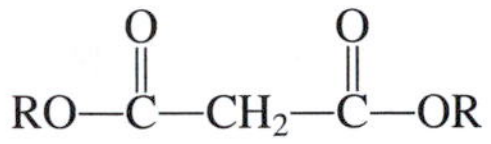

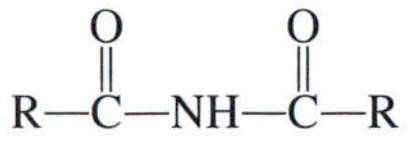

Both phenols and carboxylic acids produce resonance-stabilized conjugate bases. Thus, bases of the appropriate strength may easily remove their acidic protons to form the sodium salts.

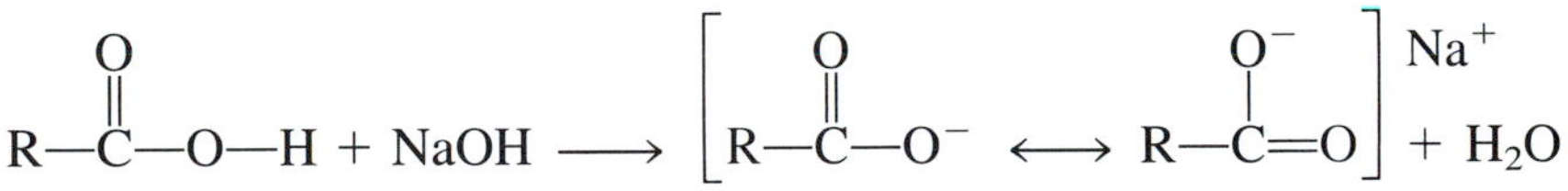

$$R—\overset{\overset{\textstyle O}{\|}}{C}—O—H + NaOH \longrightarrow \left[R—\overset{\overset{\textstyle O}{\|}}{C}—O^- \longleftrightarrow R—\overset{\overset{\textstyle O^-}{|}}{C}=O \right] Na^+ + H_2O$$

Delocalized anion

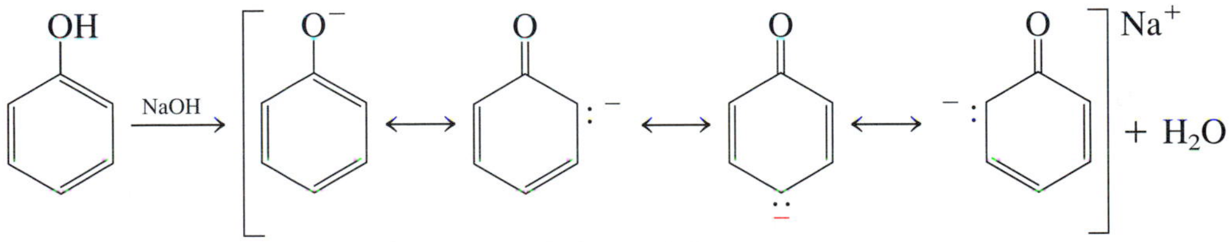

Delocalized anion

In phenols, substitution of nitro groups in the *ortho* and *para* positions of the ring increases the acidity. Nitro groups in these positions provide additional delocalization in the conjugate anion. Phenols that have two or three nitro groups in the *ortho* and *para* positions often dissolve in **both** sodium hydroxide and sodium bicarbonate solutions.

SOLUBILITY IN CONCENTRATED SULFURIC ACID

Many compounds are soluble in cold concentrated sulfuric acid. Of the compounds included in this experiment, alcohols, ketones, aldehydes, and esters are in this category. Other compounds that also dissolve include alkenes, alkynes, ethers, nitroaromatics, and amides. Since several different kinds of compounds are soluble in sulfuric acid, further chemical tests and spectroscopy will be needed to differentiate among them.

Compounds that are soluble in concentrated sulfuric acid but not in dilute acid are extremely weak bases. Almost any compound containing a nitrogen, an oxygen, or a sulfur atom can be protonated in concentrated sulfuric acid. The ions produced are soluble in the medium.

$$R-O-H + H_2SO_4 \longrightarrow R-\overset{+}{\underset{\underset{H}{|}}{O}}-H + HSO_4^- \longrightarrow R^+ + H_2O + HSO_4^-$$

$$R-\overset{\overset{O}{\|}}{C}-R + H_2SO_4 \longrightarrow R-\overset{\overset{+O-H}{\|}}{C}-R + HSO_4^-$$

$$R-\overset{\overset{O}{\|}}{C}-OR + H_2SO_4 \longrightarrow R-\overset{\overset{+O-H}{\|}}{C}-OR + HSO_4^-$$

$$\underset{R}{\overset{R}{>}}C=C\underset{R}{\overset{R}{<}} + H_2SO_4 \longrightarrow R-\overset{\overset{R}{|}}{\underset{\underset{H}{|}}{C}}-\overset{\overset{R}{|}}{\underset{+}{C}}-R + HSO_4^-$$

INERT COMPOUNDS

Compounds not soluble in concentrated sulfuric acid or any of the other solvents are said to be **inert.** Compounds not soluble in concentrated sulfuric acid include the alkanes, most simple aromatics, and the alkyl halides. Some examples of inert compounds are hexane, benzene, chlorobenzene, chlorohexane, and toluene.

Procedure 56B
Tests for the Elements (N, S, X)

$$-N- \qquad \overset{-Br}{} \qquad -C\equiv N$$
$$\qquad -NO_2$$
$$-Cl \qquad \overset{S}{\diagup\diagdown} \qquad -I$$

Except for amines (Procedure 56G), which are easily detected by their solubility behavior, all compounds issued in this experiment will contain heteroelements (N, S, Cl, Br, or I) only as **secondary** functional groups. These will be subsidiary to some other important functional group. Thus, no alkyl or aryl halides, nitro compounds, thiols, or thioethers will be issued. However, some of the unknowns may contain a halogen or a nitro group. Less frequently, they may contain a sulfur atom or a cyano group.

Consider as an example *p*-bromobenzaldehyde, an **aldehyde** that contains bromine as a ring substituent. The identification of this compound would hinge on whether the investigator could identify it as an aldehyde. It could probably be identified **without** proving the existence of bromine in the molecule. That information, however, could make the identification easier. In this experiment, methods are given for identifying the presence of a halogen or a nitro group in an unknown compound. Also given is a general method (sodium fusion) for detecting the principal heteroelements that may exist in organic molecules.

CLASSIFICATION TESTS

HALIDES	NITRO GROUPS	N, S, X(Cl, Br, I)
Beilstein test Silver nitrate Sodium iodide/acetone	Ferrous hydroxide	Sodium fusion

TESTS FOR A HALIDE

BEILSTEIN TEST

Procedure. Bend a small loop in the end of a short length of copper wire. Heat the loop end of the wire in a Bunsen burner flame. After cooling, dip the wire directly into a small sample of the unknown. Now, heat the wire in the Bunsen burner flame again. The compound will first burn. After the burning, a green flame will be produced if a halogen is present.

Test Compounds. Try this test on bromobenzene and benzoic acid.

Halogens can be detected easily and reliably by the Beilstein test. It is the simplest method for determining the presence of a halogen, but it does not differentiate among chlorine, bromine, and iodine, any one of which will give a positive test. However, when the identity of the unknown has been narrowed to two choices, of which one has a halogen and one does not, the Beilstein test will often be enough to distinguish between the two.

A positive Beilstein test results from the production of a volatile copper halide when an organic halide is heated with copper oxide. The copper halide imparts a blue-green color to the flame.

This test can be very sensitive to small amounts of halide impurities in some compounds. Therefore, one should use caution in interpreting the results of the test, especially when only a weak color has been obtained.

SILVER NITRATE TEST

Procedure. Add one drop of a liquid or five drops of a concentrated ethanolic solution of a solid unknown to 2 mL of a 2% ethanolic silver nitrate solution. If no reaction is observed after five minutes at room temperature, heat the solution on a steam bath and note whether a precipitate forms. If a precipitate forms, add two drops of 5% nitric acid and note whether the precipitate dissolves. Carboxylic acids give a false test by precipitating in silver nitrate, but they dissolve when nitric acid is added. Silver halides, on the other hand, do not dissolve in nitric acid.

Test Compounds. Apply this test to benzyl bromide (α-bromotoluene) and bromobenzene. Discard all waste reagents into a suitable waste container in the hood, since benzyl bromide is a lachrymator.

This test depends on the formation of a white or an off-white precipitate of silver halide when silver nitrate is allowed to react with a sufficiently reactive halide.

$$\text{RX} + \text{Ag}^+\text{NO}_3{}^- \longrightarrow \underset{\textbf{Precipitate}}{\textbf{AgX}} + \text{R}^+\text{NO}_3{}^- \xrightarrow{\text{CH}_3\text{CH}_2\text{OH}} \text{R—O—CH}_2\text{CH}_3$$

The test does not distinguish among chlorides, bromides, and iodides but does distinguish **labile** (reactive) halides from halides that are unreactive. Halides substituted on an aromatic ring will not usually give a positive silver nitrate test; however, alkyl halides of many types will give a positive test.

The most reactive compounds are those able to form stable carbocations in solution and those equipped with good leaving groups (X = I, Br, Cl). Benzyl, allyl, and tertiary halides give immediate reaction with silver nitrate. Secondary and primary halides do not react at room temperature but readily react when heated. Aryl and vinyl halides do not react at all, even at elevated temperatures. This pattern of reactivity fits the stability order for various carbocations quite well. Compounds that produce stable carbocations react at higher rates than those that do not.

| Benzyl | Allyl | 3° | 2° | 1° | Methyl | Vinyl | Aryl |

The fast reaction of benzylic and allylic halides is a result of the resonance stabilization that is available to the intermediate carbocations formed. Tertiary halides are more reactive than secondary halides, which are in turn more reactive than primary or methyl halides since alkyl substituents are able to stabilize the intermediate carbocations by an electron-releasing effect. The methyl carbocations has no alkyl groups and is the least stable of all the carbocations mentioned thus far. Vinyl and aryl carbocations are extremely unstable since the charge is localized on an sp^2-hybridized carbon (double-bond carbon) rather than on one that is sp^3-hybridized.

SODIUM IODIDE IN ACETONE

Procedure. This test is described in Experiment 10.

DETECTION OF NITRO GROUPS

Although nitro compounds will not be issued as distinct unknowns, many of the unknowns may have a nitro group as a secondary functional group. The presence of a nitro group, and hence nitrogen, in an unknown compound is determined most easily by infrared spectroscopy. However, many nitro compounds give a positive result in the following test.

FERROUS HYDROXIDE

Procedure. Place 1.5 mL of freshly prepared 5% aqueous ferrous ammonium sulfate in a small test tube and add about 10 mg of the unknown compound. Mix the solution well and then add first one drop of $2M$ sulfuric acid, and then 1 mL of $2M$ potassium hydroxide in methanol. Stopper the test tube and shake it vigorously. A positive test is indicated by the formation of a red-brown precipitate, usually within one minute.

Most nitro compounds oxidize ferrous hydroxide to ferric hydroxide, which is a red-brown solid. A precipitate indicates a positive test.

$$R-NO_2 + 4H_2O + 6Fe(OH)_2 \longrightarrow R-NH_2 + 6Fe(OH)_3$$

SPECTROSCOPY
INFRARED

The nitro group gives two strong bands near 1560 and 1350 cm^{-1} (6.4 and 7.4 μ).

DETECTION OF A CYANO GROUP

Although nitriles will not be given as unknowns in this experiment, the cyano group may be a subsidiary functional group whose presence or absence is important to the final identification of an unknown compound. The cyano group can be hydrolyzed in strong base, with vigorous heating, to give a carboxylic acid and ammonia gas:

$$R-C\equiv N + 2H_2O \xrightarrow[\Delta]{NaOH} R-COOH + NH_3$$

The ammonia can be detected by its odor or by moist pH paper. However, this method is somewhat difficult, and the presence of a nitrile group is confirmed most easily by infrared spectroscopy. No other functional groups (except some C≡C) absorb in the same region of the spectrum as C≡N.

SPECTROSCOPY

INFRARED

C≡N stretch is a very sharp band of medium intensity near 2250 cm^{-1} (4.5 μ).

SODIUM FUSION TESTS (OPTIONAL)
(DETECTION OF N, S, AND X)

When an organic compound containing nitrogen, sulfur, or halide atoms is fused with sodium metal, a reductive decomposition of the compound takes place, which converts these atoms to the sodium salts of the inorganic ions CN$^-$, S^{2-}, and X$^-$.

$$[\text{N, S, X}] \xrightarrow[\Delta]{\text{Na}} \text{NaCN, Na}_2\text{S, NaX}$$

When the fusion mixture is dissolved in distilled water, the cyanide, sulfide, and halide ions can be detected by standard qualitative inorganic tests.

> **CAUTION:** Always remember to manipulate the sodium metal with a knife or a forceps, and not to touch it with your fingers. Keep sodium away from water. Destroy all waste sodium with 1-butanol or ethanol. **WEAR SAFETY GLASSES.**

SODIUM FUSION

Procedure.   Using a forceps and a knife, take some sodium from the storage container, cut a small piece about the size of a small pea (3 mm on a side), and dry it on a paper towel. Place this small piece of sodium in a clean and dry small test tube (10 × 75 mm). Clamp the test tube to a ring stand and heat the bottom of the tube with a microburner until the sodium melts and its metallic vapor can be seen to rise about a third of the way up the tube. The bottom of the tube will probably have a dull red glow. Remove the burner and **immediately** drop the sample directly into the tube. About 10 mg of a solid placed on the end of a spatula or two to three drops of a liquid should be used. Be sure to drop the sample directly down the center of the tube so that it touches the hot sodium metal and does not adhere to the side of the test tube. There will usually be a flash or a small explosion if the fusion is successful. If the reaction is not successful, the tube should be heated to red heat for a few seconds to ensure complete reaction.

Allow the test tube to cool to room temperature and then carefully add 10 drops of methanol, a drop at a time, to the fusion mixture. Using a spatula or a long glass rod, reach into the test tube and stir the mixture to ensure complete reaction of any excess sodium metal. The fusion will have destroyed the test tube for other uses. Thus, the easiest way to recover the fusion mixture is to crush the test tube into a small beaker containing 5–10 mL of **distilled** water. The tube is easily crushed if it is placed in the angle of a clamp holder. Tighten the clamp until the tube is securely held near its bottom, and then—standing back from the beaker and holding the clamp at its opposite end—continue tightening the clamp until the test tube breaks and the pieces fall into the beaker. Stir the solution well, heat it to boiling, and then filter it by gravity through a fluted filter, (Figure 4–3, page 567). Portions of this solution will be used for the tests to detect nitrogen, sulfur, and the halogens.

ALTERNATIVE METHOD

Procedure. With some volatile liquids, the method given above will not work. The compounds volatilize before they reach the sodium vapors. For such compounds, place four or five drops of the pure liquid in the clean, dry test tube, clamp it, and cautiously add the small piece of sodium metal. If there is any reaction, wait until it subsides. Then heat the test tube to red heat and continue according to the instructions at the top of this page.

NITROGEN TEST

Procedure. Using pH paper and a 10% sodium hydroxide solution, adjust the pH of about 1 mL of the stock solution to pH 13. Add two drops of saturated ferrous ammonium sulfate solution and two drops of 30% potassium fluoride solution. Boil the solution for about 30 seconds. Then acidify the hot solution by adding 30% sulfuric acid dropwise until the iron hydroxides dissolve. Avoid using excess acid. If nitrogen is present, a dark blue (not green) precipitate of Prussian blue, $NaFe_2(CN)_6$, will form or the solution will assume a dark blue color.

Reagents. Dissolve 5 g of ferrous ammonium sulfate in 100 mL of water and 30 g of potassium fluoride in 100 mL of water.

SULFUR TEST

Procedure. Acidify about 1 mL of the test solution with acetic acid and add a few drops of a 1% lead acetate solution. The presence of sulfur is indicated by a black precipitate of lead sulfide, PbS.

> **CAUTION: Many compounds of lead(II) are suspected carcinogens (see p 11) and should be handled with care. Avoid contact.**

HALIDE TESTS

Procedure. Cyanide and sulfide ions interfere with the test for halides. If such ions are present, they must be removed. To accomplish this, acidify the solution with dilute nitric acid and boil it for about two minutes. This will drive off any HCN or H_2S that is formed. When the solution cools, add a few drops of a 5% silver nitrate solution. A **voluminous** precipitate indicates a halide. A faint turbidity **does not** mean a positive test. Silver chloride is white. Silver bromide is off-white. Silver iodide is yellow. Silver chloride will readily dissolve in concentrated ammonium hydroxide, whereas silver bromide is only slightly soluble.

DIFFERENTIATION OF CHLORIDE, BROMIDE, AND IODIDE

Procedure. Acidify 2 mL of the test solution with 10% sulfuric acid and boil it for about two minutes. Cool the solution and add about 0.5 mL of methylene chloride. Add a few drops of chlorine water or 2–4 mg of calcium hypochlorite.[1] Check to be sure that the solution is still acidic. Then stopper the tube, shake it vigorously, and set it aside to allow the layers to separate. An orange to brown color in the methylene chloride layer indicates bromine. Violet indicates iodine. No color or a **light** yellow indicates chlorine.

Procedure 56C
Tests for Unsaturation

The unknowns to be issued for this experiment have neither a double bond nor a triple bond as their **only** functional group. Hence, simple alkenes and alkynes can be ruled out as possible compounds. Some of the unknowns may have a double or a triple bond, however, **in addition to** another more important functional group. The tests described allow one to determine the presence of a double bond or a triple bond (unsaturation) in such compounds.

---

[1] Clorox, the commercial bleach, is a permissible substitute for chlorine water, as is any other brand of bleach, provided that it is based on sodium hypochlorite.

CLASSIFICATION TESTS

UNSATURATION	AROMATICITY
Bromine–carbon tetrachloride Potassium permanganate	Ignition test

---

TESTS FOR SIMPLE MULTIPLE BONDS

BROMINE IN CARBON TETRACHLORIDE OR METHYLENE CHLORIDE

Procedure. Dissolve 50 mg of a solid unknown or two drops of a liquid unknown in 1 mL of carbon tetrachloride (or 1,2-dimethoxyethane). Add a 2% (by volume) solution of bromine in carbon tetrachloride, dropwise with shaking, until the bromine color persists. The test is positive if more than five drops of the bromine solution are needed so that the color remains for one minute. Usually, many drops of the bromine solution will be needed if unsaturation is present. Hydrogen bromide should not be evolved. If hydrogen bromide gas is evolved, one will note a "fog" while blowing across the mouth of the test tube. The HBr can also be detected by a moistened piece of litmus or pH paper. If hydrogen bromide is evolved, the reaction is a **substitution reaction** and not an **addition reaction,** and a double or triple bond is probably not present.

Methylene Chloride. Even though carbon tetrachloride is used in very small quantities in this test, it poses certain health hazards (see p 7), and another solvent may be preferable. Methylene chloride (dichloromethane) can be substituted for carbon tetrachloride. Certain problems arise, however, because methylene chloride slowly reacts with bromine, presumably by a light-induced free-radical process, to produce HBr. After about a week, the color of a 2% solution of bromine in methylene chloride fades noticeably, and the odor of HBr can be detected in the reagent. Although the decolorization tests still work satisfactorily, the presence of HBr makes it difficult to distinguish between addition and substitution reactions. A freshly prepared solution of bromine in methylene chloride must be used to make this distinction. Deterioration of the reagent can be forestalled by storing it in a brown glass bottle. Most other substitute solvents also present problems. Ethers, for instance, react slowly in the same way as methylene chloride, and hydrocarbons, like hexane, are not general enough solvents to be able to dissolve all the possible test compounds.

Test Compounds. Try this test with cyclohexene, cyclohexane, toluene, and acetone.

A successful test depends on the addition of bromine, a red liquid, to a double or a triple bond to give a colorless dibromide:

$$\begin{array}{c}\diagup\!\!\!\diagdown \\ \diagup C = C \diagdown \end{array} + Br_2 \longrightarrow \begin{array}{c} Br \\ | \\ \diagup C - C \diagdown \\ | \\ Br \end{array}$$

Red   **Colorless**

Not all double bonds react with bromine–carbon tetrachloride solution. Only those that are electron-rich are sufficiently reactive nucleophiles to initiate the reaction. A double bond that is substituted by electron-withdrawing groups often fails to react or reacts slowly. Fumaric acid is an example of a compound that fails to give the reaction.

Fumaric acid

Aromatic compounds either do not react with bromine–carbon tetrachloride reagent or they react by **substitution.** Only the aromatic rings that have activating groups as substituents (—OH, —OR, or —NR$_2$) give the substitution reaction.

Some ketones and aldehydes react with bromine to give a **substitution** product, but this reaction is slow except for acetone and some aldehydes and ketones that have a high enol content. When substitution occurs, not only is the bromine color discharged, but hydrogen bromide gas is also evolved.

POTASSIUM PERMANGANATE (BAEYER TEST)

Procedure. Dissolve 25 mg of a solid unknown or two drops of the liquid unknown in 2 mL of water or 95% ethanol (1,2-dimethoxyethane may also be used). Slowly add a 1% aqueous solution (weight/volume) of potassium permanganate, drop by drop with shaking, to the unknown. In a positive test, the purple color of the reagent is discharged, and a brown precipitate of manganese dioxide forms, usually within one minute. If alcohol was the solvent, the solution should not be allowed to stand for more than five minutes since oxidation of the alcohol will begin slowly. Since permanganate solutions undergo some decomposition to manganese dioxide on standing, any small amount of precipitate should be interpreted with caution.

Test Compounds. Try this test on cyclohexene and toluene.

This test is positive for double and triple bonds but not for aromatic rings. It depends on the conversion of the purple ion MnO$_4^-$ to a brown precipitate of MnO$_2$ following the oxidation of an unsaturated compound.

Other easily oxidized compounds also give a positive test with potassium permanganate solution. These substances include aldehydes, some alcohols, phenols, and aromatic amines. If you suspect that any of these functional groups are present, you should interpret the test with caution.

SPECTROSCOPY

INFRARED

Double bonds (C=C)

C=C stretch usually occurs near 1680–1620 cm^{-1} (5.95–6.17 μ). Symmetrical alkenes may have no absorption.

C—H stretch of vinyl hydrogens occurs >3000 cm^{-1} (3.33 μ), but usually not higher than 3150 cm^{-1} (3.18 μ).

C—H out-of-plane bending occurs near 1000–700 cm^{-1} (10.0–14.3 μ).

Triple bonds (C≡C)

C≡C stretch usually occurs near 2250–2100 cm^{-1} (4.44–4.76 μ). The peak is usually sharp. Symmetrical alkynes show no absorption.

C—H stretch of terminal acetylenes occurs near 3310–3200 cm^{-1} (3.02–3.12 μ).

NUCLEAR MAGNETIC RESONANCE

Vinyl hydrogens have resonance near 5–7δ and have coupling values as follows: J_{trans} = 11–18 Hz, J_{cis} = 6–15 Hz, $J_{geminal}$ = 0–5 Hz. Allylic hydrogens have resonance near 2δ. Acetylenic hydrogens have resonance near 2.8–3.0δ.

TESTS FOR AROMATICITY

None of the unknowns to be issued for this experiment will be simple aromatic hydrocarbons. All aromatic compounds will have a principal functional group as a part of their structure. Nevertheless, in many cases it will be useful to be able to recognize the presence of an aromatic ring. Although spectroscopy provides the easiest method of determining aromatic systems, often they can be detected by a simple ignition test.

IGNITION TEST

Procedure. Place a small amount of the compound on a spatula and place it in the flame of a Bunsen burner. Observe whether a sooty flame is the result. Compounds giving a sooty yellow flame have a high degree of unsaturation and may be aromatic.

Test Compound. Try this test with naphthalene.

The presence of an aromatic ring or other centers of unsaturation will lead to the production of a sooty yellow flame in this test. Compounds that contain little oxygen, and have a high carbon-to-hydrogen ratio, burn at a low temperature with a yellow flame. Much carbon is produced when they are burned. Compounds that contain oxygen generally burn at a higher temperature with a clean blue flame.

SPECTROSCOPY

INFRARED

C=C aromatic ring double bonds appear in the 1650–1450 cm^{-1} (6–7 μ) region. There are often four sharp absorptions that occur in pairs near 1600 cm^{-1} (6.3 μ) and 1450 cm^{-1} (6.9 μ), which are characteristic of an aromatic ring.

Special ring absorptions: There are often weak ring absorptions around 2000–1600 cm^{-1} (5–6 μ). These are often obscured, but when they can be observed, the relative shapes and numbers of these peaks can often be used to ascertain the type of ring substitution (see Appendix 3, "Infrared Spectroscopy").

=C—H stretch, aromatic ring: The aromatic C—H stretch always occurs at a higher frequency than 3000 cm^{-1} (shorter wavelength than 3.33 μ).

=C—H out-of-plane bending peaks appear in the region 900–690 cm^{-1} (11–15 μ). The number and position of these peaks can be used to determine the substitution pattern of the ring (see Appendix 3, "Infrared Spectroscopy").

NUCLEAR MAGNETIC RESONANCE

Hydrogens attached to an aromatic ring usually have resonance near 7δ. Monosubstituted rings not substituted by anisotropic or electronegative groups usually give a single resonance for all the ring hydrogens. Monosubstituted rings with anisotropic or electronegative groups usually have the aromatic resonances split into two groups integrating either 3:2 or 2:3. A nonsymmetric, *para*-disubstituted ring has a characteristic four-peak splitting pattern (see Appendix 4, "Nuclear Magnetic Resonance").

Procedure 56D
Aldehydes and Ketones

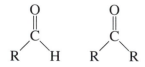

Compounds containing the carbonyl functional group, $\diagdown$C=O, where it has only hydrogen atoms or alkyl groups as substituents are called aldehydes, RCHO, or ketones, RCOR′. The chemistry of these compounds is primarily due to the chemistry of the carbonyl functional groups. These compounds are identified by the distinctive reactions of the carbonyl function.

SOLUBILITY CHARACTERISTICS	CLASSIFICATION TESTS

HCl   NaHCO$_3$ NaOH   H$_2$SO$_4$ Ether
(−)   (−)   (−)   (+)   (+)
 	
Water: <C$_5$ and some C$_6$ (+)
             >C$_5$ (−)
	Aldehydes and ketones
	   2,4-Dinitrophenylhydrazine
	Aldehydes only **Methyl ketones**
	   Chromic acid Iodoform test
	   Tollens reagent
	Compounds with high enol content
	   Ferric chloride test

CLASSIFICATION TESTS

Most aldehydes and ketones give a solid, yellow-to-red precipitate when mixed with 2,4-dinitrophenylhydrazine. However, only aldehydes will reduce chromium(VI) or silver(I). By this difference in behavior, you can differentiate between aldehydes and ketones.

2,4-DINITROPHENYLHYDRAZINE

Procedure. Place one drop of the liquid unknown in a small test tube and add 1 mL of the 2,4-dinitrophenylhydrazine reagent. If the unknown is a solid, dissolve about 10 mg (estimate) in a minimum amount of 95% ethanol or bis(2-ethoxyethyl) ether before adding the reagent. Shake the mixture vigorously. Most aldehydes and ketones will give a yellow-to-red precipitate immediately. However, some compounds will require up to 15 minutes, or even **gentle** heating, to give a precipitate. A precipitate indicates a positive test.

Test Compounds. Try this test on cyclohexanone, benzaldehyde, and benzophenone.

> **CAUTION: Many derivatives of phenylhydrazine are suspect carcinogens (see p 11) and should be handled with care. Avoid contact.**

Reagent. Dissolve 3.0 g of 2,4-dinitrophenylhydrazine in 15 mL of concentrated sulfuric acid. In a beaker mix 20 mL of water and 70 mL of 95% ethanol. With vigorous stirring, slowly add the 2,4-dinitrophenylhydrazine solution to the aqueous ethanol mixture. After thorough mixing, filter the solution by gravity through a fluted filter.

Most aldehydes and ketones give a precipitate, but esters generally do not give this result. Thus, an ester usually can be eliminated by this test. The color of the

$$\underset{\substack{\text{Aldehyde} \\ \text{or ketone}}}{\overset{R}{\underset{R'}{\diagup}}C=O} + \underset{\text{2,4-Dinitrophenylhydrazine}}{H_2N-NH-\!\!\!\underset{NO_2}{\overset{O_2N}{\bigcirc}}\!\!\!-NO_2} \xrightarrow{H^+}$$

$$\underset{\text{2,4-Dinitrophenylhydrazone}}{\overset{R}{\underset{R'}{\diagup}}C=N-NH-\!\!\!\underset{}{\overset{O_2N}{\bigcirc}}\!\!\!-NO_2 + H_2O}$$

2,4-dinitrophenylhydrazone (precipitate) formed is often a guide to the amount of conjugation in the original aldehyde or ketone. Unconjugated ketones, such as cyclohexanone, give yellow precipitates, whereas conjugated ketones, such as benzophenone, give orange-to-red precipitates. Compounds that are highly conjugated give red precipitates. However, the 2,4-dinitrophenylhydrazine reagent is itself orange-red, and the color of any precipitate must be judged cautiously. Occasionally, compounds that are either strongly basic or strongly acidic precipitate the unreacted reagent.

Some allylic and benzylic alcohols give this test because the reagent can oxidize them to aldehydes and ketones, which subsequently react. Some alcohols may be contaminated with carbonyl impurities, either because of their method of synthesis (reduction) or because they have become air-oxidized. A precipitate formed from small amounts of impurity in the solution will be formed in small amount. With some caution, a test that gives only a slight amount of precipitate can usually be ignored. The infrared spectrum of the compound should establish its identity and identify any impurities present.

CHROMIC ACID TEST

Procedure. Dissolve one drop of a liquid or 10 mg (approximate) of a solid aldehyde in 1 mL of **reagent-grade** acetone. Add several drops of the chromic acid reagent, a drop at a time with shaking. A positive test is indicated by a green precipitate and a loss of the orange color in the reagent. With aliphatic aldehydes, RCHO, the solution turns cloudy within five seconds and a precipitate appears within 30 seconds. With aromatic aldehydes, ArCHO, it generally takes 30 to 120 seconds for a precipitate to form; but with some, it may take even longer.

In a negative test, usually there is no precipitate. In some cases, however, a precipitate forms, but the solution remains orange.

In performing this test, one should make quite sure that the acetone used for the solvent does not give a positive test with the reagent. Add several drops of the chromic acid reagent to a few drops of the reagent acetone contained in a small test tube. Allow this mixture to stand for three to five minutes. If no reaction has occurred by this time, the acetone is pure enough to use as a solvent for the test. If a positive test resulted, try another bottle of acetone, or distill some acetone from potassium permanganate to purify it.

Test Compounds. Try this test on benzaldehyde, butanal (butyraldehyde), and cyclohexanone.

> **CAUTION: Many compounds of chromium(VI) are suspected carcinogens (see p 11) and should be handled with care. Avoid contact.**

Reagent. Dissolve 1.0 g of chromic oxide, CrO_3, in 1 mL of concentrated sulfuric acid. Then dilute this mixture carefully with 3 mL of water.

This test has as its basis the fact that aldehydes are easily oxidized to the corresponding carboxylic acid by chromic acid. The green precipitate is due to chromous sulfate.

$$2CrO_3 + 2H_2O \xrightleftharpoons{H^+} 2H_2CrO_4 \xrightleftharpoons{H^+} H_2Cr_2O_7 + H_2O$$

$$3RCHO + H_2Cr_2O_7 + 3H_2SO_4 \longrightarrow 3RCOOH + Cr_2(SO_4)_3 + 4H_2O$$

Orange   Green

Primary and secondary alcohols are also oxidized by this reagent (see Procedure 56H). Therefore, this test is not useful in identifying aldehydes **unless** a positive identification of the carbonyl group has already been made. Aldehydes give a 2,4-dinitrophenyl-hydrazine test, whereas alcohols do not.

There are numerous other tests used to detect the aldehyde functional group. Most are based on an easily detectible oxidation of the aldehyde to a carboxylic acid. The most common tests are the Tollens, Fehling, and Benedict tests. The Benedict test is described in Experiment 57. Only the Tollens test is described here.

TOLLENS TEST

Procedure. The reagent must be prepared immediately before use. To prepare the reagent, mix 1 mL of Tollens solution A with 1 mL of Tollens solution B. A precipitate of silver oxide will form. Add enough dilute (10%) ammonia solution (dropwise) to the mixture to **just** dissolve the silver oxide. The reagent so prepared can be used immediately for the test below.

Dissolve one drop of a liquid aldehyde or 10 mg (approximate) of a solid aldehyde in the minimum amount of bis(2-ethoxyethyl) ether. Add this solution, a little at a time, to the 2–3 mL of reagent contained in a small test tube. Shake the solution well. If a mirror of silver is deposited on the inner walls of the test tube, the test is positive. In some cases it may be necessary to warm the test tube in a bath of warm water.

> **CAUTION:** The reagent should be prepared immediately before use and all residues disposed of immediately after use. Wash any residues down a sink with a large quantity of water. On standing, the reagent tends to form silver fulminate, a *very explosive* substance. Solutions containing the mixed Tollens reagent should never be stored.

Test Compounds. Try the test on acetone and benzaldehyde.

Reagents. Solution A: Dissolve 3.0 g of silver nitrate in 30 mL of water. Solution B: Prepare a 10% sodium hydroxide solution.

Most aldehydes reduce ammoniacal silver nitrate solution to give a precipitate of silver metal. The aldehyde is oxidized to a carboxylic acid:

$$RCHO + 2Ag(NH_3)_2OH \longrightarrow 2Ag + RCOO^-NH_4^+ + H_2O + NH_3$$

Ordinary ketones do not give a positive result in this test. The test should be used only if it has already been shown that the unknown compound is either an aldehyde or a ketone.

IODOFORM TEST

Procedure. Using a Pasteur pipet, add four drops of a liquid unknown to a large test tube (20 × 150-mm). Alternatively, 0.1 g of a solid unknown may be used. Add to the test tube 2 mL of bis(2-ethoxyethyl) ether, 5 mL of water, and 1 mL of 10% sodium hydroxide solution. Obtain 3 mL of the iodine-potassium iodide test solution, and add it with a dropper in about six portions to the solution in the test tube. Shake the tube after each addition. During the early additions, the intense iodine color is decolorized rapidly, but near the end the color may not be discharged as rapidly. In any case, **cork** the test tube and shake it vigorously until the intensely colored solution has decolorized completely to give a pale yellow solution. It may be necessary to heat the solution slightly on a steam bath or in a water bath at 60 °C to aid in the discharge of the color. Again, shake the stoppered test tube vigorously after each heating period.

After the solution has been decolorized or nearly so, fill the test tube with water, cork the tube, and shake it vigorously. Allow the tube to stand for 15 minutes. A yellow precipitate of iodoform will form if the unknown was a methyl ketone or a compound easily oxidized to a methyl ketone. Other ketones will also decolorize the iodine solution, but they will not give a precipitate of iodoform **unless** there is an impurity of a methyl ketone in the unknown. To prove the identity of the yellow precipitate as iodoform, one should collect and dry the solid and determine its melting point (iodoform, mp 119–121 °C).

Test Compound. Try the test on acetone.

Reagent. Dissolve 20 g of potassium iodide and 10 g of iodine in 100 mL of water.

The basis of this test is the ability of certain compounds to form a precipitate of iodoform when treated with a basic solution of iodine. Methyl ketones are the most common type of compounds that give a positive result in this test. However, acetaldehyde, CH_3CHO, and alcohols with the hydroxyl group at the 2-position of the chain also give a precipitate of iodoform. Alcohols of the type described are easily oxidized to methyl ketones under the conditions of the reaction. The other product of the reaction, besides iodoform, is the potassium salt or the sodium salt of a carboxylic acid.

$$R\text{—}\underset{\underset{\text{OH}}{|}}{CH}\text{—}CH_3 \xrightarrow[\text{NaOH}]{I_2} R\text{—}\underset{\underset{O}{\|}}{C}\text{—}CH_3 \xrightarrow[\text{NaOH}]{I_2} R\text{—}\underset{\underset{O}{\|}}{C}\text{—}CI_3 \xrightarrow{\text{OH}^-} R\text{—}\underset{\underset{O}{\|}}{C}\text{—}O^- + HCI_3$$

Alcohol   **Methyl ketone**   **Iodoform**

FERRIC CHLORIDE TEST

Procedure. Some aldehydes and ketones, those that have a high **enol content,** give a positive ferric chloride test, as described for phenols in Procedure 56F.

SPECTROSCOPY

INFRARED

The carbonyl group is usually one of the strongest-absorbing groups in the infrared spectrum, with a very broad range: 1800–1650 cm^{-1} (5.85–6.20 μ). The aldehyde functional group has **very characteristic** CH stretch absorptions: two sharp peaks that lie **far outside** the usual region for —C—H, =C—H, or ≡C—H.

Aldehydes
C=O stretch at approximately 1725 cm^{-1} (5.80 μ) is normal. 1725–1685 cm^{-1} (5.80–5.95 μ).[2]
C—H stretch (aldehyde —CHO). Two weak bands at about 2750 cm and 2850 cm^{-1} (3.65 μ and 3.50 μ).

Ketones
C=O stretch at approximately 1715 cm^{-1} (5.85 μ) is normal. 1780–1665 cm^{-1} (5.62–6.01 μ).[3]

NUCLEAR MAGNETIC RESONANCE

Hydrogens alpha to a carbonyl group have resonance in the region between 2 and 3δ. The hydrogen of an aldehyde group has a characteristic resonance between 9 and 10δ.

---

[2] **Conjugation** moves the absorption to lower frequencies (higher wavelength). **Ring strain** (cyclic ketones) moves the absorption to higher frequencies (lower wavelength).

[3] See Footnote 2.

In aldehydes, there is coupling between the aldehyde hydrogen and any alpha hydrogens (J = 1–3 Hz).

DERIVATIVES

The most common derivatives of aldehydes and ketones are the 2,4-diphenyl-hydrazones, oximes, and semicarbazones. Procedures for preparing these derivatives are given in Appendix 2 at the end of the book.

$$\begin{array}{c}R \\ \diagdown \\ \diagup \\ R\end{array} C{=}O \; + \; H_2N{-}NH{-}\overset{O_2N}{\underset{}{\bigcirc}}{-}NO_2 \; \longrightarrow \; \begin{array}{c}R \\ \diagdown \\ \diagup \\ R\end{array} C{=}N{-}NH{-}\overset{O_2N}{\underset{}{\bigcirc}}{-}NO_2$$

2,4-Dinitrophenylhydrazine   2,4-Dinitrophenylhydrazone

$$\begin{array}{c}R \\ \diagdown \\ \diagup \\ R\end{array} C{=}O \; + \; H_2N{-}OH \; \longrightarrow \; \begin{array}{c}R \\ \diagdown \\ \diagup \\ R\end{array} C{=}N{-}OH$$

Hydroxylamine   Oxime

$$\begin{array}{c}R \\ \diagdown \\ \diagup \\ R\end{array} C{=}O \; + \; H_2N{-}NH\overset{O}{\overset{\|}{C}}NH_2 \; \longrightarrow \; \begin{array}{c}R \\ \diagdown \\ \diagup \\ R\end{array} C{=}N{-}NH\overset{O}{\overset{\|}{C}}NH_2$$

Semicarbazide   Semicarbazone

Procedure 56E
Carboxylic Acids

$$\overset{O}{\overset{\|}{\underset{R \qquad OH}{C}}}$$

Carboxylic acids are detectable mainly by their solubility characteristics. They are soluble in **both** dilute sodium hydroxide and sodium bicarbonate solutions.

SOLUBILITY CHARACTERISTICS	CLASSIFICATION TESTS
HCl NaHCO$_3$ NaOH H$_2$SO$_4$ Ether (−) (+) (+) (+) (+) Water: <C$_6$ (+) >C$_6$ (−)	pH of an aqueous solution Sodium bicarbonate Silver nitrate Neutralization equivalent

CLASSIFICATION TESTS

pH OF AN AQUEOUS SOLUTION

Procedure. If the compound is soluble in water, simply prepare an aqueous solution and check the pH with pH paper. If the compound is an acid, the solution will have a low pH.

Compounds that are insoluble in water can be dissolved in ethanol (or methanol) and water. First dissolve the compound in the alcohol and then add water until the solution **just** becomes cloudy. Clarify the solution by adding a few drops of the alcohol, and then determine its pH using pH paper.

SODIUM BICARBONATE

Procedure. Dissolve a small amount of the compound in a 5% aqueous sodium bicarbonate solution. Observe the solution carefully. If the compound is an acid, bubbles of carbon dioxide will be seen to form.

$$RCOOH + NaHCO_3 \longrightarrow RCOO^-Na^+ + H_2CO_3 \text{ (unstable)}$$

$$H_2CO_3 \longrightarrow CO_2 + H_2O$$

SILVER NITRATE

Procedure. Acids give a false silver nitrate test, as described in Procedure 56B.

NEUTRALIZATION EQUIVALENT

Procedure. Accurately weigh (three significant figures) approximately 0.2 g of the acid into a 125-mL Erlenmeyer flask. Dissolve the acid in about 50 mL of water or aqueous ethanol (the acid need not dissolve completely, since it will dissolve as it is titrated). Titrate the acid, using a solution of sodium hydroxide of known molarity (about 0.1M) and a phenolphthalein indicator.

Calculate the neutralization equivalent (NE) from the equation

$$NE = \frac{mg\ acid}{molarity\ of\ NaOH \times mL\ of\ NaOH\ added}$$

The NE is identical to the equivalent weight of the acid. If the acid has only one carboxyl group, the neutralization equivalent and the molecular weight of the acid are identical. If the acid has more than one carboxyl group, the neutralization equivalent equals the molecular weight of the acid divided by the number of carboxyl groups, that is, the equivalent weight. The NE can be used much like a derivative to identify a specific acid.

Many phenols are acidic enough to behave much like carboxylic acids. This is especially true of those substituted with electron-withdrawing groups at the *ortho* and *para* ring positions. These phenols, however, can usually be eliminated either by the ferric chloride test (Procedure 56F) or by spectroscopy (phenols have no carbonyl group).

SPECTROSCOPY

INFRARED

C=O stretch is very strong and often broad in the region between 1725 cm^{-1} and 1690 cm^{-1} (5.8–5.9 μ).

O—H stretch is a very broad absorption in the region between 3300 cm^{-1} and 2500 cm^{-1} (3.0–4.0 μ); it usually overlaps the CH stretch region.

NUCLEAR MAGNETIC RESONANCE

The acid proton of a —COOH group usually has resonance near 12.0δ.

DERIVATIVES

Derivatives of acids are usually amides. They are prepared via the corresponding acid chloride:

$$R\overset{\overset{\displaystyle O}{\|}}{-C}-OH + SOCl_2 \longrightarrow R\overset{\overset{\displaystyle O}{\|}}{-C}-Cl + SO_2 + HCl$$

The most common derivatives are the amides, the anilides, and the *p*-toluidides.

$$R\overset{\overset{\displaystyle O}{\|}}{-C}-Cl + 2NH_4OH \longrightarrow R\overset{\overset{\displaystyle O}{\|}}{-C}-NH_2 + 2H_2O + NH_4Cl$$

Ammonia (aq.)   **Amide**

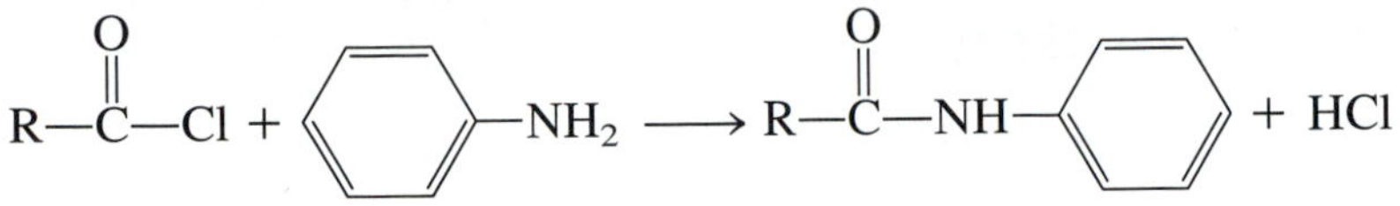

Aniline → Anilide

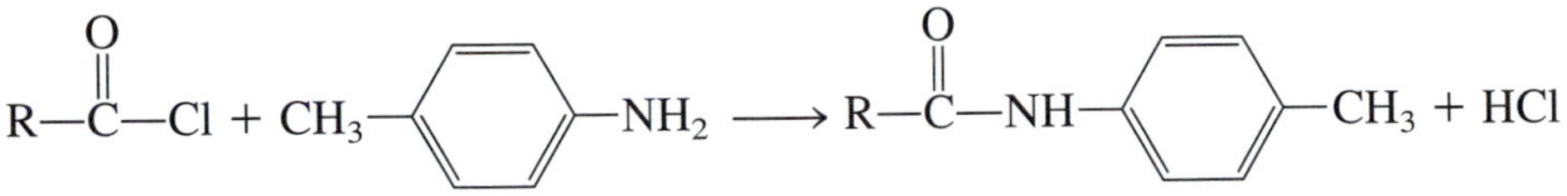

p-Toluidine → *p*-Toluidide

Procedures for the preparation of these derivatives are given in Appendix 2.

Procedure 56F
Phenols

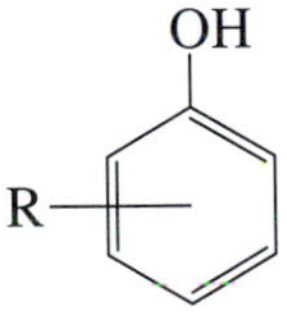

Like carboxylic acids, phenols are acidic compounds. However, except for the nitrosubstituted phenols (discussed in the section covering solubilities), they are not as acidic as the carboxylic acids. The pK_a of a typical phenol is 10, whereas the pK_a of a carboxylic acid is usually near 5. Hence, phenols are generally not soluble in the weakly basic sodium bicarbonate solution but do dissolve in sodium hydroxide solution, which is more strongly basic.

SOLUBILITY CHARACTERISTICS	CLASSIFICATION TESTS

HCl   NaHCO$_3$ NaOH   H$_2$SO$_4$ Ether
(−)   (−)   (+)   (+)   (+)
Water: Most are insoluble, although
    phenol itself and the nitrophenols
    are soluble
	Colored phenolate anion
	Ferric chloride
	Bromine/water

CLASSIFICATION TESTS

SODIUM HYDROXIDE SOLUTION

Procedure. With phenols that have a high degree of conjugation possible in their conjugate base (phenolate ion), the anion is often colored. To observe the color, it is necessary only to dissolve a small amount of the phenol in 10% aqueous sodium hydroxide solution. Some phenols do not give a color. Others have an insoluble anion and give a precipitate. The more acidic phenols, like the nitrophenols, tend more toward colored anions.

FERRIC CHLORIDE

Procedure 1 (Water-Soluble Phenols). Add several drops of a 2.5% aqueous solution of ferric chloride to 1 mL of a dilute aqueous solution (about 1–3% by weight of the phenol). Most phenols produce an intense red, blue, purple, or green color. Some colors are transient, and it may be necessary to observe the solution carefully just as the solutions are mixed. The formation of a color is usually immediate, but the color may not be permanent over any great period. Some phenols do not give a positive result in this test, so a negative test must not be taken as significant without other adequate evidence.

Test Compound. Try this test on phenol.

Procedure 2 (Water-Insoluble Phenols). Many phenols do not give a positive result when Procedure 1 is used. Often the procedure now to be described will give a positive result with such phenols. Dissolve or suspend 20 mg of a solid phenol or one drop of a liquid phenol in 1 mL of methylene chloride. Add one drop of pyridine and three to five drops of a 1% (weight/volume) solution of ferric chloride in methylene chloride.

The colors observed in this test result from the formation of a complex of the phenols with Fe(III) ion. Carbonyl compounds that have a high enol content also give a positive result in this test.

BROMINE/WATER

Procedure. Prepare a 1% aqueous solution of the unknown, and then add a saturated solution of bromine in water to it, drop by drop with shaking, until the bromine color is no longer discharged. A positive test is indicated by the precipitation of a substitution product at the same time that the bromine color of the reagent is discharged.

Test Compound. Try this test on phenol.

Aromatic compounds with ring-activating substituents give a positive test with bromine in water. The reaction is an aromatic substitution reaction, which introduces bromine atoms into the aromatic ring at the positions *ortho* and *para* to the hydroxyl group. All available positions are usually substituted. The precipitate is the brominated phenol, which is generally insoluble because of its large molecular weight.

Other compounds that give a positive result with this test include aromatic compounds that have activating substituents other than hydroxyl. These compounds include anilines and alkoxyaromatics.

SPECTROSCOPY

INFRARED

O—H stretch is observed near 3600 cm^{-1} (2.8 μ).
C—O stretch is observed near 1200 cm^{-1} (8.3 μ).
The typical aromatic ring absorptions between 1650 cm^{-1} and 1400 cm^{-1} (6–7 μ) are
   also found.
Aromatic CH is observed near 3100 cm^{-1} (3.2 μ).

NUCLEAR MAGNETIC RESONANCE

Aromatic protons are observed near 7δ. The hydroxyl proton has a resonance position that is concentration-dependent.

DERIVATIVES

Phenols form the same derivatives as alcohols do (Procedure 56H). They form urethanes by reaction with isocyanates, but whereas phenylurethanes are used for alcohols, the α-naphthylurethanes are more useful for phenols. Like alcohols, phenols also yield 3,5-dinitrobenzoates.

α-**Naphthyl isocyanate**   **An α-naphthylurethane**

3,5-Dinitrobenzoyl
chloride   A 3,5-dinitrobenzoate

The bromine-water reagent yields solid bromo derivatives of phenols in several cases. These solid derivatives can be used to characterize an unknown phenol. Procedures for preparing these derivatives are given in Appendix 2.

Procedure 56G
Amines

Amines are detected best by their solubility behavior and their basicity. They are the only basic compounds that will be issued for this experiment. Hence, once the compound has been identified as an amine, the main problem that remains is to decide whether it is primary (1°), secondary (2°), or tertiary (3°). This can usually be decided by the Hinsberg and nitrous acid tests.

SOLUBILITY CHARACTERISTICS	CLASSIFICATION TESTS
HCl NaHCO$_3$ NaOH H$_2$SO$_4$ Ether (+) (−) (−) (+) (+) Water: <C$_6$ (+) >C$_6$ (−)	pH of an aqueous solution Hinsberg test Nitrous acid test Acetyl chloride

CLASSIFICATION TESTS

HINSBERG TEST

Procedure. Place 0.1 mL of a liquid amine or 0.1 g of a solid amine, 0.2 g of p-toluenesulfonyl chloride, and 5 mL of 10% potassium hydroxide solution in a small test tube. Stopper the test tube tightly and shake it intermittently for three to five minutes. Remove the stopper and warm the test tube, with shaking, on a steam bath for one minute. Cool the solution and test a drop of it with pH paper to see whether it is still basic; if it is not, add more potassium hydroxide. If a precipitate has formed, dilute the basic mixture with 5 mL of water and shake it well. If the precipitate is insoluble, a disubstituted sulfonamide is probably present, which indicates that the unknown was a 2° amine.

> **NOTE: The precipitate may also be unreacted p-toluenesulfonyl chloride, leading to confusing results.**

If no precipitate remains after you dilute the mixture, or if none formed initially, carefully add 5% hydrochloric acid until the solution is just acidic to litmus (avoid excess acid). If a precipitate forms at this point, it should be the monosubstituted sulfonamide, indicating that the original compound was a 1° amine. If no reaction was apparent during the test, the original compound was probably a 3° amine.

If the above procedure gives confusing results(!), the procedure can be repeated, using 0.2 mL of benzenesulfonyl chloride instead of the p-toluenesulfonyl chloride. However, this reagent is likely to lead to the production of oils instead of solids.

Test Compounds. Try this test on aniline, N-methylaniline, and N,N-dimethylaniline. All three tests should be run simultaneously to allow easy comparison of the results.

This test is based on the production of monosubstituted and disubstituted sulfonamides from primary and secondary amines, respectively. Monosubstituted sulfonamides are soluble in base, while disubstituted sulfonamides are not soluble since they have no acidic hydrogens that can be removed to form a soluble salt.

$$CH_3-\langle\ \rangle-SO_2Cl + RNH_2 \longrightarrow$$

p-Toluenesulfonyl chloride (pTsCl)   **1° Amine**

$$CH_3-\langle\ \rangle-SO_2\ddot{N}HR \underset{HCl}{\overset{NaOH}{\rightleftharpoons}} CH_3-\langle\ \rangle-SO_2\ddot{N}R\ \overset{..}{\underset{..}{^-}}Na^+$$
$$\qquad\qquad\qquad |$$
$$\qquad\qquad\quad H$$

Insoluble
(but converted to soluble salt in base)   **Soluble salt**

$$CH_3\!-\!\langle\text{benzene ring}\rangle\!-\!SO_2Cl + R\!-\!\underset{\underset{R}{|}}{N}H \longrightarrow$$

p-Toluenesulfonyl   2° Amine
chloride (*p*TsCl)

$$CH_3\!-\!\langle\text{benzene ring}\rangle\!-\!SO_2\!-\!N\!\underset{R}{\overset{R}{<}} \xrightarrow{\;\text{NaOH}\;}$$

Will not dissolve since there is no hydrogen on N to be removed

Insoluble

Some sulfonamides of primary amines form insoluble sodium salts. This may lead to the mistaken assumption that the amine is secondary.

    Tertiary amines seem to be unreactive under these conditions, presumably because a tertiary amine has **no** amino hydrogens to be replaced. Actually, tertiary amines do react with benzene- and *p*-toluenesulfonyl chlorides; in most cases, however, they yield an overall result that gives the impression that no reaction has occurred. Two types of behavior can be observed for tertiary amines, one characteristic of tertiary alkylamines, R_3N, and the other characteristic of tertiary arylamines (that is, anilines).

    Most tertiary **alkylamines** react according to the following pattern:

$$R_3N: + R'\!-\!\langle\text{ring}\rangle\!-\!\overset{\overset{O}{\|}}{\underset{\underset{O}{\|}}{S}}\!-\!Cl \longrightarrow \left[R'\!-\!\langle\text{ring}\rangle\!-\!\overset{\overset{O}{\|}}{\underset{\underset{O}{\|}}{S}}\!-\!NR_3^+ \; Cl^-\right]$$

$$\text{OH}^-\;\swarrow$$

$$R_3N: + R'\!-\!\langle\text{ring}\rangle\!-\!\overset{\overset{O}{\|}}{\underset{\underset{O}{\|}}{S}}\!-\!OH$$

This leads to an observation of "no reaction." However, reaction has definitely occurred. In addition, many tertiary alkylamines give complex precipitates when left standing in the reaction medium. Therefore, the reaction time allowed for the test must be relatively short, or a reaction may really be observed.

    Tertiary **arylamines** are usually not very soluble in the reaction medium and often form an "oil" in the bottom of the test tube. Under these conditions, hydroxide ions of the reaction medium react more rapidly with the sulfonyl chloride than the insoluble amines can, and this also leads to an observation of "no reaction."

$$(ArNR_2) + R'\!-\!\langle\text{ring}\rangle\!-\!\overset{\overset{O}{\|}}{\underset{\underset{O}{\|}}{S}}\!-\!Cl \xrightarrow{\;\text{OH}^-\;} (ArNR_2) + R'\!-\!\langle\text{ring}\rangle\!-\!\overset{\overset{O}{\|}}{\underset{\underset{O}{\|}}{S}}\!-\!OH$$

Insoluble oil   **Insoluble oil**

When tertiary arylamines do dissolve in the medium, complicated secondary reactions may occur, especially if there is either excess amine or sulfonyl chloride present or if the reaction is heated. Often these secondary products include intensely colored dyes.[4]

The following table summarizes the results generally observed when the traditional Hinsberg test is used.

HINSBERG TEST SUMMARY

Primary amine $+$ pTsCl $\xrightarrow{\text{NaOH}}$ solution $\xrightarrow{\text{HCl}}$ precipitate

Secondary amine $+$ pTsCl $\xrightarrow{\text{NaOH}}$ precipitate

Tertiary amine $+$ pTsCl $\xrightarrow{\text{NaOH}}$ no reaction $\xrightarrow{\text{HCl}}$ clear solution
(apparent)

Two cautions should be observed when you are performing the Hinsberg test. First, these tests work well with reagent-grade amines; however, practical grades often contain impurities. For instance, secondary amines are often made from primary amines that may be contaminants. Similarly, tertiary amines may contain traces of secondary amines. Therefore, trace precipitates should not be considered as definitive results. Second, reaction times should be short, and any heating should be gentle—many tertiary amines will react under more vigorous conditions.

NITROUS ACID TEST

Procedure. Dissolve 0.1 g of an amine in 2 mL of water to which eight drops of concentrated sulfuric acid have been added. Use a large test tube. Cool the solution to 5 °C or less in an ice bath. Also cool 2 mL of 10% aqueous sodium nitrite in another test tube. In a third test tube, prepare a solution of 0.1 g β-naphthol in 2 mL of aqueous sodium hydroxide, and place it in an ice bath to cool. Add the cold sodium nitrite solution, drop by drop with shaking, to the cooled solution of the amine. Look for bubbles of nitrogen gas. Be careful not to confuse the evolution of the **colorless** nitrogen gas with an evolution of **brown** nitrogen oxide gas. Substantial evolution of gas at 5 °C or below indicates a primary aliphatic amine, RNH_2. The formation of a yellow oil or a yellow solid usually indicates a secondary amine, R_2NH. Either tertiary amines do not react, or they behave like secondary amines.

If little or no gas evolves at 5 °C, take **half** the solution and warm it gently to about room temperature. Nitrogen gas bubbles at this elevated temperature indicate that the original compound was a primary **aromatic** $ArNH_2$. Take the remaining solution and drop by drop add the solution of β-naphthol in base. If a red dye precipitates, the unknown has been conclusively shown to be a primary aromatic amine, $ArNH_2$.

---

[4]A complete discussion of this chemistry is in C. R. Gambill, T. D. Roberts, and H. Shechter, "Benzenesulfonyl Chloride Does React with Tertiary Amines," *Journal of Chemical Education, 49*:4 (April 1972), p 287. These authors also give an alternative version of the Hinsberg test, which may help students who have difficulty in interpreting their results using the tests described above.

Test Compounds. Try this test with aniline, *N*-methylaniline, and butylamine.

> **CAUTION: The products of this reaction may include nitrosamines. Nitrosamines are suspected carcinogens. Avoid contact and dispose of all residues in designated waste containers.**

Before you make this test, it should definitely be proved by some other method that the unknown is an amine. Many other compounds react with nitrous acid (phenols, ketones, thiols, amides), and a positive result with one of these could lead to an incorrect interpretation.

The test is best used to distinguish **primary** aromatic and **primary** aliphatic amines from secondary and tertiary amines. It also differentiates aromatic and aliphatic primary amines. It cannot distinguish between secondary and tertiary amines. Primary aliphatic amines lose nitrogen gas at low temperatures under the conditions of this test. Aromatic amines yield a more stable diazonium salt and do not lose nitrogen until the temperature is elevated. In addition, aromatic diazonium salts produce a red azo dye when β-naphthol is added. Secondary and tertiary amines produce yellow nitroso compounds, which may be soluble or may be oils or even solids. Many nitroso compounds have been shown to be carcinogenic. Avoid contact and immediately dispose of all such solutions in an appropriate waste container.

pH OF AN AQUEOUS SOLUTION

Procedure. If the compound is soluble in water, simply prepare an aqueous solution and check the pH with pH paper. If the compound is an amine, it will be basic, and the solution will have a high pH. Compounds that are insoluble in water can be dissolved in ethanol-water or 1,2-dimethoxyethane–water.

ACETYL CHLORIDE

Procedure. Amines give a positive acetyl chloride test (liberation of heat). This test is described for alcohols in Procedure 56H. When the test mixture is diluted with water, primary and secondary amines often give a solid acetamide derivative; tertiary amines do not.

SPECTROSCOPY

INFRARED

N—H stretch. Both aliphatic and aromatic **primary** amines show two absorptions (doublet due to symmetric and asymmetric stretches) in the region $3500–3300$ cm^{-1} ($2.86–3.03$ μ).
   Secondary amines show a single absorption in this region.
   Tertiary amines have no N—H bonds.
N—H bend. **Primary** amines have a strong absorption at $1640–1560$ cm^{-1} ($6.10–6.41$ μ).

Secondary amines have an absorption at $1580–1490$ cm^{-1} ($6.33–6.71$ μ). Aromatic amines show bands typical for the aromatic ring in the region $1650–1400$ cm^{-1} ($6–7$ μ). Aromatic CH is observed near 3100 cm^{-1} (3.2 μ).

NUCLEAR MAGNETIC RESONANCE

The resonance position of amino hydrogens is extremely variable. The resonance may also be very broad (quadrupole broadening). Aromatic amines give resonances near 7δ due to the aromatic ring hydrogens.

DERIVATIVES

The derivatives of amines that are most easily prepared are the acetamides and the benzamides. These derivatives work well for both primary and secondary amines but not tertiary amines.

$$CH_3-\overset{\overset{\textstyle O}{\|}}{C}-Cl + RNH_2 \longrightarrow CH_3-\overset{\overset{\textstyle O}{\|}}{C}-NH-R + HCl$$

Acetyl chloride   **An acetamide**

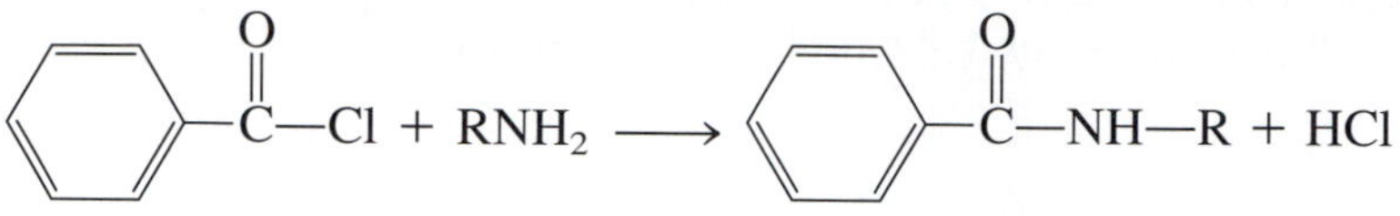

Benzoyl chloride + RNH₂ ⟶ A benzamide + HCl

Benzoyl chloride   **A benzamide**

In some cases, the solid *p*-toluenesulfonamides and benzenesulfonamides prepared from primary and secondary amines in the Hinsberg test (see above) can be used as derivatives.

The most general derivative that can be prepared is the picric acid salt, or picrate, of an amine. This derivative can be used for primary, secondary, and tertiary amines.

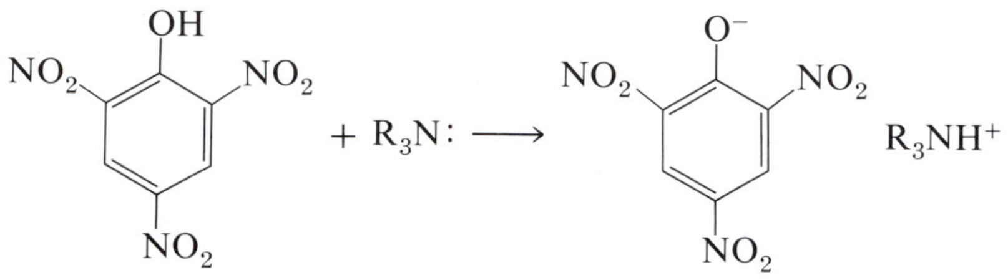

Picric acid + R₃N: ⟶ A picrate R₃NH⁺

Picric acid   **A picrate**

For tertiary amines, the methiodide salt is often useful.

$$CH_3I + R_3N: \longrightarrow CH_3-NR_3{}^+I^-$$

A methiodide

Procedures for preparing derivatives from amines can be found in Appendix 2.

Procedure 56H
Alcohols

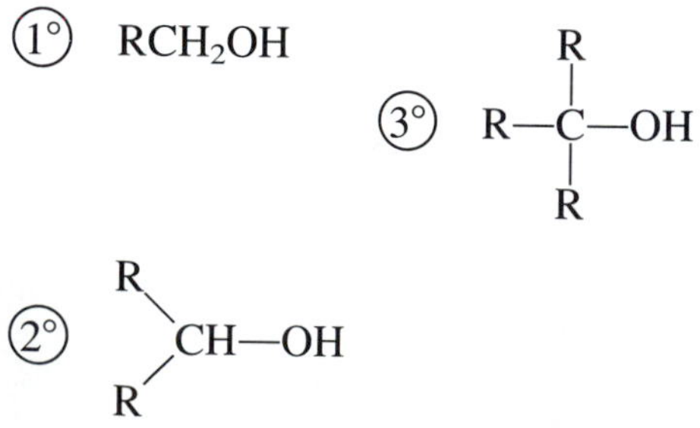

(1°) RCH_2OH

(3°) $R-\overset{\overset{\textstyle R}{|}}{\underset{\underset{\textstyle R}{|}}{C}}-OH$

(2°) $\overset{\textstyle R}{\underset{\textstyle R}{>}}CH-OH$

Alcohols are neutral compounds. The only other classes of neutral compounds used in this experiment are the aldehydes and ketones and the esters. Alcohols and esters usually do not give a positive 2,4-dinitrophenylhydrazine test, whereas aldehydes and ketones do. Esters do not react with acetyl chloride or with Lucas reagent, as alcohols do, and they are easily distinguished from alcohols on this basis. Primary and secondary alcohols are easily oxidized, whereas esters and tertiary alcohols are not. A combination of the Lucas test and the chromic acid test will differentiate among primary, secondary, and tertiary alcohols.

SOLUBILITY CHARACTERISTICS					CLASSIFICATION TESTS
HCl $(-)$	NaHCO$_3$ $(-)$	NaOH $(-)$	H$_2$SO$_4$ $(+)$	Ether $(+)$	Acetyl chloride
Water: <C$_6$ $(+)$					Lucas test
>C$_6$ $(-)$					Chromic acid test
					Iodoform test

CLASSIFICATION TESTS

ACETYL CHLORIDE

Procedure. Cautiously add about 10–15 drops of acetyl chloride, drop by drop, to about 0.5 mL of the alcohol contained in a small test tube. Evolution of heat and hydrogen chloride gas indicates a positive reaction. Addition of water will sometimes precipitate the acetate.

Acid chlorides react with alcohols to form esters. Acetyl chloride forms acetate esters.

$$CH_3-\overset{\overset{\displaystyle O}{\|}}{C}-Cl + ROH \longrightarrow CH_3-\overset{\overset{\displaystyle O}{\|}}{C}-O-R + HCl$$

Usually the reaction is exothermic, and the heat evolved is easily detected. Phenols react with acid chlorides somewhat as alcohols do. Hence, phenols should be eliminated as possibilities before this test is attempted. Amines also react with acetyl chloride to evolve heat (see Procedure 56G).

LUCAS TEST

Procedure. Place 2 mL of Lucas reagent in a small test tube and add three to four drops of the alcohol. Stopper the test tube and shake it vigorously. Tertiary (3°), benzylic, and allylic alcohols give an immediate cloudiness in the solution as the insoluble alkyl halide separates from the aqueous solution. After a short time, the immiscible alkyl halide will form a separate layer. Secondary (2°) alcohols produce a cloudiness after two to five minutes. Primary (1°) alcohols dissolve in the reagent to give a clear solution. Some secondary alcohols may have to be heated slightly to encourage reaction with the reagent.

This test works only for alcohols that are soluble in the reagent. This often means that alcohols with more than six carbon atoms cannot be tested.

Test Compounds. Try this test with 1-butanol (*n*-butyl alcohol), 2-butanol (*sec*-butyl alcohol), and 2-methyl-2-propanol (*t*-butyl alcohol).

Reagent. Cool 10 mL of concentrated hydrochloric acid in a beaker, using an ice bath. While still cooling, and with stirring, dissolve 16 g of anhydrous zinc chloride in the acid.

This test depends on the appearance of an alkyl chloride as an insoluble second layer when an alcohol is treated with a mixture of hydrochloric acid and zinc chloride (Lucas reagent):

$$R\text{—}OH + HCl \xrightarrow{ZnCl_2} R\text{—}Cl + H_2O$$

Primary alcohols do not react at room temperature; therefore, the alcohol is seen simply to dissolve. Secondary alcohols react slowly, whereas tertiary, benzylic, and allylic alcohols react instantly. These relative reactivities are explained on the same basis as the silver nitrate reaction, which is discussed in Procedure 56B. Primary carbocations are unstable and do not form under the conditions of this test. Hence, no results are observed for primary alcohols.

CHROMIC ACID TEST

Procedure. Dissolve one drop of a liquid or about 10 mg of a solid alcohol in 1 mL of **reagent-grade** acetone. Add one drop of the chromic acid reagent and note the result that occurs within 2 seconds. A positive test for a primary or a secondary alcohol is the appearance of a blue-green color. Tertiary alcohols do not give the test within two seconds, and the solution remains orange. To make sure that the acetone solvent is pure and does not give a positive test, add one drop of chromic acid to 1 mL of acetone that does not have an unknown dissolved in it. The orange color of the reagent should persist for **at least** three seconds. If it does not, a new bottle of acetone should be used.

Test Compounds. Try this test with 1-butanol (*n*-butyl alcohol), 2-butanol (*sec*-butyl alcohol), and 2-methyl-2-propanol (*t*-butyl alcohol).

> **CAUTION: Many compounds of chromium(VI) are suspected carcinogens (see p 11) and should be handled with care. Avoid contact.**

Reagent. Dissolve 1 g of chromic oxide (CrO_3) in 1 mL of concentrated sulfuric acid. Carefully add the mixture to 3 mL of water.

This test is based on the reduction of chromium(VI), which is orange, to chromium(III), which is green, when an alcohol is oxidized by the reagent. A change in color of the reagent from orange to green represents a positive test. Primary alcohols are oxidized by the reagent to carboxylic acids; secondary alcohols are oxidized to ketones.

$$2CrO_3 + 2H_2O \xrightarrow{H^+} 2H_2CrO_4 \xrightarrow{H^+} H_2Cr_2O_7 + H_2O$$

$$\underset{\text{Primary alcohol}}{R-\overset{\displaystyle H}{\underset{\displaystyle OH}{C}}-H} \xrightarrow{Cr_2O_7{}^{2-}} R-\overset{\displaystyle}{\underset{\displaystyle O}{C}}-H \xrightarrow{Cr_2O_7{}^{2-}} R-\overset{\displaystyle}{\underset{\displaystyle O}{C}}-OH$$

$$\underset{\text{Secondary alcohol}}{R-\overset{\displaystyle H}{\underset{\displaystyle OH}{C}}-R} \xrightarrow{Cr_2O_7{}^{2-}} R-\overset{\displaystyle}{\underset{\displaystyle O}{C}}-R$$

Although primary alcohols are first oxidized to aldehydes, the aldehydes are further oxidized to carboxylic acids. The ability of chromic acid to oxidize aldehydes but not ketones is taken advantage of in a test that uses chromic acid to distinguish between aldehydes and ketones (Procedure 56D). Secondary alcohols are oxidized to ketones, but no further. Tertiary alcohols are not oxidized at all by the reagent. Hence, this test can be used to distinguish primary and secondary alcohols from tertiary alcohols. Unlike the Lucas test, this test can be used with all alcohols regardless of molecular weight and solubility.

IODOFORM TEST

Alcohols with the hydroxyl group at the 2-position of the chain give a positive iodoform test. See the discussion in Section 56D, Aldehydes and Ketones.

SPECTROSCOPY

INFRARED

O—H stretch. A medium to strong, and usually broad, absorption occurs in the region 3600–3200 cm^{-1} (2.8–3.1 μ). In dilute solutions or with little hydrogen bonding, it is a sharp absorption near 3600 cm^{-1} (2.8 μ). In more concentrated solutions, or with considerable hydrogen bonding, it is a broad absorption near 3400 cm^{-1} (2.9 μ). Sometimes both bands appear.

C—O stretch. There is strong absorption in the region 1200–1050 cm^{-1} (9.5–8.3 μ). Primary alcohols absorb nearer 1050 cm^{-1} (9.5 μ), tertiary alcohols and phenols nearer 1200 cm^{-1} (8.3 μ). Secondary alcohols absorb in the middle of this range.

NUCLEAR MAGNETIC RESONANCE

The hydroxyl resonance is extremely concentration-dependent, but it is usually found between 1δ and 5δ. Under normal conditions, the hydroxyl proton does not couple with protons on adjacent carbon atoms.

DERIVATIVES

The most common derivatives for alcohols are the 3,5-dinitrobenzoate esters and the phenylurethanes. Occasionally, the α-naphthylurethanes (Procedure 56F) are also prepared, but these latter derivatives are more often used for phenols.

3,5-Dinitrobenzoyl chloride   A 3,5-dinitrobenzoate

Phenyl isocyanate   A phenylurethane

Procedures for preparing these derivatives are given in Appendix 2.

Procedure 56I
Esters

$$R-\overset{\overset{\displaystyle O}{\|}}{C}\diagdown_{\displaystyle O-R'}$$

Esters are formally considered "derivatives" of the corresponding carboxylic acid. They are frequently synthesized from the carboxylic acid and the appropriate alcohol:

$$R-COOH + R'-OH \underset{}{\overset{H^+}{\rightleftharpoons}} R-COOR' + H_2O$$

Thus, esters are sometimes referred to as though they were composed of an acid part and an alcohol part.

Although esters, like aldehydes and ketones, are neutral compounds that have a carbonyl group, they do not usually give a 2,4-dinitrophenylhydrazine test. The two most common tests for identifying esters are the basic hydrolysis and ferric hydroxamate tests. The **saponification equivalent** is also used. However, it usually requires a difficult and time-consuming procedure and will not be considered here. Procedures for determining the saponification equivalent can be found in the references at the beginning of this experiment.

SOLUBILITY CHARACTERISTICS					CLASSIFICATION TESTS
HCl	NaHCO$_3$	NaOH	H$_2$SO$_4$	Ether	Ferric hydroxamate test
($-$)	($-$)	($-$)	($+$)	($+$)	Basic hydrolysis
Water: <C$_4$ ($+$)					
>C$_5$ ($-$)					

CLASSIFICATION TESTS

FERRIC HYDROXAMATE TEST

Procedure. Before starting, you must determine whether the compound to be tested already has enough enolic character in acid solution to give a positive ferric chloride test. Dissolve one drop of a liquid unknown or a few crystals of a solid unknown in 1 mL of 95% ethanol and add 1 mL of 1M hydrochloric acid. Add a drop or two of 5% ferric chloride solution. If a definite color, except yellow, appears, the ferric hydroxamate test (described below) cannot be used.

If the compound did not show enolic character, continue as follows. Dissolve two or three drops of a liquid ester, or about 40 mg of a solid ester, in a mixture of 1 mL of

0.5M hydroxylamine hydrochloride (dissolved in 95% ethanol) and 0.2 mL of 6M sodium hydroxide. Heat the mixture to boiling for a few minutes. Cool the solution and then add 2 mL of 1M hydrochloric acid. If the solution becomes cloudy, add 2 mL of 95% ethanol to clarify it. Add a drop of 5% ferric chloride solution and note whether a color is produced. If the color fades, continue to add ferric chloride until the color persists. A positive test should give a deep burgundy or magenta color.

On being heated with hydroxylamine, esters are converted to the corresponding hydroxamic acids:

$$R-\underset{\substack{\| \\ O}}{C}-O-R' + H_2N-OH \longrightarrow R-\underset{\substack{\| \\ O}}{C}-NH-OH + R'-OH$$

Hydroxylamine   A hydroxamic acid

The hydroxamic acids form strong, colored complexes with ferric ion.

$$R-\underset{\substack{\| \\ O}}{C}-NH-OH + FeCl_3 \longrightarrow \left(R-C \underset{NH-O}{\overset{O}{\diagup}} \right)_3 Fe + 3HCl$$

BASIC HYDROLYSIS

Procedure. Place 0.7 g of the ester in a 10-mL flask with 7 mL of 25% aqueous sodium hydroxide. Add a boiling stone and attach the water condenser. Boil the mixture for about 30 minutes. Stop the heating and observe the solution to determine whether the oily ester layer has disappeared or whether the odor of the ester (usually pleasant) has disappeared. Low-boiling esters (below 110 °C) usually dissolve within 30 minutes if the alcohol part has a low molecular weight. If the ester has not dissolved, reheat the mixture to reflux for one to two hours. After that time, the oily ester layer should have disappeared along with the characteristic odor. Esters boiling up to 200 °C should hydrolyze during this time. Compounds remaining after this extended period of heating either are unreactive esters or are **not** esters.

For esters derived from solid acids, the acid part can, if desired, be recovered after hydrolysis. Extract the basic solution with ether to remove any unreacted ester (even if it appears to be gone), acidify the basic solution with hydrochloric acid, and extract the acidic phase with ether to remove the acid. Dry the ether layer over anhydrous sodium sulfate, decant, and evaporate the solvent to obtain the parent acid from the original ester. The melting point of the parent acid can provide valuable information in the identification process.

This procedure converts the ester to its separate acid and alcohol parts. The ester dissolves because the alcohol part (if small) is usually soluble in the aqueous medium, as is the sodium salt of the acid. Acidification produces the parent acid:

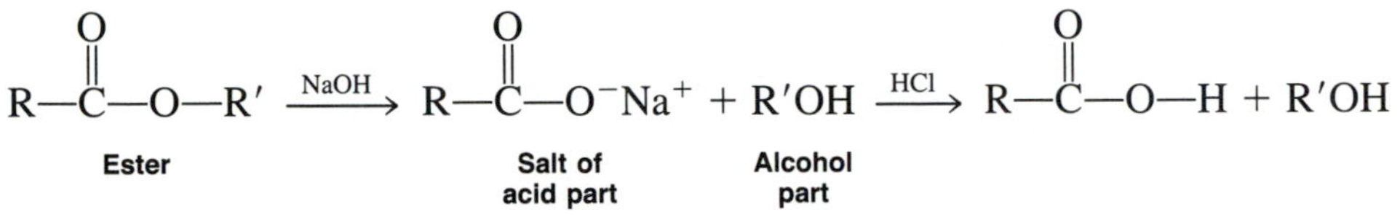

$$R-\overset{\overset{\displaystyle O}{\|}}{C}-O-R' \xrightarrow{\text{NaOH}} R-\overset{\overset{\displaystyle O}{\|}}{C}-O^-Na^+ + R'OH \xrightarrow{\text{HCl}} R-\overset{\overset{\displaystyle O}{\|}}{C}-O-H + R'OH$$

Ester   **Salt of**   **Alcohol**
                                    acid part   **part**

All derivatives of carboxylic acids are converted to the parent acid on basic hydrolysis. Thus, amides, which are not covered in this experiment, would also dissolve in this test, liberating the free amine and the sodium salt of the carboxylic acid.

SPECTROSCOPY

INFRARED

The ester-carbonyl group (C=O) peak is usually a strong absorption, and so is the absorption of the carbonyl-oxygen link (C—O) to the alcohol part. C=O stretch at approximately 1735 cm^{-1} (5.75 μ) is normal.[5] C—O stretch usually gives two or more absorptions, one stronger than the others, in the region 1280–1050 cm^{-1} (7.8–9.5 μ).

NUCLEAR MAGNETIC RESONANCE

Hydrogens that are alpha to an ester carbonyl group have resonance in the region 2–3δ. Hydrogens alpha to the alcohol oxygen of an ester have resonance in the region 3–5δ.

DERIVATIVES

Esters present a double problem when one is trying to prepare derivatives. To characterize an ester completely, one needs to prepare derivatives of **both** the acid part and the alcohol part.

ACID PART

The most common derivative of the acid part is the *N*-benzylamide derivative.

---

[5]Conjugation with the carbonyl group moves the carbonyl absorption to lower frequencies (longer wavelength). Conjugation with the alcohol oxygen raises the carbonyl absorption to higher frequencies (shorter wavelength). Ring strain (lactones) moves the carbonyl absorption to higher frequencies (shorter wavelength).

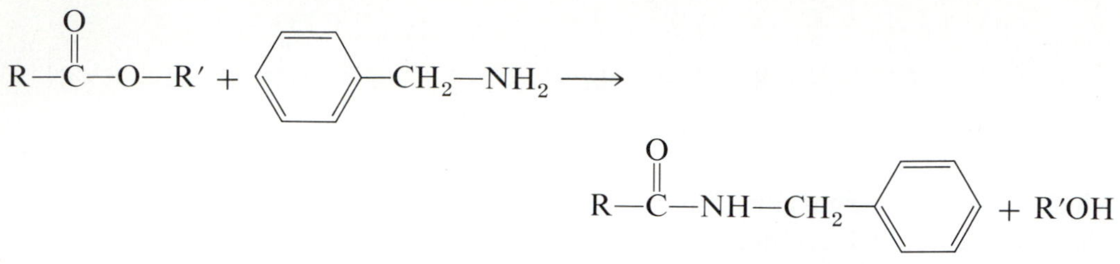

$$R-\overset{\overset{\text{O}}{\|}}{C}-O-R' + \langle\!\!\bigcirc\!\!\rangle-CH_2-NH_2 \longrightarrow$$

$$R-\overset{\overset{\text{O}}{\|}}{C}-NH-CH_2-\langle\!\!\bigcirc\!\!\rangle + R'OH$$

An *N*-benzylamide

The reaction does not proceed well unless R′ is methyl or ethyl. For alcohol portions that are larger, the ester must be transesterified to a methyl or an ethyl ester before preparing the derivative.

$$R-\overset{\overset{\text{O}}{\|}}{C}-OR' + CH_3OH \xrightarrow{\text{H}^+} R-\overset{\overset{\text{O}}{\|}}{C}-O-CH_3 + R'OH$$

Hydrazine also reacts well with methyl and ethyl esters to give acid hydrazides.

$$R-\overset{\overset{\text{O}}{\|}}{C}-OR' + NH_2NH_2 \longrightarrow R-\overset{\overset{\text{O}}{\|}}{C}-NHNH_2 + R'OH$$

An acid hydrazide

The saponification equivalent is also sometimes used. This value gives the molecular weight of the ester divided by the number of its ester groups.

ALCOHOL PART

The best derivative of the alcohol part of an ester is the 3,5-dinitrobenzoate ester, which is prepared by an acyl interchange reaction:

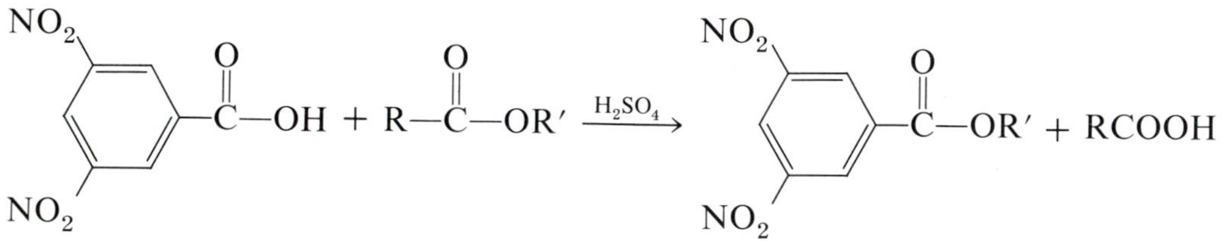

$$\begin{array}{c} NO_2 \\ \langle\!\!\bigcirc\!\!\rangle-\overset{\overset{\text{O}}{\|}}{C}-OH \\ NO_2 \end{array} + R-\overset{\overset{\text{O}}{\|}}{C}-OR' \xrightarrow{\text{H}_2\text{SO}_4} \begin{array}{c} NO_2 \\ \langle\!\!\bigcirc\!\!\rangle-\overset{\overset{\text{O}}{\|}}{C}-OR' \\ NO_2 \end{array} + RCOOH$$

**A 3,5-dinitrobenzoate
ester**

Most esters are composed of very simple acid and alkyl portions. For this reason, spectroscopy is usually a better method of identification than is the preparation of derivatives. Not only is it necessary to prepare two derivatives with an ester, but all esters with the same acid portion, or all those with the same alcohol portion, give identical derivatives of those portions.

Experiment 57
Carbohydrates

In this experiment, you perform tests that distinguish among various carbohydrates. The carbohydrates included and the classes they represent are as follows:

> Aldopentoses: xylose and arabinose
> Aldohexoses: glucose and galactose
> Ketohexoses: fructose
> Disaccharides: lactose and sucrose
> Polysaccharides: starch and glycogen

The structures of these carbohydrates can be found in your lecture textbook. The tests are classified in the following groups:

A. Tests based on the production of furfural or a furfural derivative: Molisch's test, Bial's test, and Seliwanoff's test
B. Tests based on the reducing property of a carbohydrate (sugar): Benedict's test and Barfoed's test
C. Osazone formation
D. Iodine test for starch
E. Hydrolysis of sucrose
F. Mucic acid test for galactose and lactose
G. Tests on unknowns

REQUIRED READING

New: Read the sections in your lecture textbook that give the structures and describe the chemistry of aldopentoses, aldohexoses, ketohexoses, disaccharides, and polysaccharides.

SPECIAL INSTRUCTIONS

All the procedures in this experiment involve simple test-tube reactions. Most of the tests are short; however, Seliwanoff's test, osazone formation, and the mucic acid test take relatively longer to complete. You will need a minimum of 10 test tubes (15 × 125 mm) numbered in order. Clean them carefully each time they are used. The 1% solutions of carbohydrates and the reagents needed for the tests have been prepared in advance by the laboratory instructor. Be sure to shake the starch solution before using it.

A. TESTS BASED ON PRODUCTION OF FURFURAL OR A FURFURAL DERIVATIVE

Under acidic conditions, aldopentoses and ketopentoses **rapidly** undergo dehydration to give furfural (Equation 1). Ketohexoses **rapidly** yield 5-hydroxymethylfurfural (Equation 2). Disaccharides and polysaccharides can first be hydrolyzed in an acid medium to produce monosaccharides, which then react to give furfural or 5-hydroxymethylfurfural.

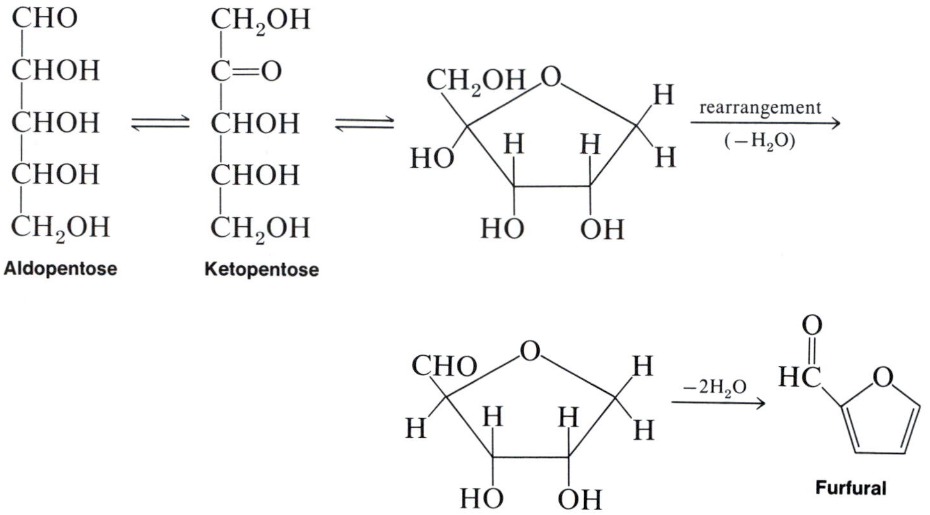

(1)

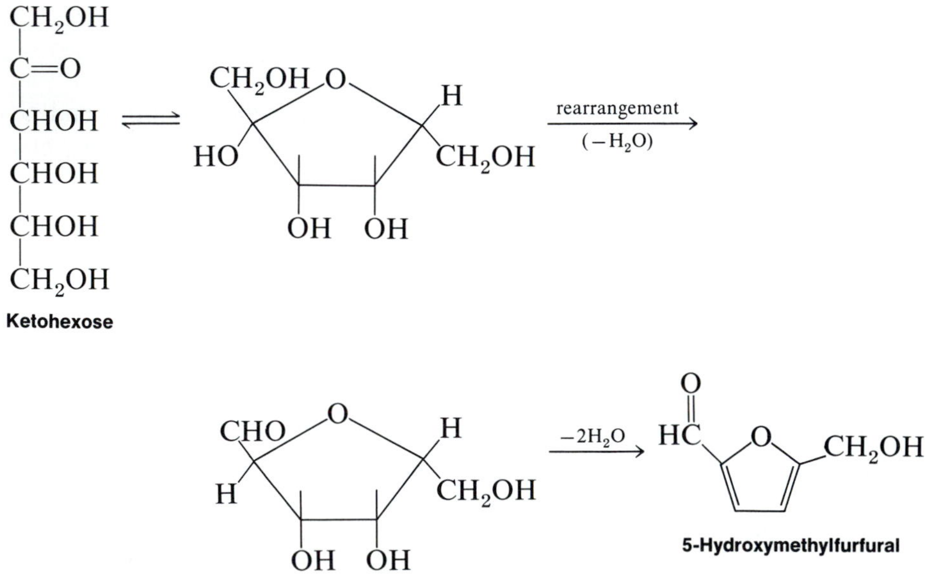

(2)

Aldohexoses are **slowly** dehydrated to 5-hydroxymethylfurfural. One possible mechanism is shown in Equation 3. The mechanism is different from that given in

Equations 1 and 2 in that dehydration is at an early step and the rearrangement step is absent.

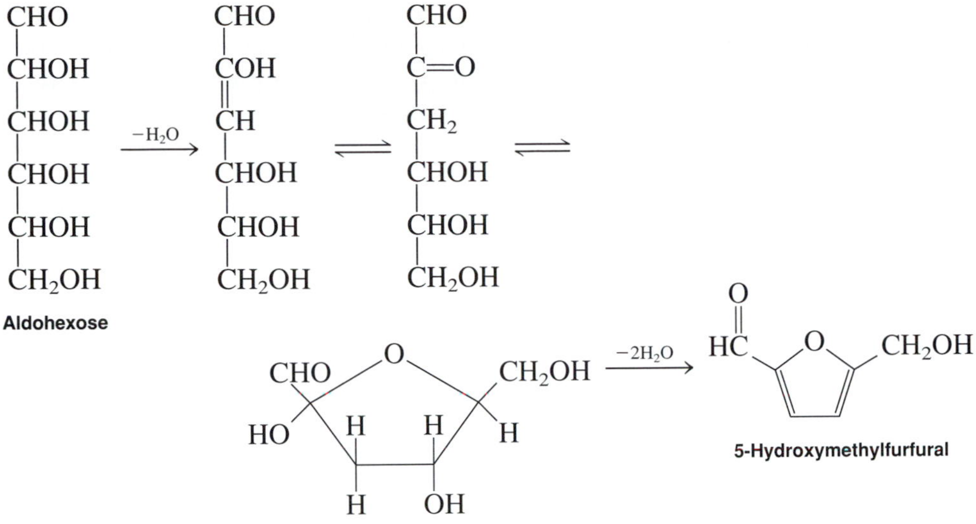

Aldohexose

(3)

5-Hydroxymethylfurfural

Once furfural or 5-hydroxymethylfurfural is produced by Equations 1, 2, or 3, either will then react with a phenol to produce a colored condensation product. The substance α-naphthol is used in the Molisch test, orcinol in Bial's test, and resorcinol in Seliwanoff's test.

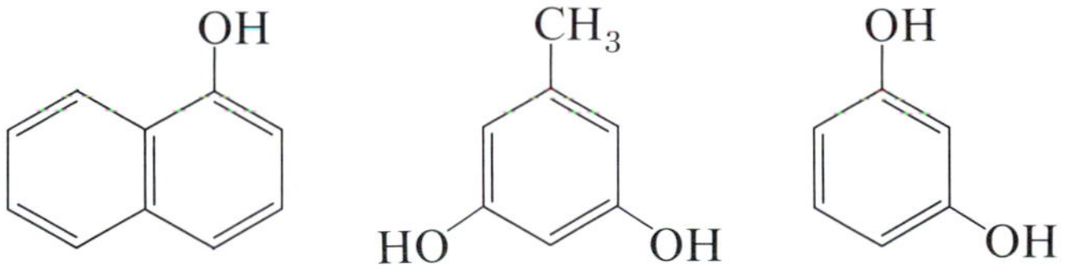

α-Naphthol   **Orcinol**   **Resorcinol**
(Molisch test)   **(Bial test)**   **(Seliwanoff test)**

The colors and the rates of formation of these colors are used to differentiate between the carbohydrates. The various color tests are discussed in Sections 1, 2, and 3. A typical colored product formed from furfural and α-naphthol (Molisch's test) is the following (Equation 4):

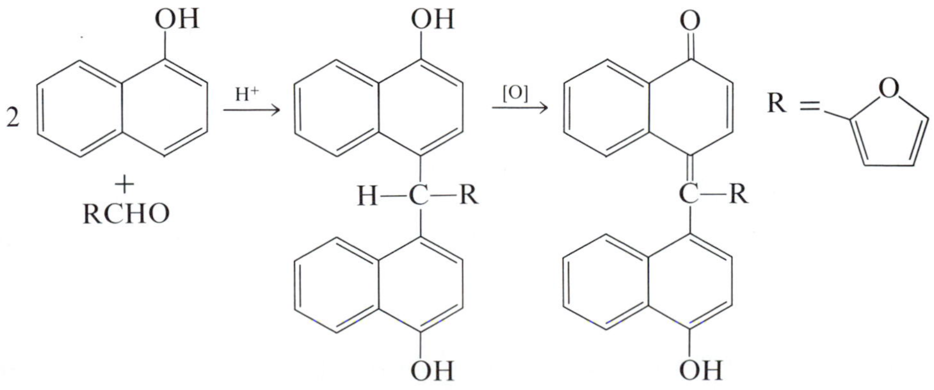

(4)

Purple

1. MOLISCH TEST FOR CARBOHYDRATES

The Molisch test is a **general** test for carbohydrates. Most carbohydrates are dehydrated with concentrated sulfuric acid to form furfural or 5-hydroxyfurfural. These furfurals react with the α-naphthol in the test reagent to give a purple product. Compounds other than carbohydrates may react with the reagent to give a positive test. A negative test usually indicates that there is no carbohydrate.

Procedure for the Molisch Test. Place 1 mL of each of the following 1% carbohydrate solutions in nine separate test tubes: xylose, arabinose, glucose, galactose, fructose, lactose, sucrose, starch (shake it), and glycogen. Also add 1 mL of distilled water to another tube to serve as a control.

Add two drops of the Molisch reagent[1] to each test tube and thoroughly mix the contents of the tube. Tilt each test tube slightly and cautiously add 1 mL of concentrated sulfuric acid down the sides of the tubes. An acid layer forms at the bottom of the tubes. Note and record the color at the interface between the two layers in each tube. A purple color constitutes a positive test.

2. BIAL TEST FOR PENTOSES

The Bial test is used to differentiate pentose sugars from hexose sugars. Pentose sugars yield furfural on dehydration in acidic solution. Furfural reacts with orcinol and ferric chloride to give a blue-green condensation product. Hexose sugars give 5-hydroxymethylfurfural, which reacts with the reagent to yield colors such as green, brown, and reddish brown.

Procedure for Bial's Test. Place 1 mL of each of the following 1% carbohydrate solutions in separate test tubes: xylose, arabinose, glucose, galactose, fructose, lactose, sucrose, starch (shake it), and glycogen. Also add 1 mL of distilled water to another tube to serve as a control.

Add 1 mL of Bial's reagent[2] to each test tube. Carefully heat each tube over a Bunsen burner flame until the mixture just begins to boil. Note and record the color produced in each test tube. If the color is not distinct, add 2.5 mL of water and 0.5 mL of 1-pentanol to the test tube. After shaking them, again observe and record the color. The colored condensation product will be concentrated in the 1-pentanol layer.

3. SELIWANOFF TEST FOR KETOHEXOSES

The Seliwanoff test depends on the relative rates of dehydration of carbohydrates. A ketohexose reacts rapidly by Equation 2 to give 5-hydroxymethylfurfural, whereas an aldohexose reacts more slowly, by Equation 3, to give the same product. Once 5-hydroxymethylfurfural is produced, it reacts with resorcinol to give a dark red condensation product. If the reaction is followed for some time, it will be found that

---

[1] Dissolve 2.5 g of α-naphthol in 50 mL of 95% ethanol.

[2] Dissolve 3 g of orcinol in 1 L of concentrated hydrochloric acid and add 3 mL of 10% aqueous ferric chloride.

sucrose hydrolyzes to give fructose, which eventually reacts to produce a dark red color.

Procedure for Seliwanoff's Test.   Prepare a boiling-water bath for this experiment. Place 0.5 mL of each of the following 1% carbohydrate solutions in separate test tubes: xylose, arabinose, glucose, galactose, fructose, lactose, sucrose, starch (shake it), and glycogen. Add 0.5 mL of distilled water to another tube to act as a control.

Add 2 mL of Seliwanoff's reagent[3] to each test tube. Place all 10 tubes in a beaker of boiling water for **60 seconds.** Remove them and note the results in the notebook.

For the remainder of Seliwanoff's test, it is convenient to place a group of three or four tubes in the boiling water bath and to complete the observations before going on to the next group of tubes. Place three or four tubes in the boiling-water bath. Observe the color in each of the tubes at one-minute intervals for five minutes beyond the original minute. Record the results at each one-minute interval. Leave the tubes in the boiling-water bath during the entire five-minute period. After the first group has been observed, remove that set of test tubes, and place the next group of three or four tubes in the bath. Follow the color changes as before. Finally, place the last group of tubes in the bath and follow the color changes over the five-minute period.

B. TESTS BASED ON THE REDUCING PROPERTY OF A CARBOHYDRATE (SUGAR)

Monosaccharides, and those disaccharides that have a potential aldehyde group, will reduce reagents such as Benedict's solution to produce a red precipitate of cuprous oxide:

$$RCHO + 2Cu^{2+} + 4OH^- \longrightarrow RCOOH + Cu_2O + 2H_2O$$

Red precipitate

Glucose, for example, is a typical aldohexose, showing reducing properties. The two diastereomeric α- and β-D-glucoses are in equilibrium with each other in aqueous solution. The α-D-glucose opens at the anomeric carbon atom (hemiacetal) to produce the free aldehyde. This aldehyde rapidly closes to give β-D-glucose, and a new hemiacetal is produced. It is the presence of this free aldehyde that makes glucose a reducing carbohydrate (sugar). It reacts with Benedict's reagent to produce a red precipitate, the basis of the test. Carbohydrates that have the hemiacetal functional group show reducing properties.

---

[3]Dissolve 0.5 g of resorcinol in 1 L of dilute hydrochloric acid (one volume of concentrated hydrochloric acid and two volumes of distilled water).

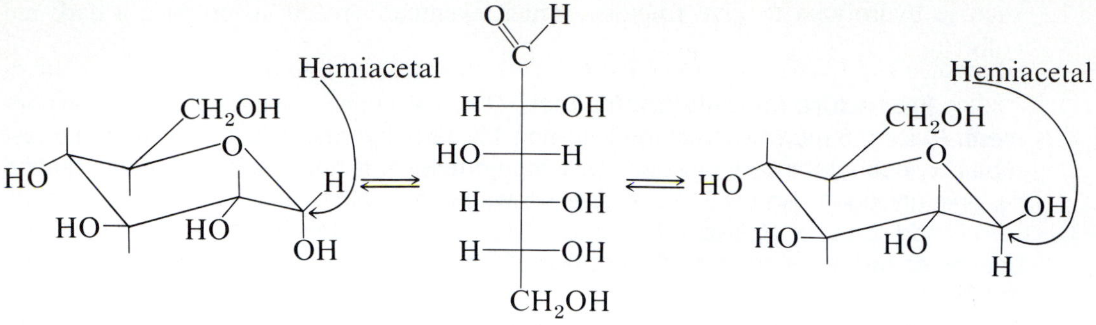

α-D-Glucose   D-Glucose   β-D-Glucose

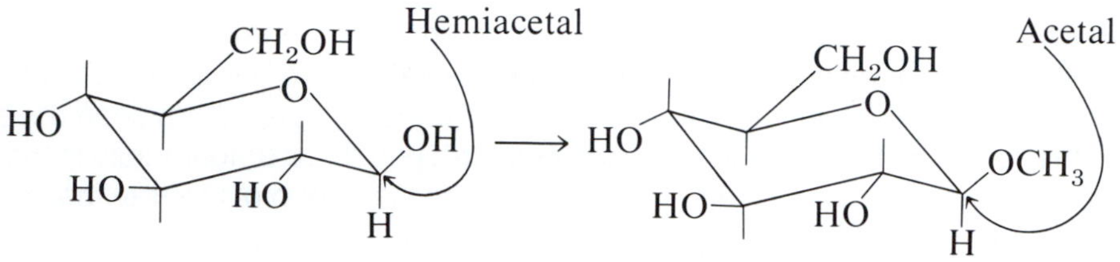

If the hemiacetal is converted to an acetal by methylation, the carbohydrate (sugar) will no longer reduce Benedict's reagent.

With disaccharides, two situations may arise. If the anomeric carbon atoms are bonded (head to head) to give an acetal, then the sugar will not reduce Benedict's reagent. If, however, the sugar molecules are joined head to tail, then one end will still be able to equilibrate through the free aldehyde form (hemiacetal). Examples of a reducing and a nonreducing disaccharide are shown below.

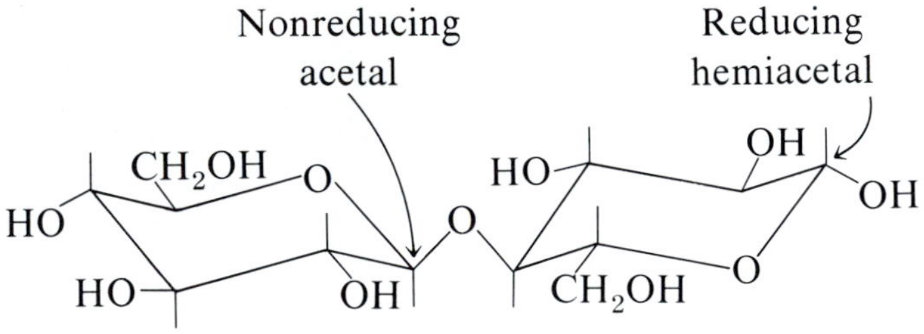

Cellobiose
(reducing sugar)

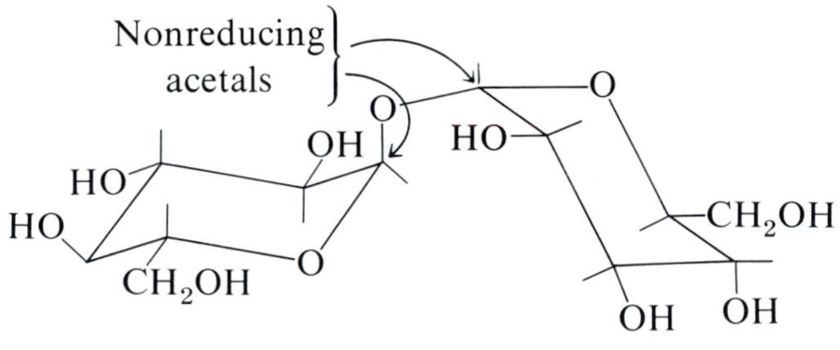

Trehalose
(nonreducing sugar)

1. BENEDICT'S TEST FOR REDUCING SUGARS

Benedict's test is performed under mildly basic conditions. The reagent reacts with all reducing sugars to produce the red precipitate cuprous oxide, as shown on p 501. It also reacts with water-soluble aldehydes that are not sugars. Ketoses, such as fructose, also react with Benedict's reagent. Benedict's test is considered one of the classical tests for determining the presence of an aldehyde functional group.

Procedure for Benedict's test. Prepare a boiling-water bath for this experiment. Place 0.5 mL of each of the following 1% carbohydrate solutions in separate test tubes: xylose, arabinose, glucose, galactose, fructose, lactose, sucrose, starch (shake it), and glycogen. Add 0.5 mL of distilled water to another tube to serve as a control.

Add 2 mL of Benedict's reagent[4] to each test tube. Place the test tubes in a boiling-water bath for two to three minutes. Remove the tubes and note the results in a notebook. A red, brown, or yellow precipitate indicates a positive test for a reducing sugar. Ignore a change in color of the solution. A precipitate must form for the test to be positive.

2. BARFOED'S TEST FOR REDUCING MONOSACCHARIDES

Barfoed's test distinguishes reducing monosaccharides and reducing disaccharides by a difference in rate of reaction. The reagent consists of cupric ions, like Benedict's reagent. In this test, however, Barfoed's reagent reacts with reducing monosaccharides to produce cuprous oxide faster than with reducing disaccharides.

$$RCHO + 2Cu^{2+} + 2H_2O \longrightarrow RCOOH + Cu_2O + 4H^+$$

**Red
precipitate**

Procedure for Barfoed's Test. Place 0.5 mL of each of the following 1% carbohydrate solutions in separate test tubes: xylose, arabinose, glucose, galactose, fructose, lactose, sucrose, starch (shake it), and glycogen. Add 0.5 mL of distilled water to another tube to function as a control.

Add 2 mL of Barfoed's reagent[5] to each test tube. Place the tubes in a boiling-water bath for 10 minutes. Remove the tubes and note the results in a notebook.

C. OSAZONE FORMATION

Carbohydrates react with phenylhydrazine to form crystalline derivatives, called **osazones.**

---

[4]Dissolve 173 g of hydrated sodium citrate and 100 g of anhydrous sodium carbonate in 800 mL of distilled water, with heating. Filter the solution. Add to it a solution of 17.3 g of cupric sulfate ($CuSO_4 \cdot 5H_2O$) dissolved in 100 mL of distilled water. Dilute the combined solutions to 1 L.

[5]Dissolve 66.6 g of cupric acetate in 1 L of distilled water. Filter the solution, if necessary, and add 9 mL of glacial acetic acid.

$$
\begin{array}{ccccc}
\text{CHO} & & \text{HC}\!=\!\text{NNHPh} & & \text{HC}\!=\!\text{NNHPh} \\
| & & | & & | \\
\text{CHOH} & & \text{CHOH} & & \text{C}\!=\!\text{NNHPh} \\
| & \xrightarrow{\text{PhNHNH}_2} & | & \xrightarrow{\text{2PhNHNH}_2} & | \\
\text{CHOH} & & \text{CHOH} & & \text{CHOH} \qquad +\ \text{NH}_3\ +\ \text{PhNH}_2 \\
| & & | & & | \\
\text{CHOH} & & \text{CHOH} & & \text{CHOH} \\
| & & | & & | \\
\text{CHOH} & & \text{CHOH} & & \text{CHOH} \\
| & & | & & | \\
\text{CH}_2\text{OH} & & \text{CH}_2\text{OH} & & \text{CH}_2\text{OH}
\end{array}
$$

An osazone

An osazone can be isolated as a derivative and its melting point determined. However, some of the monosaccharides give **identical** osazones (glucose, fructose, and mannose). Also, the melting points of different osazones are often in the same range. This limits the usefulness of an isolation of the osazone derivative.

A good experimental use for the osazone is to observe its rate of formation. The rates of reaction vary greatly even though the **same** osazone may be produced from different sugars. For example, fructose forms a precipitate in about two minutes, whereas glucose forms a precipitate about five minutes later. The osazone is the same in each case. The crystal structure of the osazone is often distinctive. Arabinose, for example, produces a fine precipitate, whereas glucose produces a coarse precipitate.

> **CAUTION: Phenylhydrazine is a sus-
> pected carcinogen. Handle with gloves.**

Procedure for Osazone Formation. A boiling-water bath is needed for this experiment. Place 0.5 mL of each of the following **10%** carbohydrate solutions in separate test tubes: xylose, arabinose, glucose, galactose, fructose, lactose, sucrose, starch (shake it), and glycogen. Add 2 mL of phenylhydrazine reagent[6] to each tube. Place the tubes in a boiling-water bath simultaneously. Watch for a precipitate, or in some cases, cloudiness. Note the time at which the precipitate begins to form. After 30 minutes, cool the tubes and record the crystalline form of the precipitates. Reducing disaccharides will not precipitate until the tubes are cooled. Nonreducing disaccharides will hydrolyze first, and then the osazones will precipitate.

D. IODINE TEST FOR STARCH

Starch forms a typical blue color with iodine. This color is due to the absorption of iodine into the open spaces of the amylose molecules (helices) present in starch. Amylopectins, which are the other types of molecules present in starch, form a red to purple color with iodine.

---

[6]Dissolve 50 g of phenylhydrazine hydrochloride and 75 g of sodium acetate trihydrate in 500 mL of distilled water. The reagent deteriorates over time and should be prepared fresh.

Procedure for the Iodine Test. Place 1 mL of each of the following 1% carbohydrate solutions in three separate test tubes: glucose, starch (shake it), and glycogen. Add 1 mL of distilled water to another tube to act as a control.

Add one drop of iodine solution to each test tube and observe the results.[7] Add a few drops of sodium thiosulfate to the solutions and note the results.[8]

E. HYDROLYSIS OF SUCROSE

Sucrose can be hydrolyzed in acid solution to its component parts, fructose and glucose. The component parts can then be tested with Benedict's reagent.

Procedure for the Hydrolysis of Sucrose. Place 1 mL of a 1% solution of sucrose in a test tube. Add two drops of concentrated hydrochloric acid and heat the tube in a boiling water bath for 10 minutes. Cool the tube and neutralize the contents with 10% sodium hydroxide solution until the mixture is just basic to litmus (about 12 drops are needed). Test the mixture with Benedict's reagent (Part B). Note the results and compare them with the results obtained on sucrose that has not been hydrolyzed.

F. MUCIC ACID TEST FOR GALACTOSE AND LACTOSE

Procedures are given in Experiment 53 for the oxidation of galactose and lactose to mucic acid. This test confirms the presence of galactose or a galactose unit in a carbohydrate (sugar).

G. TESTS ON UNKNOWNS

Obtain an unknown solid carbohydrate from the laboratory instructor or assistant. The unknown will be one of the following carbohydrates: xylose, arabinose, glucose, galactose, fructose, lactose, sucrose, starch, or glycogen. Carefully dissolve part of the unknown in distilled water to prepare a 1% solution (0.060 g carbohydrate in 6 mL water). Also prepare a 10% solution by dissolving 0.1 g of carbohydrate in 1 mL of water. Save the remainder of the solid for the mucic acid test. Apply whatever tests are necessary to identify the unknown.

At the instructor's option, the optical rotation can be determined as part of the experiment. Experimental details are given in Experiment 54 and Technique 16. Optical rotation data and decomposition points for carbohydrates and osazones are given in the standard reference works on identification of organic compounds (Experiment 56).

---

[7]The iodine solution is prepared as follows. Dissolve 1 g of potassium iodide in 25 mL of distilled water. Add 0.5 g of iodine and shake the solution until the iodine dissolves. Dilute the solution to 50 mL.

[8]The sodium thiosulfate solution is prepared as follows. Dissolve 1.25 g of sodium thiosulfate in 50 mL of water.

QUESTIONS

1. Find the structures for the following carbohydrates (sugars) in a reference work or a textbook, and decide whether they are reducing or nonreducing carbohydrates (sugars): sorbose, mannose, ribose, maltose, raffinose, and cellulose.

2. Mannose gives the same osazone as glucose. Explain.

3. Predict the results of the following tests with the carbohydrates listed in Question 1: Molisch, Bial, Seliwanoff (after one minute and six minutes), Barfoed, and mucic acid tests.

4. Give a mechanism for the hydrolysis of the acetal linkage in sucrose.

5. The rearrangement in Equations 1 and 2 can be considered a type of pinacol rearrangement. Give a mechanism for that step.

6. Give a mechanism for the acid-catalyzed condensation of furfural with two moles of α-naphthol, shown in Equation 4.

Experiment 58
Paper Chromatography of Amino Acids

Paper chromatography

In this experiment, you determine the composition of proteins by paper chromatographic analysis of the hydrolysates.

Proteins can be hydrolyzed in acidic or basic solution or with enzymes. During the hydrolysis, the peptide bonds break to give shorter polymers (polypeptides), which in turn are further degraded to amino acids. Total hydrolysis of protein can be achieved in 20% hydrochloric acid solution at 100 °C for 12 to 48 hours. However, adequate hydrolysis for purposes of this experiment can be achieved in refluxing acid in less than one hour.

$$-NH-\overset{R}{\underset{|}{CH}}-\overset{O}{\overset{\|}{C}}-NH-\overset{R}{\underset{|}{CH}}-\overset{O}{\overset{\|}{C}}-NH-\overset{R}{\underset{|}{CH}}-\overset{O}{\overset{\|}{C}}- \longrightarrow n\ NH_2\overset{R}{\underset{|}{CH}}COOH$$

Protein   **Amino acids**

These hydrolysates can be analyzed for their amino acid contents by chromatographic methods, such as paper chromatography. The most common amino acids are listed in the table on page 507. They are listed in the order of increasing R_f values. The amino acid contents for the proteins casein, gelatin, silk, and hair are also listed. The chief amino acid constituents in each of the proteins are indicated by an asterisk. Note particularly the large differences in the amino acid content of the various proteins. Since there are variations in amino acid content among samples, the values in the table are to be considered **approximate**.

Amino Acids, R$_f$ Values, and Approximate Compositions of Casein, Gelatin, Silk, and Hair

AMINO ACID	FORMULA	R$_f$ VALUE	APPROXIMATE PERCENTAGE COMPOSITION OF PROTEINS			
			Casein	*Gelatin*	*Silk*	*Hair*
Cystine	$$\begin{array}{c}NH_2\\ |\\ S-CH_2CHCOOH\\ |\\ S-CH_2CHCOOH\\ |\\ NH_2\end{array}$$	0.16	0.4	0.1		18.0*
Aspartic acid	$HOOCCH_2\overset{\overset{\displaystyle NH_2}{|}}{C}HCOOH$	.32	6.8*	6.7*		3.9
Glutamic acid	$HOOCCH_2CH_2\overset{\overset{\displaystyle NH_2}{|}}{C}HCOOH$	.40	22.4*	11.5*		13.1*
Glycine	H_2NCH_2COOH	.42	2.6	25.5*	42.3*	4.1
Serine	$HOCH_2\overset{\overset{\displaystyle NH_2}{|}}{C}HCOOH$	.43	7.4*	0.4	12.6*	10.6*
Threonine	$CH_3\overset{\overset{\displaystyle }{|}}{C}H\underset{\underset{\displaystyle }{|}}{{}}-\overset{\overset{\displaystyle }{|}}{C}HCOOH$ (OH NH$_2$)	.51	4.7	1.9	1.5	8.5*
Lysine (cation)	$\overset{+}{H_3}NCH_2CH_2CH_2CH_2\overset{\overset{\displaystyle NH_2}{|}}{C}HCOOH$	.53	7.9*	4.1	0.4	1.9
Alanine	$CH_3\overset{\overset{\displaystyle NH_2}{|}}{C}HCOOH$	.59	2.9	8.7*	24.5*	2.8
Arginine (cation)	$H_2\overset{+}{N}=CNHCH_2CH_2CH_2\overset{\overset{\displaystyle NH_2}{|}}{C}HCOOH$ (NH$_2$)	.60	3.9	8.0*	1.1	8.9*
Tyrosine	$HO-\langle\bigcirc\rangle-CH_2\overset{\overset{\displaystyle NH_2}{|}}{C}HCOOH$	.62	6.1*	0.4	10.6*	2.2
Valine	$CH_3CH-CHCOOH$ (CH$_3$ NH$_2$)	.75	6.9*	2.5	3.2	5.5
Methionine	$CH_3SCH_2CH_2\overset{\overset{\displaystyle NH_2}{|}}{C}HCOOH$	.77	3.3	1.0		0.7
Leucine	$CH_3CHCH_2CHCOOH$ (CH$_3$ NH$_2$)	.79	8.8*	4.6	0.8	11.2*
Phenylalanine	$\langle\bigcirc\rangle-CH_2\overset{\overset{\displaystyle NH_2}{|}}{C}HCOOH$	.82	4.8	2.2		2.4
Proline	$$\begin{array}{c}\text{pyrrolidine ring}\\N-COOH\\ |\\ H\end{array}$$	.85	10.9*	18.0*	1.5	4.3

*Principal constituents

The individual amino acids on a chromatogram are made visible with ninhydrin. Ninhydrin reacts with amino acids to produce characteristic deep blue colors. A few amino acids produce a different color, however; proline, for example, produces a pale yellow color with ninhydrin. The reactions involved in the production of the color are as follows:

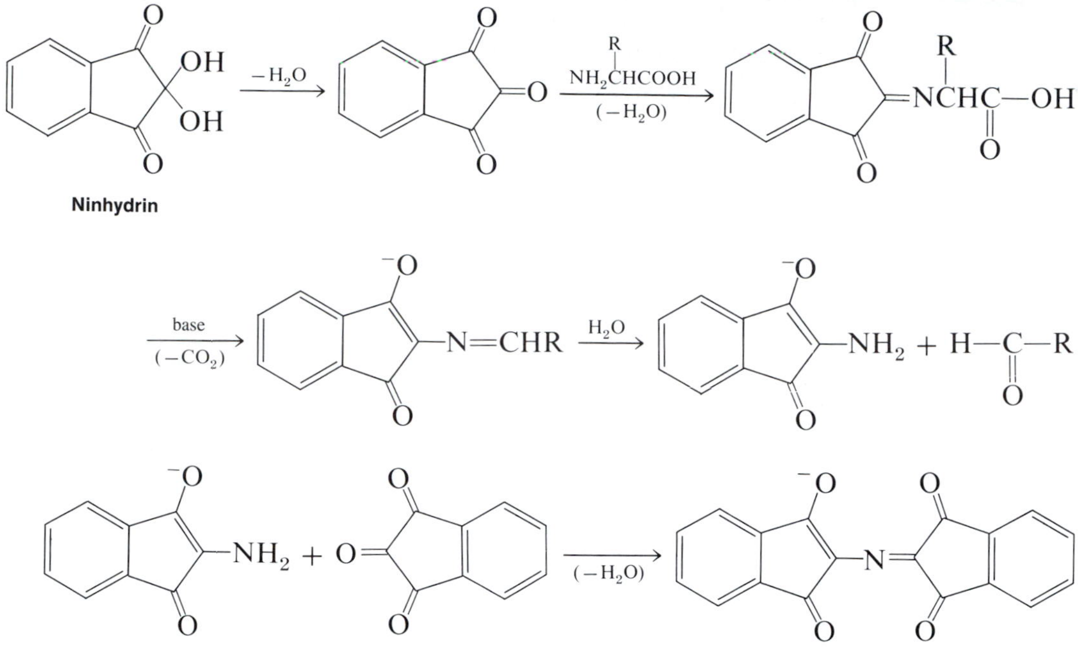

Ninhydrin

Blue

REQUIRED READING

Review: Technique 13   Sections 13.4, 13.5, 13.7, 13.9, and 13.11
          Paper chromatography is similar to thin-layer chromatography

New: Read the sections in your lecture textbook that describe the structures of
          amino acids and proteins.

SPECIAL INSTRUCTIONS

The hydrolysates of casein, gelatin, silk, and hair are best provided by the laboratory instructor. The procedure for preparing them is given below. The casein isolated in Experiment 53 can be used. The chromatographic development takes about four hours to complete. The laboratory instructor may need to be responsible for removing the chromatogram at the end of the four-hour development period.

The purpose of this experiment is to identify individual, unknown amino acids and to identify the constituent amino acids in the hydrolysates of some common proteins. This identification is accomplished by comparing R_f values of known amino acids with R_f values of the unknown. If the hydrolysates are provided by the laboratory instructor, go directly to the paper chromatography part of the experiment.

> **NOTE TO THE INSTRUCTOR:** Tests on amino acids and proteins are given in the instructor's manual.

PROCEDURES

HYDROLYSIS OF A PROTEIN

Assemble a large-scale reflux apparatus, using a 100-mL round-bottomed flask. Place 20 mL of 19% hydrochloric acid (equal volumes of concentrated hydrochloric acid and distilled water), 0.5 g of the protein (casein, gelatin, hair, or silk), and a boiling stone in the flask. For casein or gelatin, carefully heat the mixture under reflux for 35 minutes using a heating mantle or a very small flame. For hair or silk, heat the mixture under reflux for 50 minutes. After the reflux period is completed, the hydrolysate must be decolorized. Add about 0.5 g of decolorizing carbon to the hot hydrolysate and swirl the mixture. Gravity-filter the hydrolysate into a 50-mL Erlenmeyer flask. The filtrate should be colorless or pale yellow. Check the hydrolysate with the biuret test to see whether the reaction is complete.[1] The test should be negative. If hydrolysis is not complete (a positive biuret test), add 3–5 mL of 19% hydrochloric acid and heat the mixture for an additional 15 minutes. Again check the hydrolysate with the biuret test to see whether it is negative. After the hydrolysis is complete, cool the hydrolysate, stopper it, and save the solution for the analysis by paper chromatography.

SEPARATION OF AMINO ACIDS BY PAPER CHROMATOGRAPHY

Obtain a 24 × 15.5-cm rectangle of Whatman No. 1 filter paper. Handle this paper by the top edge only. If the paper is handled carelessly, fingerprints may appear as colored spots on the chromatogram. Place the rectangle on a clean paper towel or notebook paper with the "x" in the upper right corner (see page 510). This "x" was placed on the rectangle when the paper was cut. Since the thickness of the paper is uneven, this marking procedure ensures that the direction of the solvent flow on each piece of chromatography paper is the same for all students. The R_f values are more reproducible when the chromatograms are developed in the same direction.

---

[1] Place five drops of the hydrolysate and 10 drops of 10% sodium hydroxide in a test tube. Check the solution with litmus paper to make sure that it is definitely alkaline. Add more 10% sodium hydroxide if necessary. Add four or five drops of 2% copper sulfate solution. A blue color indicates that the hydrolysis is complete. A pink or violet color indicates that the hydrolysis is not complete.

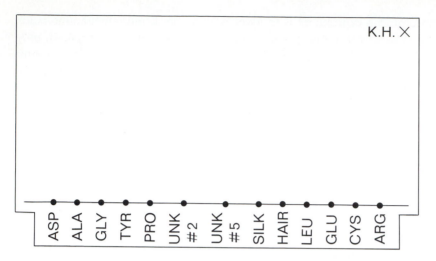

Cut a 1-cm square from the two lower corners. About 2 cm from the bottom of the paper, pencil a line (do not use a pen) across the paper. Also pencil 13 dots, evenly spaced 1.5 cm from each other along this line, with the first dot 3 cm from the left-hand edge. Nine standard amino acids, two hydrolysates, and two unknown amino acids will be placed on the dots. We suggest you use the sixth and seventh dots for the unknowns, the eighth and ninth dots for the hydrolysates, and the rest of the dots for the individual standard amino acids. One possible sequence is shown in the figure. The standard amino acids that will be used are 0.1M solutions of aspartic acid (ASP), alanine (ALA), glycine (GLY), tyrosine (TYR), proline (PRO), leucine (LEU), glutamic acid (GLU), cystine (CYS), and arginine (ARG). Each of these 0.1M solutions has been acidified with 10 drops of 19% hydrochloric acid for each 10 mL of solution. Record the sequence in the laboratory notebook. Place your name or initials in the upper right-hand corner of the chromatogram.

Prepare capillary micropipets for applying the solutions (Technique 13, Section 13.4, p 728) or use the micropipets inserted in the standard amino acids, unknowns, or hydrolysate samples. It may be wise to practice the spotting technique on a small piece of Whatman No. 1 paper before trying to spot the actual chromatogram. The correct method of spotting is described in Technique 13, Section 13.4, p 728. It is important that the spots should be made as small as possible and that the paper should not be overloaded. If either of these conditions is not observed, the spots will tail and overlap after development. The applied spot should be 1–2 mm in diameter. Apply the appropriate samples to the dots with the micropipets, taking extreme care to avoid contaminating the samples. After applying the samples to each dot, allow the spots to dry completely and then make a second application at the same point. The hydrolysates should be spotted a third time. Allow the spots to dry completely before the chromatogram is developed.

CAUTION: Phenol will cause burns if it touches the skin. Immediately wash any affected area with copious quantities of soap and water. Clean up all spills.

Use a 32-oz wide-mouthed screw-cap jar for a developing chamber. Insert a pipet **below** the protective layer of ligroin and remove 20 mL of an 80% aqueous phenol solution with a bulb.[2] Carefully transfer the phenol solution to the developing chamber so that it is not splattered on the sides of the jar. Check to see that the depth of the phenol solution does not exceed 1.5 cm. If it does, carefully pipet enough of the solution so that the depth is less than 1.5 cm. Curl the paper into a large cylinder so that the line of spots is on the bottom and on the inner surface. Overlap the top about 0.5 cm and make sure that the bottom edge is even. Hold the chromatogram together with a paper clip or a staple. Insert the cylinder into the jar so that the bottom end of the paper is immersed in the solvent. Tighten the lid securely and allow the development to proceed for four hours without disturbing the jar.

Remove the chromatogram (no fingers), mark the solvent front with a pencil, and allow the solvent to evaporate in air.[3] In the next laboratory period, spray the paper uniformly with ninhydrin spray and place the paper in a 110 °C oven. Colored spots will begin to appear within five minutes. Remove the chromatogram and outline all the spots with a pencil. With a millimeter ruler, measure the distance each spot has traveled from the point of origin to the **front (top)** of the spot. Also measure the distance that the solvent traveled. Calculate the R_f value for each of the spots (Technique 13, Section 13.9, p 733). In some cases, the separation may not be complete enough to calculate the R_f value accurately. Record the calculated values.

From the R_f values and colors of the standard amino acids, identify the two unknown amino acids and the principal constituents of the two protein hydrolysates. The table of protein composition may prove useful as you analyze the hydrolysates. Remember that amino acids with similar R_f values may not separate well enough to be seen on the chromatogram. You may have to place the amino acids in groups. Record your findings. Submit the chromatogram with your laboratory report.

[2]The 80% aqueous phenol is prepared by mixing 80 g of phenol per 20 mL of distilled water. Heat the mixture until it dissolves completely. Add a protective layer of ligroin so that air is excluded from the aqueous phenol. If the ligroin is not added, the solution should be used as soon as possible.

[3]This part of the procedure may have to be done by the laboratory instructor or assistant if there is not enough time.

Part Four

Macroscale Experiments

Essay
HOW TO SCALE UP A REACTION: MACROSCALE METHODS

Most of the techniques used in Parts One and Two involved the use of 0.1 gram or less of starting material. You have obtained a great deal of skill in doing these **microscale** experiments. There are, however, a few important techniques that you may not have used in the laboratory. These latter techniques are employed with **gram** quantities of starting materials used in **macroscale** experimentation. The important macroscale techniques, figures, and page references are listed in the following table. You should review these figures in advance of the laboratory period.

MACROSCALE METHOD	FIGURE NUMBERS	PAGE REFERENCES
Separatory funnel	7–8 and 7–9	pages 627 and 628
Simple distillation	8–11	page 650
Fractional distillation	10–11	page 680
Refluxing mixtures	3–2B	page 550
Crystallization	4–6, 5–3, and 5–4	page 571, 581 and 583

These macroscale techniques are often used when an organic chemist must prepare a compound that involves several sequential steps (a multistep synthesis). The chemist must start out with gram quantities of material so that an adequate amount of product may be obtained. It simply may be unacceptable to start with 0.1 g of starting material. The techniques listed above are used in Experiments 59 through 63.

How to Scale Up a Reaction. You should first determine the scaling factor for the reaction of interest. For example, consider a given microscale procedure that yields 0.1 g of product. You are asked to prepare 5 grams of this material. You will need to multiply each of the quantities given in the microscale procedure by 50 to determine the quantities needed for the macroscale experiment. You will need to scale up all quantities given in the procedure, including the amount of solvent needed.

Since the concentrations of reactants are the same in both the microscale and macroscale experiments, the reaction time should be the same for both experiments. In practice, however, you should allow more time than given in a microscale procedure because of the extra time needed to bring the larger mass of material up to its boiling point.

In most cases, you will need to use a round-bottom flask rather than a conical vial. Determine the size of the flask needed by keeping in mind that it should not be more than one-half full. Round-bottom flasks are available in the following sizes (volume in mL): 10, 25, 50, 100, 250, and 500. You may need to obtain the required large-scale equipment from your instructor or chemistry stockroom.

Reactions Involving Reflux. Chemists often use heating mantles to heat mixtures under reflux in macroscale experiments. Boiling stones are required to prevent bumping. In addition, you should use a water-cooled condenser rather than an air-cooled one to help prevent the loss of vapor into the laboratory. The apparatus is shown in Technique 3, Figure 3–2B, p 550.

Separatory Funnels. The separatory funnel is used to separate an aqueous layer from the organic phase in macroscale experiments. You should review Technique 7, Section 7.7, p 627, especially Figures 7–8 and 7–9.

Simple Distillation. With larger amounts of material, it is impractical to distill with a Hickman head because of its limited capacity to hold liquid (about 1 mL). Instead, the simple distillation apparatus shown in Technique 8, Figure 8–11, p 650, is used. Since the boiling point range is obtained during the distillation simply by reading the thermometer, it is usually not necessary to obtain a microboiling point. This special distillation equipment may be obtained from the instructor or the stockroom.

Fractional Distillation. The macroscale fractional distillation equipment is shown in Technique 10, Figure 10–11, p 680. Commonly, one condenser is packed with some stainless steel sponge to create the distillation column shown in the figure. You must not run water through the jacket of this column. Another, water-jacked, condenser is used to condense the vapors. It is convenient to use a heating mantle or a steam bath as the heat source, depending on the boiling point of the mixture. The progress of the fractionation may be monitored by observing the temperature on the thermometer. Fractions are obtained by changing receiving flasks when the temperature reaches the desired value.

Crystallization of Large Quantities of Solid. When larger amounts of a solid are to be crystallized, it is impractical to use a Craig tube. Instead, you will need to collect a solid by vacuum filtration using a Büchner or Hirsch funnel (Technique 4, Figure 4–6, p 571). The crystallization procedure is shown in Technique 5, Figure 5–3, p 581. You should use this procedure when there is no insoluble material present in the hot solution. If there is some insoluble material present, you should remove it by decantation or by passing the solution through a fluted filter as shown in Figure 5–4A or 5–4C, p 583.

Weighing and Measuring. With larger amounts of solids, it is usually sufficient to weigh to the nearest 0.01 gram. You may use a watch glass, weighing paper, or a piece of notebook paper on the balance pan to hold the solid.

In general, graduated and automated pipets are not used in macroscale experiments. Graduated cylinders are used for handling liquids. However, it may still be necessary to weigh the liquid to obtain an accurate weight for a limiting reagent.

Laboratory Safety and Disposal of Wastes. You should take even greater care with reagents and solvents when doing a macroscale experiment because of the larger amounts of chemicals used. Spillage is often a potential problem, and you should use funnels and spatulas when transferring liquids and solids in order to minimize this problem. Be sure to wear your safety goggles at all times.

Disposal of reagents and solvents is also a more serious problem with macroscale experiments. You should be even more concerned about proper disposal of materials because of the larger amounts of waste generated in a macroscale experiment. Place the waste in the proper container. As with microscale experimentation, do not pour waste solvents into a sink.

QUESTIONS

1. Below is a list of microscale experiments which a chemist might wish to scale upwards *(make macroscale)*. Assuming an average yield of about 70%, describe how **you** would scale each of them to yield about 5 g of final product.

Exp 13A	Synthesis of n-Butyl Bromide
Exp 27B	Benzoic Acid
Exp 28	Benzoin Condensation
Exp 42	Benzocaine
Exp 43	Methyl Salicylate

Pay attention to the following considerations.
(a) Amounts of each reagent, solvent, and catalyst, and what devices you would use to measure them.
(b) Reaction time(s) and temperature(s).
(c) Sketch the apparatus to be used, paying attention to:
    (i)   type of flask (include size) or equipment to use
    (ii) heating device or bath
    (iii) trapping any dangerous gases evolved
    (iv) how to effect a controlled rate of addition if necessary
    (v) protection from moisture
(d) Describe your workup procedure, sketching the equipment you would use at each stage to do any
    (i)   extractions or separations
    (ii) filtrations
    (iii) crystallizations
    (iv) distillations

2. The macroscale preparation of salicylic acid by hydrolysis of methyl salicylate is described in Experiment 63. Describe how you would downscale *(make microscale)* the procedure, paying attention to the same considerations outlined in the question above, so that it would yield about 100 mg of product.

Experiment 59
Isopentyl Acetate (Banana Oil)

Esterification
Macroscale technique
Heating under reflux
Extraction
Simple distillation

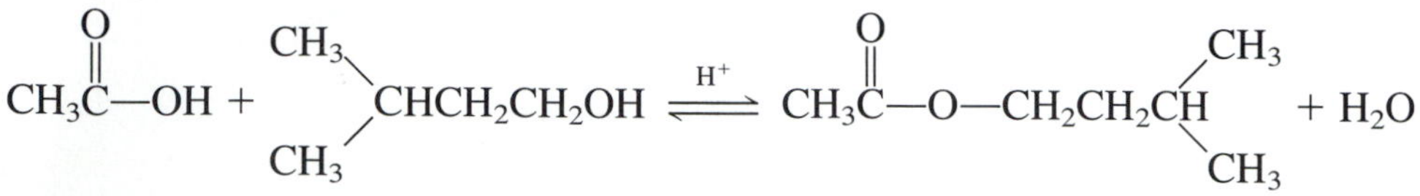

$$CH_3\overset{\displaystyle O}{\overset{\|}{C}}{-}OH + \underset{CH_3}{\overset{CH_3}{\diagup}}CHCH_2CH_2OH \underset{}{\overset{H^+}{\rightleftharpoons}} CH_3\overset{\displaystyle O}{\overset{\|}{C}}{-}O{-}CH_2CH_2CH\underset{CH_3}{\overset{CH_3}{\diagdown}} + H_2O$$

Acetic acid   Isopentyl alcohol   Isopentyl acetate
(excess)

REQUIRED READING

Review: Introductory material in Experiment 6
                Techniques 1 and 2
                Technique 18

New: Essay   How To Scale Up a Reaction: Macroscale Methods
             Technique 3   Reaction Methods, Sections 3.2 and 3.4
             Technique 7   Extractions, Separations, and Drying Agents, Sections
                                     7.7 and 7.8
             Technique 8   Simple Distillation, Section 8.4

SPECIAL INSTRUCTIONS

Since a one-hour reflux is involved, this experiment should be started at the very beginning of the laboratory period. During the reflux period, other experimental work may be performed. Be careful in handling concentrated sulfuric acid. It will cause extreme burns if it is spilled on the skin. This procedure is written for use with glassware kits equipped with ⊺ 19/22 standard-taper joints. If your kit uses ⊺ 14/20 joints you can adapt this procedure simply by reducing all of the quantities by half.

PROCEDURE

Pour 15 mL (12.2 g, 0.138 mole) of isopentyl alcohol (also called isoamyl alcohol or 3-methyl-1-butanol) and 20 mL (21 g, 0.35 mole) of glacial acetic acid into a 100-mL round-bottomed flask. Carefully add 4 mL of concentrated sulfuric acid to the contents of the flask, with swirling. Add several boiling stones to the mixture.

> **CAUTION:** **Extreme care must be exercised to avoid contact with concentrated sulfuric acid. It will cause serious burns if it is spilled on the skin. If it comes in contact with the skin or clothes, it must be washed off immediately with excess water. In addition, sodium bicarbonate may be used to neutralize the acid. Clean up all spills immediately.**

Assemble a reflux apparatus as shown in Technique 3, Figure 3–2B, p 550. Bring the mixture to a boil with a suitable heating source, such as a heating mantle or an oil bath. Heat the mixture under reflux for 1 hour (Technique 3, Section 3.2, p 549). Remove the heating source and allow the mixture to cool to room temperature. Pour the cooled mixture into a separatory funnel and carefully add 55 mL of cold water. Rinse the reaction flask with 10 mL of cold water and pour the rinsings into the separatory funnel. Use a stirring rod to mix the materials somewhat. Stopper the separatory funnel and shake it several times (Technique 7, Section 7.7, p 627). Separate the lower aqueous layer from the upper organic layer (density 0.87 g/mL). Discard the aqueous layer after making certain that the correct layer has been saved.

The crude ester in the organic layer contains some acetic acid, which can be removed by extraction with 5% aqueous sodium bicarbonate solution. Carefully add 25 mL of the 5% base to the organic layer contained in the separatory funnel. Swirl the separatory funnel gently until carbon dioxide gas is no longer evolved. Stopper and gently shake the funnel once or twice, and then vent the vapors. Shake the funnel until no vapors are evolved when the separatory funnel is vented. Remove the lower layer, and repeat the above extraction with 25 mL of 5% sodium bicarbonate solution. Remove the lower layer and check to see whether it is basic to litmus. If it is not basic, repeat the procedure with additional 25-mL portions of 5% base until the aqueous layer is basic. Discard the basic washings and extract the organic layer with one 25-mL portion of water. Add 5 mL of saturated aqueous sodium chloride to aid in layer separation. Stir the mixture gently; do not shake it. Carefully separate the lower aqueous layer and discard it. When the water has been removed, pour the ester from the top of the separatory funnel into a flask. Add about 2 g of anhydrous magnesium sulfate to dry the ester (Technique 7, Section 7.8, p 629). Stopper the flask and swirl it gently. Allow the crude ester to stand until the liquid is clear. About 15 minutes will be needed to complete drying. If the solution is still cloudy after this period, decant the solution, and add to it a fresh 0.5-g quantity of drying agent.

Assemble a simple distillation apparatus as shown in Technique 8, Figure 8–11, p 650. Dry all glassware thoroughly before use. Carefully decant the ester into the distilling flask so that the drying agent is excluded. Add several boiling stones and distill the ester (Technique 8, Section 8.4, p 648). The receiver must be cooled in an ice bath. Collect the fraction boiling between 134 and 143 °C in a dry flask. Weigh the product and calculate the percentage yield.

At your instructor's option, obtain an infrared spectrum (see Technique 18). Compare the spectrum with the one found on page 86. Include it with your report to the instructor. Submit the prepared sample to the instructor in a labeled vial.

Experiment 60
Ethanol from Sucrose

Fermentation
Macroscale technique
Simple distillation
Fractional distillation
Azeotropes

Sucrose

+ H_2O
invertase

Fructose

α-D-(+)-Glucose
(β-D-(+)-glucose is also present,
-OH equatorial)

zymase

$$4CH_3CH_2OH + 4CO_2$$

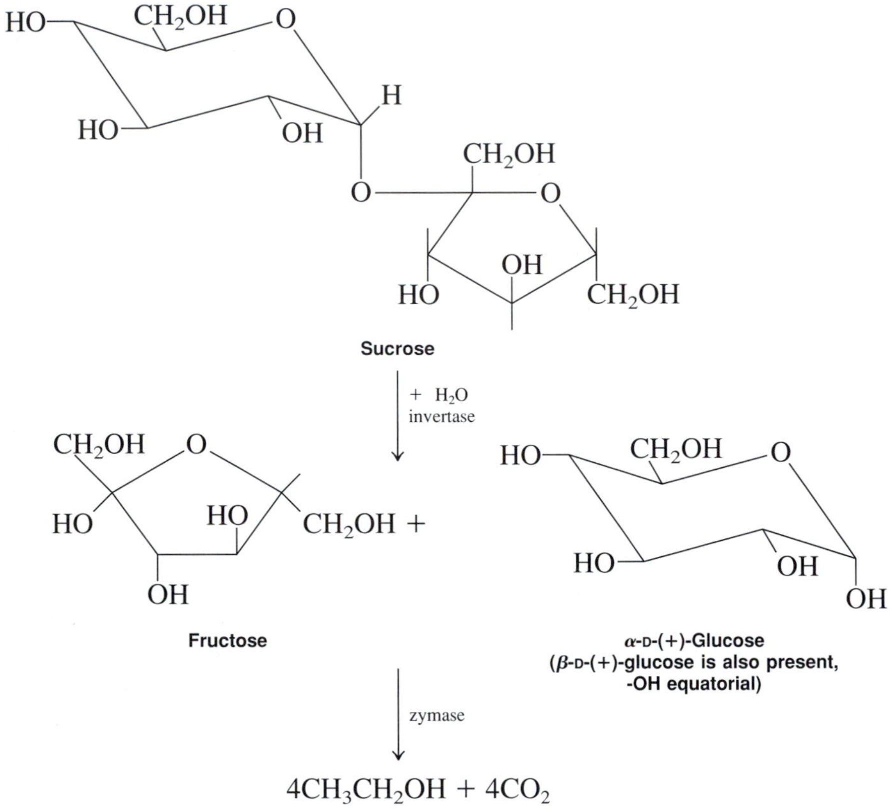

REQUIRED READING

Review: Introductory material in Experiment 20
          Techniques 1 and 2

New: Essay   How To Scale Up a Reaction: Macroscale Methods
          Technique 3   Reaction Methods, Sections 3.2 and 3.4
          Technique 4   Filtration, Sections 4.3–4.5
          Technique 8   Simple Distillation, Section 8.4
          Technique 10   Fractional Distillation, Azeotropes, Sections 10.5–10.8

520

SPECIAL INSTRUCTIONS

The fermentation must be started at least one week before ethanol is actually to be isolated. When the aqueous ethanol is to be separated from the yeast cells, it is important to siphon carefully as much of the clear, supernatant liquid as possible, without agitating the mixture. This procedure is written for use with glassware kits equipped with ℑ 19/22 standard-taper joints. If your kits uses ℑ 14/20 joints you can adapt this procedure simply by reducing all of the quantities by half.

PROCEDURE

Place 40 g of sucrose (common granulated sugar) in a 500-mL Erlenmeyer flask. Add 350 mL water, warmed to room temperature; 35 mL Pasteur's salts; and half a package of dried baker's yeast or 15 g of cake yeast.[1] Shake vigorously and fit the flask with a one-hole rubber stopper with a glass tube leading to a beaker or a test tube containing a solution of barium hydroxide. Protect the barium hydroxide from air by adding some mineral oil or xylene to form a layer above the barium hydroxide. The figure on page 223 depicts the apparatus for this experiment. A precipitate of barium carbonate will form, indicating that CO_2 is being evolved. Alternatively, a balloon may be substituted for the barium hydroxide trap assembly. The gas will cause the balloon to expand as the fermentation continues. Oxygen from the atmosphere is excluded by these techniques. If oxygen were allowed to continue in contact with the fermenting solution, the ethanol could be further oxidized to acetic acid or even all the way to carbon dioxide and water. So long as carbon dioxide continues to be liberated, ethanol is being formed.

Allow the mixture to stand at about 25 °C until fermentation is complete, as indicated by the cessation of gas evolution. Usually about 1 week is required. After this time, carefully move the flask to a desk and remove the stopper. Siphon the liquid out of the flask so that as little of the sediment as possible is removed. A siphon is easily started by filling a short section of rubber tubing with water, pinching one end closed, and placing the other end in the liquid in the flask. Release the end of the tubing that is not in the flask and hold it over the edge of the desk top. Allow the ethanol-water solution to run into a large beaker or flask. When the level of the liquid approaches the sediment, slow the siphoning by pinching the rubber tubing slightly. It is better to leave some liquid behind than to draw some sediment out of the flask.

If the siphoned liquid is not clear, clarify it as follows. Place about 2 tablespoons of Filter Aid (Johns-Manville Celite) in a beaker with about 200 mL water. Stir the mixture vigorously and then pour the contents into a Büchner funnel (with filter paper) while applying a vacuum, as in a vacuum filtration (Technique 4, Section 4.3, p 571). This procedure will cause a thin layer of Filter Aid to be deposited on the filter paper (Technique 4, Section 4.4, p 572). Discard the water that passes through this filter. The siphoned liquid containing the ethanol is then passed through this filter under gentle suction. The extremely tiny yeast particles are trapped in the pores of the Filter Aid. The

---

[1] See footnote on p 203.

liquid contains ethanol in water plus smaller amounts of dissolved metabolites (fusel oils) from the yeast.

Add about 46 g of anhydrous potassium carbonate to the filtered solution for each 100 mL of liquid. The solution, after becoming saturated with potassium carbonate, will be subjected to fractional distillation. Transfer the solution to a distillation apparatus equipped with a fractionating column packed with a metal sponge (Technique 10, Figure 10–11, p 680, and Section 10.6, p 678). Distill the liquid slowly through the fractionating column to get the best possible separation. Collect the fraction boiling between 78 and 88 °C and discard the residue in the distilling flask. The extent of purification of the ethanol is limited, since ethanol and water form a constant-boiling mixture, or an azeotrope, with a composition of 95% ethanol and 5% water (bp 78.1 °C). No amount of distillation will remove the last 5% of water (Technique 10, Section 10.7, p 680).

Calculate the percentage yield of alcohol, assuming that the product is 85% alcohol and 15% water, and submit the ethanol to the instructor in a labeled vial.[2]

---

[2] A careful analysis by flame-ionization gas chromatography on a typical student-prepared ethanol sample provided the following results:

Acetaldehyde	0.060%
Diethylacetal of acetaldehyde	0.005
Ethanol	88.3 (by hydrometer)
1-Propanol	0.031
2-Methyl-1-propanol	0.092
5-Carbon and higher alcohols	0.140
Methanol	0.040
Water	11.3 (by difference)

Experiment 61
Synthesis of *t*-Pentyl Chloride

Synthesis of alkyl halides
Macroscale techniques
Extraction
Simple distillation

$$
\underset{\substack{\text{*t*-Pentyl alcohol}}}{CH_3CH_2\overset{\displaystyle CH_3}{\underset{\displaystyle OH}{C}}CH_3} + HCl \longrightarrow \underset{\substack{\text{*t*-Pentyl chloride}}}{CH_3CH_2\overset{\displaystyle CH_3}{\underset{\displaystyle Cl}{C}}CH_3} + H_2O
$$

REQUIRED READING

Review: Introductory material in Experiment 13
          Techniques 1 and 3

New:	Essay	How To Scale Up a Reaction: Macroscale Methods
	Technique 7	Extractions, Separations, and Drying Agents, Sections 7.7–7.9
	Technique 8	Simple Distillation, Section 8.4

SPECIAL INSTRUCTIONS

Exercise caution when using concentrated hydrochloric acid; it is corrosive. It can cause burns if it is spilled on the skin. This procedure is written for use with glassware kits equipped with ℸ 19/22 standard-taper joints. If your kit uses ℸ 14/20 joints you can adapt this procedure simply by reducing all of the quantities by half.

PROCEDURE

In a 125-mL separatory funnel, place 22 mL (density 0.805 g/mL) of *t*-pentyl alcohol (*t*-amyl alcohol or 2-methyl-2-butanol) and 50 mL (density 1.18 g/mL; 37.3% HCl) of concentrated hydrochloric acid. Do not stopper the funnel. Gently swirl the mixture in the separatory funnel for about 1 minute. After this period of swirling, stopper the separatory funnel and carefully invert it. Without shaking the separatory funnel, immediately open the stopcock to release the pressure. Close the stopcock, shake the funnel several times, and again release the pressure through the stopcock (Technique 7, Section 7.7, p 627). Shake the funnel for 2 to 3 minutes, with occasional venting. Allow the mixture to stand in the separatory funnel until the two layers have completely separated. The *t*-pentyl chloride has a density of 0.865 g/mL. Which layer contains the alkyl halide? Separate the layers.

    The operations in this paragraph should be done as rapidly as possible since the *t*-pentyl chloride is unstable in water and sodium bicarbonate solution. Wash (swirl and shake) the organic layer with one 25-mL portion of water. Again, separate the layers and discard the aqueous phase after making certain that the proper layer has been saved (Technique 7, Section 7.7, p 627). Wash the organic layer with a 25-mL portion of 5% aqueous sodium bicarbonate. Gently swirl the funnel (unstoppered) until the contents are thoroughly mixed. Stopper the funnel, and carefully invert it. Release the excess pressure through the stopcock. Gently shake the separatory funnel, with frequent release of pressure. Following this, vigorously shake the funnel, again with release of pressure, for about 1 minute. Allow the layers to separate, and drain the lower aqueous bicarbonate layer. Wash (swirl and shake) the organic layer with one 25-mL portion of water, and again drain the lower aqueous layer.

    Transfer the organic layer to a small dry Erlenmeyer flask. Pour it from the top of the separatory funnel. Dry the crude *t*-pentyl chloride over anhydrous calcium chloride until it is clear (Technique 7, Section 7.8, p 629). Swirl the alkyl halide with the drying agent to aid the drying. Decant the **clear** material into a small **dry** distilling flask. Add a boiling stone and distill the crude *t*-pentyl chloride in a **dry** apparatus (Technique 8,

Section 8.4, Figure 8–11, p 650), using a steam bath. Collect the pure *t*-pentyl chloride in a receiver cooled in ice. Collect the material that boils between 79 and 84 °C. Weigh the product and calculate the percentage yield. Submit the sample to the instructor in a labeled vial.

Experiment 62
Nitration of Methyl Benzoate

Aromatic substitution
Macroscale technique
Crystallization

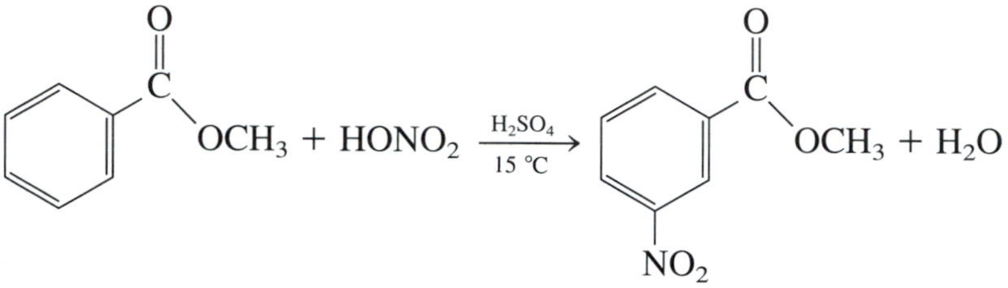

Methyl benzoate   Methyl *m*-nitrobenzoate

REQUIRED READING

Review: Introductory material in Experiment 38
             Techniques 1, 2, and 3
             Technique 18

New: Essay   How To Scale Up a Reaction: Macroscale Methods
             Technique 5   Crystallization: Purification of Solids, Sections 5.3, 5.6,
                                   and 5.7

SPECIAL INSTRUCTIONS

It is important that the temperature of the reaction mixture be maintained at or below 15 °C. Nitric acid and sulfuric acid, especially when mixed, are very corrosive substances. Be careful not to get these acids on your skin. If you do get some of these acids on your skin, flush the affected area liberally with water. This procedure is written for use with glassware kits equipped with ℸ 19/22 standard-taper joints. If your kit uses

℉ 14/20 joints you can adapt this procedure simply by reducing all of the quantities by half.

PROCEDURE

In a 150-mL beaker cool 12 mL of concentrated sulfuric acid to about 0 °C and add 6.1 g of methyl benzoate. Using an ice-salt bath (see Technique 2, Section 2.6, p 545), cool the mixture to 0 °C or below and add, VERY SLOWLY, using a Pasteur pipet, a cool mixture of 4 mL of concentrated sulfuric acid and 4 mL of concentrated nitric acid. During the addition of the acids, stir the mixture continuously and maintain the temperature of the reaction below 15 °C. If the mixture rises above this temperature, the formation of by-product increases rapidly, bringing about a decrease in the yield of the desired product.

After all the acid has been added, warm the mixture to room temperature. After 15 minutes, pour the acid mixture over 50 g of crushed ice in a 250-mL beaker. After the ice has melted, isolate the product by vacuum filtration through a Büchner funnel and wash it with two 25-mL portions of cold water and then with two 10-mL portions of ice-cold methanol. Weigh the product and recrystallize it from an equal weight of methanol (Technique 5, Section 5.3, p 58). The melting point of the recrystallized product should be 78 °C. Determine the infrared spectrum of the product as a Nujol mull (Technique 18, Section 18.6, p 779). Submit the product to your instructor in a labeled vial, along with your infrared spectrum. Compare the spectrum with the one found on p 309.

Experiment 63
Hydrolysis of Methyl Salicylate

Hydrolysis of an ester
Macroscale technique
Heating under reflux
Filtration
Crystallization
Melting-point determination

Esters can be hydrolyzed to their constituent carboxylic acid and alcohol parts under either acidic or basic conditions. In this experiment, **methyl salicylate,** an ester known as **oil of wintergreen** because of its natural source, is treated with aqueous base. The immediate product of this hydrolysis, besides methanol and water, is the sodium salt of salicylic acid. The reaction mixture is acidified with sulfuric acid, which converts the sodium salt to the free acid. The overall organic products of the reaction, therefore, are salicylic acid and methanol. The salicylic acid is a solid, which can be isolated and purified by crystallization. The chemical equations that describe this experiment are

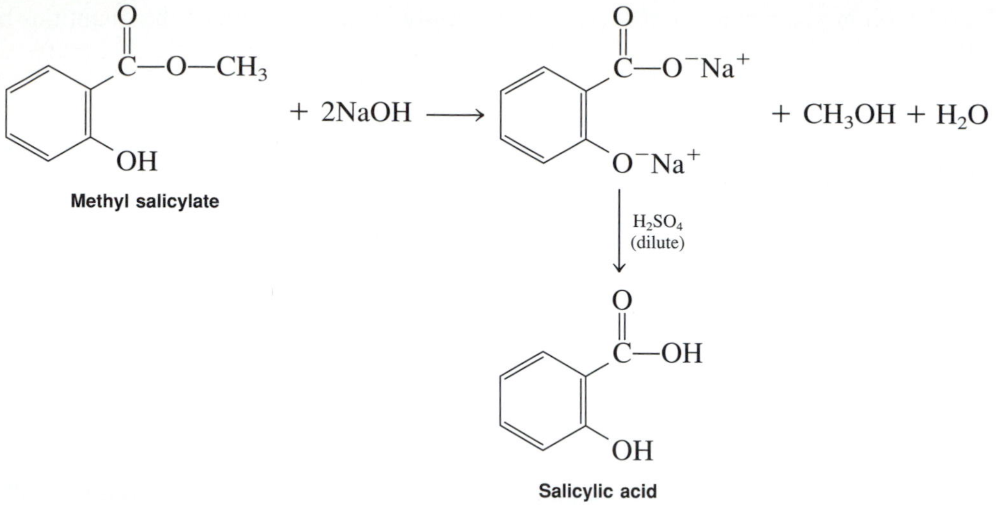

Methyl salicylate

Salicylic acid

Because the phenolic hydroxyl group is acidic, it is also converted to the corresponding sodium salt during the basic hydrolysis. In the subsequent acidification, this group becomes reprotonated.

REQUIRED READING

Review: Techniques 1, 2, and 6

New: Essay   How To Scale Up a Reaction: Macroscale Methods
          Technique 3   Reaction Methods, Sections 3.2 and 3.4
          Technique 4   Filtration, Sections 4.1, 4.3, and 4.5
          Technique 5   Crystallization: Purification of Solids, Sections 5.3 and
                            5.7

SPECIAL INSTRUCTIONS

This experiment can be scheduled conveniently with other experiments. It can be stopped at nearly any point. This procedure is written for use with glassware kits equipped with ⊤ 19/22 standard-taper joints. If your kit uses ⊤ 14/20 joints you can adapt this procedure simply by reducing all of the quantities by half.

PROCEDURE

Dissolve 10 g of sodium hydroxide in 50 mL of water. When the solution has cooled, place it, along with 5.0 g (0.033 mole) of methyl salicylate, in a 250-mL round-bottomed flask. A white solid may form at this point, but it will dissolve on heating. Attach a reflux condenser to the flask, according to the instructions given in Technique 3, Section 3.2, p 549. The standard taper joints should be greased lightly.

Add one or two boiling stones to the reaction mixture to prevent bumping when the solution is heated. Heat the solution at its boiling point for about 20 minutes with a heating mantle. After heating the mixture, allow it to cool to room temperature. When the solution is cool, transfer it to a 250-mL beaker, and carefully add enough 1M sulfuric acid to make the solution acidic to litmus paper (blue litmus turns pink). As much as 150 mL of 1M sulfuric acid may have to be added at this stage. When the litmus turns pink, add an extra 15 mL of the sulfuric acid, thus causing the salicylic acid to precipitate from the solution. Cool the mixture in an ice-water bath to about 0 °C. Allow this cold mixture to settle. Collect the product by vacuum filtration, using a Büchner funnel with filter paper. See Technique 4, Section 4.3, p 571, for details of this method. The filtration can be conducted most easily by decanting most of the supernatant liquid through the Büchner funnel before adding the mass of crystals.

Recrystallize the crude salicylic acid from water in a 125-mL Erlenmeyer flask. Add 100 mL of hot water and a boiling stone, and heat the mixture to boiling to dissolve the solid.

If the solid does not dissolve on boiling, add enough extra water to dissolve the solid. Filter the hot solution by gravity filtration through a fluted filter paper (Technique 4, Section 4.1, p 565, and Figure 4–3, p 567), using a fast filter paper, and set the solution aside to cool.

This gravity filtration must be carried out carefully. Filter the hot solution using only a small quantity at a time. Use a short-stemmed funnel for the filtration to reduce the probability that crystals might form in the stem, clogging the funnel. The filtration assembly should be placed on a steam bath. If salicylic acid begins to crystallize in the funnel, add to the filter the **minimum** amount of boiling water needed to redissolve the crystals.

After the filtered solution has cooled, place the flask in an ice-water bath to aid crystallization. When the crystals of salicylic acid have formed, collect them by vacuum filtration (Technique 4, Section 4.3, p 571). Allow the crystals to dry overnight on a watch glass. When the crystals are thoroughly dry, weigh them and determine the percentage yield. Determine the melting point of the pure material (Technique 6, Sections 6.5–6.7, p 600–603). The melting point of pure salicylic acid is 159–160 °C. Place the sample of product in a labeled vial and submit it to the instructor.

Part Five

The Techniques

Technique 1
MEASUREMENT OF VOLUME AND WEIGHT

Special care must be taken when working with small amounts of liquid or solids. In the typical microscale experiment, a student will use from 10–200 mg of a liquid or solid. Specially designed microscale equipment will be used for these small scale reactions. You may not be used to working with such small quantities, but after a while, you will adjust to "thinking small."

Liquids should be transferred using one of the following devices: automatic pipets, graduated pipets, Pasteur pipets, or syringes. In most cases, volumes and densities will be provided for liquids used as reactants in the experimental procedures. Liquids should be measured carefully using an automatic pipet or graduated pipet. You may calculate the weight from the following relationship:

$$\text{Weight (g)} = \text{Density (g/mL)} \times \text{Volume (mL)}$$

For limiting reagents, you should always check the weight after transferring the liquid to the preweighed (tared) container. In cases where a reagent is not limiting, it is not necessary to weigh the sample, and it will suffice to calculate the weight using the expression given above. In some cases, solutions of solids will be used in the experiments. They should be measured carefully. Sometimes the volume of liquid does not need to be measured precisely. In this case, a calibrated Pasteur pipet works well.

Solids must be weighed on a balance that reads to the nearest milligram (0.001 g) or tenth of a milligram (0.0001 g). You should use a spatula to transfer a solid. Never pour, dump, or shake a material from a bottle.

1.1 AUTOMATIC PIPETS

When available, an automatic pipet increases the speed of transfer of liquids from reagent bottles. These pipets are very expensive and must be shared by the entire laboratory. A number of different types of units are available commercially. We will describe the use of the continuously adjustable automatic pipet. This type of pipet can be adjusted for any volume within its defined range using a three- or four-digit readout. Several types of adjustable automatic pipets are shown in Figure 1–1. The typical laboratory may have several units available: one 10–100-μL (0.01–0.10-mL) pipet for smaller volumes, and two 100–1000-μL (0.10–1.00-mL) pipets for larger volumes. Disposable tips are available for each of these units and are color coded: yellow and blue for the small and large units, respectively. The automatic pipet is very accurate with aqueous solutions, but it is not quite as accurate with organic liquids.

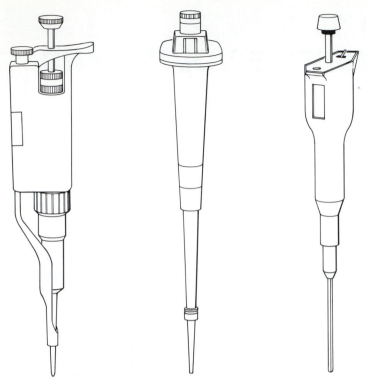

FIGURE 1–1. The adjustable automatic pipet

      In most cases, the instructor will adjust the pipet so that it will deliver the desired volume. It will be placed in a convenient location near the reagent bottle, usually in a hood, and students will reuse the tip. Your instructor will give directions for the correct use of the automatic pipet. Students must practice using the automatic pipet by following the instructions given on p 40. Remember that the automatic pipet is very expensive and must be handled carefully. To protect the unit, you must always use a tip on the end of the pipet. Liquid must be drawn only into this plastic tip and never up into the unit itself. Keep the pipet upright and immerse the tip just below the surface of the liquid.

1.2 GRADUATED PIPETS

A suitable alternative to an automatic pipet is the graduated serological pipet. These **glass** pipets are available commercially in a number of sizes. "Disposable" pipets may be used many times and discarded only when the graduations become too faint to be seen. A good assortment of these pipets consists of the following:

      0.50-mL pipets calibrated in 0.01-mL divisions (5/10 in 1/100 mL)
      1.00-mL pipets calibrated in 0.01-mL divisions (1 in 1/100 mL)
      2.00-mL pipets calibrated in 0.01-mL divisions (2 in 1/100 mL)

Liquids may be measured and transferred using a graduated pipet and a pipet pump as illustrated on p 43. The style of pipet pump shown in Figure 1–2A is available in four different sizes. The 2-mL size (blue) works well with the range of pipets indicated above. To fill the pipet, one simply rotates the knurled wheel forward so that the piston moves upward. The liquid is discharged by slowly turning the wheel backwards until the proper amount of liquid has been expelled. The top of the pipet must be inserted securely into the pump and held there with one hand to obtain an adequate seal. The other hand is used to load and release the liquid.

The pipet pump shown in Figure 1–2B may also be used with graduated pipets. The knob is turned counterclockwise to draw in the liquid, and then the liquid is released by turning the knob clockwise. With this style of pipet, the top of the pipet is held securely by a rubber O-ring and it is easily handled with one hand. You should be certain that the pipet is held securely by this O-ring before using it. Disposable pipets may not fit tightly in the O-ring because they often have smaller diameters than non-disposable pipets.

Excellent results may be obtained with graduated pipets if you transfer by difference between marked calibrations and avoid transferring the entire contents of the pipet. When expelling the liquid, be sure to touch the tip of the pipet to the inside of the container before withdrawing the pipet.

Pipets may be obtained in a number of different styles, but only three types will be described here (Figure 1–3). One type of graduated pipet is calibrated ''to deliver'' (TD) its total capacity when the last drop is blown out. This style of pipet, shown in Figure 1–3A is probably the most common type of graduated pipet in use in the laboratory; it is designated by two rings at the top. Of course, one does not need to transfer the entire volume to a container. In order to deliver a more accurate volume,

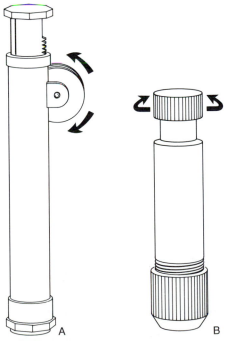

FIGURE 1–2. Pipet pumps

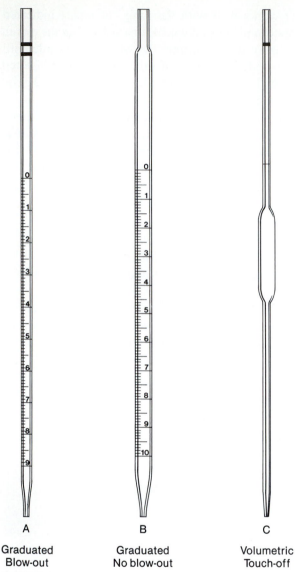

A
Graduated
Blow-out

B
Graduated
No blow-out

C
Volumetric
Touch-off

FIGURE 1–3. Pipets

you should transfer an amount less than the total capacity of the pipet using the graduations on the pipet as a guide.

Another type of graduated pipet is shown in Figure 1–3B. This pipet is calibrated to deliver its total capacity when the meniscus is located on the last graduation mark near the bottom of the pipet. For example, the pipet shown in the Figure 1–3B delivers 10.0 mL of liquid when it has been drained to the point where the meniscus is located on the 10.0-mL mark. With this type of pipet, you must not drain the entire pipet or blow it out. In contrast, notice that the pipet discussed above (Figure 1–3A) has its last graduation at 9.0 mL. The last 1.0-mL volume is blown out to give the 10.0-mL volume.

A non-graduated volumetric pipet is shown in Figure 1–3C. It is easily identified by the large bulb in the center of the pipet. This pipet is calibrated so that it will retain its last drop after the tip is touched on the side of the container. It must not be blown out. These pipets often have a single colored band at the top that identifies it as a "touch-off" pipet. The color of the band is keyed to its total volume. This type of pipet is commonly used in analytical chemistry.

1.3 PASTEUR PIPETS

The Pasteur pipet is shown in Figure 1–4A with a 2-mL rubber bulb attached. There are two sizes of pipets: a long one (9 inch) and a short one (5¾ inch). It is important that the pipet bulb fit securely. You should not use a medicine dropper bulb because of its small capacity. A Pasteur pipet is an indispensable piece of equipment that you will use often for routine transfer of liquids. It is also used for separations (Technique 7). Pasteur pipets may be packed with cotton for use in gravity filtration (Technique 4) or packed with an adsorbent for small scale column chromatography (Technique 12). Although they are considered to be disposable, you should be able to clean them for reuse as long as the tip remains unchipped.

Pipets may be calibrated for use in operations where the volume does not need to be known precisely. Examples include measurement of solvents needed for extraction and for washing a solid obtained following crystallization. It is suggested that you

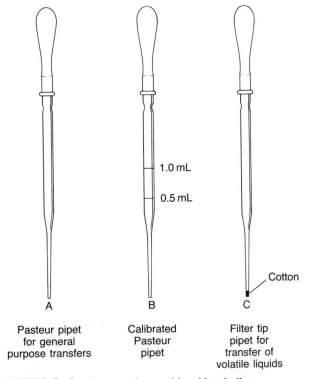

1.0 mL

0.5 mL

Cotton

A	B	C
Pasteur pipet for general purpose transfers	Calibrated Pasteur pipet	Filter tip pipet for transfer of volatile liquids

FIGURE 1–4. Pasteur pipets with rubber bulbs

calibrate several 5¾ inch pipets following the procedure given on p 44. A calibrated Pasteur pipet is shown in Figure 1–4B.

In general, Pasteur pipets should not be used to measure volumes of reagents needed for organic reactions as they are not accurate enough for this purpose. In some cases, however, your instructor may have available a calibrated pipet for transferring non-limiting reagents that may damage an automatic pipet. For example, a calibrated Pasteur pipet may be used with concentrated acids.

> **CAUTION:** You should not assume that a certain number of drops equals a 1-mL volume. The common rule that 20 drops equal 1 mL does not hold true for a Pasteur pipet!

A Pasteur pipet may be packed with cotton to create a filter tip pipet shown in Figure 1–4C. This pipet is prepared by the instructions given in Technique 4, Section 4.6, p 574. Pipets of this type are very useful in transferring volatile solvents during extractions and in filtering small amounts of solid impurities from solutions.

1.4 SYRINGES

Syringes may be used to add a pure liquid or a solution to a reaction mixture. They are especially useful when anhydrous conditions must be maintained. The needle is inserted through a septum and the liquid is added to the reaction mixture. Although syringes come in a number of different sizes, we will use a 1-mL unit in this textbook. Caution should be used with disposable syringes as they often use solvent-soluble rubber gaskets on the plungers. A syringe should be cleaned carefully after each use by drawing acetone or another volatile solvent into it and expelling the solvent with the plunger. Repeat this procedure several times to clean the syringe thoroughly. Draw air through the barrel with an aspirator to dry the syringe.

Syringes are usually supplied with volume graduations inscribed on the barrel. Large volume syringes are not accurate enough to be used for measuring liquids in microscale experiments. A small microliter syringe, such as that used in gas chromatography, delivers a very precise volume.

1.5 GRADUATED CYLINDERS

Graduated cylinders are used to measure relatively large volumes of liquids where accuracy is not required. For example, you could use a 10-mL graduated cylinder to obtain about 2 mL of a solvent for a crystallization procedure. You should use an automatic pipet or a graduated pipet for accurate transfer of liquids in microscale work.

1.6 MEASURING VOLUMES WITH CONICAL VIALS, BEAKERS, AND ERLENMEYER FLASKS

Conical vials, beakers and Erlenmeyer flasks all have graduations inscribed on them. Beakers and flasks can only be used to give a crude approximation of the volume. They are much less precise than graduated cylinders for measuring volume. In some cases, a conical vial may be used to estimate volumes. For example, the graduations are sufficiently accurate for measuring a solvent needed to wash a solid obtained on a Hirsch funnel after a crystallization. You should use an automatic pipet or graduated transfer pipet for accurate measurement of liquids.

1.7 BALANCES

Solids and some liquids will need to be weighed on a balance that reads to at least the nearest milligram (0.001 g). A top-loading balance (see Figure 1–5) works well if the balance pan is covered with a plastic draft shield. The shield has a flap that opens to allow access to the balance pan. An analytical balance (see Figure 1–6) may also be used. This type of balance will weigh to the nearest tenth of a milligram (0.0001 g) when provided with a glass draft shield.

Modern electronic balances have a tare device that automatically subtracts the weight of a container or a piece of paper from the combined weight to give the weight of the sample. With solids, it is easy to place a piece of paper on the balance pan, press the tare device so that the paper appears to have zero weight, and then add your solid

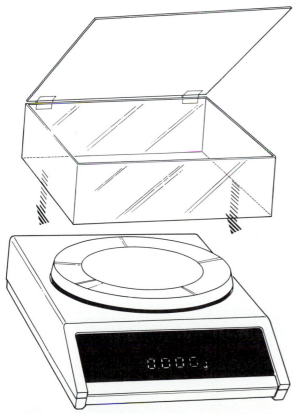

FIGURE 1–5. A top-loading balance with plastic draft shield

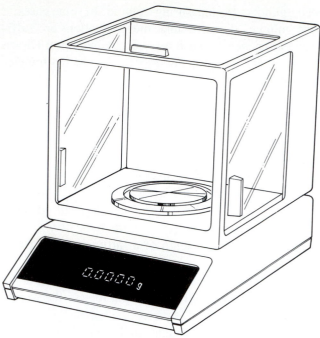

FIGURE 1–6. An analytical balance with glass draft shield

until the balance gives the weight you desire. You can then transfer the weighed solid to a container. You should always use a spatula to transfer a solid and never pour material from a bottle. In addition, solids must be weighed on paper and not directly on the balance pan. Remember to clean up any spills that occur.

With liquids, you should weigh the conical vial to determine the tare weight, transfer the liquid with a automatic pipet or graduated pipet into the vial, and then reweigh it. With liquids, usually it is only necessary to weigh the limiting reagent. The other liquids may be transferred using an automatic pipet or graduated pipet. Their weights can be calculated by knowing the volumes and densities of the liquids.

PROBLEMS

1. What measuring device would you use to measure the volume under each of the conditions described below? In some cases, there may be more than one answer to the question.
   (a) 5 mL of a solvent needed for a crystallization
   (b) 0.18 mL of a liquid needed for a reaction
   (c) 0.56 mL of a liquid needed for a reaction
   (d) 1 mL of a solvent needed for an extraction
2. Assume that the liquids used in parts **(b)** and **(c)** are limiting reagents for their respective reactions. What should you do after measuring the volumes?
3. Calculate the weight of a 0.25-mL sample of each of the following liquids.
   (a) Diethyl ether (ether)
   (b) Methylene chloride (dichloromethane)
   (c) Acetone
4. A laboratory procedure calls for 0.146 g of acetic anhydride. Calculate the volume of this reagent needed in the reaction.

Technique 2
HEATING AND COOLING METHODS

Most organic reaction mixtures need to be heated in order to complete the reaction. In general chemistry you used a Bunsen burner for heating because non-flammable aqueous solutions were used. In an organic laboratory, however, the student must heat non-aqueous solutions that may contain **highly flammable** solvents. You **should not heat organic mixtures with a Bunsen burner** unless you are directed by your laboratory instructor. Open flames present a potential fire hazard.

2.1 SAND BATH WITH HOT PLATE/STIRRER

The sand bath is the most widely used device for heating organic mixtures in microscale experiments. Sand provides a clean and quick way of distributing heat to a reaction mixture. You should never heat a conical vial directly on a hot plate. To prepare a sand bath, place about a 1-cm depth of sand in a crystallizing dish or a Petri dish and then set the dish on a hot plate/stirrer unit. The apparatus is shown in Figure 2–1. It is a good idea to use the same hot plate each time. Record the number printed on the unit that you are using in your notebook. Clamp the thermometer into position in the sand bath. You should calibrate the sand bath as described on p 36 so that you have an approximate idea where to set the dial to achieve a desired temperature. If you need a sand bath, one of the first things you should do when you come into the laboratory is to place your sand bath on the hot plate and turn on the heater. By using the hot plate settings from your calibration curve, you can save considerable time by preheating the bath to the desired temperature in advance of when it will be needed.

As a practical note, you will rarely heat a mixture above 200 °C. Do not heat the sand bath much above 200 °C or you may break the dish. If you need to heat at very high temperatures, you should use an aluminum block rather than a sand bath (Section 2.3). With sand baths, it may be necessary to cover the dish with aluminum foil to achieve a temperature near 200 °C. It should be noted that there may be several different types of hot plates in use in the laboratory, and they may not give the same temperature for a given setting. Do not rely on the values obtained by someone else. In fact, two hot plates of the same type may give different sand bath temperatures with an identical setting.

Keep in mind that the temperature obtained at a particular setting on the hot plate may vary for several reasons. First, you may place the thermometer at a different depth from time to time. Second, because of the relatively poor heat conduction of sand, you may obtain a different temperature in the conical vial depending on the depth

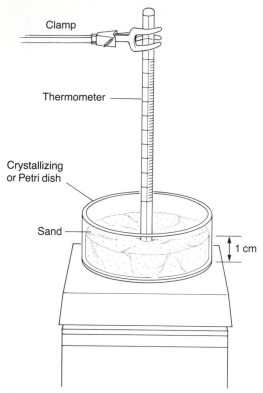

Clamp

Thermometer

Crystallizing
or Petri dish

Sand

1 cm

FIGURE 2–1. The sand bath

of the vial in the sand bath. Because of the poor heat conductivity of sand, a temperature gradient is established within the sand bath. It is warmer near the bottom of the sand bath and cooler near the top for a given setting on the hot plate. To make use of this gradient, you may find it convenient to bury the vial or flask in the sand to heat a mixture more rapidly. Once the mixture is boiling, you can then slow the rate of heating by raising the vial or flask. These adjustments may be made easily and do not require a change in the setting on the hot plate.

Although we provide sand bath temperatures in most of the experiments in this textbook, they should be taken as **approximate** values. You may need to adjust the temperature of the sand bath appropriately to achieve the conditions you require. When a sand bath temperature is not given in the procedure and the liquid needs to be brought to a boil, you can determine the approximate setting from the boiling point of the liquid. Because the temperature inside the vial is lower than the sand bath temperature, you should add 10–20 °C to the boiling point of liquid and set the sand bath at this higher temperature.

Many organic mixtures need to be stirred to achieve satisfactory results. To stir a mixture, place a magnetic spin vane (Technique 3, Figure 3–4A, p 552) into a conical vial containing the reaction mixture. If the mixture is to be heated as well as stirred, a water condenser or an air condenser should be attached as shown in Figure 2–2. With the combination stirrer/hot plate unit, it is possible to stir and heat a mixture simultaneously. Many reactions in this textbook are stirred continuously during the course of the

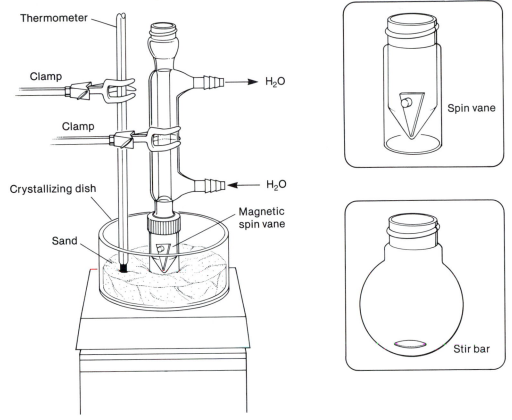

FIGURE 2–2.   Heating a mixture, with stirring

reaction. With round-bottomed flasks, a magnetic stir bar must be used to stir mixtures (Technique 3, Figure 3–4, p 552). More uniform stirring will be obtained if the vial or flask is placed near the bottom of the sand bath (closer to the magnet). Mixing may also be achieved by boiling the mixture. A boiling stone (Technique 3, Section 3.4, p 552) should be added when a mixture is boiled without magnetic stirring.

2.2   SAND BATH WITH HEATING MANTLE

A sand bath may also be prepared using a small heating mantle as the heat source as shown in Figure 2–3. When preparing a sand bath using a mantle, it is important to use one that has a solid ceramic core. A mantle consisting of a woven blanket of spun fiberglass will not work because of the possible penetration of sand into the fabric. Place sand in the heating mantle to a depth of 1–2 cm. Clamp the thermometer into position in the sand bath so that the thermometer bulb is nearly covered by the sand. The temperature in the heating mantle is regulated by a temperature controller unit. The primary disadvantage of using a heating mantle for a sand bath is that the reaction mixture cannot be stirred magnetically. There is also a risk that the heating mantle may be damaged by overheating the unit. Avoid temperatures in excess of 250 °C.

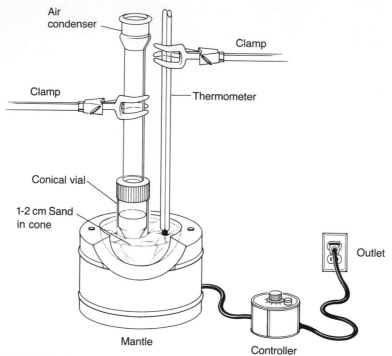

Air condenser

Clamp

Clamp

Thermometer

Conical vial

1-2 cm Sand in cone

Outlet

Mantle

Controller

FIGURE 2–3. Sand bath prepared with a heating mantle

2.3 ALUMINUM BLOCK WITH HOT PLATE/STIRRER

A reaction mixture may also be heated using an aluminum block placed on the same hot plate used in Section 2.1. This device is illustrated in Figure 2–4. The holes in the block are drilled so that different sized conical vials will fit into the holes. In addition, there is a place for a thermometer. This aluminum block may also be used in crystallizations using a Craig tube. A magnetic spin vane can be added to a vial so that the mixture may be stirred as well as heated. Also shown in Figure 2–4 is an aluminum auxiliary collar which may be used when required for very high boiling liquids. It has the same effect as ''pushing the vial deeper into a sand bath.'' The collar is split to facilitate easy placement around a 5-mL conical vial.

There are several advantages to heating with an aluminum block. First, the metal will heat faster than a sand bath. Second, one can obtain a higher temperature than with a sand bath. Third, one can cool the metal rapidly by removing it with beaker tongs and immersing it in cold water. The aluminum block may have to be used when temperatures significantly higher than 200 °C are needed. Higher temperatures are often needed when distilling high boiling liquids at atmospheric pressure or under vacuum. The calibrations that you obtained previously for the sand bath (p 36) will not apply when using an aluminum block.

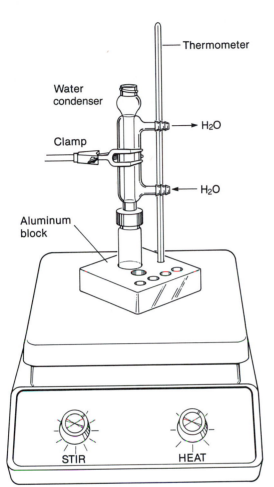

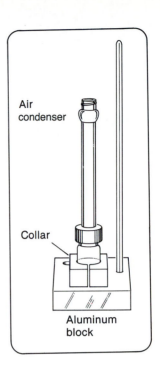

FIGURE 2–4. Heating with an aluminum block

2.4 WATER BATH WITH HOT PLATE/STIRRER

The methods described above may be used over a range of about 50 °C to over 200 °C. A hot water bath, however, may be a suitable alternative for temperatures below about 80 °C. A beaker (100 mL or 250 mL) is partially filled with water and heated on a hot plate. A thermometer is clamped into position in the water bath. You may need to cover the water bath with aluminum foil to prevent evaporation, especially at higher temperatures. The water bath is illustrated in Figure 2–5. A mixture can be stirred with a magnetic spin vane (Technique 3, Section 3.3, p 551). A hot water bath has some advantage over a sand bath in that the temperature in the bath is uniform. In addition, it is sometimes easier to establish a lower temperature with a water bath than with a sand bath. Finally, the temperature of the reaction mixture will be closer to the temperature of the water, which allows for a more precise control of the reaction conditions.

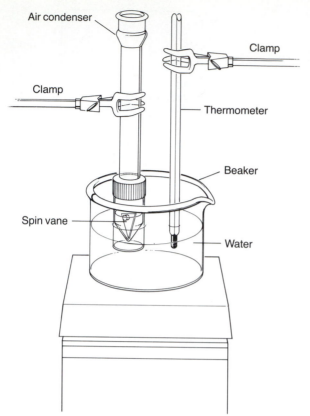

Air condenser

Clamp

Clamp

Thermometer

Beaker

Spin vane

Water

FIGURE 2–5. Water bath

2.5 FLAMES

The simplest technique for heating mixtures is to use a Bunsen burner. Because of the high danger of fires, however, the use of the Bunsen burner should be strictly limited to those cases for which the danger of fire is low or for which no reasonable alternative source of heat is available. A flame should generally be used only to heat aqueous or very-high-boiling solutions. You should always check with your instructor about using a burner. If you use a burner at your bench, great care should be taken to ensure that persons in the vicinity are not using flammable solvents.

In heating a flask with a Bunsen burner, you will find that using a wire gauze can produce more even heating over a broader area. The wire gauze, when placed under the object being heated, spreads the flame to keep the flask from being heated in one small area only.

Bunsen burners may be used to prepare capillary micropipets for thin layer chromatography or to prepare other pieces of glassware requiring an open flame. Burners should be used in designated areas in the laboratory and not at your laboratory bench for these purposes.

2.6   COLD BATHS

At times, one may need to cool a conical vial or flask below room temperature. A cold bath is used for this purpose. The most common cold bath is an **ice bath,** which is a highly convenient source of 0 °C temperatures. An ice bath requires water along with ice to work well. If an ice bath is made up of only ice, it is not a very efficient cooler since the large pieces of ice do not make good contact with the flask or vial. Enough water should be present with ice so that the flask is surrounded by water but not so much that the temperature is no longer maintained at 0 °C. In addition, if too much water is present, the buoyancy of a flask resting in the ice bath may cause it to tip over. There should be enough ice in the bath to allow the flask to rest firmly.

For temperatures somewhat below 0 °C, one may add some solid sodium chloride to the ice-water bath. The ionic salt lowers the freezing point of the ice, so that temperatures in the range of 0 to -10 °C can be reached. The lowest temperatures are reached with ice-water mixtures that contain relatively little water.

A temperature of -78.5 °C can be obtained with solid carbon dioxide or dry ice. Large chunks of dry ice do not provide uniform contact with a flask being cooled. A liquid such as isopropyl alcohol is mixed with small pieces of dry ice to provide an efficient cooling mixture. Acetone and ethanol are also used in place of isopropyl alcohol. One should be cautious when handling dry ice since it can inflict severe frostbite. Extremely low temperatures can be obtained with liquid nitrogen (-195.8 °C).

2.7   STEAM BATHS

The steam cone or steam bath is a good source of heat when temperatures around 100 °C are needed. Steam baths are used to heat reaction mixtures and to heat solvents needed for crystallization. A steam cone and a portable steam bath are shown in Figure 2–6. These methods of heating have the disadvantage that water vapor may be introduced, through condensation of steam, into the mixture being heated. A slow flow of steam may minimize this difficulty.

Because water condenses in the steam line when it is not in use, it is necessary to purge the line of water before the steam will begin to flow. This purging should be accomplished before the flask is placed on the steam bath. The steam flow should be started with a high rate to purge the line; then, the flow should be reduced to the desired rate. When using a portable steam bath, be certain that condensate (water) is drained into a sink. Once the steam bath or cone is heated, a slow steam flow will maintain the temperature of the mixture being heated. There is no advantage to having a Vesuvius on your desk! An excessive steam flow may cause problems with condensation in the flask being heated. This condensation problem can often be avoided by selecting the correct place to locate the flask on top of the steam bath.

The top of the steam bath consists of several flat concentric rings. The amount of heat delivered to the flask being heated can be controlled by selecting the correct

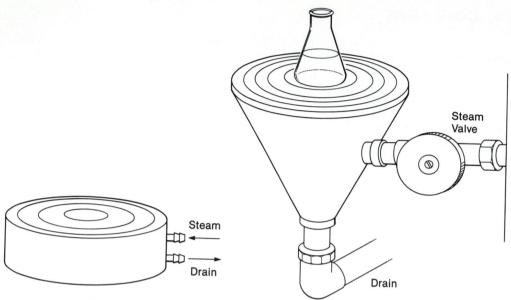

FIGURE 2–6.   Steam bath and steam cone

sizes of these rings. Heating is most efficient when the largest opening that will still support the flask is used. Heating large flasks on a steam bath while using the smallest opening provides slow heating and wastes laboratory time.

PROBLEMS

1. What would be the preferred heating device in each of the following situations?
   (a) Reflux a solvent with a 56 °C boiling point
   (b) Reflux a solvent with a 110 °C boiling point
   (c) Distillation of a substance that boils at 220 °C
2. Obtain the boiling points for the following compounds by making use of a handbook (Technique 19, Section 19.1, p 787). In each case, suggest a heating device(s) that should be used for refluxing the substance.
   (a) Butyl benzoate
   (b) 1-Pentanol
   (c) 1-Chloropropane
3. What type of bath would you use to get a temperature of −10 °C?
4. Obtain the melting point and boiling point for benzene and ammonia from a handbook (Technique 19, Section 19.1, p 787) and answer the following questions.
   (a) A reaction was conducted in benzene as the solvent. Because the reaction was very exothermic, the mixture was cooled in a salt-ice bath. This was a bad choice. Why?
   (b) What bath should be used for a reaction that is conducted in liquid ammonia as the solvent?

Technique 3
REACTION METHODS

The successful completion of an organic reaction requires the chemist to be familiar with a variety of laboratory methods. These methods include operating safely, choosing and handling solvents correctly, heating reaction mixtures, adding liquid reagents, maintaining anhydrous conditions in the reaction, and collecting gaseous products. Several techniques which are used in bringing a reaction to a successful conclusion are treated in this chapter.

3.1 SOLVENTS

Organic solvents must be used with care to ensure that they are handled safely. Always remember that organic solvents are all at least mildly toxic and that many are flammable. You should become thoroughly familiar with the introductory chapter, "Laboratory Safety."

Read "Laboratory Safety," pp 4–12.

The most common organic solvents are listed in Table 3–1 along with their boiling points. Solvents marked in boldface type will burn. Ether, pentane, and hexane are especially dangerous; if they are combined with the correct amount of air they will explode.

TABLE 3–1. Common Organic Solvents

SOLVENT	BP (°C)	SOLVENT	BP (°C)
HYDROCARBONS		**ETHERS**	
Pentane	36	**Ether** (Diethyl)	35
Hexane	69	**Dioxane***	101
Benzene*	80	**1,2-Dimethoxyethane**	83
Toluene	111	**OTHERS**	
HYDROCARBON MIXTURES		Acetic acid	118
Petroleum ether	30–60	Acetic anhydride	140
Ligroin	60–90	**Pyridine**	115
CHLOROCARBONS		**Acetone**	56
Methylene chloride	40	**Ethyl acetate**	77
Chloroform*	61	Dimethylformamide	153
Carbon tetrachloride*	77	Dimethylsulfoxide	189
ALCOHOLS			
Methanol	65		
Ethanol	78		
Isopropyl alcohol	82		

NOTE: Boldface type indicates flammability.
*Suspect carcinogen (see p 11).

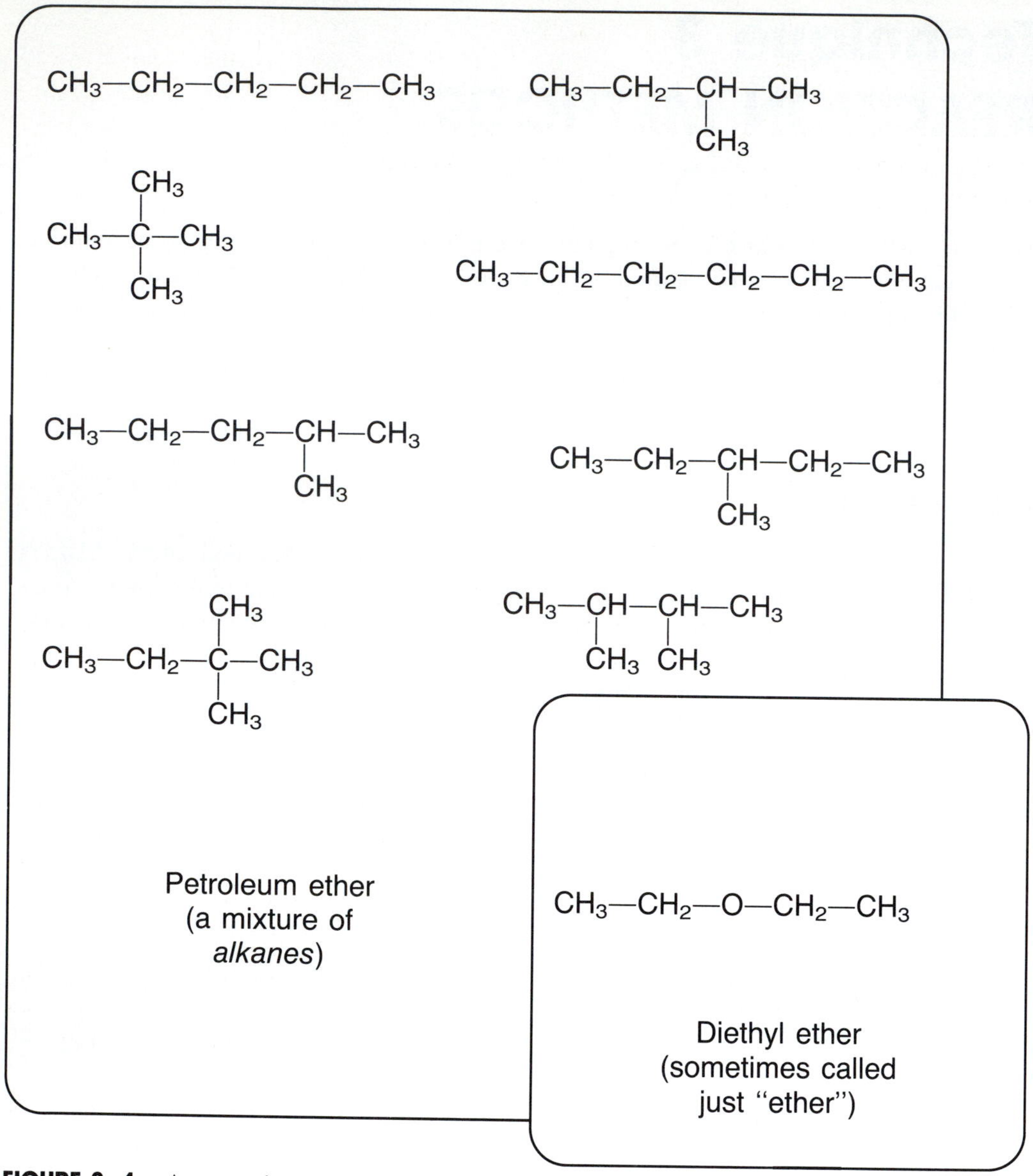

$$CH_3-CH_2-CH_2-CH_2-CH_3$$

$$CH_3-CH_2-\underset{\underset{CH_3}{|}}{CH}-CH_3$$

$$CH_3-\underset{\underset{CH_3}{|}}{\overset{\overset{CH_3}{|}}{C}}-CH_3$$

$$CH_3-CH_2-CH_2-CH_2-CH_2-CH_3$$

$$CH_3-CH_2-CH_2-\underset{\underset{CH_3}{|}}{CH}-CH_3$$

$$CH_3-CH_2-\underset{\underset{CH_3}{|}}{CH}-CH_2-CH_3$$

$$CH_3-CH_2-\underset{\underset{CH_3}{|}}{\overset{\overset{CH_3}{|}}{C}}-CH_3$$

$$CH_3-\underset{\underset{CH_3}{|}}{CH}-\underset{\underset{CH_3}{|}}{CH}-CH_3$$

Petroleum ether
(a mixture of
alkanes)

$$CH_3-CH_2-O-CH_2-CH_3$$

Diethyl ether
(sometimes called
just "ether")

FIGURE 3–1. A comparison between "ether" (diethyl ether) and "petroleum ether"

The terms **petroleum ether** and **ligroin** are often confusing. Petroleum ether is a mixture of hydrocarbons with isomers of formulas C_5H_{12} and C_6H_{14} predominating. Petroleum ether is not an ether at all since there are no oxygen-bearing compounds in the mixture. In organic chemistry, an ether is usually a compound containing an oxygen atom to which two alkyl groups are attached. Figure 3–1 shows some of the hydrocarbons that appear commonly in petroleum ether. Use special care when instructions call for either **ether** or **petroleum ether;** the two must not become accidentally confused. Confusion is particularly easy when one is selecting a container of solvent from the supply shelf.

Ligroin, or high-boiling petroleum ether, is like petroleum ether in composition except that compared with petroleum ether, ligroin generally includes higher-boiling alkane isomers. Depending on the supplier, ligroin may have different boiling ranges. While some brands of ligroin have boiling points ranging from about 60 °C to about 90 °C, other brands have boiling points ranging from about 60 °C to about 75 °C. The boiling point ranges of petroleum ether and ligroin are often included on the labels of the containers.

3.2   HEATING UNDER REFLUX

Often we wish to heat a mixture for a long time and to leave it untended. A **reflux apparatus** (see Figure 3–2) allows such heating. It also keeps the solvent from being lost by evaporation. A condenser is attached to the reaction vial or boiling flask.

Choice of Condenser.   The condenser used in a reflux apparatus can be either of two types. An **air condenser** is simply a long tube. The surrounding air removes heat from the vapors within the tube and condenses them to liquid. A **water-jacketed condenser** consists of two concentric tubes with the outer cooling tube sealed onto the inner tube. The vapors rise within the inner tube and water circulates through the outer tube. The circulating water removes heat from the vapors and condenses them. The air condenser is suitable for use with high-boiling liquids or with small quantities of material which are being heated gently. The water-jacketed condenser must be used when the vapors are difficult to condense, usually because the substance is very volatile, or when vigorous boiling action is desired. In either case, the condenser prevents the vapors from escaping. Glassware assemblies using both air and water-jacketed condensers are shown in Figure 3–2A. The figure also shows a typical macroscale apparatus for heating large quantities of material under reflux (Figure 3–2B).

When using a water-jacketed condenser, the direction of the water flow should be such that the condenser will fill with cooling water. The water should enter the bottom of the condenser and leave from the top. The water should flow fast enough to withstand any changes in pressure in the water lines, but it should not flow any faster than absolutely necessary. An excessive flow rate greatly increases the chance of a flood, and high water pressure may force the hose from the condenser. Cooling water should be flowing before heating is begun! If the water is to remain flowing overnight, it is advisable to fasten the rubber tubing securely with wire to the condenser. If a flame is used as a source of heat, it is convenient to use a wire gauze beneath the flask to provide an even distribution of heat from the flame. In most cases, a sand bath, water bath, heating mantle, or steam bath is preferred over a flame.

Stirring.   When heating a solution, always use a magnetic stirrer or a boiling stone (see Sections 3.3 and 3.4) to keep the solution from "bumping."

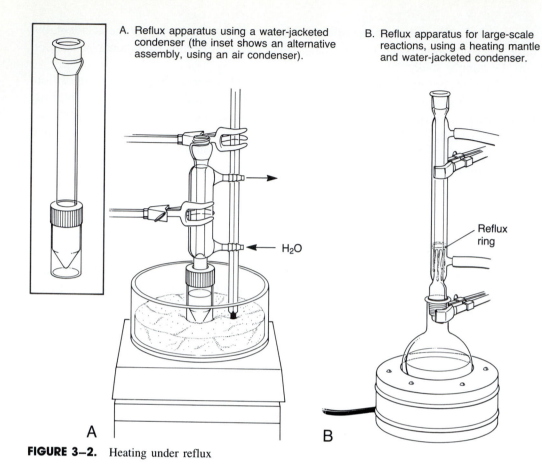

A. Reflux apparatus using a water-jacketed condenser (the inset shows an alternative assembly, using an air condenser).

B. Reflux apparatus for large-scale reactions, using a heating mantle and water-jacketed condenser.

H₂O

Reflux ring

A

B

FIGURE 3–2. Heating under reflux

Rate of Heating. If the heating rate has been correctly adjusted, the liquid being heated under reflux will travel only part way up the condenser tube before condensing. Below the condensation point, solvent will be seen running back into the flask; above it, the interior of the condenser will appear dry. The boundary between the two zones will be clearly demarcated, and a **reflux ring** or a ring of liquid will appear there. The reflux ring can be seen in Figure 3–2B. In heating under reflux, the rate of heating should be adjusted so that the reflux ring is no higher than a third to a half the distance to the top of the condenser. With microscale experiments the quantities of vapor rising in the condenser frequently are so small that a clear reflux ring cannot be seen. In those cases, the heating rate must be adjusted so that the liquid boils smoothly but not so rapidly that solvent can escape the condenser. With such small volumes, the loss of even a small amount of solvent can affect the reaction. With large-scale reactions, the reflux ring is much easier to see, and one can adjust the heating rate more easily.

Tended Reflux. It is possible to heat small amounts of a solvent under reflux in an Erlenmeyer flask. With gentle heating, the evaporated solvent will condense in

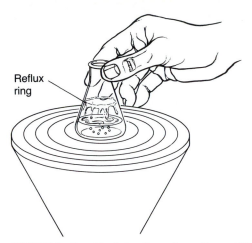

FIGURE 3–3.   Tended reflux of small quantities on a steam cone

the relatively cold neck of the flask and return to the solution. This technique (see Figure 3–3) requires constant attention. The flask must be swirled frequently and removed from the heating source for a short period if the boiling becomes too vigorous. When heating is in progress, the reflux ring should not be allowed to rise into the neck of the flask.

3.3   MAGNETIC STIRRERS

When a solution is heated, there is a danger that it may become superheated. When this happens, very large bubbles sometimes erupt violently from the solution; this is called **bumping.** Bumping must be avoided since it brings with it the risk that material may be lost from the apparatus, that a fire might start, or that the apparatus may break.

Magnetic stirrers are used to prevent bumping since they produce turbulence in the solution. The turbulence breaks up the large bubbles that form in boiling solutions. An additional purpose for using a magnetic stirrer is to stir the reaction to ensure that all of the reagents are thoroughly mixed. A magnetic stirring system consists of a magnet that is rotated by an electric motor. The rate at which this magnet rotates can be adjusted by a potentiometric control. One places a small magnet, which is coated with a non-reactive material such as Teflon or glass, into the flask. The magnet within the flask rotates in response to the rotating magnetic field caused by the motor-driven magnet. The result is that the inner magnet stirs the solution as it rotates. A very common type of magnetic stirrer includes the stirring system within a hot plate. This type of hot plate-stirrer permits one to heat the reaction and stir it simultaneously.

Magnetic stirring bars are available in several sizes and shapes. For microscale apparatus, a magnetic spin vane is often used. It is designed to contain a tiny bar magnet and to have a shape which conforms to the conical bottom of a reaction vial. A small Teflon-coated magnetic stirring bar, often called a "flea bar," works well with very small round-bottom boiling flasks. For larger flasks, longer stirring bars are used. A variety of magnetic stirring bars is illustrated in Figure 3–4.

A. Microscale magnetic   B. Small magnetic stirring   C. Standard-sized magnetic
   spin vane   bar ("flea bar")   stirring bars

FIGURE 3–4.   Magnetic stirring bars

3.4   BOILING STONES

A boiling stone, also known as a boiling chip or Boileezer, is a small lump of porous material that produces a steady stream of fine air bubbles when it is heated in a solvent. This stream of bubbles and the turbulence that accompanies it breaks up the large bubbles of gases in the liquid. In this way it reduces the tendency of the liquid to become superheated, and it promotes the smooth boiling of the liquid. The boiling stone decreases the chances for bumping. Boiling stones are generally made from pieces of pumice, carborundum, or marble.

Because boiling stones act to promote the smooth boiling of liquids, one should always make certain a boiling stone has been placed in a liquid **before** heating is begun. If one waits until the liquid is hot, it may have become superheated. Adding a boiling stone to a superheated liquid will cause all the liquid to try to boil at once. The liquid, as a result, would erupt entirely out of the flask, or froth violently.

As soon as boiling ceases in a liquid containing a boiling stone, the liquid is drawn into the pores of the boiling stone. When this happens, the boiling stone no longer can produce a fine stream of bubbles; it is spent. You may have to add a new boiling stone if you have allowed boiling to stop for a long period.

Wooden applicator sticks are used in some applications. They function in the same manner as boiling stones. Glass beads also find occasional application. Their presence also causes sufficient turbulence in the liquid to prevent bumping.

3.5   ADDITION OF LIQUID REAGENTS

Liquid reagents and solutions are added to a reaction by several means, some of which are shown in Figure 3–5. For microscale experiments, the simplest approach is simply to add the liquid to the reaction by means of a Pasteur pipet. This method is seen in Figure 3–5A. In this technique, the system is open to the atmosphere. A second microscale method, shown in Figure 3–5B, is suitable for experiments where the reaction should be kept isolated from the atmosphere. In this approach, the liquid is kept in a hypodermic syringe. The syringe needle is inserted through a rubber septum, and the liquid is added dropwise from the syringe. The septum seals the apparatus from the atmosphere, which makes this technique useful for reactions which are conducted under an atmosphere of inert gas or where anhydrous conditions must be maintained. As an alternative, the rubber septum may be replaced by a cap and Teflon insert or liner. A disadvantage of the Teflon insert, however, is that the insert may no longer form an effective seal after being punctured by the needle.

The most common type of assembly for macroscale experiments is shown in Figure 3–5C. In this apparatus, a separatory funnel is attached to the sidearm of a three-necked round-bottom flask. The separatory funnel must be equipped with a standard-taper, ground-glass joint to be used in this manner. The liquid is stored in the

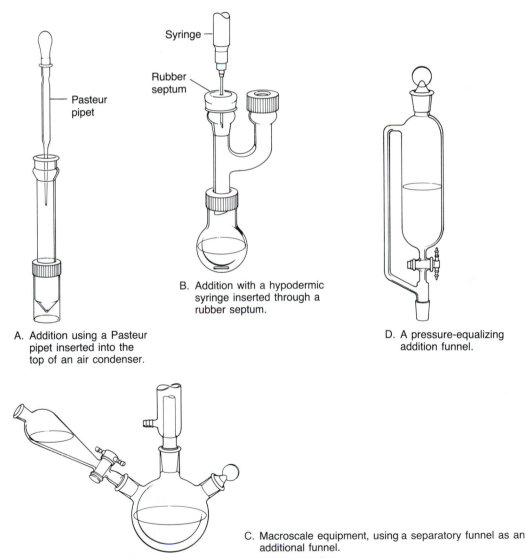

Syringe

Rubber septum

Pasteur pipet

A. Addition using a Pasteur pipet inserted into the top of an air condenser.

B. Addition with a hypodermic syringe inserted through a rubber septum.

D. A pressure-equalizing addition funnel.

C. Macroscale equipment, using a separatory funnel as an additional funnel.

FIGURE 3–5.   Methods for adding liquid reagents to a reaction

separatory funnel (which is called an **addition funnel** in this application) and is added to the reaction. The rate of addition is controlled by adjusting the stopcock. When it is being used as an addition funnel, the upper opening must be kept open to the atmosphere. If the upper hole is stoppered, a vacuum will develop in the funnel and will prevent the liquid from passing into the reaction vessel. Since the funnel is open to the atmosphere, there is a danger that atmospheric moisture can contaminate the liquid reagent as it is being added. To prevent this outcome, a drying tube (see Section 3.6) is attached to the upper opening of the addition funnel. The drying tube allows the funnel to maintain atmospheric pressure without allowing the passage of water vapor into the reaction.

Figure 3–5D shows an alternative type of addition funnel that is useful for reactions which must be maintained under an atmosphere of inert gas. This is the **pressure-equalizing addition funnel.** With this glassware, the upper opening is stoppered. The sidearm allows the pressure above the liquid in the funnel to be in equilibrium with the pressure in the rest of the apparatus, and it allows the inert gas to flow over the top of the liquid as it is being added.

With either type of macroscale addition funnel, one can control the rate of addition of the liquid by carefully adjusting the stopcock. Even after careful adjustment, changes in pressure can occur, causing the flow rate to change. In some cases the stopcock can become clogged. It is important, therefore, to monitor carefully the addition rate and to refine the adjustment of the stopcock as needed to maintain the desired rate of addition.

3.6 DRYING TUBES

With certain reactions, atmospheric moisture must be prevented from entering the reaction vessel. A **drying tube** can be used to maintain anhydrous conditions within the apparatus. Two types of drying tubes are shown in Figure 3–6. The typical drying tube

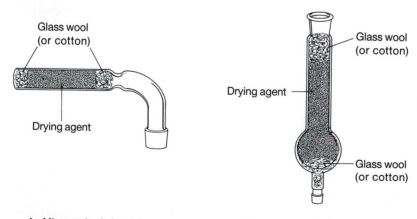

A. Microscale drying tube   B. Standard-sized drying tube

FIGURE 3–6. Drying tubes

is prepared by placing a small, loose plug of glass wool or cotton into the constriction at the end of the tube nearest the ground-glass joint or hose connection. The plug is tamped gently with a glass rod or piece of wire to place it in the correct position. A drying agent, typically calcium sulfate (''Drierite'') or calcium chloride (see Technique 7, Section 7.8, p 629), is poured on top of the plug to the approximate depth shown in Figure 3–6. Another loose plug of glass wool or cotton is placed on top of the drying agent to prevent the solid material from falling out of the drying tube. The drying tube is then attached to the flask or condenser.

Air which enters the apparatus must pass through the drying tube. The drying agent absorbs any moisture from air passing through it, so that air entering the reaction vessel has had the water vapor removed from it.

3.7 CAPTURING NOXIOUS GASES

Many organic reactions involve the production of a noxious gaseous product. The gas may be corrosive, such as hydrogen chloride, hydrogen bromide, or sulfur dioxide; or it may be toxic, such as carbon monoxide. The safest way to avoid exposure to these gases is to conduct the reaction in a ventilated hood where the gases can be safely drawn away by the ventilation system.

In many instances, however, it is quite safe and efficient to conduct the experiment on the laboratory bench, away from the hood. This is particularly true when the gases are soluble in water. Some techniques for capturing noxious gases are presented in this section.

A. Drying Tube Method

Microscale experiments have the advantage that the amounts of gases produced are very small. Hence, it is easy to trap them and prevent them from escaping into the laboratory room. One can take advantage of the water-solubility of corrosive gases such as hydrogen chloride, hydrogen bromide, and sulfur dioxide. A simple technique is to attach the drying tube (see Figure 3–6A) to the top of the reaction vial or condenser. The drying tube is filled with moistened glass wool. The moisture in the glass wool absorbs the gas, preventing its escape. To prepare this type of gas trap, fill the drying tube with glass wool and then add water dropwise to the glass wool until it has been moistened to the desired degree. Moistened cotton can also be used, although cotton will absorb so much water that it is easy to plug the drying tube.

When using glass wool in a drying tube, moisture from the glass wool must not be allowed to drain from the drying tube into the reaction. It is best to use a drying tube that has a constriction between the part where the glass wool is placed and the neck, where the joint is attached. The constriction acts as a partial barrier preventing the water from leaking into the neck of the drying tube. Make certain not to make the glass wool too moist.

B. External Gas Traps

Another approach to capturing gases is to prepare a trap which is separate from the reaction apparatus. The gases are carried from the reaction to the trap by means of tubing. There are several variations on this type of trap. One method which works well for microscale experiments is to place a thermometer adapter (Technique 8, Figure 8–9, p 648) into the opening in the reaction apparatus. A Pasteur pipet is inserted upside-down through the adapter, and a piece of fine flexible tubing is fitted over the narrow tip. It might be helpful to break the Pasteur pipet before using it for this purpose, so that only the narrow tip and a short section of the barrel is used. The other end of the flexible tubing is placed through a large plug of moistened glass wool in a test tube. The water in the glass wool serves to absorb the water-soluble gases. This method is shown in Figure 3–7.

A variation on the Pasteur pipet method uses a hypodermic syringe needle inserted upside-down (from the inside) through a rubber septum, which has been fitted over the opening at the top of the reaction apparatus. Flexible tubing, fitted over the syringe needle, leads to a trap such as the one using wet glass wool described above. This variation is also shown in Figure 3–7.

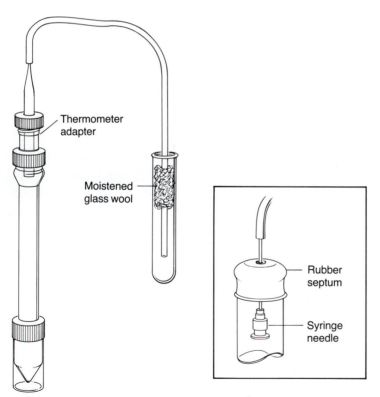

FIGURE 3–7. Microscale external gas trap. (The inset shows an expanded view of an alternative fitting, using a syringe needle and a rubber septum.)

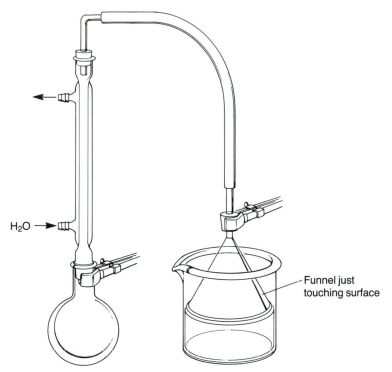

Funnel just
touching surface

FIGURE 3–8. Inverted-funnel gas trap

     With large-scale reactions, a trap using an inverted funnel placed in a beaker of water is used. A piece of glass tubing, inserted through a thermometer adapter attached to the reaction apparatus, is connected to flexible tubing. The tubing is attached to a conical funnel. The funnel is clamped in place inverted over a beaker of water. The funnel is clamped so that its lip is **slightly** beneath the water surface, but not so deep that water can be sucked back into the reaction, if the pressure in the reaction vessel changes suddenly. This type of trap can also be used in microscale applications. An example of the inverted funnel type of gas trap is shown in Figure 3–8.

C. Removal of Noxious Gases using an Aspirator

An aspirator can be used to remove noxious gases from the reaction. The simplest approach is to clamp a disposable Pasteur pipet so that its tip is placed well into the condenser atop the reaction vial. An inverted funnel clamped over the apparatus can also be used. The pipet or funnel is attached to an aspirator with flexible tubing. A trap should be placed between the pipet or funnel and the aspirator. As gases are liberated from the reaction, they rise into the condenser. The vacuum draws the gases away from the apparatus. Both types of systems are shown in Figure 3–9. In the special case where the noxious gases are soluble in water, connecting a water aspirator to the pipet or funnel would remove the gases from the reaction and trap them in the flowing water without the need for a separate gas trap.

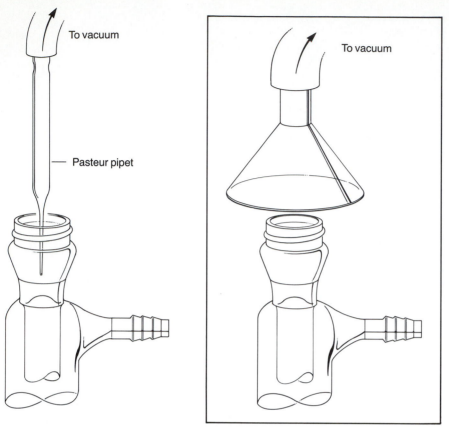

FIGURE 3–9. Removal of noxious gases under vacuum. (The inset shows an alternative assembly, using an inverted funnel in place of the Pasteur pipet.)

3.8 COLLECTING GASEOUS PRODUCTS

In Section 3.7, means of removing unwanted gaseous products from the reaction system were examined. Some experiments produce gaseous products that one wishes to collect and analyze. Methods to collect gaseous products are all based on the same principle. The gas is carried through tubing from the reaction to the opening of a flask or a test tube, which has been filled with water and is inverted in a container of water. The gas is allowed to bubble into the inverted collection tube (or flask). As the collection tube fills with gas, the water is displaced into the water container. If the collection tube is graduated, as in a graduated cylinder or a centrifuge tube, one may monitor the quantity of gas produced in the reaction.

If the inverted gas collection tube is constructed from a piece of glass tubing, a rubber septum can be used to close the upper end of the container. This type of collection tube is shown in Figure 3–10. A sample of the gas can be removed using a syringe equipped with a needle. The gas which is removed can be analyzed by gas chromatography (see Technique 14).

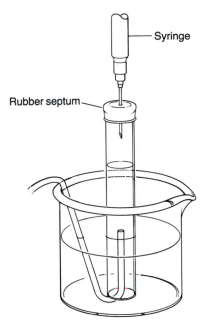

FIGURE 3–10.   Gas collection tube, with rubber septum

Many of the glassware kits for microscale experiments contain a special, all-glass, capillary gas delivery tube. The tube is attached to the top of the reaction apparatus by means of a ground-glass joint, and the open end of the capillary tubing is placed into an inverted, water-filled flask or test tube, clamped over a water bath. An example of a microscale kit gas delivery tube is shown in Figure 3–11. This type of tube is an

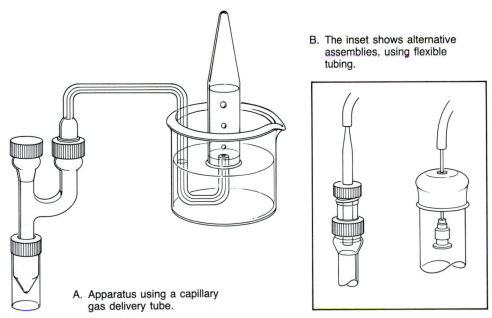

B. The inset shows alternative assemblies, using flexible tubing.

A. Apparatus using a capillary gas delivery tube.

FIGURE 3–11.   Gas delivery tubes

efficient means of collecting gases. A disadvantage, however, is that it is expensive and relatively easy to break.

A simpler, less expensive approach is to use flexible tubing of a fine diameter to lead the gases from the reaction vessel to the collecting container. One method is to place a hypodermic syringe needle, point upwards, through a rubber septum. The septum is attached to the top of the reaction apparatus, and a piece of fine flexible tubing is fitted over the end of the needle. The free end of the tubing is placed in the water bath, underneath the opening of the water-filled collection container. The gases will bubble into the container, where they will be collected. This alternative apparatus is shown as an inset in Figure 3–11 and is also depicted in Figure 3–10.

Another alternative, which may also be used with larger-scale experiments, is to place a piece of glass tubing or the tip of a Pasteur pipet through a thermometer adapter. The thermometer adapter is attached to the top of the reaction apparatus and flexible tubing is attached to the piece of glass tubing. The free end of the tubing is positioned in the opening of the water-filled collection vessel, as described above. This variation is also shown as an inset in Figure 3–11. As an option, one may attach a second piece of glass tubing to the free end of the flexible hose. This piece of glass tubing sometimes makes it easier to fix the open end in the proper position in the opening of the collection flask.

3.9 EVAPORATION OF SOLVENTS

In many experiments, it is necessary to remove excess solvent from a solution. An obvious approach is to allow the container to stand unstoppered in the hood for several hours until the solvent has evaporated. This method is generally not practical, however, and a quicker, more efficient means of evaporating solvents must be used. Figures 3–12 and 3–13 show several methods of removing solvents by evaporation. Figure 3–12 depicts microscale methods, while Figure 3–13 is devoted to large-scale procedures.

> **It is good laboratory practice to evaporate solvents in the hood.**

Microscale Methods.   A simple means of evaporating a solvent is to place the reaction vial in a warm sand bath. The heat from the sand bath will warm the solvent to a temperature where it can evaporate within a short time. The heat from the sand bath can be adjusted to provide the best rate of evaporation, but the liquid should not be allowed to boil vigorously. The evaporation rate can be increased by allowing a stream of dry air or nitrogen to be directed into the vial. The moving gas stream will sweep the vapors from the vial and accelerate evaporation. As an alternative, a vacuum can be applied over the vial to draw away solvent vapors.

During a crystallization procedure, often one must remove excess solvent from the solution. The excess solvent can be removed in a Craig tube (see Technique 5, Section 5.4, p 586). The Craig tube is placed in a warm sand bath. A microspatula is placed into the Craig tube, and it is twirled rapidly as the solvent evaporates. The

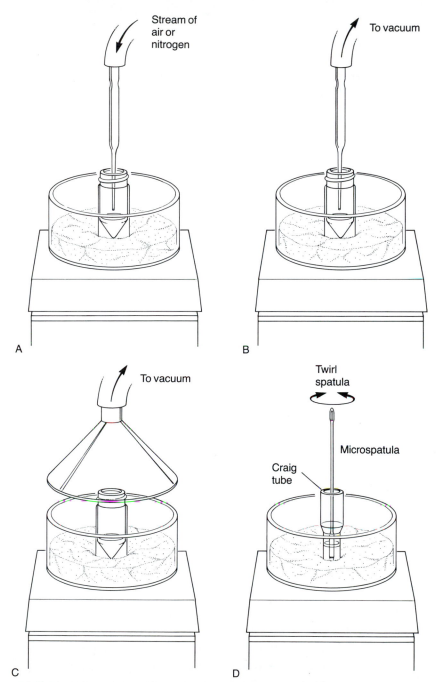

FIGURE 3–12. Evaporation of solvents (microscale methods)

twirling microspatula acts in the same manner as a boiling stone; it prevents bumping and accelerates the evaporation.

Larger-scale Methods. On a large scale, these evaporation methods can also be applied to standard-size glassware. Solvents can be evaporated from solutions in

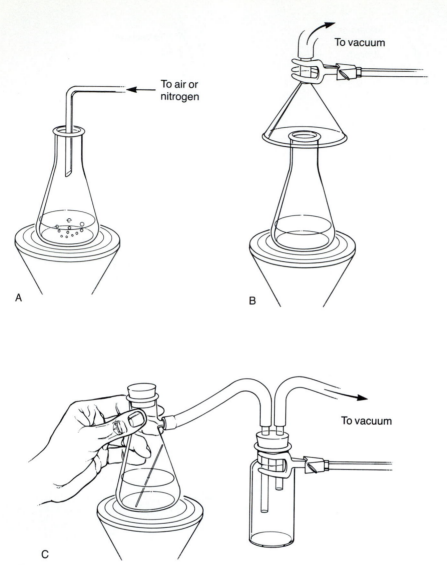

FIGURE 3–13. Evaporation of solvents (standard-sized methods)

Erlenmeyer flasks by adapting the techniques described previously. An Erlenmeyer flask can be placed on a source of heat, and the solvent can be removed by evaporation under a gas stream or a vacuum. Besides a sand bath, other sources of heat that can be used with Erlenmeyer flasks include steam baths and hot plates. A solution can also be placed in a sidearm test tube or a filter flask which is attached to a source of vacuum. A wooden stick is often placed in the solution, and the flask or test tube is swirled over the source of heat to reduce the possibility of bumping.

3.10   ROTARY EVAPORATOR

In the research laboratory, solvents are evaporated under reduced pressure using a **rotary evaporator.** This is a motor-driven device, which is designed for rapid evapo-

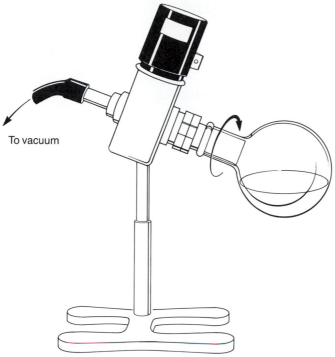

FIGURE 3—14. Rotary evaporator

ration of solvents, with heating, while minimizing the possibility of bumping. A vacuum is applied to the flask, and the motor spins the flask. The rotation of the flask spreads a thin film of the liquid over the surface of the glass. This accelerates evaporation. The rotation also agitates the solution sufficiently to reduce the problem of bumping. A water bath can be placed under the flask to warm the solution and increase the vapor pressure of the solvent. One can select the speed at which the flask is rotated and the temperature of the water bath to attain the desired evaporation rate. A rotary evaporator is shown in Figure 3–14.

PROBLEMS

1. What is the difference between
   (a) **ether** and **petroleum ether?**
   (b) **ether** and **diethyl ether?**
   (c) **ligroin** and **petroleum ether?**
2. What would be the appropriate condenser to use in order to heat a reaction under reflux, when the solvent is
   (a) methylene chloride?
   (b) toluene?
3. What is the best type of stirring device to use for stirring a reaction which takes place in
   (a) a conical vial?
   (b) a 10-mL round-bottomed flask?
   (c) a 250-mL round-bottomed flask?

4. Should you use a drying tube for the following reaction? Explain.

$$CH_3-\overset{\overset{\displaystyle O}{\|}}{C}-OH + CH_3-\overset{\overset{\displaystyle}{|}}{\underset{\underset{\displaystyle CH_3}{|}}{CH}}-CH_2-CH_2-OH \rightleftharpoons CH_3-\overset{\overset{\displaystyle O}{\|}}{C}-O-CH_2-CH_2-\overset{\overset{\displaystyle}{|}}{\underset{\underset{\displaystyle CH_3}{|}}{CH}}-CH_3 + H_2O$$

5. For which of the following reactions should you use a trap to collect noxious gases?

(a)

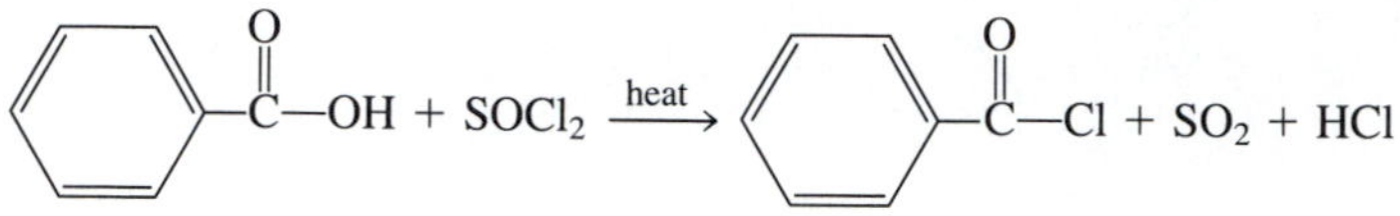

$$\text{(benzene ring)}-\overset{\overset{\displaystyle O}{\|}}{C}-OH + SOCl_2 \xrightarrow{heat} \text{(benzene ring)}-\overset{\overset{\displaystyle O}{\|}}{C}-Cl + SO_2 + HCl$$

(b)

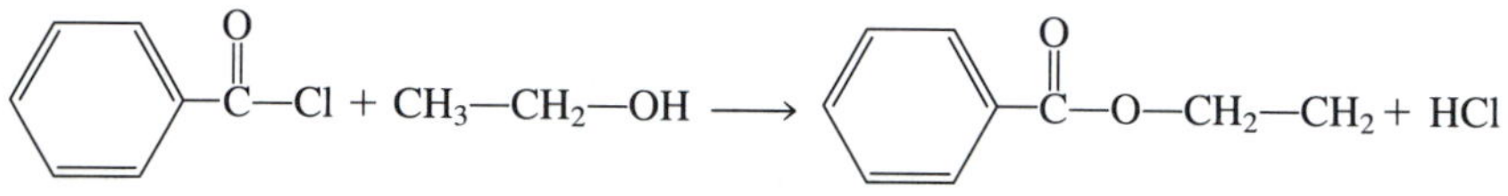

$$\text{(benzene ring)}-\overset{\overset{\displaystyle O}{\|}}{C}-Cl + CH_3-CH_2-OH \longrightarrow \text{(benzene ring)}-\overset{\overset{\displaystyle O}{\|}}{C}-O-CH_2-CH_2 + HCl$$

(c) $C_{12}H_{22}O_{11} + H_2O \longrightarrow 4CH_3-CH_2-OH + 4CO_2$
   (Sucrose)

(d) $CH_3-\overset{\overset{\displaystyle}{|}}{\underset{\underset{\displaystyle H}{|}}{C}}{=}NH + H_2O \xrightarrow[heat]{base} CH_3-\overset{\overset{\displaystyle}{|}}{\underset{\underset{\displaystyle H}{|}}{C}}{=}O + NH_3$

Technique 4
FILTRATION

Filtration is a technique used for two main purposes. The first is to remove solid impurities from a liquid. The second is to collect a desired solid from the solution from which it was precipitated or crystallized. Several different kinds of filtration are commonly used: two general methods include gravity filtration and vacuum (or suction) filtration. Two techniques specific to the microscale laboratory are filtration with a filter tip pipet and filtration with a Craig tube. The various filtration techniques and their applications are summarized in Table 4–1. These techniques will be discussed in more detail in the following sections.

TABLE 4–1. Filtration Methods

METHOD	APPLICATION	SECTION
GRAVITY FILTRATION filter cones	The volume of liquid to be filtered is about 10 mL or greater and the solid collected in the filter is saved.	4.1A
fluted filters	The volume of liquid to be filtered is greater than about 10 mL and solid impurities are removed from a solution; often used in crystallization procedures.	4.1B
filtering pipets	Used with volumes less than about 10 mL to remove solid impurities from a liquid.	4.1C
VACUUM FILTRATION Hirsch funnels	Primarily used to collect a desired solid from a relatively small volume of liquid (1–10 mL); used frequently to collect the crystals obtained from crystallizations.	4.3
Büchner funnels	Used in the same way as Hirsch funnels, except the volume of liquid is usually greater.	4.3
FILTERING MEDIA	Used to remove finely divided impurities.	4.4
FILTER TIP PIPETS	May be used to remove a small amount of solid impurities from a small volume (1–2 mL) of liquid; also useful for pipetting volatile liquids, especially in extraction procedures.	4.6
CRAIG TUBES	Used to collect a small amount of crystals resulting from crystallizations in which the volume of the solution is less than 2 mL.	4.7

4.1 GRAVITY FILTRATION

The most familiar filtration technique is probably filtration of a solution through a paper filter held in a funnel, allowing gravity to draw the liquid through the paper. Because even a small piece of filter paper will absorb a significant volume of liquid in most microscale procedures requiring filtration, this technique is only useful when the volume of mixture to be filtered is greater than about 10 mL. For many microscale procedures a more suitable technique, which also makes use of gravity, is to use a Pasteur (or disposable) pipet with a cotton or glass wool plug called a filtering pipet.

A. Filter Cones

This filtration technique is most useful when the solid material being filtered from a mixture is to be collected and used later. The filter cone, because of its smooth sides, can easily be scraped free of collected solids. Because of the many folds, fluted filter paper, described in the next section, cannot be scraped easily. The filter cone is likely to be used in microscale experiments only when a relatively large volume (greater than about 10 mL) is being filtered and when a Hirsch funnel (Section 4.3) is not appropriate.

The filter cone is prepared as indicated in Figure 4–1. It is then placed into a funnel of an appropriate size. With filtrations using a simple filter cone, solvent may form seals between the filter and the funnel and the funnel and the lip of the receiving

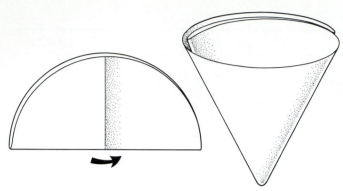

FIGURE 4–1.   Folding a filter cone

flask. When a seal forms, the filtration stops because the displaced air has no possibility of escaping. To avoid the solvent seal, you can insert a small piece of paper, a paper clip, or some other bent wire between the funnel and the lip of the flask to let the displaced air escape. As an alternative, you can support the funnel by a clamp fixed **above** the flask, rather than by placing it on the neck of the flask. A gravity filtration using a filter cone is shown in Figure 4–2.

B. Fluted Filters

This filtration method is also most useful when filtering a relatively large amount of liquid. Because a fluted filter is used when the desired material is expected to remain in solution, this filter is used to remove undesired solid materials such as dirt particles, decolorizing charcoal, and undissolved impure crystals. A fluted filter is often used to filter a hot solution saturated with a solute during a crystallization procedure.

The technique for folding a fluted filter paper is shown in Figure 4–3. An advantage of a fluted filter is that it increases the speed of filtration, which occurs for

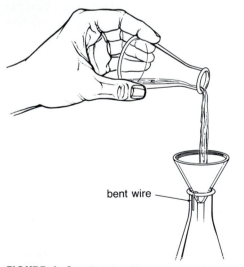

bent wire

FIGURE 4–2.   Gravity filtration with a filter cone

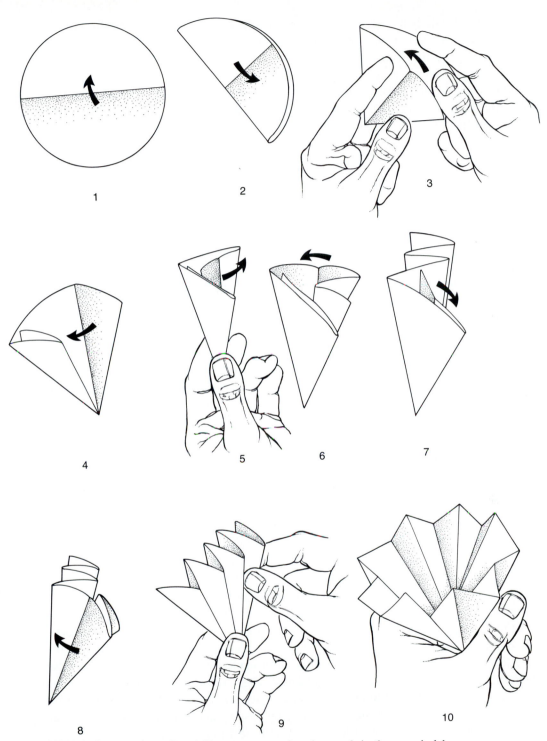

FIGURE 4–3. Folding a fluted filter paper, or origami at work in the organic lab

two reasons. First, it increases the surface area of the filter paper through which the solvent seeps; second, it allows air to enter the flask along its sides to permit rapid pressure equalization. If pressure builds up in the flask from hot vapors, filtering slows down. This problem is especially pronounced with filter cones. The fluted filter tends to reduce this problem considerably, but it may be a good idea to clamp the funnel above the receiving flask or to use a piece of paper, paper clip, or wire between the funnel and the lip of the flask as an added precaution against solvent seals.

Filtration with a fluted filter is relatively easy to perform when the mixture is at room temperature. However, when it is necessary to filter a hot solution saturated with a dissolved solute, a number of steps must be taken to ensure that the filter doesn't become clogged by solid material accumulated in the stem of the funnel or in the filter paper. When the hot saturated solution comes in contact with a relatively cold funnel (or a cold flask, for that matter), the solution is cooled and may become supersaturated. If crystallization then occurs in the filter, either the crystals will fail to pass through the filter paper or they will clog the stem of the funnel.

Four measures are feasible for preventing clogging of the filter. The first is to use a short-stemmed or a stemless funnel. In these funnels, there is less likelihood that the stem of the funnel will become clogged by solid material. The second measure is to keep the liquid to be filtered at or near its boiling point at all times. The third way is to preheat the funnel by pouring hot solvent through it before the actual filtration. This keeps the cold glass from causing instantaneous crystallization. And fourth, it is helpful to keep the **filtrate** (filtered solution) in the receiver hot enough to continue boiling **slightly** (by setting it on a hot plate, for example). The refluxing solvent heats the receiving flask and the funnel stem and washes them clean of solids. This boiling of the filtrate also keeps the liquid in the funnel warm.

C. Filtering Pipets

A filtering pipet is a microscale technique most often used to remove solid impurities from a liquid with a volume less than about 10 mL. It is important that the mixture being filtered be at or near room temperature since it is difficult to prevent premature crystallization in a hot solution saturated with a solute.

To prepare this filtration device, a small piece of cotton is inserted into the top of a Pasteur (disposable) pipet and pushed down to the beginning of the lower constriction in the pipet, as shown in Figure 4–4. It is important that enough cotton is used to collect all the solid being filtered; however, the amount of cotton used should not be so large that the flow rate through the pipet is significantly restricted. For the same reason, the cotton should not be packed too tightly. The cotton plug can be conveniently pushed down with a long thin object such as a glass stirring rod or a wooden applicator stick. It is advisable to wash the cotton plug by passing about 1 mL of solvent (usually the same solvent that is to be filtered) through the filter.

In some cases, such as when filtering a strongly acidic mixture or when one wants to make a very rapid filtration to remove dirt or impurities of large particle size from a solution, it may be better to use glass wool in place of the cotton. The disadvan-

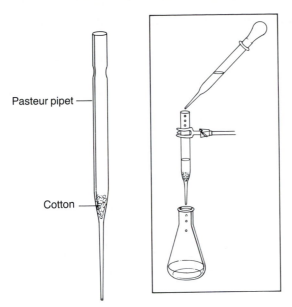

FIGURE 4–4. Filtering pipet

tage in using glass wool is that the fibers do not pack together as tightly, and small particles will pass through the filter more easily.

To carry out a filtration (with either a cotton or glass wool plug), the filtering pipet is clamped so that the filtrate will drain into an appropriate container. The mixture to be filtered is usually transferred to the filtering pipet with another Pasteur pipet. If a small volume of liquid is being filtered (less than 1 mL or 2 mL), it is advisable to rinse the filter and plug with a small amount of solvent after the last of the filtrate has passed through the filter. The rinse solvent is then combined with the original filtrate. If

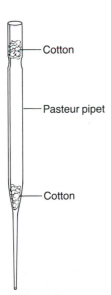

— Cotton

— Pasteur pipet

— Cotton

FIGURE 4–5. Double-filtering pipet

desired, the rate of filtration can be increased by gently applying pressure to the top of the pipet using a pipet bulb.

Depending on the amount of solid being filtered and the size of the particles (small particles are more difficult to remove by filtration), it may sometimes be necessary to put the filtrate through a second filtering pipet. This should be done with a new filtering pipet rather than the one already used.

An alternative to using two filtering pipets is to use two cotton plugs in the same Pasteur pipet, one at the lower constriction as indicated above and a second plug at the upper constriction of the pipet, as shown in Figure 4–5. A double-filtering pipet is particularly useful when filtering mixtures at room temperature in which the solid particles are very fine. However, this is not a useful technique with hot mixtures in which the solution is saturated above room temperature. Because the filtration process with a double-filtering pipet is fairly slow, it would be difficult to prevent cooling and premature crystallization. It is also not advisable to try to increase the flow rate by applying pressure with a pipet bulb since this is likely to dislodge the top cotton plug.

4.2   FILTER PAPER

Many kinds and grades of filter paper are available. The paper must be correct for a given application. In choosing, one should be aware of the various properties of filter paper. **Porosity** is a measure of the size of the particles that can pass through the paper. Highly porous paper does not remove small particles from solution; paper with low porosity removes very small particles. **Retentivity** is a property that is the opposite of porosity. Paper with low retentivity does not remove small particles from the filtrate. The **speed** of filter paper is a measure of the time it takes a liquid to drain through the filter. Fast paper allows the liquid to drain quickly; with slow paper, it takes much longer to complete the filtration. Since all these properties are related, fast filter paper usually has a low retentivity and high porosity, and slow filter paper usually has high retentivity and low porosity.

Table 4–2 compares some commonly available qualitative filter paper types and ranks them according to porosity, retentivity, and speed. Eaton-Dikeman (E&D), Schleicher and Schuell (S&S), and Whatman are the most common brands of filter paper. The numbers in the table refer to the grades of paper used by each company.

TABLE 4–2.   Some Common Qualitative Filter Paper Types and Approximate Relative Speeds and Retentivities

				TYPE (by number)		
			SPEED	E&D	S&S	Whatman
			Very slow	610	576	5
			Slow	613	602	3
			Medium	615	597	2
			Fast	617	595	1
			Very fast	. . .	604	4

Fine — High — Slow (top)
Porosity — Retentivity — Speed
Coarse — Low — Fast (bottom)

4.3 VACUUM FILTRATION

Vacuum, or suction, filtration is more rapid than gravity filtration and is most often used to collect solid products resulting from precipitation or crystallization. This technique is used primarily when the volume of liquid being filtered is more than 1–2 mL. With smaller volumes, use of the Craig tube (Section 4.7) is the preferred technique.

In a vacuum filtration, a receiver flask with a sidearm, a **filter flask,** is used. For microscale laboratory work, the most useful size is a 25-mL filter flask. The sidearm is connected by **heavy-walled** rubber tubing to a source of vacuum. Thin-walled tubing will collapse under vacuum, due to atmospheric pressure on its outside walls, and will seal the vacuum source from the flask. A **Hirsch funnel** (see Figure 4–6) is sealed to the filter flask by a rubber stopper or a filter (Neoprene) adapter. A better seal can be made by using both a #1 and a #2 Neoprene adapter. Because this apparatus is unstable and can easily tip over, it should be clamped, as shown in Figure 4–6.

It is essential that the filter flask be clamped.

The flat bottom of the Hirsch funnel, which should be 1–2 cm in diameter, is covered with an unfolded piece of circular filter paper. To prevent the escape of solid materials from the Hirsch funnel, one must be certain that the filter paper fits the Hirsch

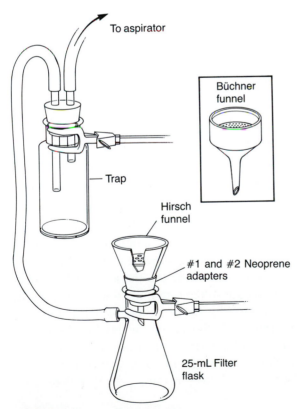

FIGURE 4–6. Vacuum filtration

funnel exactly. The paper must be neither too big nor too small. It must cover all the holes in the bottom of the funnel but not extend up the sides of the funnel. Before beginning the filtration, it is advisable to moisten the paper with a small amount of solvent. The moistened filter paper adheres more strongly to the bottom of the Hirsch funnel and prevents unfiltered mixture from passing around the edges of the filter paper.

Since the filter flask is attached to a source of vacuum, a solution poured into the Hirsch funnel is literally "sucked" rapidly through the filter paper. For this reason a vacuum filtration is generally not used to separate fine particles such as decolorizing charcoal since the small particles would likely be pulled through the filter paper. However, this problem can be alleviated when desired by the use of specially prepared filter beds (see Section 4.4).

Two types of funnels are useful for vacuum filtration. The Hirsch funnel, which has already been considered, is used for filtering smaller amounts of solid from solution. The **Büchner** funnel, which is also shown in Figure 4–6, operates on the same principle as the Hirsch funnel, but it is usually larger and its sides are vertical rather than sloped. It is sealed to the filter flask with a rubber stopper or a Neoprene adapter. In the Büchner funnel, the filter paper must also cover all the holes in the bottom but must not extend up the sides.

4.4 FILTERING MEDIA

It is occasionally necessary to use specially prepared filter beds to separate fine particles when using vacuum filtration. Often, very fine particles either pass right through a paper filter or they clog it so completely that the filtering stops. This is avoided by using a substance called Filter Aid, or Celite. This material is also called **diatomaceous earth** because of its source. It is a finely divided inert material derived from the microscopic shells of dead diatoms (a type of phytoplankton that grows in the sea).

> **WARNING: LUNG IRRITANT**
> **When using Filter Aid, take care not to breathe the dust.**

Filter Aid will not clog the fiber pores of filter paper. It is **slurried,** mixed with a solvent to form a rather thin paste, and filtered through a Hirsch or Büchner funnel (with filter paper in place) until a layer of diatoms about 2–3 mm thick is formed on top of the filter paper. The solvent in which the diatoms were slurried is poured from the filter flask, and, if necessary, the filter flask is cleaned before the actual filtration is begun. Finely divided particles can now be suction-filtered through this layer and will be caught in the Filter Aid. This technique is used for removing impurities and not for collecting a product. The filtrate (filtered solution) is the desired material in this procedure. If the material caught in the filter were the desired material, one would have to try to separate the product from all those diatoms! Filtration with Filter Aid is not appropriate when the desired substance is likely to precipitate or crystallize from solution.

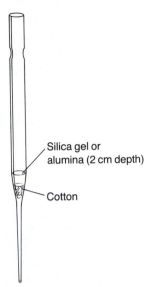

FIGURE 4–7. Pasteur pipet with filtering media

In microscale work, it may sometimes be more convenient to use a column prepared with a Pasteur pipet to separate fine particles from a solution. The Pasteur pipet is packed with alumina or silica gel, as shown in Figure 4–7.

4.5 THE ASPIRATOR

The most common source of vacuum (approximately 10–20 mmHg) in the laboratory is the water aspirator, or "water pump," illustrated in Figure 4–8. This device passes

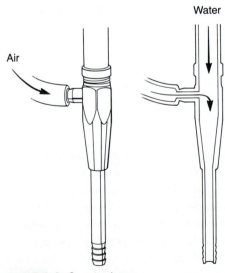

FIGURE 4–8. Aspirator

water rapidly past a small hole to which a sidearm is attached. The water pulls air in through the sidearm. This phenomenon, called the Bernoulli effect, causes a reduced pressure along the side of the rapidly moving water stream and creates a partial vacuum in the sidearm. The aspirator works most effectively when the water is turned on to the fullest extent.

A water aspirator can never lower the pressure beyond the vapor pressure of the water used to create the vacuum. Hence, there is a lower limit to the pressure (on cold days) of 9–10 mmHg. A water aspirator does not provide as high a vacuum in the summer as in the winter, due to this water-temperature effect.

A trap must be used with an aspirator. One type of trap is illustrated in Figure 4–6; however, in some laboratories it may be more convenient to use a second filter flask as a trap. If the water pressure in the laboratory line drops suddenly, the pressure in the filter flask may suddenly become lower than the pressure in the water aspirator. This would cause water to be drawn from the aspirator stream into the filter flask and contaminate the filtrate. The trap stops this reverse flow. A similar flow will occur if the water flow at the aspirator is stopped before the tubing connected to the aspirator sidearm is disconnected.

> **Always disconnect the tubing before stopping the aspirator.**

If a "back-up" begins, disconnect the tubing as rapidly as possible before the trap fills with water. Some workers like to fit a stopcock into the stopper on top of the trap. A three-hole stopper is required for this purpose. With a stopcock in the trap, the system can be vented before the aspirator is shut off. Then water cannot back up into the trap.

Aspirators do not work well if too many people use the water line at the same time since the water pressure is lowered. Also, the sinks at the ends of the lab benches or the lines that carry away the water flow may have a limited capacity for draining the resultant water flow from too many aspirators. Care must be taken to avoid floods.

4.6 FILTER TIP PIPET

There are two common uses for a filter tip pipet illustrated in Figure 4–9. The first is to remove a small amount of solid, such as dirt or filter paper fibers, from a small volume of liquid (1–2 mL). It can also be helpful when using a Pasteur pipet to transfer a highly volatile liquid, especially during an extraction procedure (see Technique 7, Section 7.2, p 617).

Preparing a filter tip pipet is similar to preparing a filtering pipet, except that a much smaller amount of cotton is used. A very tiny piece of cotton is loosely shaped into a ball and placed into the large end of a Pasteur pipet. Using a wire with a diameter slightly smaller than the inside diameter of the narrow end of the pipet, the ball of cotton is pushed to the bottom of the pipet. If it becomes difficult to push the cotton, you have probably started with too much cotton; if the cotton slides through the narrow end with little resistance, you have probably not used enough.

To use a filter tip pipet as a filter, the mixture is drawn up into the Pasteur pipet using a pipet bulb and then expelled. With this procedure, a small amount of solid will

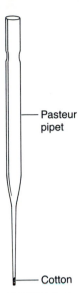

FIGURE 4–9. Filter tip pipet

be captured by the cotton. However, very fine particles such as activated charcoal cannot be removed efficiently with a filter tip pipet, and this technique is not effective in removing more than a trace amount of solid from a liquid.

Transferring many organic liquids with a Pasteur pipet can be a somewhat difficult procedure for two reasons. First, the liquid may not adhere well to the glass. Second, as a student handles the Pasteur pipet, the temperature of the liquid in the pipet increases slightly and the increased vapor pressure may tend to ''squirt'' the liquid out the end of the pipet. This problem can be particularly troublesome when separating two liquids during an extraction procedure. The purpose of the cotton plug in this situation is to slow the rate of flow through the end of the pipet so that one can control the movement of liquid in the Pasteur pipet more easily.

4.7 CRAIG TUBES

The **Craig tube,** illustrated in Figure 4–10, is used primarily to separate crystals from a solution after a microscale crystallization procedure has been performed (Technique 5, Section 5.4, p 586). Although it may not be a filtration procedure in the traditional sense, the outcome is similar. The outer part of the Craig tube is similar to a test tube, except that the diameter of the tube becomes wider part of the way up the tube, and the glass is ground at this point so that the inside surface is rough. The inner part (plug) of the Craig tube may be made of Teflon or glass. If this part is glass, the end of the plug is also ground. With either a glass or a Teflon inner plug, there is only a partial seal where the plug and the outer tube come together. Liquid may pass through, but solid will not. This is where the solution is separated from the crystals.

After crystallization has been completed in the outer Craig tube, replace the inner plug (if necessary) and connect a thin copper wire to the narrow part of the inner plug, as indicated in Figure 4–11A. While holding the Craig tube in an upright posi-

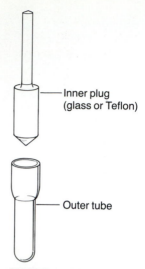
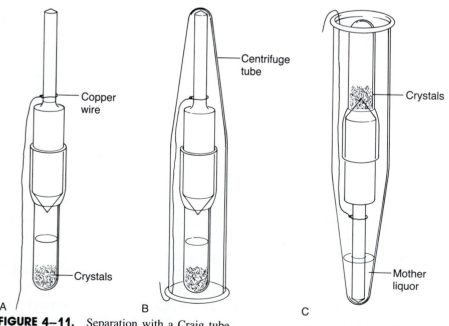

FIGURE 4–10. Craig tube (2-mL)

tion, a centrifuge tube is placed over the Craig tube so that the bottom of the centrifuge tube rests on top of the inner plug, as shown in Figure 4–11B. The copper wire should extend just below the lip of the centrifuge tube and is now bent upward around the lip of the centrifuge tube. This apparatus is then turned over so that the centrifuge tube is in an upright position. The Craig tube is spun in a centrifuge (be sure it is balanced by placing another tube filled with water on the opposite side of the centrifuge) for several minutes until the **mother liquor** (solution from which the crystals grew) goes to the bottom of the centrifuge tube and the crystals collect on the end of the inner plug (see

FIGURE 4–11. Separation with a Craig tube

Figure 4–11C). Depending on the consistency of the crystals and the speed of the centrifuge, the crystals may spin down to the inner plug, or (if you are unlucky) they may remain at the other end of the Craig tube. If the latter situation occurs, it may be helpful to centrifuge the Craig tube longer, or, if this problem is anticipated, stirring the crystal and solution mixture with a spatula or stirring rod before centrifugation may prevent this from occurring.

Using the copper wire, the Craig tube is then pulled out of the centrifuge tube. If the crystals collected on the end of the inner plug, it is now a simple procedure to remove the plug and scrape the crystals with a spatula onto a watch glass, a clay plate, or a piece of smooth paper. Otherwise, it will be necessary to scrap the crystals from the inside surface of the outer part of the Craig tube.

PROBLEMS

1. In each of the following situations, what type of filtration device would you use?
   (a) Remove powdered decolorizing charcoal from 20 mL of solution.
   (b) Collect crystals obtained from crystallizing a substance from about 1 mL of solution.
   (c) Remove a very small amount of dirt from 1 mL of liquid.
   (d) Isolate 0.2 g of crystals from about 5 mL of solution after performing a crystallization.
   (e) Remove dissolved colored impurities from about 3 mL of solution.
   (f) Remove solid impurities from 5 mL of liquid at room temperature.

Technique 5
CRYSTALLIZATION: PURIFICATION OF SOLIDS

Organic compounds that are solid at room temperature are usually purified by crystallization. The general technique involves dissolving the material to be crystallized in a **hot** solvent (or solvent mixture) and cooling the solution slowly. The dissolved material has a decreased solubility at lower temperatures and will separate from the solution as it is cooled. This phenomenon is called either **crystallization** if the crystal growth is relatively slow and selective or **precipitation** if the process is rapid and nonselective. Crystallization is an equilibrium process and produces very pure material. A small seed crystal is formed initially, and it then grows layer by layer in a reversible manner. In a sense, the crystal "selects" the correct molecules from the solution. In precipitation, the crystal lattice is formed so rapidly that impurities are trapped within the lattice. Therefore, any attempt at purification with too rapid a process should be avoided.

In microscale organic work, two methods are commonly used to perform crystallizations. The first method, which is carried out with an Erlenmeyer flask to dissolve the material and a Hirsch funnel to filter the crystals, is normally used when the weight of solid to be crystallized is more than about 0.1 g. This technique, called **semi-microscale crystallization,** will be discussed in Section 5.3. The second method is performed with a Craig tube and is used with lesser amounts of solid. Referred to as **microscale crystallization,** this technique will be discussed in Section 5.4. The weight of solid to be crystallized, however, is not the only factor to consider when choosing a method for crystallization. Since the solubility of a substance in a given solvent must also be taken into account, the weight, 0.1 g, should not be adhered to rigidly in determining which method to use. In this textbook, you will usually be advised which method to use in the experimental procedure.

The method described here for semi-microscale crystallizations is nearly identical to that used for crystallizing larger amounts of materials than those encountered in this textbook. Therefore, this technique can also be used to perform crystallizations at the macroscale level (more than several grams).

5.1 SOLUBILITY

The first problem in performing a crystallization is selecting a solvent in which the material to be crystallized shows the desired solubility behavior. In the ideal case, the material should be sparingly soluble at room temperature and yet quite soluble at the boiling point of the solvent selected. The solubility curve should be steep, as can be seen in line A of Figure 5–1. A curve with a low slope (line B, Figure 5–1) would not cause significant crystallization when the temperature of the solution was lowered. A

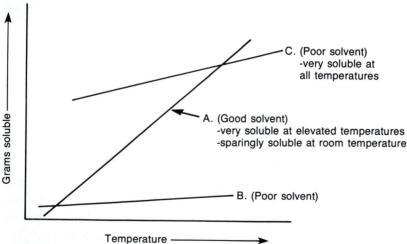

FIGURE 5–1.   Graph of solubility versus temperature

solvent in which the material was very soluble at all temperatures (line C, Figure 5–1) would also not be a suitable crystallization solvent. The basic problem in performing a crystallization is to select a solvent (or mixed solvent) that will provide a steep solubility-versus-temperature curve for the material to be crystallized. A solvent that allows the behavior shown in line A is an ideal crystallization solvent.

The solubility of organic compounds is a function of the polarities of both the solvent and the **solute** (dissolved material). A general rule is ''like dissolves like.'' If the solute is very polar, a very polar solvent will be needed to dissolve it; if it is nonpolar, a nonpolar solvent will be needed. Usually compounds having functional groups that can form hydrogen bonds (for example, —OH, —NH—, —COOH, —CONH—) will be more soluble in hydroxylic solvents such as water or methanol than in hydrocarbon solvents such as toluene or hexane. However, if the functional group is not a major part of the molecules, this solubility behavior may be reversed. For instance, dodecyl alcohol, $CH_3(CH_2)_{10}CH_2OH$, is almost insoluble in water; its 12-carbon chain causes it to behave more like a hydrocarbon than an alcohol. The list found in Table 5–1 gives an approximate order for decreasing polarity of organic functional groups.

TABLE 5–1. Solvents, in Decreasing Order of Polarity

DECREASING POLARITY (APPROXIMATE)		
	H_2O	Water
	RCOOH	Organic acids (acetic acid)
	$RCONH_2$	Amides (N,N-dimethylformamide)
	ROH	Alcohols (methanol, ethanol)
	RNH_2	Amines (triethylamine, pyridine)
	RCOR	Aldehydes, ketones (acetone)
	RCOOR	Esters (ethyl acetate)
	RX	Halides ($CH_2Cl_2 > CHCl_3 > CCl_4$)
	ROR	Ethers (diethyl ether)
	ArH	Aromatics (benzene, toluene)
	RH	Alkanes (hexane, petroleum ether)

The stability of the crystal lattice also affects solubility. Other things being equal, the higher the melting point (the more stable the crystal), the less soluble the compound. For instance, p-nitrobenzoic acid (mp 242 °C) is, by a factor of 10, less soluble in a fixed amount of ethanol than the *ortho* (mp 147 °C) and *meta* (mp 141 °C) isomers.

5.2 THEORY OF CRYSTALLIZATION

A successful crystallization depends on a large difference in the solubility of a material in a hot solvent and its solubility in the same solvent when it is cold. When the impurities in a substance are equally soluble in both the hot and the cold solvent, an effective purification is not easily achieved through crystallization. A material can be purified by crystallization when both the desired substance and the impurity have similar solubilities, but only when the impurity represents a small fraction of the total solid. The desired substance will crystallize on cooling, but the impurities will not.

For example, consider a case in which the solubilities of substance A and its impurity B are both 10 mg/mL of solvent at 20 °C and 100 mg/mL of solvent at 100 °C. In an impure sample of A, the composition is given to be 90 mg of A and 20 mg of B for this particular example. At 20 °C, this total amount of material will not dissolve in 1 mL of solvent. However, if the solvent is heated to 100 °C, all 110 mg dissolve. The solvent has the capacity to dissolve 100 mg of A **and** 100 mg of B at this temperature. If the solution is cooled to 20 °C, only 10 mg of each solute can remain dissolved, so 80 mg of A and 10 mg of B crystallize, leaving 20 mg of material in the solution. This crystallization is shown in Figure 5–2. The solution that remains after a crystallization is called the **mother liquor.** If the process is now repeated by treating the crystals with 1 mL of fresh solvent, 70 mg of A will crystallize again, leaving 10 mg of A and 10 mg of B in the mother liquor. As a result of these operations, 70 mg of pure A are obtained, but with the loss of 40 mg of material. Again, this second crystallization step is illustrated in Figure 5–2. The final result illustrates an important aspect of crystallization—it is wasteful. Nothing can be done to prevent this waste; some A must be lost along with the impurity B for the method to be successful. Of course, if the impurity B were **more** soluble than A in the solvent, the losses would be reduced. Losses could also be reduced if the impurity were present in **much smaller** amounts than the desired material.

It should be noticed that for the above case, the method operated successfully because A was present in substantially larger quantity than its impurity B. If there had been a 50/50 mixture of A and B initially, no separation would have been achieved. In general, a crystallization is successful only if there is a **small** amount of impurity. As the amount of impurity increases, the loss of material must also increase. Two substances with nearly equal solubility behavior, present in equal amounts, cannot be separated. If the solubility behavior of two components present in equal amounts is different, however, a separation or purification is frequently possible.

In the above example, two crystallization procedures were performed. Normally this will not be necessary; however, when it is, the second crystallization is more

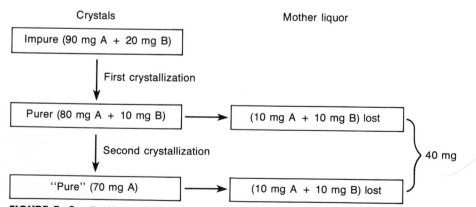

FIGURE 5–2. Purification of a mixture by crystallization

appropriately called **recrystallization.** As illustrated in this example, a second crystallization results in purer crystals, but the yield is lower.

5.3 SEMI-MICROSCALE CRYSTALLIZATIONS—HIRSCH FUNNELS

The crystallization technique described in this section is used when the weight of solid to be crystallized is more than about 0.1 g. The four main steps in a semi-microscale crystallization are

1. Dissolving the solid
2. Removing insoluble impurities (when necessary)

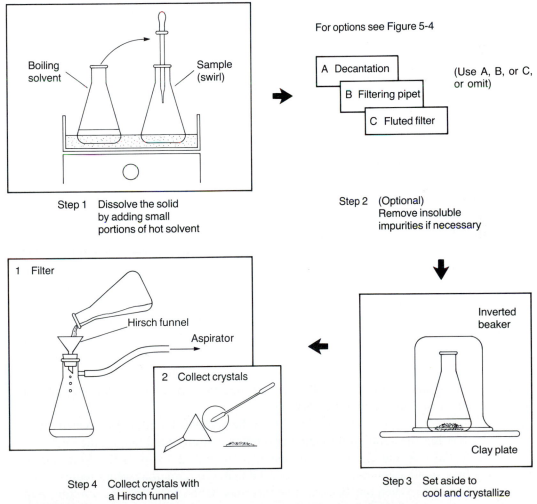

FIGURE 5—3. Steps in a semi-microscale crystallization (no decolorization)

3. Crystallization
4. Isolation of crystals

These steps are illustrated in Figure 5–3. It should be pointed out that a microscale crystallization with a Craig tube involves the same four steps, although the apparatus and procedures are somewhat different (see Section 5.4).

A. Dissolving the Solid

To minimize losses of material to the mother liquor, it is desirable to **saturate** the boiling solvent with solute. This solution, when cooled, will return the maximum possible amount of solute as crystals. To achieve this high return, the solvent is brought to its boiling point and the solute is dissolved in the **minimum amount(!) of boiling solvent.** For this procedure, it is advisable to maintain a container of boiling solvent (either on a hot plate or a sand bath). From this container, a small portion (about 0.5 mL) of the solvent is added to the flask (usually a 10 mL or 25 mL Erlenmeyer flask) containing the solid to be crystallized, and this mixture is heated with occasional swirling until it resumes boiling. If the solid does not dissolve in the first portion of boiling solvent, then another small portion of boiling solvent is added to the flask. The mixture is heated again until it resumes boiling. If the solid dissolves, no more solvent is added. But if the solid has not dissolved, another portion of boiling solvent is added, as before, and the process is repeated until the solid dissolves. (If the solid totally dissolves in less than 2 mL of solvent, a Craig tube should be used for crystallization.) It is important to stress that the portions of solvent added each time are small, so that only the **minimum** amount of solvent necessary for dissolving the solid is added. It is also important to emphasize that the procedure requires the addition of solvent to solid. One must never add portions of solid to a fixed quantity of boiling solvent. By this latter method, it may be impossible to tell when saturation has been achieved.

In many of the experiments in this textbook, a specified amount of solvent for a given weight of solid will be recommended. This is done to give you a rough idea how much solvent will be required; however, the procedure described above should still be used to determine the actual amount of solvent needed to dissolve the sample.

Occasionally, one encounters an impure solid that contains small particles of insoluble impurities, pieces of dust, or paper fibers that will not dissolve in the hot crystallizing solvent. A common error is to add too much of the hot solvent in an attempt to dissolve these small particles without realizing that they are not soluble. In such cases, one must be careful not to add too much solvent. It is probably better to add too little solvent and not dissolve all the desired solid than to add too much solvent and lower the yield of solid returned as crystals.

It is sometimes necessary to decolorize the solution by adding activated charcoal or by passing the solution through a column containing alumina or silica gel (see Section 5.6, Parts A and C, and Technique 12, Section 12.13, p 717). (Note: Often there may be a small amount of colored material that will remain in solution during the crystallization step. When one believes that this may be the case, the decolorizing step should be omitted.)

B. *Removing Insoluble Impurities*

It is necessary to use one of the three methods described below only if insoluble material remains in the hot solution or if decolorizing charcoal has been used.

> **Indiscriminate use of the procedure can lead to needless loss of your product.**

Decantation is the easiest method of removing solid impurities and should be considered first. A filtering pipet is used when the volume of liquid to be filtered is less

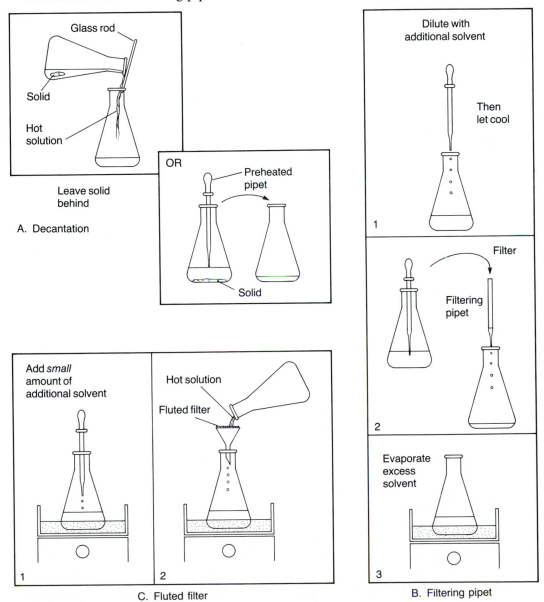

FIGURE 5–4. Methods for removing insoluble impurities in a semi-microscale crystallization

than about 10 mL (see Technique 4, Section 4.1, Part C, p 568), and you should use gravity filtration through a fluted filter when the volume is about 10 mL or greater (see Technique 4, Section 4.1, Part B, p 566). These three methods are illustrated in Figure 5–4.

Decantation.   If the solid particles are relatively large in size or they easily settle to the bottom of the flask, it may be possible to separate the hot solution from the impurities by carefully pouring off the liquid, leaving the solid behind. This is done most easily be holding a glass stirring rod along the top of the flask and tilting the flask so that the liquid pours out along one end of the glass rod into another container. A technique similar in principle to decantation, which may be easier to perform with smaller amounts of liquid, is to use a **preheated Pasteur pipet** to remove the hot solution. With this method, it may be helpful to place the tip of the pipet against the bottom of the flask when removing the last portion of solution. The small space between the tip of the pipet and the inside surface of the flask prevents solid material from being drawn into the pipet. An easy way to preheat the pipet is to draw up a small portion of hot **solvent** (not the **solution** being transferred) into the pipet and expel the liquid. Repeat this process several times.

Filtering Pipet.   If the volume of solution after dissolving the solid in hot solvent is less than about 10 mL, gravity filtration with a filtering pipet may be used to remove solid impurities. However, using a filtering pipet to filter a hot solution saturated with solute can be difficult to perform without premature crystallization. The best way to prevent this from occurring is to add enough solvent to dissolve the desired product at room temperature (be sure not to add too much solvent) and perform the filtration at room temperature, as described in Technique 4, Section 4.1, Part C, p 568. After filtration, the excess solvent is evaporated by boiling until the solution is saturated at the boiling point of the mixture (see Technique 3, Section 3.9, p 560). If powdered decolorizing charcoal was used, it will probably be necessary to perform two filtrations with a filtering pipet, or else the method described next can be used.

Fluted Filter.   This is the best method to remove solid impurities when the volume of liquid is greater than about 10 mL or when decolorizing charcoal has been used (see Technique 4, Section 4.1, Part B, p 566). One should add a small amount of extra solvent to the hot mixture. This procedure helps to prevent crystal formation in the filter paper or the stem of the funnel during the filtration. The funnel is fitted with a fluted filter and installed at the top of the Erlenmeyer flask to be used for the actual filtration. It is advisable to place a small piece of wire between the funnel and the mouth of the flask to relieve any increase in pressure caused by hot filtrate.

The Erlenmeyer flask containing the funnel and fluted paper is placed on top of a sand bath or hot plate (low setting). The liquid to be filtered is brought to its boiling point and poured through the filter in portions. (If the volume of the mixture is less than 10 mL, it may be more convenient to transfer the mixture to the filter with a preheated Pasteur pipet.) It is necessary to keep the solutions in both flasks at their boiling temperatures to prevent premature crystallization. The refluxing action of the filtrate keeps the funnel warm and reduces the chance that the filter will clog with crystals that may have formed during the filtration. With low-boiling solvents, be aware that some solvent may be lost through evaporation. Consequently, extra solvent must be added to make up for this loss. If crystals begin to form in the filter during filtration, a minimum amount of boiling solvent is added to redissolve the crystals and to allow the solution to pass through the funnel. If the volume of liquid being filtered is less than 10 mL, a small amount of hot solvent should be used to rinse the filter after all the filtrate has been collected. The rinse solvent is then combined with the original filtrate.

After the filtration, it may be necessary to remove extra solvent by evaporation until the solution is saturated at the boiling point of the solvent (see Technique 3, Section 3.9, p 560).

C. Crystallization

An Erlenmeyer flask, not a beaker, should be used for crystallization. The large open top of a beaker makes it an excellent dust catcher. The narrow opening of the Erlenmeyer flask reduces contamination by dust and allows the flask to be stoppered if it is to be set aside for a long period. Mixtures set aside for long periods must be stoppered after cooling to room temperature to prevent evaporation of solvent. If all the solvent evaporates, no purification is achieved, and the crystals originally formed become coated with the dried contents of the mother liquor. Even if the time required for crystallization to occur is relatively short, it is advisable to cover the top of the Erlenmeyer flask with a small watch glass or inverted beaker to prevent evaporation of solvent while the solution is cooling to room temperature.

The chances of obtaining pure crystals is improved if the solution cools to room temperature slowly. When the volume of solution is about 10 mL or less, the solution is likely to cool more rapidly than is desired. This can be prevented by placing the flask on a surface that is a poor heat conductor and covering the flask with a beaker to provide a layer of insulating air. Appropriate surfaces include a clay plate or several pieces of filter paper on top of the laboratory bench. It may also be helpful to use a clay plate that has been warmed slightly on a hot plate or in an oven.

After crystallization has occurred, it is sometimes desirable to cool the flask in an ice-water bath. Because the solute is less soluble at lower temperatures, this will increase the yield of crystals.

If a cooled solution does not crystallize, it will be necessary to induce crystallization. Several techniques are described in Section 5.7, Part A.

D. Isolation of Crystals

After the flask has been cooled, the crystals are collected by vacuum filtration through a Hirsch (or Büchner) funnel (see Technique 4, Section 4.3, p 571, and Figure 4–6). The crystals should be washed with a small amount of **cold** solvent to remove any mother liquor adhering to their surface. Hot or warm solvent will dissolve some of the crystals. The crystals should then be left for a short time in the funnel, where air, as it passes, will dry them free of most of the solvent. It is often wise to cover the Hirsch funnel with an oversize filter paper or towel during this air-drying. This prevents accumulation of dust in the crystals. When the crystals are nearly dry, they should be gently scraped (so paper fibers are not removed with the crystals) off the filter paper onto a watch glass or clay plate for further drying (see Section 5.8).

5.4 MICROSCALE CRYSTALLIZATIONS—CRAIG TUBES

In most microscale experiments, the amount of solid to be crystallized is small enough (generally less than 0.1 g) that a **Craig tube** (see Figure 4–10, p 576) is the preferred method for crystallization. The main advantage of the Craig tube is that it minimizes the number of transfers of solid material, thus resulting in a greater yield of crystals. Also, the separation of the crystals from the mother liquor with the Craig tube is very efficient, and little time is required for drying the crystals. The steps involved are, in principle, the same as those performed when a crystallization is done with an Erlenmeyer flask and a Hirsch funnel. The steps in a microscale crystallization using a Craig tube are illustrated in Figure 5–5.

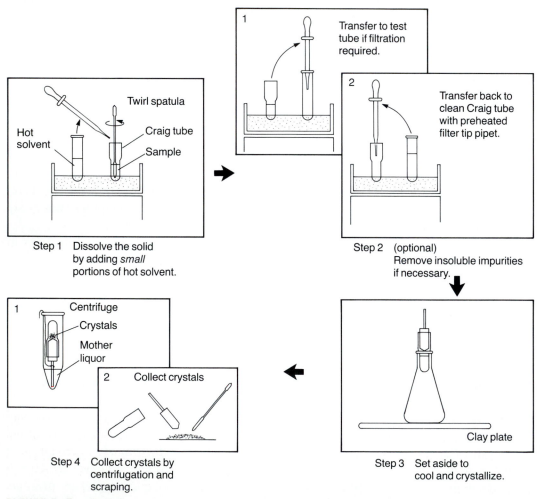

FIGURE 5—5.   Steps in a microscale crystallization (no decolorization)

A. Dissolving the Solid

In crystallizations in which a filtration step is not required to remove insoluble impurities such as dirt or activated charcoal, this first step can be done directly in the Craig tube. Otherwise, use a small test tube. The solid is transferred to the Craig tube and the appropriate solvent contained in a test tube is heated to boiling in a sand bath. A small portion (several drops) of hot solvent is added to the Craig tube, which is subsequently heated in the sand bath. The hot mixture should be stirred continuously with a microspatula using a twirling motion. Stirring not only helps to dissolve the solute, but also prevents the boiling liquid from bumping. Additional portions of hot solvent are added until all the solid has dissolved. In order to obtain the maximum yield, it is important not to add too much solvent, although any excess solvent can be evaporated later.

If the mixture is **highly** colored and it is clear that the color is due to impurities and not to the actual color of the substance being crystallized, it will be necessary to decolorize the liquid. If decolorization is necessary, it should be done before the filtration step described below. Decolorizing charcoal may be used or the mixture may be passed through an alumina or silica gel column (see Section 5.6, Parts B and C, and Technique 12, Section 12.13, p 717).

B. Removing Insoluble Impurities

You should be alert for the presence of insoluble impurities that will not dissolve in the hot solvent, no matter how much solvent is added. If it appears that most of the solid has dissolved and the remaining solid has no tendency to dissolve, or if the liquid has been decolorized with charcoal, it will be necessary to remove the solid particles. Two methods will be discussed.

If the impurities are relatively large or concentrated in one part of the mixture, it may be possible to use a Pasteur pipet preheated with hot solvent to draw up the liquid without removing any solid. One way to do this is to expel the air from the pipet and then place the end of the pipet on the bottom of the tube, being careful not to trap any solid in the pipet. The small space between the pipet and the bottom of the tube should allow you to draw up the liquid without removing any solid.

When filtration is necessary, a preheated Pasteur pipet is used to transfer the mixture to a test tube. After making this transfer, the Craig tube is rinsed with a few drops of solvent, which are also added to the test tube. The Craig tube is then washed and dried. The test tube containing the mixture is also heated in the sand bath. An additional 5–10 drops of solvent are added to the test tube to ensure that premature crystallization doesn't occur during the filtration step. To filter the mixture, take up the mixture in a filter tip pipet (see Technique 4, Section 4.6, p 574), which has been preheated with hot solvent, and quickly transfer the liquid to the clean Craig tube. Passing the liquid through the cotton plug in the filter tip pipet should remove the solid impurities. **If this doesn't occur,** it may be necessary to add more solvent (to prevent crystallization) and filter the mixture through a filtering pipet (Technique 4, Section 4.1, Part C, p 568). In either case, once the filtered solution has been returned to the Craig tube, it will be necessary to evaporate some solvent until the solution is saturated near the boiling point of the liquid. This is most conveniently done by placing the Craig tube in the sand bath, and, with rapid stirring using a microspatula (twirling is most effective), the solution is boiled. When one begins to observe a trace of solid material coating the spatula just above the level of the liquid, the solution is near saturation and evaporation should be stopped.

C. Crystallization

The hot solution is cooled slowly in the Craig tube to room temperature. Recall that slow cooling is important in the formation of pure crystals. When the volume of solution is 2 mL or less and the mass of glassware is relatively small, slow cooling is somewhat difficult to achieve. One method of increasing the cooling time is to insert the inner plug into the outer part of the Craig tube and place the Craig tube into a 10-mL Erlenmeyer flask. The layer of air in the flask will help insulate the hot solution as it cools. The Erlenmeyer flask is placed on a surface such as a clay plate (warmed slightly, if desired) or several pieces of paper. Another method is to fill a 10-mL Erlenmeyer flask with 8–10 mL of hot water at a temperature below the boiling point of the solvent. The assembled Craig tube is placed in the Erlenmeyer flask that is set on an appropriate surface. Be careful not to put so much water in the Erlenmeyer flask that the Craig tube floats. After crystallization at room temperature is complete, the Craig tube can be placed in an ice-water bath to maximize the yield.

If crystals haven't formed after the solution has cooled to room temperature, it will be necessary to induce crystallization. Several techniques are described in Section 5.7.

A common occurrence with crystallizations using a Craig tube is to obtain a seemingly solid mass of very small crystals. This may not be a problem, but if there is very little mother liquor present or the crystals are impure, it may be necessary to repeat the crystallization. This situation may have resulted either because the cooling process occurred too rapidly, or because the solubility-versus-temperature curve was so steep for a given solvent that very little mother liquor remained after the crystallization. In either case, it is possible that you will want to repeat the crystallization to obtain a better (purer) yield of crystals. Three measures may be taken to avoid this problem. A small amount of extra solvent may be added before heating the mixture again and allowing it to cool. A second measure is to cool the solution more slowly. Finally, it may be helpful to try to induce crystallization **before** the solution has cooled to room temperature.

D. Isolation of Crystals

When the crystals have formed and the mixture has cooled in an ice-water bath (if desired), the Craig tube is placed in a centrifuge tube and the crystals are separated from the mother liquor by centrifugation (see Technique 4, Section 4.7, p 575). The crystals are then scraped off the end of the inner plug or from inside the Craig tube onto a watch glass or piece of paper. Minimal drying will be necessary (see Section 5.8).

5.5   SELECTING A SOLVENT

A solvent that dissolves little of the material to be crystallized when it is cold but a great deal of the material when it is hot is a good solvent for the crystallization. With compounds that are well known, such as the compounds that are either isolated or prepared in this textbook, the correct crystallization solvent is already known through the experiments of earlier workers. In such cases, the chemical literature can be consulted to determine which solvent should be used. Sources such as handbooks or tables will frequently provide this information. Quite often, the correct crystallization solvents are indicated in the experimental procedures in this textbook.

When the appropriate solvent is not known, one selects a solvent for crystallization by experimenting with various solvents and a very small amount of the material to be crystallized. Experiments are conducted on a small test-tube scale before the entire quantity of material is committed to a particular solvent. Such trial-and-error methods are common when one is trying to purify a solid material that has not been previously studied.

When choosing a crystallization solvent, care should be taken not to pick one whose boiling point is higher than the melting point of the substance to be crystallized. If the boiling point of the solvent is high, the solid may melt in the solvent rather than dissolve. In such a case, the solid may **"oil out."** Oiling out occurs when the solid substance melts and forms a liquid that is not soluble in the solvent. On cooling, the liquid refuses to crystallize; rather, it becomes a supercooled liquid, or oil. Oils may solidify if the temperature is lowered, but often they will not crystallize. A solidified oil becomes an amorphous solid or a hardened mass—a condition that does not result in the purification of the substance. It can be very difficult to deal with oils when trying to obtain a pure substance. One must try to redissolve them and hope that they will precipitate as crystals with slow, careful cooling. During the cooling period, it may be helpful to scratch the glass container where the oil is present with a glass stirring rod which has not been fire-polished. Seeding the oil as it cools with a small sample of the original solid is another technique sometimes helpful in working with difficult oils. Other methods of inducing crystallization are discussed in Section 5.7.

One additional criterion for selecting the correct crystallization solvent is the **volatility** of that solvent. Volatile solvents are those that have low boiling points or evaporate easily. A solvent with a low boiling point may be removed from the crystals through evaporation without much difficulty. It will be difficult to remove a solvent with a high boiling point from the crystals without heating them under vacuum.

Table 5–2 lists common crystallization solvents. The solvents used most commonly are listed first in the table.

5.6   DECOLORIZATION

Small amounts of highly colored impurities may make the original crystallization solution appeared colored; this color can often be removed by **decolorization,** either by using activated charcoal (often called Norit) or by passing the solution through a col-

TABLE 5–2. Common Solvents for Crystallization

	BOILS (°C)	FREEZES (°C)	SOLUBLE IN H₂O	FLAMMABILITY
Water	100	0	+	−
Methanol	65	*	+	+
95% Ethanol	78	*	+	+
Ligroin	60–90	*	−	+
Toluene	111	*	−	+
Chloroform†	61	*	−	−
Acetic acid	118	17	+	+
Dioxane†	101	11	+	+
Acetone	56	*	+	+
Diethyl ether	35	*	Slightly	++
Petroleum ether	30–60	*	−	++
Methylene chloride	41	*	−	−
Carbon tetrachloride†	77	*	−	−

*Lower than 0 °C (ice temperature).
†Suspected carcinogen.

umn packed with alumina or silica gel. A decolorizing step should be done only if the color is due to impurities, not to the color of the desired product, and if the color is significant. Small amounts of colored impurities will remain in solution during crystallization, making the decolorizing step unnecessary. The use of activated charcoal will be described separately for semi-microscale and microscale crystallizations, and then the column technique, which can be used with both crystallization techniques, will be described.

A. Semi-microscale—Powdered Charcoal

As soon as the solute is dissolved in the minimum amount of boiling solvent, the solution is allowed to cool slightly and a small amount of Norit is added to the mixture. The Norit adsorbs the impurities. When doing a crystallization in which the filtration will be performed with a fluted filter, powdered Norit should be added since it has a larger surface area and can remove impurities more effectively. A reasonable amount of Norit would be what could be held on the end of a microspatula, or about 0.01–0.02 g. If too much Norit is used, it will adsorb product as well as impurities. A small amount of Norit should be used, and its use should be repeated if necessary. (It is difficult to determine if the initial amount added is sufficient until after the solution is filtered, because the suspended particles of charcoal will obscure the color of the liquid.) Caution should be exercised so that the solution does not froth or erupt when the finely divided charcoal is added. The mixture is boiled for several minutes and then filtered by gravity, using a fluted filter (see Section 5.3 and Technique 4, Section 4.1, Part B, p 566), and the crystallization is carried forward as described in Section 5.3.

The Norit preferentially adsorbs the colored impurities and removes them from the solution. The technique seems to be most effective with hydroxylic solvents. In using Norit, one should be careful not to breathe the dust. Normally, small quantities are used so that little risk of lung irritation exists.

B. Microscale—Pelletized Norit

If the crystallization is being done in a Craig tube, it is advisable to use pelletized Norit. Although this is not as effective in removing impurities as powdered Norit, it is easier to carry out the subsequent filtration, and the amount of pelletized Norit required is more easily determined since one can see the solution as it is being decolorized. Again, the Norit is added to the hot solution (the solution should not be boiling) after the solid has dissolved. This should be done in a test tube rather than in a Craig tube. About 0.02 g is added and the mixture is boiled for a minute or so to see if more Norit is required. More Norit is added, if necessary, and the liquid is boiled again. It is important not to add too much pelletized Norit because the Norit will also adsorb some of the desired material, and it is possible that not all the color can be removed no matter how much is added. The decolorized solution is then removed with a preheated filter tip pipet (see Section 5.4 and Technique 4, Section 4.6, p 574) to filter the mixture and transferred to a Craig tube for crystallization as described in Section 5.4.

C. Decolorization on a Column

The other method for decolorizing a solution is to pass the solution through a column containing alumina or silica gel. The adsorbent removes the colored impurities while allowing the desired material to pass through (see Figure 4–7, p 573, and Technique 12, Section 12.13, p 717). If this technique is used, it will be necessary to dilute the solution with additional solvent to prevent crystallization from occurring during the process. The excess solvent must be evaporated after the solution is passed through the column (Technique 3, Section 3.9, p 560), and the crystallization procedure is continued as described in Sections 5.3 and 5.4.

5.7 INDUCING CRYSTALLIZATION

If a cooled solution does not crystallize, several techniques may be used to induce crystallization. Although identical in principle, the actual procedures vary slightly when doing semi-microscale and microscale crystallizations.

A. Semi-microscale

In the first technique, one should try vigorous scratching of the inside surface of the flask with a glass rod that **has not been** fire-polished. The motion of the rod should be vertical (in an out of the solution) and should be vigorous enough to produce an audible scratching. Such scratching often induces crystallization, although the effect is not well understood. The high-frequency vibrations may have something to do with initiating crystallization; or perhaps—a more likely possibility—small amounts of solution dry by evaporation on the side of the flask, and the dried solute is pushed into the solution. These small amounts of material provide ''seed crystals,'' or nuclei, on which crystallization may begin.

A second technique that can be used to induce crystallization is to cool the solution in an ice bath. This decreases the solubility of the solute.

A third technique is useful when small amounts of the original material to be crystallized are saved. The saved material can be used to "seed" the cooled solution. A small crystal dropped into the cooled flask often will start the crystallization—this is called **seeding.**

If all these measures fail to induce crystallization, it is likely that too much solvent was added. The excess solvent must then be evaporated (Technique 3, Section 3.9, p 560), and the solution is again allowed to cool.

B. Microscale

The strategy is basically the same as described for semi-microscale crystallizations. Vigorous scratching with a glass rod **should be avoided,** however, since the Craig tube is fragile and expensive. **Gentle** scratching may be used.

Another measure is to dip a spatula or glass stirring rod into the solution and allow the solvent to evaporate so that a small amount of solid will form on the surface of the spatula or glass rod. When placed back into the solution, the solid will seed the solution. A small amount of the original material, if some was saved, may also be used to seed the solution.

A third technique is to cool the Craig tube in an ice-water bath. This method may also be combined with either of the previous suggestions.

If none of these measures are successful, it is possible that too much solvent is present and it may be necessary to evaporate some of the solvent (Technique 3, Section 3.9, p 560) and allow the solution to cool again.

5.8 DRYING CRYSTALS

The most common method of drying crystals involves placing them on a watch glass, a clay plate, or a piece of paper and allowing them to dry in air. While the advantage of this method is that heat is not required, thus reducing the danger of decomposition or melting, exposure to atmospheric moisture may cause the hydration of strongly **hygroscopic** materials. A hygroscopic substance is one that absorbs moisture from the air.

Another method of drying crystals is to place the crystals on a watch glass, a clay plate, or a piece of absorbent paper in an oven. Although this method is simple, some possible difficulties deserve mention. Crystals that sublime readily should not be dried in an oven since they might vaporize and disappear. Care should be taken that the temperature of the oven does not exceed the melting point of the crystals. One must remember that the melting point of crystals is lowered by the presence of solvent, and one must allow for this melting-point depression when selecting a suitable oven temperature. Some materials decompose on exposure to heat, and they should not be dried in an oven. Finally, when many different samples are being dried in the same oven, crystals might be lost due to confusion or reaction with another person's sample. It is important to label the crystals when they are placed in the oven.

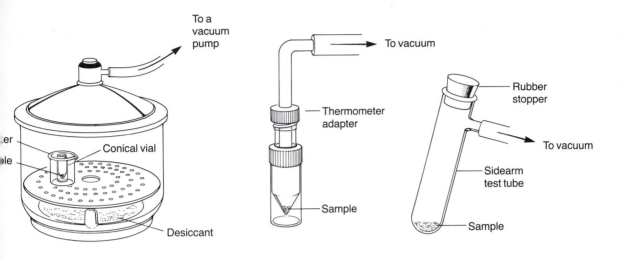

A. Desiccator

B. Conical vial or sidearm test tube

FIGURE 5–6. Methods for drying crystals in a vacuum

A third method, which requires neither heat nor exposure to atmospheric moisture, is drying *in vacuo*. Two procedures are illustrated in Figure 5–6.

Procedure A. In this method a desiccator is used. The sample is placed under vacuum in the presence of a drying agent. Two potential problems must be noted. The first deals with samples that sublime readily. Under vacuum, the likelihood of sublimation is increased. The second problem deals with the vacuum desiccator itself. Since the surface area of glass that is under vacuum is large, there is some danger that the desiccator could implode. A vacuum desiccator should never be used unless it has been placed within a protective metal container (cage). If a cage is not available, the desiccator can be wrapped with electrical or duct tape. If you use an aspirator as a source of vacuum, you should use a water trap (see Figure 4–6, p 571).

Procedure B. This method can be accomplished with a conical vial and a thermometer adapter equipped with a short piece of glass tubing, as illustrated in Figure 5–6B. The glass tubing is connected by vacuum tubing to either an aspirator or a vacuum pump. A convenient alternative utilizing a sidearm test tube is also shown in Figure 5–6B. With either apparatus, install a water trap when an aspirator is used.

5.9 MIXED SOLVENTS

Often the desired solubility characteristics for a particular compound are not found in a single solvent. In these cases a mixed solvent may be used. One simply selects a first solvent in which the solute is soluble and a second solvent, miscible with the first, in which the solute is relatively insoluble. The compound is dissolved in a minimum amount of the boiling solvent in which it is soluble. Following this, the second hot solvent is added to the boiling mixture, dropwise, until the mixture barely becomes cloudy. The cloudiness indicates precipitation. At this point, more of the first solvent

TABLE 5–3. Common Solvent Pairs for Crystallization

Methanol–Water	Ether–Acetone
Ethanol–Water	Ether–Petroleum ether
Acetic acid–Water	Toluene–Ligroin
Acetone–Water	Methylene chloride–Methanol
Ether–Methanol	Dioxane*–Water

*Suspected carcinogen.

should be added. Just enough is added to clear the cloudy mixture. At that point the solution is saturated, and as it cools, crystals should separate. Common solvent mixtures are listed in Table 5–3.

It is important not to add an excess of the second solvent or to cool the solution too rapidly. Either of these actions may cause the solute to oil out, or separate as a viscous liquid. It this happens, one should reheat the solution and add more of the first solvent.

STEPS IN A CRYSTALLIZATION

A. Dissolving the Solid

1. Find a solvent with a steep solubility-vs-temperature characteristic. (Done by trial and error using small amounts of material or by consulting a handbook.)
2. Heat the desired solvent to its boiling point.
3. Dissolve the solid in a **minimum** of boiling solvent (either in a flask or a Craig tube).
4. If necessary, add decolorizing charcoal or decolorize the solution on a silica gel or alumina column.

B. Removing Insoluble Impurities

1. Decant or remove the solution with a Pasteur pipet, or
2. Filter the hot solution through a fluted filter, a filtering pipet, or a filter tip pipet to remove insoluble impurities or charcoal.

> **NOTE: If no decolorizing charcoal has been added or if there are no undissolved particles, Part B should be omitted.**

C. Crystallizing

1. Allow the solution to cool.
2. If crystals appear, cool the mixture in an ice-water bath (if desired) and go to Part D. If crystals do not appear, go to the next step.

3. Inducing crystallization.
   (a) Scratch the flask with a glass rod; or, if using a Craig tube, dip a glass rod or spatula into the solution, let the liquid evaporate, and place the glass rod or spatula back into the solution to seed it.
   (b) Seed the solution with original solid, if available.
   (c) Cool the solution in an ice-water bath.
   (d) Evaporate excess solvent and allow the solution to cool again.

D. Collecting and Drying

1. Collect crystals by vacuum filtration using a Hirsch funnel or by centrifugation using a Craig tube.
2. If using a Hirsch funnel, rinse crystals with a small portion of **cold** solvent.
3. Continue suction until crystals are nearly dry, if using vacuum filtration.
4. Drying.
   (a) Air-dry the crystals, or
   (b) Place the crystals in a drying oven, or
   (c) Dry the crystals *in vacuo*.

PROBLEMS

1. Listed below are solubility-vs-temperature data for an organic substance A dissolved in water.

TEMPERATURE (°C)	SOLUBILITY OF A IN 100 mL OF WATER
0	1.5 g
20	3.0 g
40	6.5 g
60	11.0 g
80	17.0 g

   (a) Graph the solubility of A vs temperature. Use the data given above. Connect the data points with a smooth curve.
   (b) Suppose 0.1 g of A and 1.0 mL of water were mixed and heated to 80 °C. Would all the substance A dissolve?
   (c) The solution prepared in (b) is cooled. At what temperature will crystals of A appear?
   (d) Suppose the cooling described in (c) were continued to 0 °C. How many grams of A would come out of solution? Explain how you obtained your answer.
2. What would be likely to happen if a hot saturated solution were filtered by vacuum filtration using a Hirsch funnel?
3. A compound you have prepared is reported in the literature to have a pale yellow color. When dissolving the substance in hot solvent to purify it by crystallization, the resulting solution is yellow. Should you use decolorizing charcoal before allowing the hot solution to cool? Explain your answer.
4. After dissolving a crude product in 1.5 mL of hot solvent, the resulting solution is a dark brown color. Since the pure compound is reported in the literature to be colorless, it is necessary to perform a decolorizing procedure. Should you use pelletized Norit or powdered activated charcoal to decolorize the solution? Explain your answer.

5. While performing a crystallization, you obtain a light tan solution after dissolving your crude product in hot solvent. A decolorizing step is determined to be unnecessary, and there are no solid impurities present. Should you perform a filtration before allowing the solution to cool? Why or why not?

6. (a) Draw a graph of a cooling curve (temperature vs time) for a solution of a solid substance that shows no supercooling effects. Assume that the solvent does not freeze.

    (b) Repeat the above instructions for a solution of a solid substance that shows some supercooling behavior but eventually yields crystals if the solution is cooled sufficiently.

7. A solid substance A is soluble in water to the extent of 10 mg/mL of water at 25 °C and 100 mg/mL of water at 100 °C. You have a sample that contains 100 mg of A and an impurity B.

    (a) Assuming that 2 mg of B is present along with 100 mg of A, describe how you could purify A if B is completely insoluble in water.

    (b) Assuming that 2 mg of the impurity B is present along with 100 mg of A, describe how you could purify A if B had the same solubility behavior as A. Would one crystallization produce absolutely pure A?

    (c) Assume that 25 mg of the impurity B is present along with 100 mg of A. Describe how you could purify A if B had the same solubility behavior as A. Each time, use the correct amount of water to just dissolve the solid. Would one crystallization produce absolutely pure A? How many crystallizations would be needed to produce pure A? How much A would have been recovered when the crystallizations had been completed?

Technique 6
PHYSICAL CONSTANTS: MELTING POINTS, BOILING POINTS, DENSITY

6.1 INTRODUCTION

The physical properties of a compound are those properties that are intrinsic to a given compound when it is pure. Often a compound may be identified simply by determining a number of its physical properties. The most commonly recognized physical properties of a compound include its color, melting point, boiling point, density, refractive index, molecular weight, and optical rotation. Modern chemists would include the various types of spectra (infrared, nuclear magnetic resonance, mass, and ultraviolet-visible) among the physical properties of a compound. A compound's spectra do not vary from one pure sample to another. In this chapter, we look at methods of determining the melting point, boiling point, and density of compounds. Refractive index, optical rotation, and spectra are considered separately in their own technique chapters.

Many reference books list the physical properties of substances. Useful works for finding lists of values for the non-spectroscopic physical properties include:

The Merck Index
The CRC Handbook of Chemistry and Physics
The Dictionary of Organic Compounds
Lange's Handbook of Chemistry
Tables for the Identification of Organic Compounds

Complete literature citations for these references may be found in Technique 19 (Guide to the Chemical Literature, p 787). While the *CRC Handbook* has very good tables, it adheres strictly to IUPAC nomenclature. For this reason, it may be easier to use one of the other references, particularly *The Merck Index,* in your first attempt to locate information. *The Dictionary of Organic Compounds* is a multi-volume work. A trip to the reference shelves of your library will be required for you to use it, but it is a very complete source book.

PART A. MELTING POINTS

6.2 THE MELTING POINT

The melting point of a compound is used by the organic chemist not only to identify it, but also to establish its purity. A small amount of material is heated **slowly** in a special apparatus equipped with a thermometer or thermocouple, a heating bath or heating coil, and a magnifying eyepiece for observing the sample. Two temperatures are noted. The first is the point at which the first drop of liquid forms among the crystals; the second is the point at which the whole mass of crystals turns to a **clear** liquid. The melting point is recorded by giving this range of melting. One might say, for example, that the melting point of a substance is 51–54 °C. That is, the substance melted over a three degree range.

The melting point indicates purity in two ways. First, the purer the material, the higher its melting point. Second, the purer the material, the narrower the melting point range. Adding successive amounts of an impurity to a pure substance generally causes its melting point to decrease in proportion to the amount of impurity. Looking at it another way, adding impurities will cause a depression of the freezing point. The freezing point, a colligative property, is simply the melting point (solid → liquid) approached from the opposite direction (liquid → solid).

Figure 6–1 is a graph of the usual melting-point behavior of mixtures of two substances, A and B. The two extremes of the melting range (the low and high temperatures) are shown for various mixtures of the two. The upper curves indicate the temperature at which all of the sample has melted. The lower curves indicate the temperature at which melting is observed to begin. With pure compounds, melting is sharp and without any range. This is shown at the left and right hand edges of the graph. If one begins with pure A, the melting point decreases as impurity B is added. At some point, a minimum temperature, or **eutectic,** is reached, and the melting point begins to increase to that of substance B. The vertical distance between the lower and upper curves represents the melting range. Notice that for mixtures that contain relatively small amounts of impurity (<15%) and are not close to the eutectic, the melting

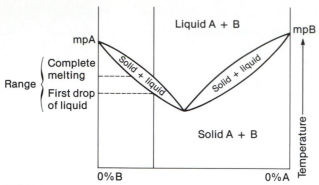

FIGURE 6–1. Melting point-composition curve

range increases as the sample becomes less pure. The range indicated by the lines in Figure 6–1 represents the typical behavior.

　　We can generalize the behavior shown in Figure 6–1. Pure substances melt with a narrow range of melting. With impure substances, the melting range will become wider and the entire melting range will be lowered. Be careful to note, however, that at the minimum point of the melting point-composition curves, the mixture often forms a eutectic, which also melts sharply. Not all binary mixtures form eutectics, and some caution must be exercised in assuming that every binary mixture follows the behavior described above. Some mixtures may form more than one eutectic, others might not form even one. In spite of these variations, both the melting point and its range are useful indications of purity, and they are easily determined by simple experimental methods.

6.3 MELTING POINT THEORY

Figure 6–2 is a phase diagram describing the usual behavior of a two component mixture (A + B) on melting. The behavior on melting depends on the relative amounts

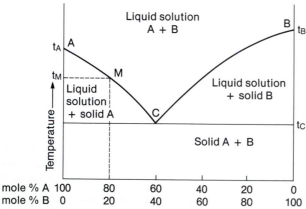

FIGURE 6–2. Phase diagram for melting in a two-component system

of A and B in the mixture. If A is a pure substance (no B), then A melts sharply at its melting point, t_A. This is represented by Point A on the left side of the diagram. When B is a pure substance, it melts at t_B; its melting point is represented by Point B on the right side of the diagram. At either Point A or Point B, the pure solid passes cleanly, with a narrow range, from solid to liquid.

In mixtures of A and B, the behavior is different. Using Figure 6–2, consider a mixture of 80% A and 20% B on a mole-per-mole basis (that is, mole percentage). The melting point of this mixture is given by t_M at Point M on the diagram. That is, adding B to A has lowered the melting point of A from t_A to t_M. It has also expanded the melting range. The temperature t_M corresponds to the **upper limit** of the melting range.

Lowering the melting point of A by adding impurity B comes about in the following way. Substance A has the lower melting point in the phase diagram shown, and with heating, it begins to melt first. As A begins to melt, solid B begins to dissolve in the liquid A that is formed. When solid B dissolves in liquid A, the melting point is depressed. To understand this, consider the melting point from the opposite direction. When a liquid at a high temperature cools, it reaches a point at which it solidifies, or "freezes." The temperature at which a liquid freezes is identical to its melting point. It will be recalled that the freezing point of a liquid can be depressed by adding an impurity. Since the freezing point and the melting point are identical, lowering the freezing point corresponds to lowering the melting point. Therefore, as more impurity is added to a solid, its melting point becomes lower. There is, however, a limit to how far the melting point can be depressed. One cannot dissolve an infinite amount of the impurity substance in the liquid. At some point the liquid will become saturated with the impurity substance. The solubility of B in A has an upper limit. In Figure 6–2, the solubility limit of B in liquid A is reached at point C, the **eutectic point.** The melting point of the mixture cannot be depressed below t_C, the melting temperature of the eutectic.

Now consider what happens when the melting point of a mixture of 80% A and 20% B is approached. As the temperature is increased, A begins to "melt." This is not really a visible phenomenon in the beginning stages; it happens before liquid is visible. It is a softening of the compound to a point at which it can begin to mix with the impurity. As A begins to soften, it dissolves B. As it dissolves B, the melting point is lowered. The lowering continues until all B is dissolved or until the eutectic composition (saturation) is reached. When the maximum possible amount of B has been dissolved, actual melting begins, and one can observe the first appearance of liquid. The initial temperature of melting will be below t_A. The amount below t_A at which melting begins is determined by the amount of B dissolved in A, but will never be below t_C. Once all B has been dissolved, the melting point of the mixture begins to rise as more A begins to melt. As more A melts, the semisolid solution is diluted by more A, and its melting point rises. While all of this is happening, one can observe **both** solid and liquid in the melting point-capillary. Once all A has begun to melt, the composition of the mixture M becomes uniform and will reach 80% A and 20% B. At this point, the mixture finally melts sharply, giving a clear solution. The maximum melting point will be t_M, since t_A is depressed by the impurity B that is present. The lower end of the melting range will always be t_C; however, melting will not always be observed at this temperature. An observable melting at t_C only comes about when a large amount of B is present. Otherwise, the amount of liquid formed at t_C will be too small to observe. Therefore, the melting behavior that is actually observed will have a smaller range, as shown in Figure 6–1.

6.4 MIXED MELTING POINTS

The melting point can be used as supporting evidence in identifying a compound in two different ways. Not only may the melting points of the two individual compounds be compared, but a special procedure called a **mixed melting point** may also be performed. The mixed melting point requires that an authentic sample of the same compound be available from another source. In this procedure, the two compounds (authentic and suspected) are finely pulverized and mixed together in equal quantities. Then the melting point of the mixture is determined. If there is a melting-point depression, or if the range of melting is expanded by a large amount, compared to the individual substances, one may conclude that one compound has acted as an impurity toward the other and they are not the same compound. If there is no depression of the melting point for the mixture (the melting point is identical with those of pure A and pure B), then A and B are almost certainly the same compound.

6.5 PACKING THE MELTING POINT TUBE

Melting points usually are determined by heating the sample in a piece of thin-walled capillary tubing (1 mm × 100 mm) that has been sealed at one end. To pack the tube, one presses the open end gently into a **pulverized** sample of the crystalline material. Crystals will stick in the open end of the tube. The amount of solid pressed into the tube should correspond to a column no more than 1–2 mm high. To transfer the crystals to the closed end of the tube, one drops the capillary tube, closed end first, down a ⅔-m length of glass tubing, which is held upright on the desk top. When the capillary tube hits the desk top, the crystals will pack down into the bottom of the tube. This procedure is repeated if necessary. Tapping the capillary on the desk top with fingers is not recommended, since it is easy to drive the small tubing into a finger if the tubing should break.

Some commercial melting point instruments have a built-in vibrating device that is designed to pack capillary tubes. With these instruments, the sample is pressed into the open end of the capillary tube and the tube is placed in the vibrator slot. The action of the vibrator will transfer the sample to the bottom of the tube and pack it tightly.

6.6 DETERMINING THE MELTING POINT—THE THIELE TUBE

There are two principal types of melting-point apparatus available: the Thiele tube and commercially available, electrically heated instruments. The Thiele tube, shown in

Figure 6–3, is the simpler device, and it is widely available. It is a glass tube designed to contain a heating oil (mineral oil or silicone oil) and a thermometer to which a capillary tube containing the sample is attached. The shape of the Thiele tube allows convection currents to form in the oil when it is heated. These currents maintain a uniform temperature distribution throughout the oil in the tube. The sidearm of the tube is designed to generate these convection currents and thus transfer the heat from the flame evenly and rapidly throughout the oil. The sample, which is in a capillary tube attached to the thermometer, is held by a rubber band or a thin slice of rubber tubing. It is important that this rubber band be above the level of the oil (allowing for expansion of the oil on heating), so that the oil does not soften the rubber and allow the capillary tubing to fall into the oil. If a cork or a rubber stopper is used to hold the thermometer, a triangular wedge should be sliced in it to allow pressure equalization.

The Thiele tube is usually heated by a microburner. During the heating, the rate of temperature increase should be regulated. One usually holds the burner by its cool base, and, using a gentle flame, moves the burner slowly back and forth along the bottom of the arm of the Thiele tube. If the heating is too fast, the burner is removed for a few seconds, and then heating is resumed. The rate of heating should be **low** near the melting point (about 1 °C min.) to ensure that the temperature increase is not faster than

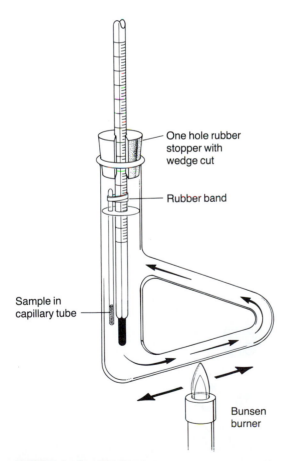

Labels:
- One hole rubber stopper with wedge cut
- Rubber band
- Sample in capillary tube
- Bunsen burner

FIGURE 6–3. Thiele tube

the rate at which heat can be transferred to the sample being observed. At the melting point, it is necessary that the mercury in the thermometer and the sample in the capillary tube be at temperature equilibrium.

6.7 DETERMINING THE MELTING POINT—ELECTRICAL INSTRUMENTS

Two types of electrically heated melting point instruments are illustrated in Figure 6–4. In each case, the melting point tube is filled as described in Section 6.5 and placed in a holder located just behind the magnifying eyepiece. The apparatus is operated by moving the switch to the ON position, adjusting the potentiometric control dial for the desired rate of heating, and observing the sample through the magnifying eyepiece. The temperature is read from a thermometer, or in the most modern instruments, from a digital display attached to a thermocouple. Your instructor will demonstrate and explain the type used in your laboratory.

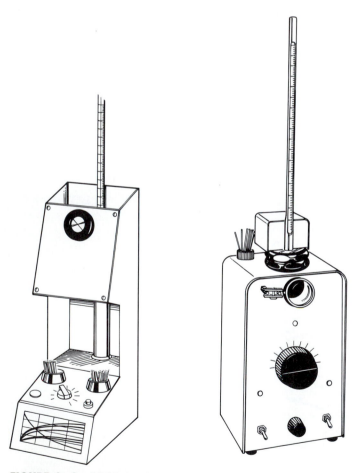

FIGURE 6–4.   Melting point apparatus

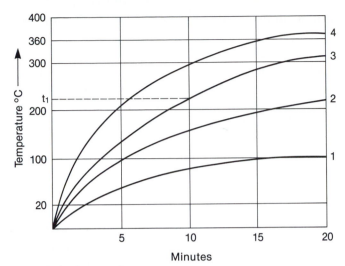

FIGURE 6–5. Heating-rate curves

Most electrically heated instruments do not heat or increase the temperature of the sample linearly. Although the rate of increase may be linear in the early stages of heating, it usually decreases and leads to a constant temperature at some upper limit. The upper-limit temperature is determined by the setting of the heating control. Thus, a family of heating curves is usually obtained for various control settings, as shown in Figure 6–5. The four hypothetical curves shown (1–4) might correspond to different control settings. For a compound melting at temperature t_1, the setting corresponding to Curve 3 would be ideal. In the beginning of the curve, the temperature is increasing too rapidly to allow determination of an accurate melting point, but after the change in slope, the temperature increase will have slowed to a more usable rate.

If the melting point of the sample is unknown, one can often save time by preparing two samples for melting-point determination. With one sample, one rapidly determines a crude melting-point value. Then the experiment is repeated more carefully using the second sample. For the second determination, one already has an approximate idea of what the melting point temperature should be and a proper rate of heating can be chosen.

When measuring temperatures above 150 °C, thermometer errors can become significant. For an accurate melting point with a high melting solid, you may wish to apply a **stem correction** to the thermometer as described in Section 6.13. An even better solution is to calibrate the thermometer as described in Section 6.12.

6.8 DECOMPOSITION, DISCOLORATION, SOFTENING, SHRINKAGE, AND SUBLIMATION

Many solid substances undergo some degree of unusual behavior before melting. At times it may be difficult to distinguish these other types of behavior from actual melting. One should learn, through experience, how to recognize melting and how to distinguish it from decomposition, discoloration, and particularly softening and shrinkage.

Some compounds decompose on melting. This decomposition is usually evidenced by discoloration of the sample. Frequently this decomposition point is a reliable physical property to be used in lieu of an actual melting point. Such decomposition points are indicated in tables of melting points by placing the symbol **d** immediately after the listed temperature. An example of a decomposition point is thiamine hydrochloride, whose melting point would be listed as 248 ° d, indicating that this substance melts with decomposition at 248 °C. When decomposition is a result of reaction with the oxygen in air, it may be avoided by determining the melting point in a sealed, evacuated melting point tube.

Figure 6–6 shows two simple methods of evacuating a packed tube. Method A uses an ordinary melting point tube, while Method B constructs the melting point tube from a disposable Pasteur pipet. Before using Method B, be sure to determine that the tip of the pipet will fit into the sample holder in your melting point instrument.

Method A. In Method A, a hole is punched through a rubber septum using a large pin or a small nail, and the capillary tube is inserted from the inside, sealed end first. The septum is placed over a piece of glass tubing connected to a vacuum line. After evacuating the tube, the upper end of the tube may be sealed by heating and pulling it closed.

Method B. In Method B, the thin section of a 9-inch Pasteur pipet is used to construct the melting point tube. Carefully seal the tip of the pipet using a flame. Be

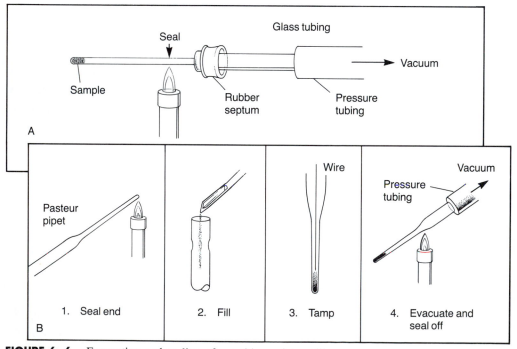

FIGURE 6–6. Evacuation and sealing of a melting point capillary

sure to hold the tip **upward** as you seal it. This will prevent water vapor from condensing inside the pipet. When the sealed pipet has cooled, the sample may be added through the open end using a microspatula. A small wire may be used to compress the sample into the closed tip. (If your melting point apparatus has a vibrator it may be used in place of the wire to simplify the packing.) When the sample is in place, the pipet is connected to the vacuum line with tubing and evacuated. The evacuated sample tube is sealed by heating it with a flame and pulling it closed.

Some substances begin to decompose **below** their melting points. Thermally unstable substances may undergo elimination reactions or anhydride formation reactions during heating. The decomposition products formed represent impurities in the original sample, so the melting point of the substance may be lowered due to their presence.

It is normal for many compounds to soften or shrink immediately before melting. Such behavior represents not decomposition but a change in the crystal structure or a mixing with impurities. Some substances "sweat," or release solvent of crystallization, before melting. These changes do not indicate the beginning of melting. Actual melting begins when the first drop of liquid becomes visible, and the melting range continues until the temperature at which all the solid has been converted to the liquid state. With experience, one soon learns to distinguish between softening, or "sweating," and actual melting. If you wish, the temperature of the onset of softening or sweating may be reported as a part of your melting point range: 211 °C (softens), 223–225 °C (melts).

Some solid substances have such a high vapor pressure that they sublime at or below their melting points. In many handbooks, the sublimation temperature is listed along with the melting point. The symbols **sub, subl,** and sometimes **s,** are used to designate a substance that sublimes. In such cases, the melting-point determination must be performed in a sealed capillary tube to avoid loss of the sample. The simplest way to accomplish sealing a packed tube is to heat the open end of the tube in a flame and pull it closed with a tweezers or forceps. A better way, although more difficult to master, is to heat the center of the tube in a small flame, rotating it about its axis, keeping the tube straight, until the center collapses. If this is not done quickly, the sample may melt or sublime while you are working. With the smaller chamber, the sample will not be able to migrate to the cool top of the tube that may be above the viewing area. Figure 6–7 illustrates the method.

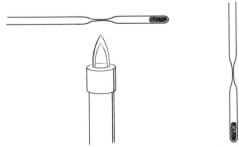

FIGURE 6–7.   Sealing a tube for a substance that sublimes

PART B. BOILING POINTS

6.9   THE BOILING POINT

As a liquid is heated, the vapor pressure of the liquid increases to the point where it just equals the applied pressure (usually atmospheric pressure). At this point the liquid will be observed to boil. The normal boiling point is measured at 760 mmHg (760 torr) or 1 atm. At a lower applied pressure, the vapor pressure needed for boiling is also lowered, and the liquid boils at a lower temperature. The relation between applied pressure and temperature of boiling for a liquid is determined by its vapor pressure-temperature behavior. Figure 6–8 is an idealization of the typical vapor pressure-temperature behavior of a liquid.

Since the boiling point is sensitive to pressure, it is important to record the barometric pressure when determining a boiling point if the determination is being conducted at an elevation significantly above or below sea level. Normal atmospheric variations may affect the boiling point, but they are usually of minor importance. However, if a boiling point is being monitored during the course of a vacuum distillation (Technique 9) that is being performed with an aspirator or a vacuum pump, the variation from the atmospheric value will be especially marked. In these cases it is quite important to know the pressure as accurately as possible.

As a rule of thumb, the boiling point of many liquids drops about 0.5 °C for a 10-mm decrease in pressure when in the vicinity of 760 mmHg. At lower pressures, a 10 °C drop in boiling point is observed for each halving of the pressure. For example, if the observed boiling point of a liquid is 150 °C at 10-mm pressure, then the boiling point would be about 140 °C at 5 mmHg.

A more accurate estimate of the change in boiling point with a change of pressure can be made by using a **nomograph.** In Figure 6–9, a nomograph is given and a method is described for using it to obtain boiling points at various pressures when the boiling point is known at some other pressure.

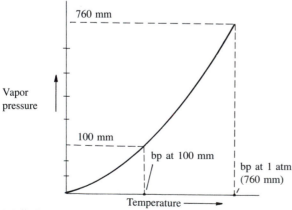

FIGURE 6–8.   The vapor pressure-temperature curve for a typical liquid

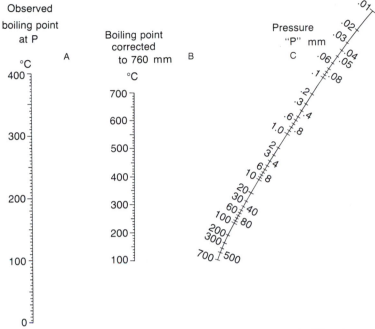

FIGURE 6–9.　Pressure–temperature alignment nomograph. **How to use the nomograph:** Assume a reported boiling point of 100 °C at 1 mm. To determine the boiling point at 18 mm, connect 100 °C (column A) to 1 mm (column C) with a transparent plastic rule and observe where this line intersects column B (about 280 °C). This value would correspond to the normal boiling point. Next, connect 280 °C (column B) with 18 mm (column C) and observe where this intersects column A (151 °C). The approximate boiling point will be 151 °C at 18 mm. Reprinted by courtesy of MC/B Manufacturing Chemists, Inc.

6.10　DETERMINING THE BOILING POINT—MICROSCALE METHODS

Two experimental methods of determining boiling points are easily available. When you have large quantities of material, you can simply record the boiling point (or boiling range) as viewed on a thermometer while performing a simple distillation (see Technique 8). With smaller amounts of material, you can carry out a microscale or a semi-microscale determination of the boiling point by using the apparatus shown in Figure 6–10.

Semi-Microscale Method.　To carry out the semi-micro determination, a piece of 5-mm glass tubing sealed at one end is attached to a thermometer with a rubber band or a thin slice of rubber tubing. The liquid whose boiling point is being determined is introduced with a Pasteur pipet into this piece of tubing and a short piece of melting point capillary (sealed at one end) is dropped in with the open end down. The whole unit is then placed in a Thiele tube. The rubber band should be placed above the level of the oil in the Thiele tube. If it is not, the band may soften in the hot oil. When

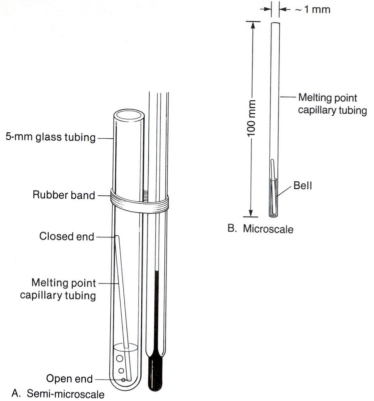

FIGURE 6—10.   Boiling point determinations

positioning the band, keep in mind that the oil will expand when heated. Next, the Thiele tube is heated in the same fashion as described in Section 6.6 for determining a melting point. Heating is continued until a rapid and continuous stream of bubbles emerges from the inverted capillary. At this point, heating is stopped. Soon the stream of bubbles slows down and stops. When the bubbles stop, the liquid enters the capillary tube. The moment at which the liquid enters the capillary tube corresponds to the boiling point of the liquid, and the temperature is recorded.

Microscale Method.   When performing microscale experiments, there is often too little product available to perform the semi-microscale method described above. However, the method described can be scaled down in the following manner. The liquid is placed in a 1-mm melting point capillary tube to a depth of about 4–6 mm. Use a syringe or a Pasteur pipet that has had its tip drawn thinner to transfer the liquid into the capillary tube. It may be necessary to use a centrifuge to transfer the liquid to the bottom of the tube. Next, prepare an appropriately-sized inverted capillary, or **bell,** in the following way. A piece of 1-mm open-end capillary tubing (same size as a melting point capillary) is rotated along its axis in a flame while being held horizontally. Use your index fingers and thumbs to rotate the tube and don't change the distance between your two hands while rotating. When the tubing is soft, it is removed from the flame and pulled to a thinner diameter. When pulling, keep the tube straight

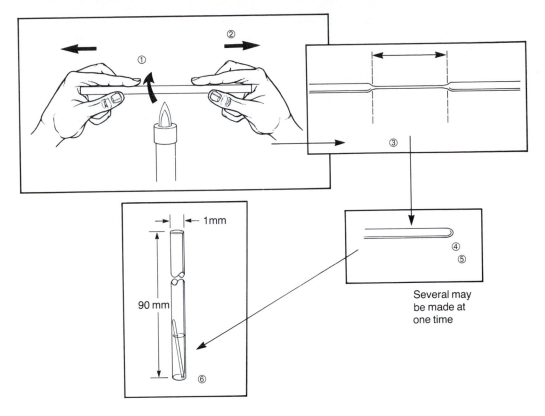

1. Rotate in flame until soft;
2. Remove from flame and pull;
3. Break pulled section out;
4. Seal one end;
5. Break to length;
6. Place bell in tube.

FIGURE 6–11.   Construction of microcapillary bell for microscale boiling point determination

by **moving both your hands and your elbows outward** by about four inches. Hold the pulled tube in place a few moments until it cools. Using the edge of a file or your fingernail, break out the thin center section. Seal one end of the thin section in the flame, then break it to a length that is about one and one-half times the height of your sample liquid (6–9 mm). Be sure the break is done squarely. Invert the bell (open end down) and place it in the capillary tube containing the sample liquid. It may be necessary to push the bell to the bottom with a fine copper wire if it adheres to the side of the capillary tube. A centrifuge may be used if your prefer. Figure 6–11 shows the construction method for the bell along with the final assembly.[1]

Place the microscale assembly in a standard melting point apparatus (or a Thiele tube if an electrical apparatus is not available) to determine the boiling point. Heating is continued until a rapid and continuous stream of bubbles emerges from the inverted capillary. At this point, heating is stopped. Soon the stream of bubbles slows down and

---

[1] A larger size of tubing (4–5 mm) than the suggested 1-mm capillary tubing may be used to construct bells. Although it is more difficult to draw the larger tubing out to form a long straight section, there is the advantage that several bells may be made at once.

stops. When the bubbles stop, the liquid enters the capillary tube. The moment at which the liquid enters the capillary tube corresponds to the boiling point of the liquid, and the temperature is recorded.

Explanation of the Method.   During the initial heating, the air trapped in the inverted bell expands and leaves the tube, giving rise to a stream of bubbles. When the liquid begins boiling, most of the air has been expelled and the bubbles of gas are due to the boiling action of the liquid. Once the heating is stopped, most of the vapor pressure left in the bell comes from the vapor of the heated liquid that seals its open end. There is always vapor in equilibrium with a heated liquid. If the temperature of the liquid is above its boiling point, the pressure of the trapped vapor will either exceed or equal the atmospheric pressure. As the liquid cools, its vapor pressure decreases. When the vapor pressure drops just below atmospheric pressure (just below the boiling point), the liquid is forced into the capillary tube.

Difficulties.   Three problems are common to this method. The first arises when the liquid is heated so strongly that it evaporates or boils away. The second arises when the liquid is not heated above its boiling point before heating is discontinued. If the heating is stopped at any point below the actual boiling point of the sample, the liquid will enter the bell **immediately** giving an apparent boiling point that is too low. Be sure that a continuous stream of bubbles, too fast for individual bubbles to be distinguished, is observed before lowering the temperature. Also be sure that the bubbling action decreases slowly before the liquid enters the bell. If your melting point apparatus has fine enough control and fast response, you can actually begin heating again and force the liquid out of the bell before it becomes completely filled with the liquid. This allows a second determination to be performed on the same sample. The third problem is that the bell may be so light that the bubbling action of the liquid causes the bell to move up the capillary tube. This problem can sometimes be solved by using a longer (heavier) bell or by sealing the bell so that a larger section of solid glass is formed at the sealed end of the bell.

When measuring temperatures above 150 °C, the thermometer errors can become significant. For an accurate boiling point with a high boiling liquid, you may wish to apply a **stem correction** to the thermometer as described in Section 6.13, or to calibrate the thermometer as described in Section 6.12

6.11 DETERMINING BOILING POINTS—OTHER METHODS

With some liquids it is difficult to obtain an accurate boiling point by using the inverted capillary methods described above. In these cases (provided enough material is available) it may be necessary to use one of the more direct methods shown in Figure 6–12. With these methods, the bulb of the thermometer can be immersed in vapor from the boiling liquid for a period of time long enough to allow it to equilibrate and give a good temperature reading. The values obtained are very reliable.

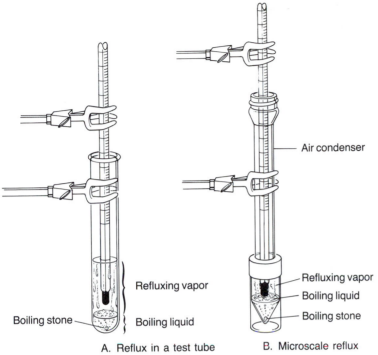

Air condenser

Refluxing vapor

Boiling stone

Boiling liquid

Refluxing vapor

Boiling liquid

Boiling stone

A. Reflux in a test tube

B. Microscale reflux

FIGURE 6–12.   Other methods of determining the boiling point

PART C. THERMOMETER CALIBRATION AND CORRECTION

6.12 THERMOMETER CALIBRATION

When a melting point or boiling point determination has been completed, one expects to obtain a result that exactly duplicates the result recorded in a handbook or in the original chemical literature. It is not infrequent, however, that there will be a discrepancy of several degrees from the literature value. Such a discrepancy does not necessarily indicate that the experiment was incorrectly performed or that the material is impure; rather it may indicate that the thermometer used for the determination was slightly in error. Most thermometers used in the laboratory do not measure the temperature with perfect accuracy.

To determine accurate values, one must calibrate the thermometer that is used. This calibration is done by determining the melting points of a variety of standard substances with the thermometer. A plot is drawn of the observed temperature versus the published value of each standard substance. A smooth line is drawn through the points to complete the chart. A correction chart prepared in this way is shown in Figure 6–13. This chart is used to correct any melting point determined with that particular thermometer. Each thermometer will require its own calibration curve. A list of suitable standard substances for calibrating thermometers is given in Table 6–1. These standards, of course, must be pure in order for the corrections to be valid.

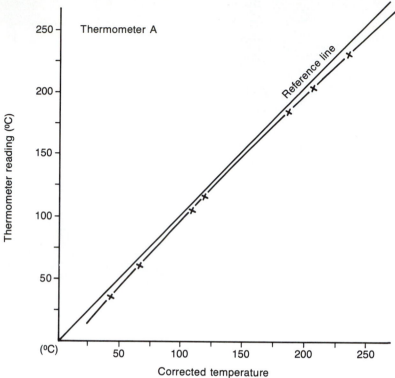

FIGURE 6–13. Thermometer calibration curve

TABLE 6–1. Melting-Point Standards

COMPOUND	MELTING POINT (°C)
Ice (solid-liquid water)	0
Acetanilide	115
Benzamide	128
Urea	132
Succinic acid	189
3,5-Dinitrobenzoic acid	205

6.13 THERMOMETER STEM CORRECTIONS

Three types of thermometers are available: bulb immersion, stem immersion (partial immersion), and total immersion. **Bulb immersion** thermometers are calibrated by the manufacturer to give the correct temperature reading when only the bulb (not the rest of the thermometer) is placed in the medium to be measured. **Stem immersion** thermometers are calibrated to give a correct temperature reading when they are immersed to a specified depth in the medium to be measured. Stem immersion thermometers are easily recognized since the manufacturer always scores a mark, or immersion ring, completely around the stem at the specified depth of immersion. The immersion ring is normally found below any of the temperature calibrations. **Total immersion** thermometers are calibrated when the entire thermometer is immersed in the medium to be measured. The three types of thermometer are often marked on the back (opposite side from the calibrations) by the words **bulb, immersion,** or **total,** but this may vary from one manufacturer to another. Since total immersion thermometers are less expensive than the other types, they are the type you are most likely to encounter in the laboratory.

Manufacturers design total immersion thermometers to read correctly only when they are immersed totally in the medium to be measured. The entire mercury thread must be covered. Since this situation is rare, a **stem correction** should be added to the observed temperature. This correction, which is positive, can be fairly large when high temperatures are being measured. Keep in mind, however, that if your thermometer has been calibrated for its desired use (such as described in Section 6.12 for a melting point apparatus), a stem correction should not be necessary for any temperature within the calibration limits. You are most likely to want a stem correction when you are performing a distillation. If you determine a melting point or boiling point using an uncalibrated, total-immersion thermometer you will also want to use a stem correction.

When you wish to make a stem correction for a total immersion thermometer, the formula given below may be used. It is based on the fact that the portion of the mercury thread in the stem is cooler than the portion immersed in the vapor or the heated area around the thermometer. The mercury will not have expanded in the cool stem to the same extent as in the warmed section of the thermometer. The equation used is

$$(0.000154)(T - t_1)(T - t_2) = \text{correction to be added to T observed.}$$

1. The factor 0.000154 is a constant, the coefficient of expansion for the mercury in the thermometer.

2. The term $(T - t_1)$ corresponds to the length of the mercury thread not immersed in the heated area. It is convenient to use the temperature scale on the thermometer itself for this measurement rather than an actual length unit. T is the observed temperature, and t_1 is the **approximate** place where the heated part of the stem ends and the cooler part begins.

3. The term $(T - t_2)$ corresponds to the difference between the temperature of the mercury in the vapor, T, and the temperature of the mercury in the air outside of the heated area (room temperature). The term T is the observed temperature, and t_2 is measured by hanging another thermometer so that the bulb is close to the stem of the main thermometer.

Figure 6–14 shows how to apply this method for a distillation. By the formula given above, it can be shown that high temperatures are more likely to require a stem correction and that low temperatures need not be corrected. The calculations given on p 614 illustrate this point.

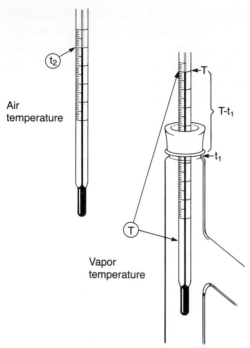

Air
temperature

Vapor
temperature

FIGURE 6—14. Measurement of a thermometer stem correction during distillation

EXAMPLE 1	EXAMPLE 2
T = 200 °C t_1 = 0 °C t_2 = 35 °C (0.000154)(200)(165) = 5.1 ° stem correction 200 °C + 5 °C = 205 °C corrected temp	T = 100 °C t_1 = 0 °C t_2 = 35 °C (0.000154)(100)(65) = 1.0 ° stem correction 100 °C + 1 °C = 101 °C corrected temp

PART D. DENSITY

6.14 DENSITY

Density is defined as mass per unit volume and is generally expressed in units of g/mL for a liquid and g/cm^3 for a solid.

$$\text{Density} = \text{mass/volume} \quad \text{or} \quad D = M/V$$

In organic chemistry, density is most commonly used in converting the weight of liquid to a corresponding volume, or vice versa. It is often easier to measure a volume of a liquid rather than to weigh it. As a physical property, density is also useful for identifying liquids in much the same way that boiling points are used.

While precise methods have been developed that allow the measurement of the densities of liquids at the microscale level, they are often difficult to perform. An approximate method for measuring densities can be found in using disposable micropipets (see Figure 6–15). These micropipets (''Microcaps'') are often used in biochemical laboratories. They are made of short lengths of precision capillary tubing which hold the indicated volume when they are completely filled end-to-end (they are TC or ''to contain'' pipets). They are filled by capillary action. It is only necessary to touch one end of them to the surface of the liquid to be measured. In those cases where there is not sufficient surface tension to fill them by capillary action, one may use the simple filling device that is supplied with them. A 50 λ or 100 λ (μL) size is convenient for this measurement. With either of these sizes, the type of top loading balance that is common in the microscale laboratory ($+/-0.001$ g) will give two- or three-figure accuracy, provided the liquid being measured does not have a low boiling point. This degree of accuracy is often quite adequate to distinguish two unknowns of different density.

To use this method, the empty micropipet should be weighed while held in a small, empty Erlenmeyer flask (10 mL). Do not touch either the Erlenmeyer flask or the micropipet with your fingers. Fingerprints will add weight. Handle the flask with a disposable tissue and the micropipet with a tweezers or forceps. Pick up the micropipet with the tweezers and touch it to the liquid to fill it. When the micropipet is filled, return it to the Erlenmeyer flask and reweigh it. The difference between the two weighings will yield the weight of the indicated volume of liquid. Table 6–2 compares some literature values with those obtained by this experimental method.

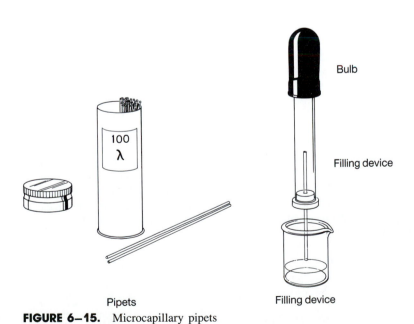

FIGURE 6–15. Microcapillary pipets

TABLE 6–2. Densities determined by the MicroCap method (g/mL)

Substance	bp	lit	50 μL	100 μL
Water	100	1.000	1.02	1.01
Hexane	69	0.660	0.65	0.66
Acetone	56	0.788	0.76	0.77
Dichloromethane	40	1.33	1.25	1.27
Diethyl ether	35	0.713	0.615	0.665

PROBLEMS

1. Two substances, A and B, have the same melting point. How can one determine if they are the same without using any form of spectroscopy? Explain in detail.

2. Using Figure 6–5, determine which heating curve would be most appropriate for a substance with a melting point of about 150 °C.

3. What steps can you take to determine the melting point of a substance that sublimes before it melts?

4. Using the temperature-pressure alignment chart in Figure 6–9, answer the following questions.
    (a) What is the normal boiling point (at 760 mmHg) for a compound that boils at 150 °C at 10 mmHg pressure?
    (b) Where would the compound in (a) boil if the pressure were 40 mmHg?
    (c) A compound was distilled at atmospheric pressure and had a boiling point of 285 °C. What would be the approximate boiling range for this compound at 15 mmHg?

5. Calculate the corrected boiling point for nitrobenzene by using the method given in Section 6.13. The boiling point was determined using an apparatus similar to that shown in Figure 6–12A. The observed boiling point was 205 °C. The reflux ring in the test tube just reached up to the 0 °C mark on the thermometer. A second thermometer suspended alongside the test tube, at a slightly higher level than the one inside, gave a reading of 35 °C.

6. Suppose you had calibrated the thermometer in your melting point apparatus against a series of melting point standards. After reading the temperature and converting it using the calibration chart, should you also apply a stem correction? Explain.

7. The density of a liquid was determined by the microcapillary method. A 100 μL microcapillary pipet was used. The liquid had a mass of 0.082 g. What was the density in g/mL of the liquid?

8. A compound melting at 134 °C was suspected to be either aspirin (mp 135 °C) or urea (mp 133 °C). Explain how you could determine whether one of these two suspected compounds was identical to the unknown compound without using any form of spectroscopy.

9. An unknown compound gave a melting point of 230 °C. When the molten liquid solidified, the melting point was redetermined and found to be 131 °C. Give a possible explanation for this discrepancy.

10. During the micro boiling point determination of an unknown liquid, heating was discontinued at 154 °C and the liquid immediately began to enter the inverted bell. Heating was begun again at once, and the liquid was forced out of the bell. Heating was again discontinued at 165 °C, at which time a very rapid stream of bubbles emerged from the bell. On cooling, the rate of bubbling gradually diminished until the liquid reached a temperature of 161 °C, and entered and filled the bell. Explain this sequence of events. What was the boiling point of the liquid?

Technique 7
EXTRACTIONS, SEPARATIONS, AND DRYING AGENTS

7.1 EXTRACTION

Transferring a solute from one solvent into another is called **extraction,** or more precisely liquid-liquid extraction. The solute is extracted from one solvent into the other because the solute is more soluble in the second solvent than in the first. The two solvents must not be miscible (mix freely), and they must form two separate phases or layers, in order for this procedure to work. Extraction is used in many ways in organic chemistry. Many natural products (organic chemicals that exist in nature) are present in animal and plant tissues having high water content. Extracting these tissues with a water-immiscible solvent is useful for isolating natural products. Often diethyl ether (commonly referred to as ''ether'') is used for this purpose. Sometimes, alternative water-immiscible solvents such as hexane, petroleum ether, ligroin, and methylene chloride are used. For instance, caffeine, a natural product, can be extracted from an aqueous tea solution by shaking it successively with several portions of methylene chloride. Water can be used to extract or ''wash'' water soluble impurities from an organic reaction mixture.

A generalized extraction process is illustrated in Figure 7–1 using a conical vial. The first solvent contains a mixture of black and white molecules (Figure 7–1A). A second solvent is added that is not miscible with the first. After capping the vial and shaking it, the layers separate. In this example, the second solvent is less dense, so it becomes the top layer (Figure 7–1B). Because of differences in physical properties, the white molecules are more soluble in the second solvent, while the black molecules are more soluble in the first solvent. Most, but not all, of the white molecules are in the upper layer but there are some black molecules there, too. Likewise, most of the black molecules are in the lower layer. However, there are a few white molecules in this lower phase. A Pasteur pipet may be used to remove the lower layer (Figure 7–1C). In this way, a partial separation of black and white molecules has been achieved. In this example, notice that it was not possible to effect a complete separation with one extraction. This is a common occurrence in organic chemistry. Many organic substances are soluble in both water and organic solvents.

7.2 DISTRIBUTION COEFFICIENT

When a solution (solute A in solvent 1) is shaken with a second solvent (solvent 2) with which it is not miscible, the solute distributes itself between the two liquid phases. When the two phases have separated again into two distinct solvent layers, an equilib-

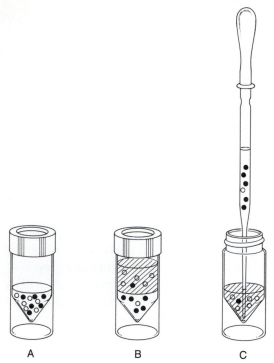

A. Solvent 1 contains a mixture of molecules (black and white).

B. After shaking with solvent 2 (shaded), most of the white molecules have been extracted into the new solvent. The white molecules are more soluble in the second solvent, while the black molecules are more soluble in the original solvent.

C. With removal of the lower phase with a Pasteur pipet, the black and white molecules have been partially separated.

A   B   C

FIGURE 7–1. The extraction process

rium will have been achieved such that the ratio of the concentrations of the solute in each layer defines a constant. The constant, called the **distribution coefficient** (or partition coefficient) K, is defined by

$$K = \frac{C_2}{C_1}$$

where C_1 and C_2 are the concentrations at equilibrium, in grams per liter or mg per mL, of solute A in solvent 1 and in solvent 2, respectively. This relationship is a ratio of two concentrations and is independent of the actual amounts of the two solvents mixed. The distribution coefficient has a constant value for each solute considered and depends on the nature of the solvents used in each case.

It is apparent that not all the solute will be transferred to solvent 2 in a single extraction unless K is very large. Usually several extractions are needed to remove all the solute from solvent 1. In extracting a solute from a solution, it is always better to use several small portions of the second solvent than to make a single extraction with a large portion. Suppose, as an illustration, a particular extraction proceeds with a distribution coefficient of 10. The system consists of 50 mg of organic compound dissolved in 1.00 mL of water (solvent 1). In this illustration, the effectiveness of three 0.50-mL extractions with ether (solvent 2) is compared with one 1.50-mL extraction with ether. In the first 0.50-mL extraction, the amount extracted into the ether layer is given by the following calculation. The amount of compound remaining in the aqueous phase is given by x.

$$K = 10 = \frac{C_2}{C_1} = \frac{\left(\dfrac{50.0 - x}{0.50} \dfrac{mg}{mL \text{ ether}}\right)}{\left(\dfrac{x}{1.00} \dfrac{mg}{mL \text{ water}}\right)}; \; 10 = \frac{(50.0 - x)(1.00)}{0.50x}$$

$$5.0x = 50.0 - x$$
$$6.0x = 50.0$$
$$x = 8.3 \text{ mg remaining in the aqueous layer}$$
$$50.0 - x = 41.7 \text{ mg in the ether layer}$$

As a check on the calculation, it is possible to substitute the value 8.3 mg for x in the original equation and demonstrate that the concentration in the ether phase divided by the concentration in the water phase equals the distribution coefficient.

$$\frac{\left(\dfrac{50.0 - x}{0.50} \dfrac{mg}{mL \text{ ether}}\right)}{\left(\dfrac{x}{1.00} \dfrac{mg}{mL \text{ water}}\right)} = \frac{\dfrac{41.7}{0.50}}{\dfrac{8.3}{1.00}} = \frac{83 \text{ mg/mL}}{8.3 \text{ mg/mL}} = 10 = K$$

The second extraction with another 0.50-mL portion of fresh ether is performed on the aqueous phase, which now contains 8.3 mg of the solute. The amount of solute extracted is given by the calculation shown in Figure 7–2. Also shown in the figure is

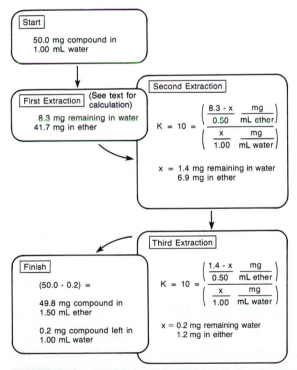

FIGURE 7–2. The result of extraction of 50.0 mg of compound in 1.00 mL of water by three successive 0.50-mL portions of ether. Compare this result with that of Figure 7–3.

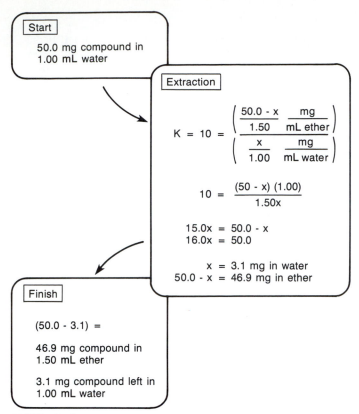

FIGURE 7–3.   The result of extraction of 50.0 mg of compound in 1.00 mL of water with one 1.50-mL portion of ether. Compare this result with that of Figure 7–2.

a calculation for a third extraction with another 0.50-mL portion of ether. This third extraction will transfer 1.2 mg of solute into the ether layer, leaving 0.2 mg of solute remaining in the water layer. A total of 49.8 mg of solute will be extracted into the combined ether layers, and 0.2 mg will remain in the aqueous phase.

Figure 7–3 shows the result of a **single** extraction with 1.50 mL of ether. As shown there, 46.9 mg of solute was extracted into the ether layer, leaving 3.1 mg of compound in the aqueous phase. One can see that three successive 0.50-mL ether extractions (Figure 7–2) succeeded in removing 2.9 mg more solute from the aqueous phase than using one 1.5-mL portion of ether (Figure 7–3). This differential represents 5.8% of the total material.

7.3   CHOOSING AN EXTRACTION METHOD AND A SOLVENT

Three different types of apparatus are used for extractions: conical vials, centrifuge tubes, and separatory funnels. These are shown in Figure 7–4. Conical vials may be used with volumes of less than 4 mL, while volumes of up to 10 mL may be handled in

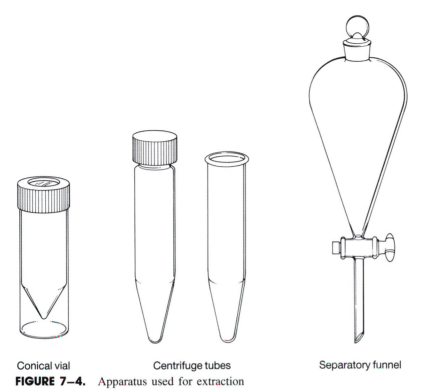

Conical vial   Centrifuge tubes   Separatory funnel
FIGURE 7—4.   Apparatus used for extraction

centrifuge tubes. The separatory funnel is used in large-scale reactions. Each of these types of equipment will be discussed in a separate section.

Most extractions consist of an aqueous phase and an organic phase. In order to extract a substance from an aqueous phase, an organic solvent that is not miscible with water must be used. Table 7—1 lists a number of the common organic solvents that are not miscible with water and are used for extraction.

Those solvents which have a density less than that of water (1.00 g/mL) will separate as the top layer when shaken with water. Those solvents which have a greater density than water will separate into the lower layer. For instance, diethyl ether (d = 0.71 g/mL) when shaken with water will form the upper layer, whereas methylene chloride (d = 1.33g/mL) will form the lower layer. When performing an extraction,

TABLE 7—1.   Densities of Common Extraction Solvents

SOLVENT	DENSITY (g/mL)
Ligroin	0.67–0.69
Diethyl ether	0.71
Toluene	0.87
Water	1.00
Methylene chloride	1.33

slightly different methods are used when you wish to separate the lower layer (whether or not it is the aqueous layer or the organic layer) than when you wish to separate the upper layer.

7.4   THE CONICAL VIAL—SEPARATING THE LOWER LAYER

The 5-mL conical vial is the most useful piece of equipment for carrying out extractions on a microscale level. In this section we will consider the method used when you wish to remove the lower layer. A concrete example would be the extraction of a desired product from an aqueous layer using methylene chloride ($d = 1.33$ g/mL) as the extraction solvent. Methods for removal of the upper layer are discussed in the next section.

> **Always place a conical vial in a small beaker to prevent the vial from falling over.**

Removing the Lower Layer.   Suppose we extract an aqueous solution with methylene chloride. This solvent is more dense than water and will settle to the bottom of the conical vial. Use the following procedure, which is illustrated in Figure 7–5, to remove the lower layer.

1. Place the aqueous phase containing the dissolved product into a 5-mL conical vial (Figure 7–5A).

2. Add about 1 mL of methylene chloride, cap the vial, and shake the mixture vigorously. Vent or unscrew the cap slightly to release the pressure in the vial. Allow the phases to separate completely so that you can detect two distinct layers in the vial. The organic phase will be the lower layer in the vial (Figure 7–5B). If necessary, tap the vial with your finger or stir the mixture gently if some of the organic phase is suspended in the aqueous layer.

3. Prepare a Pasteur filter tip pipet (Technique 4, Section 4.6, p 574) using a 5¾ inch pipet. Attach a 2-mL rubber bulb to the pipet, depress the bulb, and insert the pipet into the vial so that the tip touches the bottom (Figure 7–5C). The filter tip pipet will give you better control in removing the lower layer. In some cases, however, you may be able to use a Pasteur pipet (no filter tip) but considerably more care must be taken to avoid losing liquid from the pipet during the transfer operation. With experience, you should be able to judge how much to squeeze the bulb to draw in the desired volume of liquid.

4. Slowly draw the lower layer (methylene chloride) into the pipet in such a way that you exclude the aqueous layer and any emulsion (Section 7.9) that might be at the interface between the layers (Figure 7–5D). Be sure to keep the tip of the pipet squarely in the V at the bottom of the vial.

5. Transfer the withdrawn organic phase into a **dry** test tube or another **dry** conical vial if one is available. It is best to have the test tube or vial located next to the extraction vial. (If you wish, hold them both in the same hand between your index finger and thumb.) This avoids messy and disastrous transfers. The aqueous layer (upper layer) is left in the original conical vial (Figure 7–5E).

In performing an actual extraction in the laboratory, you would extract the aqueous phase with a second 1-mL portion of fresh methylene chloride to achieve a more complete extraction. Steps 2–5 above would be repeated, and the organic layers from both extractions would be combined. In some cases, you may need to extract a third time with yet another 1-mL portion of methylene chloride. Again, the methylene chloride would be combined with the other extracts. The overall process would use three 1-mL portions of methylene chloride to transfer the product from the water layer into methylene chloride. Sometimes you will see the statement, ''extract the aqueous phase with three 1-mL portions of methylene chloride,'' in an experimental procedure. This statement describes in a shorter fashion the process described above. Finally, the methylene chloride extracts will contain some water, and they need to be dried with a drying agent as indicated in Section 7.8.

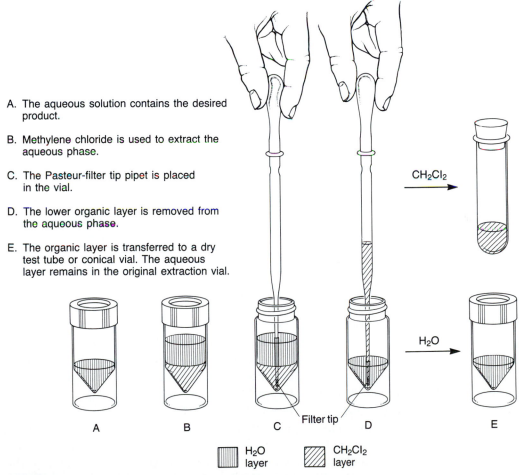

A. The aqueous solution contains the desired product.

B. Methylene chloride is used to extract the aqueous phase.

C. The Pasteur-filter tip pipet is placed in the vial.

D. The lower organic layer is removed from the aqueous phase.

E. The organic layer is transferred to a dry test tube or conical vial. The aqueous layer remains in the original extraction vial.

CH_2Cl_2

H_2O

Filter tip

A   B   C   D   E

H_2O layer CH_2Cl_2 layer

FIGURE 7–5. Extraction of an aqueous solution using a solvent more dense than water: Methylene chloride

In this example we extracted water with the heavy solvent methylene chloride and removed it as the lower layer. If you were extracting a light solvent (for instance, diethyl ether) with water, and you wished to keep the water layer, the water would be the lower layer and would be removed using the same procedure. You would not dry the water layer, however.

7.5 THE CONICAL VIAL—SEPARATING THE UPPER LAYER

In this section we will consider the method used when you wish to remove the upper layer. Two different methods are possible. We will consider both methods. A concrete example would be the extraction of a desired product from an aqueous layer using diethyl ether (d = 0.71 g/mL) as the extraction solvent. Methods for removal of the lower layer were discussed above.

> **Always place a conical vial in a small beaker to prevent the vial from falling over.**

Removing the Upper Layer. Suppose we extract an aqueous solution with diethyl ether (ether). This solvent is less dense than water and will rise to the top of the conical vial. Use one of the following procedures, which are illustrated in Figure 7–6 and Figure 7–7, to remove the upper layer.

Method A. In this method, only one layer is removed from the conical vial. In Method B, both layers are removed.

1. Place the aqueous phase containing the dissolved product in a 5-mL conical vial (Figure 7–6A).

2. Add about 1 mL of ether, cap the vial, and shake the mixture vigorously. Vent or unscrew the cap slightly to release the pressure in the vial. Allow the phases to separate completely so that you can detect two distinct layers in the vial. The ether phase will be the upper layer in the vial (Figure 7–6B).

3. Prepare a Pasteur filter tip pipet (Technique 4, Section 4.6, p 574) using a 5¾ inch pipet. Attach a 2-mL rubber bulb to the pipet, depress the bulb, and insert the pipet into the vial so that the tip touches the bottom. The filter tip pipet will give you better control in removing the lower layer. In some cases, however, you may be able to use a Pasteur pipet (no filter tip) but considerably more care must be taken to avoid losing liquid from the pipet during the transfer operation. With experience, you should be able to judge how much to squeeze the bulb to draw in the desired volume of liquid. Slowly draw the lower **aqueous** layer into the pipet. Be sure to keep the tip of the pipet squarely in the V at the bottom of the vial (Figure 7–6C).

4. Transfer the withdrawn aqueous phase into a test tube or another conical vial for temporary storage. It is best to have the test tube or vial located next to the extraction vial. This avoids messy and disastrous transfers. (If you wish, hold them both in the same hand between your index finger and thumb.) The ether layer is left behind in the vial. (Figure 7–6D)

5. The ether phase remaining in the origin conical vial should be transferred with a Pasteur pipet into a test tube for storage and the aqueous phase returned to the original conical vial (Figure 7–6E).

In performing an actual extraction, you would extract the aqueous phase with another 1-mL portion of fresh ether to achieve a more complete extraction. Steps 2–5 above would be repeated, and the organic layers from both extractions would be combined in the test tube. In some cases, you may need to extract the aqueous layer a third time with yet another 1-mL portion of ether. Again, the ether would be combined with the other two layers. This overall process uses three 1-mL portions of ether to transfer the product from the water layer into ether. The ether extracts contain some water and must be dried with a drying agent as indicated in Section 7.8.

Method B. In this method, shown in Figure 7–7, **both** layers are removed from the conical vial with a Pasteur filter tip pipet (Technique 4, Section 4.6, p 574).

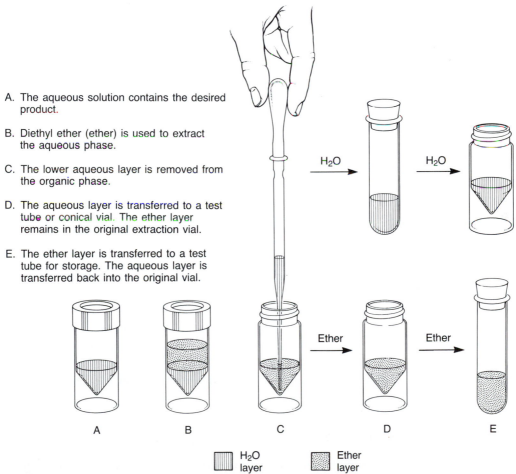

A. The aqueous solution contains the desired product.

B. Diethyl ether (ether) is used to extract the aqueous phase.

C. The lower aqueous layer is removed from the organic phase.

D. The aqueous layer is transferred to a test tube or conical vial. The ether layer remains in the original extraction vial.

E. The ether layer is transferred to a test tube for storage. The aqueous layer is transferred back into the original vial.

H₂O

H₂O

Ether

Ether

A   B   C   D   E

⬚ H₂O layer   ⬚ Ether layer

FIGURE 7–6. Extraction of an aqueous solution using a solvent less dense than water: Diethyl ether (Method A)

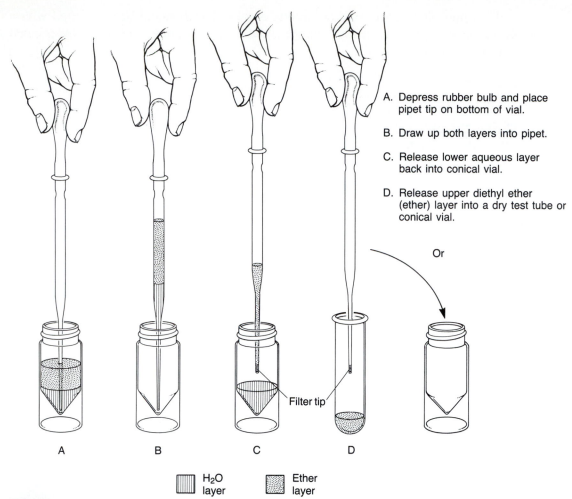

A. Depress rubber bulb and place pipet tip on bottom of vial.

B. Draw up both layers into pipet.

C. Release lower aqueous layer back into conical vial.

D. Release upper diethyl ether (ether) layer into a dry test tube or conical vial.

Or

Filter tip

A   B   C   D

| H_2O layer | Ether layer |

FIGURE 7–7. Extraction of an aqueous solution using a solvent less dense than water: Diethyl ether (Method B)

This method is best applied to a separation when the total volume of both the aqueous and ether phase is less than 2 mL, and when phases are **clearly** separated. After both phases have been drawn up into a pipet, the lower aqueous layer is returned to the conical vial. The upper ether phase is transferred to a **dry** test tube or another **dry** conical vial. To carry out this procedure successfully, avoid drawing air into the pipet so as to prevent mixing of the layers. If this procedure can be used, considerable time can be saved in separating layers over the procedure described in Method A.

7.6   THE CENTRIFUGE TUBE

A centrifuge tube may be employed instead of a conical vial for separations (Figure 7–4B). You should use the same extraction and separation procedures described in Sections 7.4 and 7.5. It is most convenient to use a centrifuge tube that is equipped with a screw cap. One may also use a "regular" centrifuge tube, but it will be necessary to cork the tube when shaking it. Alternatively, a vortex mixer may be used to mix the phases without the need of shaking the tube. A vortex mixer works well with a variety of containers, including small flasks, test tubes, conical vials, and centrifuge tubes. If an emulsion has formed after mixing or shaking, you can use a centrifuge to aid in the separation of the layers (Section 7.9). Once the layers have separated, it is easy to use a Pasteur pipet to withdraw the lower layer from the tapered bottom of the centrifuge tube.

7.7 THE SEPARATORY FUNNEL

The separatory funnel is widely used in large scale reactions. This apparatus is illustrated in Figure 7–8. To fill the separatory funnel, one usually supports it in an iron ring attached to a ring stand. Pieces of rubber tubing should be cut and attached to the iron

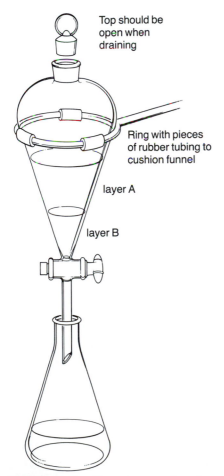

Top should be open when draining

Ring with pieces of rubber tubing to cushion funnel

layer A

layer B

FIGURE 7–8. The separatory funnel

ring to cushion the separatory funnel as shown in the figure. This protects the funnel against possible breakage.

When beginning an extraction, the first step is to close the stopcock. (Don't forget!) Both the solution and the extraction solvent are poured into the funnel. The separatory funnel is swirled gently by holding it by its upper neck, and then it is stoppered. The separatory funnel is picked up with two hands and held as shown in Figure 7–9. It is essential to hold the stopper in place firmly because the two immiscible solvents build up pressure when they mix, and this pressure may force the stopper out of the separatory funnel. To release this pressure, the funnel is vented by holding it upside-down (hold the stopper securely) and slowly opening the stopcock. Usually the rush of vapors out of the opening can be heard. Shaking and frequent venting should be continued until the ''whoosh'' is no longer audible. The funnel is then placed in the iron ring and the top stopper is removed immediately. The two immiscible solvents separate into two layers after a short time, and they can be separated from one another by draining most of the lower layer through the stopcock. A few minutes are allowed to pass so that any of the lower phase adhering to the inner glass surfaces of the separatory funnel can drain down. The stopcock is again opened and the remainder of the lower layer is allowed to drain until the interface between the upper and lower phases just begins to enter the bore of the stopcock. At this moment, the stopcock is closed. The remaining upper layer is removed by pouring it from the top opening of the separatory funnel.

When methylene chloride is used as the extracting solvent with an aqueous phase, it will settle to the bottom and will be removed through the stopcock. The

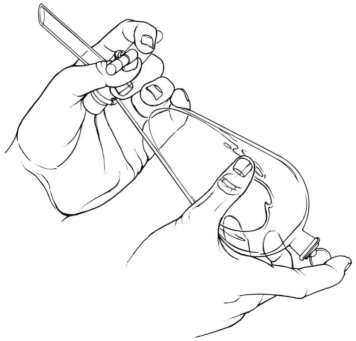

FIGURE 7–9.   Correct way of shaking and venting the separatory funnel

aqueous layer remains in the funnel. A second extraction of the remaining aqueous layer with fresh methylene chloride may be needed.

With a diethyl ether (ether) extraction of an aqueous phase, the organic layer will form on top. The lower aqueous layer is removed through the stopcock and the upper ether layer is poured from the top of the separatory funnel. The aqueous phase may be poured back into the separatory funnel and extracted a second time with fresh ether. The combined organic phases must be dried using a suitable drying agent (Section 7.8) before the solvent is removed.

7.8 DRYING AGENTS

After an organic solvent has been shaken with an aqueous solution, it will be ''wet''; that is, it will have dissolved some water even though its miscibility with water is not great. The amount of water dissolved varies from solvent to solvent; diethyl ether represents a solvent in which a fairly large amount of water dissolves. To remove water from the organic layer, a **drying agent** is used. A drying agent is an **anhydrous** inorganic salt that acquires waters of hydration when exposed to moist air or a wet solution.

The following **anhydrous** salts are commonly used in microscale and large scale experiments: sodium sulfate, magnesium sulfate, calcium chloride, calcium sulfate (Drierite), and potassium carbonate. These salts vary in their properties and applications. For instance, not all will absorb the same amount of water for a given weight, nor will they dry the solution to the same extent. **Capacity** is a term used to refer to the amount of water a drying agent absorbs per unit weight. Sodium and magnesium sulfates absorb a large amount of water (high capacity), but magnesium sulfate dries a solution more completely. **Completeness** refers to a compound's effectiveness in removing all the water from a solution by the time equilibrium has been reached. Magnesium ion, a strong Lewis acid, has the disadvantage since it sometimes causes rearrangements of compounds such as epoxides. Calcium chloride is a good drying agent but cannot be used with many compounds containing oxygen or nitrogen since it forms complexes. Calcium chloride absorbs methanol and ethanol in addition to water, so it is useful for removing these materials when they are present as impurities along with water. Potassium carbonate is a base and is used for drying solutions of basic substances, such as amines. Calcium sulfate dries a solution completely but has a low total capacity to absorb water.

Anhydrous sodium sulfate is the most widely used drying agent. The granular variety is recommended because it is easier to remove the dried solution from it than from the powdered variety. Sodium sulfate is mild and effective. It will remove water from most common solvents, with the possible exception of diethyl ether where a prior drying with saturated salt solution may be advised (see p 630). Sodium sulfate must be used at room temperature to be effective; it cannot be used with boiling solutions. Table 7–2 compares the various common drying agents.

TABLE 7–2. Common Drying Agents

	ACIDITY	HYDRATED	CAPACITY*	COMPLETE-NESS†	RATE‡	USE
Magnesium sulfate	Neutral	$MgSO_4 \cdot 7H_2O$	High	Medium	Rapid	General
Sodium sulfate	Neutral	$Na_2SO_4 \cdot 7H_2O$ $Na_2SO_4 \cdot 10H_2O$	High	Low	Medium	General
Calcium chloride	Neutral	$CaCl_2 \cdot 2H_2O$ $CaCl_2 \cdot 6H_2O$	Low	High	Rapid	Hydro-carbons Halides
Calcium sulfate (Drierite)	Neutral	$CaSO_4 \cdot \frac{1}{2}H_2O$ $CaSO_4 \cdot 2H_2O$	Low	High	Rapid	General
Potassium carbonate	Basic	$K_2CO_3 \cdot 1\frac{1}{2}H_2O$ $K_2CO_3 \cdot 2H_2O$	Medium	Medium	Medium	Amines, esters bases, ketones
Potassium hydroxide	Basic	···	···	···	Rapid	Amines only
Molecular sieves (3 Å or 4 Å)	Neutral	···	High	Extremely high	···	General

*Amount of water removed per given weight of drying agent.
†Refers to amount of H_2O still in solution at equilibrium with drying agent.
‡Refers to rate of action (drying).

Microscale Reactions. Add one spatulaful of granular anhydrous sodium sulfate (or other drying agent) from the V-grooved end of a microspatula into a solution contained in a conical vial or test tube. If all of the drying agent "clumps," add another spatulaful of sodium sulfate. The solution should be allowed to dry for at least 15 minutes. Stir the mixture occasionally with a spatula during that period. The mixture will be dry if there are no visible signs of water, and the drying agent flows freely in the container when stirred with a microspatula. The solution should not be cloudy. Add more drying agent if necessary. You should not add more drying agent if a "puddle" (water layer) forms or if drops of water are visible. Instead, you will need to transfer the organic layer to a dry container before adding fresh drying agent. When dry, use a **dry** Pasteur pipet or a **dry** filter tip pipet (Technique 4, Section 4.6, p 574) to remove the solution from the drying agent and transfer the solution to a **dry** conical vial. Rinse the drying agent with a small amount of fresh solvent and transfer this solvent to the vial containing the solution. The solvent is removed by evaporation using heat and a stream of air or nitrogen (Technique 3, Section 3.9, p 560).

An alternative method of drying an organic phase is to pass it through a filtering pipet (Technique 4, Section 4.1C, p 568) that has been packed with a small amount (ca. 2 cm) of drying agent. Again, the solvent is removed by evaporation.

Large Scale Reactions. To dry a large amount of solution (about 10–20 mL) you should add enough granular anhydrous sodium sulfate to give a 1–3-mm layer on the bottom of the flask, depending on the volume of the solution. Dry the solution for at least 15 minutes. The mixture will be dry if there are no visible signs of water and the drying agent flows freely in the container when stirred or swirled. The solution should not be cloudy. Add more drying agent if necessary. You should not add more drying agent if a ''puddle'' (water layer) forms. Instead, you will need to transfer the organic layer to a dry container before adding fresh drying agent. When dry, the drying agent should be removed by using decantation or it should be transferred with a Pasteur pipet. With large volumes, you can use gravity filtration (see Technique 4, Section 4.1B, p 566) to remove the drying agent. The solvent is removed by distillation (Technique 8, Section 8.4, p 648) or evaporation (Technique 3, Section 3.9, p 560).

      Saturated Salt Solution. At room temperature, diethyl ether (ether) dissolves 1.5% by weight of water, and water dissolves 7.5% of ether. Ether, however, dissolves a much smaller amount of water from a saturated aqueous sodium chloride solution. Hence, the bulk of water in ether, or ether in water, can be removed by shaking it with a saturated aqueous sodium chloride solution. A solution of high ionic strength is usually not compatible with an organic solvent and forces separation of it from the aqueous layer. The ether phase (organic layer) will be on top and the saturated sodium chloride solution will be on the bottom (density 1.2 g/mL). After removing the organic phase from the aqueous sodium chloride, dry the organic layer completely with sodium sulfate or with one of the other drying agents listed in Table 7–2.

7.9 EMULSIONS

An **emulsion** is a colloidal suspension of one liquid in another. Minute droplets of an organic solvent often are held in suspension in an aqueous solution when the two are mixed or shaken vigorously; these droplets form an emulsion. This is especially true if any gummy or viscous material was present in the solution. Emulsions are encountered often in performing extractions. Emulsions may require a long time to separate into two layers and are a nuisance to the organic chemist.

      Fortunately, several techniques may be used successfully for breaking a difficult emulsion once it has formed.

1. Often an emulsion will break up if it is allowed to stand for a period of time. Patience is important here. Gentle stirring with a stirring rod or spatula may also be useful.
2. If one of the solvents is water, adding a saturated aqueous sodium chloride solution will help destroy the emulsion. This makes the aqueous and organic layers less compatible, thereby forcing separation.
3. With microscale experiments, the mixture may be transferred to a centrifuge tube. The emulsion will often break during centrifugation. Remember to place another tube filled with water on the opposite side of the centrifuge to balance it.
4. Adding a very small amount of a water-soluble detergent may also help. This method has been used in the past for combating oil spills. The detergent helps to solubilize the tightly-bound oil droplets.
5. Gravity filtration (see Technique 4, Section 4.1, p 565) may help to destroy an emulsion by removing gummy polymeric substances. With large scale reactions, you might try filtering the mixture through a fluted filter (Technique 4, Section 4.1B, p 566) or a piece of cotton. With small scale reactions, a filtering pipet may work

(Technique 4, Section 4.1C, p 568). In many cases, once the gum is removed, the emulsion breaks up rapidly.

6. If you are using a separatory funnel, you might try to use a gentle swirling action in the funnel to help break an emulsion. Gentle stirring with a stirring rod may also be useful.

When it is known, through prior experience, that some mixtures may form difficult emulsions, you should avoid vigorous shaking. When using conical vials for extractions, it may be better to use a magnetic spin vane for mixing and not shake the mixture at all. When using separatory funnels, extractions should be performed with gentle swirling instead of shaking or with several gentle inversions of the separatory funnel. The separatory funnel must not be shaken vigorously in these cases. It is important to stress that you must use a longer extraction period if the more gentle techniques described in this paragraph are being employed. Otherwise, you will not transfer all of the material from the first phase to the second one.

7.10 PURIFICATION AND SEPARATION METHODS

In nearly all the synthetic experiments undertaken in this textbook, a series of operations involving extractions are used after the actual reaction has been concluded. These extractions form an important part of the purification. Using them, the desired product is separated from unreacted starting materials or from undesired side products in the reaction mixture. These extractions may be grouped into three categories, depending on the nature of the impurity they are designed to remove.

The first category involves extracting or ''washing'' an organic mixture with water. Water washes are designed to remove highly polar materials such as inorganic salts, strong acids or bases, and low-molecular-weight, polar substances including alcohols, carboxylic acids, and amines. Many organic compounds containing fewer than five carbons are water-soluble. Water extractions are also used immediately following extractions of a mixture with either acid or base to ensure that all traces of acid or base have been removed.

The second category concerns extraction of an organic mixture with a dilute acid, usually 5% or 10% hydrochloric acid. Acid extractions are intended to remove basic impurities, especially such basic impurities as organic amines. The bases are converted to their corresponding cationic salts by the acid used in the extraction. If an amine is one of the reactants, or if pyridine or another amine is a solvent, such an extraction might be used to remove any excess amine present at the end of a reaction.

$$RNH_2 + HCl \longrightarrow RNH_3^+ Cl^-$$
(Water-soluble salt)

Cationic salts are usually soluble in the aqueous solution, and they are thus extracted from the organic material. A water extraction may be used immediately following the acid extraction to ensure that all traces of the acid have been removed from the organic material.

The third category is extraction of an organic mixture with a dilute base, usually 5% sodium bicarbonate, although extractions with dilute sodium hydroxide can also be used. Such basic extractions are intended to convert acidic impurities, such as organic acids, to their corresponding anionic salts. For example, in the preparation of an ester, a sodium bicarbonate extraction might be used to remove any excess carboxylic acid that is present.

$$\text{RCOOH} + \text{NaHCO}_3 \longrightarrow \ \text{RCOO}^-\text{Na}^+ \ + \text{H}_2\text{O} + \text{CO}_2$$

$\text{(pK}_a \sim 5)$ **(Water-soluble salt)**

Anionic salts, being highly polar, are soluble in the aqueous phase. As a result, these acidic impurities are extracted from the organic material into the basic solution. A water extraction may be used after the basic extraction to ensure that all the base has been removed from the organic material.

Occasionally, phenols may be present in a reaction mixture as impurities, and removing them by extraction may be desired. Because phenols, although they are acidic, are about 10^5 times less acidic than carboxylic acids, basic extractions may be used to separate phenols from carboxylic acids by a careful selection of the base. If sodium bicarbonate is used as a base, carboxylic acids are extracted into the aqueous base, but phenols are not. Phenols are not sufficiently acidic to be deprotonated by the weak base, bicarbonate. Extraction with sodium hydroxide, on the other hand, extracts both carboxylic acids and phenols into the aqueous basic solution, since hydroxide ion is a sufficiently strong base to deprotonate phenols.

$\text{(pK}_a \sim 10)$ **(Water-soluble salt)**

It would be a useful exercise for you to examine the experimental instructions for some of the preparative experiments given in the textbook. While you are examining these procedures, you should try to identify which impurities are being removed at each extraction step. Mixtures of acidic, basic, and neutral compounds are easily separated by extraction techniques. One such example is shown in Figure 7–10.

Materials that have been extracted can be regenerated by neutralizing the extraction reagent. If an acidic material has been extracted with aqueous base, the material can be regenerated by acidifying the extract until the solution becomes acidic to blue litmus. The material will separate from the acidified solution. A basic material can be recovered from an acidic extract by adding base to the extract. These substances can then be removed from the neturalized aqueous solutions by extraction with an organic solvent such as ether. After the ether phase is dried with a drying agent, evaporation of the ether yields the isolated compounds.

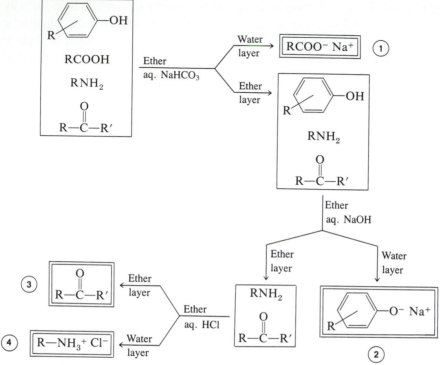

FIGURE 7–10. Separating a four-component mixture by extraction

7.11 CONTINUOUS SOLID-LIQUID EXTRACTION

The technique of liquid-liquid extraction was described in Sections 7.1–7.7. In this section, solid-liquid extraction will be described. Solid-liquid extraction is often used to extract a solid natural product from a natural source, such as a plant. A solvent is chosen that will selectively dissolve the desired compound, but that will leave behind the undesired insoluble solid. A continuous solid-liquid extraction apparatus, called a Soxhlet extractor, is commonly used in a research laboratory (Figure 7–11).

As shown in the figure, the solid to be extracted is placed in a thimble made from filter paper, and the thimble is inserted into the central chamber. A low boiling solvent such as diethyl ether is placed in the round-bottomed distilling flask, and the solvent is heated to reflux. The vapor rises through the left sidearm into the condenser where it liquifies. The condensate (liquid) drips into the thimble containing the solid. The hot solvent begins to fill the thimble and extracts the desired compound from the solid. Once the thimble is filled with solvent, the sidearm on the right acts as a siphon, and the solvent, which now contains the dissolved compound, drains back into the distillation flask. The vaporization, condensation, extraction, siphoning process is repeated hundreds of time, and the desired product is concentrated in the distilling flask. The product is concentrated in the flask because it has a boiling point higher than that of the solvent or because it is a solid.

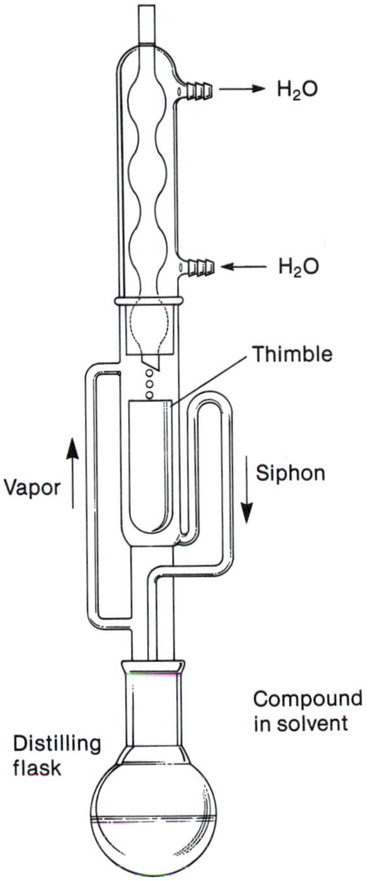

FIGURE 7–11. Continuous solid-liquid extraction using a Soxhlet extractor

7.12 CONTINUOUS LIQUID-LIQUID EXTRACTION

When a product is very soluble in water, it is often difficult to extract it using the techniques given in Sections 7.4–7.7 because of an unfavorable distribution coefficient. In this case, you would need to extract the aqueous solution numerous times with fresh batches of an immiscible organic solvent to remove the desired product from water. A less labor-intensive technique involves the use of a continuous liquid-liquid extraction apparatus. One type of extractor, used with solvents that are less dense than water, is shown in Figure 7–12. Diethyl ether is the typical solvent of choice.

The aqueous phase is placed in the extractor which is then filled with diethyl ether up to the sidearm. The round-bottom distilling flask is partially filled with ether. The ether is heated to reflux in the round-bottom flask, and the vapor is liquified in the water cooled condenser. The ether drips into the central tube, passes through the porous sintered glass tip, and flows through the aqueous layer. The solvent extracts the desired compound from the aqueous phase and the ether is recycled back into the round-bottom flask. The product is concentrated in the flask. The extraction is rather

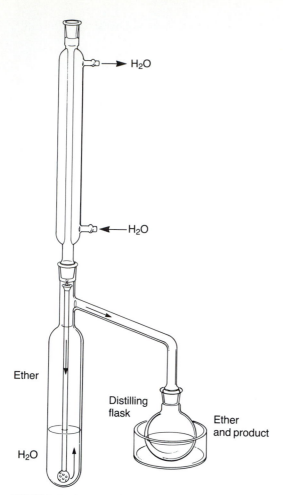

FIGURE 7–12. Continuous liquid-liquid extraction using a solvent less dense than water

inefficient and must be placed in operation for a period of at least 24 hours to remove the compound from the aqueous phase.

PROBLEMS

1. Suppose solute A has a distribution coefficient of 1.0 between water and diethyl ether. Demonstrate that if 1.0 mL of a solution of 50 mg of A in water were extracted with two 0.25-mL portions of ether, a smaller amount of A would remain in the water than if the solution were extracted with one 0.50-mL portion of ether.

2. Write an equation to show how you could recover the parent compounds from their respective salts (1, 2, and 4) shown in Figure 7–10.

3. Aqueous hydrochloric acid was used **after** the sodium bicarbonate and sodium hydroxide extractions in the separation scheme shown in Figure 7–10. Is it possible to use this reagent earlier in the separation scheme so as to achieve the same overall result? If so, explain where you would do this extraction.

4. Using aqueous hydrochloric acid, sodium bicarbonate or sodium hydroxide solutions, devise a separation scheme using the style shown in Figure 7–10 to separate the following two-component mixtures.

All of the substances are soluble in ether. Also indicate how you would recover each of the compounds from their respective salts.

(a) Give two different methods for separating this mixture.

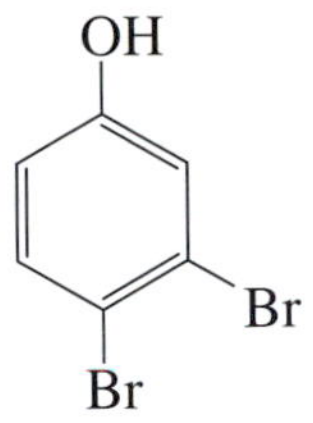

$(CH_3CH_2CH_2CH_2)_3N$

(b) Give two different methods for separating this mixture.

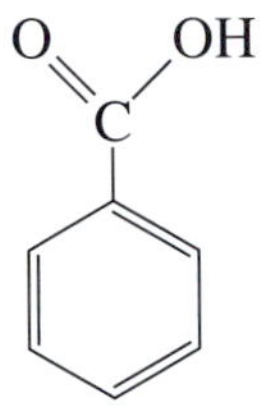

$CH_3CH_2CH_2CH_2CH_2CH_2OH$

(c) Give one method for separating this mixture.

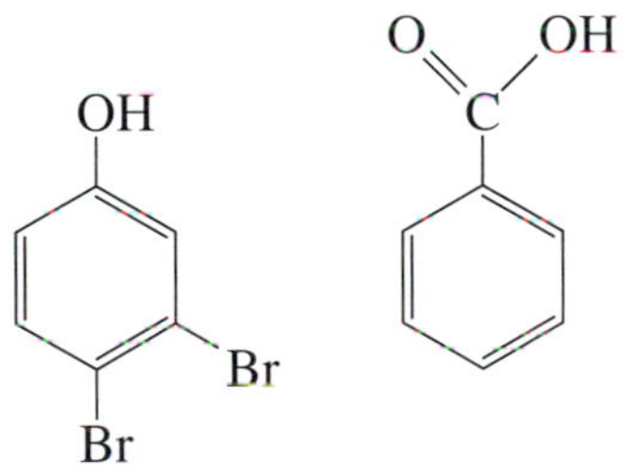

5. Solvents other than those shown in Table 7–1 may be used for extractions. Determine the relative positions of the organic layer and the aqueous layer in a conical vial or separatory funnel after shaking each of the following solvents with an aqueous phase. You will need to find the densities for each of these solvents in a handbook (Technique 19, Section 19.1, p 787).

(a) 1,1,1-Trichloroethane

(b) Hexane

6. A student prepares ethyl benzoate by the reaction of benzoic acid with ethanol using a sulfuric acid catalyst. The following compounds are found in the crude reaction mixture: ethyl benzoate (major component), benzoic acid, ethanol, and sulfuric acid. Using a handbook, obtain the solubility properties in water for each of these compounds (Technique 19, Section 19.1, p 787). Indicate how you would remove benzoic acid, ethanol, and sulfuric acid from ethyl benzoate. At some point in the purification, you will also need to use an aqueous sodium bicarbonate solution.

7. Calculate the weight of water that could be removed from a wet organic phase using 50.0 mg of magnesium sulfate. Assume that it gives the hydrate listed in Table 7–2.

Technique 8
SIMPLE DISTILLATION

Distillation is the process of vaporizing a liquid, condensing the vapor, and collecting the condensate in another container. This technique is very useful for separating a liquid mixture when the components have different boiling points, or when one of the components will not distill. It is one of the principal methods of purifying a liquid. Four basic distillation methods are available to the chemist: simple distillation, vacuum distillation (distillation at reduced pressure), fractional distillation, and steam distillation. This technique chapter discusses simple distillation. Vacuum distillation is discussed in Technique 9. Fractional distillation is discussed in Technique 10, and steam distillation is discussed in Technique 11.

8.1 INTRODUCTION

There are probably more types and styles of distillation apparatus known than exist for any other technique in chemistry. Over the centuries, chemists have devised just about every design that is conceivable. The earliest known types of distillation apparatus were the **alembic** and the **retort** (Figure 8–1). They were used by alchemists in the Middle Ages and the Renaissance, and probably even earlier by Arabic chemists. Most other distillation equipment has evolved as variations on these designs.

Figure 8–1 shows several stages in the evolution of distillation equipment as it relates to the organic laboratory. It is not intended to be a complete history; rather, it is representative. Up until recent years, equipment based on the retort design was common in the laboratory. Although the retort itself was still in use early in this century, it had evolved by that time into the distilling flask and water-cooled condenser combination which was taking its place. This early equipment was connected together with corks. By 1958, most introductory laboratories were beginning to use "organic lab kits" that included glassware connected by standard-taper glass joints. The original lab kits contained large ͲͲ 24/40 joints. Within a short time they became smaller with ͲͲ 19/22 and even ͲͲ 14/20 joints. These later kits are still being used today in many organic courses. Small scale variations of these kits are also used today by chemical researchers, but they are too expensive to use in an introductory lab. Instead, the "microscale" equipment you are using in this course is coming into common use. This equipment has ͲͲ 14/10 or ͲͲ 7/10 standard-taper joints, threaded outer joints with screwcap connectors, and its uses an internal O-ring. The distillation apparatus in microscale kits is designed for work with small amounts of material, and it is different from the more traditional larger scale equipment. It is perhaps more closely related to the alembic design than that of the retort. Because both types of equipment are in use today, after we describe microscale equipment we will also show the equivalent large scale apparatus used to perform distillation.

638

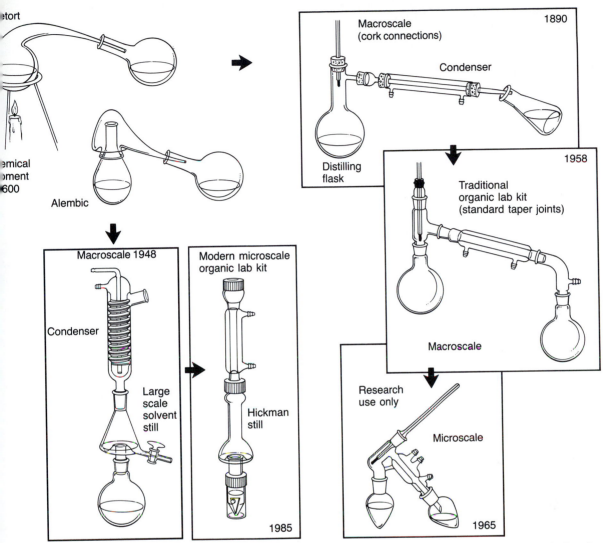

FIGURE 8–1. Some stages in the evolution of microscale distillation equipment from alchemical equipment (dates represent time of approximate popular use)

8.2 DISTILLATION THEORY

In the traditional distillation of a pure substance, vapor rises from the distilling flask and comes in contact with a thermometer that records its temperature. The vapor then passes through a condenser, which reliquifies the vapor and passes it into the receiving flask. The temperature observed during the distillation of a **pure substance** will remain constant throughout the distillation so long as both vapor **and** liquid are present in the system (see Figure 8–2A). When a **liquid mixture** is distilled, often the temperature does not remain constant but increases throughout the distillation. The reason for this is that the composition of the vapor that is distilling varies continuously during the distillation (see Figure 8–2B).

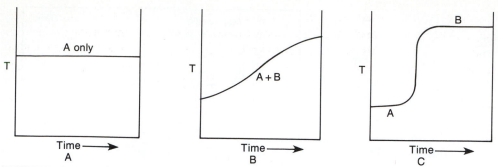

FIGURE 8–2. Three types of temperature behavior during a simple distillation. A. A single pure component. B. Two component of similar boiling points. C. Two components with widely-differing boiling points. Good separations are achieved in A and C.

For a liquid mixture, the composition of the vapor in equilibrium with the heated solution is different from the composition of the solution itself. This is shown in Figure 8–3, which is a phase diagram of the typical vapor-liquid relation for a two-component system (A + B).

On this diagram, horizontal lines represent constant temperatures. The upper curve represents vapor composition, and the lower curve represents liquid composition. For any horizontal line (constant temperature), like that shown at t, the intersections of the line with the curves give the compositions of the liquid and the vapor that are in equilibrium with each other at that temperature. In the diagram, at temperature t, the intersection of the curve at x indicates that liquid of composition w will be in equilibrium with vapor of composition z, which corresponds to the intersection at y. Composition is given as a mole percentage of A and B in the mixture. Pure A, which boils at temperature t_A, is represented at the left. Pure B, which boils at temperature t_B, is represented on the right. For either pure A or pure B, the vapor and liquid curves meet at the boiling point. Thus, either pure A or pure B will distill at a constant temperature (t_A or t_B). Both the vapor and the liquid must have the same composition in either of these cases. This is not the case for mixtures of A and B.

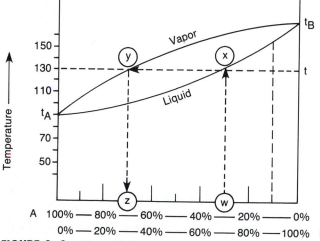

FIGURE 8–3. Phase diagram for a typical liquid mixture of two components

A mixture of A and B of composition w will have the following behavior when heated. The temperature of the liquid mixture will increase until the boiling point of the mixture is reached. This corresponds to following line wx from w to x, the boiling point of the mixture (t). At temperature t the liquid begins to vaporize, which corresponds to line xy. The vapor will have the composition corresponding to z. In other words, the first vapor obtained in distilling a mixture of A and B does not consist of pure A. It is richer in A than the original mixture, but still contains a significant amount of the higher boiling component B, **even from the very beginning of the distillation.** The result of this is that it is never possible to separate a mixture completely by a simple distillation. However, in two cases, it is possible to get an acceptable separation into relatively pure components. In the first case, if the boiling points of A and B differ by a large amount (>100 °), and if the distillation is carried out carefully, it will be possible to get a fair separation of A and B. In the second case, if A contains a fairly small amount of B (<10%), a reasonable separation of A from B can be achieved. When the boiling-point differences are not large, and when highly pure components are desired, it is necessary to do a **fractional distillation.** Fractional distillation is described in Technique 10, where the behavior during a simple distillation is also considered in detail. Note only that as vapor distills from the mixture of composition w (Figure 8–3), it is richer in A than is the solution. Thus, the composition of the material left behind in the distillation becomes richer in B (moves to the right from w toward pure B in the graph). A mixture of 90% B (dotted line on the right side in Figure 8–3) has a higher boiling point than at w. Hence, the temperature of the liquid in the distilling flask will increase during the distillation, and the composition of the distillate will change (as is shown in Figure 8–2, part B).

When two components that have a large boiling-point difference are distilled, the temperature remains constant while the first component distills. If the temperature remains constant, a relatively pure substance is being distilled. After the first substance distills, the temperature of the vapors rises, and the second component distills, again at a constant temperature. This is shown in Figure 8–2C. A typical application of this type of distillation might be an instance of a reaction mixture containing the desired component A (bp 140 °C) contaminated with a small amount of undesired component B (bp 250 °C) and mixed with a solvent such as diethyl ether (bp 36 °C). The ether is removed easily at low temperature. Pure A is removed at a higher temperature and collected in a separate receiver. Component B could then be distilled, but it usually is left as a residue and not distilled. This separation would not be difficult and would represent a case where simple distillation might be used to advantage.

8.3   MICROSCALE EQUIPMENT

Most large scale distillation equipment requires the distilled liquid to travel a large distance from the distilling flask, through the condenser, to the receiving flask. When working at the microscale level, a long distillation path must be avoided. With small quantities of liquid, there are too many opportunities to lose all the sample. The liquid will adhere to, or **wet,** surfaces and get lost in every little nook and cranny of the

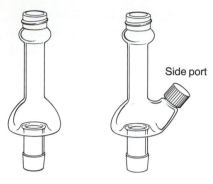

A. Unported   B. Ported
FIGURE 8–4.   The Hickman head

system. A system with a long path also has a large volume, and a small amount of liquid may not produce enough vapor to fill it. Small scale distillation requires a ''short path'' distillation. In order to make the distilling path as short as possible, the **Hickman head** has been adopted as the principal receiving device for most microscale distilling operations.

The Hickman Head.   Two types of Hickman head (also called a Hickman ''still'') are shown in Figure 8–4. One of these variations has a convenient opening, or port, in the side, making removal of liquid which has collected in it easier. In operation, the liquid to be distilled is placed in a flask or vial attached to the bottom joint of the Hickman head and heated. If desired, a condenser is attached to the top joint. Either a magnetic spin vane stirrer or a boiling stone is used to prevent bumping. Some typical assemblies are shown in Figure 8–5 and Figure 8–7. The vapors of the heated liquid rise upward and are cooled and condensed on either the walls of the condenser, or, if no condenser is used, on the inside walls of the Hickman head itself. As liquid drains downward, it collects in the circular well at the bottom of the still.

Collecting Fractions.   The liquid which distills is called the **distillate.** Portions of the distillate collected during the course of a distillation are called **fractions.** If a small fraction (usually discarded) is collected before the distillation is begun in earnest, it is called a **forerun.** The well in a Hickman head can contain anywhere from 1–2 mL of liquid. In the style with the side port, fractions may be removed by opening the port and inserting a Pasteur pipet (Figure 8–6C). The unported head works equally well, but the head is emptied from the top by using a Pasteur pipet (Figure 8–6A). If a condenser or an internal thermometer is used, the distilling apparatus must be partially disassembled to remove liquid when the well fills. In some stills the inner diameter of the head is small, and it is difficult to reach in at an angle with the pipet and make contact with the liquid. To remedy this problem, you may be able to use the longer (9-inch) Pasteur pipet instead of the shorter (5¾ inch) one. The longer pipet has a much longer narrow section (tip) and can more effectively adapt to the required angle. The disadvantage of the longer tip is that you are more likely to break it off inside the still. You may prefer to modify a short pipet by bending its tip slightly in a flame (Figure 8–6B).

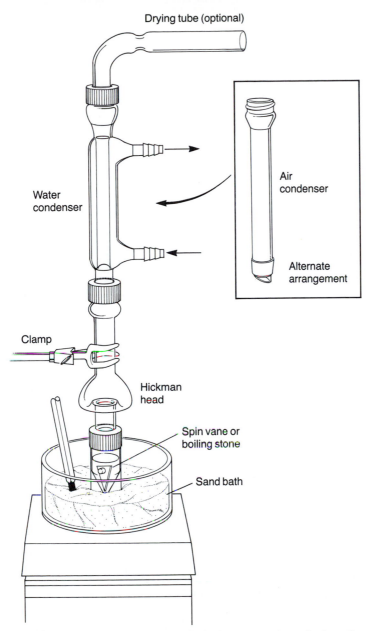

FIGURE 8–5. Basic microscale distillation (external monitoring of temperature)

Choice of Condenser. If you are careful (slow heating), or if the liquid to be distilled has a high boiling point, it may not be necessary to use a condenser with the Hickman head (Figure 8–7). In this case, the liquid being distilled must condense on the cooler sides of the head itself without any being lost through evaporation. If the liquid is low-boiling or very volatile, a condenser must be used. With very volatile liquids, a water-cooled condenser must be used; however, an air-cooled condenser may suffice for less demanding cases. Both types are shown as alternatives in Figure 8–5.

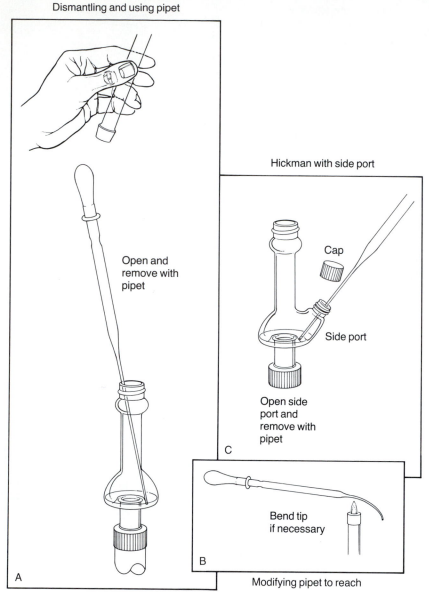

FIGURE 8–6.   Removing fractions

When using a water condenser, remember that water should enter the lower opening and exit from the upper one. If the hoses carrying the water in and out are connected in reverse fashion, the water jacket of the condenser will not fill completely.

Sealed Systems.   Whenever you perform a distillation, be sure that the system you are heating is not sealed off completely from the outside atmosphere. During a distillation, the air and vapors inside the system will both expand and contract. If pressure builds up inside a sealed system, the apparatus may explode. In performing a distillation, one usually leaves a small opening at the far end of the system. If water

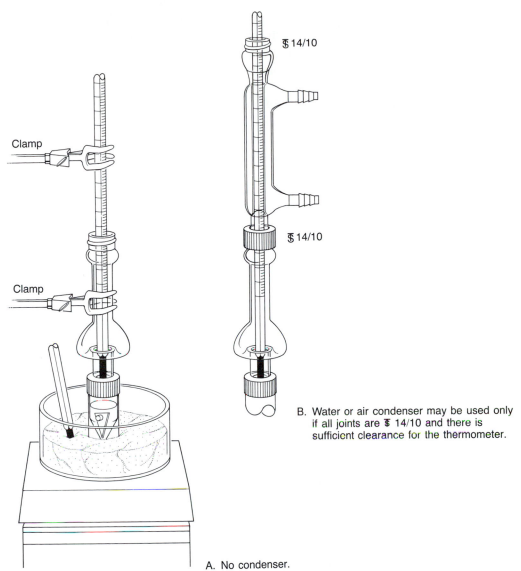

B. Water or air condenser may be used only if all joints are ⊤ 14/10 and there is sufficient clearance for the thermometer.

A. No condenser.

FIGURE 8–7. Basic microscale distillation (internal monitoring of temperature)

vapor could be harmful to the substances being distilled, a calcium chloride drying tube may be used to protect the system from moisture. Carefully examine each of the systems discussed to see how an opening to the outside is provided.

External Monitoring of Temperature. The simple assembly using the Hickman head shown in Figure 8–5 does not monitor the temperature inside the apparatus. Instead, the temperature is monitored externally with a thermometer placed in the sand bath. External monitoring of the temperature has the disadvantage that the exact temperature at which liquid distills is never known. In many cases this does not matter or is unavoidable, and the boiling point of the distilled liquid can be checked later by performing a micro boiling point determination (Technique 6, Section 6.10, p 607).

External monitoring of temperature is necessary when a distillation requiring an air or water condenser is performed with **earlier versions of the microscale kit.** The air and water condensers in the earlier style kits have ᛏ 7/10 joints at one or both ends. Most thermometers have a diameter of about 8 mm or 9 mm and will not pass through the bore of ᛏ 7/10 joints.

As a rule, there is about a 15 ° difference in temperature between the temperature of the sand bath and that of the liquid in the heated distilling vial or flask. However, the magnitude of this difference cannot be relied upon. Keep in mind that the liquid in the vial or flask may be at a different temperature than the vapor which is distilling. In many of the procedures in this text, the **approximate** temperature of the sand bath will be given instead of the boiling point of the liquid which is involved. Because this method of monitoring the temperature is rather approximate, it will be necessary for you to make the actual heater setting based on what is supposed to be occurring in the vial or flask.

Internal Monitoring of Temperature. When you wish to monitor the actual temperature of a distillation, a thermometer must be placed inside the apparatus. Figures 8–7 and 8–8 show distillation assemblies which use an internal thermometer. The apparatus in Figure 8–7A, represents the simplest possible distillation assembly. It does not use a condenser and the thermometer is suspended from a clamp. It is possible to add either an air or a water condenser to this basic assembly (Figure 8–7B) and maintain internal monitoring of the temperature, but only if the condensers have ᛏ 14/10 joints on both ends. A thermometer will not pass through the smaller ᛏ 7/10 joints that are used in some microscale kits. In some cases, the thermometer may fit too tightly even with ᛏ 14/10 joints. If there is too little clearance, a liquid seal may form around the thermometer effectively preventing the condensed liquid from draining into the Hickman head.

> **If the thermometer does not rotate freely (with good clearance) when placed through the bore of the condenser, internal monitoring of temperature should not be used.**

In the arrangement in Figure 8–8, a thermometer adapter is used. A thermometer adapter (Figure 8–9) provides a convenient way of holding a thermometer in place. The Claisen head is used to provide an opening to the atmosphere, thereby avoiding a sealed system. With the Claisen head, a drying tube may be used to protect the system from atmospheric moisture. If desired, an air condenser may be placed on top of the Hickman head to promote more efficient condensation. Once again, this arrangement will only work if you have an air condenser with ᛏ 14/10 joints on both ends.

Carefully notice the position of the thermometer in Figures 8–7 and 8–8. The bulb of the thermometer must be placed in the stem of the Hickman head, **just below the well,** or it will not read the temperature correctly.

> **Keep in mind that it is good practice to monitor the temperature internally whenever it is possible.**

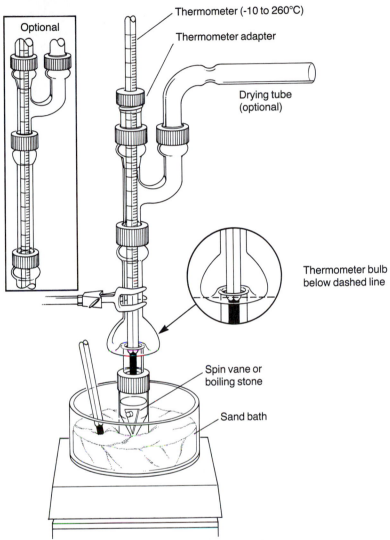

FIGURE 8—8. Basic microscale distillation using thermometer adapter (internal monitoring of temperature)

Boiling Stones or Stirring. A boiling stone should be used during distillation in order to prevent bumping. As an alternative, the liquid being distilled may be rapidly stirred. A triangular spin vane of the correct size should be used when distilling from a conical vial, whereas a stirring bar should be used when distilling from a round-bottomed flask.

Size of Distilling Flask. As a rule, the distilling flask or vial should not be filled to more than two thirds of its total capacity. This allows room for boiling and stirring action, and it prevents contamination of the distillate by bumping. A flask that is too large should also be avoided. With too large a flask, the **holdup** is excessive; the holdup is the amount of material that cannot distill since some vapor must fill the empty flask.

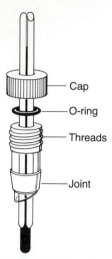

Cap

O-ring

Threads

Joint

FIGURE 8–9. The thermometer adapter

Assembling the Apparatus. You should not grease the joints when assembling the apparatus. Ungreased joints will seal well enough to allow you to perform a simple distillation. Stopcock grease can introduce a serious contaminant into your product.

Rate and Degree of Heating. You should take care not to distill too quickly. If you vaporize liquid at a rate faster than it can be recondensed, some of your product may be lost by evaporation. Carefully examine your apparatus during distillation to monitor the position of either a reflux ring or a wet appearance on the surface of the glass. Either of these indicate the place at which condensation is occurring. The position at which condensation occurs should be well inside the Hickman head. Be sure that liquid is collecting in the well. If all the surfaces are shiny (wet) and there is no distillate, you are losing material.

> **A slower rate of heating also helps to avoid bumping.**

Sometimes material is lost because the hot sand bath radiates too much heat upwards and warms the Hickman still. If you believe this to be the case, it can sometimes be remedied by placing a small square of aluminum foil over the top of the sand bath. Make a tear from one edge to the center of the foil to wrap it around the apparatus.

8.4 SEMI-MICROSCALE AND MACROSCALE EQUIPMENT

When you wish to distill quantities of liquid which are larger than about 2–3 mL, different equipment is required. Most manufacturers of microscale equipment make

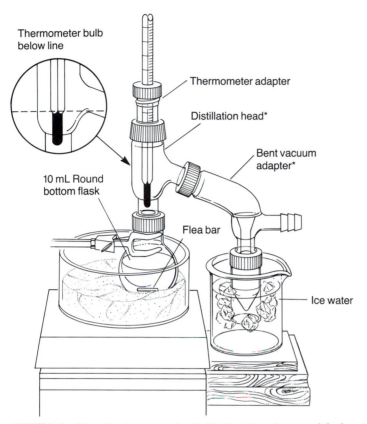

Thermometer bulb
below line

Thermometer adapter

Distillation head*

Bent vacuum
adapter*

10 mL Round
bottom flask

Flea bar

Ice water

FIGURE 8–10. Semi-microscale distillation (*requires special pieces)

two pieces of conventional distillation equipment sized so that they will work with the
℥ 14/10 microscale kit components. These two pieces, the **distillation head** and the
bent vacuum adapter, are not provided in student microscale kits, but must be purchased separately. Figure 8–10 shows a semi-microscale assembly using these components. Notice that the distilling head is designed to accommodate a thermometer. Note,
however, that the bulb of the thermometer must be placed **below the sidearm** if it is
to be bathed in vapor and give a correct temperature reading. This apparatus assumes
that a condenser is not necessary, however, one could easily be inserted between the
distilling head and the bent vacuum adapter. This insertion would give a completely
traditional distillation apparatus, but would use microscale equipment. A distillation
apparatus constructed from a ''macroscale'' organic laboratory kit is shown in Figure
8–11. This type of equipment is being used today in many organic laboratories that
have not yet converted to microscale. Electrically regulated **heating mantles** are often
used with this equipment.

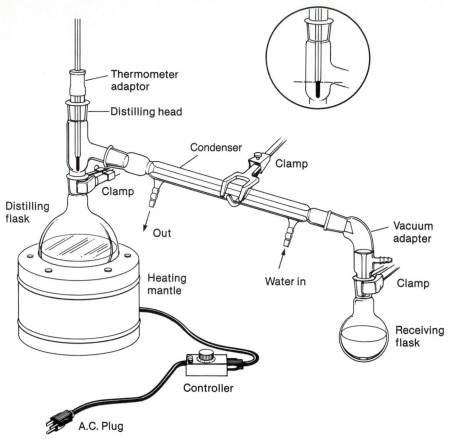

FIGURE 8–11. Distillation with the standard macroscale organic lab kit

---

PROBLEMS

1. Using Figure 8–3, answer the following questions.
   (a) What is the molar composition of the vapor in equilibrium with a boiling liquid that has a composition of 60% A and 40% B?
   (b) A sample of vapor has the composition 50% A and 50% B. What is the composition of the boiling liquid that produced this vapor?

2. Using an apparatus similar to that shown in Figure 8–10, assume that the round-bottomed flask holds 10 mL and that the Claisen head has an internal volume of about 2 mL in the vertical section. At the end of a distillation, vapor would fill this volume, but it could not be forced through the system. No liquid would remain in the distilling flask. Assuming this **holdup volume** of 12 mL, use the ideal gas law and assume a boiling point of 100 °C (760 mmHg) to calculate the number of microliters of liquid (density = 0.9 g/mL, MW 200) that would recondense into the distilling flask upon cooling.

3. Explain the significance of a horizontal line connecting a point on the lower curve with a point on the upper curve (such as line xy) in Figure 8–3.

4. Using Figure 8–3, determine the boiling point of a liquid having a molar composition of 50% A and 50% B.

5. What is the approximate difference between the temperature of a boiling liquid in a conical vial and the temperature read on an **external** thermometer when both are placed in a sand bath?

6. Give some reasons why it might not be possible to use an internal thermometer in a distillation apparatus.

7. Where should the thermometer bulb be located for internal monitoring in
(a) a distillation apparatus using a Hickman head?
(b) a large scale distillation using a Claisen head with a water condenser placed beyond it?

8. Under what conditions can a good separation be achieved with a simple distillation?

Technique 9
VACUUM DISTILLATION, MANOMETERS

Vacuum distillation (distillation at reduced pressure) is used for compounds that have high boiling points (above 200 °C). Such compounds often undergo thermal decomposition at the temperatures required for their distillation at atmospheric pressure. The boiling point of a compound is lowered substantially by reducing the applied pressure. Vacuum distillation is also used for compounds, which, when heated, might react with the oxygen present in air. It is also used when it is more convenient to distill at a lower temperature because of experimental limitations. For instance, a heating device may have difficulty heating to a temperature in excess of 250 °C.

The effect of pressure on the boiling point is discussed more thoroughly in Technique 6 (Section 6.9, p 606). A nomograph is given (Figure 6–9, p 607) that allows one to estimate the boiling point of a liquid at a pressure different from the one at which it is reported. For example, a liquid reported to boil at 200 °C at 760 mmHg would be expected to boil at 90 °C at 20 mmHg. This is a significant decrease in temperature, and it would be advantageous to use a vacuum distillation if any problems were to be expected. Counterbalancing this advantage, however, is the fact that separations of liquids of different boiling points may not be as effective with a vacuum distillation as with a simple distillation.

9.1 MICROSCALE METHODS

When working with glassware that is to be evacuated, you should wear safety glasses at all times. There is always danger of an implosion.

> **Safety glasses must be worn at all times during vacuum distillation.**

A basic apparatus similar to the one shown in Figure 9–1 (or in Figure 9–5) may be used for microscale vacuum distillations. As is the case for simple distillation, this apparatus uses the Hickman head as a means to reduce the length of the vapor path. The major difference to be found when comparing this assembly to one for simple distillation (Figure 8–8, p 647) is that the opening to the atmosphere has been replaced by a connection to a vacuum source (top right-hand side). The usual sources of vacuum are the aspirator (Technique 4, Section 4.5, p 573), a mechanical vacuum pump, or a "house" vacuum line (one piped directly to the laboratory bench). The aspirator is

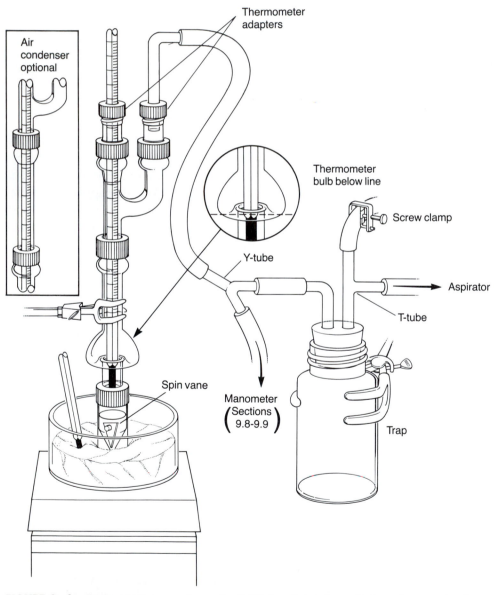

FIGURE 9–1. Reduced pressure microscale distillation (internal monitoring of temperature)

probably the simplest of these sources, and the vacuum source most likely to be available. However, if pressures below 10–20 mmHg are required, a vacuum pump will have to be used.

Assembling the Apparatus. When assembling an apparatus for vacuum distillation, it is important that all joints and connections be air tight. The joints in the newest microscale kits are standard-taper ground glass joints, with a compression cap which contains an O-ring seal. Glassware which contains this type of compression joint will hold a vacuum quite easily. Under normal conditions, it is not necessary to grease these joints.

> **Normally, you should not grease joints. It will only be necessary to grease the joints in a vacuum distillation if you cannot achieve the desired pressure without using grease.**

If you must grease joints, take care not to use too much grease. You are working with small quantities of liquid in a microscale distillation, and the grease can become a very serious contaminant if it oozes out the bottom of the joints into your system. Apply a small amount of grease (thin film) completely around the top of the **inner** joint; then, mate the joints and turn them slightly to spread the grease evenly. If you have used the correct amount of grease, it will not ooze out the bottom; rather, the entire joint will appear clear and without striations or uncovered areas.

Make doubly sure any connections to pressure tubing are tight. The pressure tubing itself should be relatively new and without cracks. If the tubing shows cracks when you stretch or bend it, it may be old and leak air into the system. Glass tubing should fit securely into any rubber stoppers. If you can move the tubing up and down with only gentle force, it is too loose, and you should obtain a larger size. Check all glassware to be sure there are not cracks in any of it and that there are no chips in the standard-taper joints. Cracked glassware may break when evacuated.

Connecting to Vacuum. In Figure 9–1, the connection to vacuum has been made using a bent piece of glass tubing and a thermometer adapter (see Figures 8–9 and 9–2A). If a thermometer adapter is not available, two alternative methods of connecting to a vacuum source may be used. These are shown in Figure 9–2B and

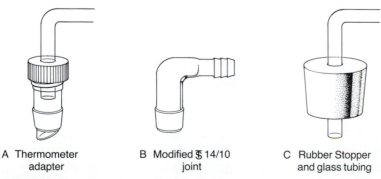

A Thermometer
    adapter

B Modified ⑤ 14/10
    joint

C Rubber Stopper
    and glass tubing

FIGURE 9–2. Alternative vacuum connections

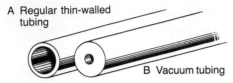

FIGURE 9–3.   Comparison of tubing

9–2C. Whichever is used, the connection to the vacuum source is made using **pressure tubing.** Pressure tubing (also called vacuum tubing), unlike the more common thin-walled tubing used to carry water or gas, has heavy walls that will not collapse inward when it is evacuated. A comparison of the two types of tubing is shown in Figure 9–3.

Water Trap.   If an aspirator is used as a source of vacuum, a water trap must be placed between it and the distillation assembly. A commonly used type of water trap is shown at the bottom right of Figure 9–1. A different type of trap, which is also common, is shown in Figure 9–5. Variations in water pressure are to be expected when using an aspirator. If the pressure drops low enough, the vacuum in the system will draw water from the aspirator into the connecting line. The trap allows you to see this happening and take corrective action (prevent water from entering the distillation apparatus). The correct action for anything but a small amount of water is to "vent the system." This can be accomplished by opening the screw clamp at the top of the trap to let air into the system. When doing a vacuum distillation, you should also realize that the system should always be vented prior to stopping the aspirator. If you turn off the aspirator while the system is still under vacuum, water will be drawn into the connecting line and trap.

Manometer Connection.   A Y-tube is shown in the line from the apparatus to the trap. This branching connection is optional, but is required if you wish to monitor the actual pressure of the system using a manometer. The operation of manometers is discussed in Sections 9.8 and 9.9.

Thermometer Placement.   If a thermometer is used, be sure that the bulb is placed in the stem of the Hickman head just below the well. If it is placed higher, it may not be surrounded by a constant stream of vapor from the material being distilled. If the thermometer is not exposed to a continuous stream of vapor, it may not reach temperature equilibrium. As a result, the temperature reading would be incorrect (low).

Preventing Bump Over.   When heating a distillation flask there is always the possibility that boiling action will become too vigorous (mainly due to superheating) and "bump" some of the undistilled liquid up into the Hickman head. The simplest way to prevent bumping is to stir the boiling liquid with a magnetic spin vane. Rapid stirring will distribute the heat evenly, keep the boiling action smooth, and prevent bumping. Boiling stones cannot be used for this purpose in a vacuum distillation; they do not work in vacuum. In a conventional vacuum distillation (macroscale), it is customary to provide smooth boiling action by using an **ebulliator tube.** The ebulliator tube agitates the boiling solution by providing a small continuous stream of air bubbles. Figure 9–4 shows how a microscale vacuum distillation may be modified to use an

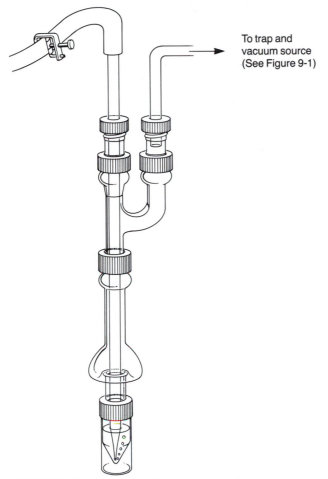

FIGURE 9—4. Use of ebulliator tube instead of thermometer

ebulliator tube. The amount of air (rate of bubbles) provided by the ebulliator is adjusted by either tightening or loosening the screw clamp at the top. A Pasteur pipet makes an excellent ebulliator tube. As Figure 9–4 shows, the ebulliator tube replaces the thermometer. Hence, the ebulliator should be used only when internal monitoring of temperature is not required. In practice, although this method works satisfactorily, better results are obtained with stirring and a slow rate of distillation.

> **A slow rate of heating helps to avoid bumping**

9.2 SIMPLIFIED MICROSCALE APPARATUS

The apparatus shown in Figure 9–5 will often give very satisfactory results when internal temperature monitoring is not required. It is the apparatus we prefer for the experiments in this textbook that require vacuum distillation. Distillation (heating) should be performed slowly with brisk stirring. As an added precaution, the stem of the

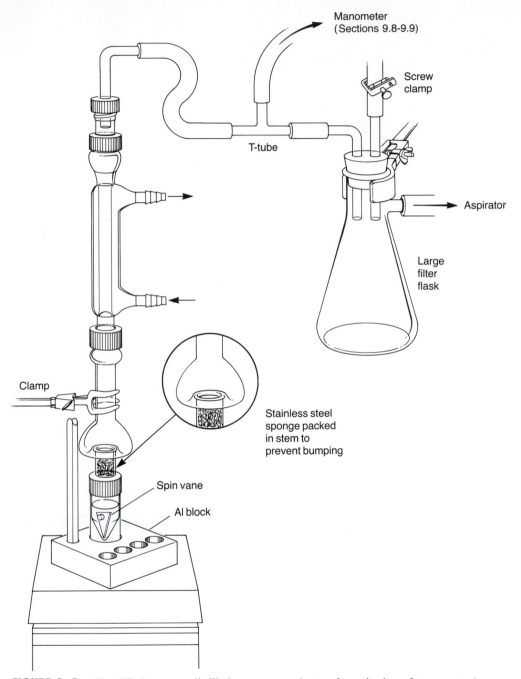

FIGURE 9–5. Simplified vacuum distillation apparatus (external monitoring of temperature)

Hickman still is loosely packed with a small portion of a stainless steel sponge. The packing very effectively prevents bump-over. Just before the well begins to fill, you will see reflux action (condensation) in the packed stem. In many cases this will occur even before there is any evidence of boiling in the heated liquid. In Figure 9–5, an

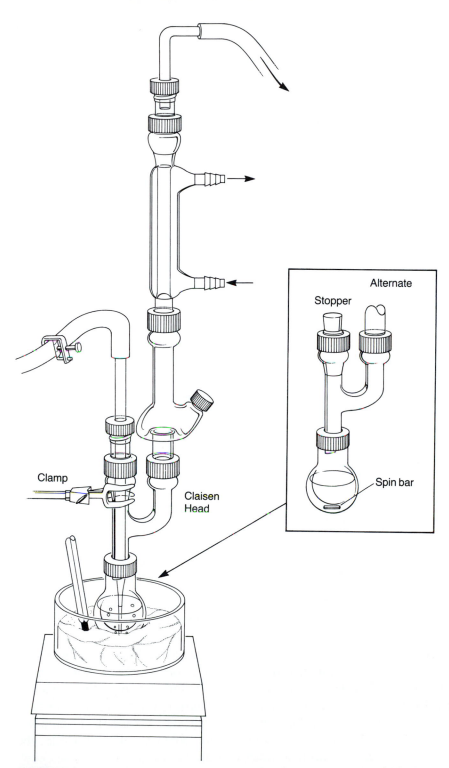

FIGURE 9–6. Apparatus for larger quantities (external temperature monitoring)

aluminum heating block is shown instead of a sand bath. The aluminum block is a more effective heat source than a sand bath whenever you want a faster heating response or a high temperature. For temperatures above 180 °C, it is the preferred heating device.

9.3 SEMI-MICROSCALE AND MACROSCALE EQUIPMENT

When you have larger quantities of liquid to distill (>5 mL), an apparatus similar to that shown in Figure 9–6 may be used. This apparatus uses a 10-mL round-bottom flask and a Claisen head. The Claisen head is used in larger scale vacuum distillations because it allows the use of an ebulliator and because the bend it provides in the distilling path prevents bump-over. Because the Claisen head increases the holdup of the system, it cannot be used with very small scale distillations. An alternative to using the ebulliator, a method based on stirring, is also shown in the figure.

A vacuum distillation apparatus using the components of the traditional organic laboratory kit is shown in Figure 9–7. It uses the ebulliator, the Claisen head, and has internal temperature monitoring with a thermometer. With high boiling liquids, this apparatus may be simplified by removing the water condenser.

9.4 STEPWISE INSTRUCTIONS FOR MICROSCALE VACUUM DISTILLATION

The following set of instructions is a step-by-step account of how to carry out a vacuum distillation. The microscale apparatus illustrated in Figure 9–1 will be used, however, the procedures apply to any vacuum distillation.

> **Safety glasses must be worn at all times during vacuum distillation.**

Evacuating the Apparatus

1. Assemble the apparatus as shown in Figure 9–1. It should be held with a clamp attached to the top of the Hickman head and placed **above** the sand bath or aluminum block.

2. If the sample contains solvent, concentrate the sample to be distilled in the conical vial (or round-bottomed flask) which you will use. Use one of the solvent removal methods discussed in Technique 3. If you have a large volume of solvent to evaporate and the sample doesn't fit in the conical vial, you will have to use an Erlenmeyer flask first, and then transfer the sample to the conical vial. (Be sure to rinse the Erlenmeyer flask with a little solvent and then reevaporate in the conical vial.) As a rule, the distilling vial or flask should be no more than two thirds full.

3. Attach the conical vial (or flask) to the apparatus and make sure all joints are sealed.

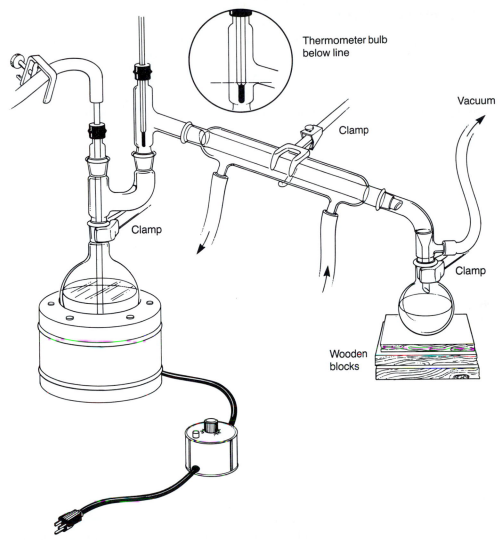

Thermometer bulb
below line

Vacuum

Clamp

Clamp

Clamp

Wooden
blocks

FIGURE 9-7. Macroscale vacuum distillation using the standard organic laboratory kit

4. Turn the aspirator on to the maximum extent.

5. Close the screw clamp on the water trap very tightly.

(If you are using an ebulliator as in Figure 9–4, you would next regulate the rate of bubbling by adjusting the tightness of the screw clamp at the top of the tube.)

6. Using the manometer, observe the pressure. It may take a few minutes to remove any residual solvent and evacuate the system. If the pressure is not satisfactory, check all connections to see whether they are tight.

(Readjust the ebulliator if necessary.)

Do not proceed until you have a good vacuum.

Beginning Distillation

7. Lower the apparatus into the sand bath (or into the aluminum block) and begin to heat. Place the external thermometer in the sand bath or block now if you wish.

8. Increase the temperature of the heat source until you begin to see distillate collect in the well of the Hickman head.

(Observe very carefully, liquid may appear almost "magically" without any sign of boiling or any obvious reflux ring.)

9. If you are using a thermometer, record the temperature and pressure when distillate begins to appear.

(If you are not using an internal thermometer, record the external temperature. If you have two thermometers, record both temperatures.)

Collecting a Fraction

10. To collect a fraction, raise the apparatus out of the sand bath or aluminum block and allow it to cool a bit before opening it.

11. Open the screw clamp on the water trap to allow air to enter the system.

(If you are using an ebulliator, you will also need to open the screw clamp at its top **immediately,** or the liquid in the distilling flask will be forced upward into it.)

12. Partially disassemble the apparatus and remove the fraction with a Pasteur pipet as shown in Figure 8–6A.

(If you have a Hickman head with a side port, you may simply open the side port to remove the fraction. This is shown in Figure 8–6C.)

> **If you do not intend to collect a second fraction, go directly to Steps 18–20.**

13. Reassemble the apparatus (or close the side port) and tighten the clamp at the top of the ebulliator tube.

14. Tighten the screw clamp on the water trap and reestablish the desired pressure. If the pressure is not satisfactory, check all connections to make sure they are sealed.

15. Lower the apparatus back into the sand bath or block and continue the distillation.

Shutdown

16. At the end of the distillation, raise the apparatus from the sand bath or block and allow it to cool. Also let the sand bath or block cool.

17. Open the screw clamp on the water trap first, and then immediately open the one at the top of the ebulliator.

18. Turn off the water at the aspirator. (Don't do this before step 17!)

19. Remove any distilled material by one of the methods shown in Figure 8–6.

20. Disassemble the apparatus and clean all glassware as soon as possible to prevent the joints from sticking.

If you used grease, thoroughly clean all grease off the joints or it will contaminate your samples used in other procedures.

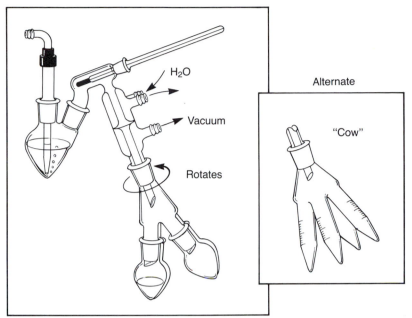

FIGURE 9—8.   Rotary fraction collector

9.5   ROTARY FRACTION COLLECTORS

With the types of apparatus we have discussed above, the vacuum must be stopped to remove fractions when a new substance (fraction) begins to distill. Quite a few steps are required in order to perform this change, and it is quite inconvenient when there are several fractions that must be collected. Two pieces of semi-microscale apparatus are shown in Figure 9–8 that are designed to alleviate the difficulty of collecting fractions while working under vacuum. The collector, which is shown to the right, is sometimes called a "cow" because of its external appearance. With these rotary fraction collecting devices, all one has to do is rotate the device to collect fractions.

9.6   BULB-TO-BULB DISTILLATION

The ultimate in microscale methods is to use a bulb-to-bulb distillation apparatus. This apparatus is shown in Figure 9–9. The sample to be distilled is placed in the glass container attached to one of the arms of the apparatus. The sample is frozen solid, usually by using liquid nitrogen, but dry ice in 2-propanol or an ice-salt-water mixture may also be used. The coolant container shown in the figure is a **Dewar flask.** The Dewar flask has a double wall with the space between the walls evacuated and sealed. A vacuum is a very good thermal insulator, and there is little heat loss from the cooling solution. It lasts a long time.

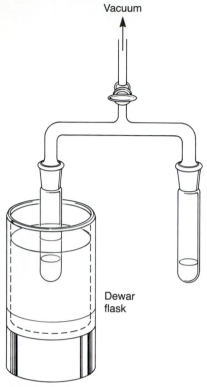

Vacuum

Dewar
flask

FIGURE 9–9.   Bulb-to-bulb distillation

After freezing the sample, the entire apparatus is evacuated by opening the stopcock. When the evacuation is complete, the stopcock is closed and the Dewar flask removed. The sample is allowed to thaw, and then it is frozen again. This freeze-thaw-freeze cycle removes any air or gases that were trapped in the frozen sample. Next, the stopcock is opened to evacuate the system again. When the second evacuation is complete, the stopcock is closed and the Dewar flask is moved to the other arm to cool the empty container. As the sample warms, it will vaporize, travel to the other side, and be frozen or liquified by the cooling solution. This transfer of the liquid from one arm to the other may take quite a while, but **no heating is required.**

The bulb-to-bulb distillation is most effective when liquid nitrogen is used as coolant, and when the vacuum system can achieve a pressure of 10^{-3} mmHg or lower. This requires a vacuum pump; an aspirator cannot be used.

9.7   THE MECHANICAL VACUUM PUMP

The aspirator is not capable of yielding pressures below about 5 mmHg. This is the vapor pressure of water at 0 °C, and water freezes at this temperature. A more realistic value of pressure for an aspirator is about 20 mmHg. When pressures below 20 mmHg are required, a vacuum pump will have to be employed. Figure 9–10 illustrates a

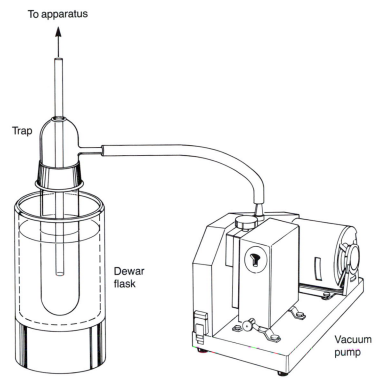

To apparatus

Trap

Dewar
flask

Vacuum
pump

FIGURE 9–10.   A vacuum pump and its trap

mechanical vacuum pump and its associated glassware. The vacuum pump operates on a principle similar to that of the aspirator, but the vacuum pump uses a high-boiling-point oil, rather than water, to remove air from the attached system. The oil used in a vacuum pump, a silicone oil or a high molecular weight hydrocarbon-based oil, has a very low vapor pressure, and very low system pressures can be achieved. A good vacuum pump, with new oil, can achieve pressures of 10^{-3} mmHg or 10^{-4} mmHg. Instead of discarding the oil as it is used, it is recycled continuously through the system.

A cooled trap is required when using a vacuum pump. This trap protects the oil in the pump from any vapors which may be present in the system. If vapors from organic solvents, or from the organic compounds being distilled, dissolve in the oil, the oil's vapor pressure will increase rendering it less effective. A special type of vacuum trap is illustrated in Figure 9–10. It is designed to fit into an insulated Dewar flask so that the coolant will last for a long period of time. At a minimum, this flask should be filled with ice water, but a dry ice-acetone mixture or liquid nitrogen will be required to achieve lower temperatures and better protect the oil. Often two traps are used; the first trap contains ice water and the second trap dry ice-acetone or liquid nitrogen. The first trap liquifies low-boiling vapors which might freeze or solidify in the second trap and block it.

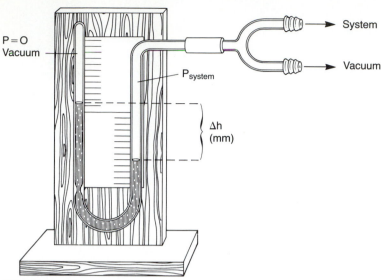

FIGURE 9–11.   A simple U-tube manometer

9.8   THE CLOSED-END MANOMETER

The principal device used to measure pressures in a vacuum distillation is the **closed-end manometer.** Two basic types are shown in Figure 9–11 and Figure 9–12. The manometer shown in Figure 9–11 is widely used since it is relatively easy to construct. It consists of a U-tube that is closed at one end and mounted on a wooden support. One

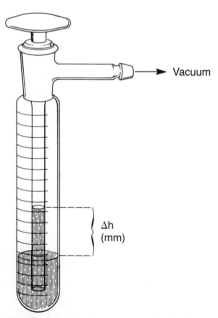

FIGURE 9–12.   Commercial "stick" manometer

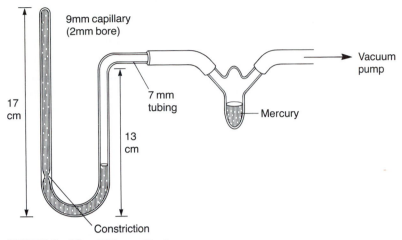

FIGURE 9–13. Filling a U-tube manometer

can construct the manometer from 9-mm glass capillary tubing and fill it as shown in Figure 9–13.

> **CAUTION. Mercury is a very toxic metal with cumulative effects. Since mercury has a high vapor pressure, it must not be spilled in the laboratory. You must not touch it with your skin. Seek immediate help from an instructor in case of a spill, or if you break a manometer. Spills must be cleaned up immediately.**

A small filling device is connected to the U-tube with pressure tubing. The U-tube is evacuated with a good vacuum pump, and then the mercury is introduced by tilting the mercury reservoir.

> **The entire filling operation should be conducted in a shallow pan in order to contain any spills that might occur.**

Enough mercury should be added to form a column about 20 cm in total length. When the vacuum is interrupted by admitting air, the mercury is forced by atmospheric pressure to the end of the evacuated tube. The manometer is then ready for use. The constriction shown in Figure 9–13 helps to protect the manometer against breakage when the pressure is released. Be sure the column of mercury in the filled manometer is long enough to pass through this constriction.

When an aspirator or any other vacuum source is used, a manometer can be connected into the system. As the pressure is lowered, the mercury rises in the right tube and drops in the left tube until Δh corresponds to the approximate pressure of the system (see Figure 9–11).

$$\Delta h = (P_{system} - P_{reference\ arm}) = (P_{system} - 10^{-3}\ mmHg) \approx P_{system}$$

A short piece of metric ruler or a piece of graph paper ruled in millimeter squares is mounted on the support board to allow Δh to be read. No addition or subtraction is necessary, since the reference pressure (created by the initial evacuation when filling) is approximately zero (10^{-3} mmHg) when referred to readings in the 10–50 mmHg range. To determine the pressure, count the number of millimeter squares beginning at the top of the mercury column on the left and continuing downward to the top of the mercury column on the right. This is the height difference, Δh, and it gives the pressure in the system directly.

A commercial counterpart to the U-tube manometer is shown in Figure 9–12. With this manometer, the pressure is given by the difference in the mercury levels in the inner and outer tubes.

The manometers described here have a range of about 1–150 mmHg in pressure. They are convenient to use when an aspirator is the source of vacuum. For high vacuum systems (pressures below 1 mmHg) a more elaborate manometer or an electronic measuring device must be used. These devices will not be discussed here.

9.9   CONNECTING AND USING A MANOMETER

The most common use of a closed-end manometer is to monitor pressure during a reduced-pressure distillation. The manometer is placed in a vacuum distillation system as shown in Figure 9–14. Generally an aspirator is the source of vacuum. Both the manometer and the distillation apparatus should be protected by a trap from possible backups in the water line. Alternative trap arrangements to that in Figure 9–14 are shown in Figure 9–1 and Figure 9–5. Notice in each case that the trap has a device (screw clamp or stopcock) for opening the system to the atmosphere. This is especially important in using a manometer since one should always make pressure changes slowly. If this is not done, there is a danger of spraying mercury throughout the system, breaking the manometer, or spurting mercury into the room. In the closed-end manometer, if the system is opened suddenly, the mercury will rush to the closed end of the U-tube. The mercury will rush with such speed and force that the end will be broken out of the manometer. Air should be admitted **slowly** by opening the valve cautiously. In a similar fashion, the valve should be closed slowly when the vacuum is being started, or mercury may be forcefully drawn into the system through the open end of the manometer.

If the pressure in a reduced-pressure distillation is lower than desired, it is possible to adjust it by means of a **bleed** valve. The stopcock can serve this function in Figure 9–14 if it is opened only a small amount. In those systems with a screw clamp on the trap (Figures 9–1 and 9–5), the screw clamp is removed from the trap valve and the base of a Tirrill-style Bunsen burner is attached. The needle valve in the base of the burner can be used to adjust precisely the amount of air that is admitted (bled) to the system and, hence, control the pressure.

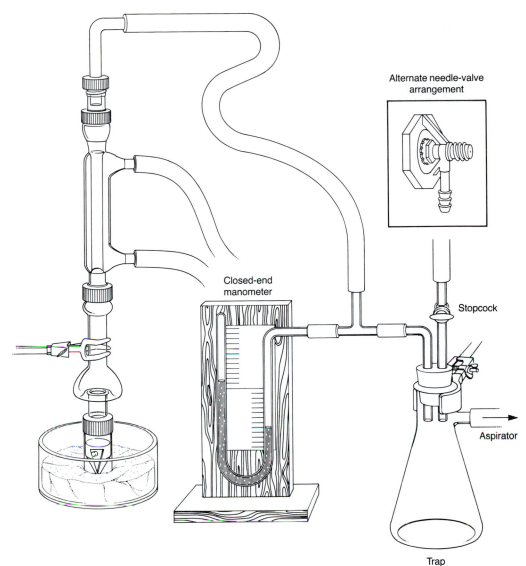

Alternate needle-valve
arrangement

Closed-end
manometer

Stopcock

Aspirator

Trap

FIGURE 9–14. Connecting a manometer to the system. To construct a ''bleed,'' the needle valve may replace the stopcock.

PROBLEMS

1. Give some reasons that would lead you to purify a liquid by using vacuum distillation rather than by using simple distillation.

2. When using an aspirator as a source of vacuum in a vacuum distillation, do you turn off the aspirator before venting the system? Explain.

3. A compound was distilled at atmospheric pressure and had a boiling range of 310–325 °C. What would be the approximate boiling range of this liquid if it was distilled under vacuum at 20 mmHg?

4. Boiling stones generally do not work when performing a vacuum distillation. What substitutes may be used?

5. What is the purpose of the trap that is used during a vacuum distillation performed with an aspirator?

Technique 10
FRACTIONAL DISTILLATION, AZEOTROPES

Simple distillation, described in Technique 8, works well for most routine separation and purification of organic compounds. When boiling-point differences of components to be separated are not large, however, **fractional distillation** must be used to achieve a good separation.

PART A. FRACTIONAL DISTILLATION

10.1 DIFFERENCES BETWEEN SIMPLE AND FRACTIONAL DISTILLATION

When an ideal solution of two liquids, such as benzene (bp 80 °C) and toluene (bp 110 °C), is distilled by simple distillation, the first vapor produced will be enriched in the lower-boiling component (benzene). When the vapor is condensed and analyzed, however, it is unlikely that the distillate will be pure benzene. The boiling point difference of benzene and toluene (30 °C) is too small to achieve complete separation by simple distillation. Likewise, the liquid remaining in the distilling flask (or vial) after collecting this first fraction will contain a larger amount of the higher-boiling toluene component than at the start of the distillation, but it also will be far from being pure.

   In principle, one could distill a solution of 50% benzene and 50% toluene by simple distillation and collect the distillate in **fractions** (portions removed separately). The first fraction would contain the largest amount of benzene and the least amount of toluene. It would also have the lowest boiling-point range. The second fraction would contain less benzene and more toluene than the first one and would have a higher boiling-point range. The trend of decreasing benzene and increasing toluene would continue until the last fraction was removed. This last fraction would have the smallest amount of benzene, the largest amount of toluene, and the highest boiling-point range. The results of this hypothetical distillation are given in Table 10–1.

   A plot of boiling point versus volume of condensate (distillate) might appear as in Figure 10–1. Clearly, separation by this method would be poor. The continuously increasing temperature observed in Figure 10–1 indicates that the composition of the vapor itself was also continuously changing. At no time did a pure substance distill.

   One could redistill each of the fractions indicated in Table 10–1. Each of the fractions would yield vapor and a resulting condensate that would contain **more** benzene than what was initially present. The residue in the distilling flask (or vial) would contain more toluene than what was initially present. The distillates and residues of

TABLE 10–1. Simple Distillation of a Mixture of Benzene and Toluene

FRACTION	BOILING RANGE (°C)	PERCENTAGE COMPOSITION	
		Benzene	*Toluene*
1	80–85	90	10
2	85–90	72	28
3	90–95	55	45
4	95–100	45	55
5	100–105	27	73
6	105–110	10	90

similar composition (similar boiling ranges) could be combined and redistilled. Eventually, one should obtain distillate that would be essentially pure benzene and a residue that would be nearly pure toluene.

Obviously the procedure described above would be very tedious; fortunately, it need not be done in usual laboratory practice. **Fractional distillation** accomplishes the same result as that described. One simply has to use a column inserted between the distilling flask and the receiver (Hickman head), as shown in Figure 10–2. This **fractionating column** is filled, or **packed,** with a suitable material such as a stainless steel sponge. This packing allows a mixture of benzene and toluene to be subjected continuously to many vaporization-condensation cycles as the material moves up the column. With each cycle within the column, the composition of the vapor is progressively enriched in the lower-boiling component (benzene). Nearly pure benzene (bp 80 °C) finally emerges from the top of the column, condenses, and passes into the receiving head or flask. This process will continue until all of the benzene is removed. The distillation must be carried out slowly to ensure that numerous vaporization-condensation cycles will occur. When nearly all of the benzene has been removed, the tempera-

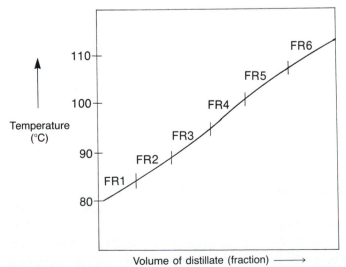

FIGURE 10–1. Temperature–distillate plot for simple distillation of a benzene–toluene mixture

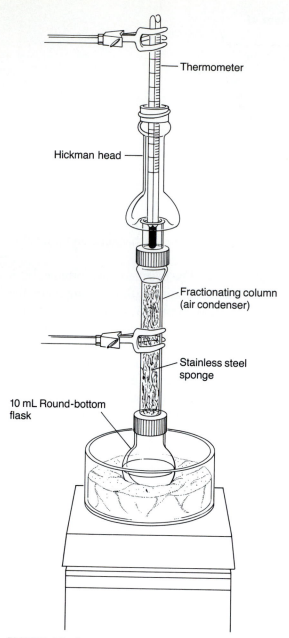

Thermometer

Hickman head

Fractionating column
(air condenser)

Stainless steel
sponge

10 mL Round-bottom
flask

FIGURE 10–2. Microscale apparatus for fractional distillation

ture will begin to rise and a small amount of a second fraction, which contains some benzene and toluene, may be collected. When the temperature reaches 110 °C, the boiling point of pure toluene, the vapor is condensed and collected as the third fraction. A plot of boiling-point versus volume of condensate (distillate) would resemble Figure 10–3. This separation would be much better than that achieved by simple distillation (Figure 10–1).

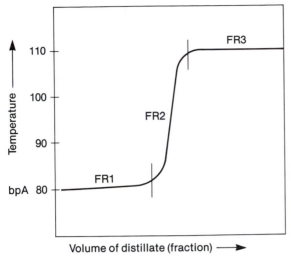

FIGURE 10–3. Temperature–distillate plot for fractional distillation of a benzene–toluene mixture

10.2 VAPOR-LIQUID COMPOSITION DIAGRAMS

A vapor-liquid composition phase diagram like the one in Figure 10–4 can be used to explain the operation of a fractionating column with an **ideal solution** of two liquids, A and B. An ideal solution is one in which the two liquids are chemically similar, miscible (mutually soluble) in all proportions, and do not interact. Ideal solutions obey **Raoult's law.** Raoult's law will be explained in detail in Section 10.3.

The phase diagram relates the compositions of the boiling liquid (lower curve) and its vapor (upper curve) as a function of temperature. Any horizontal line drawn across the diagram (a constant-temperature line) intersects the diagram in two places. These intersections relate the vapor composition to the composition of the boiling liquid that produces that vapor. By convention, composition is expressed either in **mole fraction** or in **mole percentage.** The mole fraction is defined as follows:

$$\text{mole fraction A} = N_A = \frac{\text{moles A}}{\text{moles A} + \text{moles B}}$$

$$\text{mole fraction B} = N_B = \frac{\text{moles B}}{\text{moles A} + \text{moles B}}$$

$$N_A + N_B = 1$$

$$\text{mole percentage A} = N_A \times 100$$

$$\text{mole percentage B} = N_B \times 100$$

The horizontal and vertical lines shown in Figure 10–4 represent the processes that occur during a fractional distillation. Each of the **horizontal lines** (L_1V_1, L_2V_2, etc.) represents the **vaporization** step of a given vaporization-condensation cycle and represents the composition of the vapor in equilibrium with liquid at a given temperature. For example, at 63 °C a liquid with a composition of 50% A (L₃ on the diagram) would yield vapor of composition 80% A (V₃ on diagram) at equilibrium. The vapor is richer in the lower-boiling component A than the original liquid was.

Each of the **vertical lines** (V_1L_2, V_2L_3, etc.) represents the **condensation** step of a given vaporization-condensation cycle. The composition does not change as the temperature drops on condensation. The vapor at V_3, for example, condenses to give a liquid (L_4 on the diagram) of composition 80% A with a drop in temperature from 63 to 53 °C.

In the example shown in Figure 10–4, pure A boils at 50 °C while pure B boils at 90 °C. These two boiling points are represented at the left and right-hand edges of the diagram, respectively. Now consider a solution which contains only 5% of A, but 95% of B. (Remember that these are **mole** percentages.) This solution is heated (following the dashed line) until it is observed to boil at L_1 (87 °C). The resulting vapor has composition V_1 (20% A, 80% B). The vapor is richer in A than the original liquid, but is by no means pure A. In a simple distillation apparatus, this vapor would be condensed and passed into the receiver in a very impure state. However, with a fractionating column in place, the vapor is condensed in the **column** to give liquid L_2 (20% A, 80% B). Liquid L_2 is immediately revaporized (bp 78 °C) to give a vapor of composition V_2 (50% A, 50% B), which is condensed to give liquid L_3. Liquid L_3 is revaporized (bp 63 °C) to give vapor of composition V_3 (80% A, 20% B), which is condensed to give liquid L_4. Liquid L_4 is revaporized (bp 53 °C) to give vapor of composition V_4 (95% A, 5% B). This continues to V_5 which condenses to give nearly pure liquid A. The fractionating process follows the stepped lines in the figure downward and to the left.

As this process continues all of liquid A is removed from the distilling flask or vial, leaving nearly pure B behind. If the temperature is raised, liquid B may be distilled as a nearly pure fraction. Fractional distillation will have achieved a separation

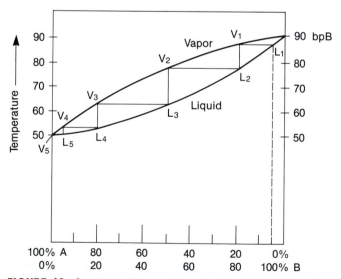

FIGURE 10–4. Phase diagram for a fractional distillation of an ideal two-component system

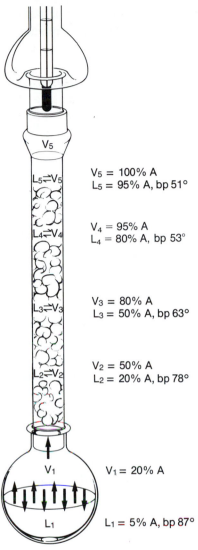

$V_5 = 100\%$ A
$L_5 = 95\%$ A, bp 51°

$V_4 = 95\%$ A
$L_4 = 80\%$ A, bp 53°

$V_3 = 80\%$ A
$L_3 = 50\%$ A, bp 63°

$V_2 = 50\%$ A
$L_2 = 20\%$ A, bp 78°

$V_1 = 20\%$ A

$L_1 = 5\%$ A, bp 87°

FIGURE 10–5. Vaporization–condensation in a fractionation column

of A and B, a separation that would have been nearly impossible with simple distillation. Notice that the boiling point of the liquid becomes lower each time that it vaporizes. Since the temperature at the bottom of a column is normally higher than the temperature at the top, successive vaporizations occur higher and higher in the column as the composition of the distillate approaches that of pure A. This process is illustrated in Figure 10–5, where the composition of the liquids, their boiling points, and the composition of the vapors present are shown alongside the fractionating column.

10.3 RAOULT'S LAW

Two liquids (A and B) that are miscible and that do not interact, form an **ideal solution** and follow Raoult's Law. The law states that the partial vapor pressure of component A in the solution, P_A, equals the vapor pressure of pure A, $P°_A$, times its mole fraction, N_A (Equation 1). A similar expression can be written for component B (Equation 2). The mole fractions, N_A and N_B, were defined in Section 10.2.

$$\text{partial vapor pressure of A in solution} = P_A = (P°_A)(N_A) \qquad (1)$$

$$\text{partial vapor pressure of B in solution} = P_B = (P°_B)(N_B) \qquad (2)$$

$P°_A$ is the vapor pressure of pure A, independent of B. $P°_B$ is the vapor pressure of B, independent of A. In a mixture of A and B, the partial vapor pressures are added to give the total vapor pressure above the solution (Equation 3). When the total pressure (sum of the partial pressures) equals the applied pressure, the solution will boil.

$$P_{total} = P_A + P_B = P°_A N_A + P°_B N_B \qquad (3)$$

The composition of A and B in the vapor produced is given by Equations 4 and 5.

$$N_A \text{ (vapor)} = \frac{P_A}{P_{total}} \qquad N_B \text{ (vapor)} = \frac{P_B}{P_{total}} \qquad (4, 5)$$

Several problems involving applications of Raoult's Law are illustrated in Figure 10–6. Note, particularly in the result from Problem 4, that the vapor is richer ($N_A = 0.67$) in the lower-boiling (higher vapor pressure) component A than it was before vaporization ($N_A = 0.50$). This proves mathematically what was described in Section 10.2.

The consequences of Raoult's Law for distillations are shown schematically in Figure 10–7. In Part A the boiling points are identical (vapor pressures the same) and no separation is attained regardless of how the distillation is conducted. In Part B a fractional distillation is required, while in Part C a simple distillation provides an adequate separation.

When a solid B (rather than another liquid) is dissolved in a liquid A, the boiling point is increased. In this extreme case, the vapor pressure of B is negligible, and the vapor will be pure A no matter how much solid B is added. Consider a solution of salt in water.

$$P_{total} = P°_{water} N_{water} + P°_{salt} N_{salt}$$

$$P°_{salt} = 0$$

$$P_{total} = P°_{water} N_{water}$$

A solution whose mole fraction of water is 0.7 will not boil at 100 °C, since $P_{total} = (760)(0.7) = 532$ mmHg and is less than atmospheric pressure. If the solution is heated to 110 °C, it will boil because $P_{total} = (1085)(0.7) = 760$ mmHg. Although the solution must be heated to 110 °C to boil it, the vapor is pure water and has a boiling point temperature of 100 °C. (The vapor pressure of water at 110 °C can be looked up in a handbook; it is 1085 mmHg.)

10.4 COLUMN EFFICIENCY

A common measure of the efficiency of a column is given by its number of **theoretical plates.** The number of theoretical plates in a column is related to the number of vapori-

Consider a solution at 100° C where $N_A = 0.5$ and $N_B = 0.5$.

1. What is the partial vapor pressure of A in the solution if the vapor pressure of pure A at 100° C is 1020 mmHg?

   Answer: $P_A = P°_A N_A = (1020)(0.5) = 510$ mmHg

2. What is the partial vapor pressure of B in the solution if the vapor pressure of pure B at 100° C is 500 mmHg?

   Answer: $P_B = P°_B N_B = (500)(0.5) = 250$ mmHg

3. Would the solution boil at 100° C if the applied pressure were 760 mmHg?

   Answer: Yes. $P_{total} = P_A + P_B = (510 + 250) = 760$ mmHg

4. What is the composition of the vapor at the boiling point?

   Answer: The boiling point is 100° C.

$$N_A \text{ (vapor)} = \frac{P_A}{P_{total}} = 510/760 = 0.67$$

$$N_B \text{ (vapor)} = \frac{P_B}{P_{total}} = 250/760 = 0.33$$

FIGURE 10–6. Sample calculations with Raoult's Law

zation-condensation cycles that occur as a liquid mixture travels through it. Using the example mixture in Figure 10–4, if the first distillate (condensed vapor) had the composition at L_2 when starting with liquid of composition L_1, the column would be said to have **one theoretical plate.** This would correspond to a simple distillation, or one vaporization-condensation cycle. A column would have two theoretical plates if the first distillate had the composition at L_3. The two-theoretical-plate column essentially carries out "two simple distillations." According to Figure 10–4, **five theoretical plates** would be required to separate the mixture that started with composition L_1. Notice that this corresponds to the number of "steps" that need to be drawn in the figure to arrive at a composition of 100% A.

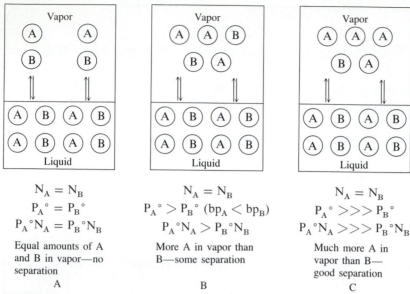

$$N_A = N_B$$
$$P_A^\circ = P_B^\circ$$
$$P_A^\circ N_A = P_B^\circ N_B$$

Equal amounts of A
and B in vapor—no
separation

A

$$N_A = N_B$$
$$P_A^\circ > P_B^\circ \ (bp_A < bp_B)$$
$$P_A^\circ N_A > P_B^\circ N_B$$

More A in vapor than
B—some separation

B

$$N_A = N_B$$
$$P_A^\circ >>> P_B^\circ$$
$$P_A^\circ N_A >>> P_B^\circ N_B$$

Much more A in
vapor than B—
good separation

C

FIGURE 10–7. Consequences of Raoult's Law. A. Boiling points (vapor pressures) are identical—**no separation;** B. Boiling point somewhat less for A than for B—**requires fractional distillation;** C. Boiling point much less for A than for B—**simple distillation will suffice.**

Most columns do not allow distillation in discrete steps, as indicated in Figure 10–4. Instead the process is **continuous,** allowing the vapors to be continuously in contact with liquid of changing composition as they pass through the column. Any material can be used to pack the column as long as it can be wetted by the liquid and as long as it does not pack so tightly that vapor cannot pass.

The approximate relationship between the number of theoretical plates needed to separate an ideal two-component mixture and the difference in boiling points is given in Table 10–2. Notice that more theoretical plates are required as the boiling-point differences between the components decrease. The entries in the table were derived

**TABLE 10–2. Theoretical Plates
Required to Separate Mixtures,
Based on Boiling-Point
Differences of Components**

BOILING-POINT DIFFERENCE	NUMBER OF THEORETICAL PLATES
108	1
72	2
54	3
43	4
36	5
20	10
10	20
7	30
4	50
2	100

from Equation 6 and rounded off to the nearest unit (up to 5) or the nearest multiple of ten (above 5). Kelvin temperatures were used, and an average boiling point of 150 °C was assumed for each mixture.

$$\text{Theoretical plates required} = \frac{T_1 + T_2}{3(T_2 - T_1)} \tag{6}$$

Equation 6 is very approximate, and most actual mixtures will require more than the calculated number of theoretical plates. Nevertheless, the equation provides a very useful guide. For instance, a mixture of A (bp 130 °C) and B (bp 166 °C) with a boiling-point difference of 36 °C would be expected to require a column with a minimum of five theoretical plates.

10.5 TYPES OF FRACTIONATING COLUMNS AND PACKINGS

Several types of fractionating columns are shown in Figure 10–8. The Vigreux column, shown in Part A, has indentations that incline downwards at angles of 45° and are in pairs on opposites sides of the column. The projections into the column provide increased possibilities for condensation and for the vapor to equilibrate with the liquid. Vigreux columns are popular in cases where only a small number of theoretical plates are required. They are not very efficient (a 20 cm column might have only 2.5 theoretical plates), but they allow for rapid distillation and have a small **holdup** (the amount of liquid retained by the column). A column packed with a stainless steel sponge is a more effective fractionating column than a Vigreux column, but not by a large margin. Glass beads, or glass helices, can also be used as a packing material, and they have a slightly

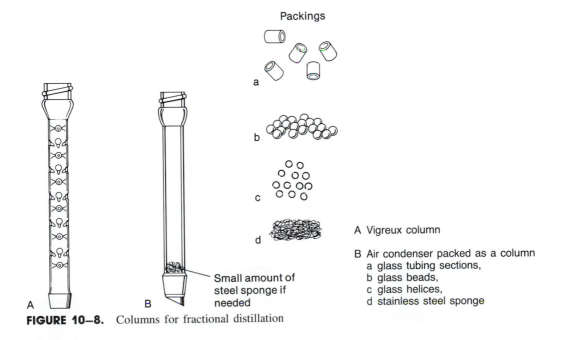

FIGURE 10–8. Columns for fractional distillation

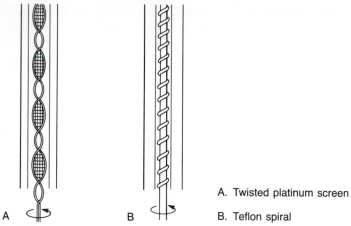

A. Twisted platinum screen

B. Teflon spiral

FIGURE 10–9.   Bands for spinning-band columns

greater efficiency yet. The air condenser or the water condenser can be used as an improvised column if an actual fractionating column is unavailable. If a condenser is packed with glass beads, glass helices, or sections of glass tubing, the packing must be held in place by inserting a small plug of stainless steel sponge into the bottom of the condenser.

The most effective type of column is the **spinning band column.** In the most elegant form of this device, a tightly-fitting twisted platinum screen or a Teflon rod with helical threads is rotated rapidly inside the bore of the column (Figure 10–9). Two different spinning band columns that are available for microscale work are shown in Figure 10–10. The larger column, shown in Part A, has a spinning band which is about 2.5 inches long, and it has about 12 theoretical plates. The smaller spinning band column, shown in part B, has a band about 2–3 cm in length and provides about 4–5 theoretical plates. It can separate 1–2 mL of a mixture with a 30° boiling-point difference. Larger research models of these spinning band columns can provide as many as 20 or 30 theoretical plates and can separate mixtures with a boiling-point difference of as little as 5–10 degrees.

Manufacturers of fractionating columns often offer them in a variety of lengths. Since the efficiency of a column is a function of its length, longer columns will have more theoretical plates than shorter ones. It is common to express efficiency of a column in a unit called **HETP,** the **H**eight of a column that is **E**quivalent to one **T**heoretical **P**late. HETP is usually expressed in units of cm/plate. When the height of the column (in cm) is divided by this value, the total number of theoretical plates is specified.

10.6 FRACTIONAL DISTILLATION: METHODS AND PRACTICE

When performing a fractional distillation, the column should be clamped in a vertical position. The distillation should be conducted as slowly as possible, but the rate of distillation should be steady enough to produce a constant temperature reading at the thermometer. Many fractionating columns must be insulated so that temperature equilibrium is established at all times. Additional insulation will not be required for columns that have an evacuated outer jacket, but those that do not can benefit from being wrapped either by glass wool or aluminum foil (shiny side in). The **reflux ratio** is defined as the ratio of the number of drops of distillate that return to the distilling flask compared to the number of drops of distillate collected. In an efficient column the reflux ratio should equal or exceed the number of theoretical plates. A high reflux ratio assures that the column will achieve temperature equilibrium and achieve its maximum efficiency. This ratio is not easy to determine, in fact, it is impossible to determine when using a Hickman head, and it should not concern a beginning student. In some cases the **throughput,** or **rate of takeoff,** of a column may be specified. This is expressed as the number milliliters of distillate that can be collected per unit time, usually as mL/min.

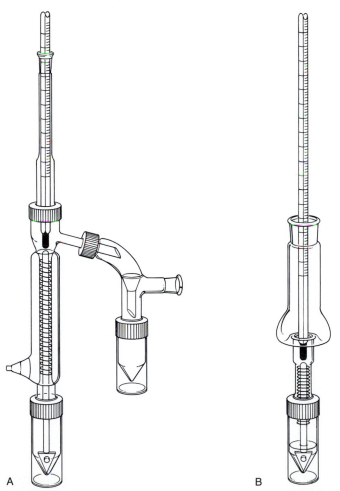

FIGURE 10–10.   Two commercially available microscale spinning-band columns

Microscale Apparatus. The apparatus shown in Figure 10–2 is the one you are most likely to use in the microscale laboratory. If your laboratory is one of the better equipped ones, you may have access to spinning band columns like those shown in Figure 10–10.

Macroscale Apparatus. Figure 10–11 illustrates a fractional distillation assembly that can be used for larger scale distillations. It has a glass-jacketed column which is packed with a stainless steel sponge. This apparatus would be common in situations where quantities of liquid in excess of 10 mL were to be distilled.

PART B. AZEOTROPES

10.7 NONIDEAL SOLUTIONS: AZEOTROPES

Some mixtures of liquids, because of attractions or repulsions between the molecules, do not show ideal behavior; they do not follow Raoult's Law. There are two types of vapor-liquid composition diagrams that result from this non-ideal behavior: **minimum-boiling-point** and **maximum-boiling-point** diagrams. The minimum or maximum points in these diagrams correspond to a constant-boiling mixture called an **azeotrope.** An azeotrope is a mixture with a fixed composition that cannot be altered by either

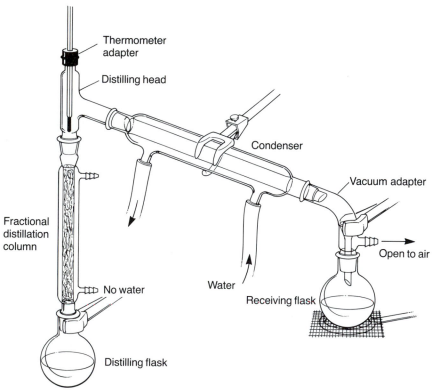

FIGURE 10–11. Large scale fractional distillation apparatus

simple or fractional distillation. An azeotrope behaves as if it were a pure compound, and it distills from the beginning to the end of its distillation at a constant temperature giving a distillate of constant (azeotropic) composition. The vapor in equilibrium with an azeotropic liquid has the same composition as the azeotrope. Because of this an azeotrope is represented as a **point** on a vapor-liquid composition diagram.

A. Minimum-Boiling-Point Diagrams

A minimum-boiling-point azeotrope results from a slight incompatibility (repulsion) between the liquids being mixed. This incompatibility leads to a higher-than-expected combined vapor pressure from the solution. This higher combined vapor pressure brings about a lower boiling point for the mixture than is observed for the pure components. The most common two-component mixture that gives a minimum-boiling-point azeotrope is the ethanol-water system shown in Figure 10–12. The azeotrope at V_3 has a composition of 96% ethanol-4% water and a boiling point of 78.1 °C. This boiling point is not much lower than that of pure ethanol (78.3 °C), but it means that it is impossible to obtain pure ethanol from the distillation of any ethanol-water mixture that contains more than 4% water. Even with the best fractionating column, one cannot obtain 100% ethanol! The remaining 4% of water can be removed by adding benzene and removing a different azeotrope, the ternary benzene-water-ethanol azeotrope (bp 65 °C). Once the water is removed, the excess benzene is removed as an ethanol-benzene azeotrope (bp 68 °C). The resulting material will be free of water and is called "absolute" ethanol.

The fractional distillation of an ethanol-water mixture of composition X can be described as follows. The mixture is heated (follow line X-L_1) until it is observed to boil at L_1. The resulting vapor at V_1 will be richer in the lower-boiling component,

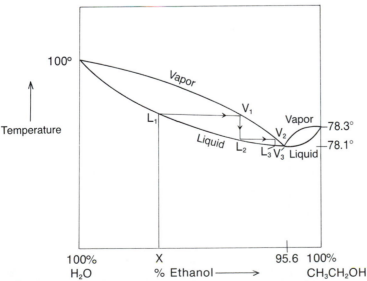

FIGURE 10–12. Ethanol–water minimum-boiling-point phase diagram

ethanol, than the original mixture. The condensate at L_2 is vaporized to give V_2. The process continues, following the lines to the right, until the azeotrope is obtained at V_3. The liquid that distills is not pure ethanol, but it has the azeotropic composition of 96% ethanol and 4% water, and it distills at 78.1 °C. The azeotrope, which is richer in ethanol than the original mixture, continues to distill. As it distills, the percentage of water left behind in the distilling flask continues to increase. When all of the ethanol has been distilled (as the azeotrope), pure water remains behind in the distilling flask, and it distills at 100 °C.

If the azeotrope obtained by the procedure above is redistilled, it distills from the beginning to the end of the distillation at a constant temperature of 78.1 °C as if it were a pure substance. There is no change in the composition of the vapor during the distillation.

Some common minimum-boiling azeotropes are given in Table 10–3. Numerous other azeotropes are formed in two- and three-component systems; such azeotropes are common. Water forms azeotropes with many substances; therefore, water must be carefully removed with **drying agents** whenever possible before compounds are distilled. Extensive azeotropic data are available in references such as the *Handbook of Chemistry and Physics*.[1]

B. Maximum-Boiling-Point Diagrams

A maximum-boiling-point azeotrope results from a slight attraction between the component molecules. This attraction leads to lower combined vapor pressure than expected in the solution. The lower combined vapor pressures cause a higher boiling point than what would be characteristic for the components. A two-component maximum-boiling-point azeotrope is illustrated in Figure 10–13. Since the azeotrope has a

TABLE 10–3.   Common Minimum-Boiling Azeotropes

AZEOTROPE	COMPOSITION (Weight percentage)	BOILING POINT (°C)
Ethanol–water	95.6% C_2H_5OH, 4.4% H_2O	78.17
Benzene–water	91.1% C_6H_6, 8.9% H_2O	69.4
Benzene–water–ethanol	74.1% C_6H_6, 7.4% H_2O, 18.5% C_2H_5OH	64.9
Methanol–carbon tetrachloride	20.6% CH_3OH, 79.4% CCl_4	55.7
Ethanol–benzene	32.4% C_2H_5OH, 67.6% C_6H_6	67.8
Methanol–toluene	72.4% CH_3OH, 27.6% $C_6H_5CH_3$	63.7
Methanol–benzene	39.5% CH_3OH, 60.5% C_6H_6	58.3
Cyclohexane–ethanol	69.5% C_6H_{12}, 30.5% C_2H_5OH	64.9
2-Propanol–water	87.8% $(CH_3)_2CHOH$, 12.2% H_2O	80.4
Butyl acetate–water	72.9% $CH_3COOC_4H_9$, 27.1% H_2O	90.7
Phenol–water	9.2% C_6H_5OH, 90.8% H_2O	99.5

[1] More examples of azeotropes, with their compositions and boiling points, can be found in the *CRC Handbook of Chemistry and Physics;* also in L. H. Horsley, ed., *Advances in Chemistry Series,* no. 116, Azeotropic Data, III, (Washington: American Chemical Society, 1973).

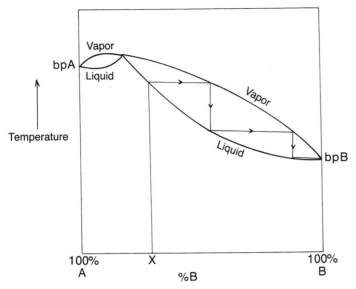

FIGURE 10-13.   A maximum-boiling-point phase diagram

higher boiling point than any of the components, it will be concentrated in the distilling flask as the distillate (pure B) is removed. The distillation of a solution of composition X would follow to the right along the lines in Figure 10–13. Once the composition of the material remaining in the flask has reached that of the azeotrope, the temperature will rise, and the azeotrope will begin to distill. The azeotrope will continue to distill until all of the material in the distillation flask has been exhausted.

Some maximum-boiling azeotropes are listed in Table 10–4. They are not nearly as common as minimum-boiling azeotropes.[2]

C. Generalizations

There are some generalizations that can be made about azeotropic behavior. They are presented here without explanation, but you should be able to verify them by thinking through each case using the phase diagrams given above. (Note that pure A is always to

TABLE 10-4.   Maximum-Boiling Azeotropes

AZEOTROPE	COMPOSITION (Weight percentage)	BOILING POINT (°C)
Acetone–chloroform	20.0% CH_3COCH_3, 80.0% $CHCl_3$	64.7
Chloroform–methyl ethyl ketone	17.0% $CHCl_3$, 83.0% $CH_3COCH_2CH_3$	79.9
Hydrochloric acid	20.2% HCl, 79.8% H_2O	108.6
Acetic acid–dioxane	77.0% CH_3COOH, 23.0% $C_4H_8O_2$	119.5
Benzaldehyde–phenol	49.0% C_6H_5CHO, 51.0% C_6H_5OH	185.6

[2] See Footnote 1.

the left of the azeotrope in these diagrams, while pure B is to the right of the azeo-trope.)

A. Minimum-Boiling Azeotropes

INITIAL COMPOSITION	EXPERIMENTAL RESULT
to left of azeotrope	azeotrope distills first, pure A second
azeotrope	unseparable
to right of azeotrope	azeotrope distills first, pure B second

B. Maximum-Boiling Azeotropes

INITIAL COMPOSITION	EXPERIMENTAL RESULT
to left of azeotrope	pure A distills first, azeotrope second
azeotrope	unseparable
to right of azeotrope	pure B distills first, azeotrope second

10.8   AZEOTROPIC DISTILLATION: APPLICATIONS

There are numerous examples of chemical reactions in which the amount of product is low because of an unfavorable equilibrium. An example is the direct acid-catalyzed esterification of a carboxylic acid with an alcohol:

$$R-\overset{\overset{\textstyle O}{\|}}{C}-OH + R-O-H \underset{}{\overset{H^+}{\rightleftarrows}} R-\overset{\overset{\textstyle O}{\|}}{C}-OR + H_2O$$

Since the equilibrium does not favor formation of the ester, it must be shifted to the right, in favor of the product, by using an excess of one of the starting materials. In most cases, the alcohol is the least expensive reagent and is the material used in excess. Isopentyl acetate (Experiment 6) and methyl salicylate (Experiment 43) are examples of esters prepared by using one of the starting materials in excess.

Another way of shifting the equilibrium to the right is to remove one of the products from the reaction mixture as it is formed. In the above example, water can be removed as it is formed by **azeotropic distillation.** A common large scale method is to use the Dean-Stark water separator shown in Figure 10–14A. In this technique, an inert solvent, commonly benzene or toluene, is added to the reaction mixture contained in the round-bottomed flask. The sidearm of the water separator is also filled with this solvent. If benzene is used, as the mixture is heated under reflux, the benzene-water azeotrope (bp 69.4 °C, Table 10–3) distills out of the flask.[3] When the vapor con-

---

[3] Actually, with ethanol, a lower-boiling three-component azeotrope distills at 64.9 °C (see Table 10–3). It consists of benzene-water-ethanol. Because some ethanol is lost in the azeotropic distillation, a large excess of ethanol is used in esterification reactions. The excess also helps to shift the equilibrium to the right.

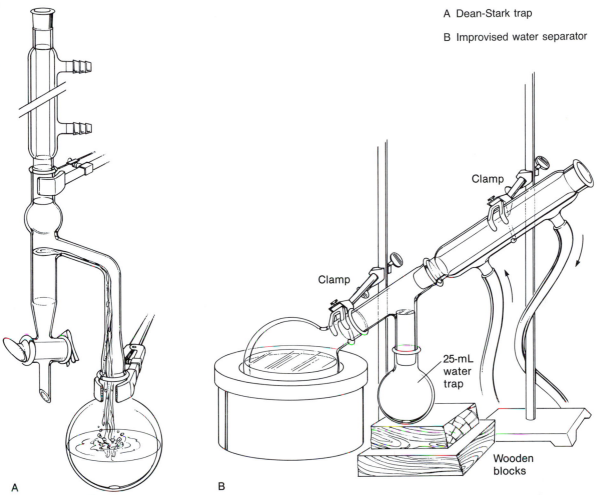

A Dean-Stark trap

B Improvised water separator

FIGURE 10–14. Large scale water separators

denses, it enters the sidearm directly below the condenser, and water separates from the benzene-water condensate; benzene and water mix as vapors, but they are not miscible as cooled liquids. Once the water (lower phase) separates from the benzene (upper phase), liquid benzene overflows from the sidearm back into the flask. The cycle is repeated continuously until no more water forms in the sidearm. One may calculate the weight of water that should theoretically be produced and compare this value with the amount of water collected in the sidearm. Since the density of water equals unity, the volume of water collected can be compared directly with the calculated amount, assuming 100% yield.

An improvised water separator, constructed from the components found in the traditional organic kit, is shown in Figure 10–14B. Although this requires the condenser to be placed in a non-vertical position, it nevertheless works quite well.

At the microscale level, water separation can be achieved using a standard distillation assembly with a water condenser and a Hickman head (Figure 10–15). The side-ported variation of the Hickman head is the most convenient one to use for this

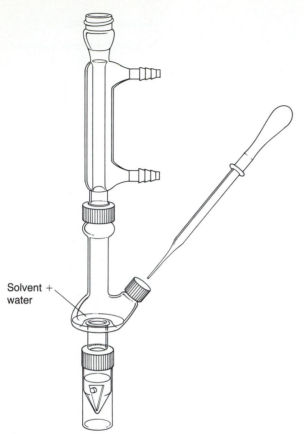

FIGURE 10–15. Microscale water separator (both layers are removed)

purpose, but it is not essential. In this variation, one simply removes all of the distillate (both solvent and water) several times during the course of the reaction. A Pasteur pipet is used to remove the distillate as shown in Technique 8 (Figure 8–6, p 644). Since both the solvent and water are removed in this procedure, it may be desirable to add additional solvent from time to time, adding it through the condenser with a Pasteur pipet.

The most important consideration in using azeotropic distillation to prepare an ester (described on p 684) is that the azeotrope containing water must have a **lower boiling point** than the alcohol used. With ethanol, the benzene-water azeotrope boils at a much lower temperature (69.4 °C) than ethanol (78.3 °C), and the technique described above works well. With higher-boiling alcohols, azeotropic distillation also works well because of the large boiling-point difference between the azeotrope and the alcohol.

However, with methanol (bp 65 °C), the boiling point of the benzene-water azeotrope is actually **higher** by about 5 °, and methanol distills first. Thus, in esterifications involving methanol, a totally different approach must be taken. For example, one can mix carboxylic acid, methanol, the acid catalyst, and **1,2-dichloroethane** in a conventional reflux apparatus (Technique 3, Figure 3–2, p 550) without a water separator. During the reaction, water separates from the 1,2-dichloroethane because it is not soluble; however, the remainder of the components are soluble, so the reaction can continue. The equilibrium is shifted to the right by the "removal" of water from the reaction mixture.

Azeotropic distillation is also used in other types of reactions, such as in ketal or acetal formation, and in enamine formation. The use of azeotropic distillation is illustrated in the formation of 2-acetylcyclohexanone (Experiment 33) via the enamine intermediate. Toluene is used in the azeotropic distillation of water. The Hickman head is used as a water separator.

Acetal Formation

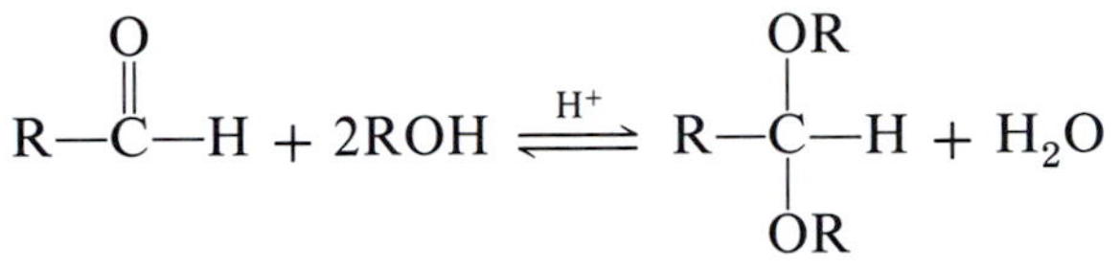

$$R-\overset{\overset{\displaystyle O}{\|}}{C}-H + 2ROH \underset{}{\overset{H^+}{\rightleftharpoons}} R-\overset{\overset{\displaystyle OR}{|}}{\underset{\underset{\displaystyle OR}{|}}{C}}-H + H_2O$$

Enamine Formation

$$RCH_2-\overset{\overset{}{\underset{\underset{\displaystyle O}{\|}}{C}}}{}-CH_2R + \underset{\underset{\displaystyle H}{|}}{\overset{\displaystyle \big\langle N \big\rangle}{}} \overset{H^+}{\rightleftharpoons} RCH=C-CH_2R + H_2O$$

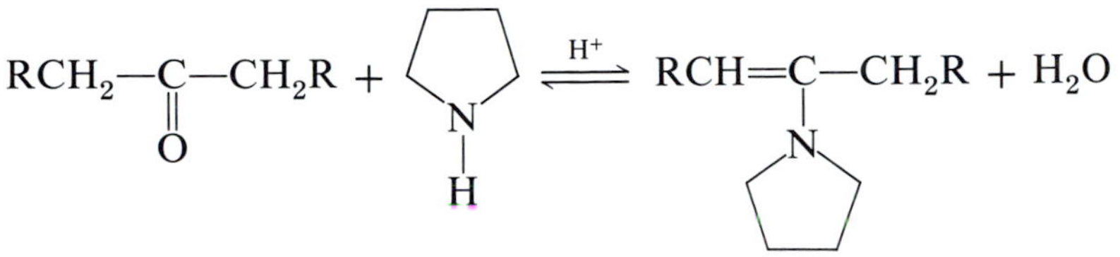

PROBLEMS

1. In the accompanying chart are approximate vapor pressures for benzene and toluene at various temperatures.

TEMP. (°C)	mmHg	TEMP. (°C)	mmHg
Benzene 30	120	Toluene 30	37
40	180	40	60
50	270	50	95
60	390	60	140
70	550	70	200
80	760	80	290
90	1010	90	405
100	1340	100	560
		111	760

(a) What is the mole fraction of each component if 3.9 g of benzene, C_6H_6, is dissolved in 4.6 g of toluene, C_7H_8?

(b) Assuming that this mixture is ideal, that is, it follows Raoult's Law, what is the partial vapor pressure of benzene in this mixture at 50 °C?

(c) Estimate to the nearest degree the temperature at which the vapor pressure of the solution equals 1 atm (bp of the solution).

(d) Calculate the composition of the vapor (mole fraction of each component) that is in equilibrium in the solution at the boiling point of this solution.

(e) Calculate the composition in weight percentage of the vapor that is in equilibrium with the solution.

2. Estimate how many theoretical plates are needed to separate a mixture that has a mole fraction of B equal to 0.70 (70% B) in Figure 10–4.

3. Two moles of sucrose are dissolved in eight moles of water. Assume that the solution follows Raoult's Law and that the vapor pressure of sucrose is negligible. The boiling point of water is 100 °C. The distillation is carried out at 1 atm (760 mmHg).
   (a) Calculate the vapor pressure of the solution when the temperature reaches 100 °C.
   (b) What temperature would be observed during the entire distillation?
   (c) What would be the composition of the distillate?
   (d) If a thermometer were immersed below the surface of the liquid of the boiling flask, what temperature would be observed?

4. Explain why the boiling point of a two-component mixture rises slowly throughout a simple distillation when the boiling-point differences are not large.

5. Given the boiling points of several known mixtures of A and B (mole fractions are known) and the vapor pressures of A and B in the pure state (P_A^0 and P_B^0) at these same temperatures, how would you construct a boiling-point-composition phase diagram for A and B. Give a stepwise explanation.

6. Describe the behavior on distillation of a 98% ethanol solution through an efficient column. Refer to Figure 10–12.

7. Construct an approximate boiling-point-composition diagram for a benzene-methanol system. The mixture shows azeotropic behavior (see Table 10–3). Include on the graph the boiling points of pure benzene and pure methanol, and the boiling point of the azeotrope. Qualitatively describe the behavior for a mixture that is initially rich in benzene (90%), and then for a mixture that is initially rich in methanol (90%).

8. Construct an approximate boiling-point-composition diagram for an acetone-chloroform system, which forms a maximum boiling azeotrope (Table 10–4). Qualitatively describe the behavior on distillation of a mixture that is initially rich in acetone (90%), then describe the behavior of a mixture that is initially rich in chloroform (90%).

9. Two compounds have boiling points of 130 °C and 150 °C. Estimate the number of theoretical plates needed to separate these substances in a fractional distillation.

10. A spinning band column has an HETP of 0.25 in/plate. If the column has 12 theoretical plates, how long is it?

Technique 11
STEAM DISTILLATION

The simple, vacuum, and fractional distillations described in Techniques 8, 9, and 10 are applicable to completely soluble (miscible) mixtures only. When liquids are **not** mutually soluble (immiscible), they can also be distilled, but with a somewhat different result. A mixture of immiscible liquids will boil at a lower temperature than the boiling points of any of the separate components as pure compounds. When steam is used to provide one of the immiscible phases, the process is called **steam distillation.** The advantage of this technique is that the desired material distills at a temperature below 100 °C. Thus, if unstable or very high-boiling substances are to be removed from a mixture, decomposition is avoided. Since all gases mix, the two substances can mix in the vapor and codistill. Once the distillate is cooled, the desired component, which is not miscible, separates from the water. Steam distillation is used widely in isolating liquids from natural sources. It is also used in removing a reaction product from a tarry reaction mixture.

11.1 DIFFERENCES BETWEEN DISTILLATION OF MISCIBLE AND IMMISCIBLE MIXTURES

$$\text{MISCIBLE LIQUIDS} \qquad P_{total} = P_A^0 N_A + P_B^0 N_B \qquad (1)$$

Two liquids A and B that are mutually soluble (miscible), and that do not interact, form an ideal solution and follow Raoult's Law, as shown in Equation 1. Note that the vapor pressures of pure liquids P_A^0 and P_B^0 are not added directly to give the total pressure P_{total}, but are reduced by the respective mole fractions N_A and N_B. The total pressure above a miscible or homogeneous solution will depend on P_A^0 and P_B^0 and also N_A and N_B. Thus, the composition of the vapor will also depend on **both** the vapor pressures and the mole fractions of each component.

$$\text{IMMISCIBLE LIQUIDS} \qquad P_{total} = P_A^0 + P_B^0 \qquad (2)$$

In contrast, when two mutually insoluble (immiscible) liquids are "mixed" to give a heterogeneous mixture, each exerts its own vapor pressure, independently of the other, as shown in Equation 2. The mole fraction term does not appear in this equation since the compounds are not miscible. One simply adds the vapor pressures of the pure liquids P_A^0 and P_B^0 at a given temperature to obtain the total pressure above the mixture. When the total pressure equals 760 mmHg, the mixture boils. The composition of the vapor from an immiscible mixture, in contrast to the miscible mixture, is determined only by the vapor pressures of the two substances codistilling. Equation 3 defines the composition of the vapor from an immiscible mixture. Calculations involving this equation are given in Section 11.2.

$$\frac{\text{Moles A}}{\text{Moles B}} = \frac{P_A^0}{P_B^0} \qquad (3)$$

A mixture of two immiscible liquids boils at a lower temperature than the boiling points of either component. The explanation for this behavior is like that given for minimum-boiling azeotropes (Technique 10, Section 10.7). Immiscible liquids behave as they do because an extreme incompatibility between the two liquids leads to higher combined vapor pressures than Raoult's Law would predict. The higher combined vapor pressures cause a lower boiling point for the mixture than for either single component. Thus, one may think of steam distillation as a special type of azeotropic distillation, in which the substance is completely insoluble in water.

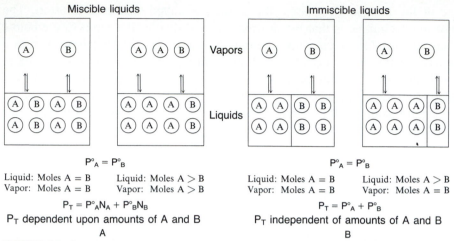

FIGURE 11–1. Total pressure behavior for miscible and immiscible liquids. A. Ideal miscible liquids follow Raoult's Law: P_T depends on the mole fractions and vapor pressures of A and B; B. Immiscible liquids do not follow Raoult's Law: P_T depends only on the vapor pressures of A and B.

The differences in behavior of miscible and immiscible liquids, where it is assumed that P_A^0 equals P_B^0, are shown in Figure 11–1. Note that with miscible liquids, the composition of the vapor depends on the relative amounts of A and B present (Figure 11–1A). Thus, the composition of the vapor must change during a distillation. In contrast, the composition of the vapor with immiscible liquids is independent of the amounts of A and B present (Figure 11–1B). Hence, the vapor composition must remain **constant** during the distillation of such liquids, as predicted by Equation 3. Immiscible liquids act as if they were being distilled simultaneously from separate compartments, as shown in Figure 11–1B, even though in practice they are "mixed" during a steam distillation. Since all gases mix, they do give rise to a homogeneous vapor and codistill.

11.2 IMMISCIBLE MIXTURES: CALCULATIONS

The composition of the distillate is constant during a steam distillation, and the boiling point of the mixture is also constant. The boiling points of steam-distilled mixtures will always be below the boiling point of water (bp 100 °C) and also below the boiling point of any of the other substances distilled. Some representative boiling points and compositions of steam distillates are given in Table 11–1. Note that the higher the boiling point of a pure substance, the more closely the temperature of the steam distillate approaches, but does not exceed, 100 °C. This is a reasonably low temperature, and it avoids the decomposition that might result at high temperatures with a simple distillation.

For immiscible liquids, the molar proportions of two components in a distillate equal the ratio of their vapor pressures in the boiling mixture, as given in Equation 3. When Equation 3 is rewritten for an immiscible mixture involving water,

TABLE 11–1. Boiling Points and Compositions of Steam Distillates

MIXTURE	BOILING POINT OF PURE SUBSTANCE (°C)	BOILING POINT OF MIXTURE (°C)	COMPOSITION (% WATER)
Benzene–water	80.1	69.4	8.9%
Toluene–water	110.6	85.0	20.2%
Hexane–water	69.0	61.6	5.6%
Heptane–water	98.4	79.2	12.9%
Octane–water	125.7	89.6	25.5%
Nonane–water	150.8	95.0	39.8%
1-Octanol–water	195.0	99.4	90.0%

Equation 4 results. Equation 4 can be modified by substituting the relation moles = (weight/molecular weight) to give Equation 5.

$$\frac{\text{moles substance}}{\text{moles water}} = \frac{P^0_{\text{substance}}}{P^0_{\text{water}}} \tag{4}$$

$$\frac{\text{wt substance}}{\text{wt water}} = \frac{(P^0_{\text{substance}})(\text{molecular weight}_{\text{substance}})}{(P^0_{\text{water}})(\text{molecular weight}_{\text{water}})} \tag{5}$$

A sample calculation using this equation is given in Figure 11–2. Notice that the result of this calculation is very close to the experimental value given in Table 11–1.

11.3   STEAM DISTILLATION: METHODS

Two methods for steam distillation are in general use in the laboratory: the **direct method** and the **live steam method.** In the first method, steam is generated *in situ* (in place) by heating a distilling flask containing the compound and water. In the second method, steam is generated outside, and it is passed into the distilling flask using an inlet tube.

A. Direct Method

Microscale.   The direct method of steam distillation is the only one suitable for microscale reactions. Steam is produced in the conical vial or distilling flask *(in situ)* by heating water to its boiling point in the presence of the compound to be distilled. This method works well for small amounts of materials. A microscale steam distillation apparatus is shown in Figure 11–3. Water and the compound to be distilled are placed in the flask and heated. A stirring bar or a boiling stone should be used to prevent bumping. An additional measure that can be used to control bumping is to place a small amount of stainless steel sponge in the stem of the Hickman head. This measure also controls foaming if it becomes a problem. The vapors of the water and the desired compound codistill when they are heated. They are condensed and collect in the Hickman head. When the Hickman head fills, the distillate is removed with a Pasteur pipet and placed in another vial for storage. For the typical microscale experiment, it

Problem   How many grams of water must be distilled to steam distill 1.55 g of 1-octanol from an aqueous solution? What will be the composition (wt %) of the distillate? The mixture distills at 99.4° C

Answer   The vapor pressure of water at 99.4° C must be obtained from the CRC Handbook (= 744 mmHg).

(a) Obtain the partial pressure of 1-octanol.

$$P°_{1\text{-octanol}} = P°_{total} - P°_{water}$$
$$P°_{1\text{-octanol}} = (760 - 744) = 16 \text{ mmHg}$$

(b) Obtain the composition of the distillate.

$$\frac{\text{wt 1-octanol}}{\text{wt water}} = \frac{(16)(130)}{(744)(18)} = 0.155 \text{ g/g-water}$$

(c) Clearly, 10 g of water must be distilled.

$$(0.155 \text{ g/g-water})(10 \text{ g-water}) = 1.55 \text{ g 1-octanol}$$

(d) Calculate the **weight** percentages.

$$1\text{-octanol} = 1.55 \text{ g}/(10 \text{ g} + 1.55 \text{ g}) = 13.4\%$$
$$\text{water} = 10 \text{ g}/(10 \text{ g} + 1.55 \text{ g}) = 86.6\%$$

FIGURE 11–2.   Sample calculations for a steam distillation

will be necessary to fill the well and remove the distillate three or four times. All of these distillate fractions are placed in the same storage container. The efficiency in collecting the distillate can sometimes be improved if the inside walls of the Hickman head are rinsed several times into the well. A Pasteur pipet is used to do the rinsing. Distillate is withdrawn from the well, and then it is used to wash the walls of the Hickman head, making sure to wash all the way around the head. After washing, and when the well is full, the distillate can be withdrawn and transferred to the storage container. It may be necessary to add additional water during the course of the distilla-

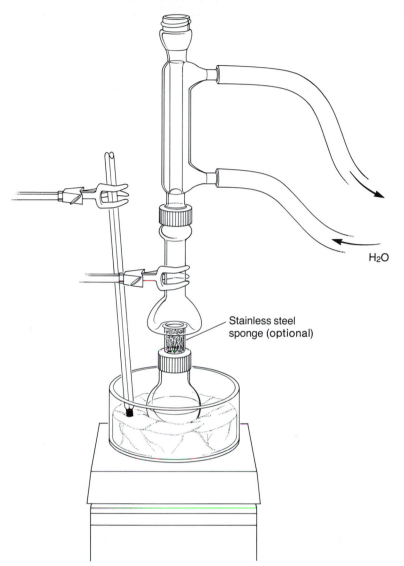

H₂O

FIGURE 11–3. Microscale steam distillation

tion. Additional water is added (remove the condenser if used) through the center of the
Hickman head by using a Pasteur pipet.

Macroscale. A larger scale direct method steam distillation is illustrated in
Figure 11–4. Although a heating mantle may be used, it is probably best to use a flame
with this method, since a large volume of water must be heated rapidly. A boiling stone
must be used to prevent bumping. The separatory funnel allows additional water to be
added during the course of the distillation.

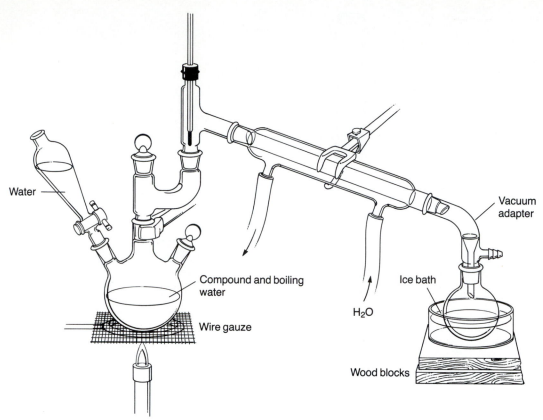

Water

Compound and boiling water

Wire gauze

Vacuum adapter

Ice bath

H_2O

Wood blocks

FIGURE 11–4. Macroscale direct steam distillation

B. Live Steam Method

Macroscale. A large scale steam distillation using the live steam method is shown in Figure 11–5. If steam lines are available in the laboratory, they may be attached directly to the steam trap (purge them first to drain water). If steam lines are not available, an external steam generator (see inset) must be prepared. The external generator usually will require a flame to produce steam at a rate fast enough for the distillation. When the distillation is first started, the clamp at the bottom of the steam trap is left open. The steam lines will have a large quantity of condensed water in them until they are well heated. When the lines become hot and condensation of steam ceases, the clamp may be closed. Occasionally, the clamp will have to be reopened to remove condensate. In this method, the steam agitates the mixture as it enters the bottom of the flask and a stirrer or boiling stone is not required.

> **Caution. Hot steam can produce very bad burns.**

Sometimes it is helpful to heat the three-necked distilling flask with a heating mantle (or flame) to prevent excessive condensation at that point. Steam must be admitted at a rate fast enough that you can see the distillate condensing as a milky white fluid in the

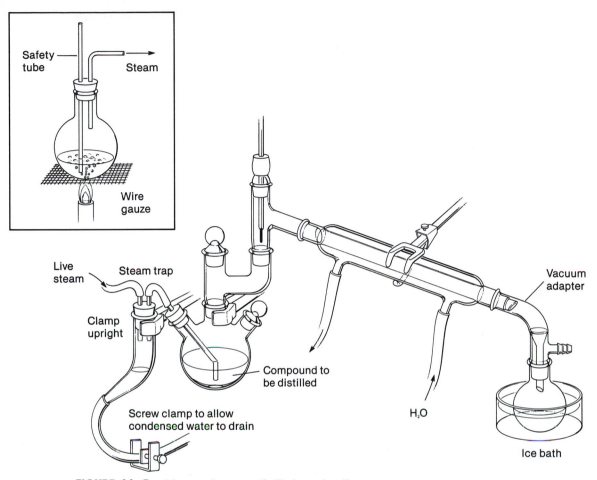

FIGURE 11-5. Macroscale steam distillation using live steam

condenser. The vapors that codistill will separate on cooling to give this cloudiness. When the condensate becomes clear, the distillation is near an end. The flow of water through the condenser should be faster than in other types of distillation to help cool the vapors. Make sure the vacuum adapter remains cool to the touch. An ice bath may be used to cool the receiving flask if desired. When the distillation is to be stopped, the screw clamp on the steam trap should be opened, and the steam inlet tube must be removed from the three-necked flask. If this is not done, liquid will back up into the tube and steam trap.

PROBLEMS

1. Calculate the weight of benzene codistilled with each gram of water and the percentage composition of the vapor produced during a steam distillation. The boiling point of the mixture is 69.4 °C. The vapor pressure of water at 69.4 °C is 227.7 mmHg. Compare the result with the data in Table 11–1.

2. Calculate the approximate boiling point of a mixture of bromobenzene and water at atmospheric pressure. A table of vapor pressures of water and bromobenzene at various temperatures is given.

TEMPERATURE (°C)	VAPOR PRESSURES (mmHg)	
	Water	*Bromobenzene*
93	588	110
94	611	114
95	634	118
96	657	122
97	682	127
98	707	131
99	733	136

3. Calculate the weight of nitrobenzene that codistills (bp 99 °C) with each gram of water during a steam distillation. You may need the data given in the previous problem.

4. A mixture of *p*-nitrophenol and *o*-nitrophenol can be separated by steam distillation. The *o*-nitrophenol is steam volatile while the *para* isomer is not volatile. Explain. Base your answer on the ability of the isomers to form hydrogen bonds internally.

Technique 12
COLUMN CHROMATOGRAPHY

The most modern and sophisticated methods of separating mixtures that the organic chemist has available all involve **chromatography.** Chromatography is defined as the separation of a mixture of two or more different compounds or ions by distribution between two phases, one of which is stationary and the other moving. Various types of chromatography are possible, depending on the nature of the two phases involved: **solid-liquid** (column, thin-layer, and paper), **liquid-liquid,** (high performance liquid chromatography), and **gas-liquid** (vapor-phase) chromatographic methods are common.

All chromatography works on much the same principle as solvent extraction (Technique 7). Basically, the methods depend on differential solubilities or adsorptivities of the substances to be separated relative to the two phases between which they are to be partitioned. In this chapter, column chromatography, a solid-liquid method is considered. A section of this chapter is also devoted to a brief examination of high performance liquid chromatography. Thin-layer chromatography is examined in Technique 13; gas chromatography, a gas-liquid method, is discussed in Technique 14.

12.1 ADSORBENTS

Column chromatography is a technique based on both adsorptivity and solubility. It is a solid-liquid phase-partitioning technique. The solid may be almost any material that does not dissolve in the associated liquid phase; those solids most commonly used are silica gel, $SiO_2 \cdot xH_2O$, also called silicic acid, and alumina, $Al_2O_3 \cdot xH_2O$. These compounds are used in their powdered or finely ground (usually 200- to 400-mesh) forms.

Most alumina used for chromatography is prepared from the impure ore bauxite, $Al_2O_3 \cdot xH_2O + Fe_2O_3$. The bauxite is dissolved in hot sodium hydroxide and filtered to remove the insoluble iron oxides; the alumina in the ore forms the soluble amphoteric hydroxide $Al(OH)_4^-$. The hydroxide is precipitated by CO_2 (which reduces the pH) as $Al(OH)_3$. When heated, the $Al(OH)_3$ loses water to form pure alumina, Al_2O_3.

$$\text{Bauxite (crude)} \xrightarrow{\text{hot NaOH}} Al(OH)_4^- \text{ (aq)} + Fe_2O_3 \text{ (insoluble)}$$

$$Al(OH)_4^- \text{ (aq)} + CO_2 \longrightarrow Al(OH)_3 + HCO_3^-$$

$$2Al(OH)_3 \xrightarrow{\text{heat}} Al_2O_3(s) + 3H_2O$$

Alumina prepared in this way is called **basic alumina,** because it still contains some hydroxides. Basic alumina cannot be used for chromatography of compounds that are base-sensitive. Therefore, it is washed with acid to neutralize the base, giving **acid-washed alumina.** This material is unsatisfactory unless it has been washed with enough water to remove **all** the acid; on being so washed, it becomes the best chromatographic material, called **neutral alumina.** If a compound is acid-sensitive, either basic or neutral alumina must be used. One should be careful to ascertain what type of alumina is being used for chromatography. Silica gel is not available in any form other than that suitable for chromatography.

12.2 INTERACTIONS

If powdered or finely ground alumina (or silica gel) is added to a solution containing an organic compound, some of the organic compound will **adsorb** onto or adhere to the fine particles of alumina. Many kinds of intermolecular forces cause organic molecules to bind to alumina. These forces vary in strength according to their type. Non-polar compounds bind to the alumina using only van der Waals forces. These are weak forces, and nonpolar molecules do not bind strongly unless they have extremely high molecular weights. The most important interactions are those typical of polar organic compounds. Either these forces are of the dipole-dipole type or they involve some direct interaction (coordination, hydrogen-bonding, or salt formation). These types of interactions are illustrated in Figure 12–1, which for convenience shows only a portion of the alumina structure. Similar interactions occur with silica gel. The strengths of such interactions vary in the approximate order:

Salt formation > coordination > hydrogen-bonding > dipole-dipole > van der Waals

Strength of interaction varies among compounds. For instance, a strongly basic amine would bind more strongly than a weakly basic one (by coordination). In fact, strong bases and strong acids often interact so strongly that they **dissolve** alumina to some extent. One can use the following rule of thumb:

> **The more polar the functional group, the stronger the bond to alumina (or silica gel).**

A similar rule holds for solubility. Polar solvents dissolve polar compounds more effectively than nonpolar solvents; nonpolar compounds are dissolved best by nonpolar solvents. Thus, the extent to which any given solvent can wash an adsorbed compound from alumina depends almost directly on the relative polarity of the solvent. For example, while a ketone adsorbed on alumina might not be removed by hexane, it might be removed completely by chloroform. For any adsorbed material, a kind of **distribution** equilibrium can be envisioned between the adsorbent material and the solvent. This is illustrated in Figure 12–2.

The distribution equilibrium is **dynamic,** with molecules constantly **adsorbing** from the solution and **desorbing** into it. The average number of molecules remaining adsorbed on the solid particles at equilibrium depends both on the particular molecule (RX) involved and the dissolving power of the solvent with which the adsorbent must compete.

12.3   PRINCIPLE OF COLUMN CHROMATOGRAPHIC SEPARATION

The dynamic equilibrium mentioned above, and the variations in the extent to which different compounds adsorb on alumina or silica gel, underlie a versatile and ingenious

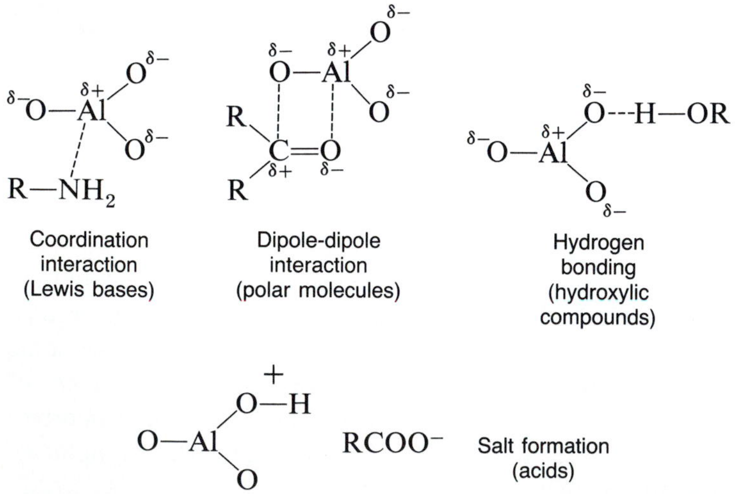

FIGURE 12–1.   Possible interactions of organic compounds with alumina

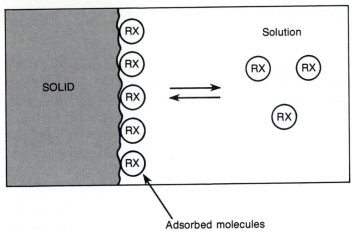

Adsorbed molecules

FIGURE 12–2. Dynamic adsorption equilibrium

method for **separating** mixtures of organic compounds. In this method, the mixture of compounds to be separated is introduced onto the top of a cylindrical glass column (Figure 12–3) **packed** or filled with fine alumina particles (stationary solid phase). The adsorbent is continuously washed by a flow of solvent (moving phase) passing through the column.

Initially, the components of the mixture adsorb onto the alumina particles at the top of the column. The continuous flow of solvent through the column **elutes,** or

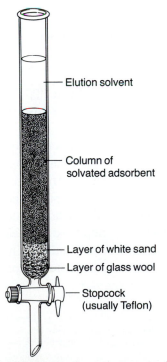

— Elution solvent

— Column of
solvated adsorbent

— Layer of white sand
— Layer of glass wool

— Stopcock
(usually Teflon)

FIGURE 12–3. Chromatographic column

washes, the solutes off the alumina and sweeps them down the column. The solutes (or materials to be separated) are called **eluates** or **elutants;** and the solvents are called **eluents.** As the solutes pass down the column to fresh alumina, new equilibria are established between the adsorbent, the solutes, and the solvent. The constant equilibration means that different compounds will move down the column at differing rates depending on their relative affinity for the adsorbent on one hand, and for the solvent on the other. Since the number of alumina particles is large, since they are closely packed, and since fresh solvent is being added continuously, the number of equilibrations between adsorbent and solvent that the solutes experience is enormous.

As the components of the mixture are separated, they begin to form moving bands (or zones), each band containing a single component. If the column is long enough and the various other parameters (column diameter, adsorbent, solvent, and flow rate) are correctly chosen, the bands separate from one another, leaving gaps of pure solvent in between. As each band (solvent and solute) passes out the bottom of the column, it can be collected before the next band arrives. If the parameters mentioned are poorly chosen, the various bands either overlap or coincide, in which case either a poor separation or no separation at all is the result. A successful chromatographic separation is illustrated in Figure 12–4.

12.4 PARAMETERS AFFECTING SEPARATION

The versatility of column chromatography results from the many factors that can be adjusted. These include:

1. Adsorbent chosen
2. Polarity of the column or solvents chosen
3. Size of the column (both length and diameter) relative to the amount of material to be chromatographed
4. Rate of elution (or flow)

By careful choice of conditions, almost any mixture can be separated. This technique has even been used to separate optical isomers. An optically active solid-phase adsorbent is used to separate the enantiomers.

Two fundamental choices for anyone attempting a chromatographic separation are the kind of adsorbent and the solvent system. In general, nonpolar compounds pass through the column faster than polar compounds since they have a smaller affinity for the adsorbent. If the adsorbent chosen binds all the solute molecules (both polar and nonpolar) strongly, they will not move down the column. On the other hand, if too polar a solvent is chosen, all the solutes (polar and nonpolar) may simply be washed through the column, with no separation taking place. The adsorbent and the solvent should be chosen so that neither is favored excessively in the equilibrium competition for solute molecules.[1]

---

[1] Often the chemist uses thin-layer chromatograph (TLC), which is described in Technique 13, to arrive at the best choices of solvents and adsorbents for the best separation. The TLC experimentation can be done quickly and with extremely small amounts (microgram quantities) of the mixture to be separated. This brings great savings of time and materials. Technique 13 describes this use of TLC.

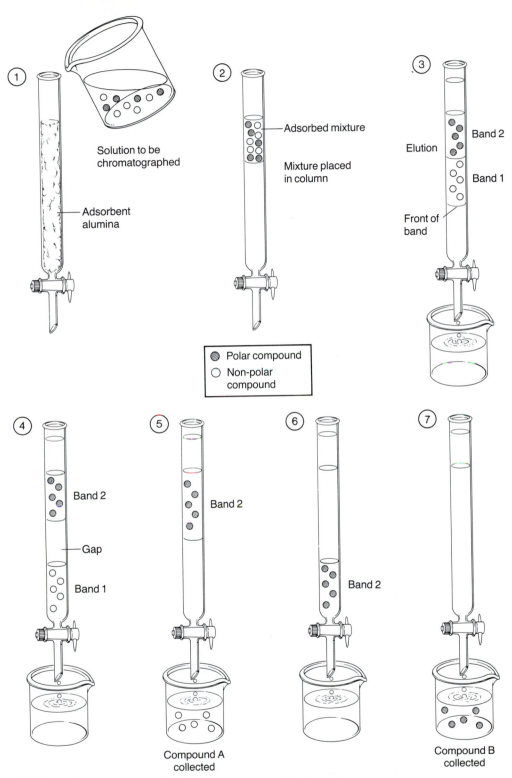

FIGURE 12—4. Sequence of steps in a chromatographic separation

A. Adsorbents

In Table 12–1 various kinds of adsorbents (solid phases) used in column chromatography are listed. The choice of adsorbent often depends on the types of compounds to be separated. Cellulose, starch, and sugars are used for polyfunctional plant and animal materials (natural products) very sensitive to acid-base interactions. Magnesium silicate is often used for separating acetylated sugars, steroids, and essential oils. Silica gel and Florisil are relatively mild toward most compounds and are widely used for a variety of functional groups—hydrocarbons, alcohols, ketones, esters, acids, azo compounds, amines. Alumina is the most widely used adsorbent and is obtained in the three forms mentioned in Section 12.1: acidic, basic, and neutral. The pH of acidic or acid-washed alumina is approximately 4. This adsorbent is particularly useful for separating acidic materials such as carboxylic acids and amino acids. Basic alumina has a pH of 10 and is useful in separating amines. Neutral alumina can be used to separate a variety of non-acidic and non-basic materials.

The approximate strength of the various adsorbents listed in Table 12–1 is also given. The order is only approximate, and therefore it may vary. For instance, the strength, or separating abilities, of alumina and silica gel largely depend on the amount of water present. Water binds very tightly to either adsorbent, taking up sites on the particles that could otherwise be used for equilibration with solute molecules. If one adds water to the adsorbent it is said to have been **deactivated.** Anhydrous alumina or silica gel are said to be highly **activated.** High activity is usually avoided with these adsorbents. Use of the highly active forms of either alumina or silica gel, or of the acidic or basic forms of alumina, can often lead to molecular rearrangement or decomposition in certain types of solute compounds.

The chemist can select the degree of activity which is appropriate to carry out a particular separation. To accomplish this, highly activated alumina is mixed thoroughly with a precisely measured quantity of water. The water partially hydrates the alumina and thus reduces its activity. By carefully determining the amount of water required, the chemist can have available an entire spectrum of possible activities.

TABLE 12–1.   Solid Adsorbents for Column Chromatography

Paper	
Cellulose	
Starch	
Sugars	
Magnesium silicate	
Calcium sulfate	**Increasing strength of**
Silicic acid	**binding interactions**
Silica gel	**toward polar compounds**
Florisil	
Magnesium oxide	
Aluminum oxide (Alumina)*	
Activated charcoal (Norit)	

*Basic, acid-washed, and neutral.

B. Solvents

In Table 12–2 some common chromatographic solvents are listed along with their relative ability to dissolve polar compounds. Sometimes a single solvent can be found that will separate all the components of a mixture. Sometimes a mixture of solvents can be found that will achieve separation. More often, one must start elution with a nonpolar solvent to remove relatively nonpolar compounds from the column and then gradually increase the solvent polarity to force compounds of greater polarity to come down the column, or elute. The approximate order in which various classes of compounds elute by this procedure is given in Table 12–3. In general, nonpolar compounds travel through the column faster (elute first), and polar compounds travel more slowly (elute last). However, molecular weight is also a factor in determining order of elution. A nonpolar compound of high molecular weight travels more slowly than a nonpolar compound of low molecular weight, and it may even be passed by some polar compounds.

TABLE 12–2. Solvents (Eluents) for Chromatography

Petroleum ether	
Cyclohexane	
Carbon tetrachloride*	
Toluene	
Chloroform*	**Increasing polarity**
Methylene chloride	**and ''solvent power''**
Diethyl ether	**toward polar functional**
Ethyl acetate	**groups**
Acetone	
Pyridine	
Ethanol	
Methanol	
Water	
Acetic acid	

*Suspected carcinogens.

TABLE 12–3. Elution Sequence for Compounds

Hydrocarbons	**Fastest (will elute with nonpolar solvent)**
Olefins	
Ethers	
Halocarbons	
Aromatics	
Ketones	**Order of elution**
Aldehydes	
Esters	
Alcohols	
Amines	
Acids, strong bases	**Slowest (need a polar solvent)**

When the polarity of the solvent has to be changed during a chromatographic separation, some precautions must be taken. Rapid changes from one solvent to another are to be avoided (especially when silica gel or alumina is involved). Usually, small percentages of a new solvent are mixed slowly into the one in use until the percentage reaches the desired level. If this is not done, the column packing often "cracks" as a result of the heat liberated when alumina or silica gel is mixed with a solvent. The solvent solvates the adsorbent, and the formation of a weak bond generates heat.

$$\text{Solvent} + \text{alumina} \longrightarrow (\text{alumina} \cdot \text{solvent}) + \text{heat}$$

Often enough heat is generated locally to evaporate the solvent. The formation of vapor creates bubbles, which force a separation of the column packing; this is called **cracking.** A cracked column does not give a good separation since it has discontinuities in the **packing.** The way in which a column is packed or filled with adsorbent is also very important in preventing cracking.

That the solvent itself has a tendency to adsorb on the alumina is an important factor in how compounds move down the column. The solvent can displace the adsorbed compound if it is more polar than the compound, and hence can move it down the column. Thus, a more polar solvent not only dissolves more compound, but also is effective in removing the compound from the alumina since it displaces the compound from its site of adsorption.

Certain solvents should be avoided with alumina or silica gel, especially with the acidic, basic, and highly active forms. For instance, with any of these adsorbents, acetone dimerizes *via* an aldol condensation to give diacetone alcohol. Mixtures of esters transesterify (exchange their alcoholic portions) when ethyl acetate or an alcohol is the eluent. Finally, the most active solvents (pyridine, methanol, water, and acetic acid) dissolve and elute some of the adsorbent itself. Generally, try to avoid going to solvents more polar than diethyl ether or methylene chloride in the eluent series (Table 12–2).

C. Column Size and Adsorbent Quantity

The column size and the amount of adsorbent must also be selected correctly to separate a given amount of sample well. As a rule of thumb, the amount of adsorbent should be 25 to 30 times, by weight, the amount of material to be separated by chromatography. Furthermore, the column should have a height-to-diameter ratio of about 8:1. Some typical relations of this sort are given in Table 12–4.

TABLE 12–4.   Size of Column and Amount of Adsorbent for Typical Sample Sizes

AMOUNT OF SAMPLE (g)	AMOUNT OF ADSORBENT (g)	COLUMN DIAMETER (mm)	COLUMN HEIGHT (mm)
0.01	0.3	3.5	30
0.10 ·	3.0	7.5	60
1.00	30.0	16.0	130
10.00	300.0	35.0	280

Note, as a caution, that the difficulty of the separation is also a factor in determining the size and length of the column to be used and in the amount of adsorbent needed. Compounds that do not separate easily may require larger columns and more adsorbent than specified in Table 12–4. For easily separated compounds, a smaller column and less adsorbent may suffice.

D. Flow Rate

The rate at which solvent flows through the column also is significant in the effectiveness of a separation. In general, the time the mixture to be separated remains on the column is directly proportional to the extent of equilibration between stationary and moving phases. Thus, similar compounds eventually separate if they remain on the column long enough. The time a material remains on the column depends on the flow rate of the solvent. If the flow is too slow, however, the dissolved substances in the mixture may diffuse faster than the rate at which they move down the column. Then the bands grow wider and more diffuse, and the separation becomes poor.

12.5   PACKING THE COLUMN: TYPICAL PROBLEMS

The most critical operation in column chromatography is packing (filling) the column with adsorbent. The **column packing** must be evenly packed and free of irregularities, air bubbles, and gaps. As a compound travels down the column, it moves in an advancing zone, or **band.** It is important that the leading edge, or **front,** of this band be horizontal, or perpendicular to the long axis of the column. If two bands are close together and do not have horizontal band fronts, it is impossible to collect one band while completely excluding the other. The leading edge of the second band begins to elute before the first band has finished eluting. This condition can be seen in Figure 12–5. There are two main reasons for this problem. First, if the top surface edge of the adsorbent packing is not level, non-horizontal bands result. Second, bands may also be non-horizontal if the column is not held in an exactly vertical position in both planes (front-to-back and side-to-side). When you are preparing a column, both of these factors must be watched carefully.

Another phenomenon, called **streaming,** or **channeling,** occurs when part of the band front advances ahead of the major part of the band. Channeling occurs if there are any cracks or irregularities in the adsorbent surface or any irregularities caused by air bubbles in the packing. A part of the advancing front moves ahead of the rest of the band by flowing through the channel. Two examples of channeling are shown in Figure 12–6.

12.6   PACKING THE COLUMN: METHODS

The following methods are used to avoid problems resulting from uneven packing and column irregularities. These procedures should be followed carefully in preparing a

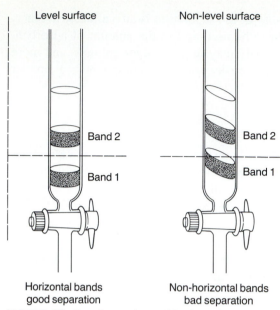

Level surface   Non-level surface

Band 2   Band 2

Band 1   Band 1

Horizontal bands good separation   Non-horizontal bands bad separation

FIGURE 12–5. Comparison of horizontal and non-horizontal band fronts

chromatography column. Failure to pay close attention to the preparation of the column may well affect the quality of the separation.

Preparation of a column involves two distinct stages. In the first stage, a support base on which the packing will rest is prepared. This must be done so that the packing, a finely divided material, does not wash out the bottom of the column. In the second stage, the column of adsorbent is deposited on top of the supporting base. Several alternatives for the second process are detailed on pp 709–711.

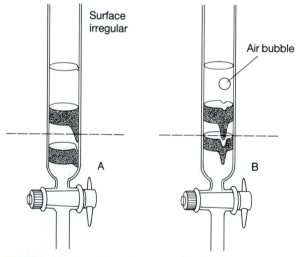

Surface irregular

Air bubble

A   B

FIGURE 12–6. Channeling complications

A. Preparing the Support Base

Microscale

For microscale applications, select a Pasteur pipet (5¾-inch) and clamp it upright in a vertical position. In order to reduce the amount of solvent which is needed to fill the column, it is convenient to break off most of the tip of the pipet. Place a small ball of cotton into the pipet and tamp it into position using a glass rod or a piece of wire. The correct position of the cotton is shown in Figure 12–7. A microscale chromatography column is packed by the "dry pack method," described in Part B of this section.

An alternative apparatus for microscale column chromatography is a commercial column, such as the one shown in Figure 12–8. This type of column is made of glass, and it has a solvent-resistant plastic stopcock at the bottom. The stopcock assembly contains a filter disc to support the adsorbent column. An optional upper fitting, also made of solvent-resistant plastic, serves as a solvent reservoir. The column shown in Figure 12–8 is equipped with the solvent reservoir. This type of column is available in a variety of lengths, ranging from 100 mm to 300 mm.

Standard Scale

For large-scale applications, clamp a chromatography column upright in a vertical position. The column (Figure 12–3) is a piece of cylindrical glass tubing with a stop-

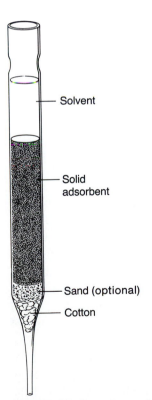

— Solvent

— Solid adsorbent

— Sand (optional)

— Cotton

FIGURE 12–7. Microscale chromatography column

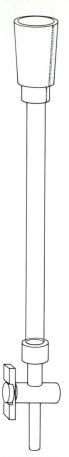

FIGURE 12–8. Commercial microscale chromatography column. (The column is shown equipped with an optional solvent reservoir.)

cock attached at one end. The stopcock usually has a Teflon plug since stopcock grease (used on glass plugs) dissolves in many of the organic solvents used as eluents. Stop-cock grease in the eluent will contaminate the eluates.

Instead of a stopcock, a piece of flexible tubing may be attached to the bottom of the column, with a screw clamp used to stop or regulate the flow (Figure 12–9). When a screw clamp is used, care must be taken that the tubing used is not dissolved by the solvents that will pass through the column during the experiment. Rubber, for instance, dissolves in chloroform, benzene, methylene chloride, toluene, or tetrahydro-furan (THF). Tygon tubing dissolves (actually, the plasticizer is removed) in many different solvents, including benzene, methylene chloride, chloroform, ether, ethyl acetate, toluene, and THF. Polyethylene tubing is the best choice for use at the end of a column, since it is inert with most solvents.

Next, the column is partially filled with a quantity of solvent, usually a non-polar solvent like hexane, and a support for the finely divided adsorbent is prepared in the following way. A loose plug of glass wool is tamped down into the bottom of the column with a long glass rod until all entrapped air is forced out as bubbles. Take care

FIGURE 12–9.   Tubing with screw clamp to regulate solvent flow on a chromatography column

not to plug the column totally by tamping the glass wool too hard. A small layer of clean white sand is formed on top of the glass wool by pouring sand into the column. The column is tapped to level the surface of the sand. Any sand adhering to the side of the column is washed down with a small quantity of solvent. The sand forms a base that supports the column of adsorbent and prevents it from washing through the stopcock. The column is packed in one of two ways: by the "slurry method" or by the "dry pack method."

B. Depositing the Adsorbent

Slurry Method

The slurry method is not recommended as a microscale method for use with Pasteur pipets. On a very small scale, it is too difficult to pack the column with the slurry without losing the solvent before the packing has been completed. Microscale columns should be packed by the "dry pack" method.

In the slurry method, the adsorbent is packed into the column as a slurry. A **slurry** is a mixture of a solvent and an undissolved solid. The slurry is prepared in a separate container by adding the solid adsorbent, a little at a time, to a quantity of the solvent. This order of addition (adsorbent to solvent) should be followed strictly since the adsorbent solvates and liberates heat. If the solvent is added to the adsorbent, it may boil away almost as fast as it is added due to heat evolved. This will be especially true if ether or another low-boiling solvent is used. When this happens, the final mixture will be uneven and lumpy. Enough adsorbent is added to the solvent, with swirling, to form a thick, but flowing, slurry. The slurry should be swirled until it is homogeneous and relatively free of entrapped air bubbles.

For a standard-sized column, the procedure is as follows. When the slurry has been prepared, the column is filled about half-full with solvent, and the stopcock is opened to allow solvent to drain slowly into a large beaker. The slurry is mixed by swirling and is then poured in portions into the top of the draining column (a wide-necked funnel may be useful here). Be sure to swirl the slurry thoroughly before each addition to the column. The column is tapped constantly and **gently** on the side, during the pouring operation, with the fingers or with a pencil fitted with a rubber stopper. A short piece of large-diameter pressure tubing may also be used for tapping. The tapping promotes even settling and mixing and gives an evenly packed column free of air bubbles. Tapping is continued until all the material has settled, showing a well-defined level at the top of the column. Solvent from the collecting beaker may be re-added to the slurry if it becomes too thick to be poured into the column at one time. In fact, the collected solvent should be cycled through the column several times to ensure that settling is complete and that the column is firmly packed. The downward flow of solvent tends to compact the adsorbent. One should take care never to let the column run dry during packing.

Dry Pack Method 1

In the first of the dry pack methods introduced here, the column is filled with solvent and allowed to drain **slowly.** The dry adsorbent is added, a little at a time, while the column is tapped constantly, as described above.

Pasteur Pipet. To fill a microscale column, fill the Pasteur pipet (with the cotton plug and the layer of sand, prepared as described above) about half full with solvent. Using a microspatula, add the solid adsorbent slowly to the solvent in the column. As the solid is added, tap the column **gently** with a pencil, a finger, or a glass rod. The tapping promotes even settling and mixing and gives an evenly packed column free of air bubbles. As the adsorbent is being added, solvent will flow out of the Pasteur pipet. Since the adsorbent must not be allowed to dry during the packing process, a means of controlling the solvent flow must be utilized. If a piece of small-diameter plastic tubing is available, it can be fitted over the narrow tip of the Pasteur pipet. The flow rate can then be controlled using a screw clamp. A simple approach to controlling the flow rate is to use a finger over the top of the Pasteur pipet, much as one controls the flow of liquid in a volumetric pipet. Continue adding the adsorbent slowly, with constant tapping, until the level of the adsorbent has reached the desired level. As the column is being packed, be careful not to let the column run dry. The final column should appear as shown in Figure 12–7.

Commercial Microscale Column. The procedure to fill a commercial microscale column is essentially the same as that used to fill a Pasteur pipet. The commercial column has the advantage that it is much easier to control the flow of solvent from the column during the filling process because the stopcock can be adjusted appropriately. It is not necessary to use a cotton plug or to deposit a layer of sand before adding the adsorbent. The presence of the fritted disc at the base of the column acts to prevent adsorbent from escaping from the column.

Macroscale Column. A plug of cotton is placed at the base of the column, and an even layer of sand is formed on top (see p 707). The column is filled about half full with solvent, and the solid adsorbent is added carefully from a beaker, while the solvent is allowed to flow slowly from the column. As the solid is added, the column is tapped as described for the "slurry method" in order to ensure that the column is packed evenly. When the column has the desired length, no more adsorbent is added. This method also gives an evenly packed column. For the same reasons as those already described, solvent should be cycled through this column (for macroscale applications) several times before each use. The same portion of solvent that has drained from the column during the packing is used to cycle through the column.

Dry Pack Method 2

Pasteur Pipet. An alternative dry pack method for microscale columns is to fill the Pasteur pipet with **dry** adsorbent, without any solvent. A plug of cotton and a thin layer of sand are positioned in the bottom of the Pasteur pipet. The desired amount of adsorbent is added slowly, with constant tapping, until the level of adsorbent has reached the desired height. Figure 12–7 can be used as a guide in order to judge the correct height of the column of adsorbent. When the column is packed, solvent is allowed to percolate through the adsorbent until the entire column is moistened. The solvent is not added until just before the column is to be used.

> **This method is not recommended for use with silica gel, nor for experiments where a very careful separation is required.**

This method is useful when the adsorbent is alumina, but it does not give satisfactory results with silica gel. Even with alumina, poor separations can arise owing to uneven packing, air bubbles, and cracking, especially if a solvent that has a highly exothermic heat of solvation is used.

Commercial Microscale Column. The method used is similar to that described for Pasteur pipets, except that the plug of cotton and the layer of sand are not required. The flow rate of solvent through the column can be controlled using the stopcock which is part of the column assembly (see Figure 12–8).

Macroscale Column. Macroscale columns can also be packed by a dry pack method which is similar to the microscale methods described above. The disadvantages described for the microscale method also apply to the macroscale method. This method is not recommended for use with silica gel or alumina since the combination leads to uneven packing, air bubbles, and cracking, especially if a solvent that has a highly exothermic heat of solvation is used.

12.7   APPLYING THE SAMPLE TO THE COLUMN

The solvent (or solvent mixture) used to pack the column is normally the least polar elution solvent one intends to use during the chromatography. The compounds to be chromatographed will not be highly soluble in the solvent. If they were, they would probably have a greater affinity for the solvent than for the adsorbent, and would pass right through the column without equilibrating with the stationary phase.

The first elution solvent, however, is generally not a good solvent to use in preparing the sample to be placed on the column. Since the compounds are not highly soluble in nonpolar solvent, it would take a large amount of the initial solvent to dissolve the compounds, and it would be difficult to get the mixture to form a narrow band on top of the column. A narrow band is ideal for an optimum separation of components. For the best separation, therefore, the compound is applied to the top of the column undiluted if it is a liquid, or in a **very small** amount of highly polar solvent if it is a solid. Water must not be used to dissolve the initial sample being chromatographed, as it reacts with the column packing.

In adding the sample to the column, the following procedure should be used. Lower the solvent level to the top of the adsorbent column by draining the solvent from the column. Add the sample (either a pure liquid or a solution) to form a small layer on top of the adsorbent. A Pasteur pipet is convenient for adding the sample to the column. Take care not to disturb the surface of the adsorbent. This is done best by touching the pipet to the inside of the glass column and slowly draining it so as to allow the sample to spread into a thin film, which slowly descends to cover the entire absorbent surface. Drain the pipet close to the surface of the adsorbent. When all the sample has been added, drain this small layer of liquid into the column until the top surface of the column **just begins** to dry. Then add a small layer of the chromatographic solvent carefully with a Pasteur pipet, again with care not to disturb the surface. Drain this small layer of solvent into the column until the top surface of the column just dries. Add another small layer of fresh solvent, if necessary, and repeat the process until it is clear that the sample is strongly adsorbed on the top of the column. If the sample is colored and the fresh layer of solvent acquires some of this color, the sample has not been properly adsorbed. Once the sample has been properly applied, the level surface of the adsorbent may be protected by carefully filling the top of the column with solvent and sprinkling clean, white sand into the column so as to form a small protective layer on top of the adsorbent. For microscale applications, this layer of sand is not required.

Separations are often better if the sample is allowed to stand a short time on the column before elution. This allows a true equilibrium to be established. In columns that stand for too long, however, the adsorbent often compacts or even swells, and the flow can become annoyingly slow. Diffusion of the sample to widen the bands also becomes a problem if a column is allowed to stand over an extended period. For small-scale chromatography, using Pasteur pipets, there is no stopcock, and it is not possible to stop the flow. In this case it is not considered necessary to allow the column to stand.

12.8 ELUTION TECHNIQUES

Solvents for analytical and preparative chromatography should be pure reagents. Commercial-grade solvents often contain small amounts of residue, which remain when the solvent is evaporated. For normal work, and for relatively easy separations that take only small amounts of solvent, the residue usually presents few problems. For large-scale work, commercial-grade solvents may have to be redistilled before use. This is especially true for hydrocarbon solvents, which tend to have more residue than other solvent types. Most of the experiments in this laboratory manual have been designed to avoid this particular problem.

The need for pure solvents is especially acute with high performance liquid chromatography (Section 12.16). The narrow bore of the column and very small particle size of the column packing require that solvents be particularly pure and free of insoluble residue. In most cases, the solvents must be filtered through ultra-fine filters and they must be **degassed** (have dissolved gases removed) before they can be used.

One usually begins elution of the products with a nonpolar solvent, like hexane or petroleum ether. The polarity of the elution solvent can be increased gradually by adding successively greater percentages of either ether or toluene (for instance, 1%, 2%, 5%, 10%, 15%, 25%, 50%, 100%) or some other solvent of greater solvent power (polarity) than hexane. The transition from one solvent to another should not be too rapid in most solvent changes. If the two solvents to be changed differ greatly in their heats of solvation in binding to the adsorbent, enough heat can be generated to crack the column. Ether is especially troublesome in this respect, as it has both a low boiling point and a relatively high heat of solvation. Most organic compounds can be separated on silica gel or alumina using hexane-ether or hexane-toluene combinations for elution, and following these by pure methylene chloride. Solvents of greater polarity are usually avoided for the various reasons mentioned above. In microscale work, the usual procedure is to use only one solvent for the chromatography. Changing solvent polarity during the experiment is relatively rare.

The flow of solvent through the column should not be too rapid, or the solutes will not have time to equilibrate with the adsorbent as they pass down the column. If the rate of flow is too low, or stopped for a period, diffusion can become a problem—the solute band will diffuse, or spread out, in all directions. In either of these cases, separation will be poor. As a general rule (and only an approximate one), most macroscale columns are run with flow rates ranging from 5 to 50 drops of effluent per minute; a steady flow of solvent is usually avoided. Microscale columns made from Pasteur pipets do not have a means of controlling the solvent flow rate, but commercial microscale columns are equipped with stopcocks. The solvent flow rate in this type of column can be adjusted in a manner similar to that used with larger columns. To avoid diffusion of the bands, do not stop the column and do not set it aside overnight.

12.9 RESERVOIRS

For a microscale chromatography, the portion of the Pasteur pipet above the adsorbent is used as a reservoir of solvent. Fresh solvent, as needed, is added by means of another Pasteur pipet. When it is necessary to change solvent, the new solvent is also added in this manner. In some cases, the chromatography may proceed too slowly; the rate of solvent flow can be accelerated by attaching a rubber dropper bulb to the top of the Pasteur pipet column and squeezing **gently.** The additional air pressure forces the solvent through the column more rapidly. If this technique is used, however, care must be taken to remove the rubber bulb from the column before releasing it. Otherwise, air may be drawn up through the bottom of the column, destroying the column packing.

When large quantities of solvent are used in a chromatographic separation, it is often convenient to use a solvent reservoir to forestall having to add small portions of fresh solvent continually. The simplest type of reservoir, a feature of many columns, is created by fusing the top of the column to a round-bottom flask (Figure 12–10A). If the column has a standard taper joint at its top, a reservoir can be created by joining a standard-taper separatory funnel to the column (Figure 12–10B). In this arrangement, the stopcock is left open, and no stopper is placed in the top of the separatory funnel. A third common arrangement is shown in Figure 12–10C. A separatory funnel is filled with solvent; its stopper is wetted with solvent and put **firmly** in place. The funnel is inserted into the empty filling space at the top of the chromatographic column, and the stopcock is opened. Solvent flows out of the funnel, filling the space at the top of the column until the solvent level is well above the outlet of the separatory funnel. As

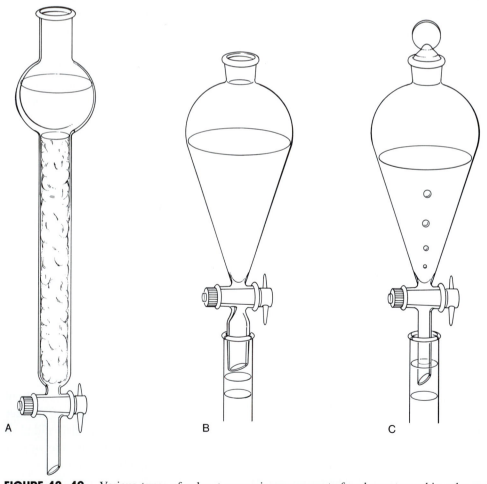

FIGURE 12–10. Various types of solvent-reservoir arrangements for chromatographic columns

solvent drains from the column, this arrangement automatically refills the space at the top of the column by allowing air to enter through the stem of the separatory funnel. Some microscale columns, such as that shown in Figure 12–8, are equipped with a solvent reservoir which fits onto the top of the column. It functions just like the reservoirs described above.

12.10 MONITORING THE COLUMN

It is a happy instance when the compounds to be separated are colored. The separation can then be followed visually and the various bands collected separately as they elute from the column. For the majority of organic compounds, however, this lucky circumstance does not exist, and other methods must be used to determine the positions of the bands. The most common method of following a separation of colorless compounds is to collect **fractions** of constant volume in preweighed flasks, to evaporate the solvent from each fraction, and to reweigh the flask plus any residue. A plot of fraction number *versus* the weight of the residues after evaporation of solvent gives a plot like that in Figure 12–11. Clearly, Fractions 2 through 7 (Peak 1) may be combined as a single compound, and so can fractions 8 through 11 (Peak 2) and 12 through 15 (Peak 3). The size of the fractions collected (1 mL, 10 mL, 100 mL, or 500 mL) depends on the size of the column and the ease of separation.

Another common method of monitoring the column is to mix an inorganic phosphor into the adsorbent used to pack the column. When the column is illuminated with an ultraviolet light, the adsorbent treated in this way fluoresces. However, many solutes have the ability to **quench** the fluorescence of the indicator phosphor. In areas in which solutes are present, the adsorbent does not fluoresce, and a dark band is visible. In this type of column, the separation can also be followed visually.

Thin-layer chromatography is often used to monitor a column. This method is described in Technique 13 (Section 13.10, p 734). Several sophisticated instrumental

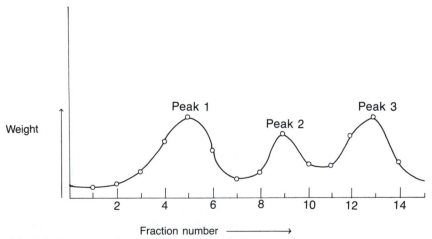

FIGURE 12–11. Typical elution graph

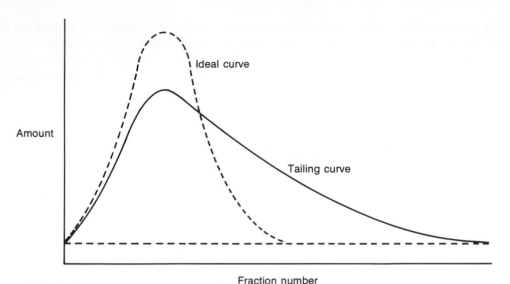

FIGURE 12–12.   Elution curves: One ideal and one that "tails"

and spectroscopic methods, which we shall not detail, can also monitor a chromatographic separation.

12.11   TAILING

Often when a single solvent is used for elution, an elution curve (weight-*vs.*-fraction) like that shown as a solid line in Figure 12–12 is observed. An ideal elution curve is shown by dashed lines. In the non-ideal curve, the compound is said to be **tailing.** Tailing can interfere with the beginning of a curve or a peak of a second component and lead to a poor separation. One way to avoid this is to increase the polarity of the solvent constantly while eluting. In this way, at the tail of the peak, where the solvent polarity is increasing, the compound will move slightly faster than at the front and allow the tail to squeeze forward, forming a more nearly ideal band.

12.12   RECOVERING THE SEPARATED COMPOUNDS

In recovering each of the separated compounds of a chromatographic separation when they are solids, the various correct fractions are combined and evaporated. If the combined fractions contain sufficient material, they may be purified by recrystallization. If the compounds are liquids, the correct fractions are combined, and the solvent is evaporated. If sufficient material has been collected, liquid samples can be purified by distillation. The combination of chromatography-crystallization or chromatography-distillation usually yields very pure compounds. For microscale applications, the amount of sample collected is too small to allow a purification by crystallization or distillation. The samples which are obtained after the solvent has been evaporated are considered to be sufficiently pure, and no additional purification is attempted.

12.13 DECOLORIZATION BY COLUMN CHROMATOGRAPHY

A common outcome of organic reactions is the formation of a product which is contaminated by highly-colored impurities. Very often these impurities are highly-polar, and they have a high molecular weight, as well as being colored. The purification of the desired product requires that these impurities be removed. Section 5.6 of Technique 5 (pp 589–591) details methods of decolorizing an organic product. In most cases, these methods involve the use of a form of activated charcoal, or Norit.

An alternative, which is applied conveniently in microscale experiments, is to remove the colored impurity by column chromatography. Owing to the polarity of the impurities, the colored components are strongly adsorbed on the stationary phase of the column, while the less-polar desired product passes through the column and is collected.

Microscale decolorization of a solution on a chromatography column requires that a column be prepared in a Pasteur pipet, using either alumina or silica gel as the adsorbent (Section 12.6). The sample to be decolorized is diluted to the point where crystallization within the column will not take place, and it is then passed through the column in the usual manner. The desired compound is collected as it exits the column, and the excess solvent is removed by evaporation (Technique 3, Section 3.9, p 560).

12.14 GEL CHROMATOGRAPHY

The stationary phase in gel chromatography consists of a cross-linked polymeric material. Molecules are separated according to their **size** by their ability to penetrate a sievelike structure. Molecules permeate the porous stationary phase as they move down the column. Small molecules penetrate the porous structure more easily than large ones. Thus, the large molecules move through the column faster than the smaller ones and elute first. The separation of molecules by gel chromatography is depicted in Figure 12–13. With adsorption chromatography using materials such as alumina or silica, the order is usually the reverse: Small molecules (of low molecular weight) pass through the column **faster** than large molecules (of high molecular weight) because large molecules are more strongly attracted to the polar stationary phase.

Equivalent terms used by chemists for the gel-chromatography technique are **gel filtration** (biochemistry term), **gel-permeation chromatography** (polymer chemistry term), and **molecular sieve chromatography. Size-exclusion chromatography** is a general term for the technique, and it is perhaps the most descriptive term for what occurs on a molecular level.

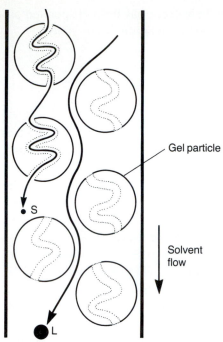

Gel particle

Solvent flow

FIGURE 12–13. Gel chromatography: Comparison of the paths of large (L) and small (S) molecules through the column during the same interval of time.

Sephadex is one of the most popular materials for gel chromatography. It is widely used by biochemists for separating proteins, nucleic acids, enzymes, and carbohydrates. Most often water or aqueous solutions of buffers are used as the moving phase. Chemically, Sephadex is a polymeric carbohydrate that has been cross-linked. The degree of cross-linking determines the size of the "holes" in the polymer matrix. In addition, the hydroxyl groups on the polymer can adsorb water, which causes the material to swell. As it expands, "holes" are created in the matrix. Several different gels are available from manufacturers, each with its own set of characteristics. For example, a typical Sephadex gel, such as G-75, can separate molecules in the molecular-weight (MW) range 3,000 to 70,000. Assume for the moment that one has a four-component mixture containing compounds with molecular weights of 10,000, 20,000, 50,000, and 100,000. The 100,000-MW compound would pass through the column first, because it cannot penetrate the polymer matrix. The 50,000-, 20,000-, and 10,000-MW compounds penetrate the matrix to varying degrees and would be separated. The molecules would elute in the order given (decreasing order of molecular weights). The gel separates on the basis of molecular size and configuration, rather than molecular weight.

Sephadex LH-20 has been developed for non-aqueous solvents. Some of the hydroxyl groups have been alkylated, and thus the material can swell under both aqueous and non-aqueous conditions (it now has "organic" character). This material can be used with several organic solvents, such as alcohol, acetone, methylene chloride, and aromatic hydrocarbons.

Another type of gel is based on a polyacrylamide structure (Bio-Gel P and Poly-Sep AA). A portion of a polyacrylamide chain is shown below.

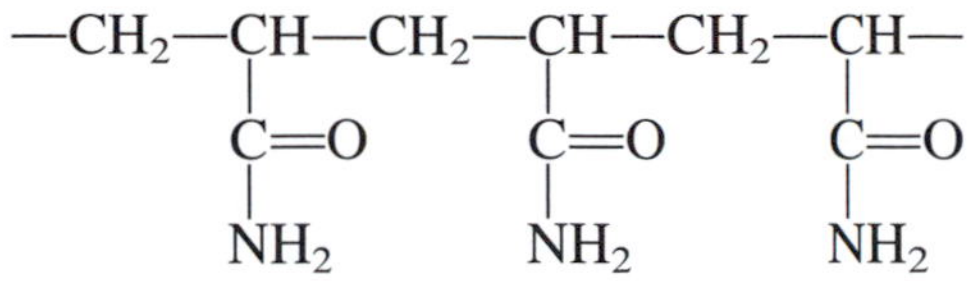

$$-CH_2-CH-CH_2-CH-CH_2-CH-$$
$$\begin{array}{ccc} | & | & | \\ C=O & C=O & C=O \\ | & | & | \\ NH_2 & NH_2 & NH_2 \end{array}$$

Gels of this type can also be used in water and some polar organic solvents. They tend to be more stable than Sephadex, especially under acidic conditions. Polyacrylamides can be used for many biochemical applications involving macromolecules. For separating synthetic polymers, cross-linked polystyrene beads (copolymer of styrene and divinylbenzene) find common application. Again the beads are swollen before use. Common organic solvents can be used to elute the polymers. As with other gels, the higher-molecular-weight compounds elute before the lower-molecular-weight compounds.

12.15 DRY-COLUMN CHROMATOGRAPHY

In dry-column chromatography, the entire column is packed dry without any solvent. The most common column for this method consists of a piece of flexible nylon tubing that has been sealed at one end (stapled). After some glass wool has been inserted into the bottom of the tubing, the bottom is perforated in several places so that air can be expelled. The column is then filled in portions with adsorbent, usually alumina. After each addition, the adsorbent is compacted by dropping the tubing repeatedly on a hard surface.

Instead of applying the sample as described in Section 12.7, the material to be separated is dissolved in a volatile solvent and mixed with the adsorbent. The solvent is removed from this adsorbent by evaporation. The resulting dry powder is introduced into the packed column. Solvent is allowed to percolate down the column slowly until the solvent front is about one centimeter from the end of the column. Only a single pure solvent or solvent mixture is used to perform the separation in dry-column chromatography.

In dry-column chromatography, in contrast to elution chromatography (Section 12.8), neither the solvent nor the separated materials actually leave the column. At the end of the chromatography, the plastic tubing is cut in segments to recover the separated compounds. Each portion of adsorbent together with the pure compound is swirled with a volatile solvent to remove the compound from the adsorbent. Once the adsorbent is removed by filtration, the solvent is evaporated to yield the purified compound. With colored compounds, one can easily decide where to cut the tubing. With colorless materials, an ultraviolet light (Technique 13, Section 13.7, p 732) may be used to detect compounds before slicing the tubing. Alternatively, one may use a new technique, in which the tubing is punctured at 1-cm intervals with glass pipets. By this technique, a small amount of adsorbent is removed with the pipet and analyzed by thin-layer chromatography, as in the method described in Technique 13, Section 13.10, p 734). Using these analyses, one can decide where the column should be cut for the best separation of components.

There is a basic similarity between dry-column chromatography and thin-layer chromatography (TLC), described in Technique 13. The separation is virtually identical to that which is obtained with TLC. One can expect the same results from both methods so long as the same adsorbent and the same solvent are used in each case. Thus, the R_f value (Technique 13, Section 13.9, p 733) obtained by TLC correlates well with the value from the dry-column chromatographic method. The main advantage of this method over conventional TLC is that much larger samples can be separated.

12.16 HIGH PERFORMANCE LIQUID CHROMATOGRAPHY

If the column packing used in column chromatography is made more dense by using an adsorbent that has a smaller particle size, the separation that can be achieved is greater. The solute molecules encounter a much larger surface area on which they can be adsorbed as they pass through the column packing. At the same time, the solvent spaces between the particles are reduced in size. As a result of this tight packing, equilibrium between the liquid and solid phases can be established very rapidly with a fairly short column, and the degree of separation is markedly improved. The disadvantage of making the column packing more dense is that the solvent flow rate becomes very slow or even stops. Gravity is not strong enough to pull the solvent through a tightly packed column.

A recently developed technique can be applied to obtain much better separations with tightly packed columns. A pump is used to force the solvent through the column packing. As a result, solvent flow rate is increased, while the advantage of better separation is retained. This technique, called **high performance liquid chromatography (HPLC),** is becoming widely applied to problems where separations by ordinary column chromatography are unsatisfactory. Because the pump often provides pressures in excess of 1000 pounds per square inch (psi), this method is also known as **high pressure liquid chromatography.** High pressures are not required, however, and satisfactory separations can be achieved with pressures as low as 100 psi.

The basic design of an HPLC instrument is shown in Figure 12–14. The instrument contains the following essential components:

1. Solvent reservoir
2. Solvent filter and degasser
3. Pump
4. Pressure gauge
5. Sample injection system
6. Column
7. Detector
8. Amplifier and electronic controls
9. Chart recorder.

There may be other variations on this simple design. Some instruments have heated ovens in order to maintain the column at a specified temperature, fraction collectors, and microprocessor-controlled data-handling systems. Additional filters for the solvent and sample may also be included. One may find it interesting to compare this schematic diagram with that shown in Technique 14 (Figure 14–2, p 740) for a gas chromatography instrument. Many of the essential components are common to both types of instruments.

The chromatography column is generally packed with silica or alumina adsorbents. Unlike column chromatography, however, the adsorbents used for HPLC have a much smaller particle size. Typically, particle size ranges from 5 to 20 micrometers in diameter for HPLC, while it is on the order of 100 micrometers for column chromatography. The actual column may be constructed of heavy glass or even stainless steel. A strong column is required to withstand the high pressures that may be used.

A flow-through **detector** must be provided to determine when a substance has passed through the column. In most applications, the detector either detects the change in index of refraction of the liquid as its composition changes, or it detects the presence of solute by its absorption of ultraviolet light. The signal generated by the detector is amplified and treated electronically in a manner similar to that found in gas chromatographs (Technique 14, Section 14.6, p 744).

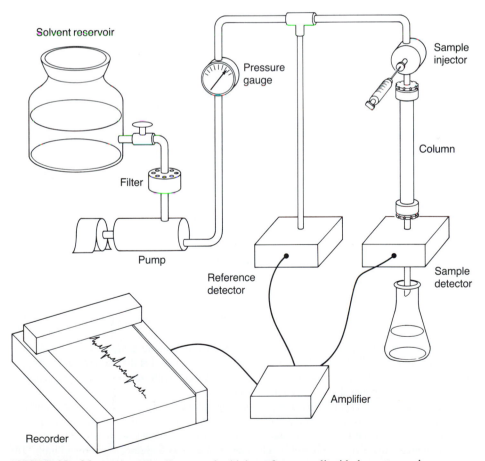

FIGURE 12–14.　Schematic diagram of a high performance liquid chromatograph

A variation unique to HPLC is to use a nonpolar column packing with a polar solvent. This is called **reversed-phase chromatography.** In this method, the silica column packing is treated with alkylating agents. As a result, nonpolar alkyl groups are bonded to the silica surface, making the adsorbent nonpolar. When a polar solvent is used, polar solutes are more strongly attracted to the mobile (solvent) phase than they are to the stationary phase (adsorbent). The order of elution, therefore, is reversed, with polar solutes passing through the column faster than nonpolar solutes. In many applications, better separations can be achieved with reversed-phase chromatography than with polar adsorbents.

If a single solvent (or solvent mixture) is used for the entire separation, the chromatogram is said to be **isochratic.** Special electronic units are available, which will allow one to program changes in the solvent composition from the beginning to the end of the chromatography. These are called **gradient elution systems.** With gradient elution, the time required for a separation may be shortened considerably.

High performance liquid chromatography is an excellent analytical technique, but the separated compounds may also be isolated. The technique can be used for preparative experiments. Just as in column chromatography, the fractions can be collected into individual receiving containers as they pass through the column. The solvents can be evaporated from these fractions, allowing one to isolate separated components of the original mixture.

REFERENCES

Deyl, Z., Macek, K., and Janák, J. *Liquid Column Chromatography*. Amsterdam: Elsevier, 1975.
Heftmann, E. *Chromatography*. 3rd ed. New York: Van Nostrand Reinhold, 1975.

PROBLEMS

1. A sample was placed on a chromatography column. Methylene chloride was used as the eluting solvent. No separation of the components in the sample was observed. What must have been happening during this experiment? How would you change the experiment in order to overcome this problem?

2. You are about to purify an impure sample of naphthalene by column chromatography. What solvent should you use to elute the sample?

3. Consider a sample which is a mixture composed of biphenyl, benzoic acid, and benzyl alcohol. Predict the order of elution of the components in this mixture. Assume that the chromatography uses a silica column and the solvent system is based on cyclohexane, with an increasing proportion of methylene chloride being added as a function of time.

4. An orange compound was added to the top of a chromatography column. Solvent was added immediately, with the result that the entire volume of solvent in the solvent reservoir turned orange. No separation could be obtained from the chromatography experiment. What went wrong?

5. A yellow compound, dissolved in methylene chloride, is added to a chromatography column. The elution is begun using petroleum ether as the solvent. After six liters of solvent had passed through the

column, the yellow band still had not traveled down the column appreciably. What should be done to make this experiment work better?

6. You have 0.50 g of a mixture which you wish to purify by column chromatography. How much adsorbent should you use to pack the column? Estimate the appropriate column diameter and height.

7. In a particular sample, you wish to collect the component with the **highest** molecular weight as the **first** fraction. What chromatographic technique should you use?

8. A colored band shows an excessive amount of tailing as it passes through the column. What can you do to rectify this problem?

9. The two isomeric 4-*tert*-butylcyclohexanols, shown below, are present in a mixture. All efforts to separate these isomers by column chromatography have failed. Describe a possible method to separate the isomers.

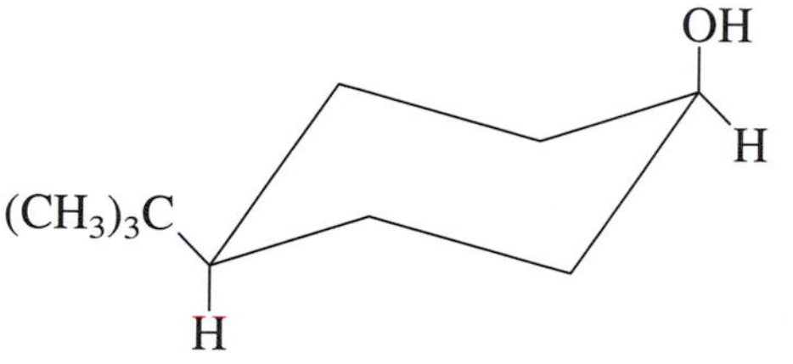

cis 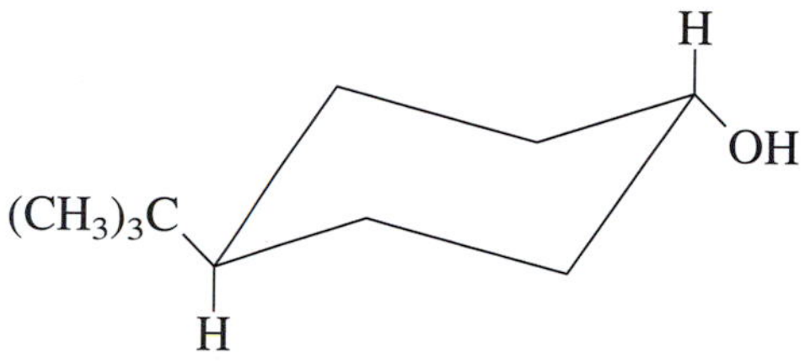   *trans*

10. How would you monitor the progress of a column chromatography when the sample is colorless? Describe at least two methods.

Technique 13
THIN-LAYER CHROMATOGRAPHY

Thin-layer chromatography (TLC) is a very important technique for the rapid separation and qualitative analysis of small amounts of material. It is ideally suited for the analysis of mixtures and reaction products in microscale experiments. The technique is closely related to column chromatography. In fact, TLC can be considered simply column chromatography **in reverse,** with the solvent ascending the adsorbent, rather than descending. Because of this close relationship to column chromatography, and because the principles governing the two techniques are similar, Technique 12, on column chromatography, should be read first.

13.1 PRINCIPLES OF THIN-LAYER CHROMATOGRAPHY

Like column chromatography, TLC is a solid–liquid partitioning technique. However, the moving, liquid phase is not allowed to percolate down the adsorbent; it is caused to **ascend** a thin layer of adsorbent coated onto a backing support. The most typical backing is a glass plate, but other materials are also used. A thin layer of the adsorbent is spread onto the plate and allowed to dry. A coated and dried plate of glass is called a **thin-layer plate** or a **thin-layer slide.** (The reference to *slide* comes about because microscope slides are often used to prepare small thin-layer plates.) When a thin-layer plate is placed upright in a vessel that contains a shallow layer of solvent, the solvent ascends the layer of adsorbent on the plate by capillary action.

In TLC, the sample is applied to the plate before the solvent is allowed to ascend the adsorbent layer. The sample is usually applied as a small spot near the base of the plate, and this technique is often referred to as **spotting.** The plate is spotted by repeated applications of a sample solution from a small capillary pipet. When the filled pipet touches the plate, capillary action delivers its contents to the plate, and a small spot is formed.

As the solvent ascends the plate, the sample is partitioned between the moving, liquid phase and the stationary, solid phase. During this process, one is said to be **developing,** or **running,** the thin-layer plate. In development, the various components in the applied mixture are separated. The separation is based on the many equilibrations the solutes experience between the moving and the stationary phases. (The nature of these equilibrations was thoroughly discussed in Technique 12, Sections 12.2 and 12.3, pp 697–700.) As in column chromatography, the least polar substances advance faster than the most polar substances. A separation results from the differences in the rates at which the individual components of the mixture advance upward on the plate. When many substances are present in a mixture, each has its own characteristic solubility and adsorptivity properties, depending on the functional groups in its structure. In general, the stationary phase is strongly polar and strongly binds polar substances. The moving, liquid phase is usually less polar than the adsorbent and most easily dissolves substances that are less polar or even nonpolar. Thus, while substances that are the most polar travel slowly upward, or not at all, nonpolar substances travel more rapidly if the solvent is sufficiently nonpolar.

When the thin-layer plate has been developed, it is removed from the developing tank and allowed to dry until it is free of solvent. If the mixture that was originally spotted on the plate was separated, there will be a vertical series of spots on the plate. Each spot corresponds to a separate component or compound from the original mixture. If the components of the mixture are colored substances, the various spots will be clearly visible after development. More often, however, the ''spots'' will not be visible because they correspond to colorless substances. If spots are not apparent, they can be made visible only if a **visualization method** is used. Often spots can be seen when the thin-layer plate is held under ultraviolet light; the ultraviolet lamp is a common visualization method. Also common is the use of iodine vapor. The plates are placed in a chamber containing iodine crystals and left to stand for a short time. The iodine reacts with the various compounds adsorbed on the plate to give colored complexes that are clearly visible. Because iodine often changes the compounds by reaction, the mixture components cannot be recovered from the plate when this method is used. (Other methods of visualization are discussed in Section 13.7.)

13.2 COMMERCIALLY PREPARED TLC PLATES

The most convenient type of TLC plate is prepared commercially and sold in a ready-to-use form. Many manufacturers supply glass plates precoated with a durable layer of silica gel or alumina. More conveniently, plates are also available that have either a flexible plastic backing or an aluminum backing. The most common types of commercial TLC plates are composed of plastic sheets that are coated with silica gel and polyacrylic acid, which serves as a binder. A fluorescent indicator may be mixed with the silica gel. The indicator renders the spots due to the presence of compounds in the sample visible under ultraviolet light (see Section 13.7). While these plates are relatively expensive when compared with plates which are prepared in the laboratory, they are far more convenient to use, and they provide more consistent results. The plates are manufactured quite uniformly. Because the plastic backing is flexible, they have the additional advantage that the coating does not flake off the plates very easily. The plastic sheets can also be cut with a pair of scissors to whatever size may be required.

13.3 PREPARATION OF THIN-LAYER SLIDES AND PLATES

The two adsorbent materials most often used for TLC are alumina G (aluminum oxide) and silica gel G (silicic acid). The G designation stands for gypsum (calcium sulfate). Calcined gypsum, $CaSO_4 \cdot \frac{1}{2}H_2O$, is better known as plaster of Paris. When exposed to water or moisture, gypsum sets in a rigid mass, $CaSO_4 \cdot 2H_2O$, which binds the adsorbent together and to the glass plates used as a backing support. In the adsorbents used for TLC, about 10–13% by weight of gypsum is added as a binder. The adsorbent materials are otherwise like those used in column chromatography; the adsorbents used in column chromatography have a larger particle size, however. The material for thin-layer work is a fine powder. The small particle size, along with the added gypsum, makes it impossible to use silica gel G or alumina G for column work. In a column, these adsorbents generally set so rigidly that solvent virtually stops flowing through the column.

A. Microscope Slide TLC Plates

For qualitative work such as identifying the number of components in a mixture or trying to establish that two compounds are identical, small TLC plates made from microscope slides are especially convenient. Coated microscope slides are easily made by dipping the slides into a container holding a slurry of the adsorbent material. Although numerous solvents can be used to prepare a slurry, methylene chloride is probably the most convenient solvent. It has the two advantages of low boiling point (40 °C) and inability to cause the adsorbent to set or form lumps. The low boiling point means that it is not necessary to dry the coated slides in an oven. Its inability to cause the gypsum binder to set means that slurries made with it are stable for several days. It has the disadvantage that the layer of adsorbent formed is fragile and must be treated carefully. For this reason, some persons prefer to add a small amount of methanol to the methylene chloride to enable the gypsum to set more firmly. The methanol solvates the calcium sulfate much as water does. More durable plates can be made by dipping plates into a slurry prepared from water. These plates must be oven-dried before use. Also, a slurry prepared from water must be used soon after its preparation. If it is not, it will begin to set and form lumps. Thus, an aqueous slurry must be prepared immediately before use; it cannot be used after it has stood for any length of time. For microscope slides, a slurry of silica gel G in methylene chloride is not only convenient, but also adequate for most purposes.

Preparing the Slurry

The slurry is most conveniently prepared in a 4-oz wide-mouthed screw-cap jar. About 3 mL of methylene chloride are required for each gram of silica gel G. For a smooth slurry without lumps, the silica gel should be added to the solvent while the mixture is being either stirred or swirled. Adding solvent to the adsorbent usually causes lumps to form in the mixture. When the addition is complete, the cap should be placed on the jar tightly and the jar shaken vigorously to ensure thorough mixing. The slurry may be stored in the tightly-capped jar until it is to be used. More methylene chloride may have to be added to replace evaporation losses.

> **CAUTION:** Avoid breathing silica dust or methylene chloride.
> Prepare and use the slurry in a hood.
> Avoid getting methylene chloride or the slurry mixture on your skin.

Preparing the Slides

If new microscope slides are available, they can be used without any special treatment. However, it is more economical to reuse or recycle used microscope slides. The slides should be washed with soap and water, rinsed with water, and then rinsed with 50% aqueous methanol. The plates should be allowed to dry thoroughly on paper towels. They should be handled by the edges, because fingerprints on the plate surface will make it difficult for the adsorbent to bind to the glass.

Coating the Slides

The slides are coated with adsorbent by dipping them into the container of slurry. Two slides can be coated simultaneously by sandwiching them together before dipping them in the slurry.

Do the coating operation in a hood.

The slurry should be shaken vigorously just before dipping the slides. Since the slurry settles on standing, it should be mixed in this way before each set of slides is dipped. The depth of the slurry in the jar should be about 3 in., and the plates should be dipped into the slurry until only about 0.25 in. at the top remains uncoated. The dipping operation should be done smoothly. The plates may be held at the top (see Figure 13–1), where they will not be coated. They are dipped into the slurry and withdrawn with a slow and steady motion. The dipping operation takes about two seconds. Some practice may be required to get the correct timing. After dipping, the cap should be replaced on the jar, and the plates should be held for a minute until most of the solvent has evaporated. The plates may then be separated and placed on paper towels to complete the drying.

The plates should have an even coating; there should be no streaks and no thin spots where glass shows through the adsorbent. The plates should not have a thick and lumpy coating.

Two conditions cause thin and streaked plates. First, the slurry may not have been mixed thoroughly before the dipping operation; the adsorbent might then have settled to the bottom of the jar, and the thin slurry at the top would not have coated the slides properly. Second, the slurry simply may not have been thick enough; more silica gel G must then be added to the slurry until the consistency is proper. If the slurry is too thick, the coating on the plates will be thick, uneven, and lumpy. To correct this, the slurry should be diluted with enough solvent to achieve the proper consistency.

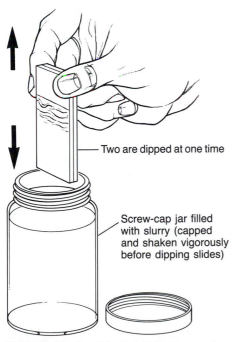

Two are dipped at one time

Screw-cap jar filled with slurry (capped and shaken vigorously before dipping slides)

FIGURE 13–1. Dipping slides to coat them

Plates with an unsatisfactory coating may be wiped clean with a paper towel and redipped. Care must be taken to handle the plates only from the top or by the sides, to avoid fingerprints on the glass surface.

B. Larger Thin-Layer Plates

For separations involving large amounts of material, or for difficult separations, it may be necessary to use larger thin-layer plates. Plates with dimensions up to 20–25 cm^2 are common. With larger plates, it is desirable to have a more durable coating, and a water slurry of the adsorbent should be used to prepare them. If silica gel is used, the slurry should be prepared in the ratio about 1 g of silica gel G to each 2 mL of water. The glass plate used for the thin-layer plate should be washed, dried, and placed on a sheet of newspaper. Along two edges of the plate are placed two strips of masking tape. More than one layer of masking tape is used if a thicker coating is desired on the plate. A slurry is prepared, shaken well, and poured along one of the untaped edges of plate.

> **Observe the precautions stated on p 726.**

A heavy piece of glass rod, long enough to span the taped edges, is used to level and spread the slurry over the plate. While the rod is resting on the tape, it is pushed along the plate from the end at which the slurry was poured toward the opposite end of the plate. This is illustrated in Figure 13–2. After the slurry is spread, the masking tape strips are removed, and the plates are dried in a 110 °C oven for about one hour. Plates of 10–25 cm^2 are easily prepared by this method. Larger plates present more difficulties. Many laboratories have a commercially manufactured spreading machine that makes the entire operation simpler.

The dry-column method (Technique 12, Section 12.15, p 719) offers an alternative for separating large quantities of material.

13.4   SAMPLE APPLICATION: SPOTTING THE PLATES

Preparing a Micropipet

To apply the sample that is to be separated to the thin-layer plate, one uses a micropipet. A micropipet is easily made from a short length of thin-walled capillary tubing

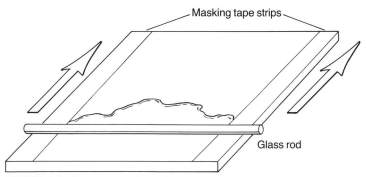

FIGURE 13–2.   Preparing a large plate

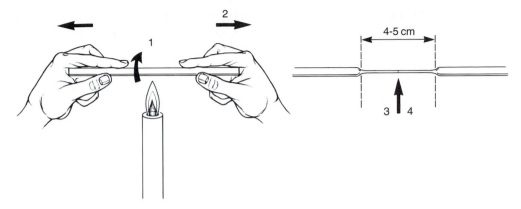

① Rotate in flame until soft.　　③ Score lightly in center of pulled section.
② Remove from flame and pull.　④ Break in half to give two pipets.

FIGURE 13—3.　Construction of two capillary micropipets

like that used for melting-point determinations. The capillary tubing is heated at its midpoint with a microburner and rotated until it is soft. When the tubing is soft, the heated portion of the tubing is drawn out until a constricted portion of tubing 4–5 cm long is formed. After cooling, the constricted portion of tubing is scored at its center with a file or scorer and broken. The two halves yield two capillary micropipets. Figure 13–3 shows how to make such pipets.

Spotting the Plate

To apply a sample to the plate, begin by placing about 1 mg of a solid test substance, or one drop of a liquid test substance, in a small container like a watch glass or a test tube. Dissolve the sample in a few drops of a volatile solvent. Acetone or methylene chloride is usually a suitable solvent. If a solution is to be tested, it can often be used directly. The small capillary pipet, prepared as described, is filled by dipping the pulled end into the solution to be examined. Capillary action fills the pipet. One empties the pipet by touching it **lightly** to the thin-layer plate at a point about 1 cm from the bottom (Figure 13–4). The spot must be high enough that it does not dissolve in the developing solvent. It is important to touch the plate very lightly and not to gouge a hole in the adsorbent. When the pipet touches the plate, solution is transferred to the plate as a

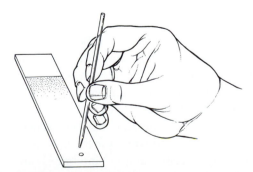

FIGURE 13—4.　Spotting the plate with a drawn capillary pipet

small spot. The pipet should be touched to the plate **very briefly** and then removed. If the pipet is held to the plate, its entire contents will be delivered to the plate. Only a small amount of material is needed. It is often helpful to blow gently on the plate as the sample is applied. This helps to keep the spot small by evaporating the solvent before it can spread out on the plate. The smaller the spot formed, the better the separation obtainable. If needed, additional material can be applied to the plate by repeating the spotting procedure. It is best to repeat the procedure with several small amounts, rather than to apply one large amount. The solvent should be allowed to evaporate between applications. If the spot is not small (about 2 mm in diameter), a new plate should be prepared. The capillary pipet may be used several times if it is rinsed between uses. It is repeatedly dipped into a small portion of solvent to rinse it and touched to a paper towel to empty it.

As many as three different spots may be applied to a microscope-slide TLC plate. Each spot should be about 1 cm from the bottom of the plate, and all spots should be evenly spaced, with one spot in the center of the plate. Due to diffusion, spots will often increase in diameter as the plate is developed. To keep spots containing different materials from merging, and to avoid confusing the samples, do not place more than three spots on a single plate. Larger plates can accommodate many more samples.

13.5 DEVELOPING (RUNNING) TLC PLATES

Preparing a Development Chamber

A convenient developing chamber for microscope-slide TLC plates can be made from a 4-oz wide-mouthed screw-cap jar. The inside of the jar should be lined with a piece of filter paper, cut so that it does not quite extend around the inside of the jar. A small vertical opening (2–3 cm) should be left for observing the development. Before development, the filter paper liner inside the jar should be thoroughly moistened with the development solvent. The solvent-saturated liner helps to keep the chamber saturated with solvent vapors, thereby speeding development. Once the liner is saturated, the level of solvent in the bottom of the jar is adjusted to a depth of about 5 mm, and the jar is capped and set aside until it is to be used. A correctly prepared development chamber (with slide in place) is shown in Figure 13–5.

Developing the TLC Plate

Once the spot has been applied to the thin-layer plate and the solvent has been selected (see Section 13.5), the plate is placed in the chamber for development. The plate must be placed in the chamber carefully so that none of the coated portion touches the filter paper liner. In addition, the solvent level in the bottom of the chamber must not be above the spot that was applied to the plate, or the spotted material will dissolve in the pool of solvent instead of undergoing chromatography. Once the plate has been placed correctly, one replaces the cap on the developing chamber and waits for the solvent to advance up the plate by capillary action. This generally occurs rapidly, and one should

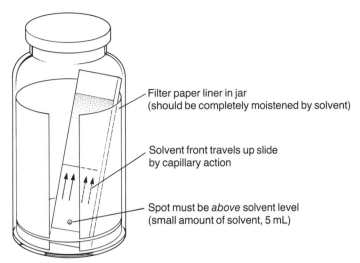

Filter paper liner in jar
(should be completely moistened by solvent)

Solvent front travels up slide
by capillary action

Spot must be *above* solvent level
(small amount of solvent, 5 mL)

FIGURE 13–5.   Development chamber with thin-layer plate undergoing development

watch carefully. As the solvent rises, the plate becomes visibly moist. When the solvent has advanced to within 5 mm of the end of the coated surface, the plate should be removed, and the position of the solvent front should be marked **immediately** by scoring the plate along the solvent line with a **pencil.** The solvent front must not be allowed to travel beyond the end of the coated surface. The plate should be removed before this happens. The solvent will not actually advance beyond the end of the plate, but spots allowed to stand on a completely moistened plate on which the solvent is not in motion expand by diffusion. Once the plate has dried, any visible spots should be outlined on the plate with a pencil. If no spots are apparent, a visualization method (Section 13.7) may be needed.

13.6   CHOOSING A SOLVENT FOR DEVELOPMENT

The development solvent used depends on the materials to be separated. One may have to try several solvents before a satisfactory separation is achieved. Since microscope slides can be prepared and developed rapidly, an empirical choice is usually not hard to make. A solvent that causes all the spotted material to move with the solvent front is too polar. One that does not cause any of the material in the spot to move is not polar enough. As a guide to the relative polarity of solvents, Table 12–2 in Technique 12 (p 703) should be consulted.

Methylene chloride and toluene are solvents of intermediate polarity and good choices for a wide variety of functional groups to be separated. For hydrocarbon materials, good first choices are hexane, petroleum ether (ligroin), or toluene. Hexane or petroleum ether with varying proportions of toluene or ether gives solvent mixtures of moderate polarity that are useful for many common functional groups. Polar materials may require ethyl acetate, acetone, or methanol.

A rapid way to determine a good solvent is to apply several sample spots to a single plate. The spots should be placed a minimum of about 1 cm apart. A capillary

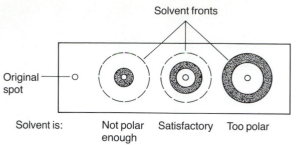

FIGURE 13—6. Concentric-ring method of testing solvents

pipet is filled with a solvent and gently touched to one of the spots. The solvent will expand outward in a circle. The solvent front should be marked with a pencil. A different solvent is applied to each spot. As the solvents expand outward, the spots expand as concentric rings. From the appearance of the rings, one can judge approximately the suitability of the solvent. Several types of behavior experienced with this method of testing are shown in Figure 13–6.

13.7 VISUALIZATION METHODS

If the compounds separated by TLC are colored, it is a fortunate result, because the separation can be followed visually. More often than not, however, the compounds are colorless. Then the separated materials must be made visible by some reagent or some method that makes the separated compounds visible. Reagents that give rise to colored spots are called **visualization reagents.** Methods of viewing that make the spots apparent are **visualization methods.**

The visualization reagent used most often is iodine. Iodine reacts with many organic materials to form complexes that are either brown or yellow. In this visualization method, the developed and dried TLC plate is placed in a 4-oz wide-mouthed screw-cap jar along with a few crystals of iodine. The jar is capped and gently warmed on a steam bath at low heat. The jar fills with iodine vapors, and the spots begin to appear. When the spots are sufficiently intense, the plate is removed from the jar, and the spots are outlined with a pencil. The spots are not permanent. Their appearance results from the formation of complexes the iodine makes with the organic substances. As the iodine sublimes off the plate, the spots fade. Hence, they should be marked immediately. Nearly all compounds except saturated hydrocarbons and alkyl halides form complexes with iodine. The intensities of the spots do not accurately indicate the amount of material present, except in the crudest way.

The second most common method of visualization is by an ultraviolet (UV) lamp. Under UV light, compounds often look like bright spots on the plate. This often suggests the structure of the compound because certain types of compounds shine very brightly under UV light since they fluoresce.

Another method with good results involves adding a fluorescent indicator to the adsorbent used to coat the plates. A mixture of zinc and cadmium sulfides is often used. When treated in this way and held under UV light, the entire plate fluoresces. However, dark spots appear on the plate where the separated compounds are seen to quench this fluorescence.

In addition to the above methods, several chemical methods are available that either destroy or permanently alter the separated compounds through reaction. Many of these methods are specific only for particular functional groups.

Alkyl halides can be visualized if a dilute solution of silver nitrate is sprayed on the plates. Silver halides are formed. These halides decompose if exposed to light, giving rise to dark spots (free silver) on the TLC plate.

Most organic functional groups can be made visible if they are charred with sulfuric acid. Concentrated sulfuric acid is sprayed on the plate, which is then heated in an oven at 110 °C to complete the charring. Permanent spots are thus created.

Colored compounds can be prepared from colorless compounds by making derivatives before spotting them on the plate. An example of this is the preparation of 2,4-dinitrophenylhydrazones from aldehydes and ketones to produce yellow and orange compounds. One may also spray the 2,4-dinitrophenylhydrazine reagent on the plate after the ketones or aldehydes have separated. Red and yellow spots form where the compounds are located. Other examples of this method are using ferric chloride for visualizing phenols and using bromocresol green for detecting carboxylic acids. Chromium trioxide, potassium dichromate, and potassium permanganate can be used for visualizing compounds that are easily oxidized. p-Dimethylaminobenzaldehyde easily detects amines. Ninhydrin reacts with amino acids to make them visible. Numerous other methods and reagents available from various supply outlets are specific for certain types of functional groups. These visualize only the class of compounds of interest.

13.8 PREPARATIVE PLATES

If large plates (Section 13.3B) are used, materials can be separated and the separated components can be recovered individually from the plates. Plates used in this way are said to be **preparative plates.** For preparative plates, a thick layer of adsorbent is generally used. Instead of being applied as a spot or a series of spots, the mixture to be separated is applied as a line of material about 1 cm from the bottom of the plate. As the plate is developed, the separated materials form bands. After development, the separated bands are observed, usually by UV light, and the zones are outlined in pencil. If the method of visualization is destructive, most of the plate is covered with paper to protect it, and the reagent is applied only at the extreme edge of the plate.

Once the zones have been identified, the adsorbent in those bands is scraped from the plate and extracted with solvent to remove the adsorbed material. Filtration removes the adsorbent, and evaporation of the solvent gives the recovered component from the mixture.

13.9 THE R_f VALUE

Thin-layer chromatography conditions include:

1. Solvent system

2. Adsorbent
3. Thickness of the adsorbent layer
4. Relative amount of material spotted

Under an established set of such conditions, a given compound always travels a fixed distance relative to the distance the solvent front travels. This ratio of the distance the compound travels to the distance the solvent travels is called the **R_f value.** The symbol **R_f** stands for "retardation factor," or "ratio-to-front," and it is expressed as a decimal fraction:

$$R_f = \frac{\text{distance traveled by substance}}{\text{distance traveled by solvent front}}$$

When the conditions of measurement are completely specified, the R_f value is constant for any given compound, and it corresponds to a physical property of that compound.

The R_f value can be used to identify an unknown compound; but like any other identification based on a single piece of data, the R_f value is best confirmed with some additional data. Many compounds can have the same R_f value, just as many different compounds have the same melting point.

It is not always possible, in measuring an R_f value, to duplicate exactly the conditions of measurement another worker has used. Therefore, R_f values tend to be of more use to a single worker in one laboratory than they are to workers in different laboratories. The only exception to this is when two workers use TLC plates from the same source, as in commercial plates, or know the **exact** details of how the plates were prepared. Nevertheless, the R_f value can be a useful guide. If exact values cannot be relied on, the relative values can provide another worker with useful information about what to expect. Anyone using published R_f values will find it a good idea to check them by comparing them with standard substances whose identity and R_f values are known.

To calculate the R_f value for a given compound, one measures the distance that the compound has traveled from the point at which it was originally spotted. For spots that are not too large, one measures to the center of the migrated spot. For large spots, the measurement should be repeated on a new plate, using less material. For spots that show tailing, the measurement is made to the "center of gravity" of the Spot. This first distance measurement is then divided by the distance the solvent front has traveled from the same original spot. A sample calculation of the R_f values of two compounds is illustrated in Figure 13–7.

13.10   THIN-LAYER CHROMATOGRAPHY APPLIED IN ORGANIC CHEMISTRY

Thin-layer chromatography has several important uses in organic chemistry. It can be used in the following applications:

1. To establish that two compounds are identical
2. To determine the number of components in a mixture

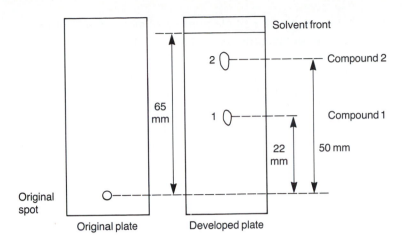

R_f (compound 1) $= \frac{22}{65} = 0.34$ R_f (compound 2) $= \frac{50}{65} = 0.77$

FIGURE 13–7.   Sample calculation of R_f values

3. To determine the appropriate solvent for a column chromatographic separation
4. To monitor a column chromatographic separation
5. To check the effectiveness of a separation achieved on a column, by crystallization, or by extraction
6. To monitor the progress of a reaction

In all these applications, TLC has the advantage that only small amounts of material are necessary. Material is not wasted. With many of the visualization methods, less than a tenth of a microgram (10^{-7} g) of material can be detected. On the other hand, samples as large as a milligram may also be used. With preparative plates that are large (about 9 in. on a side) and have a relatively thick (>500 μm) coating of adsorbent, it is often possible to separate from 0.2–0.5 g of material at one time. The main disadvantage of TLC is that volatile materials cannot be used since they evaporate from the plates.

Thin-layer chromatography can establish that two compounds suspected to be identical are in fact identical. One simply spots both compounds side by side on a single plate and develops the plate. If both compounds travel the same distance on the plate (have the same R_f value), they are probably identical. If the spot positions are not the same, the compounds are definitely not identical. It is important to spot both compounds **on the same plate.** This is especially important with hand-dipped microscope slides, since they vary widely from plate to plate, no two plates having exactly the same thickness of adsorbent. If commercial plates are used, this precaution is not necessary, although it is nevertheless a good idea.

Thin-layer chromatography can establish whether a sample is a single substance or a mixture. A single substance gives a single spot no matter what solvent is used to develop the plate. On the other hand, the number of components in a mixture can be established by trying various solvents on a mixture. A word of caution should be given. It may be difficult, in dealing with compounds of very similar properties, isomers for example, to find a solvent that will separate the mixture. Inability to achieve a separation is not absolute proof that a sample is a single pure substance. Many compounds can be separated only by **multiple developments** of the TLC slide with a fairly non-polar solvent. In this method, the plate is removed after the first development and allowed to dry. After being dried, it is placed in the chamber again and developed once more. This effectively doubles the length of the slide. At times, several developments may be necessary.

When a mixture is to be separated, TLC can be used to choose the best solvent to separate it if column chromatography is contemplated. Various solvents can be tried on a plate coated with the same adsorbent as will be used in the column. The solvent that resolves the components best will probably work well on the column. These small-scale experiments are quick, use very little material, and save time that would be wasted by attempting to separate the entire mixture on the column. Similarly, TLC plates can **monitor** a column. A hypothetical situation is shown in Figure 13–8. A solvent was found that would separate the mixture into four components (A–D). A column was run using this solvent and 11 fractions of 15 mL each were collected. Thin-layer analysis of the various fractions showed that Fractions 1–3 contained Component A; Fractions 4–7, Component B; Fractions 8–9, Component C; and Fractions 10–11, Component D. A small amount of cross-contamination was observed in Fractions 3, 4, 7, and 9.

In another TLC example, a worker found a product from a reaction to be a mixture. It gave two spots, A and B, on a TLC slide. After the product was crystallized, the crystals were found by TLC to be pure A, whereas the mother liquor was found to have a mixture of A and B. The crystallization was judged to have purified A satisfactorily.

Finally, it is often possible to monitor the progress of a reaction by TLC. At various points during a reaction, samples of the reaction mixture are taken and subjected to TLC analysis. An example is given in Figure 13–9. In this case, the desired reaction was the conversion of A to B. At the beginning of the reaction (0 hr), a TLC slide was prepared that was spotted with pure A, pure B, and the reaction mixture.

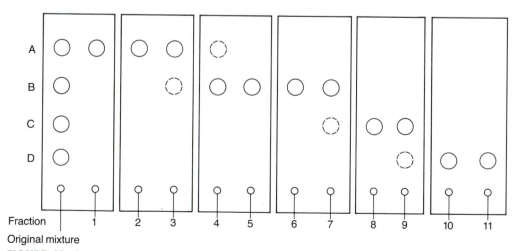

FIGURE 13–8. Monitoring a column

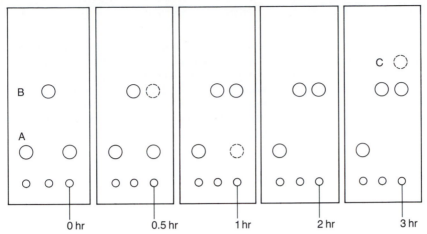

FIGURE 13–9. Monitoring a reaction

Similar slides were prepared at one-half, one, two, and three hours after the start of the reaction. The slides showed that the reaction was complete in two hours. When the reaction was run longer than two hours, a new compound, side-product C, began to appear. Thus, the optimum reaction time was judged to be two hours.

13.11 PAPER CHROMATOGRAPHY

Paper chromatography is often considered related to thin-layer chromatography. The experimental techniques are somewhat like those of TLC, but the principles are more closely related to those of extraction. Paper chromatography is actually a liquid–liquid partitioning technique, rather than a solid–liquid technique. For paper chromatography, a spot is placed near the bottom of a piece of high-grade filter paper (Whatman No. 1 is often used). Then the paper is placed in a developing chamber. The development solvent ascends the paper by capillary action and moves the components of the spotted mixture upward at differing rates. Although paper consists mainly of pure cellulose, the cellulose itself does not function as the stationary phase. Rather, the cellulose absorbs water from the atmosphere, especially from an atmosphere saturated with water vapor. Cellulose can absorb up to about 22% of water. It is this water adsorbed on the cellulose that functions as the stationary phase. To ensure that the cellulose is kept saturated with water, many development solvents used in paper chromatography contain water as a component. As the solvent ascends the paper, the compounds are partitioned between the stationary water phase and the moving solvent. Since the water phase is stationary, the components in a mixture that are most highly water-soluble, or those that have the greatest hydrogen-bonding capacity, are the ones that are held back and move most slowly. Paper chromatography applies mostly to highly polar compounds or to those that are polyfunctional. The most common use of paper chromatography is for sugars, amino acids, and natural pigments. Since filter paper is manufactured with good uniformity, R_f values can often be relied on in paper chromatographic work. However, R_f values are customarily measured from the leading edge (top) of the spot—not from its center, as is customary in TLC.

PROBLEMS

1. A student spots an unknown sample on a TLC plate and develops it in dichloromethane solvent. Only one spot, for which the R_f value is 0.95, is observed. Does this indicate that the unknown material is a pure compound? What can be done to verify the purity of the sample?

2. You and another student were each given an unknown compound. Both samples contained colorless material. You each prepared your own hand-dipped TLC plates and developed the plate using the same solvent. Each of you obtained a single spot, of $R_f = 0.75$. Were the samples that you and the other student were assigned necessarily the same substance? How could you prove unambiguously that they were identical, using TLC?

3. Consider a sample that is a mixture composed of biphenyl, benzoic acid, and benzyl alcohol. The sample is spotted on a TLC plate and developed in a dichloromethane-cyclohexane solvent mixture. Predict the **relative** R_f values for the three components in the sample. *HINT:* See Table 12–3.

4. Calculate the R_f value of a spot which travels 5.7 cm, while the solvent front travels 13 cm.

5. A student spots an unknown sample on a TLC plate and develops it in pentane solvent. Only one spot, for which the R_f value is 0.05, is observed. Is the unknown material a pure compound? What can be done to verify the purity of the sample?

6. A **colorless** unknown substance is spotted on a TLC plate and developed in the correct solvent. The spots do not appear when visualization with a UV lamp or iodine vapors is attempted. What could you do in order to visualize the spots if the compound is
   (a) an alkyl halide
   (b) a ketone
   (c) an amino acid
   (d) a sugar.

Technique 14
GAS CHROMATOGRAPHY

Gas chromatography is one of the most useful instrumental tools for separating and analyzing organic compounds that can be vaporized without decomposition. Common uses include testing the purity of a substance and separating the components of a mixture. The relative amounts of the components in a mixture may also be determined. In some cases, gas chromatography can be used to identify a compound. In microscale work, it can also be used as a preparative method to isolate pure compounds from a small amount of a mixture.

Gas chromatography resembles column chromatography in principle, but differs in three respects. First, the partitioning processes for the compounds to be separated are carried out between a **moving gas phase** and a **stationary liquid phase.** (Recall that in column chromatography the moving phase is a liquid and the stationary phase is a solid adsorbent.) A second difference is that the temperature of the gas system can be controlled since the column is contained in an insulated oven. And third, the concentration of any given compound in the gas phase is a function of its vapor pressure only. Because gas chromatography separates the components of a mixture primarily on the basis of their vapor pressures (or boiling points), this technique is also similar in principle to fractional distillation. In microscale work it is sometimes used to separate and isolate compounds from a mixture when fractional distillation would normally be used to accomplish the same task with larger amounts of material.

Gas chromatography (GC) is also known as vapor-phase chromatography (VPC) and as gas-liquid partition chromatography (GLPC). All three names, as well as their indicated abbreviations, are often found in the literature of organic chemistry. In reference to the technique, the last term, GLPC, is the most strictly correct and is preferred by most authors.

14.1 THE GAS CHROMATOGRAPH

The apparatus used to carry out a gas–liquid chromatographic separation is generally called a **gas chromatograph.** A typical student-model gas chromatograph, the GOW-MAC model 69–350, is illustrated in Figure 14–1. A schematic block diagram of a basic gas chromatograph is shown in Figure 14–2. The basic elements of the apparatus are easily seen. In short, the sample is injected into the chromatograph, and it is immediately vaporized in a heated injection chamber and introduced into a moving stream of gas, called the **carrier gas.** The vaporized sample is then swept into a column filled with particles coated with a liquid adsorbent. The column is contained in a temperature-controlled oven. As the sample passes through the column, it is subjected

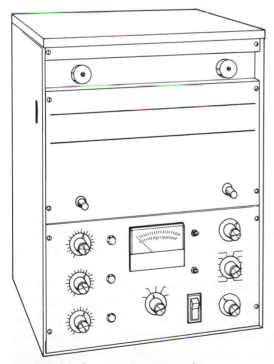

FIGURE 14–1. Gas chromatograph

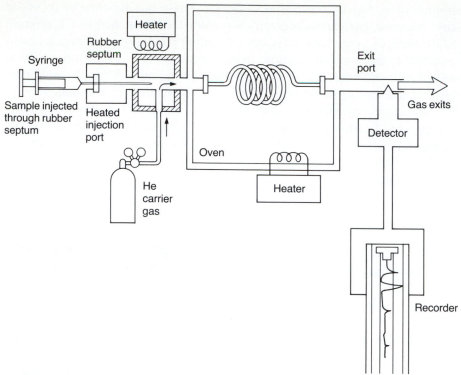

FIGURE 14–2. Schematic diagram of gas chromatograph

to many gas-liquid partitioning processes, and the components are separated. As each component leaves the column, its presence is detected by an electrical detector that generates a signal, which is recorded on a strip chart recorder.

Many modern instruments are also equipped with a microprocessor, which can be programmed to change parameters, such as the temperature of the oven, while a mixture is being separated on a column. With this capability, it is possible to optimize the separation of components and to complete a run in a relatively short time period.

14.2   THE COLUMN

The heart of the gas chromatograph is the packed column. This column is usually made of copper or stainless steel tubing, but sometimes glass is used. The most common diameters of tubing are ⅛ in. (3 mm) and ¼ in. (6 mm). To construct a column, one cuts a piece of tubing to the desired length and attaches the proper fittings on each of the two ends to connect it to the apparatus. The most common length is 4–12 ft, but some columns may be up to 50 ft in length.

The tubing (column) is then packed with the **stationary phase.** The material chosen for the stationary phase is usually a liquid, a wax, or a low-melting solid. This material should be relatively nonvolatile, that is, it should have a low vapor pressure and a high boiling point. Liquids commonly used are high-boiling hydrocarbons, silicone oils, waxes, and polymeric esters, ethers, and amides. Some typical substances are listed in Table 14–1.

TABLE 14–1. Typical Liquid Phases

Increasing polarity →

TYPE	COMPOSITION	MAXIMUM TEMPERATURE (°C)	TYPICAL USE
Apiezons (L, M, N, etc.) — Hydrocarbon greases (varying MW)	Hydrocarbon mixtures	250–300	Hydrocarbons
SE–30 — Methyl silicone rubber	Like silicone oil, but cross-linked	350	General applications
DC–200 — Silicone oil (R = CH_3)	$R_3Si-O-\left[\begin{array}{c} R \\ \mid \\ Si-O \\ \mid \\ R \end{array}\right]_n SiR_3$	225	Aldehydes, ketones, halocarbons
DC–710 — Silicone oil (R = CH_3) (R' = C_6H_5)	$\left[\begin{array}{c} R' \\ \mid \\ Si-O \\ \mid \\ R \end{array}\right]_n$	300	General applications
Carbowaxes (400–20M) — Polyethylene glycols (varying chain lengths)	Polyether $HO-(CH_2CH_2-O)_n-CH_2CH_2OH$	Up to 250	Alcohols, ethers, halocarbons
DEGS — Diethylene glycol succinate	Polyester $\left(CH_2CH_2-O-\underset{O}{\overset{\parallel}{C}}-(CH_2)_2-\underset{O}{\overset{\parallel}{C}}-O\right)_n$	200	General applications

TABLE 14–2. Typical Solid Supports

Crushed firebrick	Chromosorb T
Nylon beads	(Teflon beads)
Glass beads	Chromosorb P
Silica	(Pink diatomaceous earth,
Alumina	highly absorptive, pH 6–7)
Charcoal	Chromosorb W
Molecular sieves	(White diatomaceous earth,
	medium absorptivity, pH 8–10)
	Chromosorb G
	(like the above,
	low absorptivity, pH 8.5)

The liquid phase is usually coated onto a **support material.** A common support material is crushed firebrick. Many methods exist for coating the high-boiling liquid phase onto the support particles. The easiest is to dissolve the liquid (or low-melting wax or solid) in a volatile solvent like methylene chloride (bp 40 °C). The firebrick (or other support) is added to this solution, which is then slowly evaporated (rotary evaporator) so as to leave each particle of support material evenly coated. Other support materials are listed in Table 14–2.

In the final step, the liquid-phase-coated support material is packed into the tubing as evenly as possible. The tubing is bent or coiled so that it fits into the oven of the gas chromatograph with its two ends connected to the gas entrance and exit ports.

Selection of a liquid phase usually revolves about two factors. First, most of them have an upper temperature limit above which they cannot be used. Above the specified limit of temperature, the liquid phase itself will begin to "bleed" off the column. Second, the materials to be separated must be considered. For polar samples, it is usually best to use a polar liquid phase; for nonpolar samples, a nonpolar liquid phase is indicated. The liquid phase performs best when the substances to be separated **dissolve** in it.

Most workers today buy packed columns from commercial sources, rather than pack their own. A wide variety of types and lengths are available.

Alternatives to packed columns are Golay or glass capillary columns of diameters 0.1–0.2 mm. With these columns, no solid support is required, and the liquid is coated directly on the inner walls of the tubing. Liquid phases commonly used in glass capillary columns are similar in composition to those used in packed columns. They include DB-1 (similar to SE-30), DB-17 (similar to DC-710), and DB-WAX (similar to Carbowax 20M). The length of a capillary column is usually very long, typically 50–100 ft. Because of the length and small diameter, there is increased interaction between the sample and the stationary phase. Gas chromatographs equipped with these small diameter columns are able to separate components more effectively than instruments using packed columns.

14.3 PRINCIPLES OF SEPARATION

After a column is selected, packed, and installed, the **carrier gas** (usually helium, argon, or nitrogen) is allowed to flow through the column supporting the liquid phase. The mixture of compounds to be separated is introduced into the carrier gas stream,

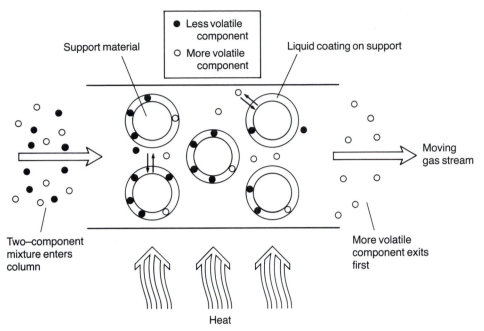

FIGURE 14–3. The separation process

where its components are equilibrated (or partitioned) between the moving gas phase and the stationary liquid phase (Figure 14–3). The latter is held stationary because it is adsorbed onto the surfaces of the support material.

The sample is introduced into the gas chromatograph by a microliter syringe. It is injected as a liquid or as a solution through a rubber septum into a heated chamber, called the **injection port,** where it is vaporized and mixed with the carrier gas. As this mixture reaches the column, which is heated in a controlled oven, it begins to equilibrate between the liquid and gas phases. The length of time required for a sample to move through the column is a function of how much time it spends in the vapor phase and how much time it spends in the liquid phase. The more time it spends in the vapor phase, the faster it will get to the end of the column. In most separations, the components of a sample have similar solubilities in the liquid phase. Therefore, the time the different compounds spend in the vapor phase is primarily a function of their vapor pressure, and the more volatile component will arrive at the end of the column first, as illustrated in Figure 14–3. By selecting the correct temperature of the oven and the correct liquid phase, the compounds in the injected mixture travel through the column at different rates and are separated.

14.4 FACTORS AFFECTING SEPARATION

Several factors determine the rate at which a given compound travels through a gas chromatograph. First of all, compounds of low boiling point will generally travel through the gas chromatograph faster than compounds of higher boiling point. This is because the column is heated, and low-boiling compounds always have higher vapor pressures than compounds of higher boiling point. If the column is heated to a temperature that is too high, however, the entire mixture to be separated is flushed through the column at the same rate as the carrier gas, and no equilibration takes place with the liquid phase. On the other hand, at too low a temperature, the mixture dissolves in the liquid phase and never revaporizes. Thus, it is retained on the column.

Second, the rate of flow of the carrier gas is important. The carrier gas must not move so rapidly that molecules of the sample in the vapor phase cannot equilibrate with those dissolved in the liquid phase. This may result in poor separation between components in the injected mixture. If the rate of flow is too slow, however, the bands will broaden significantly, leading to poor resolution (see Section 14.8).

The third factor is the choice of liquid phase used in the column. The molecular weights, functional groups, and polarities of the component molecules in the mixture to be separated must be considered when a liquid phase is being chosen. One generally uses a different type of material for hydrocarbons, for instance, than for esters. The materials to be separated should **dissolve** in the liquid. The useful temperature limit of the liquid phase selected must also be considered.

Fourth, the length of the column is important. Compounds that resemble one another closely, in general, require longer columns than dissimilar compounds. Many kinds of isomeric mixtures fit into the "difficult" category. The components of isomeric mixtures are so much alike that they travel through the column at very similar rates. A longer column, therefore, is needed to take advantage of any differences that may exist.

14.5 ADVANTAGES OF GAS CHROMATOGRAPHY

All factors that have been mentioned must be adjusted by the chemist for any mixture to be separated. Considerable preliminary investigation is often required before a mixture can be separated successfully into its components by gas chromatography. Nevertheless, the advantages of the technique are many.

First, many mixtures can be separated by this technique when no other method is adequate. Second, as little as 1–10 μL (1 μL = 10^{-6} L) of a mixture can be separated by this technique. This advantage is particularly important when working at the microscale level. Third, when gas chromatography is coupled with an electronic recording device (see below), the amount of each component present in the separated mixture can be estimated quantitatively.

The range of compounds that can be separated by gas chromatography extends from gases, such as oxygen (bp −183 °C) and nitrogen (bp −196 °C), to organic compounds with boiling points over 400 °C. The only requirement for the compounds to be separated is that they have an appreciable vapor pressure at a temperature at which they can be separated and that they be thermally stable at this temperature.

14.6 MONITORING THE COLUMN (THE DETECTOR)

To follow the separation of the mixture injected into the gas chromatograph, it is necessary to use an electrical device called a **detector.** Two types of detectors in common use are the **thermal conductivity detector (TCD),** and the **flame ionization detector (FID).**

The thermal conductivity detector is simply a hot wire placed in the gas stream at the column exit. The wire is heated by constant electrical voltage. When a steady stream of carrier gas passes over this wire, the rate at which it loses heat and its electrical conductance have constant values. When the composition of the vapor stream changes, the rate of heat flow from the wire, and hence its resistance, changes. Helium, which has a higher thermal conductivity than most organic substances, is a common carrier gas. Thus, when a substance elutes in the vapor stream, the thermal conductivity of the moving gases will be lower than with helium alone. The wire then heats up, and its resistance decreases.

A typical TCD operates by difference. Two detectors are used: one exposed to the actual effluent gas and the other exposed to a reference flow of carrier gas only. To achieve this situation, a portion of the carrier gas stream is diverted **before** it enters the injection port. The diverted gas is routed through a reference column into which no sample has been admitted. The detectors mounted in the sample and reference columns are arranged so as to form the arms of a Wheatstone bridge circuit, as shown in Figure 14–4. As long as the carrier gas alone flows over both detectors, the circuit is in balance. However, when a sample elutes from the sample column, the bridge circuit becomes unbalanced, creating an electrical signal. This signal can be amplified and used to activate a strip chart recorder. The recorder is an instrument that plots, by means of a moving pen, the unbalanced bridge current versus time on a continuously moving roll of chart paper. This record of detector response (current) versus time is called a **chromatogram.** A typical gas chromatogram is illustrated in Figure 14–5. Deflections of the pen are called **peaks.**

When a sample is injected, some air (CO_2, H_2O, N_2, and O_2) is introduced along with the sample. The air travels through the column almost as rapidly as the carrier gas; as it passes the detector, it causes a small pen response, thereby giving a

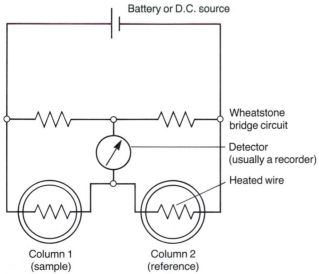

FIGURE 14–4.   Typical thermal conductivity detector

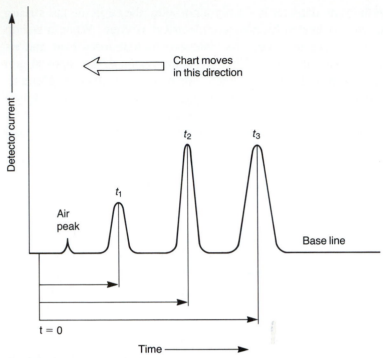

FIGURE 14–5.   Typical chromatogram

peak, called the **air peak.** At later times (t_1, t_2, t_3), the components also give rise to peaks on the chromatogram as they pass out of the column and past the detector.

In a flame ionization detector, the effluent from the column is directed into a flame produced by the combustion of hydrogen, as illustrated in Figure 14–6. As organic compounds burn in the flame, ion fragments are produced which collect on the ring above the flame. The resulting electrical signal is amplified and sent to a recorder,

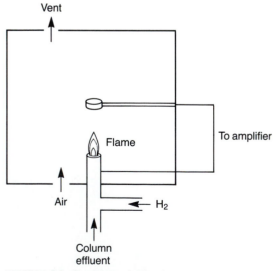

FIGURE 14–6.   Flame ionization detector

in a similar manner to a TCD, except that a FID does not produce an air peak. The main advantage of the FID is that it is more sensitive and can be used to analyze smaller quantities of sample. Also, because a FID does not respond to water, a gas chromatograph with this detector can be used to analyze aqueous solutions. Two disadvantages are that it is more difficult to operate and the detection process destroys the sample. Therefore, an FID gas chromatograph cannot be used to do preparative work, which is often desired in the microscale laboratory.

14.7   RETENTION TIME

The period following injection that is required for a compound to pass through the column is called the **retention time** of that compound. For a given set of constant conditions (flow rate of carrier gas, column temperature, column length, liquid phase, injection port temperature, carrier), the retention time of any compound is always constant (much like the R_f value in thin-layer chromatography, as described in Technique 13, Section 13.9, p 733). The retention time is measured from the time of injection to the time of maximum pen deflection (detector current) for the component being observed. This value, when obtained under controlled conditions, can identify a compound by a direct comparison of it with values for known compounds determined under the same conditions. For easier measurement of retention times, most strip chart recorders are adjusted to move the paper at a rate that corresponds to time divisions calibrated on the chart paper. The retention times (t_1, t_2, t_3) are indicated in Figure 14–5 for the three peaks illustrated.

14.8   POOR RESOLUTION AND TAILING

The peaks in Figure 14–5 are well **resolved.** That is, the peaks are separated from one another, and between each pair of adjacent peaks the tracing returns to the **base line.** In Figure 14–7, the peaks overlap and the resolution is not good. Poor resolution is often caused by using too much sample, too high a column temperature, too short a column, a liquid phase that does not discriminate well between the two components, a column with too large a diameter, or, in short, almost any wrongly adjusted parameter. When peaks are poorly resolved, it is more difficult to determine the relative amount of each component. Methods for determining the relative percentages of each component are given in Section 14.11.

Another desirable feature illustrated by the chromatogram in Figure 14–5 is that each peak is symmetrical. A common example of an unsymmetrical peak is one in which **tailing** has occurred, as shown in Figure 14–8. Tailing usually results from

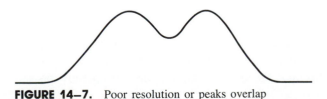

FIGURE 14–7.   Poor resolution or peaks overlap

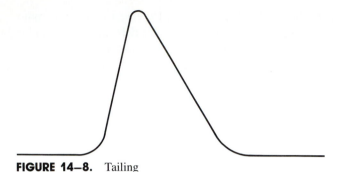

FIGURE 14–8. Tailing

injecting too much sample into the gas chromatograph. Another cause of tailing occurs with polar compounds such as alcohols and aldehydes. These compounds may be temporarily adsorbed on column walls or areas of the support material which are not adequately coated by the liquid phase. Therefore, they don't leave in a band and tailing results.

14.9 QUALITATIVE ANALYSIS

A disadvantage of the gas chromatograph is that it gives no information whatever about the identities of the substances it has separated. The little information it does provide is given by the retention time. It is hard to reproduce this quantity from day to day, however, and complete duplications of separations performed last month may be difficult to make this month. It is usually necessary to **calibrate** the column each time it is used. That is, one must run pure samples of all known and suspected components of a mixture individually, just before chromatographing the mixture, to obtain the retention time of each known compound. As an alternative, each suspected component can be added, one by one, to the unknown mixture while the operator looks to see which peak has its intensity increased relative to the unmodified mixture. Another solution is to **collect** the components individually as they emerge from the gas chromatograph. Each component can then be identified by other means, such as by infrared or nuclear magnetic resonance spectroscopy, or by mass spectrometry.

14.10 COLLECTING THE SAMPLE

For gas chromatographs with a thermal conductivity detector, it is possible to collect samples which have passed through the column. One method utilizes a gas collection tube (see Figure 14–9), which is included in most microscale glassware kits. A collection tube is joined to the exit port of the column by inserting the ＄ 5/5 inner joint into a metal adapter, which is connected to the exit port. When a sample is eluted from the column in the vapor state, it is cooled by the connecting adapter and the gas collection tube and condenses in the collection tube. The gas collection tube is removed from the adapter when the recorder indicates that the desired sample has completely passed

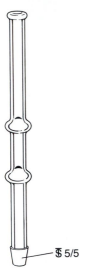

FIGURE 14–9.   Gas chromatography collection tube

through the column. After the first sample has been collected, the process can be repeated with another gas collection tube.

To isolate the liquid, the tapered joint of the collection tube is inserted into a 0.1-mL conical vial, which has a ℑ 5/5 outer joint. The assembly is placed into a test tube, as illustrated in Figure 14–10. During centrifugation, the sample is forced into the bottom of the conical vial. After disassembling the apparatus, the liquid can be

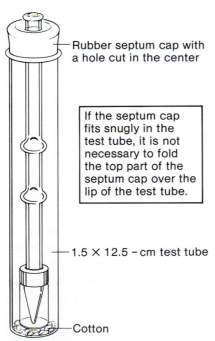

Rubber septum cap with a hole cut in the center

If the septum cap fits snugly in the test tube, it is not necessary to fold the top part of the septum cap over the lip of the test tube.

1.5 × 12.5 – cm test tube

Cotton

FIGURE 14–10.   Gas chromatography collection tube and 0.1-mL conical vial

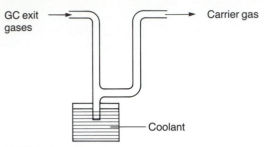

FIGURE 14–11. Collection trap

removed from the vial with a syringe for a boiling point determination or analysis by infrared spectroscopy. If a determination of the sample weight is desired, the empty conical vial and cap should be tared and reweighed after the liquid has been collected. It is advisable to dry the gas collection tube and the conical vial in an oven before use in order to prevent contamination by water or other solvents used in cleaning this glassware.

Another method for collecting samples is to connect a cooled trap to the exit port of the column. A simple trap, suitable for microscale work, is illustrated in Figure 14–11. Suitable coolants include ice water, liquid nitrogen, or dry ice-acetone. For instance, if the coolant is liquid nitrogen (bp $-196\ °C$) and the carrier gas is helium (bp $-269\ °C$), compounds boiling above the temperature of liquid nitrogen generally are condensed or trapped in the small tube at the bottom of the U-shaped tube. The small tube is scored with a file just below the point where it is connected to the larger tube, the tube is broken off, and the sample is removed for analysis. To collect each component of the mixture, one must change the trap after each sample is collected.

14.11 QUANTITATIVE ANALYSIS

The area under a gas chromatograph peak is proportional to the amount (moles) of compound eluted. Hence, the molar percentage composition of a mixture can be approximated by comparing relative peak areas. This method of analysis assumes that the detector is equally sensitive to all compounds eluted and that it gives a linear response with respect to amount. Nevertheless, it gives reasonably accurate results.

The simplest method of measuring the area of a peak is by geometrical approximation, or triangulation. In this method, one multiplies the height, h, of the peak above the base line of the chromatogram by the width of the peak at half of its height, $w_{1/2}$. This is illustrated in Figure 14–12. The base line is approximated by drawing a line between the two "sidearms" of the peak. This method works well only if the peak is symmetrical. If the peak has tailed or is unsymmetrical, it is best to cut out the peaks with scissors and weigh the pieces of paper on an **analytical** balance. Since the weight per area of a piece of good chart paper is reasonably constant from place to place, the ratio of the areas is the same as the ratio of the weights. To obtain a percentage composition for the mixture, one first adds all the peak areas (weights). Then, to

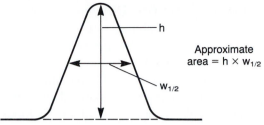

FIGURE 14–12. Triangulation of a peak

calculate the percentage of any component in the mixture, one divides its individual area by the total area and multiplies the result by 100. A sample calculation is illustrated in Figure 14–13. If peaks overlap (see Figure 14–7), either the gas chromatographic conditions must be readjusted to achieve better resolution of the peaks or the peak shape must be estimated.

There are various instrumental means, which are built into recorders, of detecting the amounts of each sample automatically. One method utilizes a separate pen that produces a trace that integrates the area under each peak. Another method employs an electronic device which automatically prints out the area under each peak and the percentage composition of the sample.

For the experiments in this textbook, one may assume that the detector is equally sensitive to all compounds eluted. Compounds with different functional groups or with widely varying molecular weights, however, produce different responses with both TCD and FID gas chromatographs. With a TCD, the responses are different

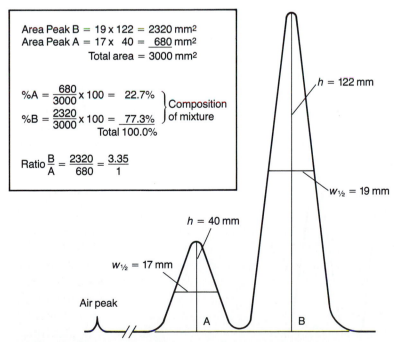

FIGURE 14–13. Sample percentage composition calculation

because not all compounds have the same thermal conductivity. Different compounds analyzed with a FID gas chromatograph also give different responses because the detector response varies with the type of ions produced. For both types of detectors, it is possible to calculate a **response factor** for each compound in a mixture. Response factors are usually determined by making up an equimolar mixture of two compounds, one of which is considered to be the reference. The mixture is separated on a gas chromatograph and the relative percentages are calculated using one of the methods described above. From these percentages, it is possible to determine a response factor for the compound being compared to the reference. If this is done for all the components in a mixture, one can then use these correction factors to make more accurate calculations of the relative percentages for the compounds in the mixture. If necessary, instructions for calculating response factors will be provided by your instructor, or you should consult a textbook on gas chromatography.

PROBLEMS

1. (a) A sample consisting of 1-bromopropane and 1-chloropropane is injected into a gas chromatograph equipped with a nonpolar column. Which compound would have the shorter retention time? Explain your answer.
   (b) If the same sample were run several days later with the conditions as nearly the same as possible, would you expect the retention times to be identical to those obtained the first time? Explain.
2. Using triangulation, calculate the percentage of each component in a mixture composed of two substances, A and B. The chromatogram is shown in Figure 14–14.
3. Make a photocopy of the chromatogram in Figure 14–14. Cut out the peaks and weigh them on an analytical balance. Use the weights to calculate the percentage of each component in the mixture. Compare your answer to what you calculated in Problem #2.
4. What would happen to the retention time of a compound if the following changes were made?
   (a) Decrease the flow rate of the carrier gas
   (b) Increase the temperature of the column
   (c) Increase the length of the column

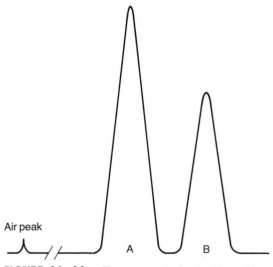

FIGURE 14–14. Chromatogram for Problem #2

Technique 15
SUBLIMATION

In Technique 6, the influence of temperature on the change in vapor pressure of a liquid was considered (see Figure 6–8, p 606). It was shown that the vapor pressure of a liquid increases with temperature. Since the boiling point of a liquid occurs when its vapor pressure is equal to the applied pressure (normally atmospheric pressure), the vapor pressure of a liquid equals 760 mmHg at its boiling point. The vapor pressure of a solid also varies with temperature. Because of this behavior, some solids can readily pass directly into the vapor phase without going through a liquid phase. This process is called **sublimation.** Since the vapor can be resolidified, the overall vaporization-solidification cycle can be used as a purification method. The purification can be successful only if the impurities have significantly lower vapor pressures than the material being sublimed.

15.1 VAPOR PRESSURE BEHAVIOR OF SOLIDS AND LIQUIDS

In Figure 15–1, vapor pressure curves for solid and liquid phases for two different substances are shown. Along lines **AB** and **DF,** the sublimation curves, the solid and

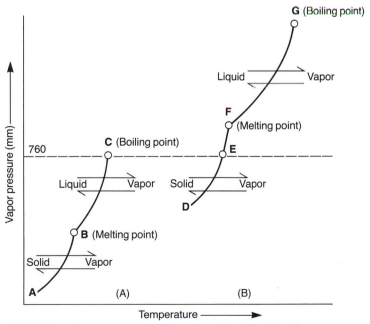

FIGURE 15—1.   Vapor pressure curves for solids and liquids. A. Substance shows normal solid to liquid to gas transitions at 760 mmHg pressure; B. substance shows a solid to gas transition at 760 mmHg pressure.

vapor are at equilibrium. To the left of these lines, the solid phase exists, and to the right of these lines the vapor phase is present. Along lines **BC** and **FG,** the liquid and vapor are at equilibrium. To the left of these lines, the liquid phase exists, and to the right, the vapor is present. The two substances vary greatly in their physical properties, as seen in Figure 15–1.

In the first case (Figure 15–1A), the substance shows normal change-of-state behavior on being heated, going from solid to liquid to gas. The dashed line, which represents an atmospheric pressure of 760 mmHg, is located **above** the melting point, **B,** in Figure 15–1A. Thus, the applied pressure, 760 mmHg, is **greater** than the vapor pressure of the solid-liquid phase at the melting point. Starting at **A,** as the temperature of the solid is raised, the vapor pressure increases along **AB** until the solid is observed to melt at **B.** At **B,** the vapor pressures of **both** the solid and liquid are identical. As the temperature continues to rise, the vapor pressure will increase along **BC** until the liquid is observed to boil at **C.** The description given is for the ''normal'' behavior expected for a solid substance. All three states (solid, liquid, and gas) are observed sequentially during the change in temperature.

In the second case (Figure 15–1B), the substance develops enough vapor pressure to vaporize completely at a temperature below its melting point. The substance shows a solid-to-gas transition only. The dashed line is now located **below** the melting point, **F,** of this substance. Thus, the applied pressure, 760 mmHg, is **less** than the vapor pressure of the solid-liquid phase at the melting point. Starting at **D,** the vapor pressure of the solid rises as the temperature increases along line **DF.** However, the vapor pressure of the solid reaches atmospheric pressure (point **E**) **before** the melting point at **F** is attained. Therefore, sublimation occurs at **E.** No melting behavior will be observed at atmospheric pressure for this substance. For a melting point to be reached and the behavior along line **FG** to be observed, an applied pressure greater than the vapor pressure of the substance at point **F** would be required. This could be achieved by using a sealed pressure apparatus.

The sublimation behavior just described is relatively rare for substances at atmospheric pressure. Several compounds exhibiting this behavior—carbon dioxide, perfluorocyclohexane, and hexachloroethane—are listed in Table 15–1. Notice that these compounds have vapor pressures **above** 760 mmHg at their melting point. In

TABLE 15–1. Vapor Pressures of Solids at their Melting Points

COMPOUND	VAPOR PRESSURE OF SOLID AT MP (mmHg)	MELTING POINT (°C)
Carbon dioxide	3876 (5.1 atm)	−57
Perfluorocyclohexane	950	59
Hexachloroethane	780	186
Camphor	370	179
Iodine	90	114
Naphthalene	7	80
Benzoic acid	6	122
p-Nitrobenzaldehyde	0.009	106

other words, their vapor pressures reach 760 mmHg below their melting points and they sublime rather than melt. Anyone trying to determine the melting point of hexachloroethane at atmospheric pressure will see vapor pouring from the end of the melting point tube! With a sealed capillary tube, the melting point of 186 °C is observed.

15.2   SUBLIMATION BEHAVIOR OF SOLIDS

Sublimation is usually a property of relatively nonpolar substances that also have a highly symmetrical structure. Symmetrical compounds have relatively high melting points and high vapor pressures. The ease with which a substance can escape from the solid is determined by the strength of intermolecular forces. Symmetrical molecular structures have a relatively uniform distribution of electron density and a small dipole moment. A smaller dipole moment means a higher vapor pressure because of lower electrostatic attractive forces in the crystal.

Solids sublime if their vapor pressures are greater than atmospheric pressure at their melting points. Some compounds are listed in Table 15–1 with the vapor pressures at their melting points. The first three entries in the table were discussed in Section 15.1. At atmospheric pressure they would sublime rather than melt, as shown in Figure 15–1B.

The next four entries in Table 15–1 (camphor, iodine, naphthalene, and benzoic acid) exhibit typical change-of-state behavior (solid, liquid, and gas) at atmospheric pressure, as shown in Figure 15–1A. These compounds sublime readily under reduced pressure, however. Vacuum sublimation is discussed in Section 15.3.

Compared with many other organic compounds, camphor, iodine, and naphthalene all have relatively high vapor pressures at relatively low temperatures. For example, they have a vapor pressure of 1 mmHg at 42°, 39°, and 53°, respectively. While this vapor pressure does not seem very large, it is high enough to lead, after a time, to **evaporation** of the solid from an open container. Mothballs (naphthalene and 1,4-dichlorobenzene) show this behavior. When iodine is allowed to stand in a closed container over a period of time, one observes movement of crystals from one part of the container to another.

Although chemists often refer to any solid-vapor transition as sublimation, the process described for camphor, iodine, and naphthalene is really an **evaporation** of a solid. Strictly speaking, a sublimation point is like a melting point or a boiling point. It is defined as the point at which the vapor pressure of the solid **equals** the applied pressure. Many liquids readily evaporate at temperatures far below their boiling points. It is, however, much less common for solids to evaporate. Solids that readily sublime (evaporate) must be stored in sealed containers. When the melting point of such a solid is being determined, some of the solid may sublime and collect toward the open end of the melting-point tube while the rest of the sample melts. To solve the sublimation problem, one seals the capillary tube or rapidly determines the melting point. It is possible to use the sublimation behavior to purify a substance. For example, at atmospheric pressure, camphor can be readily sublimed, just below its melting point, at 175 °C. At 175 °C, the vapor pressure of camphor is 320 mmHg. The vapor solidifies on a cool surface.

15.3   VACUUM SUBLIMATION

Many organic compounds sublime readily under reduced pressure. When the vapor pressure of the solid equals the applied pressure, sublimation occurs, and the behavior is identical to that shown in Figure 15–1B. The solid phase passes directly into the vapor phase. From the data given in Table 15–1, one expects camphor, naphthalene, and benzoic acid to sublime at or below the respective applied pressures of 370 mmHg, 7 mmHg, and 6 mmHg. In principle, one could sublime *p*-nitrobenzaldehyde (last entry in the table), but it would not be practical because of the low applied pressure required.

15.4   SUBLIMATION METHODS

Sublimation can be used to purify solids. The solid is warmed until its vapor pressure becomes high enough for it to vaporize and condense as a solid on a cooled surface placed closely above. Several types of apparatus are illustrated in Figure 15–2. The upper surface provided for collection may be cooled by a continuous flow of water

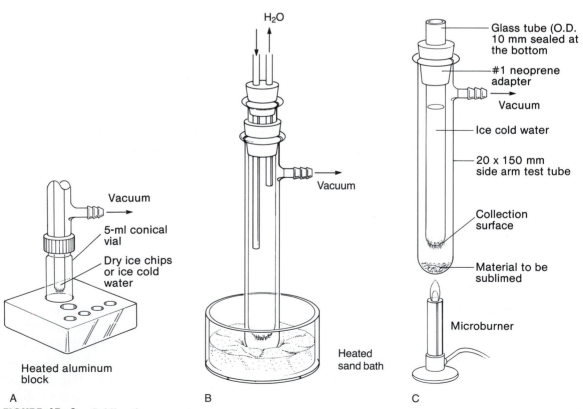

FIGURE 15–2.   Sublimation apparatus

using a ''cold-finger'' condenser (Figure 15–2B); or by dry ice chips or ice cold water (Figure 15–2A and 15–2C). The water hoses must be securely attached to the inlet and outlet of the cold-finger condenser. Otherwise, the connections may leak and allow water to pass into the sublimation apparatus. Many solids do not develop enough vapor pressure at 760 mmHg to be purified by this method, but can be sublimed at reduced pressure. Thus, most sublimation equipment has provision for connection to an aspirator or a vacuum pump. Reduction of the pressure is also advantageous in preventing thermal decomposition of substances that require high temperatures to sublime at ordinary pressures. Various means of heating the apparatus are shown in Figure 15–2.

The assembly shown in Figure 15–2C can be prepared from easily available equipment. The centrifuge tube provides a convenient surface for collecting the solid. Since the surface area at the bottom is small, the purified sample will collect on a relatively small area of glass. A complete description of the assembly of this apparatus is given in Experiment 5, p 76. Detailed procedures for using this apparatus are also given in that experiment. The sublimation apparatus shown in Figure 15–2A can be used with a 5-mL conical vial or a 10-mL round-bottom flask.[1]

It should be remembered that while performing a sublimation, it is important to keep the temperature below the melting point of the solid. After sublimation, the material that has collected on the cooled surface is recovered by removing the central tube (cold-finger) from the apparatus. Care must be used in removing this tube to avoid dislodging the crystals that have collected. The deposit of crystals is scraped from the inner tube with a spatula. If reduced pressure has been used, the pressure must be released carefully to keep a blast of air from dislodging the crystals.

15.5 ADVANTAGES OF SUBLIMATION

One advantage of sublimation is that no solvent is used and therefore none needs to be removed later. Sublimation also removes occluded material, like molecules of solvation, from the sublimed substance. For instance, caffeine (sublimes at 178 °C, melts at 236 °C) absorbs water gradually from the atmosphere to form a hydrate. During sublimation, this water is lost and anhydrous caffeine is obtained. If too much solvent is present in a sample to be sublimed, however, instead of becoming lost, it condenses on the cooled surface and thus interferes with the sublimation.

Sublimation is a faster method of purification than crystallization but not as selective. Similar vapor pressures are often a factor in dealing with solids that sublime; consequently, little separation can be achieved. For this reason, solids are far more often purified by crystallization. Sublimation is most effective in removing a volatile substance from a nonvolatile compound, particularly a salt or other inorganic material. Sublimation is also effective in removing highly volatile bicyclic or other symmetrical molecules from less volatile reaction products. Examples of volatile bicyclic compounds are borneol, isoborneol, and camphor.

---

[1] This microscale sublimator was designed by Professor Anthony Winston (University of West Virginia) and is marketed by Ace Glass Incorporated. See: Winston, A. ''Design of a Microscale Sublimator,'' Journal of Chemical Education (in press).

PROBLEMS

1. Why is solid carbon dioxide called dry ice? How does it differ from solid water in behavior?

2. Under what conditions can one have **liquid** carbon dioxide?

3. A solid substance has a vapor pressure of 800 mmHg at its melting point (80 °C). Describe how the solid behaves as the temperature is raised from room temperature to 80 °C, while the atmospheric pressure is held constant at 760 mmHg.

4. A solid substance has a vapor pressure of 100 mmHg at the melting point (100 °C). Assuming an atmospheric pressure of 760 mmHg, describe the behavior of this solid as the temperature is raised from room temperature to its melting point.

5. A substance has a vapor pressure of 50 mmHg at the melting point (100 °C). Describe how one would experimentally sublime this substance.

Technique 16
POLARIMETRY

16.1 NATURE OF POLARIZED LIGHT

Light has a dual nature since it shows properties of waves and also of particles. The wave nature of light can be demonstrated by two experiments: polarization and interference. Of the two, polarization is the more interesting to organic chemists since they can take advantage of polarization experiments to learn something about the structure of an unknown molecule.

Ordinary white light consists of wave motion in which the waves have a variety of wavelengths and vibrate in all possible planes perpendicular to the direction of propagation. Light can be made to be **monochromatic** (of one wavelength or color) by the use of filters or special light sources. Frequently, a sodium lamp (sodium D line = 5893 Å) is used. Although the light from this lamp consists of waves of only one wavelength, the individual light waves still vibrate in all possible planes perpendicular to the beam. If we imagine that the beam of light is aimed directly at the viewer, ordinary light can be represented by showing the edges of the planes oriented randomly around the path of the beam, as in the left part of Figure 16–1.

FIGURE 16–1.   Ordinary versus plane-polarized light

A Nicol prism, which consists of a specially prepared crystal of Iceland spar (or calcite), has the property of serving as a screen that can restrict the passage of light waves. Waves that are vibrating in one plane are transmitted, while those in all other planes are rejected (either refracted in another direction or absorbed). The light that passes through the prism is called **plane-polarized light,** and it consists of waves that vibrate only in one plane. A beam of plane-polarized light aimed directly at the viewer can be represented by showing the edges of the plane oriented in one particular direction, as in the right portion of Figure 16–1.

Iceland spar has the property of **double refraction,** that is, it can split, or doubly refract, an entering beam of ordinary light into two separate emerging beams of light. Each of the two emerging beams (labeled A and B in Figure 16–2) has only a single plane of vibration, and the plane of vibration in Beam A is perpendicular to the plane of Beam B. In other words, the crystal has separated the incident beam of ordinary light into two beams of plane-polarized light, with the plane of polarization of Beam A perpendicular to the plane of Beam B.

To generate a single beam of plane-polarized light, one can take advantage of the double-refracting property of Iceland spar. A Nicol prism, invented by the Scottish physicist William Nicol, consists of two crystals of Iceland spar cut to specified angles and cemented by Canada balsam. This prism transmits one of the two beams of plane-polarized light while reflecting the other at a sharp angle so that it does not interfere with the transmitted beam. Plane-polarized light can also be generated by a Polaroid filter, a device invented by E. H. Land, an American physicist. Polaroid filters consist of certain types of crystals, embedded in transparent plastic and capable of producing plane-polarized light.

After passing through a first Nicol prism, plane-polarized light can also pass through a second Nicol prism, but only if the second prism has its axis oriented so that it is **parallel** to the incident light's plane of polarization. Plane-polarized light is **absorbed** by a Nicol prism that is oriented so that its axis is **perpendicular** to the incident light's plane of polarization. These situations can be illustrated by the picket-fence analogy, as shown in Figure 16–3. Plane-polarized light can pass through a fence whose slats are oriented in the proper direction but is blocked out by a fence whose slats are oriented perpendicularly.

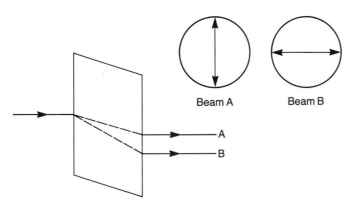

Beam A   Beam B

A

B

FIGURE 16–2. Double refraction

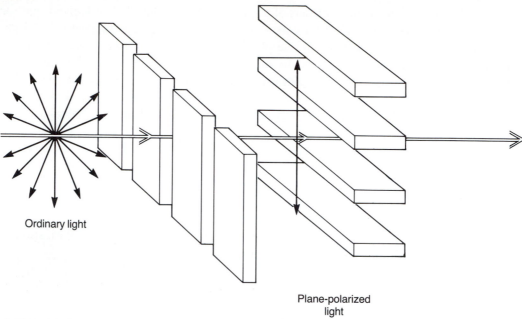

Ordinary light

Plane-polarized
light

FIGURE 16—3.   The picket-fence analogy

An **optically active** substance is one that interacts with polarized light to rotate the plane of polarization through some angle α. Figure 16—4 illustrates this phenomenon.

16.2   THE POLARIMETER

An instrument called a **polarimeter** is used to measure the extent to which a substance interacts with polarized light. A schematic diagram of a polarimeter is shown in Figure 16—5. The light from the source lamp is polarized by being passed through a fixed Nicol prism, called a polarizer. This light passes through the sample, with which it may or may not interact to have its plane of polarization rotated in one direction or the other. A second, rotatable Nicol prism, called the analyzer, is adjusted to allow the maximum amount of light to pass through. The number of degrees and the direction of rotation required for this adjustment are measured to give the **observed rotation** α.

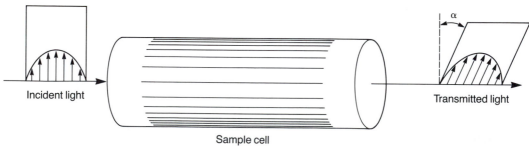

Incident light

α

Transmitted light

Sample cell

FIGURE 16—4.   Optical activity

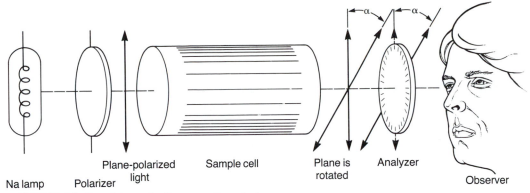

FIGURE 16–5. Schematic diagram of a polarimeter

So that data determined by several persons under different conditions can be compared, a standardized means of presenting optical rotation data is necessary. The most common way of presenting such data is by recording the **specific rotation** $[\alpha]_\lambda^t$, which has been corrected for differences in concentration, cell path length, temperature, solvent, and wavelength of the light source. The equation defining the specific rotation of a compound in solution is

$$[\alpha]_\lambda^t = \frac{\alpha}{cl}$$

where

α = observed rotation in degrees
c = concentration in grams per milliliter of solution
l = length of sample tube in decimeters
λ = wavelength of light (usually indicated as ''D,'' for the sodium D line)
t = temperature in degrees Celsius

For pure liquids, the density d of the liquid in grams per milliliter replaces c in the above formula. Occasionally one wants to compare compounds of different molecular weights, so a **molecular rotation,** based on moles instead of grams, is more convenient than a specific rotation. The molecular rotation M_λ^t is derived from the specific rotation $[\alpha]_\lambda^t$ by

$$M_\lambda^t = \frac{[\alpha]_\lambda^t \times \text{molecular weight}}{100}$$

Usually measurements are made at 25 °C with the sodium D line as a light source, so specific rotations are reported as $[\alpha]_D^{25°}$.

16.3 THE SAMPLE CELLS

It is important for the solution whose optical rotation is to be determined to contain no suspended particles of dust or dirt that might disperse the incident polarized light.

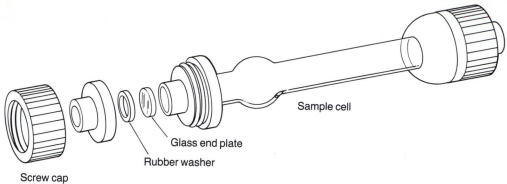

Sample cell

Glass end plate

Rubber washer

Screw cap

FIGURE 16–6. Polarimeter cell assembly

Therefore it is necessary to clean the sample cell carefully and to make certain that there are no air bubbles trapped in the path of the light. The sample cells contain an enlarged ring near one end, in which the air bubbles may be trapped. The sample cell, shown in Figure 16–6, is tilted upward and tapped until the air bubbles move into the enlarged ring. It is important not to get fingerprints on the glass endplate in reassembling the cell.

The sample is generally prepared by dissolving 0.1–0.5 g of the substance to be studied in 25 mL of solvent, usually water, ethanol, or methylene chloride; chloroform was used in the past. If the specific rotation of the substance is very high or very low, it may be necessary to make the concentration of the solution respectively lower or higher, but usually this is determined after first trying a concentration range such as that suggested above.

Sample cells are available in various lengths, with 0.5 dm and 1.0 dm being the most common. Since the shorter cells require a smaller volume of liquid, their use in microscale experiments would be more logical. Even the smallest cells, however, require a sample size which is much larger than the quantities normally produced in microscale experiments. The experiments in this textbook that require the use of a polarimeter have been written to require a larger scale in order to provide enough sample to allow the determination of optical rotation.

16.4 OPERATION OF THE POLARIMETER

The procedures given here for preparing the cells and for operating the instrument are appropriate for the Zeiss polarimeter with the circular scale; other models of polarimeter are operated similarly. It is necessary, before beginning the experiments, to turn the power switch to the ON position and wait 5 to 10 minutes until the sodium lamp is properly warmed.

The instrument should be checked initially by making a zero reading with a sample cell filled only with solvent. If the zero reading does not correspond with the zero-degree calibration mark, then the difference in readings must be used to correct all

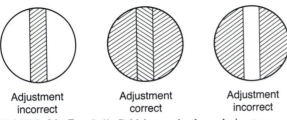

Adjustment   Adjustment   Adjustment
incorrect   correct   incorrect

FIGURE 16–7.   Split-field image in the polarimeter

subsequent readings. The reading is determined by laying the sample tube in the cradle, enlarged end up (making sure that there are no air bubbles in the light path), closing the cover, and turning the knob until the proper angle of the analyzer is reached. Most instruments, including the Zeiss polarimeter, are of the double-field type, in which the eye sees a split field whose sections must be matched in light intensity. The value of the angle through which the plane of polarized light has been rotated (if any) is read directly from the scale that can be seen through the eyepiece directly below the split-field image. Figure 16–7 shows how this split field might appear.

The cell containing the solution of the sample is then placed in the polarimeter, and the observed angle of rotation is measured in the same way. Be sure to record not only the numerical value of the angle of rotation in degrees but also the direction of rotation. Rotations clockwise are due to **dextrorotatory** substances and are indicated by the sign ''+.'' Rotations counterclockwise are due to **levorotatory** substances and are indicated by the sign ''−.'' It is best, in making a determination, to take several readings, including readings for which the actual value was approached from both sides. In other words, where the actual reading might be +75°, first approach this reading upward from a reading near zero, then on the next measurement approach this reading downward from an angle greater than +75°. Duplicating readings and approaching the observed value from both sides reduce errors. The readings are then averaged to get the observed rotation α. This rotation is then corrected by the appropriate factors, according to the formulas in Section 16.2, to provide the specific rotation. The specific rotation is always reported as a function of temperature, indicating the wavelength by ''D'' if a sodium lamp is used, and reporting the concentration and solvent used. For example, $[\alpha]_D^{20} = +43.8°$ ($c = 7.5$ g/100 mL in absolute ethanol).

16.5   OPTICAL PURITY

When one prepares a sample of an enantiomer by a resolution method, the sample is not always 100.0% pure enantiomer. It frequently is contaminated by residual amounts of the opposite stereoisomer. To determine the amount of the desired enantiomer in the sample, one calculates the **optical purity,** or the **excess** of one enantiomer in a mixture expressed as a percentage of the total. In a racemic (±) mixture, there is no excess enantiomer and the optical purity is zero; in a completely resolved material, the excess enantiomer equals the total material in weight, and the optical purity is 100%. Although the following is not the most precise equation for determining the optical purity, it should prove useful in most simple applications:

$$\text{Optical purity} = \frac{\text{observed specific rotation}}{\text{specific rotation of pure substance}} \times 100$$

A compound that is $x\%$ optically pure contains $x\%$ of one enantiomer and $(100 - x)\%$ of a **racemic mixture.**

If the optical purity is given, the relative percentages of the enantiomers can be calculated easily. If the predominant form in the impure, optically active mixture is assumed to be the $(+)$ enantiomer, the percentage of the $(+)$ enantiomer is

$$\left[x + \left(\frac{100 - x}{2} \right) \right]\%$$

and the percentage of the $(-)$ enantiomer is $[(100 - x)/2]\%$. The relative percentages of $(+)$ and $(-)$ forms in a partially resolved mixture of enantiomers can be calculated as shown below. Consider a partially resolved mixture of camphor enantiomers. The specific rotation for pure $(+)$-camphor is $+43.8°$ in absolute ethanol, but the mixture shows a specific rotation of $+26.3°$.

$$\text{Optical purity} = \frac{+26.3°}{+43.8°} \times 100 = 60\% \text{ optically pure}$$

$$\% \ (+) \text{ enantiomer} = 60 + \left(\frac{100 - 60}{2} \right) = 80\%$$

$$\% \ (-) \text{ enantiomer} = \left(\frac{100 - 60}{2} \right) = 20\%$$

PROBLEMS

1. Calculate the specific rotation of a substance that is dissolved in a solvent (0.4 g/mL) and that has an observed rotation of $-10°$ as determined with a 0.5-dm cell.
2. Calculate the observed rotation for a solution of a substance (2.0 g/mL) that is 80% optically pure. A 2-dm cell is used. The specific rotation for the optically pure substance is $+20°$.
3. What is the optical purity of a partially racemized product if the calculated specific rotation is $-8°$ and the pure enantiomer has a specific rotation of $-10°$? Calculate the percentage of each of the enantiomers in the partially racemized product.

Technique 17
REFRACTOMETRY

The **refractive index** is a useful physical property of liquids. Often a liquid can be identified from a measurement of its refractive index. The refractive index can also provide a measure of the purity of the sample being examined. This is done by comparing the experimentally measured refractive index with the value reported in the literature for an ultrapure sample of the compound. The closer the measured sample's value to the literature value, the purer the sample.

17.1 THE REFRACTIVE INDEX

The refractive index has as its basis the fact that light travels at a different velocity in condensed phases (liquids, solids) than in air. The refractive index (n) is defined as the ratio of the velocity of light in air to the velocity of light in the medium being measured:

$$n = \frac{V_{air}}{V_{liquid}} = \frac{\sin \theta}{\sin \phi}$$

It is not difficult to measure experimentally the ratio of the velocities. It corresponds to $\sin \theta / \sin \phi$, where θ is the angle of incidence for a beam of light striking the surface of the medium and ϕ is the angle of refraction of the beam of light **within** the medium. This is illustrated in Figure 17−1.

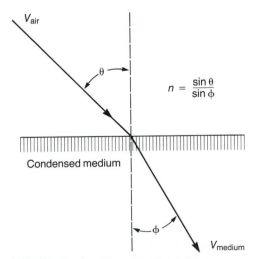

FIGURE 17−1. The refractive index

The refractive index for a given medium depends on two variable factors. First, it is **temperature**-dependent. The density of the medium changes with temperature; hence, the speed of light in the medium also changes. Second, the refractive index is **wavelength**-dependent. Beams of light with different wavelengths are refracted to different extents in the same medium and give different refractive indices for that medium. It is usual to report refractive indices measured at 20 °C, with a sodium discharge lamp as the source of illumination. The sodium lamp gives off yellow light of 589-nm wavelength, the so-called sodium D line. Under these conditions, the refractive index is reported in the following form.

$$n_D^{20} = 1.4892$$

The superscript indicates the temperature, and the subscript indicates that the sodium D line was used for the measurement. If another wavelength is used for the determination, the D is replaced by the appropriate value, usually in nanometers (1 nm = 10^{-9} m).

Notice that the hypothetical value reported above has four decimal places. It is easy to determine the refractive index to within several parts in 10,000. Therefore, n_D is a very accurate physical constant for a given substance and can be used for identification. However, it is sensitive to even small amounts of impurity in the substance measured. Unless the substance is purified **extensively,** it will not usually be possible to reproduce the last two decimal places given in a handbook or other literature source. Typical organic liquids have refractive index values between 1.3400 and 1.5600.

17.2   THE ÅBBÉ REFRACTOMETER

The instrument used to measure the refractive index is called a **refractometer.** Although many styles of refractometer are available, by far the most common instrument is the Åbbé refractometer. This style of refractometer has the following advantages:

1. White light may be used for illumination; but the instrument is compensated, so that the index of refraction obtained is actually that for the sodium D line.
2. The prisms can be temperature-controlled.
3. Only a small sample is required (a few drops of liquid using the standard method, or about 5 μL using a modified technique).

A common type of Åbbé refractometer is shown in Figure 17–2.

The optical arrangement of the refractometer is very complex; a simplified diagram of the internal workings is given in Figure 17–3. The letters **A, B, C,** and **D** label corresponding parts in both Figures 17–2 and 17–3. A complete description of refractometer optics is too difficult to attempt here, but Figure 17–3 gives a simplified diagram of the essential operating principles.

Using the standard method, the sample to be measured is introduced between the two prisms. If it is a "free-flowing" liquid, it may be introduced into a channel along the side of the prisms, injected from a Pasteur pipet. If it is a viscous sample, the prisms must be opened (they are hinged) by lifting the upper one; a few drops of liquid are applied to the lower prism with a Pasteur pipet or a wooden applicator. If a Pasteur

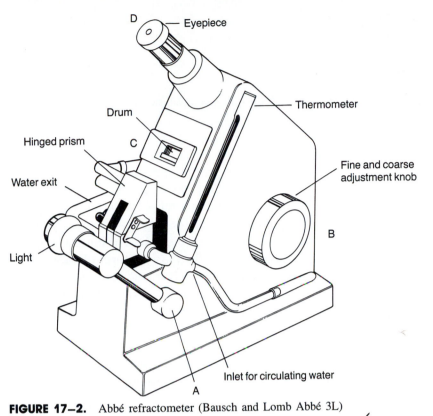

FIGURE 17–2. Abbé refractometer (Bausch and Lomb Abbé 3L)

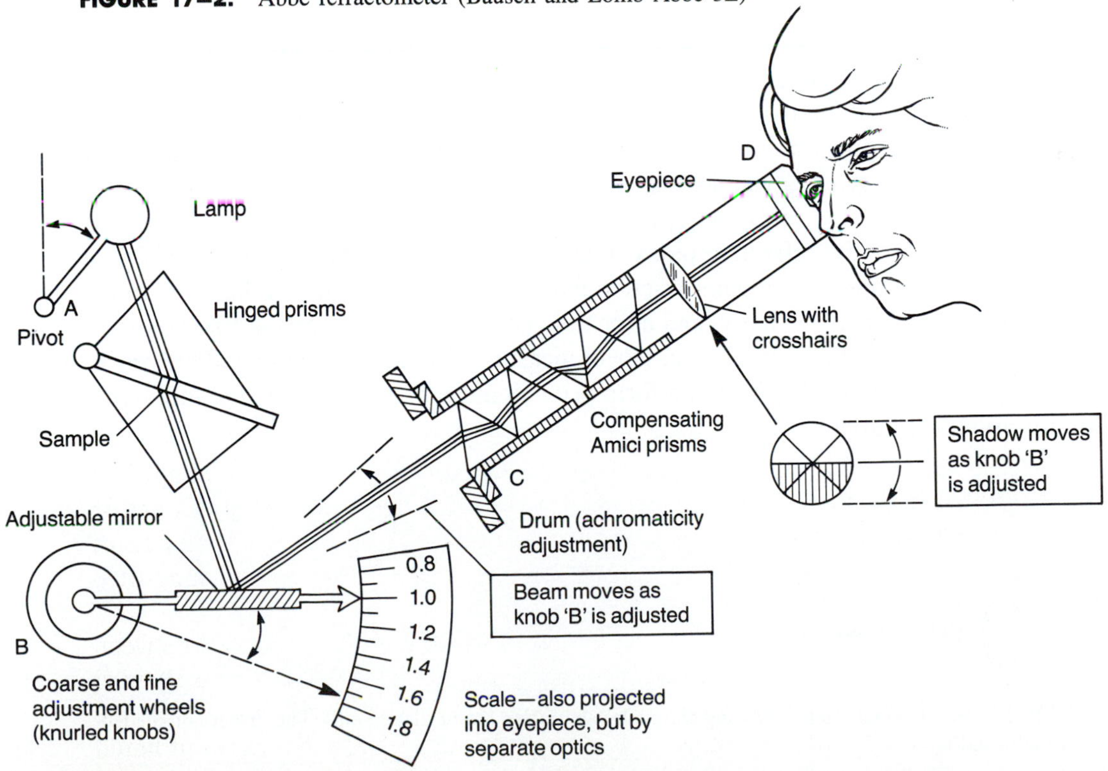

FIGURE 17–3. Simplified diagram of a refractometer

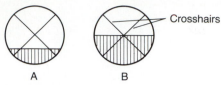

FIGURE 17–4. A. Refractometer incorrectly adjusted; B. correct adjustment

pipet is used, care must be taken not to touch the prisms since they become scratched easily. When the prisms are closed, the liquid should spread evenly to make a thin film. With highly volatile samples, the remaining operations must be performed rapidly. Even when the prisms are closed, evaporation of volatile liquids can readily occur.

Next, one turns on the light and looks into the eyepiece **D**. The hinged lamp and the coarse adjustment knob at **B** are adjusted to give the most uniform illumination to the visible field in the eyepiece (no dark areas). The light rotates at pivot **A**.

Once a uniform field is found, one rotates the coarse and fine adjustment knobs at **B** until the dividing line between the light and dark halves of the visual field coincides with the center of the cross hairs (Figure 17–4). If the cross hairs are not in sharp focus, it will be necessary to adjust the eyepiece to focus them. If the horizontal line dividing the light and dark areas appears as a colored band, as in Figure 17–5, the refractometer is showing **chromatic aberration** (color dispersion). This can be adjusted with the knob labeled **C.** This knurled knob rotates a series of prisms, called Amici prisms, that color-compensate the refractometer and cancel out dispersion. The knob should be adjusted to give a sharp, uncolored division between the light and dark segments. When one has adjusted everything correctly (as in Figure 17–4B), the refractive index is read. In the instrument described here, a small button on the left side of the housing is pressed, and the scale becomes visible in the eyepiece. In other refractometers, the scale is visible at all times, frequently through a separate eyepiece.

Occasionally the refractometer will be so far out of adjustment that it may be difficult to measure the refractive index of an unknown. When this happens, it is wise to place a pure sample of known refractive index in the instrument, set the scale to the correct value of refractive index, and adjust the controls for the sharpest line possible. Once this has been done, it will be easier to measure an unknown sample. This procedure may be especially helpful to perform prior to measuring the refractive index of a highly volatile sample.

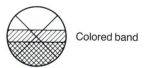

FIGURE 17–5.   Refractometer showing chromatic aberration (color dispersion). The dispersion is incorrectly adjusted.

> **There are many styles of refractometer, but most have adjustments similar to those described here.**

In the procedure described above, several drops of liquid are required to obtain the refractive index. In some microscale experiments, you may not have enough sample to use this standard method. It is possible to modify the procedure so that a reasonably accurate refractive index can be obtained on about 5 μL of liquid. Instead of placing the sample directly onto the prism, the sample is applied to a small piece of lens paper. The lens paper can be conveniently cut with a hand-held paper punch,[1] and the paper disc (0.6 cm diameter) is placed in the center of the bottom prism of the refractometer. To avoid scratching the prism, a forceps or tweezers with plastic tips should be used to handle the disc. About 5 μL of liquid is carefully placed on the lens paper using a microliter syringe. After closing the prisms, the refractometer is adjusted as described above and the refractive index is read. With this method, the horizontal line dividing the light and dark areas may not be as sharp as it is in the absence of the lens paper. It may also be impossible to eliminate color dispersion completely. Nonetheless, the refractive index values determined by this method are usually within 10 parts in 10,000 of the values determined by the standard procedure.

17.3 CLEANING THE REFRACTOMETER

You should always remember, in using the refractometer, that if the prisms are scratched the instrument will be ruined.

> **Do not touch the prisms with any hard object.**

This admonition includes Pasteur pipets and glass rods.

When measurements are completed, the prisms should be cleaned with ethanol or petroleum ether. **Soft** tissues are moistened with the solvent, and the prisms are wiped **gently.** When the solvent has evaporated from the prism surfaces, they should be locked together. The refractometer should be left with the prisms closed to avoid collection of dust in the space between them. The instrument should also be turned off when it is no longer in use.

17.4 TEMPERATURE CORRECTIONS

If the refractive index is not determined in a room in which the temperature is 20 °C, or if 20 °C cooling water is not used to circulate through the instrument, a temperature correction must be made. Although the magnitude of the temperature correction may vary from one class of compound to another, a value of 0.00045 per degree Celsius is

---

[1] In order to cut the lens paper more easily, it is helpful to place several sheets of the paper between two pieces of heavier paper, such as the paper used for file folders.

a useful approximation for most substances. The index of refraction of a substance **decreases** with **increasing** temperature. Therefore, one adds the correction to the observed n_D value for temperatures higher than 20 °C and subtracts it for temperatures lower than 20 °C. As an example, the reported n_D value for nitrobenzene is 1.5529. One would observe a value at 25 °C of 1.5506. The temperature correction would be made as follows:

$$n_D^{20} = 1.5506 + 5(0.00045) = 1.5529$$

PROBLEMS

1. A solution consisting of isobutyl bromide and isobutyl chloride is found to have a refractive index of 1.3931 at 20 °C. The refractive indices at 20 °C of isobutyl bromide and isobutyl chloride are 1.4368 and 1.3785, respectively. Determine the molar composition (in percent) of the mixture by assuming a linear relation between the refractive index and the molar composition of the mixture.
2. The refractive index of a compound at 16 °C is found to be 1.3982. Correct this refractive index to 20 °C.

Technique 18
PREPARATION OF SAMPLES FOR SPECTROSCOPY

Modern organic chemistry requires sophisticated scientific instruments. Most important among these instruments are the two spectroscopic instruments: the infrared (IR) and nuclear magnetic resonance (NMR) spectrometers. These instruments are indispensable to the modern organic chemist in proving the structures of unknown substances, in verifying that reaction products are indeed the predicted ones, and in characterizing organic compounds. The theory underlying these instruments can be found in most standard lecture textbooks in organic chemistry. Additional information, including correlation charts, to help in interpreting spectra are found in this textbook in Appendix 3 (Infrared Spectroscopy), Appendix 4 (Proton Nuclear Magnetic Resonance Spectroscopy), and Appendix 5 (Carbon-13 Nuclear Magnetic Resonance Spectroscopy). This technique chapter concentrates on the preparation of samples for these spectroscopic methods. Part A covers techniques used in infrared spectroscopy and Part B describes sample preparation for nuclear magnetic resonance spectroscopy.

PART A. INFRARED SPECTROSCOPY

18.1 INTRODUCTION

To determine the infrared spectrum of a compound, one must place it in a sample holder or cell. In infrared spectroscopy this immediately poses a problem. Glass, quartz, and plastics absorb strongly throughout the infrared region of the spectrum (any compound with covalent bonds usually absorbs) and cannot be used to construct sample cells. Ionic substances must be used in cell construction. Metal halides (sodium chloride, potassium bromide, silver chloride) are commonly used for this purpose.

Sodium Chloride Cells. Single crystals of sodium chloride are cut and polished to give plates that are transparent throughout the infrared region. These plates are then used to fabricate cells which can be used to hold **liquid** samples. Since sodium chloride is water-soluble, samples must be **dry** before a spectrum can be obtained. Sodium chloride plates cleave easily (break) when too much pressure is applied. In general, sodium chloride plates are preferred for most applications involving liquid samples.

Silver Chloride Cells. Cells may be constructed of silver chloride. These plates may be used for **liquid** samples that contain small amounts of water because silver chloride is water-insoluble. However, since water absorbs in the infrared region, as much water as possible should be removed, even when using silver chloride. Silver chloride plates must be stored in the dark, and they cannot be used with compounds that have an amino functional group. Amines react with silver chloride.

A **solid** sample is usually held in place by making a potassium bromide pellet that contains a small amount of dispersed compound. A solid sample may also be suspended in mineral oil, which absorbs only in specific regions of the infrared spectrum. Another method is to dissolve the solid compound in an appropriate solvent and place the solution between two sodium chloride or silver chloride plates.

18.2 LIQUID SAMPLES—NaCl PLATES

The simplest method of preparing the sample, if it is a liquid, is to place a thin layer of the liquid between two sodium chloride plates that have been ground flat and polished. This is the method of choice when you need to determine the infrared spectrum of a pure liquid. A spectrum determined by this method is referred to as a **neat** spectrum. No solvent is used. The polished plates are expensive because they are cut from a large, single crystal of sodium chloride. Salt plates break easily and they are water-soluble.

Preparing the Sample. Obtain two sodium chloride plates and a holder from the desiccator where they are stored. Moisture from fingers will mar and occlude the polished surfaces. Samples that contain water will destroy the plates.

> **The plates should be touched only on their edges. Be certain to use a sample that is dry or free from water.**

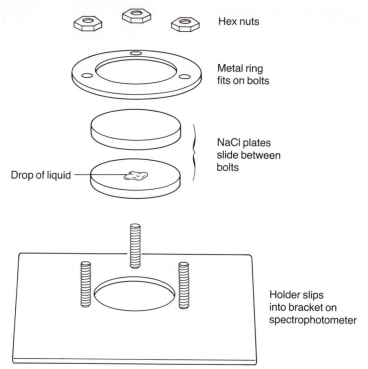

FIGURE 18–1. Salt plates and holder

Add a drop of the liquid to the surface of one plate and then place the second plate on top. The pressure of this second plate causes the liquid to spread out and form a thin, capillary film between the two plates. As shown in Figure 18–1, set the plates between the bolts in a holder and place the metal ring carefully on the salt plates. Use the hex nuts to hold the salt plates in place.

> **Do not overtighten the nuts or the salt plates will cleave or split.**

Tighten the nuts firmly, but do not use any force at all to turn them. Spin them with the fingers until they stop; then turn them just another fraction of a full turn, and they will be tight enough. If the nuts have been tightened carefully, you should observe a transparent film of sample (a uniform wetting of the surface). If a thin film has not been obtained, you should loosen one or more of the hex nuts and adjust them so that a uniform film is obtained, or add more sample.

The thickness of the film obtained between the two plates is a function of two factors: (1) the amount of liquid placed on the first plate (one drop, two drops, etc.), and (2) the pressure used to hold the plates together. If more than one or two drops of liquid have been used, it will probably be too much, and the resulting spectrum will show strong absorptions that are off the scale of the chart paper. Only enough liquid is needed to wet both surfaces.

If the sample has a very low viscosity, you may find that the capillary film is too thin to give a good spectrum. Another problem that you may find is that the liquid is so volatile that the sample evaporates before the spectrum can be determined. In these cases, you may need to use the silver chloride plates discussed in Section 18.3, or a solution cell described in Section 18.5. Often, you can obtain a reasonable spectrum by assembling the cell quickly and running the spectrum before the sample runs out of the salt plates or evaporates.

Determining the Infrared Spectrum. Slide the holder into the slot in the sample beam of the spectrophotometer. Determine the spectrum according to the instructions provided by your instructor. In some cases, your instructor may ask you to calibrate your spectrum. If this is the case, refer to Section 18.8.

Cleaning and Storing the Salt Plates. Once the spectrum has been determined, the holder should be demounted, and the salt plates should be washed with methylene chloride or **dry** acetone (Keep the plates away from water!). Use a soft tissue, moistened with the solvent, to wipe the plates. If some of your compound remains on the plates, you may observe a shiny surface. Continue to clean the plates with solvent until no more compound remains on the surfaces of the plates.

> **Avoid direct contact with methylene chloride. Return the salt plates and holder to the desiccator for storage.**

18.3 LIQUID SAMPLES—AgCl PLATES

The mini-cell[1] shown in Figure 18–2 may also be used with liquids. The cell assembly consists of a two-piece threaded body, an O-ring, and two silver chloride plates. The plates are flat on one side, and there is a circular depression (0.025 mm or 0.10 mm deep) on the other side of the plate. The advantages of using silver chloride plates are that they may be used with wet samples or solutions. A disadvantage is that silver chloride darkens when exposed to light for extended periods. They also scratch more easily than salt plates and react with amines.

Preparing the Sample. Silver chloride plates are handled in the same way as salt plates. Unfortunately, they are smaller and thinner (about like a contact lens) than salt plates and care must be taken to not lose them! They should be removed from the light-tight container with care. One side of the plate is flat and the other side has a very slight circular depression, which is very difficult to see. Your instructor may have etched a letter on each plate to indicate which side is the flat one. If you are going to determine the infrared spectrum of a pure liquid (neat spectrum), you should select the

---

[1] The Wilks Mini-Cell liquid sample holder is available from the Foxboro Company, 151 Woodward Avenue, South Norwalk, CT 06856. We recommend the AgCl cell windows with a 0.10 mm depression, rather than the 0.025 mm depression.

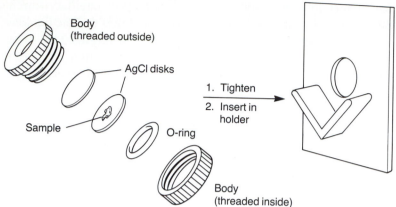

FIGURE 18–2. AgCl mini liquid cell and V-mount holder

flat side of each silver chloride plate. Insert the O-ring into the cell body as shown in Figure 18–2, place the plate into the cell body with the flat surface up, and add one drop of liquid or less to the plate.

> **Do not use amines with AgCl plates.**

Place the second plate on top of the first with the flat side down. The orientation of the silver chloride plates is shown in Figure 18–3A. This arrangement is used to obtain a capillary film of your sample. Screw the top of the mini-cell into the body of the cell so the silver chloride plates are held firmly together. A tight seal will form because AgCl deforms under pressure.

Other combinations may be used with these plates. For example, you may vary the sample path length by using the orientations shown in Figure 18–3B and C. If you add your sample to the 0.10 mm depression of one plate and cover it with the flat side of the other one, you obtain a path length of 0.10 mm (Figure 18–3B). This arrangement is useful for analyzing volatile or low viscosity liquids. Placement of the two plates with their depressions toward each other gives a path length of 0.20 mm (Figure 18–3C). This orientation may be used for a solution of a solid (or liquid) in carbon tetrachloride (Section 18.5B).

Determining the Spectrum. Slide the V-mount holder shown in Figure 18–2 into the slot on the infrared spectrophotometer. Set the cell assembly in the V-mount holder, and determine the infrared spectrum of the liquid.

Cleaning and Storing the AgCl Plates. Once the spectrum has been determined, the cell assembly holder should be demounted, and the AgCl plates washed

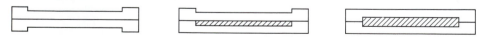

  A. Capillary film   B. 0.10–mm path length   C. 0.20–mm path length

FIGURE 18–3. Path length variations for AgCl plates

with methylene chloride or acetone. Do not use tissue to wipe the plates as they scratch easily. AgCl plates are light sensitive. Store the plates in a light-tight container.

18.4   SOLID SAMPLES—KBr PELLETS

The easiest method of preparing a solid sample is to make a potassium bromide (KBr) pellet. When KBr is placed under pressure, it flows and seals the sample into a solid ''solution'' or matrix. Since potassium bromide does not absorb in the region of the infrared spectrum that is useful to us, a spectrum can be obtained on a sample without interference.

Preparing the Sample.   Grind 1–2 mg of the solid sample for three to five minutes in an agate mortar until the particle size has become so small that the surface of the solid appears shiny. Add 100 mg of **powdered** potassium bromide (KBr) and grind the mixture for about 30 seconds with a pestle. Scrape the mixture into the middle with a spatula and grind the mixture again for about 15 seconds. This procedure assures that the sample is mixed thoroughly with the KBr. You should work as rapidly as possible because KBr absorbs water rapidly. The bottle of potassium bromide must be stored in a desiccator when it is not in use. The sample and KBr must be finely ground or it will scatter the infrared radiation excessively.

The sample and potassium bromide should be weighed on an analytical balance the first few times that a pellet is prepared. After some experience, these quantities can be estimated quite closely by eye.

Making a Pellet.   Once the powder mixture has been prepared by the method given above, transfer a **portion** of the finely ground powder (usually not more than half) into a die that compresses it into a translucent pellet. The simplest die consists of two stainless steel bolts and a threaded barrel as illustrated in Figure 18–4. The bolts have their ends ground flat. To use this die, screw one of the bolts into the barrel, but not all the way; leave one or two turns. Carefully add the powder with a spatula into the open end of the partly assembled die and tap it lightly on the bench top to give an even layer on the face of the bolt. While keeping the barrel upright, carefully screw the second bolt into the barrel until it is finger tight. Insert the head of the bottom bolt into

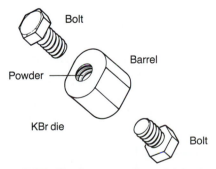

FIGURE 18–4.   KBr pellet mini-press

the hexagonal hole in a plate bolted to the bench top. This plate keeps the head of one bolt from turning. The top bolt is tightened with a torque wrench to compress the KBr mixture. Continue to turn the torque wrench until a loud click is heard (the ratchet mechanism makes softer clicks) or until the appropriate torque value (20 ft-lb) is reached. If the bolt is tighted beyond this point, the head may be twisted off one of the bolts. Leave the die under pressure for about 60 seconds and then reverse the ratchet on the torque wrench or pull the torque wrench in the opposite direction to open the assembly. When the two bolts are loose, hold the barrel **horizontally** and carefully remove the two bolts.

Determining the Infrared Spectrum. You should observe a clear or translucent KBr pellet in the center of the barrel. Even if the pellet is not totally transparent, you should be able to obtain a satisfactory spectrum as long as light will pass through the pellet. The best thing to do is to try to obtain the spectrum, and if it is unsatisfactory, prepare another pellet. To obtain the spectrum, slide the V-mount holder shown in Figure 18–2 into the slot on the infrared spectrophotometer. Set the body containing the pellet in the V-mount holder so that the sample is centered in the optical path. It may be useful to turn the barrel slightly in the V-mount holder in order to obtain a better spectrum. (Turning the barrel may provide a more transparent place in the pellet, and yield a better spectrum.) It is often possible to compensate (at least partially) for a marginal pellet by placing a wire screen or attenuator in the reference beam, thereby balancing the lowered transmittance of the pellet.

Problems with an Unsatisfactory Pellet. If the pellet is unsatisfactory (too cloudy to pass light), one of several things may have been wrong:

1. The KBr mixture may not have been ground finely enough, and the particle size may be too big. The large particle size creates too much light scattering.
2. The sample may not be dry.
3. Too much sample may have been used for the amount of KBr taken.
4. The pellet may be too thick, that is, too much of the powdered mixture was put into the die.
5. The KBr may have been ''wet'' or have acquired moisture from the air while the mixture was being ground in the mortar.
6. The die may not have been tightened enough (this is improbable when a torque wrench is used).
7. The sample may have a low melting point. Low-melting solids not only are difficult to dry but also melt under pressure. You may need to dissolve the compound in a solvent and run the spectrum in solution (Section 18.5).

Cleaning and Storing the Pellet Press. When the spectrum has been determined, punch the pellet out of the barrel with a spatula and wash both the barrel and the bolts thoroughly with water. Do not attempt to remove the pellet with one of the bolts.

> **The polished faces of the bolts must not be scratched or damaged in any way.**

Damaged bolts are useless! They are particularly prone to corrosion when the polished surfaces of the bolts are allowed to remain in contact with KBr. Once the barrel and two bolts have been thoroughly washed with water, rinse them with acetone. Allow the pieces to dry and return them to the oven or desiccator for storage.

18.5   SOLID SAMPLES—SOLUTION SPECTRA

Method A—Solution between Salt (NaCl) Plates

For substances that are soluble in carbon tetrachloride, a quick and easy method for determining the spectra of solids is available. Dissolve as much solid as possible in 0.1 mL of carbon tetrachloride. Place one or two drops of the solution between sodium chloride plates in precisely the same manner as that used for pure liquids (Section 18.2). The spectrum is determined as described for pure liquids using salt plates (Section 18.2). You will need to work as quickly as possible. If there is a delay, the solvent will evaporate from between the plates before the spectrum is recorded. Since the spectrum contains the absorptions of the solute superimposed on the absorptions of carbon tetrachloride, it is important to remember that any absorption that appears near 800 cm^{-1} (12.5 μ) may be due to the stretching of the C–Cl bond of the solvent. Information contained to the right of about 900 cm^{-1} is not usable in this method. There are no other interfering bands for this solvent (see Figure 18–6), and any other absorptions can be attributed to your sample. Chloroform solutions should not be studied by this method since the solvent has too many interfering absorptions (see Figure 18–7).

> **CAUTION: Carbon tetrachloride is a hazardous solvent. Work in the hood!**

Carbon tetrachloride, besides being toxic, is suspected of being a carcinogen. In spite of the health problems associated with its use, there is no suitable alternative solvent for Infrared spectroscopy. Other solvents have too many interfering infrared absorption bands. It is necessary for you to handle carbon tetrachloride very carefully to minimize the adverse health effects. The spectroscopic-grade carbon tetrachloride should be stored in a 1-pt stoppered bottle in a hood. A Pasteur pipet should be attached to the bottle, possibly by storing it in a test tube taped to the side of the bottle. All sample preparation should be conducted in a hood. Rubber or plastic gloves should be worn. The cells should also be cleaned in the hood. All carbon tetrachloride used in preparing samples should be disposed of in an appropriately marked waste container.

Method B—AgCl Mini-cell

The AgCl mini-cell described in Section 18.3 may be used to determine the infrared spectrum of a solid dissolved in carbon tetrachloride. Prepare a 5–10% solution (5–10 mg in 0.1 mL) in carbon tetrachloride. If it is not possible to prepare a solution of this concentration because of low solubility, dissolve as much solid as possible in the solvent. Following the instructions given in Section 18.3, position the AgCl plates as shown in Figure 18–3C to obtain the maximum possible path length of 0.20 mm. When the cell is tightened firmly, the cell will not leak.

As indicated above in Method A, the spectrum will contain the absorptions of the dissolved solid superimposed on the absorptions of carbon tetrachloride. A strong absorption appears near 800 cm^{-1} (12.5 μ) for C—Cl stretch in the solvent. No useful information may be obtained for the sample to the right of about 900 cm^{-1}, but other bands which appear in the spectrum will belong to your sample. Read the safety material given in Method A. Carbon tetrachloride is toxic, and it should be used in a hood.

> **Care should be taken in cleaning the AgCl plates. Since AgCl plates scratch easily, they should not be wiped with tissue. Rinse them with methylene chloride and keep them in a dark place. Amines will destroy the plates.**

Method C—Solution Cells (NaCl)

The spectra of solids may also be determined in a type of permanent sample cell called a **solution cell.** (The infrared spectra of liquids may also be determined in this cell.)

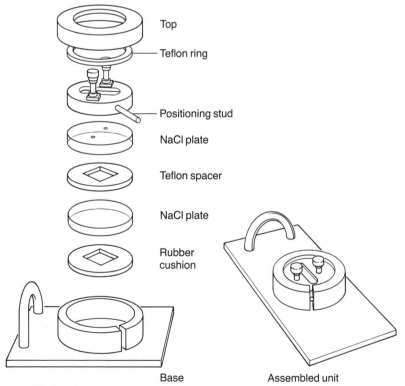

Top
Teflon ring
Positioning stud
NaCl plate
Teflon spacer
NaCl plate
Rubber cushion
Base
Assembled unit

FIGURE 18–5. Solution cell

The solution cell, shown in Figure 18–5, is made from two salt plates, mounted with a Teflon spacer between them to control the thickness of the sample. The top sodium chloride plate has two holes drilled in it so the sample can be introduced into the cavity between the two plates. These holes are extended through the face plate by two tubular extensions designed to hold Teflon plugs, which seal the internal chamber and prevent evaporation. The tubular extensions are tapered so that a syringe body (Luer lock without a needle) will fit snugly into them from the outside. The cells are thus filled from a syringe; usually they are held upright and filled from the bottom entrance port.

These cells are very expensive, and you should try either Method A or B before using solution cells. If they are needed, you should obtain your instructor's permission and receive instruction before using the cells. The cells are purchased in matched pairs, with identical path lengths. A solid is dissolved in a suitable solvent, usually carbon tetrachloride, and the solution is added to one of the cells (**sample cell**) as described in the upper paragraph. The pure solvent, identical to that used to dissolve the solid, is placed in the other cell (**reference cell**). The spectrum of the solvent is subtracted from the spectrum of the solution (not always completely), and a spectrum of the solute is thus provided. For the solvent compensation to be as exact as possible and to avoid contamination of the reference cell, it is essential that the same cell be used as a reference and that the other cell be used as a sample cell without ever being interchanged. When the spectrum is determined, it is important to clean the cells by flushing them with clean solvent. They should be dried by passing dry air through the cell.

Solvents most often used in determining infrared spectra are carbon tetrachloride, chloroform, and carbon disulfide. The spectra of these substances are shown on p 780. A 5–10% solution of solid in one of these solvents usually gives a good spectrum. Carbon tetrachloride and chloroform are suspected carcinogens. However, since there are no suitable alternative solvents, these compounds must be used in infrared spectroscopy. The procedure outlined on p 777 for carbon tetrachloride should be followed. This procedure serves equally well for chloroform.

> **Before you use the solution cells, you must obtain the instructor's permission and instruction on how to fill and clean the cells.**

18.6 SOLID SAMPLES—NUJOL MULLS

If an adequate KBr pellet cannot be obtained, or if the solid is insoluble in a suitable solvent, the spectrum of a solid may be determined as a Nujol mull. In this method, about 5 mg of the solid sample are ground finely in an agate mortar with a pestle. One or two drops of Nujol mineral oil (white) are then added, and the mixture is ground to a very fine dispersion. The solid is not dissolved in the Nujol; it is actually a suspension. This mull is then placed between two salt plates using a rubber policeman. The salt plates are mounted in the holder in the same way as for liquid samples (Section 18.2).

INFRARED SPECTRA OF SOLVENTS COMMONLY USED FOR SAMPLE PREPARATION

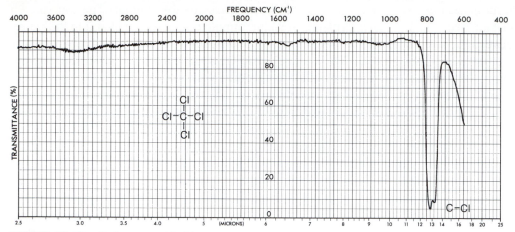

FIGURE 18—6. Carbon tetrachloride

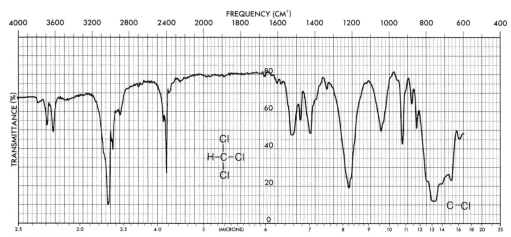

FIGURE 18—7. Chloroform

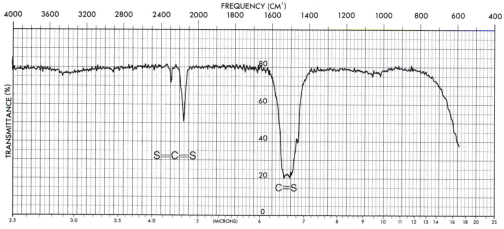

FIGURE 18—8. Carbon disulfide

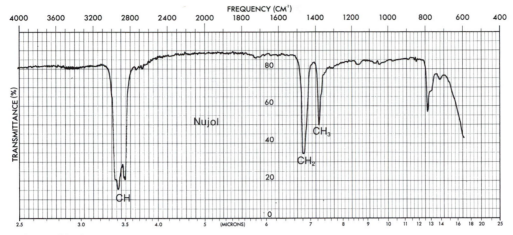

FIGURE 18—9. Nujol (mineral oil)

Nujol is a mixture of high-molecular weight hydrocarbons. Hence, it has absorptions in the C—H stretch and CH_2 and CH_3 bending regions of the spectrum (Figure 18–9). Clearly, if Nujol is used, no information can be obtained in these portions of the spectrum. In interpreting the spectrum, you must ignore these Nujol peaks. It is important to label the spectrum immediately after it was determined, noting that it was determined as a Nujol mull. Otherwise it might be forgotten that the C—H peaks belong to Nujol and not to the dispersed solid.

18.7 RECORDING THE SPECTRUM

The instructor will describe how to operate the infrared spectrophotometer since the controls vary considerably, depending on the manufacturer, model of the instrument, and type. For example, some instruments involve only pushing a few buttons, while others use a more complicated computer interface system.

In all cases, it is important that the sample, the solvent, the type of cell or method used, and any other pertinent information be written on the spectrum immediately after the determination. This information may be important, and it is easily forgotten if not recorded. You may also need to calibrate the instrument (Section 18.8).

18.8 CALIBRATION

For some instruments, the frequency scale of the spectrum must be calibrated so that you know the position of each absorption peak precisely. The calibration is accomplished by recording a very small portion of the spectrum of polystyrene over the spectrum of your sample. The complete spectrum of polystyrene is shown in Figure

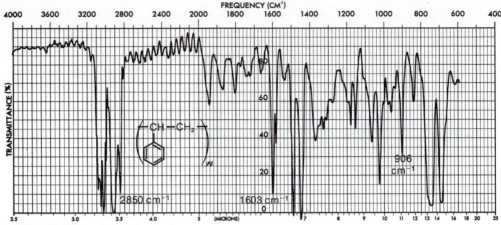

FIGURE 18–10. Infrared spectrum of polystyrene (thin film)

18–10. The most important of these peaks is at 1603 cm^{-1} (6.238 μ), while other useful peaks are at 2850 cm^{-1} (3.509 μ) and 906 cm^{-1} (11.035 μ). After the spectrum of your sample has been recorded, a thin film of polystyrene is substituted for the sample cell, and the **tips** (not the entire spectrum) of the most important peaks are recorded over the sample spectrum.

It is always a good idea to calibrate a spectrum when the instrument uses chart paper with a preprinted scale. It is difficult to align the paper properly so that the scale matches the absorption lines precisely. You often need to know the precise values for certain functional groups (for example, the carbonyl group). Calibration is essential in these cases.

With computer interfaced instruments, the instrument does not need to be calibrated. With this type of instrument, the spectrum and scale are printed on blank paper at the same time. The instrument has an internal calibration that assures that the positions of the absorptions will be known precisely, and that they will be placed at the proper positions on the scale. With this type of instrument it is often possible to print a list of the locations of the major peaks as well as obtain the complete spectrum of your compound.

PART B. NUCLEAR MAGNETIC RESONANCE

18.9   PREPARING A SAMPLE FOR PROTON NMR

The NMR sample tubes used in most instruments are approximately $^{3}/_{16} \times 7''$ in overall dimension and are fabricated of uniformly thin glass tubing. The sample tube is spun around its cylindrical axis while suspended in the magnet gap by a holder. The lower tip of the tube is positioned between the magnet pole pieces and the oscillator and detector coils by a depth gauge. To be sure that the sample will be aligned correctly, fill the tube to a minimum depth of 1 to 1.5 in. from the bottom. This usually requires a 0.5–1.0-mL quantity of the sample solution (sample dissolved in a suitable solvent). Figure 18–11 depicts an NMR sample tube, showing the correct sample level.

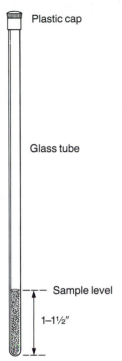

Plastic cap

Glass tube

Sample level

1–1½″

FIGURE 18–11. NMR sample tube

Liquid compounds that are not viscous can be determined "neat," that is, without solvent. Viscous compounds must be determined in solution. It is generally best to determine all samples in solution. Solutions that are 10–30% sample (weight/weight) generally give satisfactory spectra. Typically, a 20% concentration is used. Usually the sample can be prepared "by eye" if a sample of a liquid compound is being prepared. Visually divide the length of the tube required for a 1.5-inch height of sample into fifths. Fill the first fifth with sample and the remaining four fifths with solvent.

A solid sample, however, is best weighed out when preparing the solution. A deceivingly large amount (by volume), or so it seems, of solid is required to make a 20% w/w solution of a solid. One should weigh out about 150 mg of solid per 0.5 mL of solvent used.

To prepare the solution, one must of course choose the appropriate solvent *first!* There are two requirements for the solvent: (1) the sample should dissolve in it and (2) the solvent should have no NMR absorption peaks of its own, that is, no protons. Carbon tetrachloride, CCl_4, is the most widely used solvent for this purpose. Most compounds dissolve in CCl_4. It has the ability to dissolve a wide variety of functional groups and has **no protons.** If the sample does not dissolve in CCl_4, the more polar solvent deuterochloroform, or chloroform-d, $CDCl_3$, can often be used to advantage. Deuterium, ^2_1H, does not absorb in the proton region and is thus "invisible," or not seen, in the proton NMR spectrum. An even more polar solvent is deuterium oxide, D_2O. This solvent is often used for very polar compounds. Since the deuterated solvents are **expensive,** the spectrum is usually determined in carbon tetrachloride if possible.

> **CAUTION: Persons using this method must be warned that carbon tetrachloride is a hazardous solvent. See page 7.**

The chemical shift values commonly reported for various types of protons refer to the chemical shift as determined in either CCl_4 or $CDCl_3$. If another solvent is used, the chemical shift values can, and often do, change markedly, owing to the different solvent environment. When $CDCl_3$ is used, there is often a low-intensity peak in the NMR spectrum at 7.27δ due to an unavoidably small amount of $CHCl_3$ impurity. Spectra determined in D_2O often show a small peak because of OH impurity. If the sample compound has acidic hydrogens, these may **exchange** with D_2O, leading to the appearance of an OH peak in the spectrum and the **loss** of the original absorption from the acid proton, owing to the exchanged hydrogen. In many cases this will also alter the splitting patterns of a compound.

$$R-\overset{\overset{\displaystyle O}{\|}}{C}-O-H \;+\; D_2O \;\rightleftharpoons\; R-\overset{\overset{\displaystyle O}{\|}}{C}-O-D \;+\; D-OH$$

~ 12.0δ   becomes invisible   OH peak appears

$$\underset{CH_3\quad CH_3}{C=O} \;+\; D_2O \;\rightleftharpoons\; \underset{D-CH_2\quad CH_3}{C=O} \;+\; D-OH$$

$$CH_3CH_2OH + D_2O \rightleftharpoons CH_3CH_2OD + D-OH$$

Most solid carboxylic acids do not dissolve in CCl_4, $CDCl_3$, or even D_2O. In such cases one adds a small piece of sodium metal to about 1 mL of D_2O. The acid is then dissolved in this solution. The resulting basic solution enhances the solubility of the carboxylic acid. In such a case, the hydroxyl proton of the carboxylic acid cannot be observed in the NMR spectrum since it exchanges with the solvent. A large DOH peak is observed, however, due to the exchange and the H_2O impurity in the D_2O solvent.

When the above solvents fail, other special solvents can be used. Acetone, acetonitrile, dimethylsulfoxide, pyridine, benzene, and dimethylformamide can be used if one is not interested in the region or regions of the NMR spectrum in which they give rise to absorption. The deuterated (but expensive) analogues of these compounds are also used in special instances (for example, acetone-d_6, dimethylsulfoxide-d_6, dimethylformamide-d_7, and benzene-d_6). If the sample is not sensitive to acid, trifluoroacetic acid (which has no protons with $\delta < 12$) can be used. Once again, one has to be aware that these solvents often lead to different chemical shift values from those determined in CCl_4 or $CDCl_3$. Variations of as much as 0.5–1.0 ppm have been observed. In fact, it is sometimes possible, by switching to pyridine, benzene, acetone, or dimethylsulfoxide as solvents, to separate peaks that overlap when CCl_4 or $CDCl_3$ solutions are used.

Carbon tetrachloride, chloroform (and chloroform-d), and benzene (and benzene-d$_6$) are hazardous solvents. Besides being highly toxic, they are also suspected carcinogens. In spite of these health problems, these solvents are commonly used in NMR spectroscopy because there are no suitable alternatives. These solvents are used because they contain no protons and because they are excellent solvents for most organic compounds. Therefore it is necessary that one learn to handle these solvents with great care to minimize the hazard. These solvents should be stored either in a hood or in septum-capped bottles. If the bottles have screw caps, a pipet should be attached to each bottle. A recommended way of attaching the pipet is to store it in a test tube taped to the side of the bottle. Septum-capped bottles can be used only by withdrawing the solvent with a hypodermic syringe that has been designated solely for this use. All samples should be prepared in a hood, and solutions should be disposed of in an appropriately designated waste container that is stored in the hood. Rubber or plastic gloves should be worn when samples are being prepared or discarded.

> **Before using any deuterated solvent, check the solubility of the compound in its undeuterated analogue. Deuterated compounds are expensive and should not be wasted.**

18.10 PREPARING A SAMPLE FOR CARBON-13 NMR

Section 18.9 describes the technique for preparing samples for proton NMR. Much of what is described there also applies to carbon NMR. There are some differences, however, in determining a carbon spectrum. Fourier-transform (FT) instruments require a deuterium signal to stabilize (lock) the field. Therefore, the solvents must contain deuterium. Deuterated chloroform ($CDCl_3$) is used most commonly for this purpose because of its relatively low cost. Other deuterated solvents may also be used.

Because of the low natural abundance of carbon-13 in a sample, you often need to acquire multiple scans over a long period of time (Appendix 5, Section CMR.1, p 851). You can save considerable time by using a relatively concentrated sample.

Modern FT-NMR spectrometers allow chemists to obtain both the proton and carbon NMR spectra of the same sample in the same NMR tube. After changing several parameters in the program operating the spectrometer, you can obtain both spectra without removing the sample from the probe. The only real difference is that a proton spectrum may be obtained after a few scans while the carbon spectrum may require several thousand scans to obtain a suitable spectrum.

18.11 REFERENCE SUBSTANCES

Proton NMR. To provide the internal reference standard, tetramethylsilane (TMS) must be added to the sample solution. This substance has the formula $(CH_3)_4Si$. By universal convention, the protons in this substance are defined as 0.00 ppm

(0.00 δ). The spectrum should be shifted so that the TMS signal appears at this position on precalibrated paper.

The concentration of TMS in the sample should range from 1–3%. Some people prefer to add one to two drops of TMS to the sample just before determining the spectrum. Since TMS has 12 equivalent protons, not much of it needs to be added. A Pasteur pipet or a syringe may be used for the addition. It is far easier to have available in the laboratory a prepared solvent that already contains TMS. Deuterated chloroform and carbon tetrachloride often have TMS added to them. Since TMS is highly volatile (bp 26.5 °C), such solutions should be stored, tightly stoppered, in a refrigerator. Tetramethylsilane itself is best stored in a refrigerator as well.

Tetramethylsilane does not dissolve in D_2O. For spectra determined in D_2O, a different internal standard, sodium 2,2-dimethyl-2-silapentane-5-sulfonate, must be used. This standard is water-soluble and gives a resonance peak at 0.00 ppm (0.00 δ).

$$CH_3-\underset{\underset{CH_3}{|}}{\overset{\overset{CH_3}{|}}{Si}}-CH_2-CH_2-CH_2-SO_3^-Na^+$$

Sodium 2,2-dimethyl-2-silapentane-5-sulfonate (DSS)

Carbon NMR. TMS may be added as an internal reference standard where the methyl carbon is defined as 0.00 ppm. Alternatively, you may use the center peak of the $CDCl_3$ pattern, which is found at 77.0 ppm. This pattern can be observed as a small "triplet" near 77.0 ppm in a number of the spectra given in Appendix 5 (for example see Figure CMR–3 on p 854).

18.12   RECORDING THE SPECTRUM

In most instances the instructor or some qualified laboratory assistant will actually record your NMR spectrum. If you are permitted to operate the NMR spectrometer, the instructor will provide instructions. Since the controls of NMR spectrometers vary, depending on the make or model of the instrument, we shall not try to describe these controls.

> **Do not operate the NMR spectrometer unless you have been properly instructed.**

PROBLEMS

1. Comment on the suitability of running the infrared spectrum under each of the following conditions. If there is a problem with the conditions given, provide a suitable alternative method.
   (a) A neat spectrum of a liquid with a boiling point of 150 °C is determined using salt plates.
   (b) A neat spectrum of a liquid with a boiling point of 35 °C is determined using salt plates.
   (c) A KBr pellet is prepared with a compound that melts at 200 °C.

(d) A KBr pellet is prepared with a compound that melts at 30 °C.

(e) A solid aliphatic hydrocarbon compound is determined as a Nujol mull.

(f) Silver chloride plates are used to determine the spectrum of aniline.

(g) Sodium chloride plates are selected to run the spectrum of a compound that contains some water.

2. Describe the method that you should employ to determine the proton NMR spectrum of a carboxylic acid which is insoluble in **all** of the common organic solvents that your instructor is likely to make available.

3. In order to save money, a student uses chloroform instead of deuterated chloroform to run a carbon-13 NMR spectrum. Is this a good idea?

4. Look up the solubilities for the following compounds and decide whether you would select deuterated chloroform or deuterated water to dissolve the substances for NMR spectroscopy.

(a) Glycerol (1,2,3-propanetriol)

(b) 1,4-Diethoxybenzene

(c) Propyl pentanoate (propyl ester of pentanoic acid)

5. What would happen if you ran a proton NMR spectrum without any TMS in the sample?

Technique 19
GUIDE TO THE CHEMICAL LITERATURE

Often, it may be necessary to go beyond the information contained in the typical organic chemistry textbook and to use reference material in the library. At first glance, using library materials may seem formidable because of the numerous sources the library contains. If, however, one adopts a systematic approach, the task can prove rather useful. This description of various popular sources and an outline of logical steps to follow in the typical literature search should be helpful.

19.1 LOCATING PHYSICAL CONSTANTS: HANDBOOKS

To find information on routine physical constants, such as melting points, boiling points, indices of refraction, and densities, you should first consider a handbook. Examples of suitable handbooks are

J. A. Dean, ed. *Lange's Handbook of Chemistry*. 13th ed. New York: McGraw-Hill, 1985.

M. Windholz, ed. *The Merck Index*. 10th ed. Rahway, NJ: Merck & Co., 1983.

R. C. Weast, ed. *CRC Handbook of Chemistry and Physics*. 66th ed. Boca Raton, FL: CRC Press, 1985. Revised annually.

The *Handbook of Chemistry and Physics* is the handbook most often consulted. For organic chemistry, however, *The Merck Index* is probably better suited. *The Merck Index* also contains literature references on the isolation, structure determination, and synthesis of a substance, along with its molecular formula, elemental analysis, and certain properties of medicinal interest (e.g., toxicity and medicinal and veterinary uses).

A more complete handbook is

J. Buckingham, editor, *Dictionary of Organic Compounds,* Chapman & Hall/Methuen, New York: 1982.

This is a revised version of an earlier four-volume handbook edited by I. M. Heilbron and H. M. Bunbury. In its present form, it consists of five volumes with 15 supplements.

19.2 SIDE REACTIONS AND GENERAL SYNTHETIC METHODS

The easiest way of determining possible side reactions for a particular reaction is to examine all the possible reactions of the starting materials. Many standard introductory textbooks in organic chemistry provide tables that summarize most of the common reactions for a given class of compounds. These tables can be used to determine which possible side reactions the starting materials might undergo. Examples of textbooks that include tables of this type are

R. J. Fessenden and J. S. Fessenden. *Introduction to Organic Chemistry*. 3rd ed. Monterey, CA: Brooks/ Cole, 1986.

D. S. Kemp and F. Vellaccio. *Organic Chemistry*. New York: Worth Publishers, 1980.

J. McMurry. *Organic Chemistry*. 2nd ed. Monterey, CA: Brooks/Cole, 1988.

R. T. Morrison and R. N. Boyd. *Organic Chemistry*. 5th ed. Boston: Allyn and Bacon, 1987.

S. H. Pine. *Organic Chemistry*. 5th ed. New York: McGraw-Hill, 1987.

T. W. G. Solomons. *Organic Chemistry*. 4th ed. New York: John Wiley & Sons, 1988.

Similar tables may be used to identify alternative methods of preparing a particular compound. Tables of synthetic methods for each important class of compounds can be found in many of the textbooks from the above list, as well as in

J. March. *Advanced Organic Chemistry: Reactions, Mechanisms, and Structure*. 3rd ed. New York: John Wiley & Sons, 1985.

W. H. Reusch. *An Introduction to Organic Chemistry*. San Francisco: Holden-Day, 1977.

J. D. Roberts and M. C. Caserio. *Basic Principles of Organic Chemistry*. 2nd ed. Menlo Park, CA: W. H. Benjamin, 1977.

A. Streitwieser, Jr., and C. H. Heathcock. *Introduction to Organic Chemistry*. 3rd ed. New York: Macmillan, 1985.

The textbooks cited above represent only a partial list of books in which information about side reactions or other methods of preparation may be found. This information is to be found in virtually any introductory textbook on organic chemistry, although it may not be presented in a convenient series of tables.

19.3 SEARCHING THE CHEMICAL LITERATURE

If the information one is seeking is not available in any of the handbooks mentioned above, or if one is searching for more detailed information than they can provide, then a proper literature search is in order. While an examination of standard textbooks can provide some help, often one must use all the resources of the library, including journals, reference collections, and abstracts. The following sections of this chapter outline how the various types of sources should be used and what sort of information can be obtained from them.

The methods for searching the literature discussed in this chapter use printed materials. Modern methods of literature searching also make use of computerized databases. These are vast collections of data and bibliographic materials that can be scanned very rapidly from remote computer terminals. While computerized searching is becoming more widely available, its use is not usually accessible to undergraduate students.

19.4 COLLECTIONS OF SPECTRA

Collections of infrared, nuclear magnetic resonance, and mass spectra can be found in the following catalogs of spectra:

A. Cornu and R. Massot. *Compilation of Mass Spectral Data*. 2nd ed. London: Heyden and Sons, Ltd., 1975.

High-Resolution NMR Spectra Catalog. Palo Alto, CA: Varian Associates. Volume 1, 1962; Volume 2, 1963.

C. J. Pouchert. *Aldrich Library of Infrared Spectra*. 3rd ed. Milwaukee: Aldrich Chemical Co., 1981.

C. J. Pouchert. *Aldrich Library of FT-IR Spectra*. Milwaukee: Aldrich Chemical Co., 1985.

C. J. Pouchert. *Aldrich Library of NMR Spectra*. 2nd ed. Milwaukee: Aldrich Chemical Co., 1983.

Sadtler Standard Spectra. Philadelphia: Sadtler Research Laboratories. Continuing collection.

E. Stenhagen, S. Abrahamsson, and F. W. McLafferty. *Registry of Mass Spectral Data*. New York: John Wiley-Interscience, 1974. Four-volume set.

The American Petroleum Institute has also published collections of infrared, nuclear magnetic resonance, and mass spectra.

19.5 ADVANCED TEXTBOOKS

Much information about synthetic methods, reaction mechanisms, and reactions of organic compounds is available in any of the many current advanced textbooks in organic chemistry. Examples of such books are

F. A. Carey and R. J. Sundberg. *Advanced Organic Chemistry. Part A. Structure and Mechanisms; Part B. Reactions and Synthesis*. 2nd ed. New York: Plenum Press, 1983.

W. Carruthers. *Some Modern Methods of Organic Synthesis*. 3rd ed. Cambridge, U.K.: Cambridge University Press, 1986.

L. F. Fieser and M. Fieser. *Advanced Organic Chemistry*. New York: Reinhold, 1961.

I. L. Finar. *Organic Chemistry*. 6th ed. London: Longman Group, Ltd., 1986.

H. O. House. *Modern Synthetic Reactions*. 2nd ed. Menlo Park, CA: W. H. Benjamin, 1972.

J. March. *Advanced Organic Chemistry: Reactions, Mechanisms, and Structure*. 3rd ed. New York: John
    Wiley & Sons, 1985.
C. R. Noller. *Chemistry of Organic Compounds*. 3rd ed. Philadelphia: W. B. Saunders, 1965.

These books often contain references to original papers in the literature for students wanting to follow the subject further. Consequently, the student obtains not only a review of the subject from such a textbook but also a key reference that is helpful toward a more extensive literature search. The textbook by March is particularly useful for this purpose.

19.6   SPECIFIC SYNTHETIC METHODS

Anyone interested in locating information about a particular method of synthesizing a compound should first consult one of the many general textbooks on the subject. Useful ones are

D. Barton and W. D. Ollis, eds. *Comprehensive Organic Chemistry*. Oxford: Pergamon Press, 1979.
    Six-volume set.
C. A. Buehler and D. E. Pearson. *Survey of Organic Syntheses*. New York: John Wiley-Interscience, 1970
    and 1977. Two-volume set.
F. A. Carey and R. J. Sundberg. *Advanced Organic Chemistry. Part B. Reactions and Synthesis*. 2nd ed.
    New York: Plenum Press, 1983.
L. F. Fieser and M. Fieser. *Reagents for Organic Synthesis*. New York: John Wiley-Interscience, 1967–
    1988. This is a continuing series, now in thirteen volumes.
Compendium of Organic Synthetic Methods. New York: John Wiley-Interscience, 1971–1988. This is a
    continuing series, now in six volumes.
H. O. House. *Modern Synthetic Reactions*. 2nd ed. Menlo Park, CA: W. H. Benjamin, 1972.
J. March. *Advanced Organic Chemistry: Reactions, Mechanisms, and Structure*. 3rd ed. New York: John
    Wiley & Sons, 1985.
S. Patai, ed. *The Chemistry of the Functional Groups*. London: Interscience, 1964–present. This series
    consists of many volumes, each one specializing in a particular functional group.
A. I. Vogel. Revised by members of the School of Chemistry, Thames Polytechnic, *Vogel's Textbook of
    Practical Organic Chemistry, Including Qualitative Organic Analysis*. 4th ed. London: Longman
    Group, Ltd., 1978.
R. B. Wagner and H. D. Zook. *Synthetic Organic Chemistry*. New York: John Wiley & Sons, 1956.

More specific information, including actual reaction conditions, exists in collections specializing in organic synthetic methods. The most important of these is

Organic Syntheses. New York: John Wiley & Sons, 1921–present. Published annually.

One of the features of the advanced organic textbook by March is that it includes references to specific preparative methods contained in *Organic Syntheses*. While *Organic Syntheses* is published with annual volumes, older volumes are combined in groups of 10 into a series of collective volumes. Tables found at the end of each of the collective volumes classify methods according to the type of reaction, type of compound prepared, formula of compound prepared, preparation or purification of solvents and reagents, and use of various types of specialized apparatus.

More advanced material on organic chemical reactions and synthetic methods may be found in any one of a number of annual publications that review the original literature and summarize it. Examples include

Advances in Organic Chemistry: Methods and Results. New York: John Wiley & Sons, 1960–present.
Annual Reports of the Chemical Society, Section B. London: Chemical Society, 1905–present. Specifically, the section on *Synthetic Methods*.
Progress in Organic Chemistry. New York: John Wiley & Sons, 1952–1973.
Organic Reactions. New York: John Wiley & Sons, 1942–present.

Each of these publications contains a great many citations referring the reader to the appropriate articles in the original literature.

19.7 ADVANCED LABORATORY TECHNIQUES

The student who is interested in reading about more advanced techniques than those described in this textbook, or in more complete descriptions of techniques, should consult one of the advanced textbooks specializing in organic laboratory techniques. Besides focusing on apparatus construction and the performance of complex reactions, these books also provide advice on purifying reagents and solvents. Useful sources of information on organic laboratory techniques include

R. B. Bates and J. P. Schaefer. *Research Techniques in Organic Chemistry*. Englewood Cliffs, NJ: Prentice-Hall, 1971.
A. J. Krubsack. *Experimental Organic Chemistry*. Boston: Allyn and Bacon, 1973.
R. S. Monson. *Advanced Organic Synthesis: Methods and Techniques*. New York: Academic Press, 1971.
A. Weissberger et al., eds. *Technique of Organic Chemistry*. 3rd ed. New York: John Wiley-Interscience, 1959–1969. This work is in 14 volumes.
K. B. Wiberg. *Laboratory Technique in Organic Chemistry*. New York: McGraw-Hill, 1960.

Numerous works specialize in particular techniques and so do other more general textbooks. The above list is only representative of the most common books in this category. The following books deal specifically with micro and semimicro techniques.

N. D. Cheronis. "Micro and Semimicro Methods." In A. Weissberger, *Technique of Organic Chemistry*. New York: John Wiley-Interscience, Volume 6, 1954.
N. D. Cheronis and T. S. Ma. *Organic Functional Group Analysis by Micro and Semimicro Methods*. New York: John Wiley-Interscience, 1964.
T. S. Ma and V. Horak. *Microscale Manipulations in Chemistry*. New York: John Wiley-Interscience, 1976.

19.8 REACTION MECHANISMS

As with the case of locating information on synthetic methods, a great deal of information can be obtained about reaction mechanisms by consulting one of the common textbooks on physical organic chemistry. The textbooks listed here provide a general description of mechanisms, but they do not contain specific literature citations. Very general textbooks include

R. Breslow. *Organic Reaction Mechanism*. New York: Benjamin, 1966.
P. Sykes. *A Guidebook to Mechanism in Organic Chemistry*. 6th ed. London: Longman Group, Ltd., 1986.

More advanced textbooks include

F. A. Carey and R. J. Sundberg. *Advanced Organic Chemistry. Part A. Structure and Mechanisms*. 2nd ed. New York: Plenum Press, 1983.

R. D. Gilliom. *Introduction to Physical Organic Chemistry*. Reading, MA: Addison-Wesley, 1970.

L. P. Hammett. *Physical Organic Chemistry: Reaction Rates, Equilibria, and Mechanisms*. 2nd ed. New York: McGraw-Hill, 1970.

J. Hine. *Physical Organic Chemistry*. 2nd ed. New York: McGraw-Hill, 1962.

J. A. Hirsch. *Concepts in Theoretical Organic Chemistry*. Boston: Allyn and Bacon, 1974.

C. K. Ingold. *Structure and Mechanism in Organic Chemistry*. 2nd ed. Ithaca, NY: Cornell University Press, 1969.

R. A. Y. Jones. *Physical and Mechanistic Organic Chemistry*. 2nd ed. Cambridge, U.K.: Cambridge University Press, 1984.

J. B. Leffler and E. Grunwald. *Rates and Equilibria of Organic Reactions*. New York: John Wiley & Sons, 1963.

T. H. Lowry and K. S. Richardson. *Mechanism and Theory in Organic Chemistry*. 2nd ed. New York: Harper & Row, 1981.

J. W. Moore and R. G. Pearson. *Kinetics and Mechanism*. 3rd ed. New York: John Wiley & Sons, 1981.

These books include extensive bibliographies that permit the reader to delve more deeply into the subject.

Most libraries also subscribe to annual series of publications that specialize in articles dealing with reaction mechanisms. Among these are

Advances in Physical Organic Chemistry. London: Academic Press, 1963–present.

Annual Reports of the Chemical Society. Section B. London: Chemical Society, 1905–present. Specifically, the section on *Reaction Mechanisms*.

Organic Reaction Mechanisms. Chichester, U.K.: John Wiley & Sons, 1965–present.

Progress in Physical Organic Chemistry. New York: Interscience, 1963–present.

These publications provide the reader with citations from the original literature that can be very useful in an extensive literature search.

19.9   ORGANIC QUALITATIVE ANALYSIS

Experiment 56 contains a procedure for identifying organic compounds through a series of chemical tests and reactions. Occasionally, one might require a more complete description of analytical methods or a more complete set of tables of derivatives. Textbooks specializing in organic qualitative analysis should fill this need. Examples of sources for such information include

N. D. Cheronis and J. B. Entriken. *Identification of Organic Compounds: A Student's Text Using Semimicro Techniques*. New York: Interscience, 1963.

D. J. Pasto and C. R. Johnson. *Laboratory Text for Organic Chemistry: A Source Book of Chemical and Physical Techniques*. Englewood Cliffs, N.J.: Prentice-Hall, 1979.

Z. Rappoport, ed. *Handbook of Tables for Organic Compound Identification,* 3rd ed. Cleveland: Chemical Rubber Co., 1967.

R. L. Shriner, R. C. Fuson, D. Y. Curtin, and T. C. Morrill. *The Systematic Identification of Organic Compounds: A Laboratory Manual*. 6th ed. New York: John Wiley & Sons, 1980.

A. I. Vogel. *Elementary Practical Organic Chemistry*. Part 2. *Qualitative Organic Analysis*. 2nd ed. New York: John Wiley & Sons, 1966.

A. I. Vogel. Revised by members of the School of Chemistry, Thames Polytechnic. *Vogel's Textbook of Practical Organic Chemistry, Including Qualitative Organic Analysis*. 4th ed. London: Longman Group, Ltd., 1978.

19.10   BEILSTEIN AND CHEMICAL ABSTRACTS

One of the most useful sources of information about the physical properties, synthesis, and reactions of organic compounds is *Beilsteins Handbuch der Organischen Chemie*. This is a monumental work, initially edited by Friedrich Konrad Beilstein, and updated through several revisions by the Beilstein Institute in Frankfurt am Main, Germany. The original edition (the *Hauptwerk,* abbreviated H) was published in 1918 and covers completely the literature to 1909. Four supplementary series (*Ergänzungswerken*) have been published since that time. The first supplement (*Erstes Ergänzungswerk,* abbreviated E I) covers the literature from 1910–1919; the second supplement (*Zweites Ergänzungswerk,* E II) covers 1920–1929; the third supplement (*Drittes Ergänzungswerk,* E III) covers 1930–1949; and the fourth supplement (*Viertes Ergänzungswerk,* E IV) covers 1950–1959. Volumes 17–27 of supplementary series III and IV, covering heterocyclic compounds, are combined in a joint issue, E III/IV. Supplementary series III and IV are not complete, so the coverage of *Handbuch der Organischen Chemie* can be considered complete to 1929, with partial coverage to 1959.

Beilsteins Handbuch der Organischen Chemie, usually referred to simply as *Beilstein,* also contains two types of cumulative indices. The first of these is a name index (*Sachregister*) and the second is a formula index (*Formelregister*). These indices are particularly useful for a person wishing to locate a compound in *Beilstein.*

The principal difficulty in using *Beilstein* is that it is written in German. While some reading knowledge of German is useful, the beginner can obtain information from the work by learning a few key phrases. For example, *Bildung* is "formation" or "structure," *Darst.* or *Darstellung* is "preparation," K_P or *Siedepunkt* is "boiling point," and *F* or *Schmelzpunkt* is "melting point." Furthermore, the names of some compounds in German are not cognates of the English names. Some examples are *Apfelsäure* for "malic acid" (*Säure* means "acid"), *Harnstoff* for "urea," *Jod* for "iodine," and *Zimtsäure* for "cinnamic acid." If the student has access to a German-English dictionary for chemists, many of these difficulties can be overcome. The best such dictionary is

A. M. Patterson. *German-English Dictionary for Chemists*. 3rd ed. New York: John Wiley & Sons, 1959.

Beilstein is organized according to a very sophisticated and complicated system. To locate a compound in *Beilstein,* one could learn all the intricacies of this system. However, most students do not wish to become experts on *Beilstein* to this extent. A simpler, though slightly less reliable, method is to look for the compound in the formula index that accompanies the second supplement. Under the molecular formula, one will find the names of compounds that have that formula. After that name will be a series of numbers that indicate the pages and volume in which that compound is listed. Suppose, as an example, that one is searching for information on *p*-nitroaniline. This compound has the molecular formula $C_6H_6N_2O_2$. Searching for this formula in the formula index to the second supplement, one finds

4-Nitro-anilin **12** 711, **I** 349, **II** 383

This information tells us that *p*-nitroaniline is listed in the main edition (*Hauptwerk*) in volume 12, p 711. Locating this particular volume, which is devoted to isocyclic monoamines, we turn to p 711 and find the beginning of the section on *p*-nitroaniline. At the left side of the top of this page, we find "Syst. No. 1671." This is the system number given to compounds in this part of Volume 12. The system number is useful, as it can help us find entries for this compound in subsequent supplements. The organization of *Beilstein* is such that all entries on *p*-nitroaniline in each of the supplements will always be found in Volume 12. The entry in the formula index also indicated that material on this compound could be found in the first supplement on p 349 and in the second supplement on p 383. On p 349 of Volume 12 of the first supplement, there is a heading, **"XII, 710–712,"** and on the left is "Syst. No. 1671." Material on *p*-nitroaniline will be found in each supplement on a page that is headed with the volume and page of the *Hauptwerk* in which the same compound is found. On p 383 of Volume 12 of the second supplement, the heading in the center of the top of the page is **"H12, 710–712."** On the left we find "Syst. No. 1671." Again, because *p*-nitroaniline appeared in Volume 12, p 711, of the main edition, it can be located by searching through Volume 12 of any supplement until one finds a page with the heading corresponding to Volume 12, p 711. Because the third and fourth supplements are not complete, there is no comprehensive formula index for these supplements. However, one can still find material on *p*-nitroaniline by using the system number and the volume and page in the main work. In the third supplement, because the amount of information available has grown so much since the early days of Beilstein's work, Volume 12 has now expanded so that it is found in several bound parts. However, one selects the part that includes system number 1671. In this part of Volume 12, one looks through the pages until one finds a page headed "Syst. No. 1671/H 711." The information on *p*-nitroaniline is found on this page (p 1580). If Volume 12 of the fourth supplement were available, one would go on in the same way to locate more recent data on *p*-nitroaniline. This example is meant to illustrate how one can locate information on particular compounds without having to learn the Beilstein system of classification. You might do well to test your ability at finding compounds in *Beilstein* as we have described here.

Guidebooks to using *Beilstein,* which include a description of the Beilstein system, are recommended for anyone who wants to work extensively with *Beilstein.* Among such sources are

E. H. Huntress. *A Brief Introduction to the Use of Beilsteins Handbuch der Organischen Chemie.* 2nd ed. New York: John Wiley & Sons, 1938.

How to Use Beilstein. Beilstein Institute, Frankfurt am Main. Berlin: Springer-Verlag.

O. Weissbach. *The Beilstein Guide: A Manual for the Use of Beilsteins Handbuch der Organischen Chemie.* New York: Springer-Verlag, 1976.

Beilstein reference numbers are listed in such handbooks as *CRC Handbook of Chemistry and Physics* and *Lange's Handbook of Chemistry.* Additionally, Beilstein numbers are included in the *Aldrich Catalog Handbook of Fine Chemicals,* issued by the Aldrich Chemical Company. If the compound one is seeking is listed in one of these handbooks, one will find that using *Beilstein* is simplified.

Another very useful publication that can help you in finding references to research on a particular topic is *Chemical Abstracts,* published by the Chemical Abstracts Service of the American Chemical Society. *Chemical Abstracts* contains abstracts of articles appearing in more than 10,000 journals from virtually every country conducting scientific research. These abstracts list the authors, the journal in which the article appeared, the title of the paper, and a short summary of the contents of the article. Abstracts of articles that appeared originally in a foreign language are provided in English, with a notation describing the original language.

To use *Chemical Abstracts,* one must know how to use the various indices that accompany it. At the end of each volume there appears a set of indices, including a formula index, a general subject index, a chemical substances index, an author index, and a patent index. The listings in each index refer the reader to the appropriate abstract, according to the number assigned to it. There are also collective indices that combine all the indexed material appearing in a 5-year period (10-year period before 1956). In the collective indices, the listings include the volume number as well as the abstract number.

For material after 1929, *Chemical Abstracts* provides the most complete coverage of the literature. For material before 1929, use *Beilstein* before consulting *Chemical Abstracts*. *Chemical Abstracts* has the advantage that it is written entirely in English. Nevertheless, the most common requirement for a literature search by the student involves a search for a relatively simple compound. Finding the desired entry for a simple compound is much easier in *Beilstein* than in *Chemical Abstracts*. For simple compounds, the indices in *Chemical Abstracts* are likely to contain very many entries. To locate the desired information, the student will have to comb through this multitude of listings—potentially a very time-consuming task.

The opening pages of each index in *Chemical Abstracts* contain a brief set of instructions on using that index. Students who want a more complete guide to *Chemical Abstracts* may consult a textbook that contains problems designed to familiarize the reader with these abstracts and indices. An example of such a textbook is

CAS Printed Access Tools: A Workbook. Washington: Chemical Abstracts Service, American Chemical Society, 1977.

Chemical Abstracts Service maintains a computerized database that permits users to search through *Chemical Abstracts* very rapidly and thoroughly. This service is called *CAS On-Line,* and it is available in most major university and industrial libraries. The database can be searched from a remote terminal using telephone lines to connect to the main computer in Columbus, OH. The database extends from 1965 to the present, and it is being expanded currently. A modest charge is made for each search.

19.11 SCIENTIFIC JOURNALS

Ultimately, someone wanting information about a particular area of research will be required to read articles from the scientific journals. These journals are of two basic types: review journals and primary scientific journals. Journals that specialize in review articles summarize all the work that bears on the particular topic. These articles may focus on the contributions of one particular researcher but often consider the contributions of many different workers to the subject. These articles also contain extensive bibliographies, which refer the reader to the original research articles. Among the important journals devoted, at least partly, to review articles are

Accounts of Chemical Research
Angewandte Chemie (International Edition, in English)
Chemical Reviews
Chemical Society Reviews (formerly known as *Quarterly Reviews*)
Nature
Science

The details of the research of interest appear in the primary scientific journals. While there are thousands of different journals published in the world, a few important journals specializing in articles dealing with organic chemistry might be mentioned here. These are

Annalen der Chemie
Canadian Journal of Chemistry
Chemische Berichte
Journal of Organic Chemistry
Journal of the American Chemical Society
Journal of the Chemical Society, Perkin Transactions (Parts I and II)
Tetrahedron
Tetrahedron Letters

19.12 TOPICS OF CURRENT INTEREST

The journals and magazines listed below are good sources for topics of educational and current interest. They specialize in news articles and focus on current events in chemistry or in science in general. Articles in these journals (magazines) can be useful in keeping the reader abreast of developments in science that are not part of his or her normal specialized scientific reading. You may want to consult these sources as part of your search for a ''Pet Molecule'' Project (Experiment 55).

American Scientist
Chemical and Engineering News
Chemistry and Industry
Chemistry in Britain
Chemtech
Discover
Journal of Chemical Education
Nature
Omni
Science
Science Digest
Scientific American
SciQuest (formerly *Chemistry*)

Other sources for information for your ''Pet Molecule'' include the following:

Encyclopedia of Chemical Technology, also called *Kirk-Othmer Encyclopedia of Chemical Technology* (24
    volumes plus index and supplements)
McGraw-Hill Encyclopedia of Science and Technology (20 volumes and supplements)

19.13   HOW TO CONDUCT A LITERATURE SEARCH

The easiest method to follow in searching the literature is to begin with secondary
sources and then go on to the primary sources. In other words, one would try to locate
material in a textbook, *Beilstein,* or *Chemical Abstracts.* From the results of that
search, one would then consult one of the primary scientific journals.

     A literature search that ultimately requires the reader to read one or more papers
in the scientific journals is best conducted if one can identify a particular paper central
to the study. Often this reference can be obtained from a textbook or a review article on
the subject. If this is not available, a search through *Beilstein* is required. A search
through one of the handbooks that provide *Beilstein* reference numbers (see Section
19.10) may be helpful. Searching through *Chemical Abstracts* would be considered the
next logical step after *Beilstein.* From these sources, it should be possible to identify
citations from the original literature on the subject.

     Additional citations may be found in the references cited in the journal article.
In this way, the background leading to the research can be examined. It is also possible
to conduct a search forward in time from the date of the journal article through the
Science Citation Index. This publication provides the service of listing articles and the
papers in which these articles were cited. While *Science Citation Index* consists of
several types of indices, the *Citation Index* is most useful for the purposes described
here. A person who knows of a particular key reference on a subject can examine
Science Citation Index to obtain a list of papers that have used that seminal reference in
support of the work described. The *Citation Index* lists papers by their senior author,
journal, volume, page, and date, followed by citations of papers that have referred to
that article, author, journal, volume, page, and date of each. The *Citation Index* is
published in annual volumes, with quarterly supplements issued during the current
year. Each volume contains a complete list of the citations of the key articles made
during that year. A disadvantage is that *Science Citation Index* has been available only
since 1961. An additional disadvantage is that one may miss journal articles on the
subject of interest if they failed to cite that particular key reference in their bibliographies—a reasonably likely possibility.

     One can, of course, conduct a literature search by a ''brute force'' method, by
beginning the search with *Beilstein* or even with the indices in *Chemical Abstracts.*
However, the task can be made much easier by starting with a book or an article of
general and broad coverage, which can provide a few citations for starting points in the
search.

     The following guides to using the chemical literature are provided for the reader
who is interested in going further into this subject.

C. R. Burman. *How to Find Out in Chemistry*. 2nd ed. New York: Oxford University Press, 1966.

H. M. Woodburn. *Using the Chemical Literature: A Practical Guide*. New York: Marcel Dekker, 1974.

M. G. Mellon. *Chemical Publications*. 4th ed. New York: McGraw-Hill, 1965.

R. E. Maizell. *How to Find Chemical Information: A Guide for Practicing Chemists, Teachers, and Students*. 2nd ed. New York: John Wiley-Interscience, 1987.

R. F. Gould, ed. *Advances in Chemistry Series*. No. 30. *Searching the Chemical Literature*. Washington: American Chemical Society, 1961.

R. T. Bottle, ed. *The Use of Chemical Literature*. 3rd ed. London: Butterworths, 1979.

PROBLEMS

1. Using the *Merck Index* discussed in Section 19.1, find and draw structures for the following compounds:
   (a) atropine
   (b) quinine
   (c) saccharin
   (d) benzo[a]pyrene (benzpyrene)
   (e) itaconic acid
   (f) adrenosterone
   (g) chrysanthemic acid (chrysanthemumic acid)
   (h) cholesterol
   (i) vitamin C (ascorbic acid)

2. Find the melting points for the following compounds in the *Handbook of Chemistry and Physics* or *Lange's Handbook of Chemistry* (Section 19.1):
   (a) biphenyl
   (b) 4-bromobenzoic acid
   (c) 3-nitrophenol

3. Find the boiling point for each compound in the references listed in Problem #2:
   (a) octanoic acid at reduced pressure
   (b) 4-chloroacetophenone at atmospheric and reduced pressure
   (c) 2-methyl-2-heptanol

4. Find the index of refraction (n_D) and density for the liquids listed in Problem #3.

5. Using the references given in Problem #2, give the specific rotations, $[\alpha]_D$, for the enantiomers of camphor.

6. Read the section on carbon tetrachloride in the *Merck Index* and list some of the health hazards for this compound.

7. Find the following compounds in the formula index for the *Second Supplement of Beilstein* (Section 19.10). (i) List the page numbers from the Main Work and the Supplements (First and Second). (ii) Using these page numbers, look up the System Number (Syst. No.) and the Main Work number (Hauptwerk number, H) for each compound in the Main Work, and the First and Second supplements. In some cases, a compound may not be found in all three places. (iii) Now use the System Number and Main Work number to find each of these compounds in the Third and Fourth Supplements. List the page numbers where these compounds are found.
   (a) 2,5-Hexanedione (acetonylacetone)
   (b) 3-Nitroacetophenone
   (c) 4-*tert*-Butylcyclohexanone
   (d) 4-Phenylbutanoic acid (4-phenylbutyric acid, γ-phenylbuttersäure)

8. Using the *Science Citation Index* (Section 19.13), list five research papers by complete title and journal citation for each of the following chemists who have been awarded the Nobel Prize. Use the *Five-Year Cumulation Source Index* for the years 1980–1984 as your source.

   (a) H. C. Brown
   (b) R. B. Woodward
   (c) D. J. Cram
 9. The reference book by J. March is listed in Section 19.2. Using Appendix B in this book, give two
    methods for preparing the following functional groups. You will need to provide equations.
    (a) Carboxylic acids
    (b) Aldehydes
    (c) Esters (carboxylic esters)
10. *Organic Synthesis* is described in Section 19.6. There are currently six collective volumes in the
    series, each with its own index. Find the compounds listed below and provide the equations for
    preparing each compound.
    (a) 2-Methylcyclopentane-1,3-dione
    (b) *cis*-Δ^4-Tetrahydrophthalic anhydride (listed as tetrahydrophthalic anhydride)
11. Provide four ways that may be used to oxidize an alcohol to an aldehyde. Give complete literature
    references for each method as well as equations. Use the *Compendium of Organic Synthetic Methods*
    or *Survey of Organic Syntheses* by Buehler and Pearson (Section 19.6).

Appendix 1
TABLES OF UNKNOWNS
AND DERIVATIVES

> More extensive tables of unknowns may be found in Z. Rappoport, ed. *Handbook of Tables for Organic Compound Identification*, 3rd ed. Cleveland: Chemical Rubber Co., 1967

Aldehydes

COMPOUND	BP	MP	SEMI-CARBAZONE*	2,4-DINITRO-PHENYL-HYDRAZONE*
Ethanal (acetaldehyde)	21	—	162	168
Propanal (propionaldehyde)	48	—	89	148
Propenal (acrolein)	52	—	171	165
2-Methylpropanal (isobutyraldehyde)	64	—	125	187
Butanal (butyraldehyde)	75	—	95	123
3-Methylbutanal (isovaleraldehyde)	92	—	107	123
Pentanal (valeraldehyde)	102	—	—	106
2-Butenal (crotonaldehyde)	104	—	199	190
2-Ethylbutanal (diethylacetaldehyde)	117	—	99	95
Hexanal (caproaldehyde)	130	—	106	104
Heptanal (heptaldehyde)	153	—	109	108
2-Furaldehyde (furfural)	162	—	202	212
2-Ethylhexanal	163	—	254	114
Octanal (caprylaldehyde)	171	—	101	106
Benzaldehyde	179	—	222	237
Phenylethanal (phenylacetaldehyde)	195	33	153	121
2-Hydroxybenzaldehyde (salicylaldehyde)	197	—	231	248
4-Methylbenzaldehyde (*p*-tolualdehyde)	204	—	234	234
3,7-Dimethyl-6-octenal (citronellal)	207	—	82	77
2-Chlorobenzaldehyde	213	11	229	213
4-Methoxybenzaldehyde (*p*-anisaldehyde)	248	2.5	210	253
trans-Cinnamaldehyde	250 d.	—	215	255
3,4-Methylenedioxybenzaldehyde (piperonal)	263	37	230	266 d.
2-Methoxybenzaldehyde (*o*-anisaldehyde)	245	38	215 d.	254
4-Chlorobenzaldehyde	214	48	230	254
3-Nitrobenzaldehyde	—	58	246	293
4-Dimethylaminobenzaldehyde	—	74	222	325
Vanillin	285 d.	82	230	271
4-Nitrobenzaldehyde	—	106	221	320 d.
4-Hydroxybenzaldehyde	—	116	224	280 d.
(±)-Glyceraldehyde	—	142	160 d.	167

NOTE: "d" indicates decomposition.

* See Appendix 2 Procedures for Preparing Derivatives.

Ketones

COMPOUND	BP	MP	SEMI-CARBAZONE*	2,4-DINITRO-PHENYL-HYDRAZONE*
2-Propanone (acetone)	56	—	187	126
2-Butanone (methyl ethyl ketone)	80	—	146	117
3-Methyl-2-butanone (isopropyl methyl ketone)	94	—	112	120
2-Pentanone (methyl propyl ketone)	101	—	112	143
3-Pentanone (Diethyl ketone)	102	—	138	156
Pinacolone	106	—	157	125
4-Methyl-2-pentanone (isobutyl methyl ketone)	117	—	132	95
2,4-Dimethyl-3-pentanone (diisopropyl ketone)	124	—	160	95
2-Hexanone (methyl butyl ketone)	128	—	125	106
4-Methyl-3-penten-2-one (mesityl oxide)	130	—	164	205
Cyclopentanone	131	—	210	146
2,3-Pentanedione	134	—	122 (mono) 209 (di)	209
2,4-Pentanedione (acetylacetone)	139	—	—	122 (mono) 209 (di)
4-Heptanone (Dipropyl ketone)	144	—	132	75
2-Heptanone (methyl amyl ketone)	151	—	123	89
Cyclohexanone	156	—	166	162
2,6-Dimethyl-4-heptanone (diisobutyl ketone)	168	—	122	92
2-Octanone	173	—	122	58
Cycloheptanone	181	—	163	148
2,5-Hexanedione (acetonylacetone)	191	−9	185 (mono) 224 (di)	257 (di)
Acetophenone (methyl phenyl ketone)	202	20	198	238
Phenyl-2-propanone (phenylacetone)	216	27	198	156
Propiophenone (ethyl phenyl ketone)	218	21	182	191
4-Methylacetophenone	226	—	205	258
2-Undecanone	231	12	122	63
4-Chloroacetophenone	232	12	204	236
4-Phenyl-2-butanone (benzylacetone)	235	—	142	127
4-Chloropropiophenone	—	36	176	223
4-Phenyl-3-buten-2-one	—	37	187	227
4-Methoxyacetophenone	258	38	198	228
Benzophenone	305	48	167	238
4-Bromoacetophenone	225	51	208	230
2-Acetonaphthone	—	54	235	262
Desoxybenzoin	320	60	148	204
3-Nitroacetophenone	202	80	257	228
9-Fluorenone	345	83	234	283
Benzoin	344	136	206	245
4-Hydroxypropiophenone	—	148	—	229
(±)-Camphor	205	179	237	177

*See Appendix 2 Procedures for Preparing Derivatives.

Carboxylic Acids

COMPOUND	BP	MP	*p*-TOLUIDIDE*	ANILIDE*	AMIDE*
Formic acid	101	8	53	47	43
Acetic acid	118	17	148	114	82
Propenoic acid (acrylic acid)	139	13	141	104	85
Propanoic acid (propionic acid)	141	—	124	103	81
2-Methylpropanoic acid (isobutyric acid)	154	—	104	105	128
Butanoic acid (butyric acid)	162	—	72	95	115
2-Methylpropenoic acid (methacrylic acid)	163	16	—	87	102
Trimethylacetic acid (pivalic acid)	164	35	—	127	178
Pyruvic acid	165 d.	14	109	104	124
3-Methylbutanoic acid (isovaleric acid)	176	—	109	109	135
Pentanoic acid (valeric acid)	186	—	70	63	106
2-Methylpentanoic acid	186	—	80	95	79
2-Chloropropanoic acid	186	—	124	92	80
Dichloroacetic acid	194	6	153	118	98
Hexanoic acid (caproic acid)	205	—	75	95	101
2-Bromopropanoic acid	205	24	125	99	123
Octanoic acid (caprylic acid)	237	16	70	57	107
Nonanoic acid	254	12	84	57	99
Decanoic acid (capric acid)	268	32	78	70	108
4-Oxopentanoic acid (levulinic acid)	246	33	108	102	108 d.
Dodecanoic acid (lauric acid)	299	43	87	78	100
3-Phenylpropanoic acid (hydrocinnamic acid)	279	48	135	98	105
Bromoacetic acid	208	50	—	131	91
Tetradecanoic acid (myristic acid)	—	54	93	84	103
Trichloroacetic acid	198	57	113	97	141
Hexadecanoic acid (palmitic acid)	—	62	98	90	106
Chloroacetic acid	189	63	162	137	121
Octadecanoic acid (stearic acid)	—	69	102	95	109
trans-2-Butenoic acid (crotonic acid)	—	72	132	118	158
Phenylacetic acid	—	77	136	118	156
2-Methoxybenzoic acid (*o*-anisic acid)	200	101	—	131	129
2-Methylbenzoic acid (*o*-toluic acid)	—	104	144	125	142
Nonanedioic acid (azelaic acid)	—	106	201 (di)	107 (mono) 186 (di)	93 (mono) 175 (di)
3-Methylbenzoic acid (*m*-toluic acid)	263 s.	110	118	126	94
(±)-Phenylhydroxyacetic acid (mandelic acid)	—	118	172	151	133
Benzoic acid	249	122	158	163	130
2-Benzoylbenzoic acid	—	127	—	195	165
Maleic acid	—	130	142 (di)	198 (mono) 187 (di)	172 (mono) 260 (di)
Decanedioic acid (sebacic acid)	—	133	201 (di)	122 (mono) 200 (di)	170 (mono) 210 (di)
Cinnamic acid	300	133	168	153	147

NOTE: "s" indicates "sublimation"; "d" indicates decomposition.
* See Appendix 2 Procedures for Preparing Derivatives.

Carboxylic Acids (Continued)

COMPOUND	BP	MP	*p*-TOLUIDIDE*	ANILIDE*	AMIDE*
2-Chlorobenzoic acid	—	140	131	118	139
3-Nitrobenzoic acid	—	140	162	155	143
2-Aminobenzoic acid (anthranilic acid)	—	146	151	131	109
Diphenylacetic acid	—	148	172	180	167
2-Bromobenzoic acid	—	150	—	141	155
Benzilic acid	—	150	190	175	154
Hexanedioic acid (adipic acid)	—	152	239	151 (mono) 241 (di)	125 (mono) 220 (di)
Citric acid	—	153	189 (tri)	199 (tri)	210 (tri)
4-Chlorophenoxyacetic acid	—	158	—	125	133
2-Hydroxybenzoic acid (salicylic acid)	—	158	156	136	142
5-Bromo-2-hydroxybenzoic acid (5-bromosalicylic acid)	—	165	—	222	232
Methylenesuccinic acid (itaconic acid)	—	166 d.	—	152 (mono)	191 (di)
(+)-Tartaric acid	—	169	—	180 (mono) 264 (di)	171 (mono) 196 (di)
4-Chloro-3-nitrobenzoic acid	—	180	—	131	156
4-Methylbenzoic acid (*p*-toluic acid)	—	180	160	145	160
4-Methoxybenzoic acid (*p*-anisic acid)	280	184	186	169	167
Butanedioic acid (succinic acid)	235 d.	188	180 (mono) 255 (di)	143 (mono) 230 (di)	157 (mono) 260 (di)
3-Hydroxybenzoic acid	—	201	163	157	170
3,5-Dinitrobenzoic acid	—	202	—	234	183
Phthalic acid	—	210 d.	150 (mono) 201 (di)	169 (mono) 253 (di)	144 (mono) 220 (di)
4-Hydroxybenzoic acid	—	214	204	197	162
Pyridine-3-carboxylic acid (nicotinic acid)	—	236	150	132	128
4-Nitrobenzoic acid	—	240	204	211	201
4-Chlorobenzoic acid	—	242	—	194	179
Fumaric acid	—	300	—	233 (mono) 314 (di)	270 (mono) 266 (di)

NOTE: "d" indicates decomposition.
* See Appendix 2 Procedures for Preparing Derivatives.

Phenols†

COMPOUND	BP	MP	α-NAPHTHYL-URETHANE*	BROMO DERIVATIVE*			
				Mono	*Di*	*Tri*	*Tetra*
2-Chlorophenol	176	7	120	48	76	—	—
3-Methylphenol (*m*-cresol)	203	12	128	—	—	84	—
2-Methylphenol (*o*-cresol)	191	32	142	—	56	—	—
2-Methoxyphenol (guaiacol)	204	32	118	—	—	116	—
4-Methylphenol (*p*-cresol)	202	34	146	—	49	—	198
Phenol	181	42	133	—	—	95	—
4-Chlorophenol	217	43	166	33	90	—	—
2,4-Dichlorophenol	210	45	—	68	—	—	—
4-Ethylphenol	219	45	128	—	—	—	—
2-Nitrophenol	216	45	113	—	117	—	—
2-Isopropyl-5-methylphenol (thymol)	234	51	160	55	—	—	—
3,4-Dimethylphenol	225	64	141	—	—	171	—
4-Bromophenol	238	64	169	—	—	95	—
3,5-Dimethylphenol	220	68	109	—	—	166	—
2,5-Dimethylphenol	212	75	173	—	—	178	—
1-Naphthol (α-naphthol)	278	96	152	—	105	—	—
2-Hydroxyphenol (catechol)	245	104	175	—	—	—	192
3-Hydroxyphenol (resorcinol)	281	109	275	—	—	112	—
4-Nitrophenol	—	112	150	—	142	—	—
2-Naphthol (β-naphthol)	286	121	157	84	—	—	—
1,2,3-Trihydroxybenzene (pyrogallol)	309	133	—	—	158	—	—
4-Phenylphenol	305	164	—	—	—	—	—

*See Appendix 2 Procedures for Preparing Derivatives.
†Also check:
   Salicylic acid (2-hydroxybenzoic acid)
   Esters of salicylic acid (salicylates)
   Salicylaldehyde (2-hydroxybenzaldehyde)
   4-Hydroxybenzaldehyde
   4-Hydroxypropiophenone
   3-Hydroxybenzoic acid
   4-Hydroxybenzoic acid
   4-Hydroxybenzophenone

Primary Amines†

COMPOUND	BP	MP	BENZAMIDE*	PICRATE*	ACETAMIDE*
t-Butylamine	46	—	134	198	101
Propylamine	48	—	84	135	—
Allylamine	56	—	—	140	—
sec-Butylamine	63	—	76	139	—
Isobutylamine	69	—	57	150	—
Butylamine	78	—	42	151	—
Cyclohexylamine	135	—	149	—	104
Furfurylamine	145	—	—	150	—
Benzylamine	184	—	105	194	60
Aniline	184	—	163	198	114
2-Methylaniline (*o*-toluidine)	200	—	144	213	110
3-Methylaniline (*m*-toluidine)	203	—	125	200	65
2-Chloroaniline	208	—	99	134	87
2,6-Dimethylaniline	216	11	168	180	177
2-Methoxyaniline (*o*-anisidine)	225	6	60	200	85
3-Chloroaniline	230	—	120	177	74
2-Ethoxyaniline (*o*-phenetidine)	231	—	104	—	79
4-Chloro-2-methylaniline	241	29	142	—	140
4-Ethoxyaniline (*p*-phenetidine)	250	2	173	69	137
4-Methylaniline (*p*-toluidine)	200	43	158	182	147
2-Ethylaniline	210	47	147	194	111
2,5-Dichloroaniline	251	50	120	86	132
4-Methoxyaniline (*p*-anisidine)	—	58	154	170	130
4-Bromoaniline	245	64	204	180	168
2,4,5-Trimethylaniline	—	64	167	—	162
4-Chloroaniline	—	70	192	178	179
2-Nitroaniline	—	72	110	73	92
Ethyl *p*-aminobenzoate	—	89	148	—	110
o-Phenylenediamine	258	102	301 (di)	208	185 (di)
2-Methyl-5-nitroaniline	—	106	186	—	151
2-Chloro-4-nitroaniline	—	108	161	—	139
3-Nitroaniline	—	114	157	143	155
4-Chloro-2-nitroaniline	—	118	—	—	104
2,4,6-Tribromoaniline	300	120	200	—	232 (mono) 127 (di)
2-Methyl-4-nitroaniline	—	130	—	—	202
2-Methoxy-4-nitroaniline	—	138	149	—	153
p-Phenylenediamine	267	140	128 (mono) 300 (di)	—	162 (mono) 304 (di)
4-Nitroaniline	—	148	199	100	215
4-Aminoacetanilide	—	162	—	—	304
2,4-Dinitroaniline	—	180	202	—	120

*See Appendix 2 Procedures for Preparing Derivatives.

†Also check: 4-Aminobenzoic acid and its esters.

Secondary Amines

COMPOUND	BP	MP	BENZAMIDE*	PICRATE*	ACETAMIDE*
Diethylamine	56	—	42	155	—
Diisopropylamine	84	—	—	140	—
Pyrrolidine	88	—	Oil	112	—
Piperidine	106	—	48	152	—
Dipropylamine	110	—	—	75	—
Morpholine	129	—	75	146	—
Diisobutylamine	139	—	—	121	86
N-Methylcyclohexylamine	148	—	85	170	—
Dibutylamine	159	—	—	59	—
Benzylmethylamine	184	—	—	117	—
N-Methylaniline	196	—	63	145	102
N-Ethylaniline	205	—	60	132	54
N-Ethyl-m-toluidine	221	—	72	—	—
Dicyclohexylamine	256	—	153	173	103
N-Benzylaniline	298	37	107	48	58
Indole	254	52	68	—	157
Diphenylamine	302	52	180	182	101
N-Phenyl-1-naphthylamine	335	62	152	—	115

*See Appendix 2 Procedures for Preparing Derivatives.

Tertiary Amines†

COMPOUND	BP	MP	PICRATE*	METHIODIDE*
Triethylamine	89	—	173	280
Pyridine	115	—	167	117
2-Methylpyridine (α-picoline)	129	—	169	230
3-Methylpyridine (β-picoline)	144	—	150	92
Tripropylamine	157	—	116	207
N,N-Dimethylbenzylamine	183	—	93	179
N,N-Dimethylaniline	193	—	163	228 d.
Tributylamine	216	—	105	186
N,N-Diethylaniline	217	—	142	102
Quinoline	237	—	203	133

NOTE: ''d'' indicates decomposition.
*See Appendix 2 Procedures for Preparing Derivatives.

†Also check: Nicotinic acid and its esters.

Alcohols

COMPOUND	BP	MP	3,5-DINITRO-BENZOATE*	PHENYL-URETHANE*
Methanol	65	—	108	47
Ethanol	78	—	93	52
2-Propanol (isopropyl alcohol)	82	—	123	88
2-Methyl-2-propanol (*t*-butyl alcohol)	83	26	142	136
2-Propen-1-ol (allyl alcohol)	97	—	49	70
1-Propanol	97	—	74	57
2-Butanol (*sec*-butyl alcohol)	99	—	76	65
2-Methyl-2-butanol (*t*-pentyl alcohol)	102	−8.5	116	42
2-Methyl-3-butyn-2-ol	104	—	112	—
2-Methyl-1-propanol (isobutyl alcohol)	108	—	87	86
2-Propyn-1-ol (propargyl alcohol)	114	—	—	—
3-Pentanol	115	—	101	48
1-Butanol	118	—	64	61
2-Pentanol	119	—	62	—
3-Methyl-3-pentanol	123	—	96	43
2-Methoxyethanol	124	—	—	(113)†
2-Chloroethanol	129	—	95	51
3-Methyl-1-butanol (isoamyl alcohol)	130	—	70	31
4-Methyl-2-pentanol	132	—	65	143
1-Pentanol	138	—	46	46
Cyclopentanol	140	—	115	132
2-Ethyl-1-butanol	146	—	51	—
2,2,2-Trichloroethanol	151	—	142	87
1-Hexanol	157	—	58	42
Cyclohexanol	160	—	113	82
(2-Furyl)-methanol (furfuryl alcohol)	170	—	80	45
1-Heptanol	176	—	47	60
2-Octanol	179	—	32	114
1-Octanol	195	—	61	74
3,7-Dimethyl-1,6-octadien-3-ol (linalool)	196	—	—	66
Benzyl alcohol	204	—	113	77
1-Phenylethanol	204	20	92	95
2-Phenylethanol	219	—	108	78
1-Decanol	231	7	57	59
3-Phenylpropanol	236	—	45	92
1-Dodecanol (lauryl alcohol)	—	24	60	74
3-Phenyl-2-propen-1-ol (cinnamyl alcohol)	250	34	121	90
1-Tetradecanol (myristyl alcohol)	—	39	67	74
(−)-Menthol	212	41	158	111
1-Hexadecanol (cetyl alcohol)	—	49	66	73
1-Octadecanol (stearyl alcohol)	—	59	77	79
Diphenylmethanol (benzhydrol)	288	68	141	139
Benzoin	—	133	—	165
Cholesterol	—	147	—	168
(+)-Borneol	—	208	154	138

*See Appendix 2 Procedures for Preparing Derivatives.
† α-Naphthylurethane

Esters

COMPOUND	BP	MP	COMPOUND	BP	MP
Methyl formate	34	—	Ethyl lactate	154	—
Ethyl formate	54	—	Ethyl hexanoate (ethyl caproate)	168	—
Vinyl acetate	72	—	Methyl acetoacetate	170	—
Ethyl acetate	77	—	Dimethyl malonate	180	—
Methyl propanoate (methyl propionate)	77	—	Ethyl acetoacetate	181	—
Methyl acrylate	80	—	Diethyl oxalate	185	—
2-Propyl acetate (isopropyl acetate)	85	—	Methyl benzoate	199	—
Ethyl chloroformate	93	—	Ethyl octanoate (ethyl caprylate)	207	—
Methyl 2-methylpropanoate	93	—	Ethyl cyanoacetate	210	—
(methyl isobutyrate)			Ethyl benzoate	212	—
2-Propenyl acetate (isopropenyl acetate)	94	—	Diethyl succinate	217	—
2-(2-Methylpropyl) acetate	98	—	Methyl phenylacetate	218	—
(*t*-butyl acetate)			Diethyl fumarate	219	—
Ethyl acrylate	99	—	Methyl salicylate	222	—
Ethyl propanoate (ethyl propionate)	99	—	Diethyl maleate	225	—
Methyl methacrylate	100	—	Ethyl phenylacetate	229	—
Methyl trimethylacetate (methyl pivalate)	101	—	Ethyl salicylate	234	—
Propyl acetate	102	—	Dimethyl suberate	268	—
Methyl butanoate (methyl butyrate)	102	—	Ethyl cinnamate	271	—
2-Butyl acetate (*sec*-butyl acetate)	111	—	Diethyl phthalate	298	—
Methyl 3-methylbutanoate	117	—	Dibutyl phthalate	340	—
(methyl isovalerate)			Methyl cinnamate	—	36
Ethyl butanoate (ethyl butyrate)	120	—	Phenyl salicylate	—	42
Butyl acetate	127	—	Methyl *p*-chlorobenzoate	—	44
Methyl pentanoate (methyl valerate)	128	—	Ethyl *p*-nitrobenzoate	—	56
Methyl chloroacetate	130	—	Phenyl benzoate	314	69
Ethyl 3-methylbutanoate (ethyl isovalerate)	132	—	Methyl *m*-nitrobenzoate	—	78
Pentyl acetate (*n*-amyl acetate)	142	—	Methyl *p*-bromobenzoate	—	81
3-Methylbutyl acetate (isoamyl acetate)	142	—	Ethyl *p*-aminobenzoate	—	90
Ethyl chloroacetate	143	—	Methyl *p*-nitrobenzoate	—	94

Appendix 2
PROCEDURES FOR PREPARING DERIVATIVES

> **CAUTION:** Some of the chemicals used in preparing derivatives are suspected carcinogens. The list of suspected carcinogens on p 11 should be consulted before beginning any of these procedures. Care should be exercised in handling these substances.

ALDEHYDES AND KETONES

Semicarbazones. Place 0.5 mL of a $2M$ stock solution of semicarbazide hydrochloride (or 0.5 mL of a solution prepared by dissolving 1.11 g of semicarbazide hydrochloride [MW 111.5] in 5 mL of water) in a small test tube. Add an estimated one millimole (mmol) of the unknown compound to the test tube. If the unknown does not dissolve in the solution, or if the solution becomes cloudy, add enough methanol to dissolve the solid and produce a clear solution. Using a disposable Pasteur pipet, add 10 drops of pyridine and heat the mixture gently on a steam bath for about five minutes; by that time, the product should have begun to crystallize. Collect the product by vacuum filtration. The product can be recrystallized from ethanol if necessary.

Semicarbazones (Alternative Method). Dissolve 0.25 g of semicarbazide hydrochloride and 0.38 g of sodium acetate in 1.3 mL of water. Then dissolve 0.25 g of the unknown in 2.5 mL of ethanol. Mix the two solutions together in a 25-mL Erlenmeyer flask and heat the mixture to boiling for about five minutes. After heating, place the reaction flask in a beaker of ice and scratch the sides of the flask with a glass rod to induce crystallization of the derivative. Collect the derivative by vacuum filtration and recrystallize it from ethanol.

2,4-Dinitrophenylhydrazones. Place 10 mL of a solution of 2,4-dinitrophenylhydrazine (prepared as described for the classification test in Procedure 56D) in a test tube, and add an estimated 1 mmol of the unknown compound. If the unknown is a solid, it should be dissolved in the minimum amount of 95% ethanol or 1,2-dimethoxyethane before it is added. If crystallization is not immediate, gently warm the solution for a minute on a steam bath and then set it aside to crystallize. Collect the product by vacuum filtration.

CARBOXYLIC ACIDS

Using a reflux condenser and a 5-mL conical vial, heat a mixture of 0.25 g of the acid and 1 mL of thionyl chloride on a steam bath for about 30 minutes. Allow the mixture to cool and use it for one of the following three procedures:

Amides. Working in a hood, pour the reaction mixture into a beaker containing 5 mL of ice-cold concentrated ammonium hydroxide and stir it vigorously. When the reaction is complete, collect the product by vacuum filtration and recrystallize it from water or from water–ethanol, using the mixed-solvents method (Technique 5, Section 5.9).

Anilides. Dissolve 0.5 g of aniline in 13 mL of toluene and carefully add it to the reaction mixture. Warm the mixture for an additional five minutes on a hot plate. Then transfer the toluene solution to a small separatory funnel and wash it sequentially with 2.5 mL of water, 2.5 mL of 5% hydrochloric acid, 2.5 mL of 5% sodium hydroxide, and a second 2.5-mL portion of water. Dry the toluene layer over a small amount of anhydrous sodium sulfate. Decant the toluene layer away from the drying agent into a small beaker and evaporate the toluene on a hot plate in the hood. Recrystallize the product from water or from ethanol–water, using the mixed-solvents method (Technique 5, Section 5.9).

p-Toluidides. Use the same procedure as that described for the anilide, but substitute *p*-toluidine for aniline.

PHENOLS

α-Naphthylurethanes. Follow the procedure given below for preparing phenylurethanes from alcohols but substitute α-naphthylisocyanate for phenylisocyanate.

Bromo Derivatives. First, if a stock brominating solution is not available, prepare one by dissolving 0.75 g of potassium bromide in 5 mL of water and adding 0.5 g of bromine. Dissolve 0.1 g of the phenol in 1 mL of methanol or 1,2-dimethoxyethane and then add 1 mL of water. Add 1 mL of the brominating mixture to the phenol solution and swirl the mixture vigorously. Then, continue adding the brominating solution, dropwise with swirling, until the color of the bromine reagent persists. Finally, add 3–5 mL of water and shake the mixture vigorously. Collect the precipitated product by vacuum filtration and wash it well with water. Recrystallize the derivative from methanol–water, using the mixed-solvents method (Technique 5, Section 5.9).

AMINES

Acetamides. Place an estimated 1 mmol of the amine and 0.5 mL of acetic anhydride in a small Erlenmeyer flask. Heat the mixture for about five minutes; then add 5 mL of water and stir the solution vigorously to precipitate the product and hydrolyze the excess acetic anhydride. If the product does not crystallize, it may be necessary to scratch the walls of the flask with a glass rod. Collect the crystals by vacuum filtration and wash them with several portions of cold 5% hydrochloric acid. Recrystallize the derivative from methanol–water, using the mixed-solvents method (Technique 5, Section 5.9).

Aromatic amines, or those amines that are not very basic, may require pyridine (2 mL) as a solvent and a catalyst for the reaction. If pyridine is used, a longer period of heating will be required (up to one hour), and the reaction should be carried out in an apparatus equipped with a reflux condenser. After reflux, the reaction mixture must be extracted with 5–10 mL of 5% sulfuric acid to remove the pyridine.

Benzamides. Using a test tube, suspend an estimated 1 mmol of the amine in 1 mL of 10% sodium hydroxide solution and add 0.5 g of benzoyl chloride. Stopper the test tube with a cork and shake the mixture vigorously for about 10 minutes. After shaking, add enough dilute hydrochloric acid to bring the pH of the solution to pH 7 or 8. Collect the precipitate by vacuum filtration, wash it thoroughly with cold water, and recrystallize it from ethanol–water, using the mixed-solvents method (Technique 5, Section 5.9).

Benzamides (Alternative Method). Dissolve 0.25 g of the amine in a solution of 1.2 mL of pyridine and 2.5 mL of toluene. Add 0.25 mL of benzoyl chloride to the solution and heat the mixture under reflux about 30 minutes. Pour the cooled reaction mixture into 25 mL of water and stir the mixture vigorously to hydrolyze the excess benzoyl chloride. Separate the toluene layer and wash it, first with 1.5 mL of water and then with 1.5 mL of 5% sodium carbonate. Dry the toluene over anhydrous sodium sulfate, decant the toluene into a small beaker, and remove the toluene by evaporation on a hot plate in the hood. Recrystallize the benzamide from ethanol or ethanol–water, using the mixed-solvents method (Technique 5, Section 5.9).

Picrates. Dissolve 0.2 g of the unknown in about 5 mL of ethanol and add 5 mL of a saturated solution of picric acid in ethanol. Heat the solution to boiling and then allow it to cool slowly. Collect the product by vacuum filtration and rinse it with a small amount of cold ethanol.

Methiodides. Mix equal-volume quantities of the amine and methyl iodide in a test tube (about 0.25 mL is convenient) and allow the mixture to stand for several minutes. Then heat the mixture gently under reflux on a steam bath for about five minutes. The methiodide should crystallize on cooling. If it does not, you can induce crystallization by scratching the walls of the tube with a glass rod. Collect the product by vacuum filtration and recrystallize it from ethanol or ethyl acetate.

ALCOHOLS

3,5-Dinitrobenzoates. *Liquid Alcohols.* Dissolve 0.25 g of 3,5-dinitrobenzoyl chloride[1] in 0.25 mL of the alcohol and heat the mixture for about five minutes. Allow the mixture to cool and add 1.5 mL of a 5% sodium carbonate solution and 1 mL of water. Stir the mixture vigorously and crush any solid that forms. Collect the product by vacuum filtration and wash it with cold water. Recrystallize the derivative from ethanol–water, using the mixed-solvents method (Technique 5, Section 5.9).

---

[1] This is an acid chloride and undergoes hydrolysis readily. The purity of this reagent should be checked before its use by determining its melting point. When the carboxylic acid is present, the melting point will be high.

Solid Alcohols. Dissolve 0.25 g of the alcohol in 1.5 mL of dry pyridine and add 0.25 g of 3,5-dinitrobenzoyl chloride.[2] Heat the mixture under reflux for 15 minutes. Pour the cooled reaction mixture into a cold mixture of 2.5 mL of 5% sodium carbonate and 2.5 mL of water. Keep the solution cooled in an ice bath until the product crystallizes, and stir it vigorously during the entire period. Collect the product by vacuum filtration, wash it with cold water, and recrystallize it from ethanol–water, using the mixed-solvents method (Technique 5, Section 5.9).

Phenylurethanes. Place 0.25 g of the **anhydrous** alcohol in a dry test tube and add 0.25 mL of phenylisocyanate (α-naphthylisocyanate for a phenol). If the compound is a phenol, add one drop of pyridine to catalyze the reaction. If the reaction is not spontaneous, heat the mixture on a steam bath for 5 to 10 minutes. Cool the test tube in a beaker of ice and scratch the tube with a glass rod to induce crystallization. Decant the liquid from the solid product, or, if necessary, collect the product by vacuum filtration. Dissolve the product in 2.5–3 mL of hot ligroin or hexane and filter the mixture by gravity (preheat funnel) to remove any unwanted and insoluble diphenylurea present. Cool the filtrate to induce crystallization of the urethane. Collect the product by vacuum filtration.

ESTERS

Preparing the derivatives of esters is usually complicated. We recommend that esters be characterized by spectroscopic methods whenever possible. If a derivative must be prepared, consult a comprehensive textbook. Several are listed in the first section of Experiment 56.

---

[2] See Footnote 1.

Appendix 3
INFRARED (IR) SPECTROSCOPY

Almost any compound having covalent bonds, whether organic or inorganic, will be found to absorb various frequencies of electromagnetic radiation in the infrared region of the spectrum. The infrared region of the electromagnetic spectrum lies at wavelengths longer than those associated with visible light, which includes wavelengths from approximately 400 nm–800 nm (1 nm = 10^{-9} m), but at wavelengths shorter than those associated with radio waves, which have wavelengths longer than 1 cm. For chemical purposes, we are interested in the **vibrational** portion of the infrared region. This portion is defined as that including radiations with wavelengths (λ) between 2.5 μ and 15 μ (1 μ = 1 micron = 1 μm = 10^{-6} m). Although the more technically correct unit for wavelength in the infrared region of the spectrum is micrometer, we shall follow common practice and use micron as the unit. The relation of the infrared region to others included in the electromagnetic spectrum is illustrated in Figure IR–1.

As with other types of energy absorption, molecules are excited to a higher energy state when they absorb infrared radiation. The absorption of infrared radiation is, like other absorption processes, a quantized process. Only selected frequencies (energies) of infrared radiation are absorbed by a molecule. The absorption of infrared radiation corresponds to energy changes on the order of 2–10 kcal/mol. Radiation in this energy range corresponds to the range encompassing the stretching and bending vibrational frequencies of the bonds in most covalent molecules. In the absorption process, those frequencies of infrared radiation that match the natural vibrational frequencies of the molecule in question are absorbed, and the energy absorbed increases the **amplitude** of the vibrational motions of the bonds in the molecule.

Many chemists refer to the radiation in the vibrational infrared region of the electromagnetic spectrum by units called **wavenumbers** ($\overline{\nu}$). Wavenumbers are expressed in reciprocal centimeters (cm^{-1}) and are easily computed by taking the reciprocal of the wavelength (λ) expressed in centimeters. This unit has the advantage, for

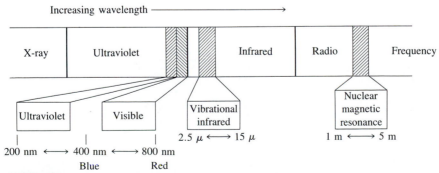

FIGURE IR–1. Portion of electromagnetic spectrum, showing relation of vibrational infrared to other types of radiation

those performing calculations, that it is directly proportional to energy. Thus, the vibrational infrared extends from about 4000–650 cm^{-1} (or wavenumbers).

Wavelengths (μ) and wavenumbers (cm^{-1}) can be interconverted by the following relationships:

$$ \text{cm}^{-1} = \frac{1}{(\mu)} \times 10{,}000 \quad \text{and} \quad \mu = \frac{1}{(\text{cm}^{-1})} \times 10{,}000 $$

IR.1 USES OF THE INFRARED SPECTRUM

Since every different type of bond has a different natural frequency of vibration, and since the same type of bond in two different compounds is in a slightly different environment, no two molecules of different structure have exactly the same infrared absorption pattern, or **infrared spectrum.** Although some of the frequencies absorbed in the two cases might be the same, in no case of two different molecules will their infrared spectra (the patterns of absorption) be identical. Thus, the infrared spectrum can be used for molecules much as a fingerprint can be used for humans. By comparing the infrared spectra of two substances thought to be identical, one can establish whether or not they are in fact identical. If the infrared spectra of two substances coincide peak for peak (absorption for absorption), in most cases, the substances are identical.

A second and more important use of the infrared spectrum is that it gives structural information about a molecule. The absorptions of each type of bond (N—H, C—H, O—H, C—X, C=O, C—O, C—C, C=C, C≡C, C≡N, and so on) are regularly found only in certain small portions of the vibrational infrared region. A small range of absorption can be defined for each type of bond. Outside this range, absorptions will normally be due to some other type of bond. Thus, for instance, any absorption in the range 3000 ± 150 cm^{-1} (around 3.33 μ) will almost always be due to the presence of a CH bond in the molecule; an absorption in the range 1700 ± 100 cm^{-1} (around 6.1 μ) will normally be due to the presence of a C=O bond (carbonyl group) in the molecule. The same type of range applies to each type of bond. The way these are spread out over the vibrational infrared is illustrated schematically in Figure IR–2. It is a good idea to try to remember this general scheme for future convenience.

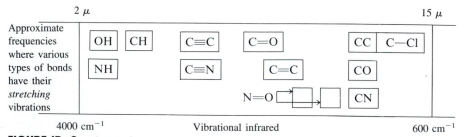

FIGURE IR–2.   Approximate regions in which various common types of bonds absorb. (Bending and twisting and other types of bond vibration have been omitted for clarity.)

IR.2 MODES OF VIBRATION AND BENDING

The simplest types, or **modes,** of vibrational motion in a molecule that are **infrared-active,** that is, give rise to absorptions, are the stretching and bending modes.

$$C—H \qquad C \overset{O}{\underset{}{\diagdown H}}$$

Stretching   Bending

Other, more complex types of stretching and bending are also active, however. To introduce several words of terminology, the normal modes of vibration for a methylene group are shown on p 816.

In any group of three or more atoms—at least two of which are identical—there are **two** modes of stretching or bending: the symmetric mode and the asymmetric mode. Examples of such groupings are —CH_3, —CH_2—, —NO_2, —NH_2, and anhydrides, $(CO)_2O$. For the anhydride, owing to asymmetric and symmetric modes of stretch, this functional group gives **two** absorptions in the C=O region. A similar phenomenon is seen for amino groups, where primary amines usually have **two** absorptions in the NH stretch region while secondary amines (R_2NH) have only one absorption peak. Amides show similar bands. There are two strong N=O stretch peaks for a nitro group, which are caused by asymmetric and symmetric stretching modes.

IR.3 WHAT TO LOOK FOR IN EXAMINING INFRARED SPECTRA

The instrument that determines the absorption spectrum for a compound is called an **infrared spectrophotometer.** The spectrophotometer determines the relative strengths and positions of all the absorptions in the infrared region and plots this information on a piece of calibrated chart paper. This plot of absorption intensity versus wavenumber or wavelength is referred to as the **infrared spectrum** of the compound. A typical infrared spectrum, that of methyl isopropyl ketone, is shown in Figure IR–3.

The strong absorption in the middle of the spectrum corresponds to C=O, the carbonyl group. Note that the C=O peak is quite intense. In addition to the characteristic position of absorption, the **shape** and **intensity** of this peak are also unique to the C=O bond. This is true for almost every type of absorption peak; both shape and intensity characteristics can be described, and these characteristics often allow one to distinguish the peak in a confusing situation. For instance, to some extent both C=O and C=C bonds absorb in the same region of the infrared spectrum:

$$C=O \ 1850–1630 \ cm^{-1} \ (5.5–6.1 \ \mu)$$
$$C=C \ 1680–1620 \ cm^{-1} \ (5.95–6.2 \ \mu)$$

However, the C=O bond is a strong absorber, whereas the C=C bond generally absorbs only weakly. Hence, a trained observer would not normally interpret a strong peak at $1670 \ cm^{-1}$ to be a carbon–carbon double bond nor a weak absorption at this frequency to be due to a carbonyl group.

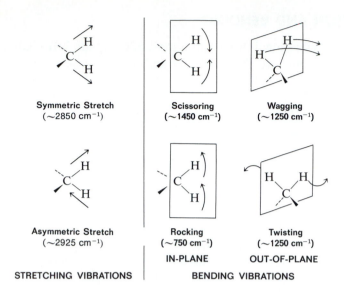

Symmetric Stretch
($\sim$2850 cm^{-1})

Scissoring
($\sim$1450 cm^{-1})

Wagging
($\sim$1250 cm^{-1})

Asymmetric Stretch
($\sim$2925 cm^{-1})

Rocking
($\sim$750 cm^{-1})

Twisting
($\sim$1250 cm^{-1})

IN-PLANE

OUT-OF-PLANE

STRETCHING VIBRATIONS

BENDING VIBRATIONS

The shape of a peak often gives a clue to its identity as well. Thus, while the NH and OH regions of the infrared overlap,

OH 3650–3200 cm^{-1} (2.75–3.12 μ)
NH 3500–3300 cm^{-1} (2.85–3.00 μ)

NH usually gives a **sharp** absorption peak (absorbs a very narrow range of frequencies), and OH, when it is in the NH region, usually gives a **broad** absorption peak. Primary amines give **two** absorptions in this region, whereas alcohols give only one.

Therefore, while you are studying the sample spectra in the pages that follow, you should also notice shapes and intensities. They are as important as the frequency at which an absorption occurs, and the eye must be trained to recognize these features. Often, in the literature of organic chemistry, one will find absorptions referred to as strong (s), medium (m), weak (w), broad, or sharp. The author is trying to convey some idea of what the peak looks like without actually drawing the spectrum.

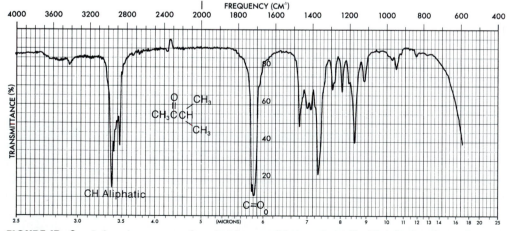

FIGURE IR–3. Infrared spectrum of methyl isopropyl ketone (neat liquid, salt plates)

IR.4 CORRELATION CHARTS AND TABLES

To extract structural information from infrared spectra, one must know the frequencies or wavelengths at which various functional groups absorb. Infrared **correlation tables** present as much information as is known about where the various functional groups absorb. The books listed at the end of this Appendix present extensive lists of correlation tables. Sometimes the absorption information is given in a chart, called a **correlation chart.** A simplified correlation table is given in Table IR–1.

TABLE IR–1. A Simplified Correlation Table

TYPE OF VIBRATION			FREQUENCY (cm^{-1})	WAVELENGTH (μ)	INTENSITY
C—H	Alkanes	(stretch)	3000–2850	3.33–3.51	s
	—CH$_3$	(bend)	1450 and 1375	6.90 and 7.27	m
	—CH$_2$—	(bend)	1465	6.83	m
	Alkenes	(stretch)	3100–3000	3.23–3.33	m
		(bend)	1700–1000	5.88–10.0	s
	Aromatics	(stretch)	3150–3050	3.17–3.28	s
		(out-of-plane bend)	1000–700	10.0–14.3	s
	Alkyne	(stretch)	ca. 3300	ca. 3.03	s
	Aldehyde		2900–2800	3.45–3.57	w
			2800–2700	3.57–3.70	w
C—C	Alkane	Not interpretatively useful			
C=C	Alkene		1680–1600	5.95–6.25	m–w
	Aromatic		1600–1400	6.25–7.14	m–w
C≡C	Alkyne		2250–2100	4.44–4.76	m–w
C=O	Aldehyde		1740–1720	5.75–5.81	s
	Ketone (acyclic)		1725–1705	5.80–5.87	s
	Carboxylic acid		1725–1700	5.80–5.88	s
	Ester		1750–1730	5.71–5.78	s
	Amide		1700–1640	5.88–6.10	s
	Anhydride		ca. 1810	ca. 5.52	s
			ca. 1760	ca. 5.68	s
C—O	Alcohols, ethers, esters, carboxylic acids		1300–1000	7.69–10.0	s
O—H	Alcohols, phenols				
	Free		3650–3600	2.74–2.78	m
	H-Bonded		3400–3200	2.94–3.12	m
	Carboxylic acids		3300–2500	3.03–4.00	m
N—H	Primary and secondary amines		ca. 3500	ca. 2.86	m
C≡N	Nitriles		2260–2240	4.42–4.46	m
N=O	Nitro (R—NO$_2$)		1600–1500	6.25–6.67	s
			1400–1300	7.14–7.69	s
C—X	Fluoride		1400–1000	7.14–10.0	s
	Chloride		800–600	12.5–16.7	s
	Bromide, iodide		< 600	>16.7	s

NOTE: s, strong; m, medium; w, weak.

TABLE IR-2.   Base Values for Absorptions of Bonds

OH	3600 cm^{-1}	2.78μ	C≡C	2150 cm^{-1}	4.65μ
NH	3500	2.86	C=O	1715	5.83
CH	3000	3.33	C=C	1650	6.06
C≡N	2250	4.44	C—O	1100	9.09

Although you may think assimilating the mass of data in Table IR-1 will be difficult, it is not if you make a modest start and then gradually increase your familiarity with the data. An ability to interpret the fine details of an IR spectrum will follow. This is most easily done by first establishing the broad visual patterns of Figure IR-2 firmly in mind. Then, as a second step, a "typical absorption value" can be memorized for each of the functional groups in this pattern. This value will be a single number that can be used as a pivot value for the memory. For instance, start with a simple aliphatic ketone as a model for all typical carbonyl compounds. The typical aliphatic ketone has a carbonyl absorption of $1715 \pm 10 \text{ cm}^{-1}$. Without worrying about the variation, memorize 1715 cm^{-1} as the base value for carbonyl absorption. Then, more slowly, learn the extent of the carbonyl range and the visual pattern of how the different kinds of carbonyl groups are arranged throughout this region. See, for instance, Figure IR-14, which gives typical values for carbonyl compounds. Also learn how factors like ring size (when the functional group is contained in a ring) and conjugation affect the base values (that is, in which direction the values are shifted). Learn the trends—always remembering the base value (1715 cm^{-1}). It might prove useful as a beginning to memorize the base values in Table IR-2 for this approach. Notice that there are only eight.

IR.5   ANALYZING A SPECTRUM (OR WHAT YOU CAN TELL AT A GLANCE)

In trying to analyze the spectrum of an unknown, you should concentrate first on trying to establish the presence (or absence) of a few major functional groups. The most conspicuous peaks are C=O, O—H, N—H, C—O, C=C, C≡C, C≡N, and NO_2. If they are present, they give immediate structural information. Do not try to analyze in detail the CH absorptions near 3000 cm^{-1} (3.33μ); almost all compounds have these absorptions. Do not worry about subtleties of the exact type of environment in which the functional group is found. A checklist of the important gross features follows:

1. Is a carbonyl group present?
   The C=O group gives rise to a strong absorption in the region $1820-1660 \text{ cm}^{-1}$ ($5.5-6.1 \mu$).
   The peak is often the strongest in the spectrum and of medium width. You can't miss it.

2. If C=O is present, check the following types. (If it is absent, go to 3.)
   ACIDS   Is OH also present?
                 Broad absorption near $3300-2500 \text{ cm}^{-1}$ ($3.0-4.0 \mu$) (usually overlaps C—H).

AMIDES	Is NH also present? Medium absorption near 3500 cm^{-1} (2.85 μ), sometimes a double peak, equivalent halves.
ESTERS	Is C—O also present? Medium intensity absorptions near 1300–1000 cm^{-1} (7.7–10 μ).
ANHYDRIDES	Have **two** C=O absorptions near 1810 cm^{-1} and 1760 cm^{-1} (5.5 and 5.7 μ).
ALDEHYDES	Is aldehyde CH present? Two weak absorptions near 2850 cm^{-1} and 2750 cm^{-1} (3.50 and 3.65 μ) on the right side of CH absorptions.
KETONES	The above five choices have been eliminated.

3. If C=O is absent

ALCOHOLS or PHENOLS	Check for OH. **Broad** absorption near 3600–3300 cm^{-1} (2.8–3.0 μ). Confirm this by finding C—O near 1300–1000 cm^{-1} (7.7–10 μ).
AMINES	Check for NH. Medium absorption(s) near 3500 cm^{-1} (2.85 μ).
ETHERS	Check for C—O (and absence of OH) near 1300–1000 cm^{-1} (7.7–10 μ).

4. Double Bonds or Aromatic Rings or Both

C=C is a **weak** absorption near 1650 cm^{-1} (6.1 μ).

Medium to strong absorptions in the region 1650–1450 cm^{-1} (6–7 μ) often imply an aromatic ring.

Confirm the above by consulting the CH region.

Aromatic and vinyl CH occur to the left of 3000 cm^{-1} (3.33 μ) (aliphatic CH occurs to the right of this value).

5. Triple Bonds

C≡N is a medium, sharp absorption near 2250 cm^{-1} (4.5 μ).

C≡C is a weak but sharp absorption near 2150 cm^{-1} (4.65 μ).

Check also for acetylenic CH near 3300 cm^{-1} (3.0 μ).

6. Nitro Groups

Two strong absorptions 1600 to 1500 cm^{-1} (6.25–6.67 μ) and 1390–1300 cm^{-1} (7.2–7.7 μ).

7. Hydrocarbons

None of the above is found.

Main absorptions are in CH region near 3000 cm^{-1} (3.33 μ).

Very simple spectrum, only other absorptions near 1450 cm^{-1} (6.90 μ) and 1375 cm^{-1} (7.27 μ).

The beginning student should resist the idea of trying to assign or interpret **every** peak in the spectrum. You simply will not be able to do this. Concentrate first on learning the principal peaks and recognizing their presence or absence. This is best done by carefully studying the illustrative spectra in the section that follows.

NOTE: In describing the shifts of absorption peaks or their relative positions, we have used the phrases "to the left" and "to the right." This was done to facilitate economy of use when using *both* microns and reciprocal centimeters. The meaning is clear, since all spectra are conventionally presented left to right from 4000 cm^{-1} to 600 cm^{-1} or from 2.5 μ–16 μ. "To the right" avoids saying each time "to lower frequency (cm^{-1})" or "to longer wavelength (μ)," which is confusing since cm^{-1} and μ have an inverse relation; as one goes up, the other goes down.

IR.6 SURVEY OF THE IMPORTANT FUNCTIONAL GROUPS

Alkanes

Spectrum is usually simple with few peaks.

C—H   Stretch occurs around 3000 cm^{-1} (3.33 μ).
     (a) In alkanes (except strained ring compounds) absorption always occurs to the right of 3000 cm^{-1} (3.33 μ).
     (b) If a compound has vinylic, aromatic, acetylenic, or cyclopropyl hydrogens, the CH absorption is to the left of 3000 cm^{-1} (3.33 μ).

CH$_2$ Methylene groups have a characteristic absorption at approximately 1450 cm^{-1} (6.90 μ).

CH$_3$ Methyl groups have a characteristic absorption at approximately 1375 cm^{-1} (7.27 μ).

C—C   Stretch—not interpretatively useful—has many peaks.

The spectrum of decane is shown in Figure IR–4.

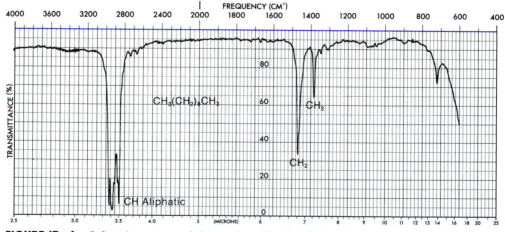

FIGURE IR–4.   Infrared spectrum of decane (neat liquid, salt plates)

Alkenes

=C—H   Stretch occurs to the left of 3000 cm^{-1} (3.33 μ).

=C—H   Out-of-plane (oop) bending at 1000–650 cm^{-1} (10–15 μ)

C=C   Stretch 1675 to 1600 cm^{-1} (5.95–6.25 μ), often weak.

Conjugation moves C=C stretch to the right.

Symmetrically substituted bonds, for example, 2,3-dimethyl-2-butene, do not absorb in the infrared region (no dipole change). Highly substituted double bonds are often vanishingly weak in absorption.

The spectrum of styrene is shown in Figure IR–5. The spectrum of cyclohexene is shown in Experiment 15.

Aromatic Rings

=C—H   Stretch is always to the left of 3000 cm^{-1} (3.33 μ).

CH   Out-of-plane (oop) bending at 900 to 690 cm^{-1} (11.0–14.5 μ)

The CH out-of-plane absorptions often allow one to determine the type of ring substitution by their numbers, intensities, and positions. The correlation chart in Figure IR–6A indicates the positions of these bands.

The patterns are generally reliable—most particularly reliable for rings with alkyl substituents, least for polar substituents.

Ring Absorptions (C=C). There are often four sharp absorptions that occur in pairs at 1600 cm^{-1} (6.25 μ) and 1450 cm^{-1} (6.90 μ) and are characteristic of an aromatic ring. See, for example, the spectra of anisole (Figure IR–10), benzonitrile (Figure IR–13) and methyl benzoate (Figure IR–17).

There are many weak combination and overtone absorptions that appear between 2000 cm^{-1} and 1667 cm^{-1} (5–6 μ). The relative shapes and numbers of these peaks can be used to tell whether an aromatic ring is monosubstituted or di-, tri-, tetra-,

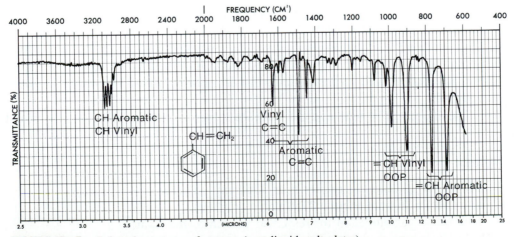

FIGURE IR–5. Infrared spectrum of styrene (neat liquid, salt plates)

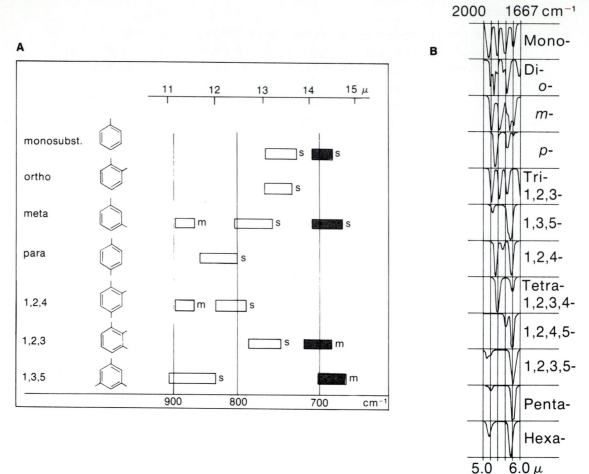

FIGURE IR–6. A. The C—H out-of-plane bending vibrations for substituted benzenoid compounds. B. The 2000–1667 cm^{-1} (5–6 μ) region for substituted benzenoid compounds. (From John R. Dyer, *Applications of Absorption Spectroscopy of Organic Compounds*. Englewood Cliffs, NJ: Prentice-Hall, 1965.)

penta-, or hexasubstituted. Positional isomers can also be distinguished. Since the absorptions are weak, these bands are best observed by using neat liquids or concentrated solutions. If the compound has a high-frequency carbonyl group, this absorption overlaps the weak overtone bands, so that no useful information can be obtained from analyzing this region. The various patterns that are obtained in this region are shown in Figure IR–6B.

The spectra of styrene and *o*-dichlorobenzene are shown in Figures IR–5 and IR–7.

Alkynes

≡C—H	Stretch is usually near 3300 cm^{-1} (3.0 μ).
C≡C	Stretch is near 2150 cm^{-1} (4.65 μ).
	Conjugation moves C≡C stretch to the right.

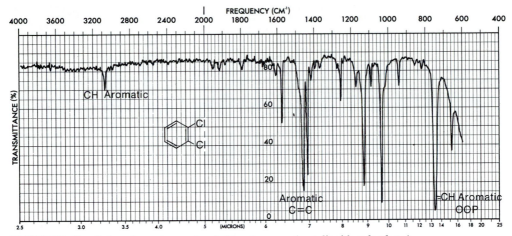

FIGURE IR—7. Infrared spectrum of *o*-dichlorobenzene (neat liquid, salt plates)

Disubstituted or symmetrically substituted triple bonds give either no absorption or weak absorption.

The spectrum of propargyl alcohol is shown in Figure IR–8.

Alcohols and Phenols

O—H Stretch is a sharp peak at 3650–3600 cm^{-1} (2.74–2.78 μ) if no hydrogen bonding takes place. (This is usually only observed in dilute solutions.)

If there is hydrogen bonding (usual in neat or concentrated solutions), the absorption is **broad** and occurs more to the right at 3500–3200 cm^{-1} (2.85–3.12 μ), sometimes overlapping C—H stretch absorptions.

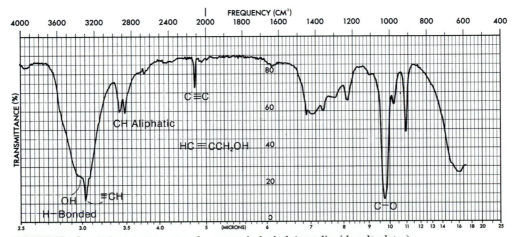

FIGURE IR—8. Infrared spectrum of propargyl alcohol (neat liquid, salt plates)

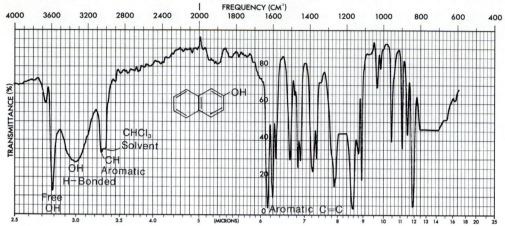

FIGURE IR–9. Infrared spectrum of 2-naphthol, showing both free and hydrogen-bonded OH (CHCl₃ solution)

C—O Stretch is usually in the range 1300–1000 cm⁻¹ (7.7–10 μ).
 Phenols are like alcohols. The 2-naphthol shown in Figure IR–9 has some molecules hydrogen-bonded and some free. The spectrum of cyclohexanol is given in Experiment 15. This alcohol, which was determined neat, would also have had a free OH spike to the left of its hydrogen-bonded band if it had been determined in dilute solution. The solution spectra of borneol and isoborneol are shown in Experiment 22.

Ethers

C—O The most prominent band is due to C—O stretch at 1300–1000 cm⁻¹ (7.7–10.0 μ). Absence of C=O and O—H bands is required to be sure C—O stretch is not due to an alcohol or ester. Phenyl and vinyl ethers are found in the left portion of the range, aliphatic ethers to the right. (Conjugation with the oxygen moves the absorption to the left.)

The spectrum of anisole is shown in Figure IR–10.

Amines

N—H Stretch occurs in the range of 3500–3300 cm⁻¹ (2.86–3.03 μ).
 Primary amines have **two** bands typically 30 cm⁻¹ (0.03 μ) apart.
 Secondary amines have one band, often vanishingly weak.
 Tertiary amines have no NH stretch.

C—N Stretch is weak and occurs in the range of 1350–1000 cm⁻¹ (7.4–10 μ).

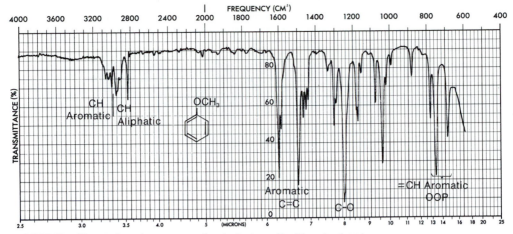

FIGURE IR-10. Infrared spectrum of anisole (neat liquid, salt plates)

N—H Scissoring mode occurs in the range of 1640–1560 cm⁻¹ (6.1–6.4 μ) (broad).

An out-of-plane bending absorption can sometimes be observed at about 800 cm⁻¹ (12.5 μ).

The spectrum of *n*-butylamine is shown in Figure IR–11.

Nitro Compounds

N=O Stretch is usually two strong bands at 1600–1500 cm⁻¹ (6.25–6.67 μ) and 1390–1300 cm⁻¹ (7.2–7.7 μ).

The spectrum of nitrobenzene is shown in Figure IR–12.

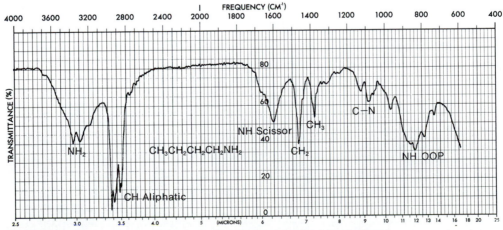

FIGURE IR-11. Infrared spectrum of *n*-butylamine (neat liquid, salt plates)

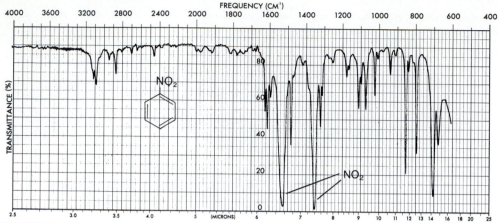

FIGURE IR–12.   Infrared spectrum of nitrobenzene, neat

Nitriles

C≡N   Stretch is a sharp absorption near 2250 cm^{-1} (4.5 μ).
Conjugation with double bonds or aromatic rings moves the absorption to the right.

The spectrum of benzonitrile is shown in Figure IR–13.

Carbonyl Compounds

The carbonyl group is one of the most strongly absorbing groups in the infrared region of the spectrum. This is mainly due to its large dipole moment. It absorbs in a variety of compounds (aldehydes, ketones, acids, esters, amides, anhydrides, and so on) in the

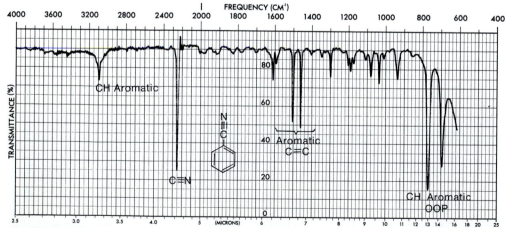

FIGURE IR–13.   Infrared spectrum of benzonitrile (neat liquid, salt plates)

5.52	5.68	5.76	5.80	5.83	5.85	5.92	μ
1810	1760	1735	1725	1715	1710	1690	cm^{-1}
Anhydride (Band 1)		Esters		Ketones		Amides	
	Anhydride (Band 2)		Aldehydes		Carboxylic acids		

FIGURE IR–14. Normal values (± 10 cm^{-1}) for various types of carbonyl groups

range of 1850–1650 cm^{-1} (5.41–6.06 μ). In Figure IR–14, the normal values for the various types of carbonyl groups are compared. In the sections that follow, each type will be examined separately.

Aldehydes

C=O Stretch at approximately 1725 cm^{-1} (5.80 μ) is normal.
Aldehydes **seldom** absorb to the left of this value.
Conjugation moves the absorption to the right.

C—H Stretch, aldehyde hydrogen (—CHO), consists of **weak** bands at about 2750 cm^{-1} (3.65 μ) and 2850 cm^{-1} (3.50 μ). Note that the CH stretch in alkyl chains does not usually extend this far to the right.

The spectrum of nonanal is shown in Figure IR–15. In addition, the spectrum of benzaldehyde is shown in Experiment 28.

Ketones

C=O Stretch at approximately 1715 cm^{-1} (5.83 μ) is normal.
Conjugation moves the absorption to the right.
Ring strain moves the absorption to the left in cyclic ketones.

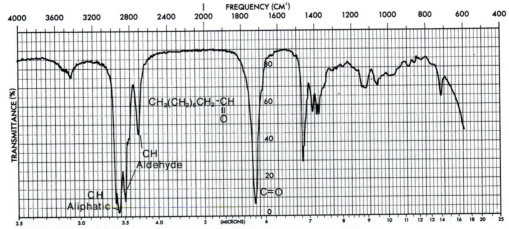

FIGURE IR–15. Infrared spectrum of nonanal (neat liquid, salt plates)

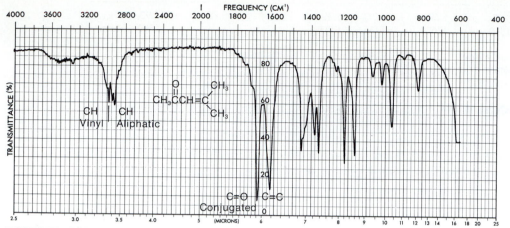

FIGURE IR–16. Infrared spectrum of mesityl oxide (neat liquid, salt plates)

The spectra of methyl isopropyl ketone and mesityl oxide are shown in Figures IR–3 and IR–16. The spectrum of camphor is shown in Experiment 22.

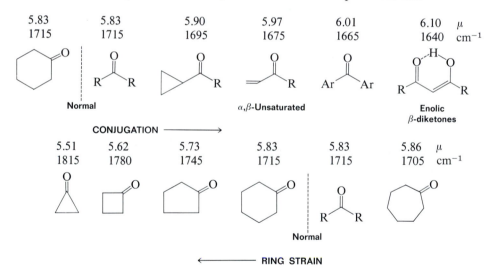

Acids

O—H   Stretch, usually **very broad** (strongly hydrogen-bonded) at 3300–2500 cm^{-1} (3.0–4.0 μ), often interferes with C—H absorptions.

C=O   Stretch, broad, 1730–1700 cm^{-1} (5.8–5.9 μ)
        Conjugation moves the absorption to the right.

C—O   Stretch, in range of 1320–1210 cm^{-1} (7.6–8.3 μ), strong.

The spectrum of benzoic acid is shown in Experiment 27B.

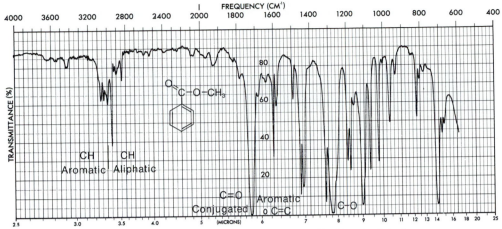

FIGURE IR-17.   Infrared spectrum of methyl benzoate (neat liquid, salt plates)

Esters (R—C—OR')

C=O   Stretch occurs at about 1735 cm^{-1} ($5.76\ \mu$) in normal esters.
  (a) Conjugation in the R part moves the absorption to the right.
  (b) Conjugation with the O in the R' part moves the absorption to the left.
  (c) Ring strain (lactones) moves the absorption to the left.

C—O   Stretch, two bands or more, one stronger than the others, is in the range of 1300–1000 cm^{-1} (7.69–$10.0\ \mu$).

The spectrum of methyl benzoate is shown in Figure IR–17. The spectra of isopentyl acetate and methyl salicylate are shown in Experiments 6 and 43.

Amides

C=O   Stretch is at approximately 1670–1640 cm^{-1} (6.0–$6.1\ \mu$).
  Conjugation and ring size (lactams) have the usual effects.

N—H   Stretch (if monosubstituted or unsubstituted) 3500–3100 cm^{-1} (2.85–$3.25\ \mu$)
  Unsubstituted amides have two bands (—NH$_2$) in this region.

N—H   Bending around 1640–1550 cm^{-1} (6.10–$6.45\ \mu$)

The spectrum of benzamide is shown in Figure IR–18.

Anhydrides

C=O   Stretch always has **two** bands: 1830–1800 cm^{-1} (5.46–$5.56\ \mu$) and 1775–1740 cm^{-1} (5.63–$5.75\ \mu$).

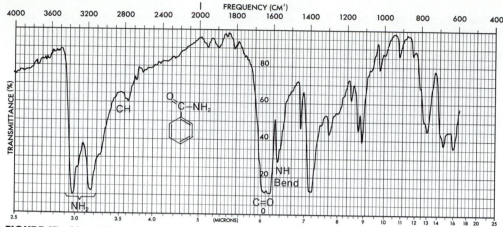

FIGURE IR–18. Infrared spectrum of benzamide (solid phase, KBr)

Unsaturation moves the absorptions to the right.
Ring strain (cyclic anhydrides) moves the absorptions to the left.

C—O Stretch is at 1300–900 cm^{-1} (7.7–11 μ). The spectrum of *cis*-norbornane-5,6-*endo*-dicarboxylic anhydride is shown in Experiment 49.

Halides

It is often difficult to determine either the presence or the absence of a halide in a compound by infrared spectroscopy. The absorption bands cannot be relied on, especially if the spectrum is being determined with the compound dissolved in CCl_4 or $CHCl_3$ solution.

C—F Stretch, 1350–960 cm^{-1} (7.41–10.4 μ)
C—Cl Stretch, 850–500 cm^{-1} (11.8–20.0 μ)
C—Br Stretch, to the right of 667 cm^{-1} (15.0 μ)
C—I Stretch, to the right of 667 cm^{-1} (15.0 μ)

The spectra of carbon tetrachloride and chloroform are shown in Technique 18, Figures 18–6 and 18–7, p 780.

REFERENCES

Bellamy, L. J. *The Infrared Spectra of Complex Molecules.* 3rd ed. New York: Methuen, 1975.

Dyer, J. R. *Applications of Absorption Spectroscopy of Organic Compounds.* Englewood Cliffs, NJ: Prentice-Hall, 1965.

Nakanishi, K., and Soloman, P. H. *Infrared Absorption Spectroscopy.* 2nd ed. San Francisco: Holden-Day, 1977.

Pasto, D. J., and Johnson, C. R. *Laboratory Text for Organic Chemistry.* Englewood Cliffs, NJ: Prentice-Hall, 1974.

Pavia, D. L., Lampman, G. M., and Kriz, G. S., Jr. *Introduction to Spectroscopy.* Philadelphia: W. B. Saunders, 1979.

Silverstein, R. M., Bassler, G. C., and Morrill, T. C. *Spectrophotometric Identification of Organic Compounds.* 4th ed. New York: John Wiley & Sons, 1981.

Appendix 4
NUCLEAR MAGNETIC RESONANCE SPECTROSCOPY

NMR.1 THE RESONANCE PHENOMENON

Nuclear magnetic resonance (NMR) spectroscopy is an instrumental technique that allows the number, type, and relative positions of certain atoms in a molecule to be determined. This type of spectroscopy applies only to those atoms that have nuclear magnetic moments because of their nuclear spin properties. Although many atoms meet this requirement, hydrogen atoms (^1_1H) are of the greatest interest to the organic chemist. Atoms of the ordinary isotopes of carbon ($^{12}_6C$) and oxygen ($^{16}_8O$) do not have nuclear magnetic moments, and ordinary nitrogen atoms ($^{14}_7N$), although they do have magnetic moments, generally fail to show typical NMR behavior for other reasons. The same is true of the halogen atoms, except for fluorine ($^{19}_9F$), which does show active NMR behavior. Atoms other than hydrogen will not be considered here.

Nuclei of NMR-active atoms placed in a magnetic field can be thought of as tiny bar magnets. In hydrogen, which has two allowed nuclear spin states ($+\frac{1}{2}$ and $-\frac{1}{2}$), either the nuclear magnets of individual atoms can be aligned with the magnetic field (spin $+\frac{1}{2}$), or they can be opposed to it (spin $-\frac{1}{2}$). A slight majority of the nuclei will be aligned with the field, as this spin orientation constitutes a slightly lower-energy spin state. If radiofrequency waves of the appropriate energy are supplied, nuclei aligned with the field can absorb this radiation and reverse their direction of spin, or become reoriented so that the nuclear magnet opposes the applied magnetic field (Figure NMR–1).

The frequency of radiation required to induce spin conversion is a direct function of the strength of the applied magnetic field. When a spinning hydrogen nucleus is placed in a magnetic field, the nucleus begins to precess with angular frequency ω, much like a child's toy top. This precessional motion is depicted in Figure NMR–2. The angular frequency of nuclear precession ω increases as the strength of the applied magnetic field is increased. The radiation that must be supplied to induce spin conversion in a hydrogen nucleus of spin $+\frac{1}{2}$ must have a frequency that just matches the

FIGURE NMR–1. The NMR absorption process

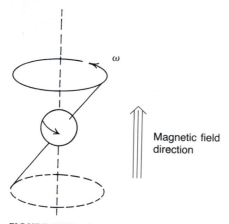

FIGURE NMR–2. Precessional motion of a spinning nucleus in an applied magnetic field

angular precessional frequency ω. This is called the resonance condition, and spin conversion is said to be a resonance process.

For the average proton (hydrogen atom), if a magnetic field of approximately 14,000 Gauss is applied, radiofrequency radiation of 60 MHz (60,000,000 cycles per second) is required to induce a spin transition. Fortunately, the magnetic-field strength required to induce the various protons in a molecule to absorb 60-MHz radiation varies from proton to proton within the molecule and is a sensitive function of the immediate **electronic** environment of each proton. The typical proton nuclear magnetic resonance spectrometer supplies a basic radiofrequency radiation of 60 MHz to the sample being

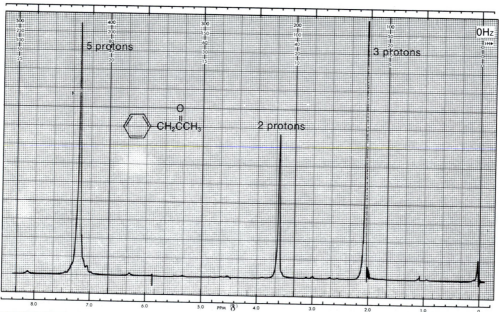

FIGURE NMR–3. Nuclear magnetic resonance spectrum of phenylacetone (the absorption peak at the far right is caused by the added reference substance TMS)

measured and **increases** the strength of the applied magnetic field over a range of several parts per million from the basic field strength. As the field increases, various protons come into resonance (absorb 60-MHz energy), and a resonance signal is generated for each proton. An NMR spectrum is a plot of the strength of the magnetic field versus the intensity of the absorptions. A typical NMR spectrum is shown in Figure NMR–3.

NMR.2 THE CHEMICAL SHIFT

The differences in the applied field strengths at which the various protons in a molecule absorb 60-MHz radiation are extremely small. The different absorption positions amount to a difference of only a few parts per million (ppm) in the magnetic field strength. Since it is experimentally difficult to measure the precise field strength at which each proton absorbs to less than one part in a million, a technique has been developed whereby the **difference** between two absorption positions is measured directly. To achieve this measurement, a standard reference substance is used and the positions of the absorptions of all other protons are measured relative to the values for the reference substance. The reference substance that has been universally accepted is tetramethylsilane, $(CH_3)_4Si$, which is also called TMS. The proton resonances in this molecule appear at a higher field strength than the proton resonances in most all other molecules do, and all the protons of TMS have resonance at the same field strength.

To give the position of absorption of a proton a quantitative measurement, a parameter called the **chemical shift** (δ) has been defined. One δ unit corresponds to a 1-ppm change in the magnetic field strength. To determine the chemical shift value for the various protons in a molecule, the operator determines an NMR spectrum of the molecule with a small quantity of TMS added directly to the sample. That is, both spectra are determined **simultaneously.** The TMS absorption is adjusted to correspond to the $\delta = 0$ position on the recording chart, which is calibrated in δ units, and the δ values of the absorption peaks for all other protons can be read directly from the chart.

Since the NMR spectrometer increases the magnetic field as the pen moves from left to right on the chart, the TMS absorption appears at the extreme right edge of the spectrum ($\delta = 0$) or at the **upfield** end of the spectrum. The chart is calibrated in δ units (or ppm), and most other protons absorb at a lower field strength (or **downfield**) from TMS.

Because the frequency at which a proton precesses, and hence the frequency at which it absorbs radiation, is directly proportional to the strength of the applied magnetic field, a second method of measuring an NMR spectrum is possible. One could hold the magnetic field strength constant and vary the frequency of the radiofrequency radiation supplied. Thus, a given proton could be induced to absorb **either** by increasing the field strength, as described earlier, or alternatively, by decreasing the frequency of the radiofrequency oscillator. A 1-ppm decrease in the frequency of the oscillator would have the same effect as a 1-ppm increase in the magnetic field strength. For reasons of instrumental design, it is simpler to vary the strength of the magnetic field than to vary the frequency of the oscillator. Most instruments operate on the former principle. Nevertheless, the recording chart is calibrated not only in δ units but in Hz as well (1 ppm = 60 Hz when the frequency is 60 MHz), and the chemical shift is customarily defined and computed using Hertz rather than Gauss:

$$\delta = \text{chemical shift} = \frac{(\text{observed shift from TMS, in Hz})}{(60 \text{ MHz})} = \frac{\text{Hz}}{\text{MHz}} = \text{ppm}$$

Although the equation defines the chemical shift for a spectrometer operating at 14,100 Gauss and 60 MHz, the chemical shift value that is calculated is **independent** of the field strength. For instance, at 23,500 Gauss the oscillator frequency would have to be 100 MHz. Although the observed shifts from TMS (in Hz) would be larger at this field strength, the divisor of the equation would be 100 MHz, instead of 60 MHz, and δ would turn out to be identical under either set of conditions.

NMR.3   CHEMICAL EQUIVALENCE—INTEGRALS

All the protons in a molecule that are in chemically identical environments often exhibit the same chemical shift. Thus, all the protons in tetramethylsilane (TMS) or all the protons in benzene, cyclopentane, or acetone have their own respective resonance values all at the same δ value. Each compound gives rise to a single absorption peak in

Molecules giving rise to one NMR absorption peak—all protons chemically equivalent

Molecules giving rise to two NMR absorption peaks—two different sets of chemically equivalent protons

its NMR spectrum. The protons are said to be **chemically equivalent.** On the other hand, molecules that have sets of protons that are chemically distinct from one another may give rise to an absorption peak from each set.

The NMR spectrum given in Figure NMR–3 is that of phenylacetone, a compound having **three** chemically distinct types of protons:

One can immediately see that the NMR spectrum furnishes a valuable type of information on this basis alone. In fact, the NMR spectrum can not only distinguish how many different types of protons a molecule has but also can reveal **how many** of each different type are contained within the molecule.

In the NMR spectrum, the area under each peak is proportional to the number of hydrogens generating that peak. Hence, in the above compound, the area ratio of the three peaks is $5:2:3$, the same as the ratio of the numbers of each type of hydrogen. The NMR spectrometer can electronically "integrate" the area under each peak. It does this by tracing over each peak a vertically rising line, which rises in height by an amount proportional to the area under the peak. Shown in Figure NMR–4 is an NMR spectrum of benzyl acetate, with each of the peaks integrated in this way.

It is important to note that the height of the integral line does not give the absolute number of hydrogens; it gives the **relative** numbers of each type of hydrogen. For a given integral to be of any use, there must be a second integral to which it is referred. The benzyl acetate case gives a good example of this. The first integral rises for 55.5 divisions on the chart paper, the second for 22.0 divisions, and the third for 32.5 divisions. These numbers are relative and give the **ratios** of the various types of protons. One can find these ratios by dividing each of the large numbers by the smallest number:

$$\frac{55.5 \text{ div}}{22.0 \text{ div}} = 2.52 \qquad \frac{22.0 \text{ div}}{22.0 \text{ div}} = 1.00 \qquad \frac{32.5 \text{ div}}{22.0 \text{ div}} = 1.48$$

Thus, the number ratio of the protons of each type is $2.52:1.00:1.48$. If we assume that the peak at 5.1δ is really caused by two hydrogens, and if we assume that the

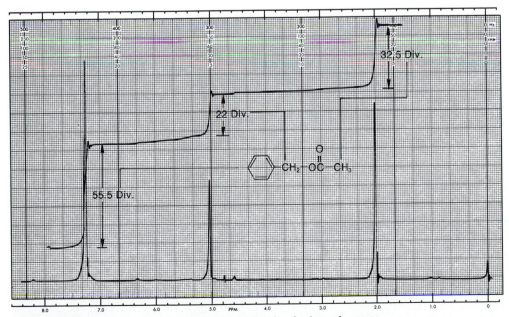

FIGURE NMR–4.   Determination of the integral ratios for benzyl acetate

integrals are slightly in error (this can be as much as 10%), then we can arrive at the true ratios by multiplying each figure by two and rounding off; we then get $5:2:3$. Clearly the peak at 7.3δ, which integrates for 5, arises from the resonance of the aromatic ring protons, and the peak at 2.0δ, which integrates for 3, is caused by the methyl protons. The two-proton resonance at 5.1δ arises from the benzyl protons. Notice then that the integrals give the simplest ratios, but not necessarily the true ratios, of the number of protons of each type.

NMR.4   CHEMICAL ENVIRONMENT AND CHEMICAL SHIFT

If the resonance frequencies of all protons in a molecule were the same, NMR would be of little use to the organic chemist. However, not only do different types of protons have different chemical shifts, but they also have a value of chemical shift that characterizes the type of proton they represent. Every type of proton has only a limited range of δ values over which it gives resonance. Hence, the numerical value of the chemical shift for a proton indicates the **type of proton** originating the signal, just as the infrared frequency suggests the type of bond or functional group. Notice, for instance, that the aromatic protons of both phenylacetone (Figure NMR–3) and benzyl acetate (Figure NMR–4) have resonance near 7.3δ and that both methyl groups attached directly to a carbonyl group have a resonance of approximately 2.1δ. Aromatic protons characteristically have resonance near $7–8\delta$, and acetyl groups (the methyl protons) have their resonance near 2δ. These values of chemical shift are diagnostic. Notice also how the resonance of the benzyl ($-CH_2-$) protons comes at a higher value of chemical shift (5.1δ) in benzyl acetate than in phenylacetone (3.6δ). Being attached to the electronegative element, oxygen, these protons are more deshielded (see Section NMR.5) than the protons in phenylacetone. A trained chemist would have readily recognized the probable presence of the oxygen by the chemical shift shown by these protons.

It is important to learn the ranges of chemical shifts over which the most common types of protons have resonance. Figure NMR–5 is a correlation chart that contains the most essential and frequently encountered types of protons. For the beginner, it is often difficult to memorize a large body of numbers relating to chemical shifts and proton types. One need actually do this only crudely. It is more important to "get a feel" for the regions and the types of protons than to know a string of factual numbers.

The values of chemical shift given in Figure NMR–5 can be easily understood in terms of two factors: local diamagnetic shielding and anisotropy. These two factors are discussed in Sections NMR.5 and NMR.6.

NMR.5   LOCAL DIAMAGNETIC SHIELDING

The trend of chemical shifts that is easiest to explain is that involving electronegative elements substituted on the same carbon to which the protons of interest are attached. The chemical shift simply increases as the electronegativity of the attached element increases. This is illustrated in Table NMR–1 for several compounds of the type CH_3X.

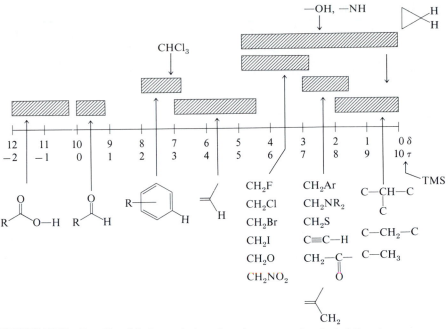

FIGURE NMR–5. Simplified correlation chart for proton chemical shift values

Multiple substituents have a stronger effect than a single substituent. The influence of the substituent, an electronegative element having little effect on protons that are more than three carbons away, drops off rapidly with distance. These effects are illustrated in Table NMR–2.

Electronegative substituents attached to a carbon atom, because of their electron-withdrawing effects, reduce the valence electron density around the protons attached to that carbon. These electrons **shield** the proton from the applied magnetic field. This effect, called **local diamagnetic shielding,** occurs because the applied magnetic field induces the valence electrons to circulate and thus to generate an induced magnetic field, which **opposes** the applied field. This is illustrated in Figure NMR–6. Electronegative substituents on carbon reduce the local diamagnetic shielding in the vicinity of the attached protons because they reduce the electron density around those protons. Substituents that produce this effect are said to **deshield** the proton. The greater the electronegativity of the substituent, the more the deshielding of the protons and, hence, the greater the chemical shift of those protons.

TABLE NMR–1. Dependence of Chemical Shift of CH₃X on the Element X

Compound CH$_3$X	CH$_3$F	CH$_3$OH	CH$_3$Cl	CH$_3$Br	CH$_3$I	CH$_4$	(CH$_3$)$_4$Si
Element X	F	O	Cl	Br	I	H	Si
Electronegativity of X	4.0	3.5	3.1	2.8	2.5	2.1	1.8
Chemical shift (δ)	4.26	3.40	3.05	2.68	2.16	0.23	0

TABLE NMR–2.   Substitution Effects

	CHCl$_3$	CH$_2$Cl$_2$	CH$_3$Cl	—C$\underline{H}_2$Br	—C$\underline{H}_2$—CH$_2$Br	—C$\underline{H}_2$—CH$_2$CH$_2$Br
δ	7.27	5.30	3.05	3.30	1.69	1.25

NMR.6   ANISOTROPY

Figure NMR–5 clearly shows that several types of protons have chemical shifts not easily explained by simple consideration of the electronegativity of the attached groups. Consider, for instance, the protons of benzene or other aromatic systems. Aryl protons generally have a chemical shift that is as large as that for the proton of chloroform! Alkenes, alkynes, and aldehydes also have protons whose resonance values are not in line with the expected magnitude of any electron-withdrawing effects. In each of these cases, the effect is due to the presence of an unsaturated system (π electrons) in the vicinity of the proton in question. In benzene, for example, when the π electrons in the aromatic ring system are placed in a magnetic field, they are induced to circulate around the ring. This circulation is called a **ring current.** Moving electrons (the ring current) generate a magnetic field much like that generated in a loop of wire through which a current is induced to flow. The magnetic field covers a spatial volume large enough to influence the shielding of the benzene hydrogens. This is illustrated in Figure NMR–7. The benzene hydrogens are deshielded by the **diamagnetic anisotropy** of the ring. An applied magnetic field is nonuniform (anisotropic) in the vicinity of a benzene molecule because of the labile electrons in the ring that interact with the applied field. Thus, a proton attached to a benzene ring is influenced by **three** magnetic fields: the strong magnetic field applied by the electromagnets of the NMR spectrometer and two

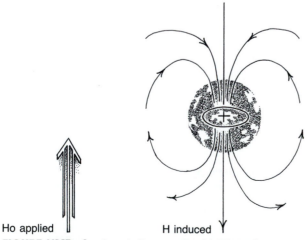

Ho applied   H induced

FIGURE NMR–6.   Local diamagnetic shielding of a proton due to its valence electrons

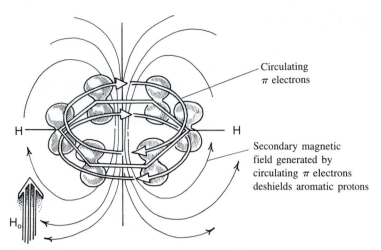

FIGURE NMR–7. Diamagnetic anisotropy in benzene

weaker fields, one due to the usual shielding by the valence electrons around the proton and the other due to the anisotropy generated by the ring system electrons. It is this anisotropic effect that gives the benzene protons a greater chemical shift than is expected. These protons just happen to lie in a **deshielding** region of this anisotropic field. If a proton were placed in the center of the ring rather than on its periphery, the proton would be shielded, since the field lines would have the opposite direction.

All groups in a molecule that have π electrons generate secondary anisotropic fields. In acetylene, the magnetic field generated by induced circulation of π electrons has a geometry such that the acetylene hydrogens are **shielded.** Hence, acetylenic hydrogens come at a higher field than expected. The shielding and deshielding regions due to the various π electron functional groups have characteristic shapes and directions; they are illustrated in Figure NMR–8. Protons falling within the cones are shielded, and those falling outside the conical areas are deshielded. Since the magnitude of the anisotropic field diminishes with distance, beyond a certain distance anisotropy has essentially no effect.

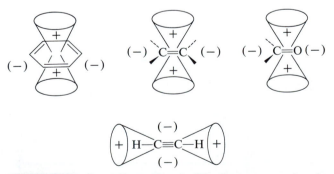

FIGURE NMR–8. Anisotropy caused by the presence of π electrons in some common multiple bond systems

NMR.7   SPIN-SPIN SPLITTING (*n* + 1 RULE)

We have already considered how the chemical shift and the integral (peak area) can give information about the number and type of hydrogens contained in a molecule. A third type of information available from the NMR spectrum is derived from spin-spin splitting. Even in simple molecules, one finds that each type of proton rarely gives a single resonance peak. For instance, in 1,1,2-trichloroethane there are two chemically distinct types of hydrogen:

$$Cl-\overset{\overset{\textstyle (H)}{|}}{\underset{\underset{\textstyle Cl}{|}}{C}}-(CH_2)-Cl$$

From information given thus far, one would predict **two** resonance peaks in the NMR spectrum of 1,1,2-trichloroethane with an area ratio (integral ratio) 2:1. In fact, the NMR spectrum of this compound has **five** peaks. A group of three peaks (called a triplet) exists at 5.77δ and a group of two peaks (called a doublet) at 3.95δ. The spectrum is shown in Figure NMR–9. The methine (CH) resonance (5.77δ) is split into a triplet, and the methylene resonance (3.95δ) is split into a doublet. The area under the three triplet peaks is **one,** relative to an area of **two** under the two doublet peaks.

This phenomenon is called spin-spin splitting. Empirically, spin-spin splitting can be explained by the "*n* + 1 rule." Each type of proton "senses" the number of equivalent protons (*n*) on the carbon atom or atoms next to the one to which it is bonded, and its resonance peak is split into *n* + 1 components.

Let's examine the case at hand, 1,1,2-trichloroethane, using the so-called *n* + 1 rule. First, the lone methine hydrogen is situated next to a carbon bearing two methylene protons. According to the rule, it has two equivalent neighbors (*n* = 2) and is split

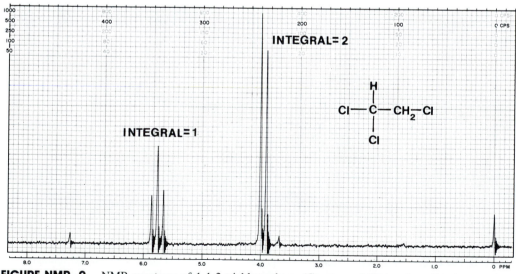

FIGURE NMR–9.   NMR spectrum of 1,1,2-trichloroethane (Courtesy of Varian Associates)

into $n + 1 = 3$ peaks (a triplet). The methylene protons are situated next to a carbon bearing only one methine hydrogen. According to the rule, they have one neighbor ($n = 1$) and are split into $n + 1 = 2$ peaks (a doublet).

Two neighbors give a triplet ($n + 1 = 3$) (area = 1)

One neighbor gives a doublet ($n + 1 = 2$) (area = 2)

Equivalent protons behave as a group

The spectrum of 1,1,2-trichloroethane can be explained easily by the interaction, or coupling, of the spins of protons on adjacent carbon atoms. The position of absorption of proton H_a may be affected by the spins of protons H_b and H_c attached to the neighboring (adjacent) carbon atom. If the spins of these protons are aligned with the applied magnetic field, the small magnetic field generated by their nuclear spin properties will augment the strength of the field experienced by the first-mentioned proton H_a. The proton H_a will thus be **deshielded.** If the spins of H_b and H_c are opposed to the applied field, they will decrease the field experienced by proton H_a. It will then be **shielded.** In each of these situations, the absorption position of H_a will be altered. Among the many molecules in the solution, one will find all the various possible spin combinations for H_b and H_c; hence, the NMR spectrum of the molecular solution will give **three** absorption peaks (a triplet) for H_a since H_b and H_c have three different possible spin combinations (Figure NMR–10). By a similar analysis, it can be seen that protons H_b and H_c should appear as a doublet.

Some common splitting patterns that can be predicted by the $n + 1$ rule and that are frequently observed in a number of molecules are shown in Figure NMR–11. Notice particularly the last entry, where **both** methyl groups (6 protons in all) function as a unit and split the methine proton into a septet ($6 + 1 = 7$).

NMR.8 THE COUPLING CONSTANT

The quantitative amount of spin-spin interaction between two protons can be defined by the coupling constant. The spacing between the component peaks in a simple multiplet

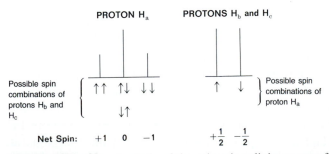

FIGURE NMR–10. Analysis of the spin-spin splitting pattern for 1,1,2-trichloroethane

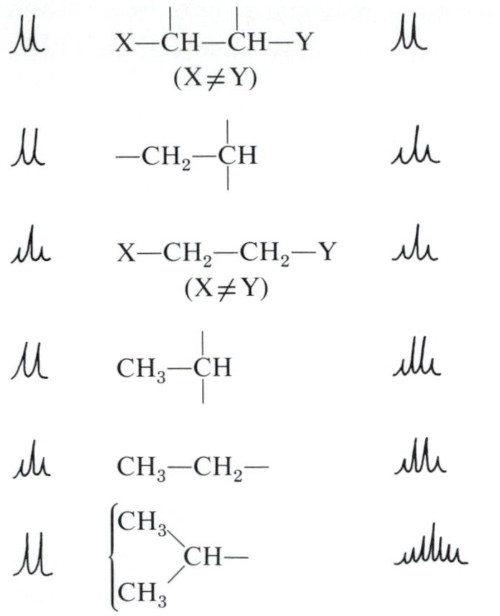

X—CH—CH—Y
(X ≠ Y)

—CH₂—CH

X—CH₂—CH₂—Y
(X ≠ Y)

CH₃—CH

CH₃—CH₂—

CH₃ \ CH—
CH₃ /

FIGURE NMR–11. Some common splitting patterns

is called the coupling constant J. This distance is measured on the same scale as the chemical shift and is expressed in Hertz (Hz).

For the interaction of most aliphatic protons in acyclic systems, the magnitudes of the coupling constants are always near 7.5 Hz. See, for instance, the NMR spectrum of 1,1,2-trichloroethane in Figure NMR–9, where the coupling constant is approximately 6 Hz. Different magnitudes of J are found for different types of protons. For instance, the *cis* and *trans* protons substituted on a double bond commonly have values where $J_{trans} \cong 17$ Hz and $J_{cis} \cong 10$ Hz are typical coupling constants. In ordinary compounds, coupling constants may range anywhere from 0–18 Hz. The magnitude of J often provides structural clues. One can usually distinguish, for example, between a *cis* olefin and a *trans* olefin on the basis of the observed coupling constants for the vinyl protons. The approximate values of some representative coupling constants are given in Table NMR–3.

NMR.9 MAGNETIC EQUIVALENCE

In the example of spin-spin splitting in 1,1,2-trichloroethane, one notices that the two protons H_b and H_c, which are attached to the same carbon atom, do not split one another. They behave as an integral group. Actually the two protons H_b and H_c **are** coupled to one another; however, for reasons we cannot explain fully here, protons that are attached to the same carbon and both of which have the **same chemical shift** do not show spin-spin splitting. Another way of stating this is that protons coupled to the same extent to **all** other protons in a molecule do not show spin-spin splitting. Protons that have the same chemical shift and are coupled equivalently to all other protons are **magnetically equivalent** and do not show spin-spin splitting. Thus, in 1,1,2-trichloro-

TABLE NMR–3. Representative Coupling Constants and Approximate Values (Hz)

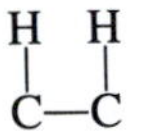 H–C–C (H, H, H)	6–8	ortho	6–10	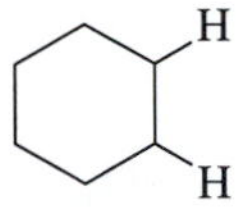 (cyclohexane a,e)	a,a 8–14 / a,e 0–7 / e,e 0–5
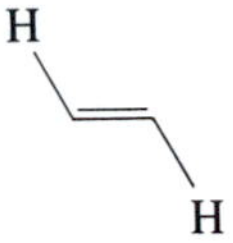 (trans vinyl)	11–18	meta	1–4	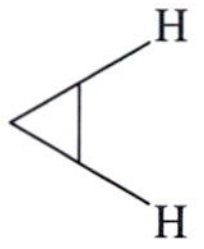 (cyclopropane)	cis 6–12 / trans 4–8
(cis vinyl)	6–15	para	0–2	(epoxide)	cis 2–5 / trans 1–3
(geminal vinyl)	0–5				
(allylic =CH–CH)	4–10	(cyclohexene)	8–11	(cyclopentene)	5–7
H–C≡C–CH	0–3	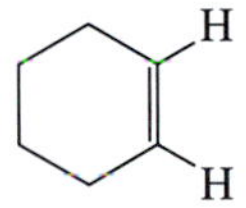		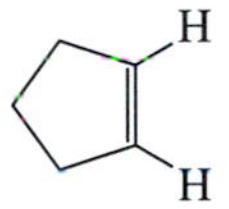	

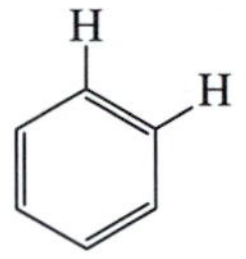

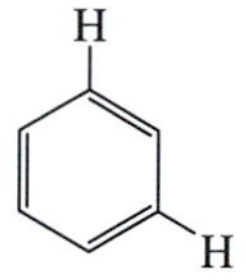

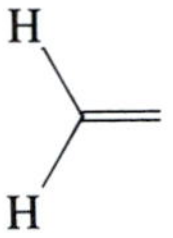

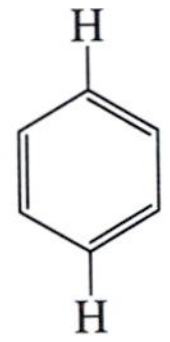

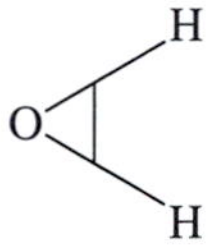

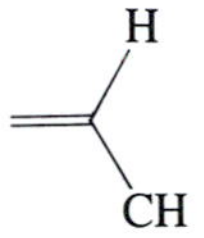

ethane, protons H_b and H_c have the same value of δ and are coupled by the same value of J to proton H_a. They are magnetically equivalent.

It is important to differentiate magnetic equivalence and chemical equivalence. Note the two compounds shown below.

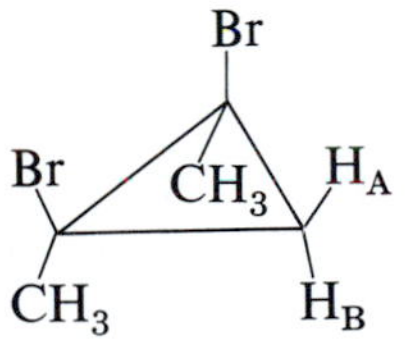

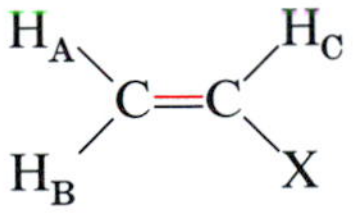

In the cyclopropane compound, the two geminal hydrogens are chemically equivalent; however, they are not magnetically equivalent. Proton H_A is on the same side of the ring as the two halogens. Proton H_B is on the same side of the ring as the two methyl groups. Protons H_A and H_B will have different chemical shifts, will couple to one another, and will show spin-spin splitting. Two doublets will be seen for H_A and H_B. For cyclopropane rings, $J_{geminal}$ is usually around 5 Hz.

Another situation in which protons are chemically equivalent but not magnetically equivalent exists in the vinyl compound. In this example, protons A and B are chemically equivalent but not magnetically equivalent. H_A and H_B have different chemical shifts. In addition, a second distinction can be made between H_A and H_B in this type of compound. Each has a different coupling constant with H_C. The constant J_{AC} is a *cis* coupling constant, and J_{BC} is a *trans* coupling constant. Whenever two protons have different coupling constants relative to a third proton, they are not mag-

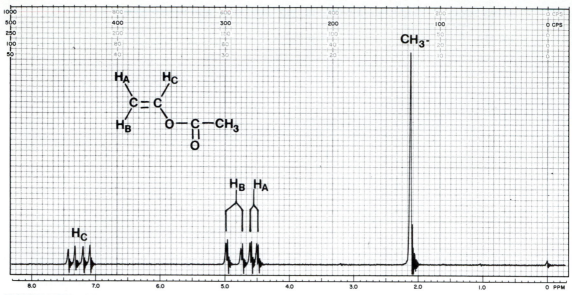

FIGURE NMR–12.   NMR spectrum of vinyl acetate (Courtesy of Varian Associates)

netically equivalent. In the vinyl compound, H_A and H_B do not act as a group to split proton H_C. Each proton acts independently. Thus H_B splits H_C with coupling constant J_{BC} into a doublet, and then H_A splits each of the components of the doublet into doublets with coupling constant J_{AC}. In such a case, the NMR spectrum must be analyzed graphically, splitting by splitting. An NMR spectrum of a vinyl compound is shown in Figure NMR–12. The graphical analysis of the vinyl portion of the NMR spectrum is in Figure NMR–13.

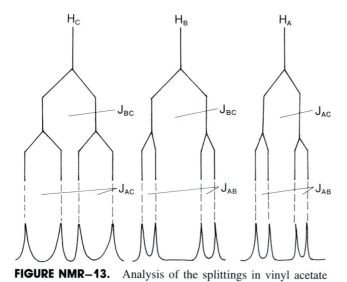

FIGURE NMR–13.   Analysis of the splittings in vinyl acetate

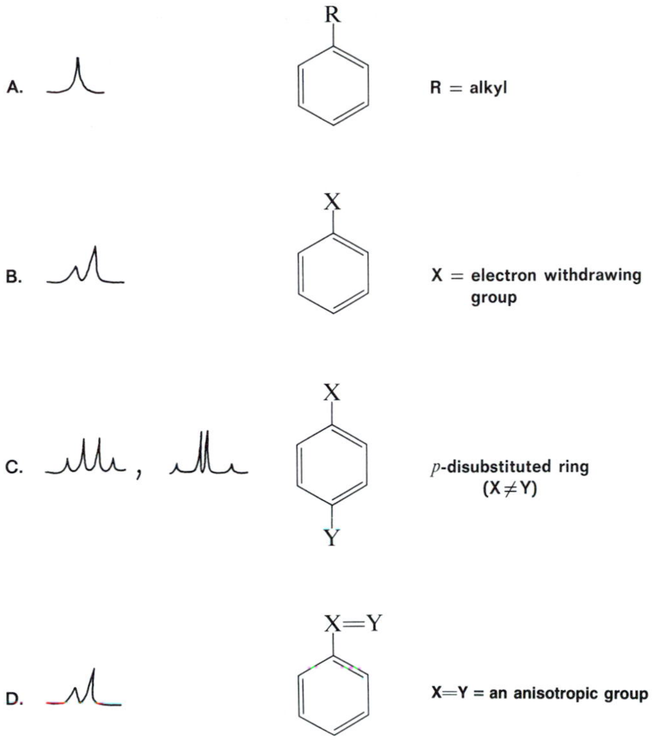

A.

R

R = alkyl

B.

X

X = electron withdrawing group

C.

X

Y

p-disubstituted ring
(X ≠ Y)

D.

X═Y

X═Y = an anisotropic group

FIGURE NMR–14. Some common aromatic patterns

NMR.10 AROMATIC COMPOUNDS

The NMR spectra of protons on aromatic rings are often too complex to explain by simple theories. However, some simple generalizations can be made that are useful in analyzing the aromatic region of the NMR spectrum. First of all, most aromatic protons have resonance near 7.0δ. In monosubstituted rings in which the ring substituent is an alkyl group, all the ring protons often have chemical shifts that are very nearly identical, and the five ring protons may appear as if they gave rise to an overly broad singlet (Figure NMR–14A). If an electronegative group is attached to the ring, all the ring protons are shifted downfield from where they would appear in benzene. However, often the *ortho* protons are shifted more than the others, as they are more affected by the group. This often gives rise to an absorption pattern like that in Figure NMR–14B. In a *para*-disubstituted ring with two substituents X and Y that are identical, all the protons in the ring are chemically and magnetically equivalent, and a singlet is observed. If X is different from Y in electronegativity, however, a pattern like that shown in the left side of Figure NMR–14C is often observed, clearly identifying a *p*-disubstituted ring. If X and Y are more nearly similar, a pattern more like the one on the right is observed. In monosubstituted rings that have a carbonyl group or a double bond attached directly to the ring, a pattern like that in Figure NMR–14D is not uncommon. In this case, the *ortho* protons of the ring are influenced by the anisotropy

TABLE NMR–4.   Typical Ranges for Groups with Variable Chemical Shift

Acids	RCOOH	$10.5-12.0\ \delta$
Phenols	ArOH	$4.0-7.0$
Alcohols	ROH	$0.5-5.0$
Amines	RNH_2	$0.5-5.0$
Amides	$RCONH_2$	$5.0-8.0$
Enols	$CH{=}CH{-}OH$	≥ 15

of the π systems that make up the CO and CC double bonds and are deshielded by them. In other types of substitution, such as *ortho* or *meta,* or polysubstituted ring systems, the patterns may be much more complicated and require an advanced analysis.

NMR.11   PROTONS ATTACHED TO ATOMS OTHER THAN CARBON

Protons attached to atoms other than carbon often have a widely variable range of absorptions. Several of these groups are tabulated in Table NMR–4. In addition, under the usual conditions of determining an NMR spectrum, protons on heteroelements normally do not couple with protons on adjacent carbon atoms to give spin-spin splitting. This is primarily because such protons often exchange very rapidly with those of the solvent medium. The absorption position is variable because these groups also undergo varying degrees of hydrogen bonding in solutions of different concentrations. The amount of hydrogen bonding that occurs with a proton radically affects the valence electron density around that proton and produces correspondingly large changes in the chemical shift. The absorption peaks for protons that have hydrogen bonding or are undergoing exchange are frequently broad relative to other singlets and can often be recognized on that basis. For a different reason, called quadrupole broadening, protons attached to nitrogen atoms often show an extremely broad resonance peak, often almost indistinguishable from the baseline.

NMR.12   SPECTRA AT HIGHER FIELD STRENGTH

Occasionally the 60-MHz spectrum of an organic compound, or a portion of it, is almost undecipherable because the chemical shifts of several groups of protons are all very similar. In these cases all of the proton resonances occur in the same area of the spectrum, and often peaks overlap so extensively that individual peaks and splittings cannot be extracted. One of the ways in which such a situation can be simplified is by the use of a spectrometer that operates at a higher frequency. Although both 60- and 90-MHz instruments are quite common, it is not unusual at a large university or an industrial research center to find instruments with operating frequencies of 100, 220, 300 MHz, or even higher.

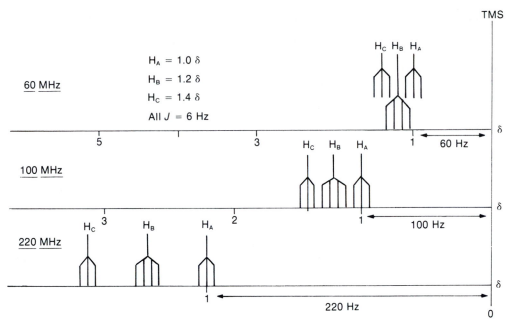

FIGURE NMR–15. A comparison of the spectrum of a compound with overlapping multiplets at 60 MHz, with spectra of the same compound also determined at 100 MHz and 220 MHz. The drawing is to scale.

Although NMR coupling constants are not dependent on the frequency or the field strength of operation of the NMR spectrometer, chemical shifts in Hz are dependent on these parameters. This circumstance can often be used to simplify an otherwise undecipherable spectrum. Suppose, for instance, that a compound contained three multiplets: a quartet and two triplets derived from groups of protons with very similar chemical shifts. At 60 MHz these peaks might overlap, as illustrated in Figure NMR–15, and simply give an unresolved envelope of absorption.

Figure NMR–15 also shows the spectrum of the same compound at two higher field strengths (frequencies). In redetermining the spectrum at higher field strengths, the coupling constants do not change, but the chemical shifts in Hz (not δ) of the proton groups (H_A, H_B, H_C) responsible for the multiplets do increase. It should be noted that at 220 MHz the individual multiplets are cleanly separated and resolved.

NMR.13 CHEMICAL SHIFT REAGENTS

It has been known for some time that interactions between molecules and solvents, such as those due to hydrogen bonding, can cause large changes in the resonance positions of certain types of protons (e.g., hydroxyl and amino). It has also been known that the resonance positions of some groups of protons can be greatly affected by changing from the usual NMR solvents such as CCl_4 and $CDCl_3$ to solvents like benzene, which impose local anisotropic effects on surrounding molecules. In many cases it was found possible to resolve partially overlapping multiplets by such a solvent change. However, the use of chemical shift reagents for this purpose is a more recent innovation, dating from about 1969. Most of these chemical shift reagents are organic complexes of paramagnetic rare earth metals from the lanthanide series of elements. When these metal complexes are added to the compound whose spectrum is being determined, profound shifts in the resonance positions of the various groups of protons are observed. The direction of the shift (upfield or downfield) depends primarily on which metal is being used. Complexes of europium, erbium, thulium, and ytterbium shift resonances to lower field; complexes of cerium, praseodymium, neodymium, samarium, terbium, and holmium generally shift resonances to higher field. The advantage of using such reagents is that shifts similar to those observed at higher field can be induced without the purchase of an expensive higher field instrument.

Of the lanthanides, europium is probably the most commonly used metal. Two of its widely used complexes are *tris*-(dipivalomethanato)europium and *tris*-(6,6,7,7,8,8,8-heptafluoro-2,2-dimethyl-3,5-octanedionato)europium. These are frequently abbreviated $Eu(dpm)_3$ and $Eu(fod)_3$, respectively.

Eu(dpm)₃
or Eu(thd)₃

Eu(fod)₃

These lanthanide complexes produce spectral simplifications in the NMR spectrum of any compound that has a relatively basic pair of electrons (unshared pair) that can coordinate with Eu^{3+}. Typically, aldehydes, ketones, alcohols, thiols, ethers, and amines will all interact:

$$2B: + Eu(dpm)_3 \longrightarrow$$

The amount of shift that a given group of protons will experience depends (1) on the distance separating the metal (Eu^{3+}) and that group of protons, and (2) on the concentration of the shift reagent in the solution. Because of the latter dependence, it is necessary when reporting a lanthanide-shifted spectrum to report the number of mole equivalents of shift reagent used or its molar concentration.

The distance factor is illustrated in the spectra of hexanol, which are given in Figures NMR–16 and NMR–17. In the absence of shift reagent, the normal spectrum

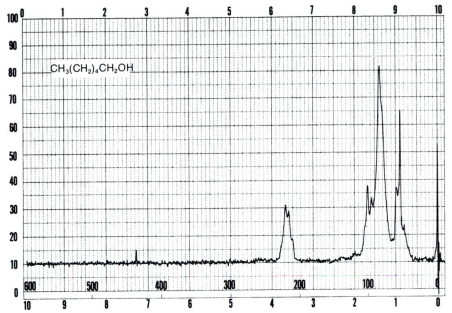

FIGURE NMR–16.   The normal 60-MHz NMR spectrum of hexanol

is obtained (Figure NMR–16). Only the triplet of the terminal methyl group and the triplet of the methylene group next to the hydroxyl are resolved in the spectrum. The other protons (aside from OH) are found together in a broad unresolved group. With shift reagent added (Figure NMR–17), each of the methylene groups is clearly separated and resolved into the proper multiplet structure. The spectrum is first order and simplified; all of the splittings are explained by the $n + 1$ rule.

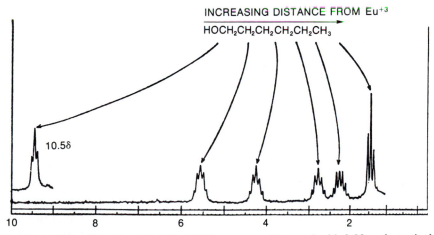

FIGURE NMR–17.   The 100-MHz NMR spectrum of hexanol with 0.29 mole equivalents of $Eu(dpm)_3$ added. (Reprinted with permission from J. K. M. Sanders and D. H. Williams, *Chem. Commun.*, 422 [1970])

One final consequence of the use of a shift reagent should be noted. Notice in Figure NMR–17 that the multiplets are not as nicely resolved into sharp peaks as one usually expects. This is due to the fact that shift reagents cause a small amount of peak broadening. At high shift reagent concentrations this problem becomes serious, but at most useful concentrations the amount of broadening experienced is tolerable.

REFERENCES

Dyer, J. R. *Applications of Absorption Spectroscopy of Organic Compounds*. Englewood Cliffs, NJ: Prentice-Hall, 1965.

Jackman, L. M., and Sternhell, S. *Applications of Nuclear Magnetic Resonance in Organic Chemistry*. 2nd ed. New York: Pergamon Press, 1969.

Pasto, D. J., and Johnson, C. R. *Laboratory Text for Organic Chemistry*. Englewood Cliffs, NJ: Prentice-Hall, 1979.

Paudler, W. W. *Nuclear Magnetic Resonance*. Boston: Allyn and Bacon, 1971.

Pavia, D. L., Lampman, G. M., and Kriz, G. S., Jr. *Introduction to Spectroscopy*. Philadelphia: W. B. Saunders, 1979.

Silverstein, R. M., Bassler, G. C., and Morrill, T. C. *Spectrometric Identification of Organic Compounds*. 4th ed. New York: John Wiley & Sons, 1981.

Appendix 5
CARBON-13 NUCLEAR MAGNETIC RESONANCE SPECTROSCOPY

CMR.1 CARBON-13 NUCLEAR MAGNETIC RESONANCE

Carbon-12, the most abundant isotope of carbon, does not possess spin ($I = 0$); it has both an even atomic number and an even atomic weight. The second principal isotope of carbon, ^{13}C, however, does have the nuclear spin property ($I = \frac{1}{2}$). ^{13}C atom resonances are not easy to observe, due to a combination of two factors. First, the natural abundance of ^{13}C is low; only 1.08% of all carbon atoms are ^{13}C. Second, the magnetic moment (μ) of ^{13}C is low. For these two reasons, the resonances of ^{13}C are about **6000 times weaker** than those of hydrogen. With special Fourier-Transform instrumental techniques, which will not be discussed here, it is possible to observe ^{13}C nuclear magnetic resonance (carbon-13) spectra on samples that contain only the natural abundance of ^{13}C.

The most useful parameter derived from carbon-13 spectra is the chemical shift. Integrals are unreliable and are not necessarily related to the relative numbers of ^{13}C atoms present in the sample. Hydrogens that are attached to ^{13}C atoms cause spin-spin splitting, but spin-spin interaction between adjacent carbon atoms is rare. With its low natural abundance (0.0108), the probability of finding two ^{13}C atoms adjacent to one another is extremely low.

CMR.2 COMPLETELY COUPLED ^{13}C SPECTRA

Figure CMR–1 shows the carbon-13 spectrum of ethyl phenylacetate. Consider first the upper trace shown in the figure. Chemical shifts, just as in proton NMR, are reported by the number of ppm (δ units) that the peak is shifted downfield from TMS. Keep in mind, however, that it is a ^{13}C atom of the methyl group of TMS that is being observed, not the 12 methyl hydrogens. Notice the extent of the scale. While the chemical shifts of protons encompass a range of only about 20 ppm, ^{13}C chemical shifts cover an extremely wide range of up to 200 ppm! Under these circumstances, even adjacent —CH_2— carbons in a long hydrocarbon chain generally have their own distinct resonance peaks, and these peaks are clearly resolved. It is unusual to find any two carbon atoms in a molecule having resonance at the same chemical shift unless these two carbon atoms are equivalent by symmetry.

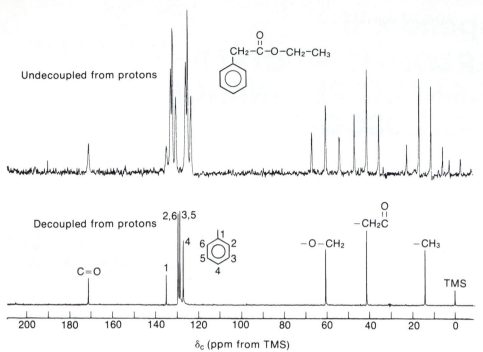

FIGURE CMR–1. Carbon-13 spectra of ethyl phenylacetate. (From Moore, J. A., and Dalrymple, D. L., *Experimental Methods in Organic Chemistry,* copyright 1976 by W. B. Saunders Co., Philadelphia, PA. Reprinted by permission of the publisher.)

Returning to the upper spectrum in Figure CMR–1, the first quartet downfield from TMS (14.2δ) corresponds to the carbon of the methyl group. It is split into a quartet ($J = 127$ Hz) by the three attached hydrogen atoms. In addition, although it cannot be seen on the scale of this spectrum, each of the quartet lines is split into a **closely spaced** triplet ($J = $ ca 1 Hz). This additional fine splitting is caused by the two protons on the adjacent —CH_2— group. These are geminal couplings (H—C—^{13}C) of a type that commonly occur in carbon-13 spectra, with coupling constants that are generally small ($J = 0$–2 Hz). The quartet is caused by **direct coupling** (^{13}C—H). Direct coupling constants are larger, usually about 100–200 Hz, and are more obvious on the scale in which the spectrum is presented.

There are two —CH_2— groups in ethyl phenylacetate. The one corresponding to the ethyl —CH_2— group is found further downfield (60.6δ), as this carbon is deshielded by the attached oxygen. It is a triplet because of the two attached hydrogens. Again, although it is not seen in this unexpanded spectrum, each of the triplet peaks is finely split into a quartet by the three hydrogens on the adjacent methyl group. The benzyl —CH_2— carbon is the intermediate triplet (41.4δ). Furthest downfield is the carbonyl group carbon (171.1δ). On the scale of presentation, it is a singlet (no directly attached hydrogens), but because of the adjacent benzyl —CH_2— group, actually it is split finely into a triplet. The aromatic ring carbons also appear in the spectrum, and they have resonances over the range from 127δ to 136δ.

CMR.3   BROAD-BAND DECOUPLED ^{13}C SPECTRA

Although the splittings in a simple molecule such as ethyl phenylacetate yield interesting structural information, namely the number of hydrogens attached to each carbon (as well as those adjacent if the spectrum is expanded), for large molecules the carbon-13 spectrum becomes very complex due to these splittings, and the splitting patterns often overlap. It is customary, therefore, to decouple **all** the protons in the molecule by irradiating them all simultaneously with a broad spectrum of frequencies in the proper range. This type of spectrum is said to be **completely decoupled.** The completely decoupled spectrum is much simpler and, for larger molecules, much easier to interpret. The decoupled spectrum of ethyl phenylacetate is presented in the lower trace of Figure CMR–1.

In the completely decoupled carbon-13 spectrum, each peak represents a different carbon atom. If two carbons are represented by a single peak, they must be equivalent by symmetry. Thus the carbons at positions 2 and 6 of the aromatic ring of ethyl phenylacetate give a single peak, and the carbons at positions 3 and 5 also give a single peak in the lower spectrum of Figure CMR–1.

CMR.4   CHEMICAL SHIFTS

Just as is the case for proton spectra, the chemical shift of each carbon indicates both its type and its structural environment. In fact, a correlation chart can be presented for ^{13}C chemical shift ranges, similar to the correlation chart for proton resonances shown in Figure NMR–5. Figure CMR–2 gives typical chemical shift ranges for the various types of carbon resonances.

Electronegativity, hybridization, and anisotropy effects all influence ^{13}C chemical shifts, just as they do for protons, but in a more complex fashion. These factors will not be discussed in any detail here, but note that the —CH_2— group carbon attached to oxygen in ethyl phenylacetate has a larger chemical shift than the —CH_2— carbon of the benzyl group. Note also that the carbonyl carbon appears relatively far downfield, probably due to an anisotropy effect.

CMR.5   SOME SAMPLE SPECTRA

The following spectra illustrate some of the effects that can be observed in carbon-13 spectra. In Figure CMR–3, the spectrum of methylcyclopentane is presented. The methyl carbon appears at highest field (20.7 ppm), as expected. The carbon atoms of the ring appear over the range 25.5–34.9 ppm. The presence of an electronegative element should deshield a carbon atom closest to it, as is illustrated in the cases of bromocyclohexane (Figure CMR–4) and cyclohexanol (Figure CMR–5). The carbon bearing the bromine in bromocyclohexane appears at 53.0 ppm; the carbon bearing the hydroxyl group of cyclohexanol appears at 70.0 ppm. In each of these cases, note that

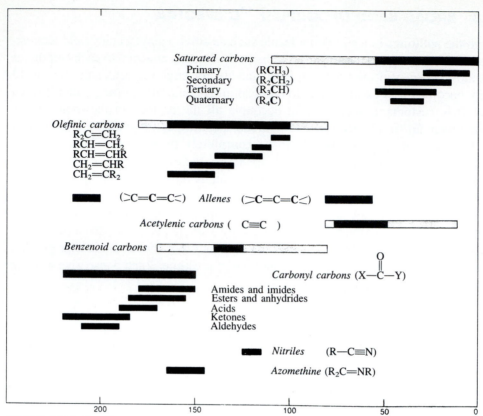

FIGURE CMR–2. ^{13}C chemical shift ranges (ppm from TMS). Extended ranges sometimes occur when polar substituents are attached to the carbons. These extended ranges are indicated by the lightly shaded areas. (From Moore, J. A., and Dalrymple, D. L., *Experimental Methods in Organic Chemistry,* copyright 1976 by W. B. Saunders Co., Philadelphia, PA. Reprinted by permission of the publisher.)

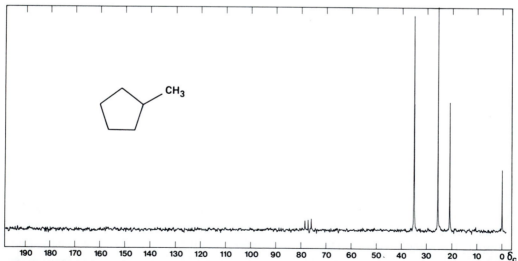

FIGURE CMR–3. Carbon-13 spectrum of methylcyclopentane. (From Johnson, L. F., and Jankowski, W. C., *Carbon-13 NMR Spectra: A Collection of Assigned, Coded, and Indexed Spectra,* copyright 1972 by John Wiley and Sons, New York, NY. Reprinted by permission of the publisher.)

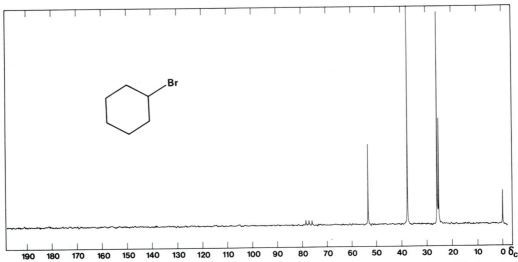

FIGURE CMR–4. Carbon-13 spectrum of bromocyclohexane. (From Johnson, L. F., and Jankowski, W. C., *Carbon-13 NMR Spectra: A Collection of Assigned, Coded, and Indexed Spectra,* copyright 1972 by John Wiley and Sons, New York, NY. Reprinted by permission of the publisher.)

as the ring carbons are located farther away from the electronegative element, their resonances appear at higher field. A carbon attached to a double bond appears deshielded, due to diamagnetic anisotropy. This effect can be seen in the spectrum of cyclohexene (Figure CMR–6). The carbon atoms of the double bond appear at 127.2 ppm. Again, it can be seen that as carbon atoms are located farther from the double bond, their resonances appear at higher field. The effect of diamagnetic anisotropy can be seen in the spectrum of toluene (Figure CMR–7), where the carbon atoms of the aromatic ring appear at low field (125.5–137.7 ppm). Finally, the strong

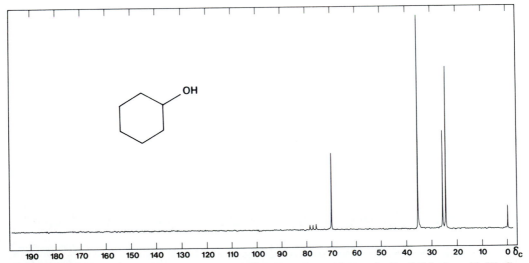

FIGURE CMR–5. Carbon-13 spectrum of cyclohexanol. (From Johnson, L. F., and Jankowski, W. C., *Carbon-13 NMR Spectra: A Collection of Assigned, Coded, and Indexed Spectra,* copyright 1972 by John Wiley and Sons, New York, NY. Reprinted by permission of the publisher.)

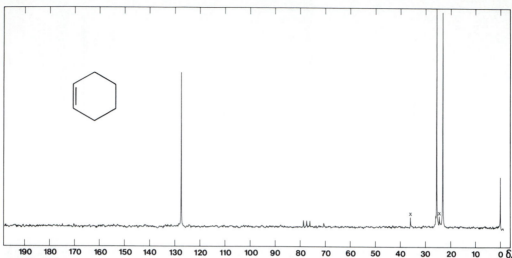

FIGURE CMR–6.   Carbon-13 spectrum of cyclohexene. (From Johnson, L. F., and Jankowski, W. C., *Carbon-13 NMR Spectra: A Collection of Assigned, Coded, and Indexed Spectra,* copyright 1972 by John Wiley and Sons, New York, NY. Reprinted by permission of the publisher.)

deshielding experienced by the carbon atom of a carbonyl group can be seen in the carbon-13 spectrum of cyclohexanone (Figure CMR–8). The carbon atom appears at a chemical shift of 211.3 ppm.

CMR.6   NUCLEAR OVERHAUSER EFFECT

As was mentioned above, integrals (areas under peaks) are not as reliable for carbon spectra as they are for hydrogen spectra. This is due in part to the **nuclear Overhauser**

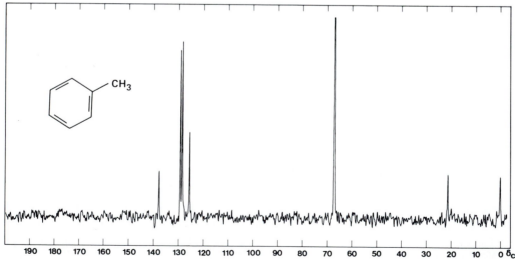

FIGURE CMR–7.   Carbon-13 spectrum of toluene. (Note: The peak at 67.4 ppm is due to the solvent, dioxane.) (From Johnson, L. F., and Jankowski, W. C., *Carbon-13 NMR Spectra: A Collection of Assigned, Coded, and Indexed Spectra,* copyright 1972 by John Wiley and Sons, New York, NY. Reprinted by permission of the publisher.)

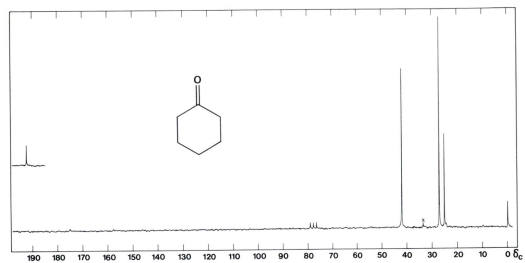

FIGURE CMR–8. Carbon-13 spectrum of cyclohexanone. (From Johnson, L. F., and Jankowski, W. C., *Carbon-13 NMR Spectra: A Collection of Assigned, Coded, and Indexed Spectra,* copyright 1972 by John Wiley and Sons, New York, NY. Reprinted by permission of the publisher.)

effect. This effect operates when two dissimilar adjacent atoms (in this case carbon and hydrogen) both exhibit spins and are NMR active. The atoms can influence the NMR absorption intensities of each other. The effect can be either positive or negative, but in the case of carbon-13 interacting with hydrogen, the effect is positive. As a result, carbon-13 NMR absorptions vary in intensity with respect to the number of hydrogen atoms that are directly attached to the carbon atom being observed. In general, the more hydrogens that are attached to a given carbon, the stronger will be its NMR absorption. Other factors also influence the absorption intensities (they are related to molecular relaxation phenomena), so the number of attached hydrogens can only be taken as a single factor influencing absorption intensity; many times this is a very helpful factor in deciding which carbon to assign to a given absorption. In Figure CMR–1 note the low intensity of the carbonyl carbon (172δ), and in Figure CMR–7 note the low intensity of the ring carbon to which the methyl group is attached (138δ). The carbonyl peak in cyclohexanone (Figure CMR–8) is also weak. None of these carbons has attached hydrogens.

CMR.7 AN EXAMPLE OF SYMMETRY

As one example of the utility of carbon-13 experiments, consider the cases of the isomers, 1,2- and 1,3-dichlorobenzene. Although these isomers could be difficult to distinguish from one another on the basis of their boiling points or their infrared spectra, each could be identified clearly by their carbon-13 spectra. 1,2-Dichlorobenzene has a plane of symmetry that gives it only three different types of carbon atoms. 1,3-Dichlorobenzene has a plane of symmetry that gives it four different types of carbon atoms. The proton-decoupled carbon-13 spectra of these two compounds are shown in Figures CMR–9 and CMR–10, respectively. It is easy to see the differences in the carbon-13 spectra of these two isomers.

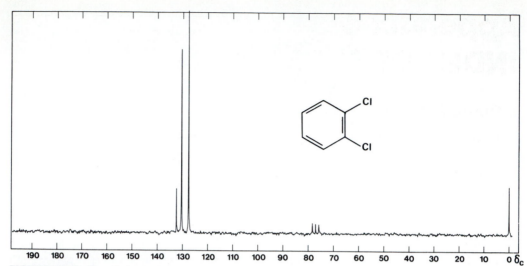

FIGURE CMR–9. Carbon-13 spectrum of 1,2-dichlorobenzene. (From Johnson, L. F., and Jankowski, W. C., *Carbon-13 NMR Spectra: A Collection of Assigned, Coded, and Indexed Spectra,* copyright 1972 by John Wiley and Sons, New York, NY. Reprinted by permission of the publisher.)

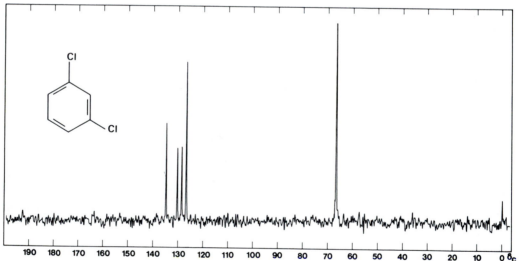

FIGURE CMR–10. Carbon-13 spectrum of 1,3-dichlorobenzene. (Note: The peak at 67.4 ppm is due to the solvent, dioxane.) (From Johnson, L. F., and Jankowski, W. C., *Carbon-13 NMR Spectra: A Collection of Assigned, Coded, and Indexed Spectra,* copyright 1972 by John Wiley and Sons, New York, NY. Reprinted by permission of the publisher.)

Appendix 6
INDEX OF SPECTRA

Infrared Spectra

¹H NMR Spectra

¹³C NMR Spectra

Ultraviolet-Visible Spectra

Mixtures

Index

862

863

864

866

867

870

873

874

876